9TH EDITION

Introductory Horticulture

9TH EDITION

Introductory Horticulture

Carroll L. Shry, Jr. • H. Edward Reiley

CENGAGE

Australia • Brazil • India • Mexico • Singapore • United Kingdom • United States

CENGAGE

Introductory Horticulture, Ninth Edition

Carroll L. Shry, Jr. and H. Edward Reiley

ISBN-13: 978-93-5350-301-7
ISBN-10: 93-5350-301-9

Registered Office:
Cengage Learning India Private Limited
T-303, 3rd Floor
DAV Complex, Plot 3
Mayur Vihar Phase 1
Delhi-110091
Email: asia.infoindia@cengage.com

Cengage Learning is a leading provider of customized learning solutions with office locations around the globe, including Andover, Melbourne, Mexico City, Stamford (CT), Toronto, Hong Kong, New Delhi, Seoul, Singapore, and Tokyo. Locate your local office at:
www.cengage.com/global

Cengage Learning Products are represented in Canada by Nelson Education, Ltd.

For product information, visit our website at **www.cengage.co.in**

Printed in India
First Indian Reprint 2020

Disclaimer:
This book contains link to resources which are a part of the US edition. The same may not be available with the Indian edition. The buyer needs to purchase a copy of the US edition to have full access to those resources. In no event the publisher shall be liable for any limitation or performance due to unavailability of these resources.

Contents

Section 11 Holiday Crafts and Floral Designs

Section 12 Business Operations

Preface

Introductory Horticulture is intended for agriculture programs in high schools offering an introduction to horticulture, agriculture, or plant science course, and presents a comprehensive review of relevant horticulture topics, activities, and tips for success within the field. Continuing in the tradition of previous editions, this all-new *Ninth Edition* offers many avenues for students to explore the horticulture field and remain on the cutting-edge of the industry with the latest technology and trends information.

Why We Wrote This Text

One of the most rapidly growing areas in the field of agriculture is that of horticulture. This interest is due in large part to an increasing awareness by the general public, as people look inwardly at a need to be closer to nature and outwardly at a world in which efficient and green methods of food production, healthier eating habits, eco-friendly aesthetic landscapes and conservation of land are increasingly important.

Consequently, starting a horticultural business that is not only profitable but also *green* and *sustainable*, is certainly a noble decision, but it also requires knowledge of the alternative methods used in biodynamic agriculture. This all-natural approach to crop production, such as the use of composted animal manures, ground dried leaves, straw, wood chips, and grass clippings for fertilization; crop rotation; mechanical cultivation; and biological pest control, allows one to produce beautiful and consumable horticultural plants that inspire and provide healthy, nutritional foods for current generations, while enabling our environment to grow and prosper for future generations. If one has the knowledge, the experience, and the drive to properly execute this effort, it will pay off in the future.

This awareness and the resulting growth of the field have generated a need for a broad-based, first-level text that explores the basic principles of horticulture and methods of practical application of these principles. *Introductory Horticulture, Ninth Edition*, offers varied multisensory applications, from group, crew, to individual, and a hands-on approach in the modalities of learning, offering myriad avenues for all students to experiment with the green industry. As agricultural educators, we must be role models to share the enthusiasm and curiosity to explore the environment through horticulture.

How This Text Is Organized

Foremost in the text is a presentation of fundamental horticulture principles, from careers in the field to the cultivation of specific crops, and from the processing and sale of ornamental plants, vegetables, fruits, nuts, flowers, and garden supplies to enhancing the living environment. Special care has been taken to treat highly technical subjects, such as plant propagation and taxonomy, in a comprehensive yet understandable manner. A section on pesticides includes up-to-date information on the subject, with special emphasis on personal safety, the protection of humans and the environment, and the use of biological control of pests whenever possible. Certain topics, such as the creation of holiday arrangements; basic floral design; the construction of dish gardens, terrariums, fairy gardens, and bonsai plantings for beginners to create a beautiful visual image; the design and care of houseplants; creating landscapes with curb appeal; and designing with basic CAD software, have been added especially for student appeal. All of these topics are arranged in a logical sequence in the following sections:

Section 1 explores the horticulture industry, reviewing career opportunities in the field and explaining plant taxonomy.

Section 2 dives into the specifics of basic plant science. This section stresses the importance of determining the nutrients in the soil, the appropriate potting mixes to use in the greenhouse, and the benefits of green practices such as high tunnels in nurseries, landscapes, and commercial and residential properties. Also covered is plant anatomy, how the environment has an effect on plant growth, and various methods for stimulating or slowing down plant growth, including a section on plant growth regulators and the use of the smartphone to calculate the correct parts per million (PPM) to ensure the optimum growth of each plant.

Section 3 covers plant propagation, including starting seeds, softwood and hardwood cuttings, grafting fruit trees, ornamentals, heirloom tomatoes, budding, and layering.

Section 4 provides an overview of the production of specific crops in the greenhouse setting: poinsettias, chrysanthemums, and Easter lilies.

Section 5 talks specifically about pest management, including important safety information in handling pesticides, insecticides, fungicides, rodenticides, molluscicides, nematocides, and herbicides.

Section 6 focuses on container-grown plants, including the increasing popularity of terrariums. Everything is covered in this section—from interior dish gardens to the artful design of fairy gardens in creating outdoor gardens with interest and appeal for all ages. Design and construction of the bonsai and interior landscapes are also addressed.

Section 7 explores the aspect of designing an exterior landscape, including an emphasis on the application of organic nutrients and compost to be more environmentally friendly. Different plants are explored, as well as pruning techniques, and the principles of landscaping, maintenance, xeriscaping, water gardens, irrigation, and green roofs. Techniques used in computer-aided landscape designs are also discussed.

Section 8 offers an overview of lawn and turfgrass practices, including establishing and maintaining a healthy lawn. It stresses the importance of soil analysis in determining the specific needs of the soil prior to establishing the lawn, as well as how eco-friendly practices, including organic applications, in maintaining the lawn can help protect the environment. This section also discusses how to renovate lawns when good maintenance practices are not followed or healthy grass is affected by environmental challenges.

Section 9 highlights the vegetable garden, including new varieties in the market and grafting heirloom tomatoes, and other vegetable plants, to increase production and disease and pest resistance. It also explains what is involved in planning and preparing the site, as well as the requirements of planting and caring for the garden. Additionally, this section provides an overview of favorite garden vegetables and herbs.

Section 10 shifts focus over to fruit and nut crops. This section includes information on planting and caring for strawberries, along with the implementation of high tunnels for extended production, blueberries, brambles, grapes, and nuts.

Section 11 inspires holiday creativity with specific units on wreaths and door swags and creating centerpieces, while also touching on special floral designs, including the design of a kissing ball for the holiday season, floral arrangements, accents, and corsages and boutonnieres.

Section 12 introduces students to the details of owning a horticulture business from how to start up a small business to financial operations, personnel management, customer sales, business communication, and the use of new technology and social media in promoting the efficiency and growth of a business. It also focuses on becoming more environmentally responsible through the use of battery-powered equipment such as chainsaws, power blowers, line trimmers and hedge trimmers, automatic mowers, and robotic plant movers that are reliable, energy efficient, and eco-friendly for a sustainable future. New technology is covered such as the use of drones in identifying deficiencies throughout a growing area to more efficiently tackle problems with crops. A unit on leadership development is also included within this section, illustrating the importance of this trait in the horticulture industry and how students have many opportunities to develop these skills through organizations such as the National FFA Organization.

Features of This Edition

LEARNING FEATURES such as **Objectives, Key Terms,** and **Competencies to Develop** frame the unit and help students identify key concepts and information. Additionally, **Review Questions** and **Student Activities** at the end of the unit encourage students to evaluate and apply what they have learned. Specific units also feature a **Materials** list that provides information on the tools and equipment required to complete the activities described in that unit.

GREEN TIPS offer practical advice on a wide range of subjects from recommendations for cultivating plants to eco-friendly and sustainable practices.

FAST FACTS provide additional engaging information that furthers understanding of the topics under discussion.

NEW! TRY THIS activities in each unit highlight a specific experiment or exercise that students can perform to apply what they have learned in a specific section.

CAUTION notes keep students safe by relaying important information about the hazards associated with using certain tools and equipment, while **SAFETY FIRST** at the end of each unit reminds students to keep apprised of the latest information from OSHA for occupational health and safety.

PROCEDURES combine visuals and step-by-step instructions to demonstrate specific techniques in the cultivation and care of plants.

NEW! SUGGESTED SUPERVISED AGRICULTURAL EXPERIENCE (SAE) ACTIVITIES at the end of each unit encourage students to seek out opportunities within the horticulture field outside of the classroom thereby broadening their knowledge and experience in various facets of the industry.

NEW! ADDITIONAL RESOURCES at the end of each unit provide a list of relevant websites that students can navigate to if they want to learn more about a specific topic.

All-New FULL-COLOR DESIGN, IMAGES, AND ILLUSTRATIONS bring the beauty of horticulture to life by engaging students in learning and inspiring them to further explore the horticulture field.

ROBUST APPENDICES provide students with valuable reference material, such as the identification of scientific and common names of plants; a guide to professional, educational, and trade organizations; a review of tools and equipment for the trade with specific care and maintenance guidelines; and listings of deer-resistance plants. Also, new to this edition, is a combined **GLOSSARY/GLOSARIO** that provides terms and definitions in both English and Spanish.

New to This Edition

In addition to the new activities and visual enhancements included in the *Ninth Edition* to facilitate learning, many revisions to the technical content have also been integrated into the text to keep students apprised of the latest information from the horticulture field:

- Statistics were thoroughly researched and current data integrated into the text where appropriate.
- New plant varieties were identified throughout the units, where applicable, so that students are well versed in the latest information related to plants.
- A focus on the use of organics and harnessing natural elements in the environment to cultivate crops and prevent disease is emphasized throughout the units as an eco-friendly, sustainable practice for both the individual and the entrepreneur.

- Cutting-edge tools and equipment, such as the driverless nursery seeder, the robotic grapevine pruner, and drones, are featured throughout to keep students apprised of the latest technology in the industry.
- The inclusion of industry-related apps, such as the PGR Mix Master calculator, demonstrates to students how technology can be useful for facilitating tasks in the field.
- The *QR code* is also investigated as a helpful marketing tool for the entrepreneur.
- An all-new section on the *agricultural educator* was added to the unit that explores career opportunities.
- An all-new section on *genetically engineered plants* and the facts surrounding this controversial topic was integrated into the information under plant science.
- An all-new section was added, including a step-by-step procedure, related to *grafting tomatoes.*
- An all-new section on *terrariums* revisits the construction and care of these miniature gardens, recognizing this as a revived popular trend.
- An all-new section on *rain gardens* to help manage the drainage of surface water illustrates another resourceful way to combat an environmental challenge through thoughtful landscaping.
- *High tunnels* are explored as an effective method to protect high-value crops from pests and environmental challenges in an economical way.
- The information on *nut crops* was moved from the Appendices, thoroughly revised, and integrated into the units.
- A thoroughly revised section on business operations covers new information on financial opportunities, marketing strategies, responsibilities to employees, current technology, best practices for business communications, and career development.
- Current information on the National FFA Organization and related activities in the unit on leadership development provides horticulture students with the tools they need to prepare for their future.

Extensive Teaching and Learning Package

Companion Site

NEW! The companion site to accompany *Introductory Horticulture, 9E,* features tools to support learning and facilitate teaching:

- *Answers to Review Questions* appearing at the end of each chapter allow teachers to track and validate student learning.
- *Lesson Plans* provide an outline of the key topics in each chapter, and correlate to the accompanying PowerPoint® presentations.
- *PowerPoint* presentations align with the Lesson Plans and include photos and illustrations to visually reinforce the key points in each chapter.
- *Testing powered by Cognero,* a flexible online system, provides chapter-by-chapter quizzes, and enables teachers to do the following:
  - Author, edit, and manage test bank content from multiple sources
  - Create multiple test versions in an instant
  - Deliver tests from teacher/school-specific learning management system (LMS) or classrooms
- *Image Gallery* offers full-color photos and illustrations from the text to enable teachers to further enhance classroom presentations.

For these instructor-specific resources, please visit CengageBrain.com at http://login.cengage.com, and follow the prompts for obtaining access to this secure site.

Lab Manual

Revised for the ninth edition, the lab manual contains a variety of exercises to help students retain and apply key concepts and information presented in the book. Labs are organized by book sections, and this edition features new exercises related to tomato grafting, designing gardens, the safe operation of power equipment, and using apps as horticultural tools.

MindTap for Introductory Horticulture, 9E

NEW! The MindTap for *Introductory Horticulture, 9E,* features an integrated course offering a complete digital experience for the student and teacher. This MindTap is highly customizable and combines assignments, videos, interactivities, and quizzing along with the enhanced ebook to enable students to directly analyze and apply what they are learning and allow teachers to measure skills and outcomes with ease.

- A Guide: Relevant interactivities combined with prescribed readings, featured multimedia, and quizzing to evaluate progress, will guide students from basic knowledge and comprehension to analysis and application.
- Personalized Teaching: Teachers are able to control course content by hiding content, rearranging existing content, or adding and creating their own content to meet the needs of their specific program.
- Promote Better Outcomes: Through relevant and engaging content, assignments, and activities, students are able to build the confidence they need to ultimately lead them to success. Likewise, teachers are able to view analytics and reports that provide a snapshot of class progress, time in course, engagement, and completion rates.

About the Authors

The authors of *Introductory Horticulture*, H. Carroll L. Shry, Jr. and Edward Reiley, were both involved in agriculture education in Maryland.

Carroll L. Shry, Jr. is a Certified Professional Horticulturist (CPH) with the Maryland Nursery and Landscape Association and is principal and owner of Brandywine Landscaping, a business that offers horticulture/landscape consulting services and snow removal. Professor Shry is presently retired after teaching for 44 years as the landscaping/nursery production instructor at the Frederick County Career and Technology Center, Maryland. He holds a bachelor of science degree in Agricultural Education/Horticulture from the University of Maryland and a master of science degree in Agricultural Education/Horticulture from West Virginia University. Mr. Shry was named Vocational Educator of Maryland in 1980 and Outstanding Vocational Educator of the Year in Region 1—American Vocational Association in 1981. He is also active in many professional organizations, including as the past president of the National Association of Agricultural Educators, and as member of the President's Council of the American Horticultural Society, the Associated Landscape Contractors of America, the American Nurserymen Association, and the National FFA Alumni Association. He is past president and currently serves as a member of FALCAN (Frederick Area Landscape Contractors and Nurserymen) and is the Eastern Region representative of the National FFA Landscape/Nursery Career Development Event. In addition, he is the past president and currently serves on the board of directors of the University of Maryland College of Agriculture & Natural Resources Agricultural Alumni, College Park, Maryland, and president of the Glade Valley Lions Club. in addition to participating as a member of the Fredrick Career and Technology Center FFA Alumni, a member of the Fredrick Memorial Hospital 1902 Club, a member of Elks Lodge 684, and serving as President and on Board of Directors of the Forest Grove Methodist Church, and on Board of Directors of the Edward F. Fry Memorial Library. Professor Shry has attended American Hort Cultivate 13 & 14, Columbus, Ohio; ICG, Washington DC & Baltimore, Maryland; MANTS Show, Baltimore, Maryland, for 15 years.

Acknowledgments

Appreciation is expressed to the following individuals for their participation and technical assistance in the revision of this and former editions, especially:

Levi Cahan, Agricultural Education Instructor Schuylerville High School, Schuylerville, New York

Frances Mary Dewey, Instructor, Junior College of Albany, Division of Russell Sage College, Albany, New York

Technical Reviewers

For those who thoroughly reviewed the text and offered their professional insight and technical expertise, we offer our gratitude:

Alex Bartel, Holton High School, Holton, Kansas

John Borst, Millbrook High School, Frederick County, Virginia

Levi Cahan, Schuylerville, High School, Schuylerville, New York

Joe Green, Pope High School, Marietta, Georgia

Eric Mayer, Ashland High School, Ashland, Ohio

Content Contributors

A special thank you to the following for their generosity in sharing of their time and expertise, and for granting us permission for the use of the exquisite imagery featured in this text:

Waverly Farm, Adamstown, Maryland

Glade Link Farm, Keymar, Maryland

SynaTek Solutions, Souderton, Pennsylvania

NaturaLawn of America, Fredrick, Maryland

Stadler Nurseries, Fredrick, Maryland; Gaithersburg, Maryland; Manassas, Virginia

United States National Arboretum, Washington, DC

Mr. Robert Sharpe of Sharpe's Flowers, Inc.

Dr. Francis Gouin, Professor, Horticulture Department, University of Maryland

Agricultural Research Service US Department of Agriculture

Shemin Nurseries, Inc., Burtonsville, Maryland

Dutch Plant Farms, Frederick, Maryland

Bluemont Nurseries, Inc., Monkton, Maryland

J. Frank Schmidt & Son Co., Boring, Oregon

Frederick County Career and Technology Center, Frederick, Maryland

Netherlands Flowerbulb Information Center Brooklyn, New York

American Nurseryman, Washington, DC

Landscape Contractors Association, Rockville, Maryland

National Landscape Association, Washington, DC

ALCA-Interior Plantscape Division, Herdon, Virginia

Prairie Nursery, Westfield, Wisconsin

Bluemels Nursery, Baltimore, Maryland

Chapel Valley Landscape, Inc., Lisbon, Maryland

Chesapeake Nurseries, Inc., Salisbury, Maryland

Kraft Gardens, Fort Pierce, Florida

Chilton Enterprises, McLean, Virginia

Tim Cleary Turf Equipment and Supply Company, Inc., Jessup, Maryland

GreenRoof LLC, A Subsidary of Hortech, Inc., Spring Lake, Michigan

Gerry Giorgio, Creative Director at MasterTag, Michigan

Kindal Marin, Director of Marketing, Bell Nursery

Anne Kemp, author and social media consultant

Craig Regelbrugge, Senior Vice President, AmericanHort

Karen Kemp Docksteader, Director of Sales, Weeks Roses

Lori Soufiane, freelance photographer

A large portion of the material in this text was classroom tested at Linganore High School and the Fredrick County Career and Technology Center, Frederick, Maryland. Unit 3 was classroom tested at Wilco Area Career Center, Lockport, Illinois. For this, we express our gratitude.

Additionally, we would like to thank Bruce Sherwin and Linda DeMasi at *Publishers Solutions, LLC,* for their professional editing expertise and the time spent in preparing the manuscript for production, and Dawn Jacobson, for her involvement in the creation of the art program and the acquisition of imagery and permissions, so that we could present a beautifully enhanced ninth edition.

AND FOR Mary Reiley and Judy Shry.

SECTION 1

Horticulture: An Introduction

1 Exploring the Horticulture Field
2 Plant Taxonomy: How Plants Are Named

UNIT 1

Exploring the Horticulture Field

Objective

To obtain an overview of the horticulture industry and jobs available in the field so that a career choice can be made.

Key Terms

Certified Professional Horticulturists (CPH)
Certified Landscape Professional (CLP)
floriculture
horticulture
landscape and nursery industry
olericulture
pharming
pomology

Competencies to Develop

After studying this unit, you should be able to:

- discuss the scope, size, and economic importance of the horticulture industry in the United States.
- describe five careers in the horticulture industry, including working conditions and educational and personal requirements.
- use the Job Analysis Chart, Self-Analysis Table, and job descriptions to make a career choice in horticulture.
- survey the local job market, and select a job that is compatible with your interests and preferences. List skills needed for the job. Describe how to prepare for the job.
- prepare a résumé for submission prior to a job interview.

The word **horticulture** is derived from a Latin word meaning "garden cultivation" (hortus [garden] + culture [cultivation]). Over the years, the horticulture industry has developed far beyond this simple definition, however. Today, horticulture includes the cultivation, processing, and sale of fruits, nuts, vegetables, ornamental plants, and flowers as well as many additional services. A market analysis by the United States Department of Agriculture (USDA) shows all horticultural sales for 2009 in the United States at $11.7 billion.

The horticulture industry consists of four major divisions:

- **Pomology.** The science and practice of growing, harvesting, handling, storing, processing, and marketing tree fruits.
- **Olericulture.** The science and practice of growing, harvesting, storing, processing, and marketing vegetables.
- **Floriculture.** The science and practice of growing, harvesting, storing, designing, and marketing flowering plants.
- **Landscape and nursery industry.** The science and practice of propagating, growing, installing, maintaining, and using grasses, annual plants, shrubs, and trees in the landscape.

Additional specialized areas include the following:

- **Seed production.** The science and practice of producing, processing, and selling high-quality seed crops for use in growing a wide range of horticultural crops that grow true from seed.
- **Sustainable horticultural occupations.** The science and practice of using an integrated system of plant production processes that makes the most efficient use of nonrenewable resources, sustains the economic viability of horticultural operations, and enhances both environmental quality and quality of life for society as a whole. Topics within this specialty include biodynamic (organic) horticulture, which uses compost for plant nutrient resources, beneficial insects, crop rotation, and low-input systems; and permaculture, which is ecological and environmental design, enhancement landscaping, and hydroponics.
- **Related occupations.** Those areas that provide products and services necessary in the production of horticultural crops. Such resources include goods (e.g., fertilizers, pesticides, and containers) and machinery (e.g., education and research).

Jobs in Horticulture

Agricultural services is an ever-expanding area of employment. Firms offering landscape and horticultural services employ almost half of the wage and salaried workers in the horticultural industry. Such firms provide landscape planning and installation, landscape architecture, lawn care, and grounds maintenance services.

Individuals trained in horticultural practices can access various sources of employment. These occupations include positions in greenhouses, nurseries, garden centers, golf courses, parks, orchards, floral design shops, and grounds maintenance operations, along with positions as vegetable and fruit growers and agricultural educators. Figure 1–1 is a Job Analysis Chart that explains positions in these areas of employment. The following job descriptions, adapted from the *Handbook of Agricultural Occupations* (published by the US Department of Agriculture), provide a more detailed account of the most common positions in the industry.

Job Title: Greenhouse Employee

JOB DESCRIPTION A greenhouse employee grows plants in a temperature-controlled glass or plastic greenhouse (Figure 1–2). The plants that are grown may be vegetables or flowers. The work involved may include the propagation of trees or ornamental shrubs from seed or cuttings. Greenhouse activities include the production of out-of-season vegetables such as tomatoes, cucumbers, or leaf lettuce; the growing of cut flowers, potted plants, and bedding plants in preparation for sale; and the starting of plants for outdoor nursery beds. When a greenhouse is

JOB TITLE	DOES IT ENTAIL YEAR-ROUND WORK?	ARE THERE REGULAR HOURS?	IS MOST OF THE WORK OUTDOORS OR INDOORS?	DOES IT OFFER VARIETY?	IS THE WORK IN ONE PLACE?	ARE THERE FRINGE BENEFITS?	DOES THE JOB INVOLVE WORKING WITH OTHERS?	WHAT ARE THE EDUCATIONAL REQUIREMENTS?	IS THERE AN OPPORTUNITY FOR PROMOTION?
greenhouse worker	yes	Generally, but some overtime is usually required.	indoors	yes	yes	yes	Yes, to some extent.	high school diploma with a course in agriculture or horticulture	yes
nursery worker	in many cases	Yes, but there are peak seasons.	mostly outdoors	yes	yes	some	yes	high school diploma with a course in agriculture or horticulture	yes
garden center employee	yes	Yes, but there are peak seasons.	both indoors and outdoors	yes	yes	some	yes	high school diploma with a course in agriculture or horticulture	yes
golf course employee	no	Yes, during the golfing season.	outdoors	yes	yes	not to a great extent	yes	high school diploma with a course in agriculture or horticulture	yes
assistant grounds keeper	yes	Yes, some overtime is required on occasion.	mostly outdoors	yes	yes	There may be some.	not necessarily	high school diploma with a course in agriculture or horticulture	yes
park employee	yes	yes	outdoors	yes	yes	not to a great extent	yes	high school diploma with a course in agriculture or horticulture	yes
vegetable grower	depends upon grower	no, seasonal	outdoors	yes	yes	some	yes	high school diploma with a course in agriculture or horticulture	yes
orchard employee	depends upon grower	no, seasonal	outdoors	yes	yes	some	yes	high school diploma with a course in agriculture or horticulture	yes
employee of small fruit grower	depends upon grower	no, seasonal	outdoors	yes	yes	some	yes	high school diploma with a course in agriculture or horticulture	yes
employee of floral design shop	yes	yes, with some overtime	indoors	no	yes	yes	yes	high school diploma with a course in agriculture or horticulture	yes
agricultural educator	yes	yes	indoors and outdoors	yes	yes	yes	yes	college education	yes

Figure 1–1 Job Analysis Chart.

▲ **Figure 1–2** A greenhouse worker grooming plants in the greenhouse.

▲ **Figure 1–3** A flower designer finishes a floral arrangement.

part of a nursery operation, a worker may have the duties of a nursery employee as well as those of a greenhouse worker. Employees in greenhouses place growing medium in containers for growing plants. They sow seed, start cuttings, and transplant seedlings and plants. They water, weed, thin, prune, fertilize, and spray growing plants. They are also responsible for maintaining the greenhouse structure and equipment.

WORKING CONDITIONS Most of the work of a greenhouse employee is done indoors. Outdoor jobs are usually done during favorable weather. This work requires a great deal of walking, stooping, and bending over plants or seedbeds. Hands and clothing may be stained from soil and plant sap or juices.

A job in a greenhouse involves a great deal of manual labor, but the labor is not usually difficult. Working hours are usually regular, with work generally steady throughout the year. Seasonal demands sometimes require working overtime.

PERSONAL AND EDUCATIONAL QUALIFICATIONS For individuals to really enjoy greenhouse work, they must be interested in watching plants grow and develop. Good health is important, but certain types of physical disabilities will not prevent an interested individual from entering the field. A high school education with a course in agricultural education or horticulture is preferred for one interested in this vocation.

Closely related to greenhouse work is flower arranging (Figure 1–3).

Job Title: Nursery Employee

JOB DESCRIPTION The nursery employee grows seedlings and plants for landscaping, fruit production, and replanting in forests (Figure 1–4). The employee may work in one of several different types of nurseries. Some nurseries specialize in producing fruit trees and small fruit transplants, some in ornamental trees and shrubs, and some in forest replanting materials. Some nursery employees operate greenhouses and produce their own seedlings and plants from cuttings. Others produce planting stock of two or more major lines of plants, trees, shrubs, and perennials.

Individuals employed in nurseries prepare seedbeds; plant seedlings; prepare cuttings for

▲ **Figure 1–4** Nursery equipment operator digging a deciduous tree with a Pizzaglia vibratory digger.

rooting; and weed, cultivate, water, and prune plants. They also perform other cultural practices such as spraying and grafting. They dig, grade, and pack plants for shipment. They may cut, lift, and lay sod. They also transplant perennials, shrubs, and trees and, in a tree nursery, gather and process forest tree seeds. Helping maintain and repair buildings and equipment is usually also a part of the job.

WORKING CONDITIONS Most of the nursery worker's time is spent outdoors. If a greenhouse is a major part of the employer's business, the employee will naturally spend a considerable amount of time in the greenhouse. Planting and cultivating must be accomplished when weather conditions are good, but much outdoor work at a nursery can be done in rather bad weather.

Most of the work is considered manual labor; some of it involves heavy lifting. Working hours are regular. Employment may be for the entire year, depending on the skills of the worker.

PERSONAL AND EDUCATIONAL QUALIFICATIONS The nursery employee should enjoy working with plants and being outdoors. A truck driver's license may be required. Employees may operate equipment, such as loaders, tree spades, spray equipment, trucks, and skidsteers. The ability to identify plants and the knowledge of how plants are used in the landscape are very desirable. A high school education with a course in agricultural education or horticulture is preferred.

Job Title: Garden Center Employee

JOB DESCRIPTION The employee of a garden center has many jobs, including caring for ornamental plants, moving plants and supplies into selling areas, arranging plants and supplies for display purposes, and selling the various products handled by the center. A garden center may be a part of a large retail store, a part of a nursery or greenhouse operation, or a retail store operated independently of other business.

Garden center employees clean, stock, and arrange garden supply displays on shelves, on counters, and in windows. They water, spray, and trim ornamental plants and control environmental conditions. They unload and unpack supplies as they arrive from wholesalers, make deliveries, and load orders onto trucks and into customers' cars. They also give information and advice to customers concerning plants and lawns and their care.

WORKING CONDITIONS Garden centers are built and arranged to attract customers. As a result, a garden center employee usually works in a clean, pleasant, and comfortable environment. Parts of the work area are usually heated during cold weather. Other areas are unheated but protected from rain, snow, and wind. Some of the work may be outdoors during the season in which ornamental shrubs and trees are sold. The work is fairly regular but has seasonal peaks. Some garden centers close completely or operate with only a small crew during the winter months.

PERSONAL AND EDUCATIONAL QUALIFICATIONS Employees of garden centers should enjoy working with people and caring for ornamental plants. They must be good salespeople and must therefore be able to talk easily and in a friendly way. A high school education with a course in agricultural education or horticulture is desirable.

Job Title: Grounds Maintenance Employee

JOB DESCRIPTION A grounds maintenance employee cares for the area surrounding an industry, business, church, school, airport, apartment building, private estate, cemetery, or shopping center. These employees plant and care for lawns and ornamental plants. The work entails mowing grass, reseeding, controlling weeds, and planting and spraying ornamental plants. They also rake grounds and dispose of leaves and other refuse. A year-round job is provided through maintenance and repair of walks, driveways, and equipment. The work may involve making minor repairs to buildings and providing for snow removal.

WORKING CONDITIONS The grounds maintenance employee works outdoors and deals mainly with ornamental plant materials. The work does provide variety but is not extremely strenuous. The environment in which the employee works is usually very pleasant, although certain jobs must be done under a variety of weather conditions. The work is steady throughout the year. Employees work regular hours, but there are some peaks in the work load.

PERSONAL AND EDUCATIONAL QUALIFICATIONS The grounds maintenance employee should not mind working alone and should enjoy working with plants, tools, and small garden equipment. The demand for high school graduates specializing in this area is increasing as job opportunities expand in the landscape industry.

Job Title: Golf Course Employee

JOB DESCRIPTION Golf course employees are responsible for the overall maintenance of golf courses. They care for the turf on both the greens and fairways. They install and use irrigation and drainage equipment, clean and maintain sand traps or bunkers, change the locations of cups, and aerate the soil. They may also prune shrubs and trees, replace soil as needed, and repair equipment and buildings.

WORKING CONDITIONS The work of the golf course employee is usually done outdoors and mostly during fairly good weather. The workday is usually 8 hours long. In the southern part of the United States, employment is steady throughout the year. In sections of the country with cold winters, employment is from March or April through October or November.

PERSONAL AND EDUCATIONAL QUALIFICATIONS Because a great deal of walking is done in golf course maintenance, the employee should have good health, but certain physical handicaps will not interfere with one's success. The employee should enjoy working outdoors and be able to get along well with others. A high school education with a course in agricultural education or horticulture is desirable.

Job Title: City, State, or National Park Employee

JOB DESCRIPTION The park employee does whatever work is needed for the proper maintenance of parks. This includes maintaining the trees, shrubs, flowers, and lawns that make up the planting area. The city park worker deals mainly with formal flower beds and lawn areas, whereas the state or national park worker is usually concerned with the care and maintenance of natural woodlands or forests.

The work of the park employee includes mowing grass; trimming the edges of walks and driveways; planting, pruning, and caring for trees, shrubs, hedges, lawns, and flower beds; controlling insects, diseases, and weeds; and caring for the soil. Other jobs include the removal of trash and snow, maintenance of swimming pools, care of boating facilities, general maintenance of buildings and equipment, and repair of roadways and drives.

WORKING CONDITIONS The park employee is outdoors a great deal of the time. Most of the work is manual labor and is accomplished in a healthy, pleasant environment. At times, park improvement jobs must be done in bad weather conditions. This occupation usually provides year-round employment, and working hours are regular. In certain types of state park work, there may be peak periods.

PERSONAL AND EDUCATIONAL QUALIFICATIONS A high school education with a course in agricultural education or horticulture is desirable.

The average hourly rate for city, state, or national park employees varies widely from region to region.

Job Title: Agricultural Educator

JOB DESCRIPTION Agricultural educators (Figure 1–5) can be found in public schools, technical schools, adult and business education programs, colleges, universities, and state-run farmer training programs. They teach a variety of subjects, such as horticulture, animal and plant science, agricultural chemicals, and agricultural technology, and they serve as advisors for Future Farmers of America (FFA). Career requirements generally include a college degree, teacher licensure and specialty certification, and agricultural experience.

WORKING CONDITIONS Agricultural educators work primarily indoors. Teachers work with a variety of people, including students,

▲ **Figure 1–5** An agricultural educator teaching students while working in the greenhouse.

colleagues, parents, and FFA members. They have the opportunity to teach a variety of subject matter relevant to a strong agricultural foundation. Educators also can remain current with agricultural education practices and FFA by being involved at the local, state, national, and global levels. Ag educators develop close working relationships with parents and students through the students' Supervised Agricultural Experience (SAE) projects, which are used to implement the "hands-on" learning.

PERSONAL AND EDUCATIONAL QUALIFICATIONS Requirements for agriculture educators vary. Most states require at least a bachelor's degree in the field of agriculture for those wanting to work at the middle school and high school levels. Coursework typically includes subjects such as agricultural mechanics, forestry/wildlife management, and agriculture management. Teacher licensure, specialty certification, and agricultural teaching are also usually required. Agriculture educators who want to work in public schools might require completion of a state-accredited teacher training program. Certificate programs in agricultural education may be an option for teachers who already possess a bachelor's degree in fields such as agriculture, forestry, or wildlife management. Master's programs in agricultural education include advanced study in areas such as agribusiness, agriculture ethics, and agricultural technology. Master's programs in agriculture education often include teaching internships and may allow students to specialize in adult education, state extension programs, or public education.

Opportunities for College Graduates

With the exception of the agricultural educator position, the majority of the jobs listed and described previously are entry-level jobs requiring only a high school diploma. However, college graduates are in great demand in the horticulture industry, and for the student interested in pursuing a degree in this field, jobs are plentiful.

What are these opportunities? *Scientific jobs* are available, such as food, soil, or water chemists; plant breeders; and forest pathologists. *Sales and service jobs* are also available, including marketing specialists, sales representatives, buyers, and landscape designers, builders, and contractors.

Other opportunities include agriculture production, farm management consulting, finance, manufacturing, processing, toxicology, engineering, entomology, communications, education, extension work, public relations, computer technology, **pharming** (genetically modifying plants for use as pharmaceuticals), and photography. The opportunity for employment is excellent for all of these jobs requiring college degrees.

If you have an interest in any of these employment opportunities, check with your teacher or school counselor to be sure you get the necessary high school math and science courses for college and a career in your chosen area.

Many of these professional careers will require more paperwork, and much of the work will be done indoors. Good fringe benefits and retirement plans are usually included with most of them, as well as high salaries during the working years. In these professions, all projects require close attention to detail, management skills, use of high-quality materials, and a strong work ethic. Horticulture professionals must also strive to ensure complete client satisfaction in every facet of each project. The college graduate in horticulture has a bright future.

Selecting a Job

To simplify the task of choosing a specific career in the horticulture industry, first study the questions posed in Figure 1–6, Self-Analysis Table. After you answer the questions, match the information with the Job Analysis Chart shown earlier in Figure 1–1. This procedure may be followed for jobs other than those listed in the Job Analysis Chart. Simply collect the information concerning the job, and match it with your self-analysis.

Career opportunities may also be broken down according to the type of work they involve: production, management, and sales (Figure 1–7). If you prefer outside work, consider an area in production. If inside work without the company of other people is appealing, consider a professional career, such as research.

There are other factors to consider before selecting a career:

- Will the job be challenging?
- What types of duties does the work involve?
- What skills does the job require?
- Are jobs in the field available?

In a career, do you desire a job that involves:
- working outside?
- working inside?
- year-round work?
- regular working hours?
- a variety of tasks?
- working in one place?
- the security of fringe benefits?
- working alone?
- working with others?
- specific educational requirements?
- the guarantee of promotion?

Figure 1–6 Self-Analysis Table.

JOBS THAT REQUIRE WORK WITH THE HANDS, TOOLS, AND EQUIPMENT. MUCH OF THE WORK IS DONE OUTSIDE, EXCEPT FOR THE GREENHOUSE WORKER.	
greenhouse worker	nursery grower
propagator	bedder
tree surgeon	pruner
turf worker	grounds keeper
greenskeeper	landscape gardener
orchardist	small fruit grower
vegetable crop grower	farm chemicals sales worker
JOBS THAT MAY REQUIRE ADVANCED TRAINING, PLANNING, AND PAPERWORK. MUCH OF THE WORK IS DONE INSIDE.	
landscape architect	consultant
teacher	researcher
plant breeder	office supervisor
salesperson	plant disease specialist
greenhouse manager	nursery manager
garden center manager	pest control specialist
pesticide specialist	landscape contractor

Note: *This represents only a partial list. It is suggested that a survey of jobs available in the local job market be made. These jobs may then be analyzed using the process outlined in Figure 1-1.*

Figure 1–7 Job list.

Using Figures 1–1 and 1–6, check several of the listed jobs to determine which comes closest to satisfying your personal job expectations. Draw up a blank form as in Figure 1–1, and fill in the information for the job you select. Research the job selected. If the job is available in your community or work area, arrange to discuss your interest with someone involved in that career, such as the manager or owner of a local business. A visit to the class by the owner or a visit to the business may be an option to discuss with your instructor.

TRY THIS Participate in the FFA Job Interview Contest. Research the contest rules and information on the FFA website, www.ffa.org. Then select a mock career area; prepare a cover letter, résumé, and follow-up letter; and practice interview skills.

The Future of the Horticulture Industry

The horticulture industry is currently in a state of rapid growth. Full of career opportunities, this industry is making use of the newest technologies in ecofriendly mechanized equipment and computers. As people become increasingly concerned about their surroundings, world resources, nature, conservation, and controlling pollution, horticulture becomes more important. More plants will be used to provide additional resources for healthy foods, to beautify surroundings, to conserve soil and control pollutants, and to protect and feed wildlife. As food costs increase, more families will plant gardens. Employment that involves working with plants or providing goods and services related to the horticulture business will be widely available.

> **GREEN TIP** The green industry is dynamic, with opportunities to promote a healthy and sustainable living environment and esthetics that enhance the quality of life.

Production of Horticultural Crops

As the population increases and people have more leisure time, more gardening of all types will be done. The demand for fruits, vegetables, flowers, ornamental plants, and landscaping services will grow as well. Increase in the demand for vegetable plants by home gardeners means a substantial increase in the production business.

Service Industries

As more horticultural goods are produced, more and more seeds, fertilizers, insecticides, fungicides, growth-regulating chemicals, hand tools, and equipment are also needed. These service industries generally seek to employ persons with a background in horticulture.

Marketing

More people are engaged in marketing and distributing horticultural products than in their production. As the volume of products increases, so does the need for employees in marketing with a background in horticulture. It is important that employees of the green industry enhance their professional expertise by becoming **Certified Professional Horticulturists (CPH)** or **Certified Professional Landscapers (CPL)**. This certification is administrated through each individual state association.

Inspection

Government agencies employ inspectors of both fresh and processed horticultural crops and products. Private processing firms interested in a quality product also employ field inspectors to spot disease and insect outbreaks as well as to determine when the crop is at peak quality for harvesting.

Teaching and Extension Work

More and more students need to be trained in horticulture at secondary and postsecondary school levels, which, in turn, requires more teachers. Quickly changing technology also calls for a field force of specialists, such as extension agents and county agricultural agents, to work with producers. The homeowners and those who garden as a hobby also require the advice of professionals. Educators often use the news media as a mass communication method to reach the public more easily.

Research

As demands for greater production are placed on the horticulture industry, researchers will play an important role (Figure 1–8) in a concerted effort to develop better varieties of plants; establish more effective insect and disease control, both through resistant varieties and through biological and chemical control; and develop better ways of growing, harvesting, preserving, packaging, storing, and marketing. In addition, the new tissue culture propagation technique still needs research for application to a wider range of plants.

> **GREEN TIP** To reduce the amount of unrecyclable or poorly recyclable materials found in stores, many companies are looking at the farmer's fields for green solutions. For example, fully recyclable plastics can be made from corn to use in the production of the plastic needed to make cell phones.

▲ **Figure 1–8** A researcher at the Agricultural Research Service, US Department of Agriculture, finds new uses for sugar, a simple carbohydrate produced by plants. By changing the molecular structure of sugar, a powerful epoxy glue can be made.

The horticulture industry has met the challenge in the past through the development of new products and procedures such as the following:

- Hybrid varieties
- Better greenhouse climate control and plant medium
- Mist propagation
- Propagation by tissue culture
- Automatic watering and fertilizing
- Control of the blooming date of plants by regulating the length of darkness or light
- Plant growth regulators
- Rooting hormones
- Weed control
- Resistant varieties of plants and biological pest control
- Extended storage life of foods
- Mechanical harvesting
- Dwarfing rootstock for tree fruits
- Better methods of packaging and processing products such as concentrated juices and frozen foods
- Improved cultural practices
- Gene splicing
- Wise water use through xeriscape gardening

As the horticulture industry becomes more complex and more technical in nature, more training will be required for the available jobs. The future holds great promise for the person willing to learn and to work in horticulture. To get started in this exciting horticultural world, you must become employed. To do this, you must convince an employer that you can help his or her business grow and become more efficient.

At this point in your career, it is important to know how to prepare a job résumé and follow up with an effective job interview. The following is brief but helpful in preparing you for your first job interview (see Figure 1–9).

The Job Résumé

The following lists some of the most important elements of a good résumé, cover letter, and application:

1. Place your name, address, e-mail address, and telephone number at the top of the first page.
2. List your educational as well as work experience. List your most recent jobs, certificates, diploma, and so forth in reverse chronological order. Include any awards for outstanding scholastic work or job-related awards. It is especially important to include anything that better prepares you for the specific job you are applying for.
3. Write in words you normally use so it will sound as though you prepared the résumé if you have an interview later.

John Doe
26 North Palm Drive
Ocean Beach
Florida 00015
e-mail address
phone number

Objective
An opportunity to use and improve my horticultural training and skills as part of a landscaping team.

Education
Graduate of Palm City Career and Technology Center, 2016

Work Experience
Landscape Associates Inc., Palm City Florida
Assistant on a Lawn Care Team, Summer of 2015
City of Ocean Beach, Park Maintenance Crew, Summer of 2014

Honors
2 Years Academic Honor Roll
Member of winning FFA State Horticulture Team

Computer Experience
Word Perfect
Photoshop
PowerPoint
Windows XP
CAD Program Word Excel

Figure 1–9 A sample résumé.

4. List your career objective if you have one.
5. List the date you are available for work: after graduation or summer vacation.
6. Research the company to which you are submitting the résumé, and state why you want to work there in the cover letter. Be positive, and do not mention money. Companies like to feel you want to become part of their team and contribute to their success.
7. Be clear and complete when filling out the job application. Follow directions explicitly.

The Job Interview

The second step after preparing and submitting a résumé is a successful job interview. Here is where you and the employer meet face to face to determine whether you would work well together.

You must be well prepared for the job interview. Dress in clean, attractive clothes that are somewhat better than the clothes you wear on the job. First impressions are very important.

You will feel more relaxed if you prepare yourself mentally for the interview and will present yourself much better. Some things you can do to prepare for a job interview include the following:

1. Write down several things you do well. Give examples of things you have done successfully.
2. Consider your interpersonal skills, and be prepared to discuss them. These skills include things such as being dependable, patient, tolerant, sympathetic, punctual, hardworking, open to suggestions, and so forth.
3. List your strengths.
4. Research the company you are applying to so you can ask appropriate questions. Do some research on how much to expect regarding salary as well.
5. Organize all your material in a folder.
6. Plan accordingly so you will be on time for the interview.

During the interview, follow these tips:

1. Let the interviewer control the meeting. Answer questions but do not brag or appear bossy.
2. Be positive. Do not say negative things about your school, teacher, principal, or any previous bosses. If you left a job, hopefully it was so you could advance or better yourself. If you got fired, learn from the situation, and do not repeat what happened if you were at fault. If you make mistakes, admit them if asked. Hasten to add that you will not make the same mistake again.
3. Anticipate questions the interviewer will ask you, such as these:
   A. *What are your greatest strengths and weaknesses?* Mention strengths that better qualify you for the job. Mention that you are working to strengthen your weaknesses and would welcome help to overcome them.
   B. *Why did you leave your previous job?* Again, be positive. If it was your fault,

say you have learned from the experience and no longer have this problem; be honest. If you and your boss could not get along, do not cut the boss down; just mention that there was a conflict.

C. *Why do you want to work for our company?* This is one of the most important questions. Know something about the company so that you can mention what you like about it. Tell the interviewer you want to grow with the company as part of its team.

D. *What specific training have you had?* Be sure you have the training skills necessary to do the job offered. If you do not, mention this and state that you are willing to train and learn whatever is necessary.

E. *How much salary do you expect?* Wait for the interviewer to mention salary. When salary is mentioned, be realistic in terms of your experience and what other companies are paying for a similar job.

F. *What are your goals?* Mention where you would like to be in the long run with this company. Mention that you are willing to learn and to advance.

G. *Why should we hire you?* Now is the time to mention your strengths and how you will use those strengths to help the company. Show your interest in the company and your desire to work and grow with it.

Helpful Hints

- Be sure both you and your clothes are clean and neat and your hair is neatly combed. Minimize jewelry, perfume/cologne, and makeup.
- Be on time or 5 minutes early for the interview.
- Do not chew gum, smoke, or bring drinks to the interview.
- Be sure you understand the questions, and think before you answer so you can give a clear, brief, honest reply.
- Make eye contact with the interviewer.
- Be enthusiastic and positive.
- Remember to let the interviewer initiate the discussion on salary and benefits.
- Thank the interviewer for the time and consideration he or she has given you, and ask how soon you should expect a decision on your employment.
- Write a thank-you note to the interviewer immediately after the interview.

All of the information provided here has been for employees. Because most of you are new to the industry, some experience working for a successful business is valuable and there is much to learn. At some point, you may want to start your own business. For an insight into small business ownership and how small businesses fit into our free-enterprise system, see Appendix B.

SAFETY FIRST

A career in horticulture offers numerous and varied opportunities, such as self-employment, greenhouse owner-operator, landscaper, lawn care specialist, teacher, and cooperative extension agent. Safety in the field of horticulture is an everyday task to maintain the health and welfare of all who work in this field. The Occupational Safety and Health Administration (OSHA) recognizes that landscaping and horticultural service workers are at risk of injury from a variety of potential hazards, including exposure to chemicals, noise, machinery, lifting, construction, and weather-related hazards. For more information on occupational safety and health in the workplace, please visit the OSHA site at www.osha.gov.

Summary

The horticultural industry is dynamic and growing rapidly. Many jobs are and will continue to be available in this field. Information in this unit should help you decide on a career. Preparing a good résumé and following through with a strong interview will help you gain employment in this ever-expanding job market with so many interesting and challenging opportunities. The keys to success in the horticultural business are honesty, integrity, pride in workmanship, continuing education, training, commitment to quality materials and service, exceeding clients' expectations, and professionalism.

Review Questions

1. What makes the horticulture industry an important component of the US economy?
2. What are the keys to success in the horticulture business? Why is each one important?
3. What are some of the specialty areas within the horticulture industry?
4. What does "sustainable horticulture" mean?
5. When working on a job, why should you always try to exceed a client's expectations?
6. When interviewing for a job, why is it important to describe your personal goals? At what point in the interview should you do this?

Student Activities

1. Conduct research online to create a list of horticulture businesses in the local area.
2. List the horticultural jobs advertised in the classified sections of online agricultural websites or journals.
3. Select one business in the community, and interview the owner as well as several employees. Create a slide show presentation, and report to the class on available jobs and the type of duties and working conditions involved in each job.
4. Surf the Web for an agricultural college in your state. Using the university website, research the types of jobs typically offered to graduates in horticulture, internship programs, and job placement services.
5. Go online or make a personal contact at a local business and complete a job application form for a horticulture-related job. Write a résumé.
6. Conduct practice interview sessions to develop and refine your interviewing skills.

Suggested Supervised Agriculture Experience (SAE) Activities

1. Interview for an SAE placement in vegetable production at a local vegetable farm.
2. Plan an SAE either at home and/or in the community by working for a local farmer's market.
3. Design, build, plant, grow, and maintain a community garden for a placement Career Development Event (CDE).

Additional Resources

www.ffa.org

http://hort.cals.cornell.edu/

www.hrt.msu.edu

www.ars.usda.gov/careers/careers.htm

UNIT 2 Plant Taxonomy: How Plants Are Named

Objective

To differentiate between scientific and common plant names and explain the binomial system of naming plants.

Key Terms

binomial system
botanical nomenclature
common name
cultivar
genus
hierarchy
Latin
macrophylla
microphylla
species
taxonomists
taxonomy

Competencies to Develop

After studying this unit, you should be able to:

- explain why scientific plant names are used.
- explain the differences among genus, species, and variety.
- list five plants by their common and scientific binomial names.
- give four examples of family names (both Latin and common).

Materials

- online nursery catalogs
- nursery catalogs
- online issues of *American Nurseryman* magazine or *Grower Talks*
- *Hortus Third*, staff of the L.H. Bailey Hortorium, Cornell University
- http://plants.usda.gov
- smartphone apps:
  - Leafsnap (plant identification)
  - Spectracide (weed identification)
  - LikeThat Garden (plant identification)
  - GRO GETTER (plant identification)
  - Pacific Northwest Early Detection Network (PNEDN) (identification of invasive plants)
  - Mid-Atlantic Early Detection Network (MAEDN) (identification of invasive plants)

Most plants have more than one **common name**, and some have several. For example, the trout lily is also known as the tiger lily, adder's-tongue, dog's-tooth violet, and yellow snowdrop. The Judas tree and redbud are the same tree but are known by these different names in different parts of the United States. Common names can be confusing because two totally different plants may have the same common name. The cowslip is one of these: in New York State, it is a marsh-loving, buttercup-like plant, whereas in England, it is a primrose-like plant found on dry, grassy slopes. Both have yellow flowers, but, apart from their color, they have little else in common. Because common names can be so misleading, it is important to use the same name when plants from different areas are spoken of—to use common names when buying or selling could be disastrous. Common names (e.g., pea, maple, tomato, and petunia) are an important part of the horticulture industry, however, because these names are a part of the everyday language used by both those in the industry and laypeople.

The Binomial System of Naming Plants

Early scholars always wrote in **Latin** or Greek, so, naturally, when they described plants or animals, they gave them scientific Latin or Latinized Greek names. However, this way of naming plants also caused problems because the names were often long and difficult. For example, *Nepeta floribus interrupte spicatis pedunculatis* was the name for catnip, and *Dianthus floribus solitariis, squamis calycinis subovatis brevissimis, corollis, crenatis* was the name for the common carnation.

The famous Swedish botanist Carolus Linnaeus (1707–1778) simplified the matter by developing the **binomial system** (*binomial* means "two-name") for naming plants, which is still used today. He gave all plants just two Latin names as their scientific name. For example, he renamed catnip *Nepeta cataria.* The first name is known as the **genus** name, which is the plant's group name. All plants having the same *genus name* are said to belong to the same genus. All plants belonging to the same genus have similar characteristics and are more closely related to each other than they are to the members of any other genus. The second name is the *specific name* or special name. All plants with the same specific name belong to the same **species.** (The Latin word *species* means "kind.") It is difficult to define exactly what a species is, but we can say that plants of the same species have the same characteristics and will consistently produce plants of the same type. Today, species are often subdivided into *varieties.* One variety of a species resembles that of another variety, but there are always one or two differences that are consistent and inherited. For example, the peach tree is known as *Prunus persica,* and the nectarine is *Prunus persica* var. *nucipersica.* Another subdivision is termed a **cultivar** (cv.), which means "cultivated variety." For example, many people consider red to be the ultimate fall color, giving the brightest, most unique display of all. The most commonly known cultivars of *Acer rubrum* (red maple) are 'Red Sunset' and 'Autumn Flame,' which are the most reliable for brilliant reds and a long-lasting display of foliage.

FAST FACT

New to the plant kingdom, *Coffea charrieriana* is the first recorded caffeine-free coffee bean species from Central Africa. The plant is named for Professor André Charrier, who managed coffee breeding research during the last 30 years of the twentieth century.

The genus name is a noun, and the species name is an adjective. Sometimes, genus names are the names of early botanists, for example, *Buddleia* was named in memory of Adam Buddle. Some common genus names include *Acer* (maple), *Chrysanthemum* (mum), *Dianthus* (pink), *Hibiscus* (mallow), *Mimulus* (monkey flower), *Sedum* (stonecrop), *Papaver* (poppy), *Pinus* (pine), and *Pelargonium* (geranium). And, in some cases, the generic name has become the common name of the plant, such as *Euonymus, Rhododendron,* and *Forsythia.*

Because the species name is an adjective, it often gives important information about the plant. Sometimes, it tells us the color of the plant; for example, *Betula lutea* is the yellow birch (*lutea* means "yellow"), *Betula alba* is the white birch (*alba* means "white"), *Quercus rubra* is the red oak (*rubra* means "red"), and *Juglans nigra* is the black walnut (*nigra* means "black"). The species name can also tell us if the plant is creeping or erect; for example, *Epigaea repens* is the scientific name for trailing arbutus.

The specific name often gives geographical information about where a plant occurs naturally; for example, *Anemone virginiana* is the Virginia anemone, and *Taxus canadensis* is the Canadian yew. *Macro* and *micro* are Greek words meaning "large" and "small," respectively. Therefore, a plant with the species name **macrophylla** could be expected to have large leaves, whereas a plant with the species name **microphylla** would have small leaves (*phyllus* means "leaf"); for example, *Philadelphus micro-phyllus* is the scientific name for the little-leaf mock orange.

When the meanings of the scientific names of plants are understood, the names are interesting and not difficult to learn. The easiest way to become familiar with them is to say them out loud; every time you transplant a seedling, dig a shrub, or sow some seeds, say the Latin name. When scientific names are used, the horticulturist can order plants from any part of the world because it is a universal language. The French, Dutch, Texan, and New Englander all use the same Latin names. Whereas common names are often misleading, the scientific name never is. For example, people often confuse the red maple with the Japanese red maple. They are two completely different trees. The red maple *(Acer rubrum)* grows to be more than 75 feet tall; it has green leaves in the summer and red foliage in the fall. The Japanese red maple *(Acer palmatum)* has red leaves throughout the year and does not grow above 25 feet tall.

TRY THIS Research the specific scientific names of several plant species. Choose *five* specific names, and *describe* the information that the name gives about the plant. It may be a geographic location, the color of the plant, or even structural characteristics. Share your findings with the class.

Plant Families

Related genera (plural of *genus*) with similar flower structures are grouped together into major units known as *families*. For example, the rose family, known as Rosaceae, consists of several genera—*Prunus* (plum), *Fragaria* (strawberry), *Rubus* (bramble), and *Malus* (apple). All members of each genus of the rose family have relatively simple flowers with separated petals (A in Figure 2–1). On the other hand, in the Solanaceae, the potato family, the petals are fused or joined to form a corolla tube (B in Figure 2–1). The Solanaceae includes the genera *Solanum* (potato), *Petunia*, and *Nicotiana*

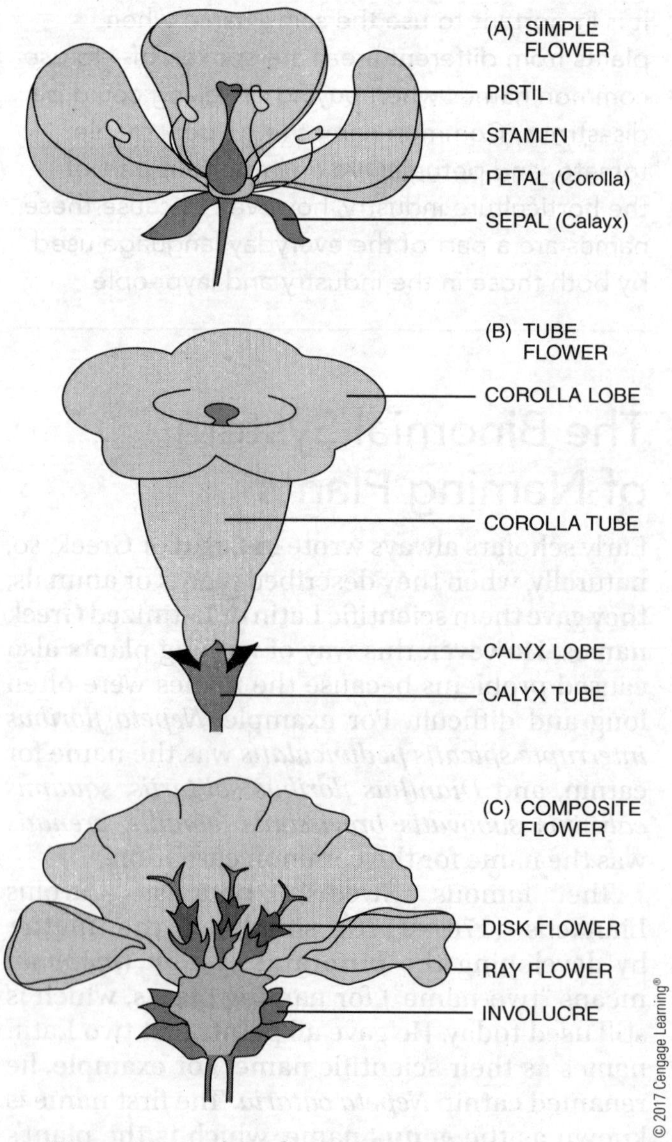

▲ **Figure 2–1** Plant families.

(tobacco). The Asteraceae (aster family) is the largest of all the plant families. Members of the Asteraceae have two kinds of flowers packed together to form a single head or "flower" (C in Figure 2–1). The outer flowers (known as ray flowers) may have large or small petals. The inner flowers (disk flowers) always have small petals. The aster family includes these genera: *Aster* (aster), *Artemisia* (silver mound), *Achillea* (yarrow), *Helianthus* (sunflower), *Chrysanthemum* (mum), *Senecio* (cineraria), *Calendula* (pot marigold), and many others. Other important families are the Brassicaceae (cabbage family), Apiaceae (carrot family or umbellifer family), Papaveraceae (poppy family), Liliaceae (lily family), and Poaceae (grass family).

Expressing Scientific Names

Why do we use the scientific name? As mentioned earlier, the common name is not specific enough and sometimes is misleading. The common name of a plant as known in one area of the country or world may be totally different in another area.

In the green industry, it is vitally important to know exact information about the plant in question. Each name is unique with precise links to describe the plant. Many people will use the common name of the plant because it is easier to learn and remember than the Latin name. But in some cases, the scientific name is the same as the common name; for example, forsythia is *Forsythia suspensa,* epimedium is *Epimedium* × *rubrum,* coreopsis is *Coreopsis verticillata* 'Moonbeam,' and sedum is *Sedum* × *telephium* 'Autumn Joy.'

Notice that when the Latin names of plants are printed, they are expressed in italics. This is because when names and phrases are written in a language other than our own, it is conventional to print them in italics or underline them if they are typewritten or handwritten. Also, by convention, the generic name is always written first, then the species name, and last the cultivar (cv.) name. The generic name always begins with a capital letter and the species name with a small letter. The cultivar name is preceded by the letters cv. or enclosed with single quotes (i.e., 'Red Sunset'), with the first letter of the cultivar name capitalized. Sometimes, when a number of species all belonging to the same genus is the subject, the generic name is abbreviated to only the first letter; for example, to express several different types of maples, the generic name for maple, *Acer,* is abbreviated to *A.* The red maple may be expressed as *A. rubrum* and the Japanese maple as *A. palmatum.* The term *cultivar* refers to a specific plant, as in *Pinus strobus* 'Pendula,' weeping white pine, or *Fagus sylvatica* 'Atropurpurea,' copper beech, whose leaves are a deep maroon color.

The following **taxonomy** chart shows the **hierarchy** of **botanical nomenclature**.

Taxonomy Chart: hierarchy of specification

Kingdom—Plant
Division or Phylum—Spermatophyta (seed plants)
Class—Angiospermae (seeds in fruit)
Order—Acerales
Family—Aceraceae
Genus—*Acer*
Species—*rubrum*
Variety or cultivar—var. 'October Glory'

The Taxonomist

Scientists who identify and classify plants are known as **taxonomists**. An international set of rules has been drawn up to ensure that every different species has a different binomial name and that the scientific name assigned to that plant is the oldest binomial name ever used for that plant. This international set of rules is known as the *International Code of Botanical Nomenclature.*

GREEN TIP

The rules that make up the *International Code of Botanical Nomenclature* give future horticulturists the opportunity to explore the world for new plants and employ plant nomenclature in an international setting.

SAFETY FIRST

In the field of horticulture, one must have the knowledge to identify plants that are poisonous because exposure to or ingestion of poisonous plants can cause bodily harm. One out of every 100 plant species of plants is poisonous to humans and animals. For example, humans can suffer ill health effects if they come in contact with plants such as poison ivy, poison oak, and poison sumac. For more information on occupational safety and health in the workplace, please visit the OSHA site at www.osha.gov.

Summary

Plants are named by scientists known as taxonomists who identify and classify the plants in a hierarchy of specification. Each taxonomist must follow the national set of rules known as the International Code of Botanical Nomenclature. The binomial system of naming plants was developed by a famous Swedish botanist, Carolus Linnaeus. Plant names are in Latin, specific with genus, species, and cultivar. Common names are expressed in English and sometimes may be confusing or misleading. Therefore, it is best to use the scientific name to be sure the correct plant is identified.

Review Questions

1. Why is it important to use a plant's scientific name when discussing it?
2. Why is the generic name of a plant a noun, and the species name an adjective?
3. Why do taxonomists follow an international set of rules when identifying and classifying plants? What are the possible consequences of not following these rules?

Student Activities

1. Diagram the seven groups of the taxonomical hierarchy of specification.
2. Access the Internet and search the taxonomy of a plant species provided by the instructor. For each species, identify the proper hierarchy of taxonomy. For each of the following, identify the correct name: kingdom, division or phylum, class, order, genus, species, and variety or cultivar.
3. Research online to identify an example of a recently discovered species of plant. Once identified, describe the taxonomy of the new species.

4. Using the following smartphone apps, identify 20 different horticultural plants:
   a) Leafsnap (plant identification)
   b) Spectracide (weed identification)
   c) LikeThat Garden (plant identification)
   d) GRO GETTER (plant identification)
5. Do an online search for "poisonous plants," and make a list of 10 poisonous plants you discover.
6. Search online to find out more about taxonomy. How does a person become a taxonomist? Where does a taxonomist work? What is a day in the life of a taxonomist like?

Suggested Supervised Agricultural Experience (SAE) Activities

1. Participate in a placement SAE specializing in agronomy.
2. Take pictures and make a CD for plant or insect identification for an exploratory SAE.

Additional Resources

https://plants.usda.gov/classification.html

www.ext.colostate.edu/mg/gardennotes/122.html

http://ibis.geog.ubc.ca/biodiversity/eflora/IntroductiontoPlantTaxonomy.html

SECTION 2

Plant Science

UNIT 3

Parts of the Plant and Their Functions

Objective

To recognize the main parts of a plant and describe the function of each.

Key Terms

adhesion
centrosome
chloroplasts
cohesion
cytoplasm
gamete
genetic engineering
genetically engineered plants
genetics
guttation
meiosis
mitosis
petals
phloem
photosynthesis
pistil
respiration
sepals
stamen
stomata
transpiration
tropism
xylem
xylem tubes

Competencies to Develop

After studying this unit, you should be able to:

- note two contributions of plants to the life cycle on earth.
- list and describe the purpose of the four main parts of plants.
- explain the process of photosynthesis.
- explain the process of respiration.
- explain the major structural difference between dicot and monocot stems and how the stems grow.
- describe the process of pollination.

Materials

- a complete plant with roots, stem, leaves, and flower or seeds
- a plant with a fibrous root system
- a plant with a taproot system
- a large flower for dissection
- knives for dissecting flower parts
- small monocot and dicot plants
- computer

The Importance of Plants

The importance of plant life on earth cannot be overemphasized. Without plants, life on earth could not exist. Directly or indirectly, plants are the primary source of food for humans and animals. Whether people eat plants or eat animals that feed on plants, plant life is vital as a food source.

> **GREEN TIP** For a career in the horticulture industry, it is necessary to develop an understanding of the importance of plants as the primary source of the food and oxygen that are essential for life on earth.

Plants play another essential role by producing oxygen, which is required for life. Plants are the major producers of oxygen on this planet. All plant life, from the smallest plankton in the ocean to the giant redwood tree, works to produce oxygen.

In addition to supplying food and oxygen, plants help keep us cool, renew the air, slow down the wind, hold soil in place, provide a home for wildlife, beautify our surroundings, perfume the air, and furnish building materials and fuel.

Parts of a Plant

Most plants are made up of four basic parts: leaves, stems, roots, and flowers, which later become fruit or seeds (Figure 3–1).

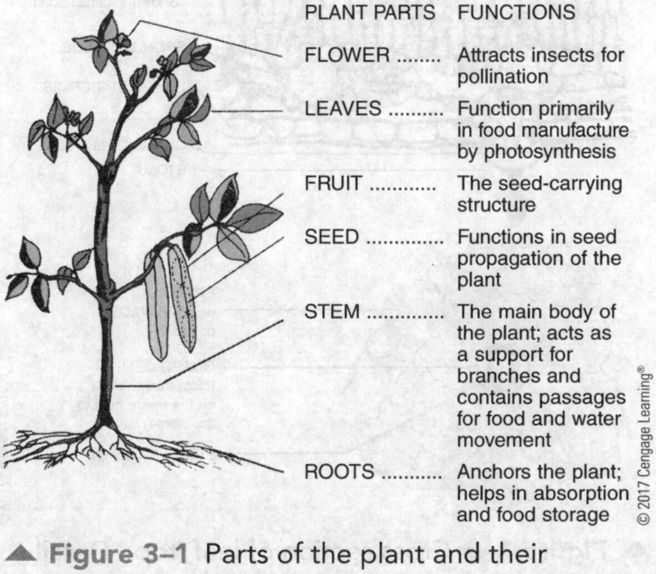

▲ **Figure 3–1** Parts of the plant and their functions.

Leaves

Leaves are the food factory of the plant, producing all food that is used by the plant and stored for later use by the plant or by animals. Leaves vary a great deal in shape and size. Most leaves are flat. Some, such as the leaves of pine trees, are needlelike, whereas others, such as onion leaves, are cylindrical. The shape and size of leaves help to identify plants (Figure 3–2).

The arrangement of leaves on plants also differs. Some plants have leaves that alternate on the stem; some are positioned opposite one another. Others are *whorled* (arranged in a circle around the stem) (Figure 3–3).

EXTERNAL LEAF STRUCTURE Leaves consist of the *petiole,* or leaf stalk, and the *blade,* a larger, usually flat part of the leaf (Figure 3–4). Notice that the leaf blade has veins and a midrib. The *midrib* is the large center vein from which all other leaf veins extend.

The *veins* of the leaf form its structural framework. Most leaves have one of the forms shown in Figure 3–2. Different leaf *margins* (edges of

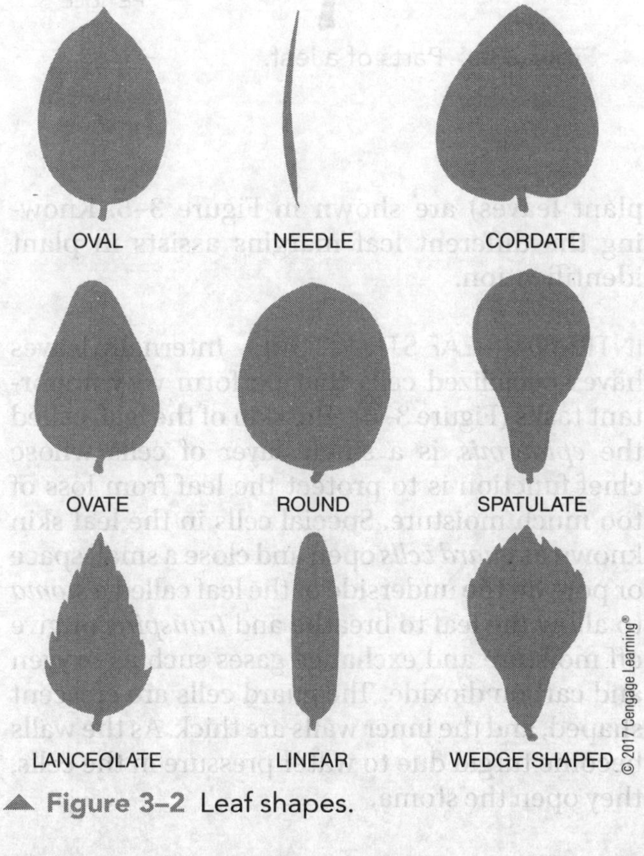

▲ **Figure 3–2** Leaf shapes.

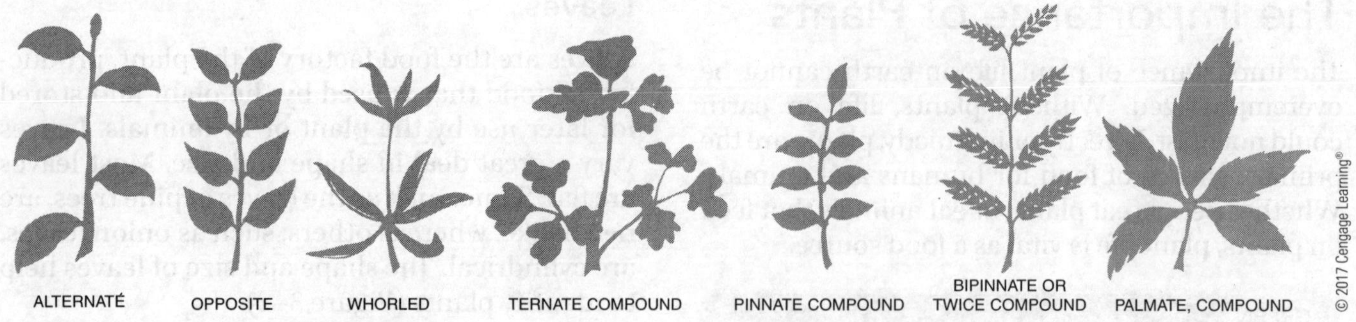

▲ **Figure 3–3** Leaf arrangements.

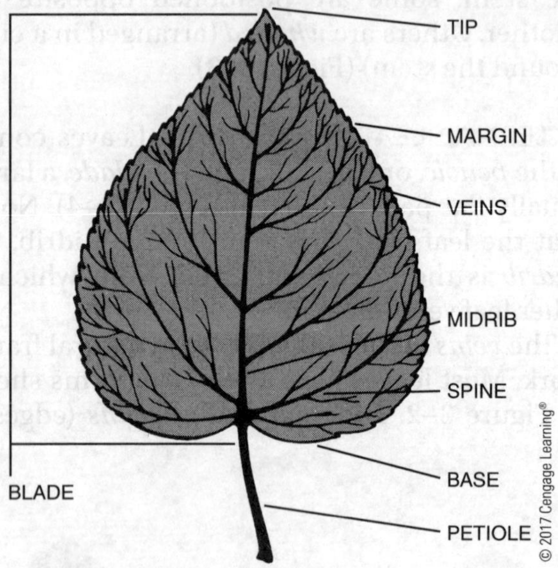

▲ **Figure 3–4** Parts of a leaf.

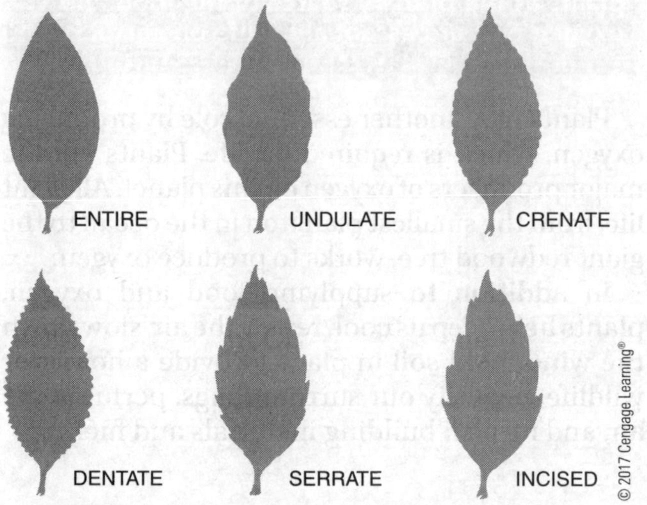

▲ **Figure 3–5** Leaf margins.

plant leaves) are shown in Figure 3–5. Knowing the different leaf margins assists in plant identification.

INTERNAL LEAF STRUCTURE Internally, leaves have specialized cells that perform very important tasks (Figure 3–6). The skin of the leaf, called the *epidermis,* is a single layer of cells whose chief function is to protect the leaf from loss of too much moisture. Special cells in the leaf skin known as *guard cells* open and close a small space or pore on the underside of the leaf called a *stoma* to allow the leaf to breathe and *transpire,* or give off moisture and exchange gases such as oxygen and carbon dioxide. The guard cells are crescent shaped, and the inner walls are thick. As the walls become turgid due to water pressure in the cells, they open the stoma.

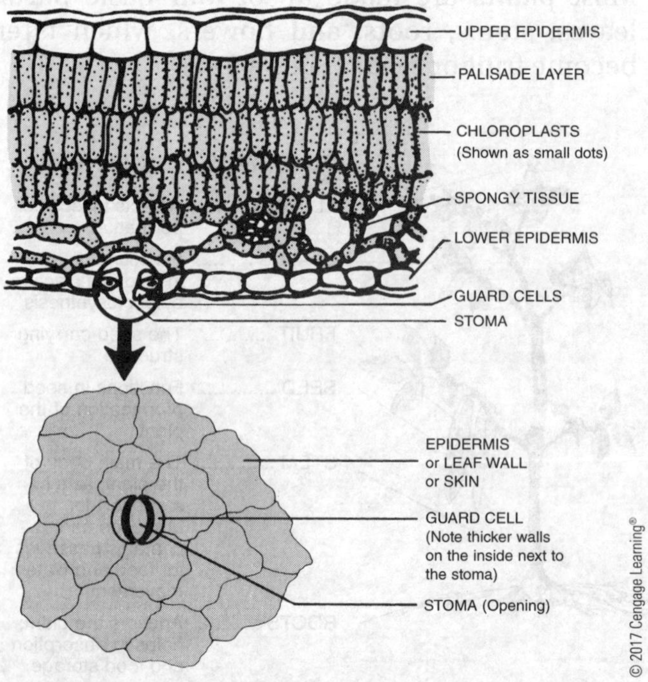

▲ **Figure 3–6** Cross section of leaf (top). Detail of guard cells and stoma (bottom).

Transpiration

Transpiration is the evaporation of water through plant leaves and stems, primarily through **stomata**. When adequate water is not available due to dry soil conditions, transpiration slows down or stops causing plants to wilt. If wilting continues in the evening, it is time to water the plant to prevent plant damage. During extreme drought, the stomata close, and photosynthesis slows or stops because no water or carbon dioxide moves through the leaf food factory.

Transpiration is necessary to a plant's survival. It both cools the plant and moves water to maintain the flow of nutrients and manufactured food from the root to the top and back again. Plants transpire more when the soil has adequate moisture, when it is windy or hot, and when the humidity is low. Water movement in plants is pulled up by the forces of cohesion and adhesion, not pushed up by the roots.

When the soil is too moist, plants force water from leaf vein tips to reduce water pressure inside the plant. When you can see the droplets of water on leaf tips and edges on a humid windless morning, this is call **guttation**.

Photosynthesis

In the center of the leaf are food-making cells that contain **chloroplasts**. The green color of the chloroplasts, which give green leaves their color, comes from the chlorophyll they contain. Through the **photosynthesis** process, these cells manufacture food. Photosynthesis is the process by which carbon dioxide and water in the presence of light are converted to sugar and oxygen. This is the beginning of the food chain for all living things on earth. The chemical process is explained here in chemist's shorthand:

LIGHT ENERGY

$$6\,CO_2 + 6\,H_2O + 672\ \text{kcal} = C_6H_{12}O_6 + 6\,O_2$$

carbon dioxide + water + light energy = glucose (sugar) + oxygen

CHLOROPHYLL

Thus, the equation states that 6 molecules of carbon dioxide plus 6 molecules of water are combined using 672 kilocalories of light energy to produce 1 molecule of glucose, also known as dextrose or corn sugar, and 6 molecules of oxygen. *Note:* One kilocalorie is the energy required to heat 1,000 grams of water 1°C or 1 quart of water approximately 2°F.

FAST FACT

Glucose, a sugar, is the main carbohydrate of photosynthesis and a primary source of energy in most living things. Glucose is one of the top candidates to replace oil as a source for fuels, plastics, and other petroleum products.

The oxygen resulting from the photosynthesis process is used directly by animals. It is the vital ingredient in all forms of *oxidation* (combining with oxygen), such as burning, rusting, and rotting.

Food manufactured in the leaves moves downward through the stem to the roots. The plant then uses this food or stores it in the stem or root in the form of sugar, starch, or protein. The leaves themselves are also used as food for various animals, including human beings. They are often the most nutritious part of the plant.

The amount of sugar produced by photosynthesis depends on an adequate supply of carbon dioxide, water, and energy in the form of light. The carbon dioxide used in photosynthesis enters the plant through the stomata, which, as stated earlier, are small openings on the undersurface of the leaf (see details in Figure 3–6). As long as temperatures are not too high (above 86°F [31°C] for most plants), or dry conditions do not cause the stomata to close to conserve water, a free flow of carbon dioxide and water can enter the leaf to be used in the manufacture of food, primarily sugar.

Under dry soil conditions, the stomata may close and reduce carbon dioxide movement, which slows down photosynthesis. Thus, we can conclude that plants manufacture more food under good growing conditions, that is, when carbon dioxide and water are readily available, and a good light source provides the energy or fuel to power the process.

Respiration

Plants also respire 24 hours a day just as animals do. In this **respiration** process, they consume oxygen and give off carbon dioxide just as animals do. Roots, stems, and leaves all use ("breathe" in) oxygen as they grow and "breathe" out carbon dioxide. An adequate supply of oxygen to all parts of the plant is necessary. Due to poor soil drainage and water pushing the air from the soil pore space, roots suffer most from short oxygen supply, and plant death is possible. Plants produce more oxygen through photosynthesis than they consume through respiration and growth processes. This excess oxygen is used by animals and by all other oxidation processes.

The following chart shows a comparison of respiration and photosynthesis.

RESPIRATION IS THE OPPOSITE OF PHOTOSYNTHESIS

Photosynthesis	Respiration
Food is manufactured.	Food is consumed.
Sun's energy is used.	Energy is released.
Carbon dioxide is used by plant.	Carbon dioxide is given off.
Oxygen is given off.	Oxygen is used.
Complex compounds form.	Simple compounds form.
Sugars and starches are produced.	Carbon dioxide and water are produced.
Light is required.	Process occurs both day and night.
Chlorophyll must be present.	Process is carried on in all cells.

Stems

Stems have two main movement-related functions: the movement of materials, such as water and minerals from roots upward to the leaves, and the movement of manufactured food from the leaves down to the roots.

Water moves through plant stems by being pulled from above, not pushed from below. Two forces, cohesion and adhesion, are responsible. Water molecules are attracted to each other like magnets with a force known as **cohesion**. As water evaporates from leaves, more water is pulled up through microscopic **xylem tubes** in a continuous column as the force of cohesion acts to keep water molecules stuck to one another. Water also clings to the walls of the xylem tubes via a force known as **adhesion**.

The second function is the support of the leaves and reproductive structures (flowers and fruit or seeds). Stems are also used for food storage, as in the Irish potato, and for reproduction methods that involve stem cuttings or grafting. Green stems manufacture food just as leaves do.

External Stem Structure

The outside of the stem consists of *lenticels,* or breathing pores, bud scale scars, and leaf scars. *Bud scale scars* indicate where a terminal bud has been located; the distance between two scars represents 1 year of growth. *Leaf scars* show where leaves were attached (Figure 3–7).

Some plants, such as the Irish potato and the gladiolus, have stems that differ greatly from the stems of other plants. The stems in these plants are used for food storage and plant reproduction. A detailed discussion of this type of stem and how it differs from others is given in Section 3, Plant Propagation.

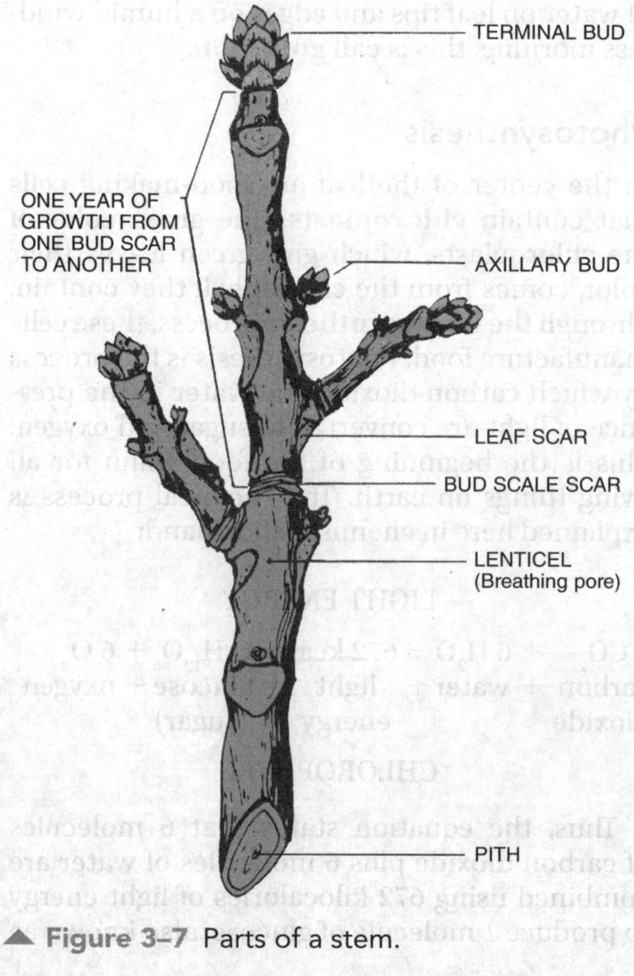

▲ **Figure 3–7** Parts of a stem.

INTERNAL STEM STRUCTURE In all stems, water and minerals travel up the **xylem**, and manufactured food travels down the **phloem**. In *dicots,* plants that have two seed leaves (cotyledons), xylem and phloem form two layers separated by the *cambium,* which produces all new cells (Figure 3–8). Dicot stems may continue to increase in diameter because the cambium builds new phloem cells on the outside and new xylem cells on the inside. Trees are good examples of dicots that have stems which continue to expand. In woody dicots, sapwood consists of new, active xylem; heartwood is old, inactive xylem; and tree bark is old, inactive phloem (plus other cells).

This theoretically unlimited continuation of growth through the production of new cells is one of the major differences between plants and animals. For example, all the members of an animal's body are differentiated before birth. A human usually stops growing at maturity. Yet, a dicotyledonous tree continues to grow new branches and roots and to grow taller as long as it lives.

Some plants, such as grasses, have a different stem structure (Figure 3–9). These plants are referred to as *monocots* because they have only one cotyledon (seed leaf). Corn is an example of a monocot. Monocots have vascular bundles that contain both phloem and xylem tissue in each small bundle. Notice that the monocot stem has no outside circling cambium. All cells are formed in the initial stage of stem growth and merely enlarge to create the size of the mature stem. There is no further enlargement of stem size by formation of new cells; however, some stems form additional vascular bundles as they grow.

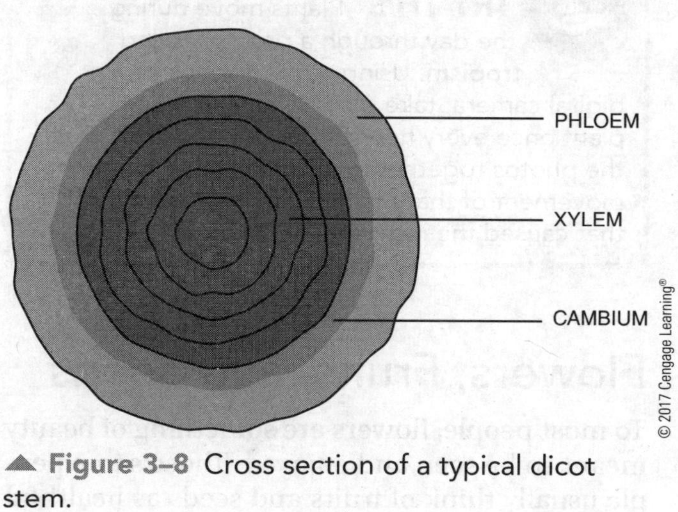

▲ **Figure 3–8** Cross section of a typical dicot stem.

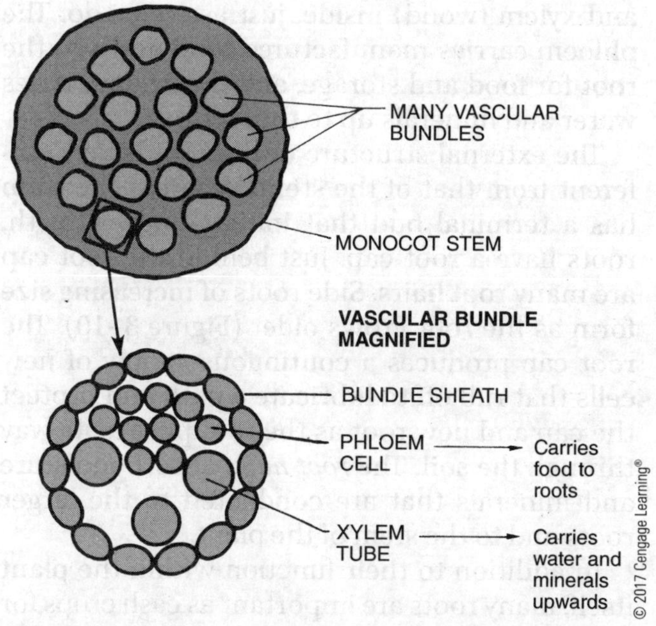

▲ **Figure 3–9** Cross section of a monocot stem.

These plants are limited in size much as animals are. Stems grow upward. They also grow toward light. This response is called *phototropism.* When plants respond to an environmental stimulus by moving, this movement is called *tropism.*

ECONOMIC IMPORTANCE The stems of some plants, such as the Irish potato and asparagus, are used as food. Others are used as building materials, such as the lumber from tree trunks.

Roots

Roots are usually underground and, therefore, not easily visible. Roots perform the following functions:

- Anchor the plant and hold it upright.
- Absorb water and minerals from the soil and conduct them to the stem.
- Store large quantities of plant food.
- Propagate or reproduce some plants.

The first three functions are essential to all plants.

STRUCTURE The internal structure of a root is much like that of a stem. Older roots of shrubs and trees have phloem (old outer layers form corklike bark) on the outside, a cambium layer, and xylem (wood) inside, just as stems do. The phloem carries manufactured food down to the root for food and storage, and the xylem carries water and minerals up to the stem.

The external structure of the root is very different from that of the stem. Whereas the stem has a terminal bud that initiates new growth, roots have a root cap. Just behind the root cap are many root hairs. Side roots of increasing size form as the root grows older (Figure 3–10). The root cap produces a continuous supply of new cells that rub off to lubricate a path and protect the cap and new root as the root pushes its way through the soil. The *root hairs* absorb moisture and minerals that are conducted to the larger roots and to the stem of the plant.

In addition to their function within the plant itself, many roots are important as cash crops for food. Vegetables such as carrots, beets, radishes, and sweet potatoes are all roots. Roots also serve in the process of propagation. Plants with tuberous roots, such as the dahlia, peony, and sweet potato, are propagated by separating the root clump or by rooting sprouts from the root, as is the case with the sweet potato.

TYPE OF ROOT SYSTEM The ease with which nursery stock is transplanted or moved depends to a great extent on the type of root system of the plant.

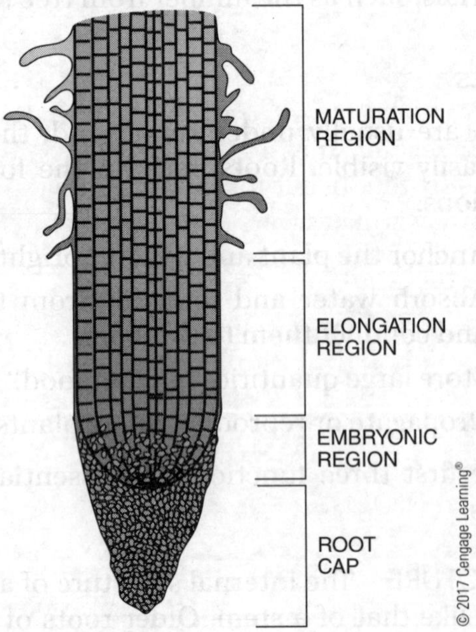

▲ **Figure 3–10** Longitudinal section of a young root (greatly enlarged).

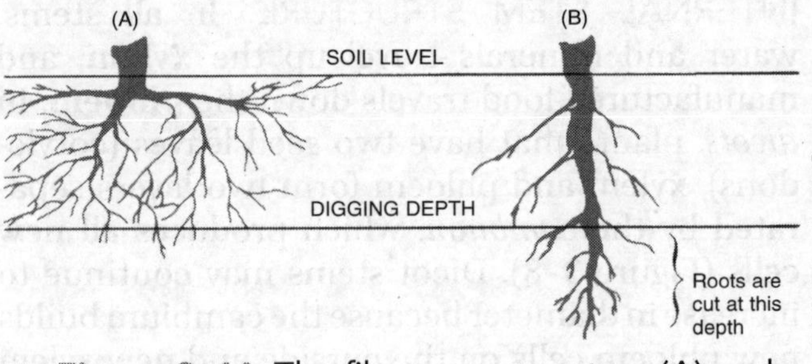

▲ **Figure 3–11** The fibrous root system (A) and the taproot system (B).

FIBROUS ROOT SYSTEM VERSUS TAPROOT SYSTEM Plants with fibrous root systems are much easier to transplant than plants that have taproot systems. As illustrated in Figure 3–11, when a plant with a fibrous root system and a plant with a taproot system are dug from the same depth in the ground, a greater percentage of the fibrous root system is saved. The roots are shorter, smaller, and more compact.

The taproot system has longer and fewer roots, which means that much of the root system is cut off when a plant is dug. The ends of the roots that are lost in the cutting contain many root hairs necessary in the absorption of water and minerals from the soil. The larger roots serve only to conduct and store water, nutrients, and food. If too many of the small roots are lost, the plant may not be able to replenish the moisture lost by the leaves, and the plant will dry out and die. Roots move toward moisture in the soil. Central roots grow downward because of the force of gravity. This is called *geotropism* and is controlled by the root tip.

TRY THIS Plants move during the day through a process called **tropism**. Using a smartphone or a digital camera, take pictures of a growing plant once every hour throughout one day. Put the photos together in a collage to show the movement of that plant. Describe the factors that caused the movement.

Flowers, Fruits, and Seeds

To most people, flowers are something of beauty meant to be seen and enjoyed. Those same people usually think of fruits and seeds as healthful foods. The parts that are admired and enjoyed by human beings, however, have an entirely different purpose for the plant.

The beauty of the flower, for example, is necessary to attract insects. In their visits for nectar or pollen, these insects fertilize the flower by means of a process called *pollination.* This is the beginning of fruit and seed formation. The fruits and seeds are made attractive to animals and birds so that they are collected, eaten, and spread. This, in turn, reproduces the plant. Some seeds have sticky coats and are carried on the animal's fur to be dropped in a new location. These devices and actions are not for the convenience of human beings or animals, but they ensure the continued existence of the plant itself.

Parts of the Flower

Flowers differ in features such as size, shape, and color, but they generally have the same basic parts, which are necessary for the production of seed. Seed is the most common way plants reproduce in nature. Seeds are produced by a sexual process with a male and a female parent involved. A *complete flower* (Monoecious) has both male and female parts, and only one parent is needed if the plant is *self-fruitful,* or can pollinate itself. The following is a description of the four main parts of a complete flower (Figure 3–12):

- **Sepals.** The green leaflike parts of the flower that cover and protect the flower bud before it opens. They form the calyx after the flowers open.
- **Petals.** Actually leaves but are generally known as the most striking part of the flower. The bright colors of the petals that are usually present on flowers act to attract insects for pollination.
- **Stamens.** Make up the male reproductive part of the flower. Each stamen consists of a short stalk called a *filament* and a saclike structure on top of the filament called an *anther.* The anther contains pollen, which is the male sex cell.
- **Pistil.** The female part of the flower located in the center. It produces the female sex cells, the eggs *(ovules).* These eggs, if fertilized, become seeds. The pistil has three main parts: a sticky *stigma* on top to catch pollen and a *style,* a tube that leads to the third part, the *ovary.* The egg cells develop in the ovary. After fertilization, the ovary grows to become a fruit or a seed coat.

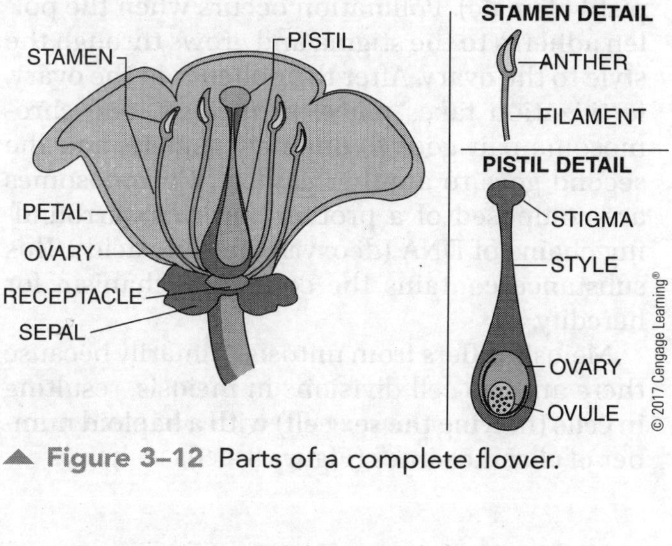

▲ **Figure 3–12** Parts of a complete flower.

The flower is constructed so that insects attracted to it for nectar must first climb over the anther and brush pollen on the hairy surface of their bodies. As they climb onto the center of the flower for nectar, part of the pollen is brushed onto the stigma of the pistil. This allows the fertilization process to begin. The pollen grain sprouts like a seed and sends a long stalk down the style to the ovary and the egg cells. The pollen sperm cell then fertilizes the egg cells, and seeds begin to develop. The ovary enlarges into a seed coat or fruit.

An *incomplete flower* (Dioecious) is missing either female or male parts. Thus, a male flower has sepals, petals, and stamens, but no pistil. A female flower has sepals, petals, and pistil, but no stamens.

Flowers play a very important role in the florist and nursery businesses. Many plants are grown solely for the beautiful flowers they produce. Plants may grow flowers only to attract insects for pollination, but people grow them for their beauty and economic value. The floriculture industry (that part of the horticulture industry that deals with flowers) is a multimillion dollar business.

The Importance of Plant Genetics

Today's plants are much improved over the plants we had a hundred years ago. They produce high-quality flower color and foliage and are more resistant to diseases and other pests. **Genetics** is the study of heredity and the ways parent plants pass their valued traits to their offspring.

Gregor Mendel, an Austrian monk, is considered to be the founder of the science of genetics. Mendel experimented with garden peas. He was able to prove that certain characteristics such as height and color are passed from the parent to the offspring. Through this basic theory, plant breeders today are doing wonders for the plant environment to breed new plants for the green industry. They are summarized as follows:

1. In every living thing, a pair of genes in every cell controls the appearance of every trait in that individual. A gene is a unit of hereditary material on a chromosome. A chromosome is a rodlike carrier for the gene.
2. Individuals receive one gene for each trait from each parent.
3. Genes are transmitted from parent to offspring as an unchanging unit.
4. In the production of reproduction cells, gene pairs separate, and one gene for each trait is contained in each gamete. A **gamete** is a reproductive cell.
5. When an individual has different genes for a trait, one usually shows while the other does not.

Cells and Cell Division

The plant structure consists of tiny units called cells (Figure 3–13), which are made of protoplasmic material with a nucleus and cell wall.

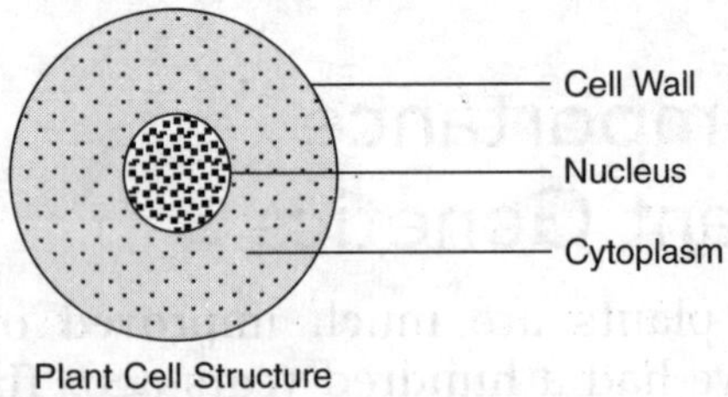

▲ **Figure 3–13** Plant cell.

Cytoplasm is the liquid that gives the plant cell inside the cell wall its shape and fosters the other cell structures.

As a basic living structure, all plant cells start as a single cell with a nucleus that contains rod-shaped bodies called chromosomes. The nucleus of the cell contains the **centrosome** and the chromosomes, with genes resting on a specific area. The simple cell division for growth is called mitosis, in which each chromosome divides in two. The wall of the nucleus disappears, and chromosomes move to opposite sides of the cell. A new nucleus forms around each group of chromosomes. Finally, the cell wall divides, resulting in two new cells complete with nuclei and pairs of chromosomes.

Mitosis

As a plant grows, the number of plant cells increases in the plant in the cell-division process called **mitosis** as mentioned previously. In mitosis, the chromosome pairs duplicate in each new cell (Figure 3–14), and each new cell is identical to the old cell.

Meiosis

Meiosis is the reproductive cell division that results in the formation of gametes. It is different from mitosis in that instead of the chromosomes dividing and moving in pairs to opposite sides of the cell, they separate and move individually to the cell wall. The gametes that are formed in meiosis are mature sex cells. The pollen is transferred from the stamen (male) to the pistil (female). Pollination occurs when the pollen adheres to the stigma and grows through the style to the ovary. After the pollen is in the ovary, fertilization takes place. In meiosis, one chromosome pair goes to one new gamete, and the second goes to another gamete. Chromosomes are composed of a protein covering surrounding chains of DNA (deoxyribonucleic acid). This substance contains the coding mechanism for heredity.

Meiosis differs from mitosis primarily because there are two cell divisions in meiosis, resulting in cells (halving the sex cell) with a haploid number of chromosomes (Figure 3–15).

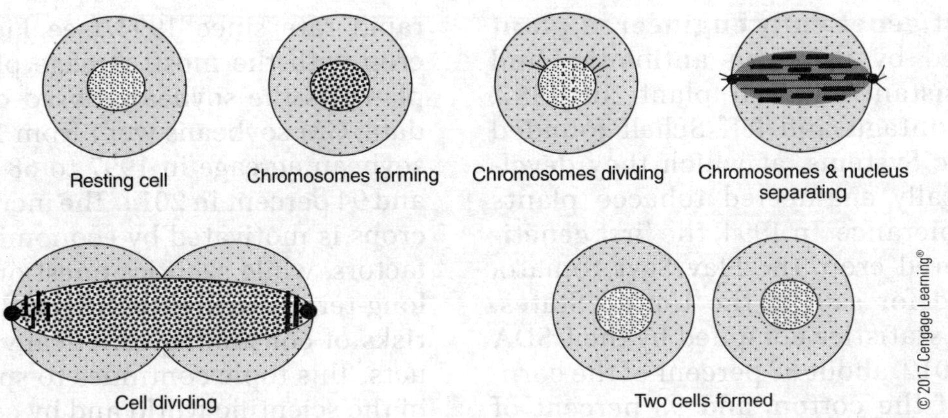

▲ **Figure 3–14** Cell division (mitosis).

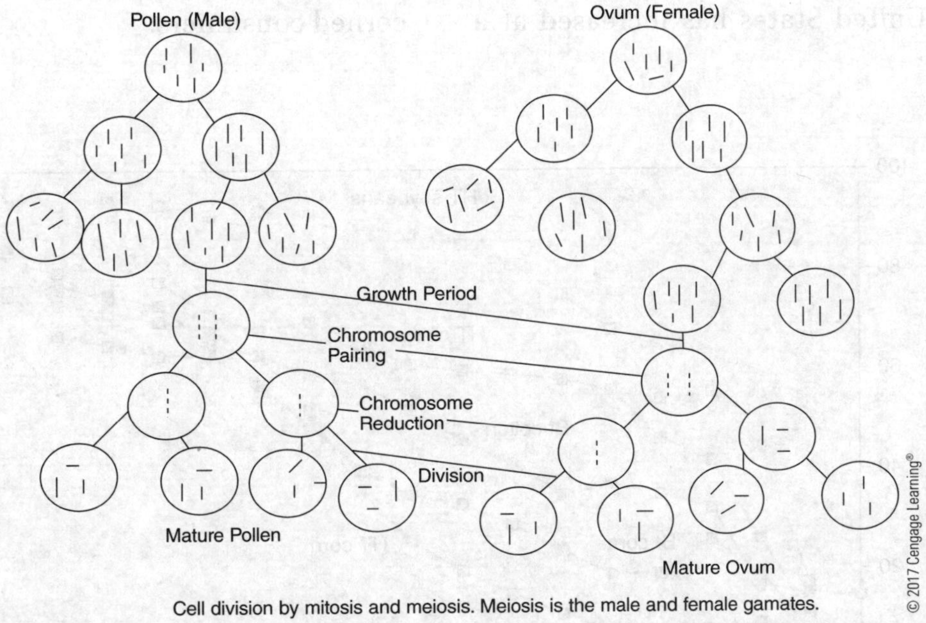

▲ **Figure 3–15** Meiosis.

The following lists the numbers of pairs of chromosomes in the cells of different plants:

Corn–20
Wheat–42
Potato–48
Radish–18
Pineapple–50
Oats–42
Mango–40
Beans–22
Carrots–18
Broccoli–18
Cabbage–18

Genetically Engineered Plants

Genetic engineering is a field of scientific study that has resulted in a myriad of new plants to enhance our environment. Genetic engineering is different from traditional breeding programs where the genes are manipulated indirectly. Genetic engineering uses the techniques of molecular cloning and transformation to alter the structure and characteristics of genes directly. Genetic engineering techniques have been successful in improving crop technology. For example, in

1982, the first **genetically engineered plant** was produced by using an antibiotic- and herbicide-resistant tobacco plant. In 1987, Marc Van Montagu and Jeff Schell founded Plant Genetic Systems, at which they developed genetically engineered tobacco plants with insect tolerance. In 1994, the first genetically engineered crop, the FlavrSavr tomato, was approved for sale in the United States. According to statistics compiled by the USDA for the year 2012, about 88 percent of the corn, 94 percent of the cotton, and 93 percent of the soybean crops grown in the United States were genetically engineered.

The adoption of genetically modified (GM) crops in the United States has increased at a rapid rate since 1994 (see Figure 3–16). The crop with the most acreage planted with GM plants is the soybean. Based on USDA survey data, GM soybeans went from 17 percent of US soybean acreage in 1997 to 68 percent in 2001 and 94 percent in 2014. The increased use of GM crops is motivated by economic and efficiency factors, while skeptics question the amount of long-term research conducted to identify the risks of consuming genetically modified products. This topic continues to spark controversy in the scientific world and by consumers in the food industry. A niche market of certified non-genetically modified organisms (non-GMO) products has emerged as an alternative for concerned consumers.

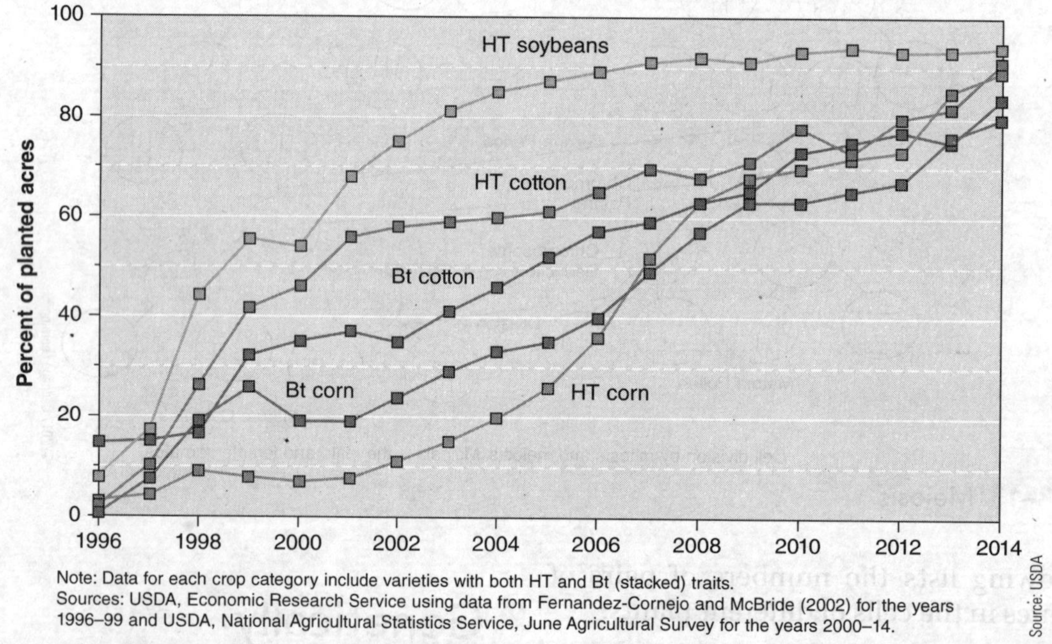

▲ **Figure 3–16** The adoption of genetically modified crops in the United States has increased dramatically in the past several years.

SAFETY FIRST

When using the knife to dissect flower parts, use caution not to cut yourself. For more information on occupational safety and health in the workplace, please visit the OSHA site at www.osha.gov.

Summary

Plants are different from animals in many ways. Plants are the primary source of food and oxygen that are absolutely essential to ensure the continuation of life on earth. Through the process of photosynthesis, a plant's leaves capture the sun's energy and use carbon dioxide and water to produce this food and oxygen. The life cycle and length of life of plants are also quite different from those of animals.

We think of plants as being nonmoving. We now know that plants are in constant motion as roots move to find water and stems move toward light so the leaves are exposed to more light, thus increasing photosynthesis. The process of photosynthesis has not been duplicated by humans and still remains a challenge as researchers look for new ways to produce food.

Review Questions

1. What are the different parts of the formula for photosynthesis? Practice stating it in words and writing it in chemist's shorthand.
2. What are some contributions plants make to the life cycle on earth? Would humans be able to survive in a world without plants? Why or why not?
3. How does the process of pollination work?
4. How would you describe a common leaf, such as a maple, by its shape, arrangement, and margin?
5. What are the pros and cons of genetically engineering plants for use as horticultural crops?

Student Activities

1. Dissect a flower. Sketch and label all parts.
2. Access an online virtual plant dissection. Show the instructor that you completed the dissection.
3. In a notebook, draw and label a cross section of the four main parts of a plant. Allow the instructor to check for accuracy.
4. Cut a cross section of a dicot stem and a monocot stem provided by the instructor. Draw a sketch, noting the differences between the two stems.

CAUTION: Exercise care when handling knives or saws.

5. Obtain two plants common to the area, and label a sketch of a leaf from each plant. Note the shape or form, type of margin, and arrangement of leaves. See Figure 3–2 through Figure 3–5 for information.
6. Obtain a cross section of a tree. Identify the cambium layer and the site of xylem and phloem.
7. In the classroom, germinate a monocot seed (corn) and a dicot seed (bean). As they develop, note that the monocot has one seed leaf, and the dicot has two seed leaves. Measure and record the growth of each.

Suggested Supervised Agricultural Experience (SAE) Activities

1. Participate in a placement SAE in the area of floriculture, nursery and landscape management, golf course management, or forest management and products.

Additional Resources

https://ag.arizona.edu/pubs/garden/mg/botany/plantparts.html

www.urbanext.illinois.edu/gpe/case1/c1facts2a.html

www.amnh.org/learn/biodiversity_counts/ident_help/Parts_Plants/parts_of_flower.htm

UNIT 4

Environmental Requirements for Good Plant Growth

Objective

To explain the basic needs of plants and the various factors that make up their environment.

Key Terms

- aggregation
- biochar
- biochemical
- capillary water
- compost
- dormancy
- field capacity
- forage radish
- gravitational water
- humus
- hydroscopic water
- limestone
- loam
- macroelements
- medium
- microelements
- nutrient holding availability
- peat moss
- perlite
- photoperiodism
- slow-release fertilizers
- soilless mix
- sphagnum moss
- texture triangle
- transpiration
- tree bark
- vermiculite

Competencies to Develop

After studying this unit, you should be able to:

- list four factors that affect the roots of plants.
- describe the differences among clay, sandy, and loamy soils and identify a sample of each.
- explain three ways to improve soil drainage and two ways to increase moisture retention of soil.
- explain what is meant by the pH value of soil.
- compose a balanced fertilizer program for one plant that is grown commercially in the area.
- list four aboveground requirements for good plant growth.
- list the three major plant food elements and two functions of each.
- list three functions of using compost in soils

Materials

- bags of both organic and synthetic garden and lawn fertilizers
- container of slow-release, water-soluble fertilizer with minor elements
- samples of sandy, clay, and loamy soils
- complete plant
- three half-gallon jars
- four potted plants for light and temperature experiments
- samples of organic matter resources to enhance soils

Environment

To properly grow into profit-making crops, plants require a certain environment. The environmental factors that have the greatest effects on plants are water, nutrition, medium, temperature, light, humidity, insects, and diseases. A few of the most important factors are discussed here.

Water

Water makes up 70 to 90 percent of the fresh weight and is the major component of green plants. Most of the water is contained in the cells. Water provides the medium for many **biochemical** reactions. Plant foods must be in soluble form in water before plants can use them. Chemical processes such as photosynthesis could not occur without water. Water also transports nutrients and manufactured food throughout the plant. Nutrients from the soil must be in a water solution before roots can absorb and transport them.

Because water is clear, it allows light from the sun to penetrate plant leaves and green stems to reach the chlorophyll and power the production of food. Water vapor evaporating from leaves cools the plant to prevent heat damage.

Inadequate water can stress plants, making them more susceptible to disease and insect attacks. Lack of water during dry soil conditions can slow the process of photosynthesis.

Nutrition

Nutrition is just as important to plants as it is to humans. Plants require at least 16 elements from the soil or from fertilizers. Good nutrition is important to plant health and good growth. In most **loam** soils, enough minor elements are available for satisfactory plant growth. In soilless mixes, minor elements may be deficient. The major elements—nitrogen, phosphorus, and potassium—are needed most often.

Temperature

Plant growth is greatly influenced by heat. Unlike animals, plants are unable to maintain their cells at a uniform temperature. There are a few exceptions; for example, plants can heat their reproduction parts (flowers) when they are ready for pollination (fertilization). Plants also cool leaves by **transpiration** of water, but this process cannot maintain a constant temperature. Roots are affected less by temperature changes because the soil moderates temperatures and slows down the rate of speed of temperature change.

It is important to select plants adapted to the area. Tropical plants do well in heat and humidity but not at temperatures below freezing. Some plants require a certain number of hours of cold temperatures below 40 to 50°F to break **dormancy** or to flower.

HIGH TEMPERATURES Plant growth increases as temperature increases up to 86°F (31°C) for most plants. High temperatures (above 86°F [31°C] for most plants) slow down or stop plant activity. Continued temperatures in the high 90°F (35°C) to 100+°F range can severely damage many plants not adapted to these temperatures. Within the favorable growth range of temperature, some plant responses double for each 10°C increase in temperature.

Some plants can reflect radiation from the sun and remain cooler. The shape and size of leaves affect the area impacted by radiation and heat buildup.

LOW TEMPERATURES The primary effect of cooling plants below their best growing temperatures is a slowdown of metabolic processes and reduced growth.

Chilling subtropical plants to near freezing causes a very rapid fall in *metabolic processes*, especially *respiration*, and can result in death in a few hours. Plants adapted to cold areas cease growth and show signs of damage only after ice forms within the cells. These plants have developed methods of moving water out of cells to concentrate sugar content and thus lower the temperature required for ice formation. Ice can form in water in spaces between cells without freeze damage to cells. Ice within cells causes damage and possible cell death.

Cold-hardy plants can survive with very dried out, *dehydrated* cells for periods of time. Very rapid drops in temperatures of several degrees per hour can even damage cold-tolerant plants because they cannot move water out of the cells fast enough to prevent freezing inside the cell.

Thus, it is important to select plants adapted to the area, which will grow well under the environmental conditions present.

Light and Day Length

The intensity of the brightness of light affects photosynthesis. Plants receiving more light produce more sugars and tend to grow and flower better. Too much direct sunlight can damage plants such as rhododendrons or hellebores that are best adapted to shade.

As days lengthen and temperatures rise in spring, plants are stimulated to start growing and flower. The shorter days of fall signal perennials, shrubs, and trees to start hardening off and preparing for dormancy and cold temperatures. This response to different periods of day and night in terms of growth and maturity of the plant is called **photoperiodism.**

Environment includes both the underground environment, in which the roots live and grow, and the aboveground environment in which the visible part of the plant exists. Figure 4–1 illustrates these two environmental systems of trees and how they relate to tree growth.

In the commercial production of plants, individuals must control the plant environment to obtain the optimum return for the investment made (Figure 4–2).

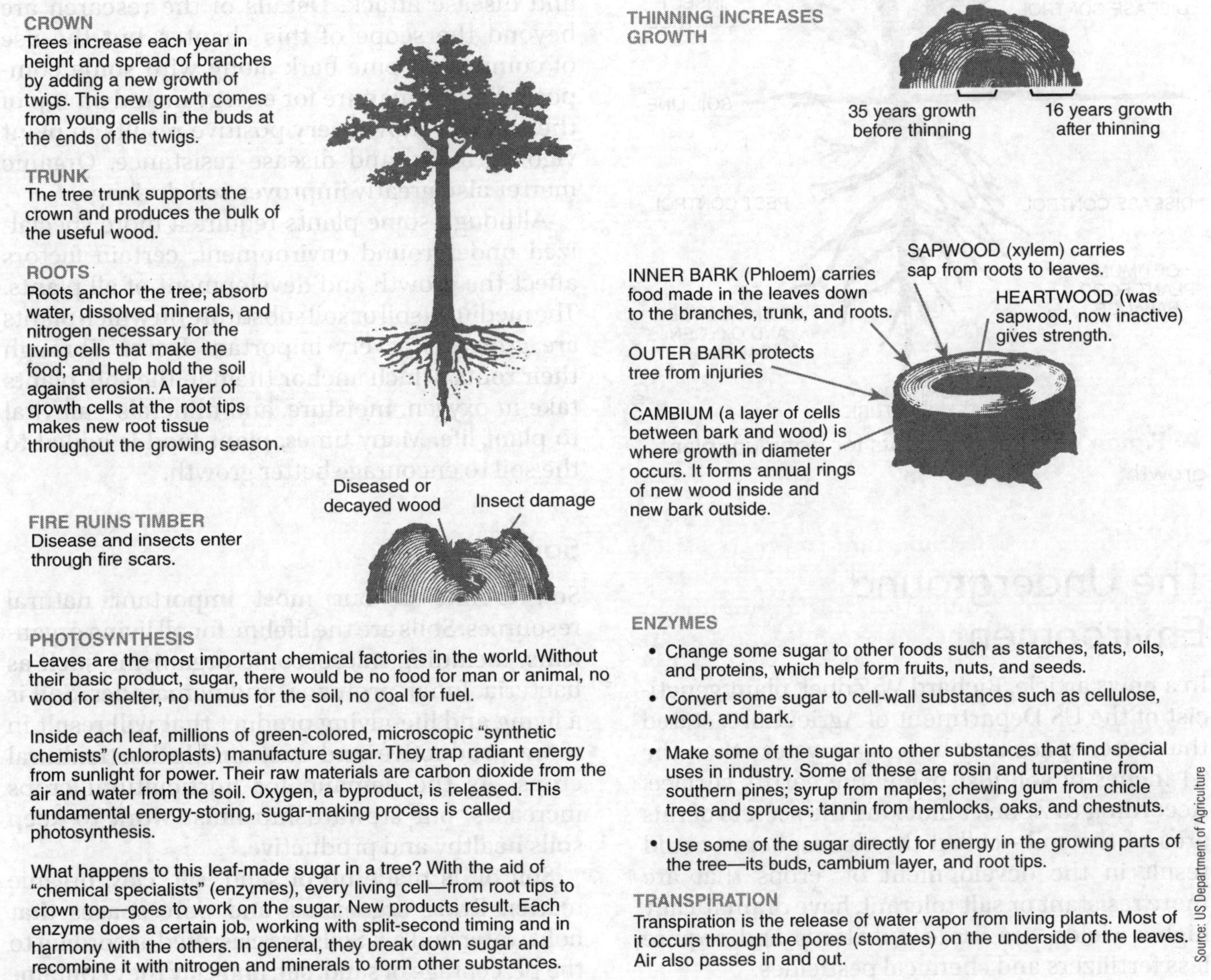

▲ **Figure 4–1** The growth of a tree.

> **TRY THIS** Using indoor horticultural grow lights, conduct several experiments to test the effect of photoperiodism on plants. The characteristics that should be altered for testing are duration of light and intensity of light (distance from the plant to the light). Record the differences for each experiment.

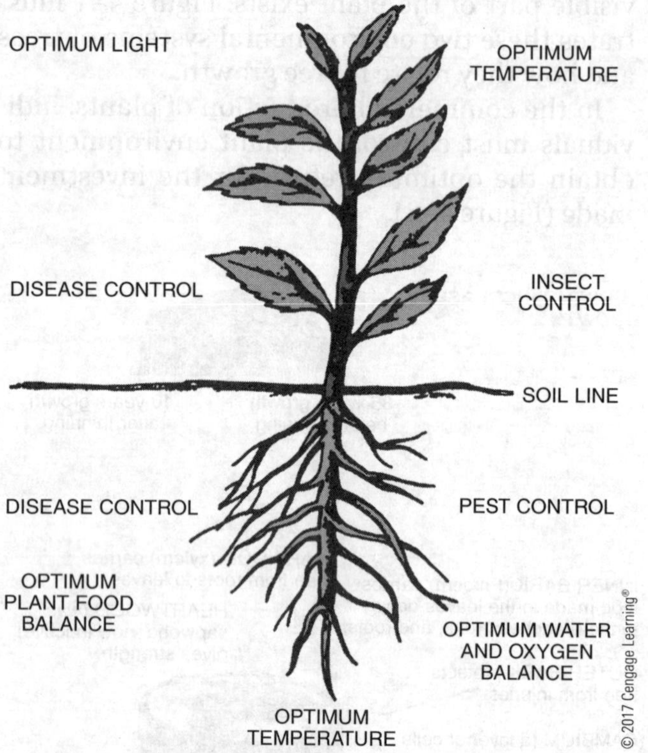

▲ **Figure 4–2** Requirements for optimum plant growth.

The Underground Environment

In a news article, Richard W. Zobel, plant geneticist of the US Department of Agriculture, stated that outer space may be less a mystery than the 24 inches of soil just below the earth's surface. According to Zobel, unlocking the secrets of this area of the soil—called the rhizosphere—could result in the development of "crops that are more resistant or salt tolerant, have dramatically higher yields, are more nutritious and require less fertilizers and chemical pesticides."

One casualty of America's agricultural revolution and the use of chemicals is the loss of native soil fungi, which allow plants to grow with less water, fewer added nutrients, and fewer pesticides. These modern practices, designed to increase production, resulted in the elimination of mycorrhizal fungi from the soil. These fungi attach to plant roots and assist them in absorbing water and nutrients more efficiently from the soil.

We have recently learned a lot about the community of both good and bad microorganisms in the soil, and we can now control some plant diseases of roots and plant foliage through proper use of compost and mulch, which favor growth of the beneficial microorganisms. This soil management technique can greatly reduce the need for chemical control of root rot diseases as well as foliar diseases. More vigorous plants in soils high in organic material are able to resist insect and disease attack. Details of the research are beyond the scope of this chapter, but the use of composted pine bark along with some composted horse manure for energy in and on top of the soil has shown very positive results in plant vigor, growth, and disease resistance. Organic matter also greatly improves soil drainage.

Although some plants require a more specialized underground environment, certain factors affect the growth and development of all plants. The medium (soil or soil substitute) in which plants are grown is a very important factor. Through their roots, which anchor them in the soil, plants take in oxygen, moisture, and minerals—all vital to plant life. Many times, plant food is added to the soil to encourage better growth.

Soil

Soil is one of our most important natural resources. Soils are the lifeline for all living organisms, including microscopic organisms such as bacteria, fungi, protozoa, and nematodes. Soil is a living and life-giving product that will result in more productive and sustainable horticultural crops. As the demand for horticultural crops increases, our stewardship must work to keep soils healthy and productive.

Soil oil is made up of sand, silt, clay, organic matter, living organisms, and pore spaces that hold water and air. Soils are classified according to the percentage of sand, silt, and clay they contain.

Soil particles vary greatly in size. A sand particle is much larger than a silt particle.

Clay particles, by far the smallest, hold moisture and plant food elements much more effectively than larger particles. A certain amount of clay in all soil is important for this reason.

Soils also differ greatly in general composition, depending on their origin. Some soils were formed as a result of rock breaking down over thousands of years; others developed as certain materials were deposited by water. A normal soil profile consists of three layers: (1) topsoil, (2) subsoil, and (3) soil bedrock or, if rock is not present, lower subsoil (Figure 4–3). Topsoil represents the depth normally plowed or tilled and contains the most organic matter or decaying plant parts. Deep-rooting plants send roots down in the subsoil, which is a well-defined layer immediately below the topsoil. If the soil is well drained, roots penetrate deeper into the subsoil because oxygen is available at greater depths. Roots may penetrate until rock, hard clay, or water prevents further growth.

The natural structure of soils is more important to the outdoor gardener, fruit grower, or nursery worker who plants outside than to the greenhouse operator or nursery worker who grows plants in containers. The worker who grows plants in containers can add ingredients to the medium to change its structure, moisture-holding ability, drainage ability, or fertility. In fact, most container plants are grown in completely soilless mixes.

An ideal soil is about 50 percent solid material (Figure 4–4) that consists mainly of minerals and a small percentage of organic matter. The other 50 percent of the ideal soil is pore space (Figure 4–5). These pores are small holes between soil particles that are filled with water and air in varying amounts. After rain or irrigation, the pores may be nearly filled with water, and the air is pushed out. As the soil dries, the amount of water decreases, and the pores gradually fill with air again. The ideal water-to-air ratio in the pores is about half and half—50 percent air to 50 percent water.

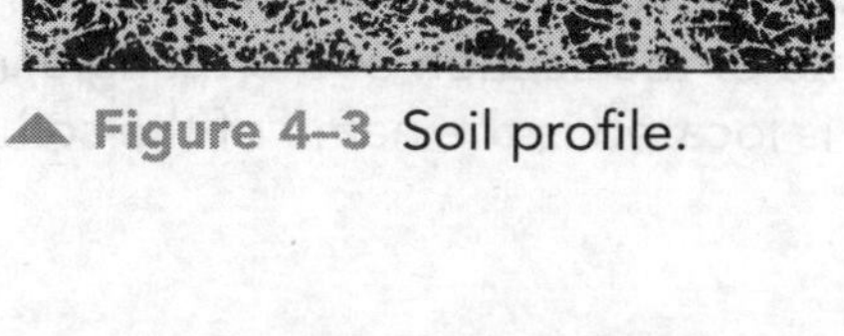

▲ **Figure 4–3** Soil profile.

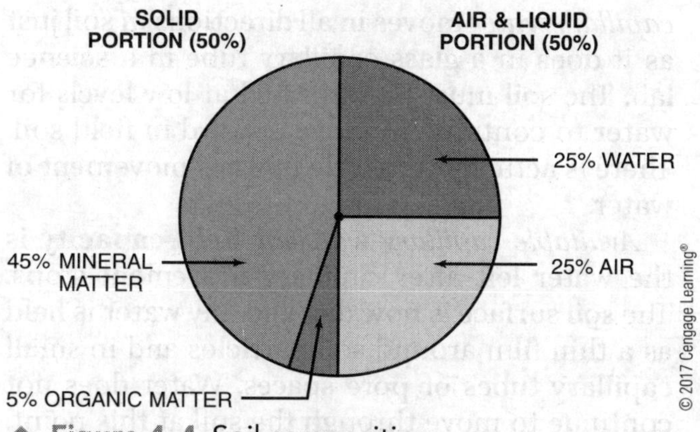

▲ **Figure 4–4** Soil composition.

The amount of moisture and air a soil will hold depends on the soil structure and type of soil. Sandy soils with large particles also have large pore spaces. Water is lost more quickly from these large pores as the force of gravity drains the water out. These are called well-drained soils. As the clay content of the soil increases, more water is held. If soils contain too much clay, they may not drain well enough to allow enough oxygen in the pore space for good plant growth.

TYPES OF WATER IN SOIL **Gravitational water** is water that the soil is unable to hold against the force of gravity. It becomes a part of groundwater or drains away in streams. Gravitational water is of little value to the plant because it drains away quickly, carrying soluble plant food elements with it. Soils with larger pore spaces lose water faster to the force of gravity than do soils with smaller pore spaces.

Capillary water is held against the force of gravity in the small pore spaces of the soil as a thin film of water around soil particles. There are three types of capillary water. *Free moving*

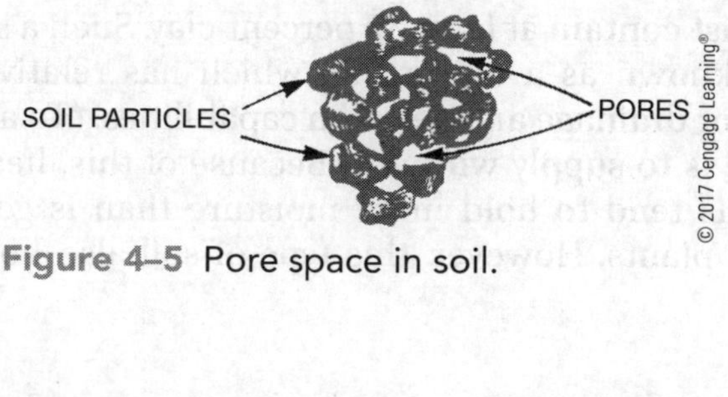

▲ **Figure 4–5** Pore space in soil.

capillary water moves in all directions in soil just as it does in a glass capillary tube in a science lab. The soil must be saturated at low levels for water to continue to move upward in field soil. There is actually very little upward movement of water.

Available capillary water or **field capacity** is the water left after capillary movement stops. The soil surface is now dry, and any water is held as a thin film around soil particles and in small capillary tubes or pore spaces. Water does not continue to move through the soil at this point. Plant roots must continue to move through the soil in the search for water as the soil immediately surrounding them is dried out from root absorption of water. Roots will not continue to grow into air-dry soil where no moisture is available. When the soil is moist with available water, they will grow again. The field capacity for water is high (water is held) in heavy soils, with clay particles providing a lot of surface area and tiny pore spaces for the water to cling. This is the most important water for plants, and they can extract and use about half of this water. As the film of water on soil particles gets thin, more force is required to absorb it. When this force becomes about 15 atmospheres or about 200 pounds per square inch (psi), most plants can no longer absorb it. At this moisture level, plants will wilt and die. Desert plants can extract water at a force greater than 200 psi.

Hydroscopic water, or *unavailable capillary water*, is not available to plants because it is held tightly as a molecular film around the soil particles and can only be moved as vapor. Soil this dry is called *air dry* because under normal drying conditions, no more moisture will move.

SANDY SOIL *Sandy* (or *light*) soils include soils in which silt and clay make up less than 20 percent of the material by weight. These soils drain well but have little capacity to hold moisture and plant food.

CLAY SOIL To be classified as a clay soil, a soil must contain at least 30 percent clay. Such a soil is known as a heavy soil, which has relatively poor drainage and aeration capabilities. (To aerate is to supply with air.) Because of this, heavy soils tend to hold more moisture than is good for plants. However, this type of soil also holds fertilizer and plant food well, which can be beneficial to plant growth.

LOAMY SOIL This is the most desirable soil for general use. *Loam* is a mixture of approximately equal parts of sand, silt, and clay. If the loam has more sand than silt or clay, it is known as a *sandy loam;* more clay, a *clay loam;* and more silt, a *silty loam*. The **texture triangle** shown in Figure 4–6 is helpful in determining the names of soils.

Soil Improvement

Soils used for outdoor plant growth may be improved through increased drainage, irrigation methods, and the addition of organic matter and plant food in the form of fertilizers.

Because it would be very expensive to change the percentage of sand, silt, or clay in a soil to

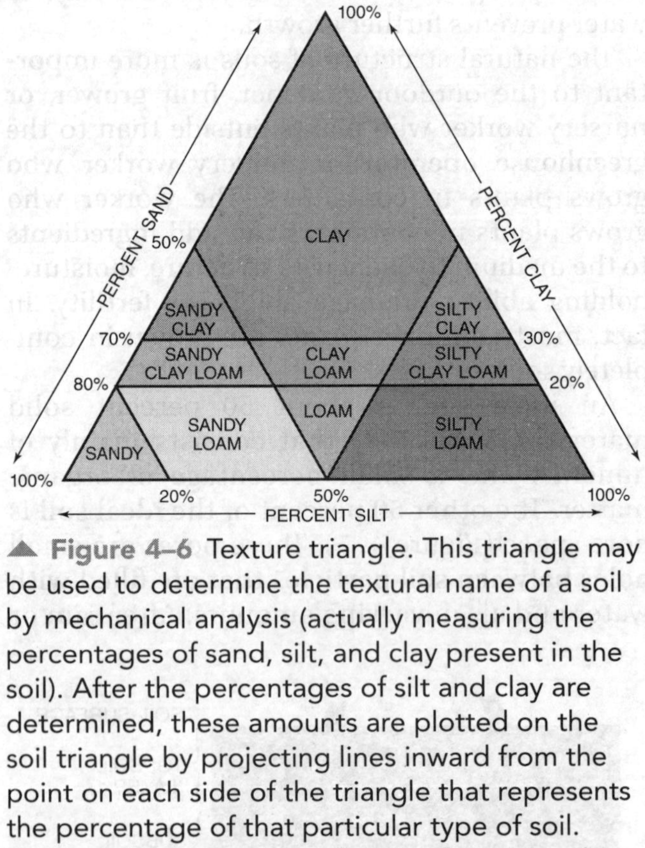

▲ **Figure 4–6** Texture triangle. This triangle may be used to determine the textural name of a soil by mechanical analysis (actually measuring the percentages of sand, silt, and clay present in the soil). After the percentages of silt and clay are determined, these amounts are plotted on the soil triangle by projecting lines inward from the point on each side of the triangle that represents the percentage of that particular type of soil. The line drawn from the silt side of the triangle is placed parallel to the sand side of the triangle. The line projected from the clay line runs parallel to the silt line. The location of the point at which these two lines intersect indicates the name of the soil. The name of the section of the triangle in which the point is located is the name of the soil.

improve the soil's drainage, aeration, or moisture-holding capacity, other practices must be used.

DRAINAGE AND AERATION Drainage and aeration can be improved by changing the soil structure. One way to change the soil structure is by adding organic matter to the soil. Organic matter is anything that was once alive and is now in or on the soil. Organic matter must decompose into **humus**, which refers to organic material that has been decomposed by microbes. The breakdown of soil organic matter is an aerobic process, meaning it needs oxygen. When the soil is tilled, aeration (adding air to the soil) occurs and speeds up the breakdown process. Therefore, less soil erosion occurs, and more nutrients are readily available to the plants for optimum growth. Maintaining organic matter content in the soil also sustains many beneficial life forms in the soil, such as earthworms, bacteria, actinomycetes, fungi, and nematodes.

Organic matter improves soil structure through **aggregation**, the clinging together of soil particles in large crumblike shapes. Lime and gypsum (calcium sulfate) also aid in soil aggregation and improve structure in some cases. In effect, aggregation increases the size of soil particles by forming larger spaces between particles.

Cover crops are commonly planted before planting trees to help improve the soil. For example, sustainable nurseries often plant **forage radish** to improve the topsoil conditions of the soil structure. Forage radishes take up nitrogen, phosphorous, and potassium (N-P-K) from both the topsoil and the subsoil layer, storing the nutrients in their shoots and root biomass. Forage radishes grow rapidly when planted in late summer or early fall and provide full leafy cover in about three weeks. The foliage of the forage radish catches the raindrops, minimizing soil-surface impact by controlling the loss of soil particles. The foliage also helps to control weed growth in the nursery, thereby reducing the use of herbicides. Even after radishes are killed by a hard freeze, a layer of (pungent) decomposing residue remains on the soil surface throughout the winter and into the early spring, providing erosion control. In addition, runoff and sediment transport are reduced due to the rapid infiltration facilitated by open root holes. These holes can capture runoff and sediment before it leaves the field. For more complete protection against erosion and to control water runoff, forage radish may be sown with a companion crop such as crimson clover.

Another drainage method uses tile drains to remove water from the soil. Raising planting beds and placing ditches between the beds are also methods used to control moisture in soil.

MOISTURE RETENTION Often, sandy soils are not able to hold sufficient water for plants; the large particles and pore spaces encourage too much drainage. Adding organic matter improves the water-holding capacity of soils. Organic material holds many times its own weight in water; for example, peat moss holds 15 times its weight in water. Organic matter also holds plant food effectively and allows the slow release of the food for plant use as the organic matter decomposes in the soil. Animal manure, green manure, peat moss, and sawdust are good sources of organic matter. (*Green manure* is a green herbaceous crop plowed under to improve soil.) *Mulches,* such as compost, wood chips, or bark, are placed on the surface of the soil to help retain soil moisture by reducing runoff, thereby allowing more rainwater to be absorbed into the soil. Mulches also keep soil cool, thus reducing evaporation loss, and add organic matter to the soil as they decompose. Irrigation methods are used to add water when rains do not supply enough.

Fertilizers should be used when necessary and according to recommendations of a soil test. Obtain directions on how to take a soil sample from a local extension agent or horticulture instructor. Send samples to the local soil test laboratory, asking for recommendations for the crop being grown.

Beyond these practices, the grower using field or nursery areas to cultivate crops has little control over the soil structure or its moisture content.

Nutritional Deficiencies

Nutritional or plant food deficiencies often appear on the leaves of plants. Yellow or pale green leaves indicate a nitrogen deficiency. A phosphorus deficiency appears as a purple color on the underside of the leaf. By the time these symptoms appear, damage has already been done to the plant. A soil test should have been used earlier to determine plant needs.

Soil Pests

Soils harbor certain diseases such as root rot and wilt. Nematodes, which are tiny animals, and insects may also damage roots. Resistant varieties of crops or natural controls such as crop rotation and compost must be used to satisfactorily control these problems. Chemicals are sometimes used as a last resort to protect plants against diseases and insects.

Soil pasteurization is needed any time soil or sand is used in a planting medium for container growing. The soil or sand should be heated thoroughly to 180°F for 30 minutes before being mixed with other materials.

Planting Media Mixes

More and more growers are using planting mixes that contain little, if any, soil. In greenhouse operations and nurseries where plants are grown in flats, pots, or other containers, this is both convenient and economical.

ADVANTAGES There are several advantages to using soilless media.

- The mix is uniform; that is, it does not vary in fertility, acidity (pH), or texture.
- The mixes are *sterile* (containing no disease organisms, insects, or weed seeds).
- **Soilless mixes** weigh less, so they are easier to handle and ship.
- Good moisture retention and drainage are possible through the proper combination of ingredients.

DISADVANTAGES Soilless mixes have some disadvantages.

- Because they are very lightweight, containers kept outdoors may be blown over.
- Because the mineral content in most mixes is low, minor plant food elements necessary for plant life may be missing, such as iron, sulfur, manganese, zinc, and calcium.
- Plants transplanted from mixes to soils often hesitate to extend roots into such a different growing medium. A problem develops when the growing medium remaining on the roots fails to blend with the soil into which the plant is transplanted. Moisture and other necessary nutrients are then unable to pass from the new soil to the roots. In addition, if the soil is heavy in clay, the plant's roots may never venture beyond the original root ball into the clayey soil. This situation may result in the death of the plant from lack of moisture. (The rhododendron and azalea, which are often grown in pinebark, peat moss, or sphagnum moss, are good examples of plants with this transplanting problem.) The problem may be solved by extending the root ends out into the soil when the plant is transplanted. Mixing sphagnum moss or pine bark into the soil at the permanent planting site also makes the transition easier.

CONTENT OF MIXES Soilless mixes contain various combinations of the following materials.

Perlite, a gray-white material of volcanic origin, is usually used to improve aeration of the media. Horticultural-grade perlite should be used because it has larger particles and provides better drainage and aeration than the finer builder's perlite. Sponge "ROK," an aluminum silicate, is used in the same manner as perlite.

Sphagnum moss is the dehydrated remains of acid bog plants, used in shredded form for seeds. Most horticulturists use shredded sphagnum moss to cover seeds because it is relatively sterile and lightweight, controls disease well, and has an excellent moisture-holding capacity. It is very acidic.

Peat moss is partially decomposed vegetation that has been preserved underwater. The peat is collected from marshes, bogs, or swamps. It has a very high moisture-holding capacity, contains about 1 percent nitrogen, and is low in phosphorus and potassium.

Vermiculite is a very light, expanded material with a neutral pH (neither acidic nor alkaline). (When vermiculite is heated, the moisture in the mineral becomes steam and causes it to expand.) It has a very high moisture-holding capacity.

Limestone refers to ground natural limestone, also known as calcium carbonate ($CaCO_3$). It tends to raise pH.

© Kaspars Porins/Shutterstock.com

▲ **Figure 4–7** Grinding tree branches to create sustainable compost.

Tree bark is usually either pine bark (conifer trees), which will influence the soil media to be more acidic, or hardwood bark (deciduous trees), which will influence the soil media to be more alkaline when broken into small pieces and used in planting mixes. Fine pine bark (1/4 inch or less in size) is now used extensively in container growing.

Slow-release fertilizers contain plant food that is gradually made available to plants over a period of time.

Compost is the result of composting—the natural decomposition of organic material such as yard waste of leaves, grass clippings, straw, and ground tree branches (see Figure 4–7). Mixing a variety of organic materials together, turning them to introduce oxygen, and adding moisture helps the materials decompose to become compost. Compost is a sustainable organic nutrient resource that supplies nitrogen (N), phosphorous (P), potassium (K), and organic matter to the soil tilth (physical condition). This is an excellent resource that can be incorporated into containers at the rate of 15 percent of the total volume of the container mix, as well as into landscape beds and in nursery field production. Compost may be applied on lawns at the rate of 2 to 3 inches over the soil surface and be tilled into the soil.

GREEN TIP
Compost fosters the use of organic material to enhance the stewardship of the productive nutrients for tomorrow's greener environment.

Biochar is a sustainable biomass that has been charred and is biologically used for the enhancement of soils and growing media. It is lightweight and porous, with efficient cation exchange capacity to increase the soil's water and **nutrient holding availability**. It also provides the food source that the microbes need to increase microbial activity to enhance plant growth. Biochar can make up 5 to 10 percent of a potting medium.

Mixes may contain two or more of the preceding ingredients. For plants that require excellent drainage, coarse materials, such as bark or perlite, should make up a high percentage of the mix. Most mixes contain an organic material such as sphagnum or peat moss for moisture retention and a material such as perlite or bark for drainage and aeration. One of the most popular commercial mixes is 50 percent shredded sphagnum moss and 50 percent vermiculite, with a slow-release plant food added. Mixes are sold bagged and ready for use.

Selecting the best soilless mix is important for growing plants in pots. Different plants have different requirements, and many special soilless mixes designed for specific plants are available. All mixes must provide a good supply of oxygen to the roots. A planting **medium** in a pot does not drain as well as outdoor soil because of the shallow column of soil in pots. Potting mixes must therefore be more coarse or open to ensure good drainage. A medium of 100 percent soil, unless it is nearly all sand, does not drain well enough to be used as a container medium. It is therefore necessary to purchase a potting medium recommended for the specific plant you are planting. For example, orchids require a very open, airy, well-draining mix composed of chunks of fir bark, coarse sphagnum peat moss, redwood or tree fern fiber, perlite, and charcoal. The African Violet Society recommends a mix of equal parts sphagnum peat moss, vermiculite, and perlite. For cactus, builder's sand and/or perlite should make up half of the mix; the other half can be a regular potting mix. The pH or acidity of the mix is also important because some plants require different pH levels to grow well.

It is much easier to purchase a prepared potting mix recommended for specific plants than to attempt to mix your own. Potting mixes are available to meet the needs of most plants.

Soilless mixes may also be used as soil conditioners by digging them into the soil in varying amounts as needed. The mixes add organic matter, which increases moisture-holding capacity and improves soil drainage.

If soil is used as part of the planting mix, be sure it is pasteurized. A mixture of one-third soil; one-third peat, shaving, or leaf mold; and one-third sand is suitable for most plants.

TRY THIS Construct a composting unit using wooden shipping pallets. Using four pallets, create a square holding area for the material to be composted. Turn the compost pile every three days until it is decomposed enough to be incorporated into the soil or growing medium.

Plant Food and Fertilizers

Plants may be grown in soil, in soilless organic substrate, or in a combination of the two. In any case, plant food must be readily available to the plant. Fertilizer must be added as the plant requires food, and the plant food elements must be in a water-soluble form to be available to the plant. Before any additional plant food is applied to the soil, to soilless organic substrate, or to any other growing medium, a soil analysis should be completed to determine the level of plant nutrients in the medium.

WATER Water is the most important plant food element by far. It makes up approximately 90 percent of the weight of plants and is the one most limiting factor in plant growth. All plant food elements are dissolved in water and move into and throughout the plant in a water solution. As noted earlier, photosynthesis uses water in the manufacture of food and could not occur without it.

Only about 1 percent of the water absorbed is actually used by the plant. The other 99 percent is lost through the leaves and stems as water vapor in a process called transpiration. From 90 to 95 percent of the water lost from the plant is lost by transpiration through stomata in the leaves. Leaf temperature affects transpiration most. As leaf temperature rises, more water is lost. For each 10°C increase in temperature, the loss is doubled. A single corn plant can transpire 2 quarts of water per day. If it were not for this loss of water through transpiration, the water needs of plants would be greatly reduced. Loss of water from transpiration faster than the roots can pick it up results in wilting and plant death. Plants are unable to stop transpiration to conserve water, but they can slow it down.

Transpiration of water is high when soils are wet, and the turgid guard cells of the stomata open wide to allow more water vapor and gas to escape. When water in the soil and plant is in short supply, the stomata close, and less water vapor is lost. There are an estimated 250,000 stomata in 1 square inch of the undersurface of an apple leaf. Stomata close at night, so transpiration is greatly reduced. They open in response to light.

Transpiration is of little value to the plant. It cools the leaf slightly, but air movement and heat radiation are most important in leaf cooling. Transpiration is not needed for the absorption or movement of minerals in the plant. Perhaps its main value is in cooling and moistening the air around the plant through evaporation. This natural air conditioning is beneficial to plants and animals.

Plant food may be divided into three groups: (1) *major elements*—(**macroelements**) nitrogen, phosphorus, and potassium (stated as *total nitrogen, available phosphoric acid*, and *soluble potash* when listed as fertilizer content); (2) *secondary nutrients*—calcium, magnesium, and sulfur; and (3) *minor nutrients*—(**microelements**) boron, copper, chlorine, iron, manganese, molybdenum, zinc, and possibly nickel. The elements are listed as major, secondary, and minor according to the relative amounts of each element needed for good plant growth. Plants require relatively large amounts of the major elements and relatively smaller amounts of the secondary and minor elements. Three elements—carbon, hydrogen, and oxygen—come from the air.

Commercial fertilizers and plant foods show the percentage or pounds per hundred weight of the three major elements in large numbers on the bag or container. This is the fertilizer analysis. If the container has the numbers 5-10-5 on the label, the mix is 5 percent elemental nitrogen, 10 percent available phosphoric acid, and 5 percent water-soluble potash. The other 80 percent is filler material that makes it easier to spread the plant food evenly. The three main plant food elements, nitrogen, phosphorus, and potassium, are always listed in the same order. It is easy to see how much actual plant food is being purchased on a percentage basis by reading the label (Figure 4–8).

Minor elements may also be listed on the container. In parts of the country that have acidic soils, the reaction of acid content to the soil may be expressed in a statement such as "Acid equivalent to ___ pounds of limestone."

Just as human beings require a balanced diet, plants need a balance of food for best growth. Only a soil test can determine the amount of various plant food elements needed. If shortages exist, the plant will often show symptoms of deficiencies.

NITROGEN Nitrogen is generally purchased in one of four forms. It is absorbed by plants in two forms—ammonium and nitrate.

▲ Figure 4–8 The fertilizer bag shown on the left is sold as garden fertilizer. The analysis, in the bottom right-hand corner (10-10-10), indicates that the mix is 10 percent nitrogen, 10 percent available phosphoric acid, and 10 percent soluble potash, with a total of 30 percent actual plant food. The other 70 percent is filler, which aids in spreading fertilizer. The bag on the right, a lawn fertilizer, has a high ratio of nitrogen compared with the other elements. Notice that the fertilizer on the left contains the same percentage of all three elements. Grasses require more nitrogen than phosphorus or potash; this accounts for the higher ratio of nitrogen in the mix on the right. The application of fertilizer must be applied at the rate recommended by the soil test analysis.

- *Nitrate of soda*, $NaNO_3$, is highly soluble, is quickly available, tends to reduce soil acidity, and contains 16 percent nitrogen.
- *Ammonium nitrate*, NH_4NO_3, is not as soluble, is available over a longer period of time, and contains 33 percent nitrogen.
- *Ammonium sulfate*, $(NH_4)_2SO_4$, becomes available more slowly, leaves the soil more acidic, is good for plants that grow well in very acidic soil, and is 21 percent nitrogen.
- *Urea formaldehyde* is an organic form of nitrogen, is more slowly available than the inorganic forms, and contains 38 percent nitrogen.

Of the three major plant food elements, nitrogen has the most noticeable effect on plants, with the effects showing soonest. Nitrogen encourages aboveground vegetative growth and gives a dark green color to leaves. It tends to produce soft, tender growth, which is a good quality for crops such as lettuce to possess. The tender growth makes the plant better tasting. Nitrogen also seems to regulate the use of the other major elements.

Because the addition of nitrogen quickly produces a visible effect, there is often a tendency to overuse it. Too much importance is often placed on this element without regard for a balanced plant food program.

Too much nitrogen may (1) lower the plant's resistance to disease; (2) weaken the stem because of long soft growth; (3) lower the quality of fruits, causing them to be too soft to ship; and (4) delay maturity or hardness of tissue and thus increase winter damage to plants. Not enough nitrogen results in a plant being (1) yellow or light green in color and (2) stunted in root and top growth.

Nitrogen is lost from the soil very easily by *leaching* (washing out). It is very soluble in water and is not held by the soil particles because of the charges of the particles involved. Soil particles have a negative charge; nitrogen also has a negative charge. The ammonium form of nitrogen has a positive charge and is held by negatively charged soil particles. Nitrate nitrogen is leached quickly because of its negative charge. However, organic matter does tend to hold insoluble nitrogen that is released slowly into the soil. Nitrogen should not be used in excess for two reasons: (1) it is quickly lost from the soil through leaching, especially in sandy soils that lose water faster, and (2) it can damage plants if applied in too great an amount. Some plants such as legumes (beans, peas, etc.) manufacture their own nitrogen.

PHOSPHORUS Phosphorus is generally purchased in the following forms:

- Superphosphate: 20 percent phosphate
- Treble superphosphate: 46 percent phosphate
- Rock phosphate: 25–35 percent phosphate
- Ammonium phosphate: 48 percent phosphate

Phosphorus is present to some extent in all soils. Unlike nitrogen, it is held tightly by soil particles and therefore is not easily leached from soil. However, because it may not be in a water-soluble form, it may not be available to plants in the amount needed. This means that additional phosphate fertilizers should be applied as determined by the use of a soil test. In the past five years, 16 US states have enacted fertilizer laws that regulate the use of phosphorus applications.

Phosphorus has positive effects on plants in several ways:

- encourages plant cell division
- enables flowers and seeds to form
- hastens maturity, thereby offsetting the quick growth caused by nitrogen
- encourages root growth and the development of strong root systems
- makes potash (potassium) more easily available
- increases the plant's resistance to disease
- improves the quality of grain, root, and fruit crops

Because phosphorus is held very tightly by soil particles, it does not usually cause damage to field-grown plants if excessive amounts are applied. However, container-grown plants can be damaged by an excess of any soluble fertilizer element because it increases the soluble salt content present in the media. Fertilizers that are high in soluble salts *dehydrate* (dry out) plant roots by pulling water from the roots.

Insufficient phosphorus results in (1) purple coloring on the undersurface of leaves; (2) reduced flower, fruit, and seed production; (3) susceptibility to cold injury; (4) susceptibility to plant diseases; and (5) poor quality fruits and seeds.

POTASSIUM The most common sources of potassium are the following:

- muriate of potash: 60 percent potash
- sulfate of potash: 49 percent potash
- nitrate of potash: 44 percent potash (also 13 percent nitrogen)

Potassium is rarely present in the soil in sufficient amounts to harm plants. It tends to modify both the fast, soft growth of nitrogen and the early maturity of phosphorus.

The presence of potassium is essential for several reasons:

- increases the plant's resistance to disease
- encourages a strong, healthy root system
- enables starch formation
- enables the development of chlorophyll
- enables tuber development
- encourages the efficient use of carbon dioxide

Because it is a major element, potash is generally added to soil. The amount is determined by soil tests.

Potassium deficiency appears as a marginal yellowing or scorch on the edges of leaves on the lower portion of the plant. This symptom is easily mistaken for moisture shortage during dry soil conditions.

LIME ($CACO_3$) Lime acts as a plant food and as a material that affects soil acidity. Soil acidity, in turn, affects the availability of other plant food elements. Lime furnishes calcium, one of the most important of the secondary food elements. Calcium is important in the formation of plant cell walls, among other functions.

Organic Fertilizers

Organic fertilizers are derived from the remains of or byproducts of organisms and have been in use since the beginning of time. Organic fertilizers depend on the microorganisms found in soil to break them down and release their essential nutrients.

These organic nutrients are resources of manure from cattle, sheep, horses, and poultry. Many byproducts of the cattle, sheep, horses, swine, and poultry such as bone meal and blood meal are excellent resources as organic fertilizer. Other sources for organic fertilizer include magnesium sulfate (Epsom salt), sulfate of potash, potassium magnesium sulfate, sodium nitrate, fish extract, seaweed, manure, guano, composts, and green manure crops. Green manure crops are crops that are planted—such as rye, wheat, soybeans, and clover—and then plowed under as a resource as green manures. Legume crops are an additional resource of nitrogen for plant growth.

Plants absorb their nutrients in the water-soluble form; therefore, the organic resources are available for the plant roots to obtain nutrients.

GREEN TIP Enhance your soil by using green manure. Legumes, such as peas, soybeans, alfalfa, clover, and bird's-foot trefoil, may be grown in the summer as preparation for your next summer season crop. The green manure plants add nitrogen along with organic matter to enrich the soil.

The use of organic fertilizers is less likely to burn the plant roots because it is naturally occurring. Organic fertilizers also enhance the environment by improving the soil structure, increasing the water-holding capacity, and releasing compatible and sustainable nutrients. The nutrients are released to the plant roots to allow the active microorganisms (such as bacteria, fungi, actinomycetes, and protozoa) to break down the nutrients into a soluble form for the plants to absorb as a food source. The mineral resources of N-P-K are vital elements to help build a healthy soil that encourages plant growth (see Figure 4–9).

Soil Acidity (pH)

Most plants grow best in soil with a pH of from 5.6 to 7.0. A pH of 7.0 is neutral; that is, the soil at pH 7.0 is neither acid nor alkaline (basic). *Alkaline* soil is the opposite of acid soil in pH rating. Hence, on a scale of 1 to 14, values lower than 7.0 indicate acidic soils, and values higher than 7.0 indicate alkaline soils (Figure 4–10). Figure 4–11 lists the pH preferences of common flowers, ornamentals, vegetables, and small fruits.

Courtesy of Lebanon Seaboard Corporation

▲ **Figure 4–9** Organic fertilizer resources for a sustainable landscape.

GREEN TIP Improve environmental stewardship of our soil to insure successful plant growth through the soil food chain. Go green!

In the United States, soils tend to be acid in areas where the parent material of the soil was acid and where the amount of rainfall exceeds evaporation of moisture from the soil. This is due to water draining through the soil and washing out the salts of sodium and calcium. This condition occurs in the eastern United States and along the West Coast down to about central California.

In areas where water evaporation from the soils is equal to or greater than the amount of

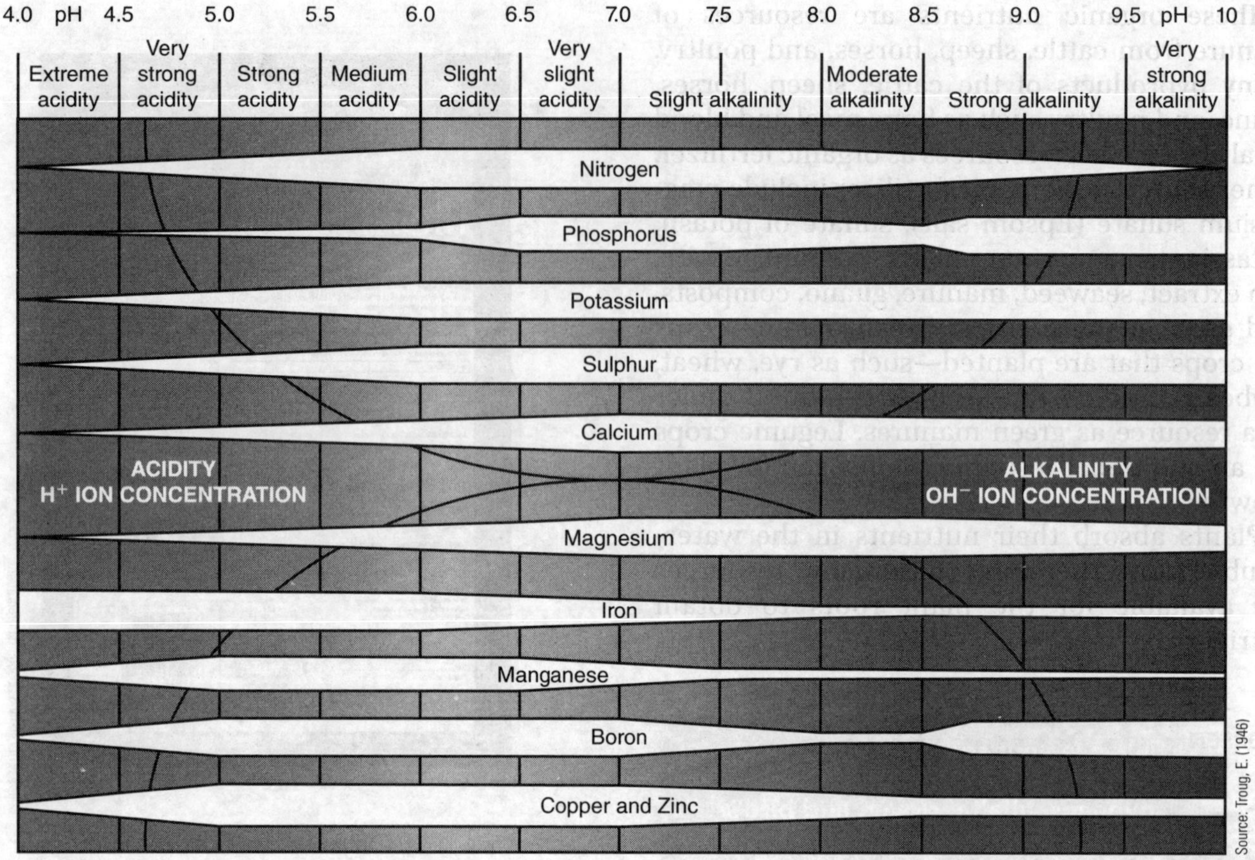

▲ **Figure 4–10** The effects of pH on plant availability.

rainfall, the salts of calcium and sodium tend to build up in the soils and increase the pH level. These salts may build up enough and thereby raise the pH so much that plants cannot grow in these soils. This occurs in many areas west of the Mississippi River, some to the point of being too alkaline to grow crops. Irrigation with low-salt-content water can wash the alkali-producing elements out of some of these soils and greatly improve growing conditions.

LOWERING pH IN ALKALI SOILS When it is necessary to lower soil pH for the best growth of plants, materials such as sulfur, iron sulfate, or aluminum sulfate may be used. This practice may be needed in alkali soil areas of the western United States. As mentioned, flushing these soils with low-salt irrigation water will also lower the pH.

LIME, pH, AND OTHER PLANT FOOD ELEMENTS Lime serves a very important function in changing soil acidity or pH. When a soil test is performed, and the soil proves too acidic, lime is added to raise the pH.

Lime also affects the availability of other plant food elements to plants. For example, if a soil is acidic with a low pH (5.5 to 6.5), phosphorus is tied up and not readily available. Adding lime releases phosphorus and makes it available to the plants. As another example, consider the fact that acidic soils release iron and aluminum into the soil. This poses a problem because aluminum may be toxic or poisonous to some plants. The application of lime decreases the availability of aluminum and iron. On the other hand, liming crops such as blueberries, azaleas, and rhododendron may cause iron deficiency if the pH is raised above 6.0.

Liming to produce the proper pH also activates soil organisms and encourages the release of plant food. Soil structure is usually improved with the addition of calcium in the form of lime or gypsum. Gypsum ($CaSO_4$) does not change soil pH, and it is most effective in improving soil structure in sodic soils (high in sodium) because it replaces and removes sodium from the soil.

FLOWERS AND ORNAMENTALS							
African violet	5.5–7.5	Chrysanthemum	6.0–7.5	Gladiolus	6.0–7.0	Mock orange	6.0–8.0
Ageratum, blue	6.0–7.5	Clematis	5.5–7.0	Gloxinia	5.5–6.5	Morning glory	6.0–7.5
Alyssum	6.0–7.5	Clivia	5.5–6.5	Heather	4.5–6.0	Narcissus	6.0–7.0
Amaryllis	5.5–6.5	Coleus	6.0–7.0	Hens and chickens	6.0–8.0	Nasturtium	5.5–7.5
Anthurium	5.0–6.0	Columbine	5.5–7.0	Hibiscus, Chinese	6.0–8.0	Pansy	5.5–6.5
Arbor vitae	6.0–7.5	Coralbells	6.0–7.0	Holly, English	4.0–5.5	Peony	6.0–7.5
Arbutus, trailing	4.0–5.0	Cosmos	5.0–8.0	American	5.0–6.0	Petunia	6.0–7.5
Aster	6.5–7.0	Coxcomb	6.0–7.5	Hollyhock	6.0–8.0	Philodendron	5.0–6.0
Azalea	4.5–6.0	Crab, flowering	5.0–6.5	Honeysuckle bush	6.5–8.0	Phlox	5.0–6.0
Baby's breath		Crocus	6.0–8.0	Hoya	5.0–6.0	Pink	6.0–8.0
(gypsophila)	6.0–7.5	Cyclamen	6.0–7.0	Hydrangea, blue	4.0–5.0	Poinsettia	6.0–7.5
Baby's tears	5.0–5.5	Daffodil	6.0–6.5	pink	6.0–7.0	Poppy	6.0–7.5
Bachelor's button	6.0–7.5	Dahlia	6.0–7.5	white	6.0–8.0	Portulaca	5.5–7.5
Barberry	6.0–7.5	Daisy, shasta	5.0–6.0	Impatiens	5.5–6.5	Primrose	5.5–6.5
Bayberry	5.0–6.0	Daphne	6.5–7.5	Iris	5.5–7.5	Rhododendron	4.5–6.0
Begonia	5.5–7.0	Delphinium	6.0–7.5	Iris, Japanese	5.5–6.5	Rose	6.0–7.0
Bird of paradise	6.0–6.5	Deutzia	6.0–7.5	Ivy, Boston	6.0–8.0	Rose of Sharon	
Bleeding heart	6.0–7.5	Dieffenbachia	5.0–6.0	English	6.0–8.0	(althea)	6.0–7.5
Bloodleaf	5.5–6.5	Dogwood	5.0–6.0	Grape	5.0–6.5	Rubber plant	5.0–6.0
Bluebell	6.0–7.5	Eucalyptus	6.0–8.0	Jasmine	5.5–7.0	Salvia	6.0–7.5
Boxwood	6.0–7.5	Fern, asparagus	6.0–8.0	Lady's slipper	5.0–6.0	Scilla	6.0–8.0
Bridal wreath	6.0–8.0	bird's nest	5.0–5.5	Larkspur	6.0–7.5	Shrimp plant	5.5–6.5
Burning bush	5.5–7.5	Boston	5.5–6.5	Laurel	4.5–6.0	Snapdragon	6.0–7.5
Butterfly bush	6.0–7.5	maidenhair	6.0–8.0	Lilac	6.0–7.5	Sweet pea	6.0–7.5
Cacti	4.5–6.0	Forget-me-not	6.0–8.0	Lily, calla	6.0–7.0	Verbena	6.0–8.0
Cactus, Christmas	5.0–6.5	Forsythia	6.0–8.0	day	6.0–8.0	Wisteria	6.0–8.0
Camellia	4.5–5.5	Foxglove	6.0–7.5	Easter	6.0–7.0	Yew, Japanese	6.0–7.0
Campanula	5.5–6.5	Freesia	6.0–7.5	madonna	6.5–7.5	Yucca	6.0–8.0
Candytuft	5.5–7.0	Fuchsia	5.5–6.5	tiger	6.0–7.0	Zinnia	5.5–7.5
Canna	6.0–8.0	Gardenia	5.0–6.0	Lily of the valley	4.5–6.0		
Carnation	6.0–7.5	Geranium	6.0–8.0	Marigold	5.5–7.0		
VEGETABLES							
Artichoke		Cauliflower	5.5–7.5	Lettuce	6.0–7.0	Radish	6.0–7.0
(Jerusalem)	6.5–7.5	Celery	5.8–7.0	Mushroom	6.5–7.5	Rhubarb	5.5–7.0
Asparagus	6.0–8.0	Chicory	5.0–6.5	Mustard	6.0–7.5	Rutabaga	5.5–7.0
Beans, lima	6.0–7.0	Chives	6.0–7.0	Okra	6.0–7.5	Sage	5.5–6.5
pole	6.0–7.5	Corn	5.5–7.5	Onion	6.0–7.0	Salsify	6.0–7.5
Beets, sugar	6.5–8.0	Cucumber	5.5–7.0	Parsley	5.0–7.0	Shallot	5.5–7.0
table	6.0–7.5	Eggplant	5.5–6.5	Parsnip	5.5–7.0	Spinach	6.0–7.5
Broccoli	6.0–7.0	Endive	5.8–7.0	Peas	6.0–7.5	Squash, crookneck	6.0–7.5
Brussels sprouts	6.0–7.5	Garlic	5.5–8.0	Peanut	5.3–6.6	Swiss chard	6.0–7.5
Cabbage	6.0–7.5	Horseradish	6.0–7.0	Pepper	5.5–7.0	Tomato	5.5–7.5
Cantaloupe		Kale	6.0–7.5	Potato	4.8–6.5	Turnip	5.5–6.8
(muskmelon)	6.0–7.5	Kohlrabi	6.0–7.5	Sweet potato	5.2–6.0	Watercress	6.0–8.0
Carrot	5.5–7.0	Leek	6.0–8.0	Pumpkin	5.5–7.5	Watermelon	5.5–6.5
SMALL FRUITS							
		Blackberries	5.0–6.0	Raspberries, black	5.0–6.5		
		Blueberries	4.0–5.5	red	5.5–7.0		
		Grapes	5.5–7.0	Strawberries	5.0–6.5		

Figure 4–11 pH preferences of plants.

The Aboveground Environment

Just as the entire plant is influenced by the underground environment of the plant roots, the entire plant is also affected by the environment surrounding the top of the plant. The aboveground environment is more changeable, and its effect on plants may be more violent. The aboveground environment may be explained in terms of the factors affecting plants: (1) temperature, (2) light, (3) humidity, (4) plant diseases, (5) insects, and (6) gases or particles in the air.

Temperature

The temperature of the air has one of the strongest effects on plant growth. Some plants, such as lettuce, cabbage, and kale, grow best in cool temperatures. Others, such as corn, beans, and tomatoes, prefer hot weather. There are temperatures above which all plant growth stops. At the other extreme, temperatures near and below freezing also stop plant growth and, in fact, kill tender crops. For the best temperatures for raising specific crops, see Figure 4–12. Generally, the plant growth rate increases as temperature increases up to temperatures of about 86°F. This varies but is a good general rule, providing that moisture is available to the plant and wilting does not occur.

Light

Light must be present for plants to manufacture food. No green plant can exist for very long without light, whether that source is sunlight or light from an artificial source. Plants vary in the amount of light they require for best growth. Some plants prefer full sunlight; others prefer varying degrees of shade.

Too much light destroys a leaf protein essential in photosynthesis. Plants have natural sun blocker compounds called *flavonoids* in their leaves. Scientists are trying to develop plants with stronger genes for making natural sun blockers. Plants have enzymes to repair leaf damage from ultraviolet light. These enzymes are activated by light and are essential for a plant's survival in light. Light also affects plants in other ways. Some plants, such as chrysanthemum, bloom

TEMPERATURE		SUNLIGHT	
Cool Temperature Plants 60–80°F	Warm Temperature Plants 75–90°F	Full Sunlight	Partial Shade
VEGETABLE CROPS	VEGETABLE CROPS	VEGETABLE CROPS	VEGETABLE CROPS
Beet, Broccoli, Cabbage, Carrot, Chard, Endive, Kale, Lettuce, Onion, Parsley, Pea, Potato, Radish, Spinach, Turnip	Bean, Corn, Cucumber, Eggplant, Melon, Pepper, Squash, Sweet potato, Tomato	Bean, Beet, Broccoli, Cabbage, Carrot, Corn, Cucumber, Eggplant, Kale, Lettuce, Melon, Onion, Pea, Pepper, Potato, Spinach, Squash, Tomato, Turnip	Beet, Broccoli, Cabbage, Kale, Lettuce, Onion, Parsley, Radish, Spinach, Turnip
FLOWERING PLANTS	FLOWERING PLANTS	FLOWERING PLANTS	FLOWERING PLANTS
Arbutus, Azalea, Deciduous azalea, Primose, Rhododendron	Aster, Marigold, Petunia (nearly all summer annual bedding plants)	Rose, Marigold (most annual bedding (plants)	Azalea, Dogwood, Lady's slipper, Mountain laurel, Rhododendron

Figure 4–12 Favorable temperature and light conditions for common vegetable crops and flowering plants.

only when the days begin to shorten. (Long nights are needed for flower buds to form.)

Flowering is one way in which plants react to varying periods of light and dark. Plants may be classified in three groups according to this flowering reaction. *Short day* plants, such as chrysanthemum and Christmas cactus, flower when days are short, and nights are long. *Long day* plants, such as lettuce and radishes, flower when days are long, and nights are short. *Indifferent* plants do not depend on certain periods of light or darkness to flower. The African violet and tomato are indifferent plants.

Plants react to the length of days in other ways as well. For example, the black raspberry roots from cane tips in 5 to 10 days if the tips are covered with soil in September or October. In contrast, a rooting in midsummer may require up to 6 weeks. Dahlias develop a fibrous root system during the relatively long days of summer but as days shorten, the roots become thick storage organs.

Plants grow toward their source of light because the plant stem produces more growth hormones on the shady side, causing the stem on that side to grow to a greater length.

Humidity

Most plants are not affected drastically by a minor change in *humidity*, the moisture level of air. Most plants grow best in the 40 to 80 percent relative humidity range. *Relative humidity* is the amount of moisture in the air as compared with the percentage of moisture that the air could hold at the same temperature if it were completely saturated.

Some plants are more sensitive to humidity than others. Provided that the roots are able to replenish moisture lost through plant leaves as fast as it is lost and that the plants do not wilt, low humidity is not a great problem for most horticultural crops. However, when hot, dry conditions cause plants to wilt, plant growth is slowed or stopped completely. If wilting is allowed to reach the extreme state of *permanent wilting*, death occurs. When the humidity is very high (80 to 100 percent relative humidity), other problems may arise. For example, high humidity may cause the spread of fungus disease.

Adaptability of Plants

The American Horticultural Society (AHS) Plant Heat-Zone Map (Figure 4–13) divides the United

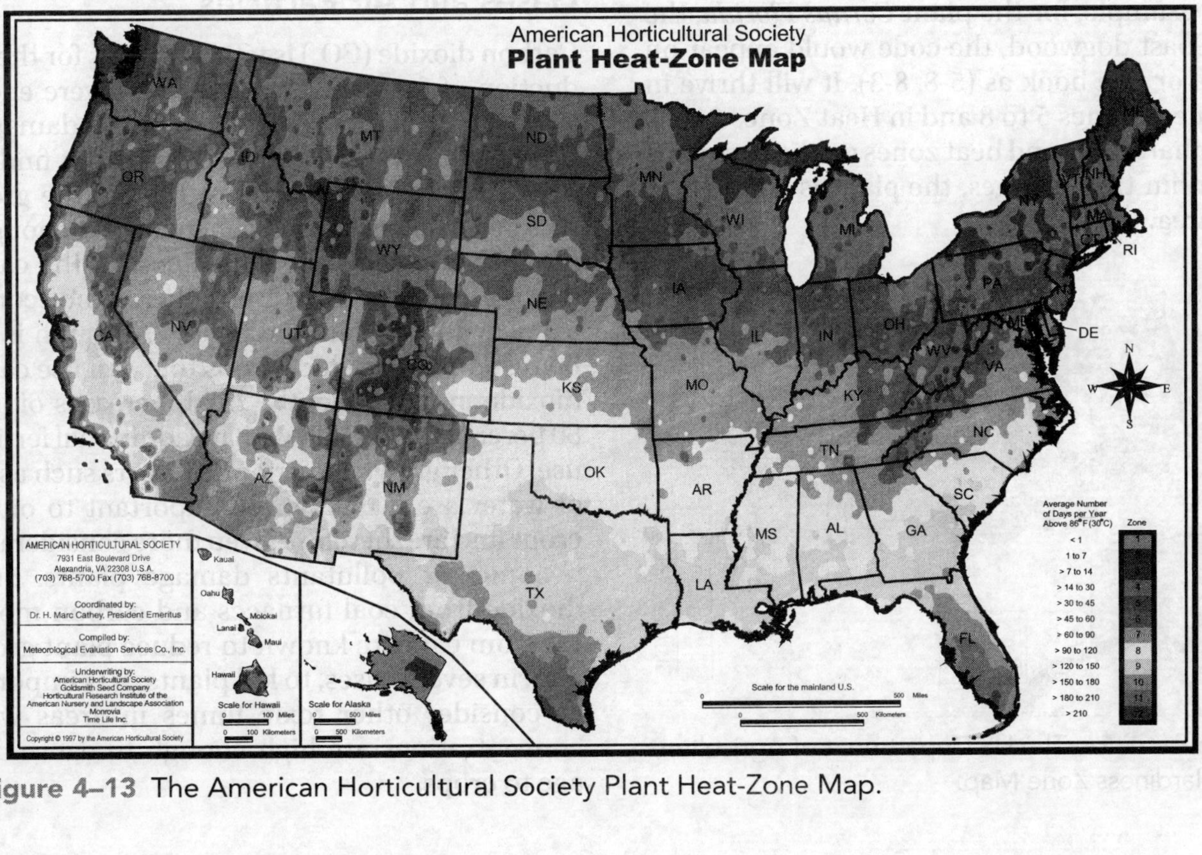

Reproduced with the permission of the American Horticultural Society (www.ahs.org)

▲ **Figure 4–13** The American Horticultural Society Plant Heat-Zone Map.

States into 12 zones based on the average number of days the daily high temperature reaches or exceeds 86°F (30°C). At 86°F, many plants begin to suffer damage from heat.

The US Department of Agriculture Plant Hardiness Zone Map (Figure 4–14) has recently been revised to reflect the earth's warming trend. Four more zones have been added (total 15) for subtropical and tropical plants.

When these two maps are used together a four-number code system can be set up to indicate where plants will grow best. The numbers are already available in some books from the AHS and on plant labels from some nurseries. In this numbering system, the first two numbers indicate cold hardiness. The first number is the coldest zone the plant will grow in, and the second number is the least cold zone the plant will thrive in. Some plants, such as apples and daffodils, need a certain amount of cold to grow and flower successfully. The second set of numbers indicates heat tolerance and requirement. The first number in the second set indicates the hottest zone the plant is adapted to. The second number in this set indicates the lowest heat zone the plant will thrive in. Some plants need a minimum number of warm days to flower and set fruit. Together, these two sets of numbers indicate the zones the plant will grow well in.

For example, for the plant *Cornus Florida*, the East Coast dogwood, the code would appear on a label or in a book as (5-8, 8-3). It will thrive in Hardiness Zones 5 to 8 and in Heat Zones 8 to 3. If your hardiness and heat zones on the two maps fall within these ranges, the plant is adapted to your area.

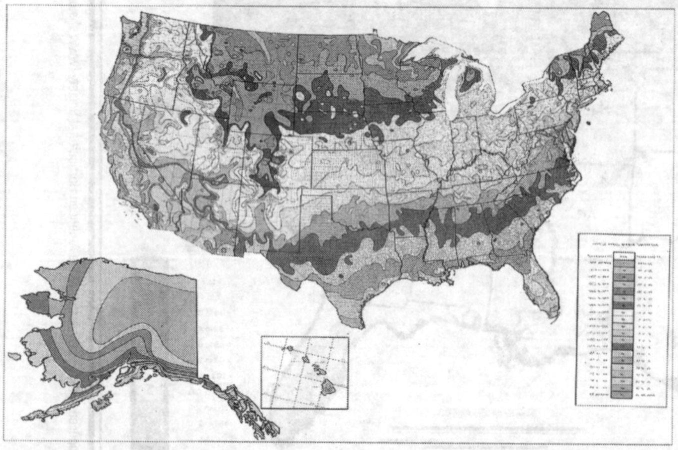

Source: The Agricultural Research Service, US Department of Agriculture

▲ **Figure 4–14** The US Department of Agriculture Plant Hardiness Zone Map.

These zones are a good place to start in determining plant adaptability for your area. Other factors such as moisture (amount of rainfall), exposure to sun and wind, humidity, and snow cover also affect how well a plant grows in a particular area. You may search on the AHS website (www.ahs.org) by zip code for databases of both hardiness and heat codes for your area.

Plant Diseases and Insects

Any time a plant is suffering from disease or insect damage, production is reduced. The amount of reduction depends on how severe the damage is and what percentage or part of the plant is infected. For example, because leaf damage reduces the area available for producing food, the more leaves that are lost, the more severely total production is reduced. Stem injury may *girdle* (circle) or clog up a stem and kill the entire plant because the supply of water and minerals to the plant top and food to the roots is completely cut off.

Some diseases and insect damage may be prevented by the use of varieties of plants that are resistant to disease and/or insects or by crop rotation or chemical sprays.

Gases and Air Particles

Carbon dioxide (CO_2) is vital to plants for the production of food. Rarely is there a severe enough shortage of carbon dioxide to cause damage to plants. However, greenhouse operators find that by adding carbon dioxide to the air, the growth rate of certain crops may be increased enough to more than pay for the added cost of the carbon dioxide. In field-grown crops, there is no economical way to add carbon dioxide. Research is now being done to produce crops that can use carbon dioxide more efficiently. Yield increases of up to 50 percent may result, with no additional fertilizer use. Other growth-restricting factors such as lack of water are usually more important to outside crops and are therefore given more consideration.

Some air pollutants damage plants. Sulfur dioxide from coal furnaces and carbon monoxide from cars are known to reduce plant growth and, in severe cases, to kill plants. It is important to consider other toxic fumes in areas where concentrations are high enough to cause damage to crops.

SAFETY FIRST

When applying plant nutrients to the soil, always follow the soil analysis test recommendations. The soil analysis gives the correct amount of nutrients to apply for optimum plant growth.

For more information on occupational safety and health in the workplace, please visit the OSHA site at www.osha.gov.

Summary

For successful plant growth, it is extremely important to consider the plant environment both underground and aboveground. Soil conditions and soilless mixes must match plant needs. The plant medium must be evaluated in terms of soil pH, drainage, and fertility.

Soilless mixes specially formulated for most plants grown in containers are now available. There is little, if any, need to mix your own anymore.

Aboveground factors such as temperature, sun, and wind exposure are also important. Select only plants adapted to the environment they will be placed into.

Review Questions

1. What USDA Plant Hardiness Zone do you live in? How does your zone affect the growing season in your area?
2. Discuss the difference between humidity and relative humidity. Why is it important to know both?
3. Using the American Horticultural Society Plant Heat-Zone Map, what heat-zone code do you live in? How does your zone affect the growing season in your area?
4. How would you define compost? What makes using compost a sustainable nursery practice?
5. Discuss the five media that are commonly used as soil additives. What enhancements does each one offer a growing medium?
6. Discuss three conservation practices that forage radish provides for our environment. What other cover crops are beneficial for enriching or protecting the soil?

Student Activities

1. Try this experiment to see how sand, silt, and clay particles are easily separated from soil. Take a half-gallon jar, fill it one-half full of soil, add 1 teaspoon of table salt (NaCl), and fill with water to a point within 2 inches of the top. Replace the lid and shake until all the soil is in suspension (mixed very well). Set the jar aside and examine it the next day.

The heavy sand particles should be on the bottom, with the silt above them, and the clay on top. (Organic matter may be floating on top of the water.) Definite layers should be visible. Classify the soil based on the percentage of sand, silt, and clay. Is it a sandy loam, silty loam, or clay loam? (Use the texture triangle shown earlier in Figure 4–6.)

2. Place sand in one jar and clay in a second jar. Pour a cup of water into each. Record how long it takes the water to drain down through the medium in each jar. For best results, place 2 inches of coarse gravel in the bottom of each jar.
3. Examine the label on a complete plant food. List all the plant food elements and record the percentage of each available. Identify major, secondary, and minor elements. Notice the small amounts of the minor elements available.
4. Experiment with the effect of temperature on the growth rate of plants by placing two plants of the same species in atmospheres with a wide difference in temperature. Note the results.
5. Place a plant in complete darkness and record its growth rate and appearance as compared with a similar plant kept in normal lighting conditions.
6. Take a soil sample from a local farm field or garden. Use a soil test kit, or send a soil sample to a commercial soil testing lab to test for pH level, phosphorus, and potassium. Make crop recommendations for which specific crops would grow best in this soil.
7. Search the Internet for the following terms: greenhouse soils; soils; agronomy; soil testing; soil texture; soil erosion. Google www.nrcs.usda.gov. Define each term, and submit the definitions to the instructor.

Suggested Supervised Agricultural Experience (SAE) Activity

1. Conduct a research SAE by working with a local college or secondary research program.

Additional Resources

http://www.extension.iastate.edu/forestry/publications/pdf_files/ctspreqs.pdf

http://aesop.rutgers.edu/~horteng/workshop/lecture2.pdf

UNIT 5 Growth Stimulants, Retardants, and Rooting Hormones

Objective

To list four types of growth regulators used in the horticulture industry.

Key Terms

allelopathy
apical dominance
auxins
blossom set
cytokinins
gibberellins
green manure
hormones
inhibitors
node
plant growth retardants (PGRs)
PGR calculator
photorespiration
retard
retardants

Competencies to Develop

After studying this unit, you should be able to:

- list one example of a substance used to stimulate plant growth.
- explain why chemical retardants are applied to floral crops, and name two commonly used retardants.
- explain the use of rooting hormones on cuttings, and list several rooting hormones.
- describe orally the source of one plant growth stimulant.
- demonstrate the proper application of a rooting hormone to cuttings.
- explain how rootstock is used in dwarfing fruit trees.
- define biostimulants.

Materials

- chrysanthemum and poinsettia plants
- rooting hormones (Hormodin #3 and #1)
- gibberellic acid
- B-Nine
- American and English holly cuttings (semihard cuttings from different shrubs may be used)
- mist system or plastic enclosed pots or flats
- knife for making cuttings
- smartphone to calculate dilution rates for growth regulators

Growth-Regulating substances in plants are called **hormones**. Hormones are organic chemicals that act and interact to affect growth rate. In the example shown in Figure 5–1, the gardenia plant on the right was treated with the growth **retardant** B-Nine to produce a more compact, lower-growing plant. Prominent hormones are listed here:

- **auxins** (Greek word meaning to increase). Accelerate growth by stimulating cell enlargement.
- **gibberellins.** Stimulate growth in stem and leaf by cell elongation. Also, they stimulate premature flowering, growth of young fruits, and breaking of dormancy.
- **cytokinins.** Stimulate cell division. They work along with auxins (will not work without auxins present).
- **inhibitors** (abscisic acid). Inhibit seed germination and stem elongation, and hasten ripening of fruit (ethylene gas).

The auxins and gibberellins promote cell enlargement, and the cytokinins stimulate cell division.

Growth hormones are organic chemicals principally produced by actively growing plant tissue such as shoot tips and young leaves. They move throughout the plant and may be found in many plant parts. These chemicals react with one another in a very complex system in the plant. In some cases, one concentration or amount of a hormone stimulates growth, and a different concentration or amount restricts growth.

Courtesy of Uniroyal Chemical Co., Inc.

▲ **Figure 5–1** The growth retardant B-Nine used on the gardenia plant on the right results in a compact, low-growing plant.

Apical Dominance

A good example of regulating growth in plants is seen in the dominance of the terminal bud or shoot. Where **apical dominance** exists, the terminal bud secretes chemicals that inhibit or prevent the growth of axillary buds on the same shoot. Axillary buds are found in the axil or angle between a leaf and the stem. This causes the plant to grow tall and not send out side branches. After the plant reaches flowering age, and the terminal bud becomes a flower bud, the chemicals are no longer secreted. At this time, the plant starts sending out side branches. It appears that this is a genetic program directing the plant to grow above competing plants in its early years. After height and access to sunlight are secured, the plant spreads over its competitors. Pinching out the terminal leaf bud of the shoot nullifies the effect on axillary buds, and branching occurs sooner than the plant would have branched if left alone. This method is commonly used by nurseries to encourage early branching of shrubs and the production of more compact plants.

For years, methods have been developed other than pruning to control the growth rate, size, and shape of plants. Some major achievements include the discovery of the following:

- chemical and natural stimulants that cause plants to grow taller or faster
- chemical retardants that cause plants to grow slower or cease growing at a certain point
- hormones that cause cuttings to root faster
- dwarfing rootstock for tree fruits

GREEN TIP Go green by using the naturally occurring chemicals in the plant to prevent, monitor, and control plant growth.

Stimulants

Applications of certain chemicals enable plants to grow taller. The most common chemical of this type is gibberellic acid. An example of the effect this chemical has on plants is shown in Figure 5–2 in which the chrysanthemum treated with gibberellic acid is much taller than normal.

Gibberellic acid causes the stems of plants to stretch out. The **nodes**, the joints at which buds, leaves, and branches grow out from the stem, are farther apart. Notice how much longer the stems that hold the blossom buds are on the treated plant. Gibberellins also occur naturally in plants.

> **GREEN TIP** By using natural Epsom salts, magnesium sulphate, 9.8 percent magnesium and 6 percent sulfur, you fortify the crown of a plant and stimulate buds to sprout new wood from the base. Epsom salts help rose plants or other flowering shrubs to increase the number of blossoms.

Courtesy of Dr. Gary Wilfret, University of Florida

▲ **Figure 5–2** A plant treated with gibberellic acid shows elongated stems. The space between nodes on the stem is lengthened, causing the plant to grow taller than normal. Gibberellic acid is naturally present in plants.

A natural growth stimulant has been found in alfalfa, and although this stimulant has been known by nurserymen for more than 30 years, interest in it has recently been renewed. The chemical that causes the stimulation of growth is the alcohol triacontanol.

Plant stimulation is brought about by mulching with alfalfa hay or watering plants with a tea made by soaking alfalfa feed pellets in water and using the tea to water the plants. The amount of triacontanol needed is very small (4 milligrams per acre of crop). Too much seems to reduce the stimulating effect. This material is also sold commercially.

Results with rhododendron seedlings are listed here:

- 55.1 percent increase in height
- 27.6 percent increase in diameter
- 27.6 percent increase in number of leaves
- 53.5 percent increase in leaf length
- 19.3 percent increase in leaf width*

This increased plant growth resulted with no added fertilizer. See Figure 5–3.

If you want to experiment with this material, see the Student Activities section at the end of this unit for details.

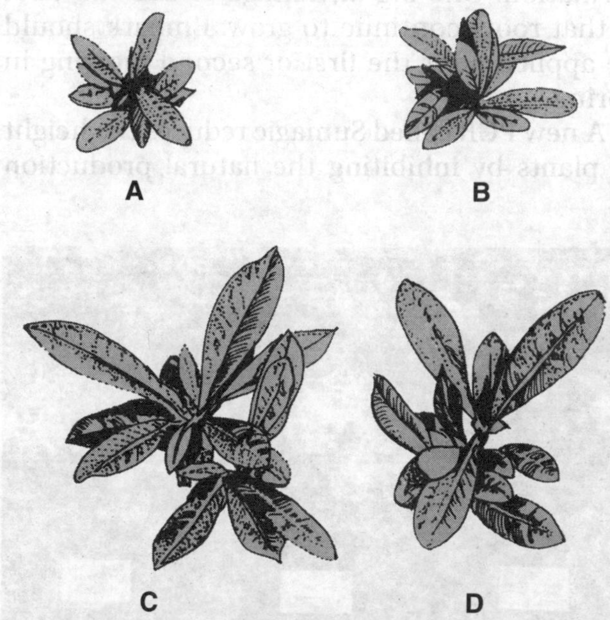

Courtesy of American Rhododendron Society Quarterly Journal, Spring 1979.

▲ **Figure 5–3** Growth increases on rhododendron using alfalfa as a stimulant. Examples A and B are untreated controls; C and D have been treated with alfalfa.

*Research done by Dr. Hinderman of Ann Arbor, Michigan

Chemical Retardants

At times, plants may grow too tall and open to be pleasing to customers. In these cases, chemicals are used to **retard** the growth of the plants, causing them to be shorter and more compact. Not only are these plants more attractive, but also the shorter stems are better able to support the flowers, thereby reducing the need for staking or tying. Today, chemical retardants (chemicals that *retard* or slow down growth) are used commercially as a normal part of the growing process of many plants. Figure 5–4 illustrates the effects of B-Nine used on azaleas.

One of the newest uses of **plant growth retardants (PGRs)** is the application of a chemical to retard lawn grass growth. If applied at first *greenup*, after the first mowing of grass in the spring, a chemical called Limit is absorbed by the roots and restricts grass growth for six to eight weeks. Another material called Embark is leaf-absorbed and also restricts grass growth.

Figure 5–5 shows two strips of turfgrass. Grass on the left side was treated with Embark, a PGR that is absorbed through the leaves and moves to the growing point where it interrupts cell division, stem elongation, and seed head formation. One big advantage of this material is that roots continue to grow. Embark should be applied after the first or second mowing in spring.

A new PGR called Sumagic reduces the height of plants by inhibiting the natural production of the plant hormone gibberellic acid. As stated previously, gibberellic acid causes stems to elongate or lengthen. This chemical is taken up by the leaves, stem, and roots and moves through the plant to the growing point where it inhibits the production of gibberellic acid.

Courtesy of PBI Gordon Corporation

▲ **Figure 5–5** Two strips of turfgrass. The strip on the left was treated with Embark. There is a tremendous reduction in growth rate compared with the untreated grass on the right.

Courtesy of Uniroyal Chemical Co., Inc.

▲ **Figure 5–4** The plant growth retardant (PGR) B-Nine was applied to the two azaleas on the right. Notice the compact growth and heavier flowering compared with the untreated plant on the left.

Plant retardants are also being used in the landscape to control growth on shrubbery and hedges. One application of the PGR Atrimmic after pruning or some regrowth is claimed to last the entire season on plants that have only one flush of growth, thus reducing or eliminating the need to prune. Plants that have a continuous growth habit would need a treatment again in eight weeks.

Atrimmic is a systemic (a material that penetrates the plant, enters the plant sap, and moves through the plant) that works by blocking the plant hormones that stimulate growth. Applied as a foliar spray, it is absorbed by the leaves and moves to the shoot tips where it temporarily stops shoot growth. Figure 5–6 shows an example of the use of Atrimmic on a hedge.

Rooting Hormones

When plants are propagated from cuttings, it is important that a large percentage of the plants root and that they root as quickly as possible. Some plants root very easily from cuttings

Courtesy of PBI Gordon Corporation

▲ **Figure 5–6** The shrubbery shown above was treated with the PGR atrimmec. The treated area on the left has made very little growth compared with the untreated area on the right.

© 2017 Cengage Learning®

▲ **Figure 5–7** The holly cutting on the left treated with 1.5% indolebutyric acid developed a good root system, whereas the untreated cutting did not.

without any chemical treatment. Geranium, azalea, and many soft, succulent houseplants require only a moist, well-aerated medium and high humidity to root. However, many other plants must have some assistance if any or very many of the cuttings are to root successfully.

Why is there a difference in ease of rooting? Early researchers discovered that some plants have more natural root-promoting chemicals than others. Indoleacetic acid is a natural plant hormone that causes roots to form on plant stems and is found in various plants in differing amounts.

Plants also root more easily or with more difficulty at different stages of maturity, which affects the hardness of the wood. Some root best when the wood is soft, others when it is almost hard.

The development of chemical rooting hormones made it possible to root certain plant cuttings that were considered impossible to root before. These chemicals also shortened the length of time required to root cuttings.

The chemical most commonly used for rooting cuttings is indolebutyric acid. Naphthaleneacetic acid and alpha-naphthaleneacetic acid are also widely used. Indoleacetic acid, the first naturally occurring plant hormone to be used, is rarely, if ever, used today because it is generally not as effective as indolebutyric acid or naphthaleneacetic acid.

Rooting hormones are either mixed with talc and used as powders or dissolved in liquid and used as a wet dip. Some of the new liquid dips give better rooting percentages than do the powders. The strength of the active chemical varies from 0.1 percent to 0.8 percent, although concentrations as great as 2 percent of indolebutyric acid have been used on certain evergreens such as camellias and yews. Whatever their concentration, all rooting hormones should contain a fungicide (a chemical that kills fungi) such as Captan to prevent rotting of the cutting. Captan also seems to help promote faster rooting.

The holly cutting on the left in Figure 5–7 was treated with indolebutyric acid; the one on the right was untreated. After 21 days, the treated cutting was rooted. The untreated cutting had developed no roots.

Dwarfing Rootstocks

As the cost of labor increased, orchardists began looking for smaller fruit trees that could be picked without the use of ladders. Research was started to discover a way to prevent apple trees from growing very tall. Pruning had been used to reduce growth, but the researchers realized that it was a time-consuming, expensive, and only partially effective method.

It was then discovered that trees growing from certain types of roots did not grow as large as other trees. These trees also bore fruit at an earlier age. Using this principle, a complete series of rootstock known as the malling rootstock was developed in England. (Rootstock is a root or piece of root that is used for grafting.) It became possible to control the size and rate of growth of apple trees by selecting the proper rooting stock. Some varieties rooted from the malling rootstock grow close to normal size, whereas others are only one-fourth to one-half the normal size.

Since the original development of the malling stock, many other types have originated, including stock for apple, peach, and pear trees. The dwarf trees themselves can be purchased commercially. Check with your local agricultural experiment station to determine which rootstock is best for grafting or budding fruit trees in your area. Consult nursery catalogs for listings of dwarf fruit trees.

The Chemical Blossom Set

One type of growth hormone is used on tomato blossoms early in the season to cause earlier development of fruit. Normally, the first blossoms on tomato plants often do not set (develop) fruit because of cool weather, lack of pollination, or for other reasons. The chemical **blossom set** causes seedless fruit to set on the blossoms and results in tomatoes ready to eat as much as 10 days earlier than normal. Experiment with some of the commercially used plant growth regulators shown in Figure 5–8.

Plant Biostimulants

Biostimulants are natural products. They are organic, meaning from living organisms, and work to both stimulate soil microbial activity and improve soil cation exchange capacity, stimulate plant growth, and promote disease resistance. Humic acid, a product of humus or rotting organic matter, is an example of a soil microbial stimulant.

Root growth biostimulants improve water and nutrient uptake by the roots and also increase the number of fibrous roots. High levels of these

TRADE NAME	EFFECT	USED ON
A-Rest	shortens stem length	chrysanthemum lily poinsettia
B-Nine	shortens stem length	chrysanthemum hydrangea petunia azalea poinsettia
Cycocel	shortens stem on poinsettia induces heavier flower bud set on azalea	poinsettia azalea geranium
Phosfon	shortens stem length	lily chrysanthemum
Florel	induces basal breaks on roses retards growth on poinsettia increases number of branches on carnation speeds ripening of tomatoes assists in faster rooting of cuttings	rose poinsettia carnation tomato cuttings
Augeo™	promoting thicker lateral branching increased number of flowers reduces apical dominance resulting in fuller and often times a more compact plant	bedding plants, herbaceous plants, perennials, woody ornamentals
Paczol	reduces plant internodal length	Bedding Plants (petunia, impatiens, dianthus, calibachoa,kale) Flowering Plants (chrysanthemums, poinsettias) Perennials (echinacea, argyanthemum, salvia 'lucantha') Woody plants (azaleas, gardeina) Bulbs (calla lily,lilium asiatic, lilium orientials, casa blanca lilium, and other orientals)

Figure 5–8 Commercially used plant growth regulators.

TRADE NAME	EFFECT	USED ON
Bonzi	shortens stem length	poinsettia geranium petunia chrysanthemum impatiens coleus pansy snapdragon celosia
Sumagic	shortens stem length	poinsettia chrysanthemum celosia coleus marigold periwinkle petunia ageratum dahlia

Growth retardants work best when the humidity is high, but plant foliage is dry. Always follow exactly the directions on the label.

(*Note:* wear protective clothing when called for on the label.)

Figure 5–8 Commercially used plant growth regulators. (*continued*)

growth stimulants are found in kelp seaweed. Fresh kelp is most effective and has been found to greatly increase growth and yields of plants. Processing such as drying, freezing, boiling, and chemical treatment decreases the stimulating effect. A product is currently available on the market that is processed to retain the stimulating effect of kelp.

Use of biostimulants can greatly reduce the need for fertilizers, especially nitrogen. This not only saves the grower money but also reduces water pollution from excess nitrogen runoff or percolation into groundwater or stream water.

Methanol (a form of alcohol) used in small amounts appears to speed up plant growth. In tests by Dr. Andrew A. Benson (University of California), methanol (methyl alcohol) increased plant yield by 36 to 100 percent. A 10 to 30 percent solution of methanol was used.

The alcohol seems to work by blocking **photorespiration**. Plants use the water for growth rather than transpiring it into the air. This material works best on plants grown in full summer sun. Water use was reduced by as much as 50 percent from some plants.

CAUTION
Stay on the low side until plants' responses or possible damage is observed.

This product is already on the market. Other plant stimulants such as triacontanol and DCPTA are being studied as possible growth regulators.

An aspirin to stimulate plants? Scientists have discovered that salicylic acid is produced by plants when attacked by a disease organism. Spraying the plants with salicylic acid stimulates them to put their defenses into action. Applying the acid before plants are attacked may prepare them to resist diseases even before they are present.

Plants' Biological Clocks

Plants have clocks that operate on a 24-hour cycle and go off each morning just as your alarm clock does. According to Autar Mattoo, a plant physiologist in Vegetable Laboratory of the Agricultural Research Service in Beltsville, Maryland, the clock is located in the nuclei of plant cells. The clock controls an enzyme that modifies a protein necessary for photosynthesis as day starts and light energy is available.

The clock also regulates plants' adjustment to the brightest, hottest part of the day by stimulating them to put on sunscreen. Too much sun destroys leaf protein, which is essential in photosynthesis. Plants use natural sun blockers called flavonoids for protection during midday in summertime.

Allelopathy

Allelopathy is the production of a chemical compound by one plant that slows down or stops the growth of another plant. This is a natural growth regulator some plants use to stop competition from other plants. Some **green manure** or cover crops grown for weed control, such as millet, have this effect on most weeds and prevent weed growth. The effect of these growth regulators may last for months or in some cases, such as the black walnut, years.

Some plants do not transplant well into beds where older plants of the same type are or recently were growing. Members of the rose family are good examples (hawthorn, apple, rose, peach, pear, etc.).

A **PGR calculator** is available as a smartphone app to assist horticultural plant growers in the calculation of the correct dilution rates and application volumes for plant growth regulators (see Figure 5–9). The user enters the desired concentration (ppm), size of application area or the number of containers to be sprayed or drenched, and the PGR calculator calculates the amount of solution and product needed.

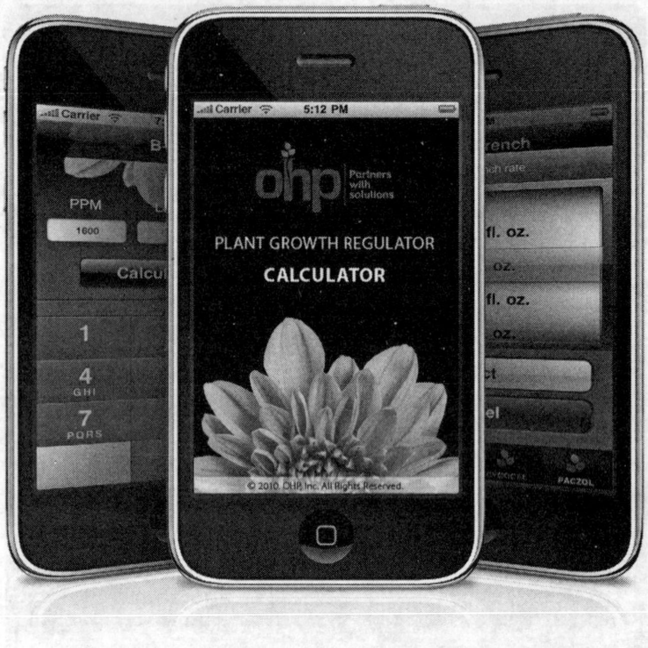

Courtesy of OHP, Inc.

▲ **Figure 5–9** Use your smartphone or tablet to assist with PGR use rates and calculations. OHP developed the OHP Plant Growth Regulator Calculator app to help growers with their rate calculations and with determining how much PGR solution to apply over a specific area with spray applications or the number of pots for drenches. The app is easy to use but is limited to OHP plant growth regulators: Augeo, B-Nine, Cycocel, and Paczol.

TRY THIS Using the Internet or a smartphone app, use a PGR calculator to figure out a mixing solution for a growth regulator provided by the instructor.

SAFETY FIRST

When working with plant growth regulators, one must adhere to the restricted entry interval (REI), or the time immediately following a pesticide application when entry into the treated area is restricted, and the signal word (REI = 24 HRS, Signal Word = Caution), required Personal Protective Equipment (PPE), and Mode of Action about how the plant growth regulator works in the plant. For more information on occupational safety and health in the workplace, please visit the OSHA site at www.osha.gov.

Summary

Plant growth is the result of many naturally occurring chemicals in the plant. The materials work together to promote natural growth rates, flowering, rooting, and other activities. Horticulturalists have learned to add greater amounts of the materials than naturally found in the plant to change growth rate (faster or slower) or to more easily root cuttings. Some manmade chemicals slow down or speed up production of growth-regulating chemicals in plants.

Review Questions

1. Discuss the different methods horticulturalists may use to control the growth rate, size, and shape of plants. Which are the advantages and disadvantages of each method?
2. Why are chemical retardants applied to floral crops? What are the advantages and disadvantages of their use?
3. Why are rooting hormones used on some plant cuttings?
4. What does it mean to dwarf a rootstock? When is this method used?
5. What are the advantages of using biostimulants?

Student Activities

1. Apply gibberellic acid to an actively growing plant. Keep an identical plant that has not been treated as a check on the results.
2. Apply B-Nine to five poinsettias and five mum plants. Keep five untreated plants of each type as check plants. Record the differences noted in the plants' growing habits. (Be sure to follow the directions on the label exactly.)
3. Start 20 holly cuttings. (If necessary, refer to Unit 7, Softwood and Semihardwood Cuttings and Micropropagation, for procedural suggestions.) Use a commercial rooting compound on 10 of the cuttings. Keep 10 untreated plants as checks. Be sure to follow the directions when using the rooting hormone. Record the results.
4. Contact a local orchardist or a state experiment station for details on dwarf fruit trees.

5. Perform an alfalfa *growth stimulus experiment*. Take some potted plants just starting their growth cycle (about a week after transplanting) and treat as follows with alfalfa:
   a) Finely chop some good-quality alfalfa hay.
   b) Mix 1/2 cup of the chopped hay with some hardwood mulch and place on top of the soil in a 6- to 10-inch pot; water it well.
   c) Compare with control plants with no alfalfa.

   An alternate method is to soak alfalfa pellets (cattle feed) in water (1 1/2 cups of pellets per 5 gallons of water) for at least a day. Water the plants with 1/2 cup of this tea every 7 to 14 days.
6. Search the Internet using the following terms: plant hormones; growth stimulants; USDA.
7. Go to www.ffa.org, and list all of the National FFA Proficiency Award areas that relate to the information in this unit.
8. Use the PGR app by OHP to calculate growth regulator concentrations for each of the products listed.

Suggested Supervised Agricultural Experience (SAE) Activities

1. Participate in a placement SAE in the area of agronomy.
2. Plan and operate an entrepreneurship SAE in the area of landscape management.

Additional Resource

https://quizlet.com/30452728/arkansas-plant-science-2-unit-1-growth-mc-book-5-flash-cards/

SECTION 3

Plant Propagation

UNIT 6 Seeds

Objective

To propagate at least one plant using seeds.

Key Terms

asexual
cultivar
dormant
embryo
endosperm
hardening off
hybrid
indirect seeding
pelleted seeds
propagation
seedling
tissue culture

Competencies to Develop

After studying this unit, you should be able to:

- identify the parts of a seed and the functions of each.
- differentiate between indirect and direct seeding methods.
- prepare a medium for seeds, sow seeds, and provide the proper conditions for germination.
- water, fertilize, and harden off seedlings before transplanting.
- transplant seedlings into flats or pots.

Materials

- planting container such as a pot or flat
- planting medium (pasteurized)
- seeds of plants commonly grown in the area
- labels and marking pen or pencil
- leveling board or skew
- watering facilities for bottom watering
- transplanting containers
- watering can
- greenhouse or other growing space

What Is Propagation?

There are many ways of propagating or reproducing plants. The most common method of reproducing flowering as well as vegetable and cereal crops is through the use of seeds. This is a sexual process that requires the union of *pollen* (the male sex cell) with the *egg* (the female sex cell) in the ovary. Male and female cells may be from the same parent (*self-pollination*) or from separate parents (*cross-pollination*).

Seeds are a means of rapidly increasing the number of a certain plant. However, not all plants *come true to seed* (reproduce exact duplicates of the parent plant from seeds). Wheat and barley are examples of plants that do come true from seed. Other plants, such as rhododendrons, azaleas, and apples, do not come true from seed, causing the offspring to differ from either parent.

Plants that are not produced directly from seeds or do not produce seeds that will grow, such as some **hybrids**, must be propagated by another method to obtain exact duplicates. Hybrids are the offspring of two different cultivars (the term **cultivar** is often interchanged with the term *variety*) of one plant, each of which possesses certain characteristics that are desired in the new plants. A female plant of one cultivar and a male plant of another cultivar are *crossed* (bred) to produce offspring with the best characteristics of each parent. However, when hybrids themselves are reproduced from seed, their offspring do not have the same characteristics as the parents; rather, they have a variety of combinations of traits possessed by the plants originally used to produce the hybrids. To produce exact duplicates of these plants, **asexual** reproduction is used. This is not a sexual process, and no seeds are used in this method. Instead, the plant is propagated from one of its parts, such as the leaf, stem, or root.

Propagation may be accomplished by division of roots; by cuttings of leaves, stems, or roots; or by budding and grafting. The strawberry reproduces itself by runners; the lily does the same by tiny bulblets. The reproduction method of the strawberry and lily occur naturally. However, over time people have developed and improved upon cuttings, budding, and grafting as reproduction methods.

Asexual reproduction is possible because each single cell of a plant contains all the characteristics of the entire plant and can regrow any missing part. Thus, a stem cutting removed from the roots of the plant develops new roots from cells along and at the base of the stem. Root cuttings develop new stem tissue in the same manner.

Plant propagation predates recorded history. Very early humans planted seeds or divided plants to increase plant numbers and to carry them over the **dormant** or resting season. The quality of plants was improved by using seed from the best plants to produce other plants, or by separating naturally propagated plants of superior quality from the parent and planting them.

One of the newest approaches to plant propagation is a method called **tissue culture** or *micropropagation*. This is an asexual method in which a growing shoot, tip, or bud of the plant is separated and grown on an agar gel or other nutrient medium. Through this method, many thousands of new plants exactly like the parent plant can be produced very quickly. This is a good way to rapidly increase the propagation of a new, superior plant and get it on the market years sooner. This method has been used for years with orchids and some other plants and is now being adopted as a commercial method in the propagation of many plants. Tissue culture is discussed in greater detail in Unit 7.

Figure 6–1 shows the various kinds of plant propagation that are discussed in this section.

TRY THIS Using stem stock provided by the instructor, prepare cuttings from a parent plant, and record the growth of new roots.

Propagation of Plants from Seeds

In the horticulture industry, many plants are started from seed because it is a quick and economical method. For successful germination, the proper environmental and cultural conditions must be provided. These conditions include temperature, moisture, light, and medium.

Seeds of annual plants (plants that complete their life cycle in 1 year) from named cultivars that are not hybrids generally come reasonably true from seed. *Named cultivars* are specific

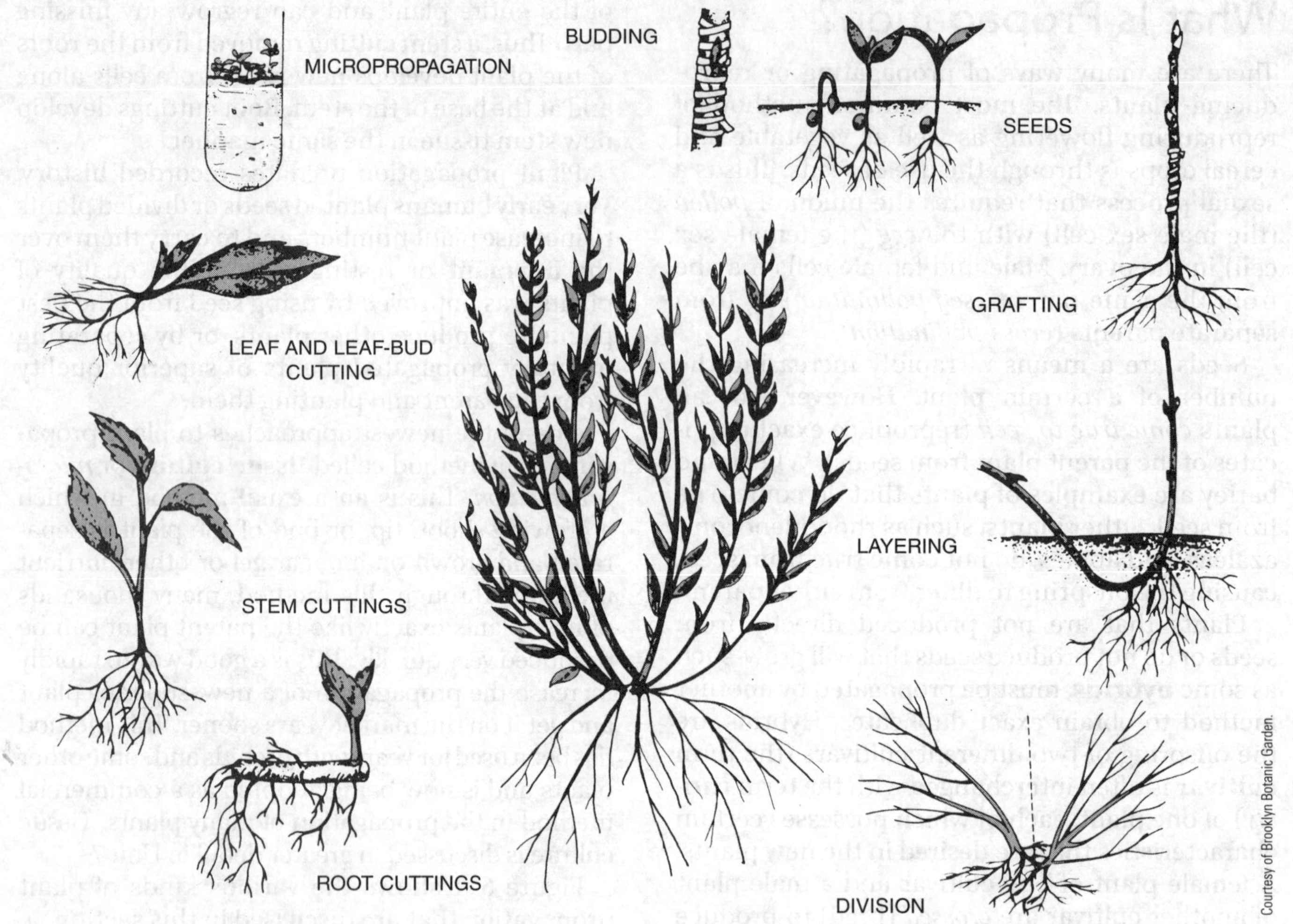

▲ **Figure 6–1** Commonly used methods of propagating plants. Except for seeds, these are all vegetative methods of propagation.

individual plants that have been named to indicate their particular traits. The tomato plant has numerous named varieties, such as 'Better Boy' and 'Marglobe,' each of which is different in size, days needed to mature, growth habits, and disease resistance. Seeds should not be saved from hybrids and planted because the resulting plants would not come true from seeds. Many perennials are grown from seed (Figure 6–2).

Special Treatment of Seeds for Germination

Some seeds have a hard seed coat that must be soaked or scratched before the seeds are able to germinate in a process called *scarification*. Examples of this type of seed are the red bud, Judas tree, and sweet clover. Sandpaper, scratching, or an acid bath may be used to weaken the seed coat.

Other seeds require a moist, cold rest period (dormant stage) at temperatures below 37°F (3°C) for 8 weeks or longer. Examples of seeds that require a rest period are the apple, peach, pear, tree peony, maple, and yew. These may be buried or stratified in moist sand and kept cool.

Still other seeds must go through alternate wetting and drying. Some must have light to germinate; others must have darkness. These specific requirements must be met when these seeds are planted. Always follow instructions on seed packets or enclosed planting directions.

Lastly, some seeds are **pelleted**. Many seeds are very small and difficult to sow evenly in a flat

PLANT	GERMINATION TIME INDOORS (DAYS)
Achillea	5–7
Alyssum saxatile (basket-of-gold)	5
Anthemis (camomile)	5
Hardy aster	7–14
Campanula	10–24
Cerastium tomentosum (snow-in-summer)	10
Chrysanthemum	5–8
Dianthus (hardy pinks)	4–6
Doronicum (leopard's bane)	8
Eupatorium (perennial ageratum)	7
Gaillardia	5
Gloriosa daisy	7
Gypsophila (baby's breath)	7
Heliopsis	4
Hollyhock	7
Iberis (perennial candytuft)	6
Lupine	3–4
Myosotis (forget-me-not)	7–12
Papaver (Oriental poppy)	6–10
Phlox paniculata	21–28
Primula (primrose)	21–30
Salvia	5–21
Shasta daisy	5–7
Verbena	10

Figure 6–2 Perennials easily grown from seed.

indoors or to direct seed into the soil outdoors. To help create a more even sowing distribution, the seeds are coated with an organic clay-based compound to create a large seed that will be easier to sow, either directly into the flat indoors or into the prepared bed outdoors for future production.

GREEN TIP Many growers are getting great germination with pelleted seed, particularly for petunias or begonias. The trick is keeping the germination trays moist, and then allowing them to dry until the first true leaves (cotyledons) come out. Petunias have a tap root that needs to branch out with drier conditions to get top growth to expand and to obtain uniform seedling growth.

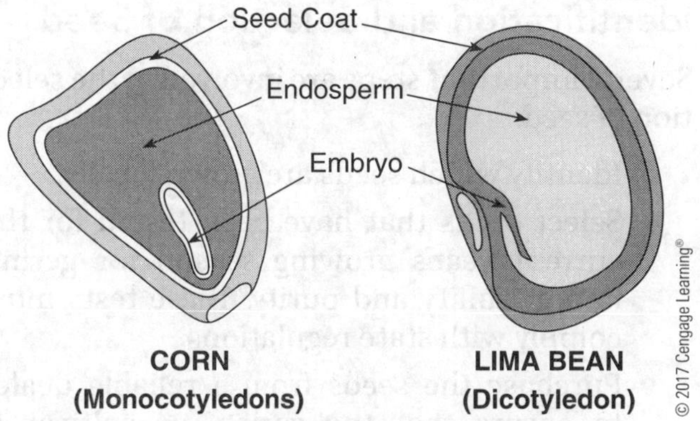

▲ **Figure 6–3** Composition of a seed.

Composition of Seeds

The basic parts of a seed are the seed coat, the **endosperm** (stored plant food), and the **embryo** (Figure 6–3).

SEED COAT The *seed coat* is the outside covering of the seed that protects the embryonic plant. The seed coat makes it possible for seeds to be transported and stored for long periods of time.

ENDOSPERM (STORED PLANT FOOD) The endosperm is the food storage tissue that nourishes the embryonic plant during germination (the first start of growth in a seed).

EMBRYO (OR EMBRYONIC PLANT) The embryo is a new plant that is developed as a result of fertilization. During germination, it extends its roots and seed leaves to form a new plant (Figure 6–4).

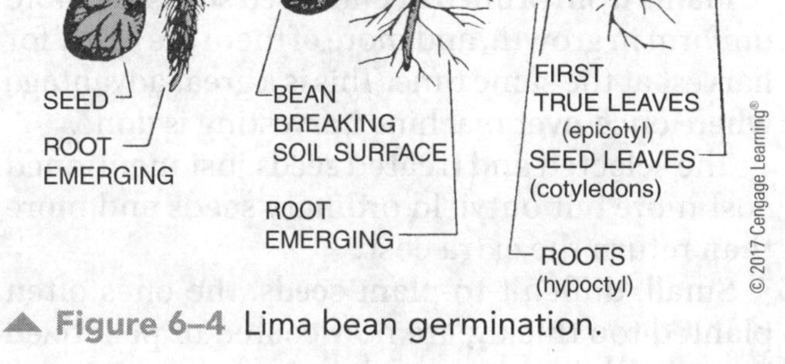

▲ **Figure 6–4** Lima bean germination.

Identification and Selection of Seed

Several important steps are involved in the selection of seed:

- Identify which seeds are grown locally.
- Select seeds that have been tested for the current year's growing season for germination ability and purity. These tests must comply with state regulations.
- Purchase the seeds from a reliable dealer to ensure that the variety or cultivar is true (pure) and that germination ability is acceptable.
- Choose hybrid varieties for greater vigor, uniformity, and flowering ability.
- Select uniform heavyweight or primed seeds.

Some new things have been done with seeds to improve germination and uniformity of growth. First, seed *quality* has been improved, and seeds have been sorted by weight. This results in all **seedlings** coming up at the same time. Heavier seeds tend to grow faster and produce larger plants. These seeds are usually worth more and are a good investment.

Second, seeds are *primed* or *enhanced* by soaking the seeds in salt solutions of potassium chloride (KCl) or ethyl alcohol. This treatment causes the growth hormones or enzymes to become active, and the seed radical or young root starts to grow. By regulating the amount of moisture and chemical, growth of the radical is stopped at a point where the seed may still be safely stored.

Seeds treated this way come up much sooner after planting. Where seeds are planted outside in the field, this is a great advantage. The sooner the seed comes up after planting, the less chance for soil insects and diseases to damage or kill the young seedling. Young plants become much more disease-resistant in just a few days, and many more survive if growth is fast.

Plants from primed or enhanced seeds are more uniform in growth, and more of them are ready for harvest at the same time. This is a great advantage where once-over machine harvesting is done.

The selected and treated seeds just mentioned cost more but outyield ordinary seeds and more than return the extra cost.

Small, difficult-to-plant seeds, the ones often planted too thickly, are now coated or pelletized to make them larger and easier to plant. Some companies even glue them to a tape with the proper spacing. Then, all you do is stretch out the tape and cover it with soil.

Germination Medium

The best medium for germination has a favorable pH level and an adequate supply of plant nutrients. It is firm, porous, uniform in texture, sterile, and free of weeds, insects, and disease organisms. A good germinating medium contains one or more of the following ingredients.

SOIL The soil should be a loam composed of 45 percent mineral matter, 5 percent organic matter, 25 percent air, and 25 percent water. Soil is seldom used as a germinating medium now.

CONSTRUCTION-GRADE SAND This is the best type of sand to use because it is more porous than some other sands, thereby allowing for better aeration and drainage. Sand particles do not hold plant nutrients in the medium as strongly as particles of soil or peat moss. Sand is seldom used now. Perlite or vermiculite is used instead.

PEAT MOSS Peat is partially decomposed vegetation that has been preserved underwater and is collected from marshes, bogs, or swamps. Peat has a very high capacity for holding water. It contains about 1 percent nitrogen and is low in phosphorus and potassium.

SPHAGNUM MOSS Sphagnum moss is the dehydrated remains of acid bog plants. It is used in shredded form in seed germination. Sphagnum moss is relatively sterile and lightweight, controls disease well, and has excellent water-holding capacity.

HORTICULTURAL-GRADE PERLITE Perlite is a gray-white material of volcanic origin that expands when it is heated. It is most commonly used to improve aeration of media. Horticultural-grade perlite consists of large particles, thereby providing good drainage and aeration.

VERMICULITE Vermiculite is a very lightweight mineral that is neutral (has a pH of 7.0) and has a very high water-holding capacity.

JIFFY MIX Jiffy Mix is composed of equal parts of shredded sphagnum moss, peat, fine-grade

Terra-Lite vermiculite, and enough nutrients to sustain initial plant growth.

Germination media are usually made up of one or more of the materials in the list. Sand or perlite should be added if the medium needs greater drainage and aeration. Peat, sphagnum moss, or vermiculite may be added to improve the moisture-holding capacity of the medium.

Soil used in any seed-germination medium should be pasteurized by heating it at 180°F for 30 minutes. Pasteurized soil may also be purchased at garden stores. It may then be mixed with other materials to form the desired medium. Special media are now commercially available for seed propagation.

Indirect Seeding

Indirect seeding is a process in which seed is sown in a place separate from where the plants will eventually grow to maturity. The seedlings are transplanted one or more times before reaching the permanent growing area.

Flats

When growing seeds, horticulturists often select a *flat*. Flats are made of plastic and come in many shapes and sizes. Size selection should be determined by the number of seeds to be sown. The medium selected is placed in the flat and leveled off to about 1/2 to 3/4 inch below the top of the flat by using a tool called a *skew*, sized to fit the flats (Figure 6–5).

After the flat is filled, rows are made in which to sow the seeds. A row marker may be used, or rows may be made one at a time using a straight board as a guide. It is best to sow seeds in rows because if disease strikes one row of seed, it can be removed without disturbing the others. Also, when several different varieties or cultivars are in the flat, it is easier to label them by rows and much easier to transplant seedlings (young plants that have been germinated several days).

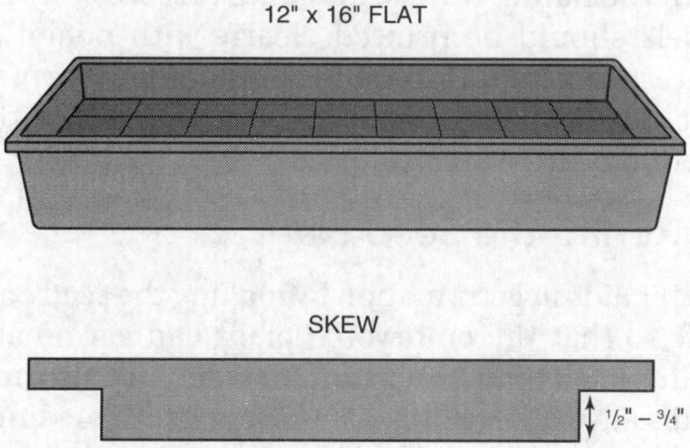

▲ Figure 6–5 Flat and skew.

Individual Cells in Cavity Seedling Trays

New seeding machines that can place one seed at a time in a series of small cell-like pots are now available. This allows the use of flats made up of many small pots. One or two seeds are planted in each cell, and are later thinned to one plant per cell. Cells are deep and narrow, making it possible to get many plants in a flat and yet have room for root growth (Figure 6–6). Plants from cells transplant with less root damage than those from solid mat flats. Figure 6-9, later in this chapter, shows a flat planted with begonia seedlings.

When to Seed

As businesspersons who must operate at a profit to survive, commercial plant growers must be sure that their plants are ready for sale at the correct time for outdoor planting and for holidays. Seeds must be planted on certain dates so that the resulting seedlings are ready for transplanting at the proper time. The charts in Figure 6–7 give planting information for seeds of various plants.

Sowing Seeds

To sow seeds properly, use the following procedure. Take the packet of seeds and shake the seeds to the bottom of the packet. Hold the seed packet with the open end slightly lower than the rest of the packet and gently tap the packet. The seeds will move out slowly and gradually, making it easy to sow the seeds properly spaced in rows. Follow directions on the package for determining the distance apart the

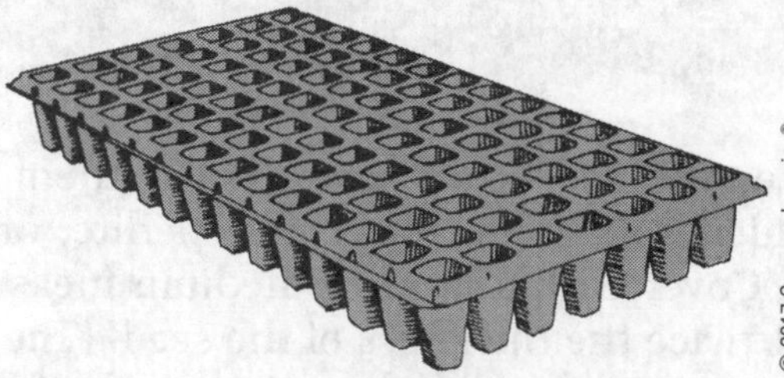

▲ Figure 6–6 Flats with 98 separate cells for seedlings.

PLANT	TIME FROM SEEDING TO MARKET	GERMINATION TIME/DAYS	OTHER OBSERVATIONS
Ageratum	8 weeks in packs 11 weeks in pots	8	Sow seed on surface Germinate at 70–80°F (21–27°C)
Annual Alyssum	7 weeks in packs 9 weeks in pots	8	Sow seed on surface Germinate at 70°F (21°C)
Begonia	14 weeks in packs 16 weeks in pots	2–3	Sow seed on surface, need light to germinate Germinate at 65–70°F (19–21°C)
Coleus	8 weeks in packs 9 weeks in pots	10	Cover seed ¼ inch Germinate at 70–80°F (21–27°C)
Impatiens	8 weeks in packs	8	Sow seed on surface, need light to germinate Germinate at 60–70°F (15–21°C)
Marigold	9 weeks in packs 10 weeks in pots	7	Cover seed ¼ inch Germinate at 70–75°F (21–24°C)
Pansy	12 weeks in packs 13 weeks in pots	10	Sow seed on surface, cover with paper to keep dark Germinate at 70°F (21°C)
Petunia	10 weeks in packs 11 weeks in pots	7–10	Sow seed on surface, need light to germinate Germinate at 70–80°F (21–27°C)
Salvia	10 weeks in packs 11 weeks in pots	14	Cover seed ⅛–¼ inch Germinate at 70°F (21°C)
Verbena	10–11 weeks in packs	14–21	Cover seed ⅛ inch Germinate at 70°F (21°C)
Vinca	13 weeks in packs 14 weeks in pots	14–21	Cover seed ⅛ inch Germinate at 75–80°F (24–27°C)
Zinnia	5 weeks in packs 9 weeks for color in pots	5	Cover seed ⅛ inch Germinate at 80°F (27°C)

Figure 6–7A Seeding information for growing plants to market size in the greenhouse.

seeds should be sown. If flats with individual cells are used, place one or two seeds in each cell.

GREEN TIP Sow the seeds of plants to create an economical, biological, and environmental niche on the land to enhance the stewardship and ways to green our good earth.

After the seeds are sown, cover them with shredded sphagnum moss, fine perlite, or fine sand. Cover with a layer of medium measuring about twice the thickness of the seed. (The seed package may have directions for depth of planting; if so, follow them for best results.) Some seeds, such as lobelia, petunia, and snapdragon, need light for germination and should not be covered when sown.

As soon as the seeds are sown, they are labeled with the name, variety, and date (Figure 6–8). All labels should be printed clearly with pencil or waterproof marking pen. Ballpoint pen should not be used because it washes out when the seedlings are watered.

Watering the Seed Flat

Water aids in germination by making the seed coat soft so that the embryonic plant can germinate. Water itself is an important nutrient and also acts to dissolve other nutrients present in the medium, making them available to the growing plants.

PLANT	TIME TO SOW SEED	DAYS FROM TRANSPLANT IN THE GARDEN TO HARVEST	GERMINATION TIME IN DAYS	OTHER OBSERVATIONS
Beet	Seed outdoors	55–68	Depends on soil temp.	Cover seed ½ inch Germinate at 50–60°F (10–15°C)
Broccoli	Seed outdoors	28–42	4–10	Cover seed ½ inch Germinate at 50–85°F (10–30°C)
Bush or Snap Bean	After soil is warm	65–90	7–10	Cover seed 1 inch
Cabbage	6–8 weeks before transplanting	80–100	4–10	Cover seed ½ inch Germinate at 50–85°F (10–30°C)
Cucumber	3–4 weeks before transplanting date	50–65	6–10	Cover seed 1 inch Germinate at 70–80°F (21–27°C)
Eggplant	8 weeks before transplanting	80–100	14	Cover seed ¼ inch Germinate at 70–80°F (21–27°C)
Lima Bean	Seed outdoors after soil is warm	65–90	Depends on soil temp.	Cover seed 1 inch
Melons	3–4 weeks before transplant date	85–120	15–21	Cover seed 1 inch Germinate at 70–80°F (21–27°C)
Onion	6–7 weeks before	100–115	7–10	Cover seed ½ inch Water regularly
Parsley	9 weeks before transplant date	60–75	15–21	Cover seed ¼ inch Germinate at 50–85°F (10–30°C)
Peppers	6–8 weeks before transplant date	57–75	8–14	Cover seed ¼ inch Germinate at 70–80°F (21–27°C)
Sweet Corn	After soil is warm	65–90 days from sowing seed	6–10	Cover seed 1½ inches Germinate at 70–75°F (21–24°C)
Tomato	4–6 weeks before transplant date	65–80	7–14	Cover seed ¼ inch Germinate at 70–80°F (21–27°C)

Figure 6–7B Seeding information for growing some vegetable crops. (See seed catalogs for others.)

To water freshly seeded flats, set the flats in a tub of water. This allows the seeds to be watered by *capillary action.* (The water is drawn up into the spaces between the soil particles to thoroughly moisten the medium.) After the medium is moistened, the flat is removed from the tub. If the flat is watered from the top, care must be taken to avoid washing the seeds out of the flat. Very small seeds should never be watered from the top because they may be washed too deeply into the soil.

▲ **Figure 6–8** Plant label.

Conditions for Germination

Seed flats containing seeds for germination should be located in a semishaded area of the greenhouse and receive a bottom heat of 65 to 70°F (18 to 21°C). (Some cool-season crops germinate well without bottom heat.) This may be accomplished in several ways. Location of the containers above the heat coil or hot water pipes or the use of a special propagating mat that is controlled with a thermostat are two possibilities. After the seeds are sown, the containers are covered with a pane of glass or clear plastic film to retain humidity. The covering should not touch the medium while the seedlings are germinating. Some seedlings, such as verbena, dusty miller, pansy, and portulaca, require a 3-day period of darkness for germination. For these plants, newspaper is used instead of glass or plastic. The paper is removed as soon as the seeds germinate so that the new seedlings receive light immediately. After germination, the medium is kept moist but never wet. The flats are watered with a gentle mist or from the bottom so that the small seedlings are not washed out.

The proper seedling medium is low in fertilizer elements. Seedlings should be fed weekly with a water-soluble fertilizer. As the seedlings approach transplanting size, a cooler temperature (55 to 60°F) (13 to 16°C) is provided to prepare the seedlings for the shock of transplanting. This is known as the **hardening-off** process. The process may also include modest withholding of water to slow active growth.

Many seedlings are now sold as plugs (Figure 6–9), which creates less transplant shock because the roots are not torn apart as they are when seedlings are transplanted from flats.

Seeds may be sown in pots, pans, or trays in a manner similar to that used for flats. The horticulturist should remember to provide drainage and uniform moisture. The containers should be watered from the bottom if possible. Plastic, newspaper, or glass is placed over the container to retain uniform moisture until the seeds germinate.

Transplanting Seedlings

After seeds germinate, they develop seed leaves or *cotyledons*, the first leaves to appear on the plant (Figure 6–10). The plant should be allowed to grow until the first true leaves are present before it is transplanted.

When handling seedlings, hold them by their true leaves using the thumb and forefinger. Do not hold by the stem; if the stem is badly bruised, the seedling could die. A bruised leaf is not nearly as serious an injury. While carefully holding the seedling, use a pot label to reach under the roots and lift, pushing the seedling out of the germinating medium. This exposes the roots. Do not shake the medium off the roots—exposure to air causes them to dry out. Plant the seedlings one by one about 2 inches apart in a flat. Using a *dibble* (a tool used to make the hole for transplanting seedlings), insert the seedlings to a depth slightly deeper than that at which they grew in the seedling flat. Gently press the medium around the roots. Water the seedlings at the soil surface with a gentle stream of water to settle the soil around the roots. The new seedlings are now ready to be grown to saleable size. (An alternate method

▲ **Figure 6–9** A flat of begonia plugs ready for transplanting to pots. Notice the small holes in the flat where some plants have been removed.

▲ **Figure 6–10** The cucumber plant on the right has only cotyledons. Plant on the left also has true leaves and is ready for transplanting.

involves using a trowel or fork when digging seedlings for transplanting.)

It is sometimes convenient to transplant seedlings into peat pots or market packs rather than flats. When a peat pot is used, one plant is placed in a pot and later transplanted directly to the garden with the plant remaining in the pot. Market packs generally hold 6 to 12 plants and are sold to the customer as such. The plants are then separated from the pack and planted individually in a permanent location.

PROCEDURE
Transplanting to Pots
(Seedlings or Rooted Cuttings)

1. Work from a large table. Place the pots in a row on your right side.
2. Place all other material to your left side.
3. Cover the drain hole with pottery or coarse sphagnum. (If using peat pots, this is not necessary.)
4. Hold the plant to be transplanted in the pot with the left hand.
5. Fill the pot to the rim with soil or soilless mix, gently holding the plant at the proper planting depth with the left hand.
6. Gently firm the planting medium around the plant by pushing down along the edge of the pot with the thumbs. Do not push hard enough to tear or bruise plant roots.
7. Label the pots or growing area according to variety, name, and date.
8. Move plants to growing area, and water until water runs through the pot and out the drain hole.

Another indirect seeding technique is to plant a single seed in a Jiffy 7 peat moss pellet. These pellets, when soaked in water, expand to seven times their original size. They contain all the necessary nutrients to feed the small seedling until it is planted in a permanent site. This eliminates the seedling transplanting stage prior to permanent planting and allows the young seedlings to be transplanted to the garden with less transplant shock. The Jiffy 7 pellets are placed in plastic trays or flats for ease of handling. Manufacturer's directions should be followed when using these pellets.

PROCEDURE
Transplanting Plants from Pots

1. Work from a large table. Spread fingers of the left hand around the plant stem and over the top of the soil surface.
2. Invert the pot and gently tap the pot edge on the edge of the table.
3. Remove the pot from the root ball as it loosens.
4. If the roots are pot bound, using your fingers, scissors, pruning shears or nursery spade, cut across the bottom of the root mass to break the matted root ball to prevent the roots from continuing to grow in a circular pattern.
5. Transplant the plant by spreading the four sections open and seating into the ground to be planted.
6. Using the soil and pine bark fines, mix evenly in the hole.
7. Water the plant with a slow stream of water from the hose. Saturate the soil to remove air pockets around the roots.
8. Mulch the plant with 2-3 inches of pine or hardwood bark.

Single seeds of squash, cucumber, and melon may also be planted in 2-inch peat pots in preparation for direct planting to the garden at a later time. These crops transplant with difficulty, and the peat pot permits transplanting without disturbing the root system.

Seedlings of nursery stock plants such as trees and shrubs are generally planted in a flat and transplanted directly to the nursery row outdoors. Seeds of these plants may also be seeded directly in the nursery row.

GREEN TIP Try to use the following seed-saving practice. Organic farmers and enthusiasts with environmentalist interests use open pollination, which happens through the natural process caused by insects, birds, or wind. Seed saving is the ultimate resource for heirloom plants.

Direct Seeding

Many seeds are planted directly in the permanent growing area. This is referred to as *direct seeding* (see Figure 6–11) and is the most economical seeding method. Plants such as corn, melons, beans, beets, peas, lettuce, carrots, and most other vegetable crops are grown by this process. Some vegetables, such as tomatoes, peppers, cabbage, and eggplant, are generally transplanted as plants to the garden. Some trees and shrubs are also grown by direct seeding.

▲ **Figure 6–11** Direct seeding in the nursery field by a robotic driverless nursery seeder.

In direct seeding, the planting medium is the soil. The soil is prepared by removing all large clods or lumps of earth so that the seeds are uniformly covered. In some cases, manure, grass clippings, or compost may be added to the soil to improve its structure. In direct seeding, it is important to plant the right cultivar at the right time in the right soil. Seeds may need treatment to prevent disease or insect damage. Weather conditions largely determine germination and initial growth.

The same conditions are needed for germination in direct seeding as when seeding in a flat: (1) there must be sufficient moisture and aeration; (2) the seedbed must allow firm contact with the seed; and (3) the temperature must be high enough to support the germination process. Requirements for germination of specific seeds are given in the directions on the label of the seed packet.

GREEN TIP Always use seed that has been tested for the current year of production. Viable seed will be labeled for the current year. Seeds not of the current year will have reduced germination.

SAFETY FIRST

For more information on occupational safety and health in the workplace, please visit the OSHA site at www.osha.gov.

Summary

Growing plants from seeds is one of the most economical ways to propagate plants, and it is used to grow many of our important food crops as well as ornamental landscaping plants. The fact that not all plants propagate true from seed must be kept in mind. Asexual methods, such as cuttings or grafting, must be used to propagate plants not growing true from seed.

Review Questions

1. Discuss the different ways plants propagate other than from seeds. What are the advantages of each method?
2. What is a primed seed?
3. What is the advantage of seeds selected to be all the same weight?
4. Why are primed or enhanced seeds best for direct field planting?

Student Activities

1. Plant seeds in a pot or flat. Care for the seeds through the transplanting stage. Keep a record of the date the seeds were planted, when they germinate, and when they are transplanted.
2. Collect various seed packets, and compile a list of information given on the packets. Discuss the importance of such information.
3. Soak a bean seed in water for a few days. Compare the soaked seed with a dry seed. What difference do you notice in the seed coat?
4. Separate a soaked bean seed, and see if you can identify the parts of the seed. Sketch and label the parts.
5. Search the Internet (www.USDA.gov) using the following terms: seed production; seed propagation; wildlife seed.
6. Participate in the FFA Floriculture Career Development Event at the state level.

Suggested Supervised Agricultural Experience (SAE) Activity

1. Research plant propagation techniques for an SAE in the area of research/agriscience.

Additional Resources

http://umaine.edu/gardening/master-gardeners/manual/propagation/plant-propagation/

http://ag.arizona.edu/pubs/garden/mg/propagation/

http://extension.missouri.edu/p/mg3

UNIT 7

Softwood and Semihardwood Cuttings and Micropropagation

Objective

To collect softwood and semihardwood cutting wood, prepare cuttings, and place them in the propagating structure for rooting.

Key Terms

asexual
clone
fungicide
herbaceous
node
tissue culture

Competencies to Develop

After studying this unit, you should be able to:

- select plants suitable for propagating by use of cuttings, and determine whether the maturity of wood is correct for optimum rooting.
- propagate at least 10 softwood or semihardwood plants from cuttings following procedures outlined in the unit, including the following:
  - collecting wood from the parent plant
  - making the actual cutting
  - treating the cutting with the proper rooting hormone when necessary
  - preparing the rooting medium and placing the cutting
  - caring for the cutting after it is placed in the medium

  (A minimum of nine of the cuttings should root for development of competency.)
- write a brief description of the tissue culture procedure.

Materials

- parent plant material from which to make cuttings
- plastic bags for collection
- labels and label marking pen
- propagating knife/or Felco hand pruners
- propagating container (flat, pot, etc.)
- rooting medium (sphagnum moss, perlite, or sand)
- rooting hormone
- watering can
- polyethylene plastic cover or sprinkling system
- smartphone apps:
  - Zapper QR Code Scanner
  - Leafsnap (to identify plants being propagated)
  - LikeThat Garden (to identify plants being propagated)

Asexual reproduction is required to produce **clones** (exact duplicates) of plants that are sterile (do not produce seeds), of hybrids as mentioned in Unit 6, and of plants that are difficult to grow from seed. The most commonly used and commercially important method of asexual reproduction is propagation via cuttings. *Cuttings* are leaves or pieces of stems or roots used for propagating plants. Most of the economically important perennials and many annuals are reproduced in this way. (*Perennials*, such as chrysanthemums, are plants that grow year after year without replanting. *Annuals*, such as petunias, normally complete their growth cycle in one year.)

Various kinds of cuttings are taken from different parts of the plant and at different stages of plant maturity (age). Cuttings may be taken from the leaf, bud, stem, or root (Figure 7–1). They may be taken when the plant tissue is soft and tender (*softwood cuttings*) or when the plant tissue is hardened and more woody (*hardwood cuttings*). Some plants root more easily from softwood; others root more easily from hardwood.

To root, cuttings require essentially the same conditions that seeds need to germinate: moisture, oxygen, and warmth (40°F [4°C] or above). In addition, cuttings need light—something that most seeds do not need to germinate. Light is essential in allowing the cuttings to produce food through the process of photosynthesis. Cuttings may begin to grow roots without light, especially in the first stages of *callus* formation (growth of tissue over a wounded plant stem), but, generally, light is needed for food production. Food production supplies energy to the rapidly developing roots. Stored hardwood cuttings are an exception because they are able to develop roots without light.

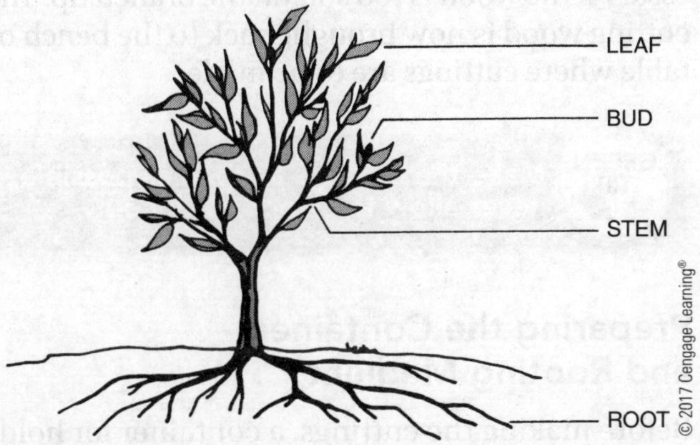

▲ **Figure 7–1** Cuttings may be taken from the bud, stem, leaf, or root.

Root formation on cuttings is stimulated due to the interruption of the downward movement of carbohydrates, hormones, and other materials from the leaves and growing tips. Because these materials cannot travel beyond the point at which the stem has been cut, they accumulate and stimulate the healing and rooting process.

Additional rooting hormones are sometimes used to aid in root formation. Some plants root more easily than others because they produce a higher level of natural rooting hormones. These plants need less synthetic rooting hormones to root satisfactorily. Cuttings with high levels of carbohydrates often root easier.

Cuttings from young plants such as seedlings and rooted cuttings often root much faster and in higher percentages. This "juvenile effect" also accounts for the better rooting qualities of cuttings that are taken near the plant's roots or crown because these shoots are often younger. Soft, newly formed shoots often root much easier due to the juvenile effect as well. This effect is stronger in some plants than in others.

Softwood and Semihardwood Stem Cuttings

One method of asexual reproduction by cuttings is softwood and semihardwood stem cuttings. These cuttings are taken after the current or present season's growth has at least partially matured or hardened. The wood for semihardwood cuttings should be able to bend but generally snaps if bent at an angle greater than 70 to 80 degrees. The softwood cutting is taken while there is still active *terminal* (tip) growth. This type of wood is very soft and breaks with a snap. The horticulturist best learns about maturity of wood through experience; it is not necessary to be exact with most plants.

Collecting Cutting Wood

The first step in making the cutting is to find a parent plant that has made at least 2 to 6 inches of new growth during the current year. This current season growth may be from the tip of a branch or a shoot that is growing from the base of the plant (Figure 7–2). Check to be sure that the wood is of proper maturity or hardness and that the plant is not wilted or dried out. If the plant appears to be suffering from lack of moisture, water it well and wait until the shoots and leaves are turgid (swollen with moisture). This may require waiting overnight. Early morning is the best time to take a cutting from the parent plant. The shoots and leaves contain more moisture at this time than later when the sun has had time to dry them out. Remember, the new cutting will have no roots with which to gather moisture, so it should be started with as much moisture as possible.

Immediately after taking the cutting from the parent plant, place it in a bucket of water or a plastic bag containing water. This prevents drying of the wood after it is cut from the plant and before the actual cutting is made from it. At this time, prepare a label with the variety name and the date. Place the label in the container with the cutting wood. Be sure the label is marked with waterproof pencil or waterproof felt-tip pen.

Softwood and semihardwood cuttings are made when plants have leaves on them. There are various degrees of tissue hardness or maturity, depending on the plant being propagated. As an example, Figure 7–3 gives the best rooting date and desired maturity of various plants for the mid-Atlantic area.

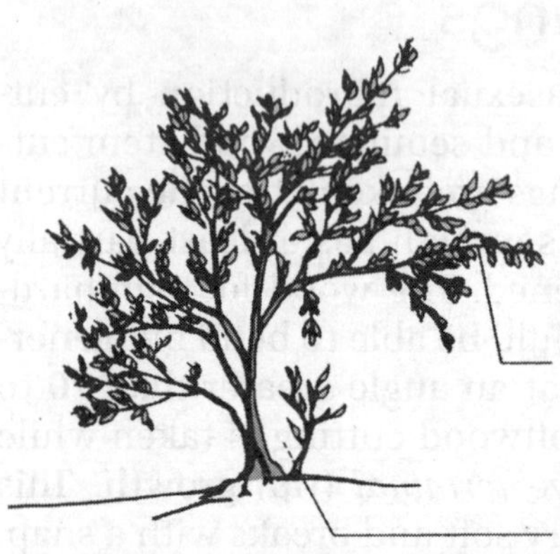

▲ **Figure 7–2** Softwood and semihardwood cuttings may be taken from the tip of a branch or from a shoot growing at the base of the plant.

PROCEDURE
Nine Steps in Taking Softwood and Semihardwood Cuttings

1. Collect the cutting wood of proper maturity from a selected parent plant.
2. Label the cutting wood.
3. Fill the propagating container with moist rooting medium.
4. Make the cutting 3 to 4 inches, and cut just below the node at a 45-degree angle.
5. Treat the cutting with rooting hormone.
6. Insert the cutting in the rooting medium.
7. Firm the medium around the cutting.
8. Insert the label in the rooting container.
9. Control the atmosphere around the cuttings by covering with plastic or placing under a mist system.

Leaves help the propagator keep the cuttings right side up—they point upward. The buds in the leaf axil are always on top on the leaf stem and toward the tip (Figure 7–4). It is necessary to know the top of the cutting because the stem must be placed in the rooting medium with the bottom end down. Stem tissue is highly polarized, and roots form much more easily on the bottom of the cutting. To avoid confusion, cut the branches from the parent plant at a 45-degree angle on the bottom end and straight across at a 90-degree angle on the top (Figure 7–5). This is important only if the tip of the branch is cut off, a process that is usually necessary if the wood is too soft on the branch tip. The cutting wood is now brought back to the bench or table where cuttings are to be made.

GREEN TIP Make the cuttings as soon as possible; never allow the wood to dry out.

Preparing the Container and Rooting Medium

Before making the cuttings, a container for holding the medium must be prepared. A flat, pot, or any container at least 4 inches deep may be used

PLANT NAME	ROOTING HORMONE	MONTH TO TAKE CUTTING IN MID-ATLANTIC STATES	CUTTING WOOD MATURITY	REMARKS
Arborvitae* *(Thuja occidentalis)* *(Thuja orientalis)*	Hormodin #2	Oct.–Dec.	hardwood	Wounding cutting helps.
Azalea (evergreen) *(Rhododendron)*	Rootone or Hormodin #1	early July	semihardwood	Use only terminal growth.
Barberry *(Berberis deciduous)*	Hormodin #2	July–Sept.	semihardwood	
Barberry (evergreen)	Hormodin #2	Nov.–Dec.	hardwood	
Boxwood* *(Buxus)*	Hormodin #2	July or Nov.–Dec.	semihardwood or hardwood	
Cotoneaster	Hormodin #2	July	semihardwood	
Dogwood *(Cornus)*	Hormodin #3	June–July	softwood	new growth after flowering
Deutzia	Rootone or	June–July	softwood	
	Hormodin #1	Nov.–Dec.	hardwood	
English ivy* *(Hedera helix)*	Rootone or Hormodin #1	Nov.–Dec.	hardwood	
Euonymus*	Rootone or	June–July–Aug.	evergreen-semihardwood	
	Hormodin #1	Dec.–March	deciduous-hardwood	
Firethorn* *(Pyracantha)*	Rootone or Hormodin #1	July–Oct.	semihardwood	
Forsythia	Rootone or	June–July	softwood in container	
	Hormodin #1	Dec.–March	hardwood, outside	
Holly *(Ilex)** (evergreen)	Hormodin #3	July–Aug.	semihardwood	terminal growth
Hydrangea	Rootone or	June–July	softwood	
	Hormodin #1	Dec.–March	hardwood	
Juniper *(Juniperus)**	Hormodin #3	Nov.–Dec.	hardwood	

Figure 7–3 Rooting date and desired maturity of various plants. *(continued)*

PLANT NAME	ROOTING HORMONE	MONTH TO TAKE CUTTING IN MID-ATLANTIC STATES	CUTTING WOOD MATURITY	REMARKS
Lilac *(Syringa)*	Hormodin #2	directly after flowering	very soft wood	Use fungicide.
Mock orange *(Philadelphus)*	Rootone or	July	softwood	
	Hormodin #1	Dec.–March	hardwood	
Pachysandra*	Rootone or	Nov.–Dec.	hardwood	
	Hormodin #1			
Privet *(Ligustrum)*	Rootone or	July	softwood	
	Hormodin #1	Dec.–March	hardwood	
Rhododendron	Hormodin #3	July and Sept.–Oct.	semihardwood	Wound cuttings; use only terminal growth.
Spirea *(Spiraea)*	Rootone or	July	softwood	species: *vanhouttei*
	Hormodin #1	Dec.–March	hardwood	
Viburnum	Rootone or	May–June	softwood	fragrant types all others
	Hormodin #1	July–Aug.	semihardwood	
Weigela	Rootone or	July–Sept.	softwood	
	Hormodin #1	Dec.–March	hardwood	
Wisteria	Rootone or	July	softwood	
	Hormodin #1			
Yew *(Taxus)*	Hormodin #3	Dec.–March	hardwood	

*May be taken in fall and midwinter.

Note: One of the new liquid dip rooting materials may be used in all cases above. Adjust the strength by diluting with water as recommended on the label. These materials penetrate plant stems better than powders. *(Wear rubber or nitrile gloves when working with all chemicals. Do not allow chemicals to get on your skin.)*

Figure 7–3 Rooting date and desired maturity of various plants. *(continued)*

▲ **Figure 7–4** Parts of the cutting.

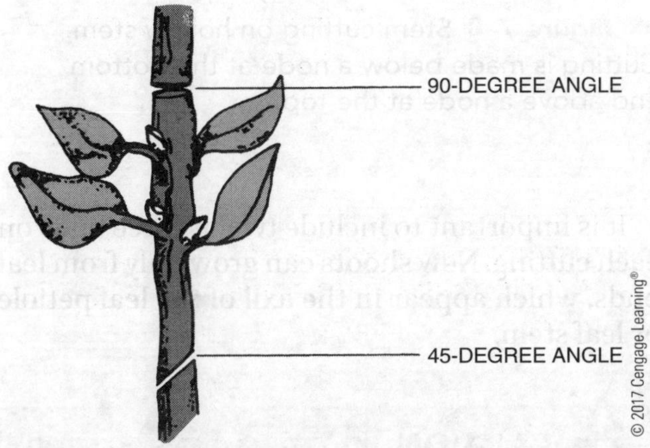

▲ **Figure 7–5** Cut branches from the parent plant at a 45-degree angle at the bottom and a 90-degree angle at the top.

(Figure 7–6). The size of the container (other than its depth) is determined by how many cuttings are to be made. The container must have holes in the bottom to allow for drainage.

GREEN TIP It is essential that the rooting medium be loose to allow circulation of air, be able to hold moisture, and be sterile.

Remember that roots need oxygen and moisture to live. The cut end of the cutting makes it easy for organisms that cause rotting to enter the plant, so the medium must be sterile or free from living organisms. A mixture of one-half horticultural perlite to help loosen the structure of the rooting medium and one-half coarse sphagnum moss provides a good medium.

The medium may depend on the type of watering system that is to be used. Excellent success can be obtained with the peat-perlite mixture

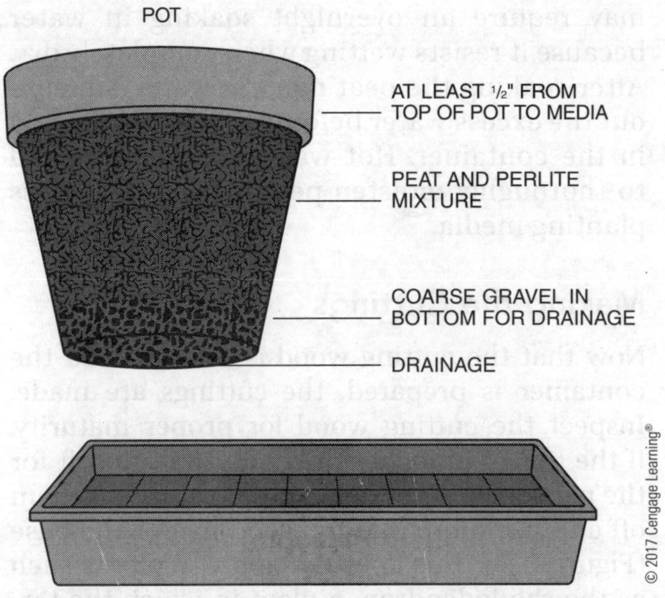

▲ **Figure 7–6** Any container at least 4 inches deep, such as a plant pot or flat, may be used to hold cuttings.

in containers that are to be covered with plastic until the cuttings root. When mist is used (a process in which fine droplets of water are sprayed on plants through nozzles), pure perlite or pure sand is satisfactory. If pure perlite or sand is not used, the percentage of peat moss should be reduced to 25 percent or less. The high percentage of peat moss recommended for pots and flats sealed with plastic would hold too much moisture under mist, thereby withholding sufficient oxygen and possibly causing the cuttings to rot. The rooting mixture should be mixed before it is wet so that a uniform mixture results.

Filling the Container

After the rooting medium is thoroughly mixed, it is placed in the container, leveled, and firmed. There should be at least 1/2 inch between the top of the container and the level of the medium surface so that water does not run off the surface as the medium is watered. The surface is leveled with a flat board, skew, or brick. The leveling is important so that water spreads evenly over the surface and all cuttings are watered equally. The medium is watered until water runs through the bottom of the container. If the peat moss used is dry, the moss may require an overnight soaking in water because it resists wetting when completely dry. After soaking the peat moss in water, squeeze out the excess water before mixing and placing in the container. Hot water is recommended to thoroughly moisten peat moss and soilless planting media.

Making the Cuttings

Now that the cutting wood is selected and the container is prepared, the cuttings are made. Inspect the cutting wood for proper maturity. If the tips of branches or shoots are too soft for the particular plant being propagated, cut them off and use more mature wood nearer the base (Figure 7–7). This does not apply to plants such as the rhododendron, a plant in which the terminal end must be used as the cutting because buds appear only near the terminal end.

Make as many 3- or 4-inch cuttings as possible from each stem or shoot. A very sharp knife or pruning shears must be used so that a clean cut is made. Remember to cut the bottom on a slant and the top of the cutting straight across to indicate which end should be placed in the rooting medium. The bottom of cuttings are cut slanted for another reason. The new cutting has no roots, and the slanted bottom allows more surface area to contact the rooting medium for moisture uptake or absorption. It is not important to cut at or near a **node** unless the plant stem is hollow. On hollow stems, cut just below the node on the bottom and just above a node at the top (Figure 7–8).

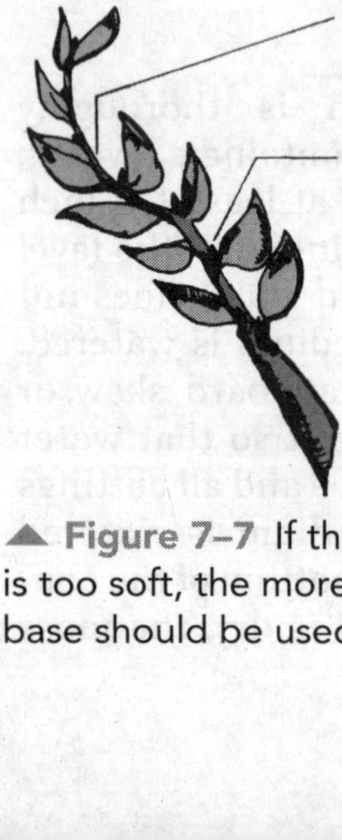

▲ **Figure 7–7** If the top of the branch or shoot is too soft, the more mature wood closer to the base should be used.

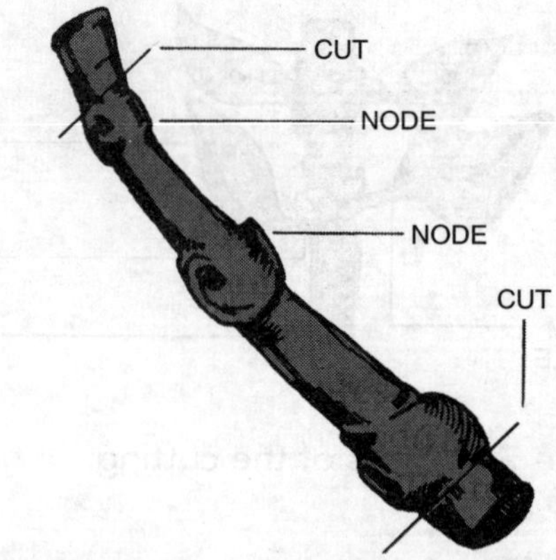

▲ **Figure 7–8** Stem cutting on hollow stem. Cutting is made below a node at the bottom and above a node at the top.

It is important to include two or three buds on each cutting. New shoots can grow only from leaf buds, which appear in the axil of the leaf petiole or leaf stem.

> **CAUTION**
> Knives used in propagation are sharp. Never cut toward yourself or another person.

Treating with Hormones and Inserting in Medium

After the cutting is made, it is treated with the proper concentration of rooting hormone. Follow directions on the label, and be sure the substance is recommended for the plants being rooted. Rooting hormones are chemicals that help cuttings grow roots more quickly and grow a larger number of roots. Hormones used for various plants are listed in Figure 7–3. If the hormone does not contain a **fungicide**, the cutting may be sprayed, dipped, or watered with a fungicide such as Captan to help prevent rotting. Captan has been reported to stimulate rooting in addition to controlling fungus infection.

The cutting is immediately placed in the rooting medium to one-half its length but not more than 2 inches deep. Cuttings should be spaced so that no leaves overlap. Large leaves may be trimmed. Press the medium around the cutting very firmly. Do not water when using powder rooting hormones.

Labeling

The cuttings must now be labeled, using either the same label that was prepared when the wood was cut from the parent plant or preparing new labels. The name of the rooting hormone should be added to the information already on the label (variety name and date, including the day, month, and year). Remember to use a waterproof pencil or pen. The label is placed in the rooting container.

Controlling the Atmosphere

The new cutting has leaves, but no roots. Because of this, these leaves lose moisture through *transpiration* (loss of water through the leaves or stems of plants) more quickly than it can be drawn up through the slanted cut end. Therefore, it is essential to keep the relative humidity around the cutting close to 100 percent to reduce moisture loss.

If a greenhouse is available, one way in which high humidity can be maintained is to place the cuttings in a misting propagation bench. This system uses a greenhouse bench equipped with misting nozzles that spray a fine mist on the cuttings from a piped water system. The system may be operated manually or set up to water the cuttings automatically at specific intervals. An alternative method for use without a greenhouse is to enclose the containers with a film of polyethylene plastic (a clear plastic that holds in moisture but lets in light and a small amount of air). Very little water is lost from the container, but there is a need to water the cuttings every 3 or 4 weeks until they have rooted.

When a pot is used as a container, a plastic food bag may be slipped over it and tied in place or tucked under the pot as shown in Figure 7–9. A flat may be covered by drilling three holes in each side and placing stiff wire over the flat. This structure is covered with polyethylene film that is tied or tucked under the bottom of the flat (Figure 7–10A and B). Containers covered with plastic must be placed out of direct sunlight by locating them against a north wall, in the shade of a tree, or under a bench. Light aids formation of carbohydrates—and thus rooting—and should be available. However, the cuttings should not receive direct sunlight because this raises the temperature

▲ **Figure 7–9** Rhododendron cutting ready to be inserted in a plastic bag. The bag is pulled over the pot and tied to seal in moisture. The pot is then placed under bright light but not direct sunlight.

▲ **Figure 7–10A** Flat with rooting medium being enclosed in a watertight framework. The plastic is pulled over the wire hoops and tied around the bottom or tucked under to form a watertight hut for propagating.

▲ **Figure 7–10B** A commercially available propagating flat similar to the framework hut.

inside the container too much, thereby killing the cuttings. If they must be placed in the sun, the plastic may be covered with cheesecloth or burlap to give shade and hold down the temperature. When the misting system is used, and cuttings are not in a tight container, direct sunlight is beneficial.

Rooting

As time nears for the cuttings to be rooted, open the plastic-covered container or go to the misting propagation bench and check the cuttings by holding each cutting and tugging gently. If the cutting does not easily slip out, roots are developing. If there are no roots, merely close the container or leave the cuttings under mist, depending on the system used, and wait another 7 to 10 days. As long as the cutting has not rotted or turned dark and is holding its leaves, it will probably root if given more time. Growth on the tip or sides of the cutting is normal but may not indicate that the cutting has rooted.

When the cuttings have a root ball measuring 2 or 3 inches across (Figure 7–11C), it is time to harden them off in preparation for transplanting (removing and planting in a new place). The hardening off must be done very slowly over a period of 7 to 14 days so that the cuttings are not exposed to a drying atmosphere too suddenly. They are accustomed to high humidity and must adapt to new, drier conditions. Open the plastic a little more each day over the 14-day period until it is totally removed. After the plastic is completely opened, care must be taken not to allow the rooting medium to dry out. The cuttings must now be watered to keep the medium moist but not wet; a water-soluble fertilizer may be used at one-fourth recommended strength every other watering. If a misting propagation bench is used, the misting cycle is changed so that the misting time becomes shorter and shorter until no mist is used and the cuttings are watered normally.

▲ Figure 7–11 (A) shows a rhododendron cutting as taken from parent plant; (B) shows the cutting after trimming, with stem and leaves trimmed and the stem slices on each side at the base; (C) illustrates a well-rooted rhododendron cutting ready to be planted in the cold frame.

After the hardening-off process, the cuttings are ready to be *lined out* (planted outside in the nursery row under shade or sprinklers) or transplanted to other containers for further growth.

Herbaceous Cuttings (Softwood)

Herbaceous cuttings are made from succulent greenhouse plants such as the geranium, chrysanthemum, coleus, and carnation. Most florist crops are propagated in this way. Cuttings are made 3 to 6 inches long with leaves left on the upper or terminal end, and several cuttings may be obtained from a plant stem. This allows the horticulturist to obtain many more cuttings than if only leaf or stem tips are used. Leaves are trimmed from the basal, or lower, end of the cutting, with three or four leaves left remaining on the top end. A fungicide should be used on all cuttings to prevent rotting. A sterile rooting medium is also essential.

Herbaceous cuttings are rooted under the same high-humidity conditions required for softwood cuttings. Bottom heat helps to speed rooting. Under proper conditions, rooting is rapid. For this reason, this type of cutting is highly recommended for the beginner. A mild concentration of rooting hormone such as Rootone #1 may be used but is not necessary. A high degree of success may be expected, with more than 90 percent of cuttings rooting. Using a flat, flower pot, or mist system in the greenhouse, try propagating at least one of the herbaceous plants listed. Care for them in the same manner as noted for the softwood cuttings.

Additional plants easily propagated as herbaceous cuttings include the lantana, Swedish ivy, wandering Jew, and begonia.

Micropropagation—Tissue Culture

As mentioned previously, one of the newest approaches to plant propagation is **tissue culture** (also called *micropropagation* or *meristem culture*). This is an asexual method using sterilized terminal shoots or leaf buds placed on a sterile (disease-free) agar gel or other nutrient-growing medium (Figure 7–12). For sterilizing, 1 part bleach to 10 parts medium is used, and the proper growth-regulating chemicals and nutrients are mixed in the growing medium. The buds or stem tips and medium are placed inside a test tube or small jar and kept sealed to keep out mold and other disease organisms.

GREEN TIP Micropropagation creates opportunities to produce clones of the parent plant that will not reproduce true from seeds and conserves endangered plant species.

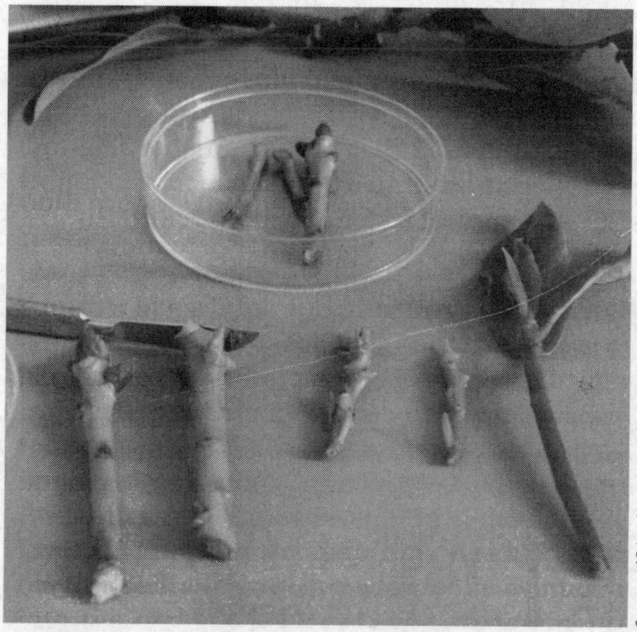
Courtesy of Briggs Nursery, Sidhu & Sons Nursery USA Inc.

▲ **Figure 7–12** Small stems being trimmed and sterilized. They will be placed in a small jar or test tube on nutrient agar gel to start sending out shoots. Shoots will emerge from the buds or growing tips of the stem.

Soon, new tiny plants sprout out from the plant tissue placed in the sealed container (Figures 7–13 and 7–14). As the tiny sprouts grow large enough to be moved, they are pulled off with sterile tweezers and placed in a new medium in another container to grow roots (Figure 7–15), or the new shoots may be placed back in a jar to produce more shoots. This process is repeated again and again as new sprouts appear. As soon as roots develop, the container may be opened, a little more each day for about a week, allowing the plants to harden off. Plants may then be transplanted to small containers

Courtesy of Briggs Nursery, Sidhu & Sons Nursery USA Inc.

▲ **Figure 7–13** Small shoots (microcuttings) growing from a stem in a jar.

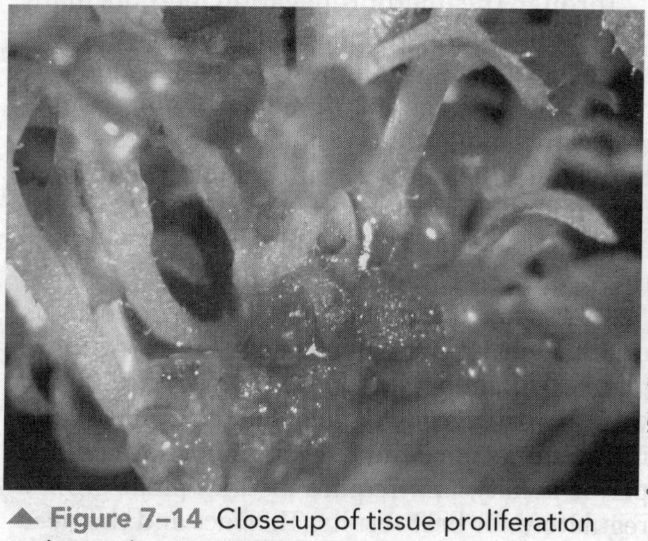
Courtesy of Briggs Nursery, Sidhu & Sons Nursery USA Inc.

▲ **Figure 7–14** Close-up of tissue proliferation and new shoots sprouting.

▲ **Figure 7–15** Shoots (microcuttings) being separated and placed in pots to root. They will root like any other softwood cutting taken from a full-sized plant. Because they are very tender, extreme care must be taken that they do not dry out.

and treated in the same manner as small seedling transplants.

GREEN TIP It is very important that all containers, growing media, hands, and so on, are sterile so that no plant disease organisms are present.

Through the micropropagation method, many thousands of new plants exactly like the parent can be produced very quickly. This is a good way to rapidly increase the numbers of a new superior plant and get it on the market years earlier than could be done with cuttings or other propagation methods. It is estimated that in 28 weeks, a single blackberry tip could produce 2.5 million tissue particles ready for rooting.

Micropropagation has been used for years on orchids but is now being used for many kinds of plants. Different plant types and even different varieties of the same type of plant often need varying amounts of chemical plant food, have unique pH or acidity, and require distinct plant growth regulators for best results. This fact and the need for absolutely sterile conditions make it hard for the untrained person to propagate plants through micropropagation. For this reason, no more detail will be given here. If you are interested in trying this newest propagating technique and have the laboratory equipment, get all the details from your local agricultural college or experiment station. Be prepared to perform almost daily observation on a strict time schedule. Also, be sure you can keep sterile growing and transplanting conditions until the small plants are stuck for rooting.

An alternative way of obtaining tissue culture plants is to purchase them at the stages shown in Figures 7–13 and 7–15 and then perform the rooting step yourself. This method does not require as much care for disease prevention.

Research done at McMaster University in Ontario, Canada, has made it possible to produce small duplicate plants from the tip of a growing plant. A plant hormone is applied to the stem and a new plant starts to form right on the plant. No expensive sterile laboratory is needed in this process.

Plant Tissue Culture

A new propagation system called *in vitro soil* (IVS), started by the Western Australia Department of Agriculture, may change existing tissue culture methods. The difference with this system is in the second step when plants develop roots. Instead of being placed in an agar medium to root, the microcuttings are planted in a 0.5-2-2 ratio of sphagnum peat moss, coarse river sand, and horticultural perlite. Results are much better than methods now used; cuttings root faster and more vigorously.

TRY THIS Research online to find a scientific article that discusses the use of *in vitro soil* (IVS) propagation. Prepare a summary of the findings in the article.

Other Types of Softwood Cuttings

Leaf Cuttings

The African violet is a good example of a plant propagated by leaf cuttings. Refer to Figure 7–16 for details, and try propagating this easy-to-root

▲ **Figure 7–16** (Left plant) African violet cutting as taken from the parent plant. (Middle plant) Rooted cutting showing the small plant growing from the base of the leaf stem. It is ready to be planted as a new plant. The old leaf is pruned off at this stage or soon after. (Right plant) Adult African violet plant in bloom.

plant. (See Figure 7–18 later in this unit for details on making mallet cuttings.)

It will be necessary to advance to another unit while the cuttings are rooting. A careful check should be made each day, however, to see if the cuttings are progressing and to be sure that no damage has been done to the containers or the cuttings.

Leaf-Bud Cuttings

Many plants normally propagated using softwood and semihardwood cuttings may also be propagated using leaf-bud cuttings. Rhododendrons root well this way and may be used as a project to test this method.

The cutting is taken by cutting a bud from the stem very much like the cut made for the bud in the T-budding procedure in Unit 11. The difference is that the leaf is left intact, and thus the cutting must be protected from moisture loss the same as with a softwood or semihardwood cutting. The protection process was explained earlier in this unit.

The first step in the cutting procedure is to cut the leaf and bud from the plant as shown in Figure 7–17(A). The wooden shield of the cutting should be about 3/4 to 1 inch long, and the leaf should be cut off crossways to get it to that size. (Refer to Figure 7–11 for leaf trimming.) This smaller leaf will require less moisture to survive. Figure 7–17(B) shows the complete leaf-bud cutting. Figure 7–17(C) shows the cutting inserted in the rooting medium. It must be firmed in, watered, and covered in a moisture-tight container. The advantage to leaf-bud cuttings is that many more cuttings can be made from the same number of parent plants than can be made using stem cuttings. This may be important in introducing new plants. Usually, fewer plants will root, and sometimes the bud is slow to break into new growth. The strength of the chemical rooting hormone should be weaker than that for stem cuttings (about 50 percent or less in strength). Rooting hormones appear to slow initial bud growth and thus new shoot development.

Mallet Cutting

The mallet cutting is made using a 2- to 4-inch piece of 2-year-old wood with a current season or 1-year-old shoot on it (Figure 7–18). The mallet is placed horizontally and covered about 1 inch with rooting medium.

Root Cuttings

Root cuttings may be made from any plant that will sprout or sucker from the root. Roots from

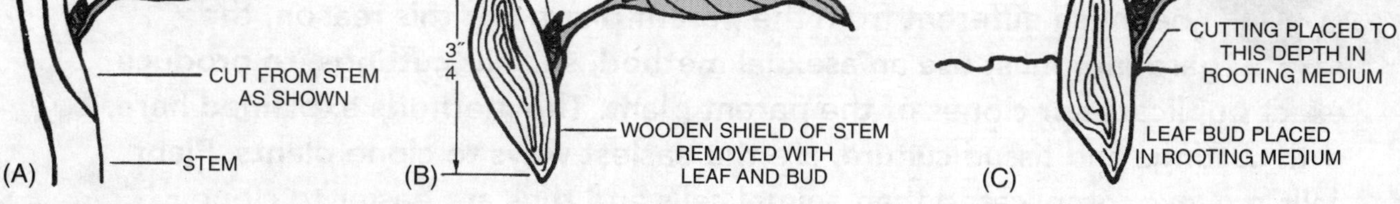

▲ **Figure 7–17** Leaf-bud cutting.

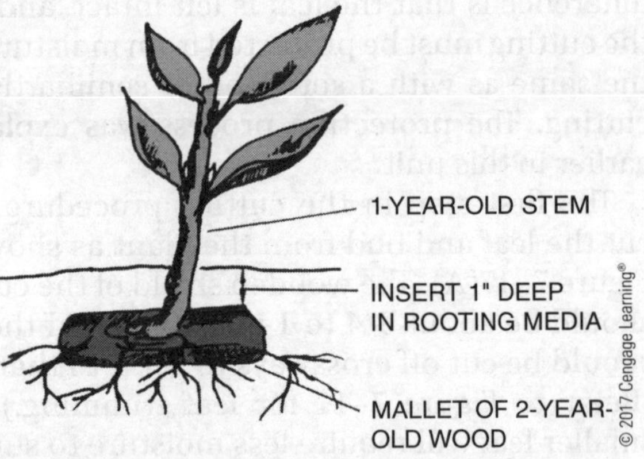

▲ **Figure 7–18** Mallet cutting. Depth of inserting the cutting is 1 inch.

1/8 to 1/2 inch in diameter cut into pieces 1 to 4 inches long are placed in a flat or other container and watered well. Glass or clear plastic over the flat helps to hold moisture. When new shoots sprout from the flat, the new plants are moved to the nursery row. Plants commonly propagated this way include the Oriental poppy, wisteria, spirea, red raspberry, and trumpet creeper. Figures 7–19 and 7–20 illustrate the propagation of plants by use of root cuttings.

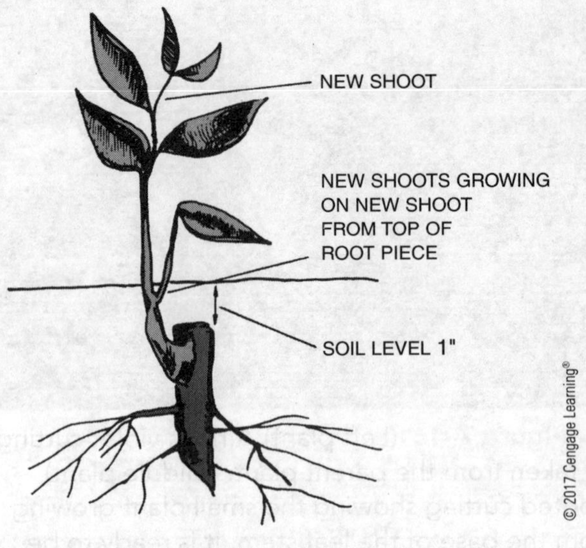

▲ **Figure 7–19** Fleshy root cutting. After cutting is placed in soil, it is given no special care.

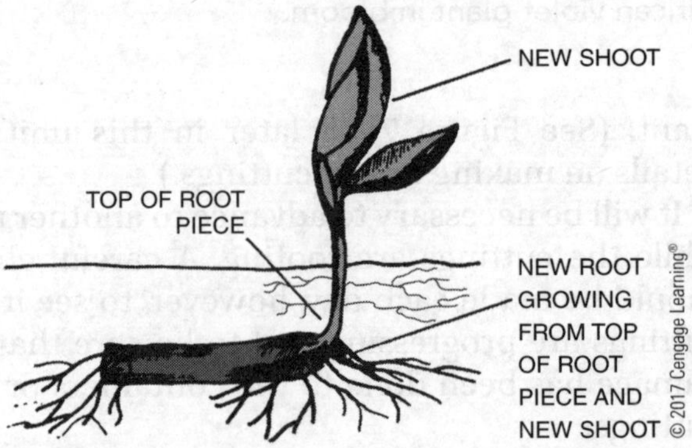

▲ **Figure 7–20** Small, nonfleshy root cutting. This cutting is placed horizontally in the soil. Depth is 3/4 inch below soil level.

GREEN TIP When obtaining propagated cuttings, use a knife or Felco pruners to make the proper cuts to enhance the rooting ability of the cutting.

SAFETY FIRST

Use caution when collecting cuttings; the knife and/or Felco pruners are sharp! For more information on occupational safety and health in the workplace, please visit the OSHA site at www.osha.gov.

Summary

Many plants do not propagate true from seeds. Seedlings will be much different from the parent plant. For this reason, the gardener must use an asexual method, such as cuttings, to produce exact duplicates or clones of the parent plant. The methods explained here, with cuttings and tissue culture, are the easiest ways to clone plants. Plant cells are less complicated than animal cells and thus are easier to clone.

Review Questions

1. What is asexual plant propagation? What type of plants are usually propagated by asexual propagation?
2. Why is it necessary to reproduce plants asexually?
3. Discuss the proper timing for taking softwood and semihardwood stem cuttings.
4. Describe the nine key steps in taking softwood or semihardwood cuttings.
5. What part of the plant is used to propagate a plant by tissue culture?

Student Activities

1. Practice making softwood and semihardwood cuttings by completing each step of this exercise:
   a) Collect cutting wood for at least 10 softwood or semihardwood cuttings. Label the cuttings and keep moist until placed in the rooting mix. Check with the instructor on the proper maturity of cutting wood.
   b) Select a container and fill it with the proper rooting mix. (Provide for drainage.) Water properly as determined by the instructor.
   c) Using a sharp knife, make 10 cuttings and insert them in the rooting medium. Treat with rooting hormone, label, and space properly. (The text may be used for reference.)
   d) Place the cuttings under a mist system or cover with plastic. Ninety percent of the cuttings should root. Check to be sure all of the previous steps in the exercise have been completed correctly. If 9 of 10 cuttings root, you are a competent propagator. If less than 90 percent root, repeat the project if time allows. If not, review the steps in the procedure and trace your errors.

> **GREEN TIP** Spray cuttings with a systemic fungicide or dip them in a fungicide when they are made. This is usually sufficient treatment to prevent rotting if a sterile medium is used.

2. Plants that are more difficult to root, such as evergreens, may be tried as an additional activity. A good choice might be the rhododendron. The following are necessary extra steps for this plant:
   a) Take a cutting with three or four leaves from terminal current season's growth in July or from September 15 to October 30. (Trim leaves if necessary.)
   b) Cut to 2 1/2 inches long.
   c) Trim a slice off each side of the cutting 1 inch from the base and extending to the base just deeply enough to cut into the wood.
   d) Treat with Hormodin #3, 0.8 percent indolebutyric acid, or a liquid such as Dip-N-Grow. (Liquid dip treatments are more effective.)
   e) Complete by following the same steps given for other cuttings.
3. Use one of the commercially available tissue culture kits to propagate.
4. Search the Internet for educational videos using the following terms: asexual propagation; stem cuttings; leaf cuttings; micropropagation; tissue culture. Find a video that demonstrates proper propagation methods and share with the class.

Suggested Supervised Agricultural Experience (SAE) Activities

1. Develop an entrepreneurship SAE by establishing a business using micropropagation techniques for reproduction.
2. Conduct a research SAE studying the effectiveness of tissue culture kits.

Additional Resources

http://aggie-horticulture.tamu.edu/tisscult/Microprop/microprop.html

www.biotech.iastate.edu/publications/lab_protocols/AV_Micropropagation.html

www.sandiegozoo.org/CF/plants/micropropagation.html

https://extension.unh.edu/resources/files/Resource000867_Rep913.pdf

UNIT 8

Hardwood Cuttings

Objective

To perform the six steps in taking hardwood cuttings so that success in rooting 60 to 80 percent of the cuttings is achieved.

Key Terms

bare root
callus
deciduous
dormant
hardwood cutting
lining out

Competencies to Develop

After studying this unit, you should be able to:

- list five plants commonly propagated by hardwood cuttings.
- select, collect, and label mature cutting wood from one of the five identified parent plants.
- make and store 10 hardwood cuttings.
- line out cuttings in the field or greenhouse at the proper planting time.
- keep accurate records of the percentage of cuttings that root.

Materials

- cutting wood
- string to tie bundles
- storage area
- propagating knife and/or Felco pruners
- labels and waterproof pen or pencil to mark labels
- smartphone apps:
  - Zapper QR Code Scanner
  - Leafsnap (to identify plants being propagated)
  - LikeThat Garden (to identify plants being propagated)

Propagation by hardwood cuttings is one of the easiest and least expensive methods of vegetative or asexual propagation. Cuttings are prepared in the **dormant** winter season when time usually is more available to the propagator. **Hardwood cuttings** may be shipped long distances or stored for long periods of time. Expensive misting equipment or rooting benches are not needed, and the necessary cold storage may be done outdoors in low-cost facilities if temperatures are 50°F (10°C) or lower. Many **deciduous** woody plants are easily propagated by hardwood cuttings. Figure 8–1 lists deciduous and narrowleaf evergreens propagated by hardwood cuttings.

Propagation by hardwood cuttings differs from softwood and semihardwood cuttings in several ways:

- the time of year in which the cuttings are taken
- the hardness, or maturity, of wood used for the cuttings
- the usual absence of leaves on the cuttings
- the storage of cuttings instead of immediate planting

SOME DECIDUOUS PLANTS GENERALLY PROPAGATED BY HARDWOOD CUTTINGS	
blueberry	mock orange
currant (ribes)	mulberry
deutzia	Persian lilac
dogwood	plum
euonymous	pomegranate
fig	privet
forsythia	rosa multiflora
gooseberry	spirea
grape	weigela
honeysuckle	wisteria
kerria	

SOME NARROWLEAF EVERGREENS PROPAGATED BY HARDWOOD CUTTINGS	
chamaecyparis	spruce
hemlock	thuja
juniper	yew
pine	

Figure 8–1 Deciduous and narrowleaf evergreens propagated by hardwood cuttings.

PROCEDURE
Six Steps in Taking Hardwood Cuttings

1. Identify the plant to be propagated, and select proper cutting wood.
2. Collect and label wood for cuttings.
3. Make the cuttings.
4. Store the cuttings.
5. Line out or plant the cuttings.
6. Determine the rooting percentage of the cuttings.

Selecting the Cutting Wood

Wood for hardwood cuttings is generally taken from the current season's growth, as is done with softwood cuttings. The wood is cut from the ends of branches or from long shoots that grow from the base of the plant (Figure 8–2). This material may be gathered soon after the plants become dormant. Plants that are dormant have lost their leaves and are preparing for the winter rest cycle. Cuttings may be taken throughout the winter months. The wood taken is current year's growth, but it is now mature or hardwood. Plants that are healthy, vigorous, and grown in full sunlight yield cutting wood with more stored carbohydrates and more vigor in rooting.

GREEN TIP Using a natural hardwood cutting program for your plantings is not only economical, but the program also provides greater control of the plants overall health, reducing the need for chemical spraying. You will have higher producing plants and protect the health of the surrounding plants.

▲ **Figure 8–2** Strong, reddish-brown young shoots growing from the base of a blueberry plant. These shoots (located to the left of the parent plant) make excellent hardwood cuttings. Notice that the leaves have all dropped off the plant because the plant is dormant.

Collecting the Cutting Wood

Using a sharp knife or hand pruner, cut the selected wood from the parent plants. Because the wood has no leaves, drying out is not as much of a problem as it is with softwood cuttings. Label the cuttings according to variety name and date collected. The wood may be made into 6- to 8-inch cuttings for immediate use or stored in a cool moist place (below 50°F [10°C]) for later use (Figure 8–3). If stored, the branches should never be allowed to dry out or become too wet. A covering of moist (but not wet) sawdust, sand, or peat moss works well to maintain moisture. There should be little variation in temperature during storage because temperature variation tends to hasten drying.

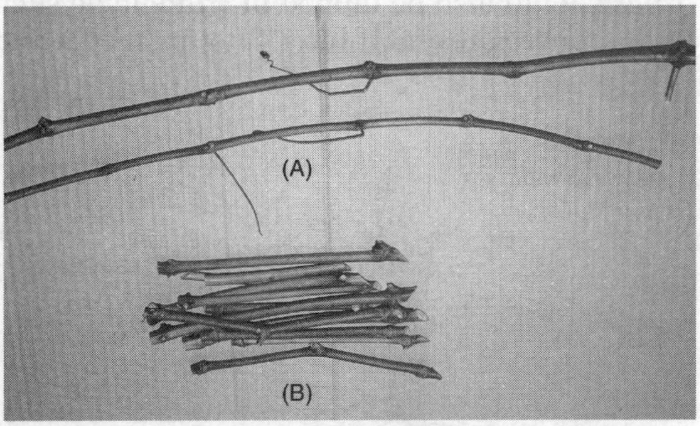

▲ **Figure 8–3** (A) illustrates grape hardwood cutting wood as collected from the plant. (Stems are longer than shown in the photograph.) (B) shows hardwood cuttings cut to the proper length and ready to be tied in bundles and stored.

Taking Cuttings

Cuttings, usually 6 to 8 inches long, are made with a sharp knife or pruning shears (Figure 8-3A). The bottom cut is made just below a node and the top cut about 1 inch above a node or bud (Figure 8–3B). Make the bottom cut (end toward the base of the plant) at a 45-degree angle and the top cut at a 90-degree angle so that cuttings can be planted bottom down. The diameter of cutting wood is not extremely important. Medium-sized wood, however, tends to survive storage and root better than wood that is very thin.

Cuttings should be made early enough so that they may be stored 6 to 8 weeks prior to planting.

Storage of Cuttings

The cuttings are treated with a rooting hormone for better growth. The finished cuttings are then tied in bundles for storage. They should be placed so that all the tops are on the same end of the bundle and buried in moist sawdust or similar material. The bundles should not be tightly sealed; the cuttings are alive and need oxygen to remain healthy. A tight seal could also cause excess moisture to collect, resulting in rotting.

The storage period allows the cut ends of the cuttings to **callus** and the rooting process to begin. Callus formation, which occurs during storage, is an advantage when cuttings are placed outdoors for rooting under drying conditions because callused cuttings root more quickly. New roots begin to grow during storage and may be visible at this point.

The cuttings may be buried outside in sand-filled containers in a well-drained area. Outside temperatures must be cold enough to prevent growth beginning at the tops of the cuttings during the storage period. During the first 4 weeks of storage, the temperature should be 50 to 55°F (10 to 13°C). This relatively high temperature is favorable for callus formation. After callus has formed, the temperature may be lowered to below 40°F (4°C) but never below 32°F (0°C). This lower temperature prevents growth on the tops before the cuttings are lined out (planted).

TRY THIS Test the impact of the storage temperature on the successful growth of cuttings. Using a laboratory freezer, store one bundle of cuttings at 40°F, one bundle at 56°F, and one bundle at 31°F. Place cuttings outside and record the differences in growth.

Lining Out Cuttings

As soon as the soil is ready in the spring, the cuttings are planted outside in a process known to nursery workers as lining out. Before **lining out** the cuttings, the soil must be prepared. A well-drained, sunny site with some wind protection is best.

Rows are dug deeply enough so that the cutting can be covered with soil with only the top bud remaining above the soil level (Figure 8–4). If they are to be left for only 1 year, the cuttings are placed 6 inches apart in rows 1 foot apart. When the cuttings are to be left for more than 1 year or when especially fast-growing plants are involved, the cuttings are spaced 9 to 10 inches apart and in rows 2 to 3 feet apart.

A small shovel or other tool is used to break the soil. The cuttings are placed in the soil at the depth described earlier. The cutting should not be pushed into the soil under pressure; this could damage the tissue or break the cutting. After the cutting is placed in the row, the soil is pressed firmly around the cutting. The soil surface may then be mulched to help hold moisture and control weeds, or it may be cultivated for weed control. The soil should never be allowed to dry out around the cuttings. Provision should be made for watering as needed.

The number of plants that will eventually root depends on the variety of the plant. However, between 60 and 80 percent of the cuttings can be expected to root and grow.

As the weather becomes warm, the cuttings develop leaves and shoots although they may not yet be rooted. However, if growth continues into the summer months, it is almost certain that the cuttings have rooted. The care required beyond this stage varies with the plant being propagated.

Narrowleaf evergreens propagated by hardwood cuttings are often very difficult and slow to root. In general, the slow-growing *Juniperus* species root easily, the yews fairly easily, and the upright junipers, spruces, and hemlocks with difficulty. Pines are extremely difficult to root. Rooting hormones with high concentrations of indolebutyric acid (0.8 to 1.0 percent actual chemical solution or powder) are usually beneficial. These cuttings are taken from late fall to late winter and are best rooted in a greenhouse under high light intensity and in high humidity. This is achieved by placing the cuttings under plastic or using a misting system as specified for softwood and semihardwood cuttings. Cuttings should also be dipped in a fungicide such as Captan to prevent rotting. Rooting media and

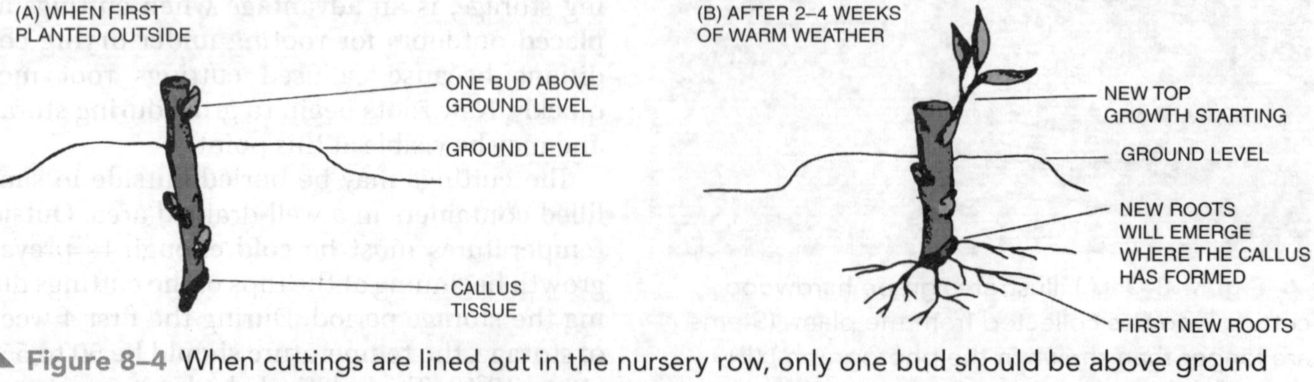

▲ **Figure 8–4** When cuttings are lined out in the nursery row, only one bud should be above ground level. At least one (preferably two) buds should be below ground level. (A) shows cutting directly after planting; (B) shows root development and initial top growth.

other environmental conditions are essentially the same as those for softwood and semihardwood cuttings.

Hardwood cuttings of deciduous plants do not require the extensive care given to softwood and semihardwood cuttings. Because no leaves are present in the initial rooting stages, the demand for moisture is not high, and these cuttings need not be placed under plastic or in a mist system. The evergreen hardwood cutting, which does have needles at the time of rooting, must be handled as softwood cuttings are.

Rooted cuttings of deciduous plants are usually dug for transplanting when they are dormant and without leaves. They may be dug **bare root**. Evergreens, however, that have needles should be dug with a ball of soil attached to the roots. Care must be taken to keep the root ball intact.

GREEN TIP When obtaining propagated cuttings, use a knife or Felco pruners to make the proper cuts to enhance the rooting ability of the cutting.

SAFETY FIRST

Use caution when collecting the cuttings; the knife and/or Felco pruners are sharp! For more information on occupational safety and health in the workplace, please visit the OSHA site at www.osha.gov.

Summary

Hardwood cuttings are an easy, inexpensive way to root some plants such as grapes. Rooting occurs outside or in a greenhouse, and much of the work can be done in winter when more time is available. This method does not work well for many plants.

Review Questions

1. Discuss how propagation by hardwood cuttings differs from propagation by semihardwood cuttings. What are some advantages of propagation by hardwood cuttings?
2. What plants are commonly propagated by hardwood cuttings?
3. Discuss the process of lining out cuttings. What are some important considerations to keep in mind during this process?

Student Activities

1. Visit a local nursery to observe a demonstration of hardwood cutting techniques.
2. a) Select at least one plant commonly propagated by hardwood cuttings. Perform the six basic steps to produce at least 10 cuttings.

b) Set up a calendar to determine the following:
   1. the storage temperature and dates to change temperature
   2. the approximate date for lining out certain cuttings
   3. inspection dates so that things such as moisture conditions, temperature, and rooting date can be noted

c) Line out the cuttings in a nursery row in the spring and follow up to determine rooting success. Keep a record of observations, including dates when shoots and roots first develop.

3. Start a small nursery on the school grounds or at home in which to propagate and grow plants for future use. This is an inexpensive way to keep a supply of laboratory materials on hand.
4. Search online for current articles about the following topics: asexual reproduction; stem cuttings. Select one article to summarize.

Suggested Supervised Agricultural Experience (SAE) Activities

1. Participate in a placement SAE by working in an orchard. Record all of the time worked, and give a description of the experience.
2. Work in a nursery that has a cutting program. Record the experience information for credit in a placement SAE.

Additional Resources

www.extension.org/pages/29211/blueberry-propagation#.VVYsSVV0yM8

https://ucanr.edu/repositoryfiles/ca1906p4-59136.pdf

www.wnps.org/education/resources/documents/Garden_Links/hardwood_cuttings_and-live-stakes.pdf

UNIT 9

Separation and Division

Objective

To propagate one plant by separation and one by division.

Key Terms

bulb
bulblets
corm
cormels
division
double-nose
herbaceous perennials
offset bulbs
rhizomes
separation
splits or slabs
tuber
tuberous roots

Competencies to Develop

After studying this unit, you should be able to:

- describe the processes of separation and division, and explain the major difference between the two.
- identify five specialized plant structures used in propagation, and explain how each is used.
- list five plants propagated by separation or division and the type of specialized structure used in each case.
- write a brief definition of bulb, corm, tuber, tuberous root, and rhizome.

Materials

- bulbs, corms, tubers, tuberous roots, rhizomes
- sharp knife and/or Felco pruners
- storage container and packing material

Many plants are propagated by one of two processes usually associated with one another—the process of separation and the process of division. In both cases, plants are propagated by use of underground plant parts. The primary function of these plant parts is to act as an organ for food storage. These plant parts are found on herbaceous perennials having shoots that die back at the end of the growing season; the underground part lives through the winter (dormant season) and sends forth a new top the following growing season. The second function of these plant parts is that of vegetative or asexual reproduction.

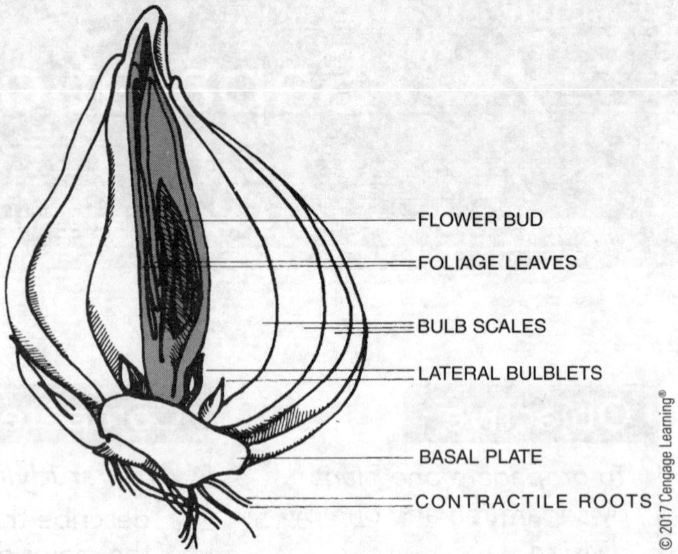

▲ **Figure 9–1** Parts of the tulip bulb. The bulb scales grow small bulblets at their base. The bulblets eventually grow larger and may be removed and grown to flowering size.

Separation is a method of propagation in which natural structures produced by certain plants are removed from the parent plant to become new plants. Bulbs and corms are examples of removable structures. These parts are usually removed while the parent plant is in a dormant or resting stage. Some bulb-producing plants propagated by separation are the tulip, amaryllis, lily, narcissus, daffodil, hyacinth, and grape hyacinth. Examples of corm-producing plants are the crocus, gladiolus, and colchicum.

Division is a method of propagation in which parts of the plant are cut into sections, each capable of developing a new plant. Division is possible with plants that grow **rhizomes**, stem **tubers**, and **tuberous roots**. Examples of plants with rhizomes are the iris, canna, and calla lily. Those with stem tubers include the caladium, peony, and Irish potato. Tuberous roots are found on the dahlia, gloxinia, and bleeding heart.

Propagation by Separation

As mentioned, plants that are propagated by separation are those that produce bulbs or corms as specialized underground plant parts that are responsible for food storage and propagation of the plant. Separation is a natural process.

Bulbs

PARTS OF THE BULB A **bulb**, such as the tulip bulb, is a plant structure containing many parts but primarily composed of leaf scales (Figure 9–1). Immediately next to and outside of the foliage leaves are bulb scales that produce small **bulblets** (tiny bulbs) at their base. These tiny bulblets grow larger and become small bulbs that may be separated and planted as individual plants. After several growing seasons, the small bulbs become large enough to bloom, which produces **offset bulbs**. The process of developing offset bulbs is a simple commercial method for propagating the tulip, daffodil, bulbous iris, and hyacinth (Figure 9–2).

Offset daffodil bulbs are called **splits or slabs** when first separated from the mother bulb. After one year of growth, the slab or split is called a *round bulb* and is capable of flowering the following season. Within the second year, a second flower bud forms. At this time, the bulb is known as a double nose and is capable of producing two flower stalks. The round and **double-nose** sizes are sold commercially. Large mother bulbs are kept to produce more splits. Splits or slabs are planted to grow for another year to reach flowering size.

TYPES OF BULBS The tulip bulb is a fairly solid structure and therefore does not require special care in handling. The lily, however, is a loosely scaled bulb that cannot withstand rough handling or drying. The tulip is a *laminate* or *tunicate* bulb; these types of bulbs have dry membranous

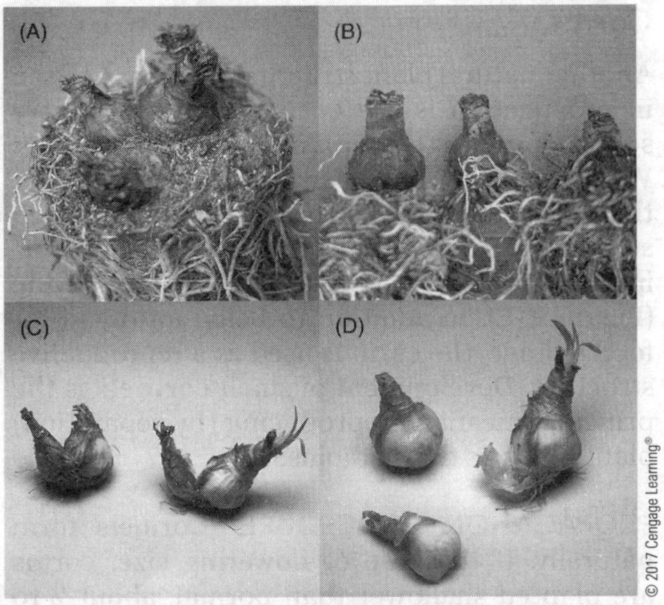

▲ **Figure 9–2** (A) An amaryllis bulb with smaller offset bulbs attached. (B) The same bulbs, after the offsets have been separated from the mother, or parent, bulb. (C) Two daffodil bulbs with smaller offset bulbs attached. (D) The same bulbs, after the offset bulbs have been separated from the parent plant. The two offset daffodils may require another year to grow to blooming size.

outer scales that protect them against drying and injury. The lily represents a *nontunicate* or *scaly* bulb, which has no tough outer cover.

PROPAGATION PROCEDURE Bulbs may be dug and separated after the foliage of the plant dies back and the plant is in a dormant state. The new bulbs are then stored at 65 to 68°F (18 to 20°C) and planted at the proper time. For tulips, planting time is in the fall. The minimum size for flowering of tulip bulbs is 10 centimeters. Separated bulbs that are smaller than this must be grown for 1 to 3 years before they reach flowering size.

Dry bulbs are stored at 55 to 60°F (13 to 16°C). Bulbs should be washed clean of all soil before storage and treated for rot.

The tops of bulb-producing plants should not be cut off until they have turned brown. Early removal of tops greatly reduces the amount of food produced and stored and thus reduces bulb size. This could result in the bulb being too small to flower the following season.

Potted plants, such as the amaryllis, should be allowed to continue to grow for 6 to 8 months after blooming before water is withheld to force them into dormancy. This allows time for the bulb to grow and store sufficient food reserves to bloom during the next growing period.

The other plants mentioned earlier, such as narcissus, hyacinth, and grape hyacinth, are propagated in a similar manner.

LILY PROPAGATION The lily (Figure 9–3) is also propagated by bulbs but at a much slower rate. Mother bulbs are split at the base to force production of new bulblets. Some lilies naturally grow small bulblets on the stems and even in leaf axils. Commercially grown lily bulbs are separated, and individual bulb scales are planted to produce new bulblets (Figure 9–4). This is a

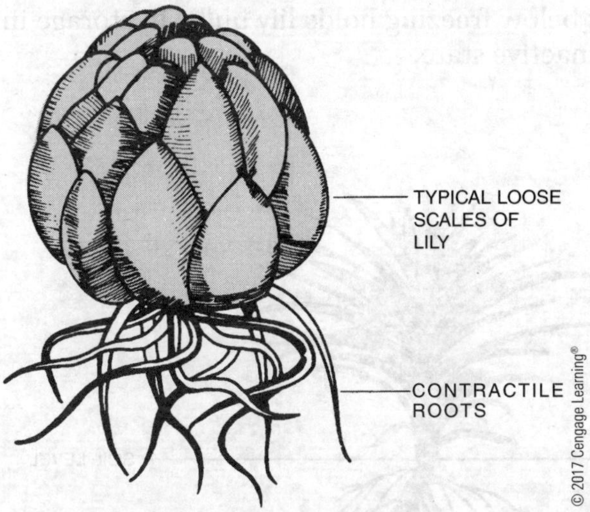

▲ **Figure 9–3** Typical lily bulb.

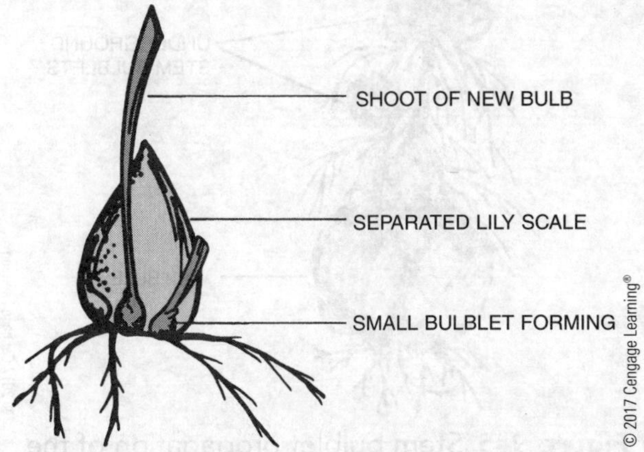

▲ **Figure 9–4** Bulb scale propagation of the lily.

more rapid method of increasing the number of bulbs because the lily bulb contains many loose, easily separated bulb scales.

The Easter lily is propagated commercially with underground stem bulblets (Figure 9–5). The flowering stems are pulled from the ground in late August through mid-September, and the small bulblets and stems are kept moist by sprinkling them. In about mid-October, the small bulblets are placed 4 inches deep and planted about 1 inch apart to grow the first season. They are moved again in September of the following season and planted about 6 inches deep, spaced 6 inches apart. By the end of the second year, they are dug and sold as flowering bulbs. At this time, bulb size should be 7 inches in circumference or larger. When dug, bulbs must be handled carefully to prevent bruising and must not be allowed to dry out. The best storage is in moist, but not wet, sphagnum moss. A temperature of just below freezing holds lily bulbs in storage in an inactive state.

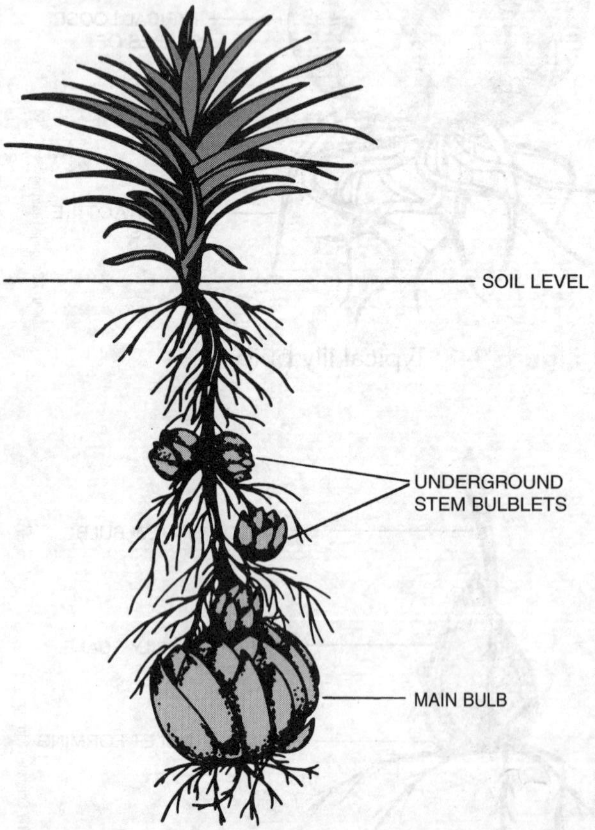

▲ **Figure 9–5** Stem bulblet propagation of the Easter lily.

Corms

Another natural plant structure that can be used in propagation is the **corm**. A corm is a very solid, compact stem with nodes and internodes. Whereas the bulb is composed of leaf scales, the corm consists of a very short specialized stem for food storage. Corms have a dry covering that protects them from injury and drying (Figure 9–6). In addition to being an organ for food storage, the corm is used as a reproductive structure. Development of small **cormels** is the principal means of reproducing (by separation) plants such as the gladiolus.

PROPAGATION PROCEDURE Cormels form naturally. If they are of flowering size, corms are planted shallower than normal, about 2 to 3 inches deep. More cormels will form at this depth than if corms are planted deeper. When the plant top dies back, the plant may be dug and the small cormels separated and grown to larger size. The plant top should be allowed to die back normally by the effects of frost or should be grown at least 3 months after blooming so that the food supply manufactured in the top and stored in the specialized stem (the corm) is sufficient to develop good-sized cormels.

After frost or pulling of plants, the corms and cormels are separated from the rest of the plant and dried for storage. They should be treated with a fungicide and stored at 40°F (5°C) in a well-ventilated area with 80 percent humidity to prevent too much drying.

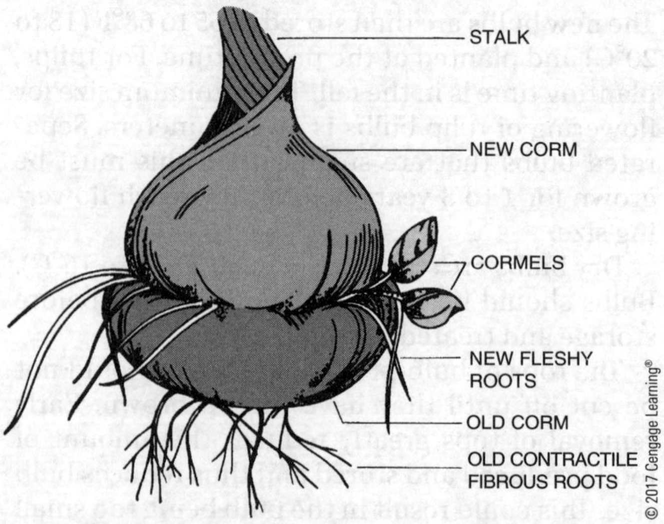

▲ **Figure 9–6** Gladiolus corm.

A hot-water treatment for disease control may also be necessary for cormels. Contact a commercial grower for information concerning the best treatment.

In the field, new cormels are planted in rows to grow larger, much as large seeds are. Two years are usually required for them to reach blooming size.

Propagation by Division

Division differs from separation in that it is not a natural process; the parts used for propagation do not separate naturally from the mother plant. Parts of the plant that are to become new plants must be cut from the mother plant with pruners or a knife. Rhizomes, stem tubers, and tuberous roots are used in propagation by division.

Rhizomes

Rhizomes are underground stems that grow horizontally and produce roots on the bottom and stems on the top. A rhizome may be thought of as a plant lying on its side with the stem covered with soil. An example of a plant that reproduces in this way is the iris.

PROPAGATION PROCEDURE When plants become crowded, they may be divided for the purpose of reproducing or increasing numbers of plants. Rhizomes generally grow very near the surface of the soil. They are removed from the soil by digging underneath the plant with a garden (spading) fork or shovel and lifting them out of the ground. All soil is washed from the plant so that the parts are clearly visible for division. Division is done by cutting the rhizome into sections. Because a rhizome is an underground stem, the same care must be taken as with a stem cutting to make sure that each section has at least one bud (eye) and preferably several.

The iris is used here to illustrate division of a rhizome. Iris are divided in late summer after bloom. Each clump is washed clean and cut apart (Figure 9–7). The tops should be cut back to about one-third of their original height to balance the root loss. After being cut apart, the new rhizomes are dusted with a fungicide, especially on the cut surfaces. They may then be planted in a new location.

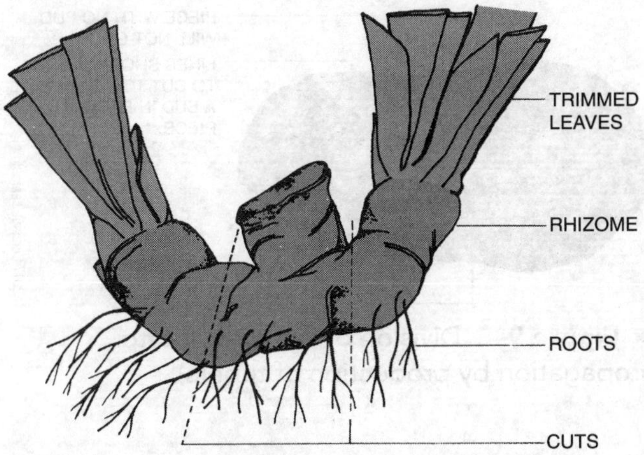

▲ **Figure 9–7** Division of the iris (rhizome structure).

> ***TRY THIS*** With the supervision of the instructor, create a clear grow box using Plexiglas construction. Fill the Plexiglas structure with growing media and plant a *laminate* or *tunicate* bulb, *nontunicate* or *scaly* bulb, corm, and rhizome in the media. Care for them and watch the *laminate* or *tunicate* bulb, *nontunicate* or *scaly* bulb, corm, rhizome, and tuber root under the medium from all sides. Record the growth of the roots.

Tubers

A tuber is a swollen end of an underground side shoot or stem. Tubers are distinguished by their eyes. Each eye produces a separate plant as it sprouts, developing a shoot with roots at the base of the new shoot. The tuber contains stored food on which the new plant feeds until new leaves take over the job of food production. One common tuber is the Irish potato.

> **GREEN TIP** **Plants are lessening the need for manufactured medicines. For example, the potato is a rich source of vitamin C, which aids in reducing cholesterol levels.**

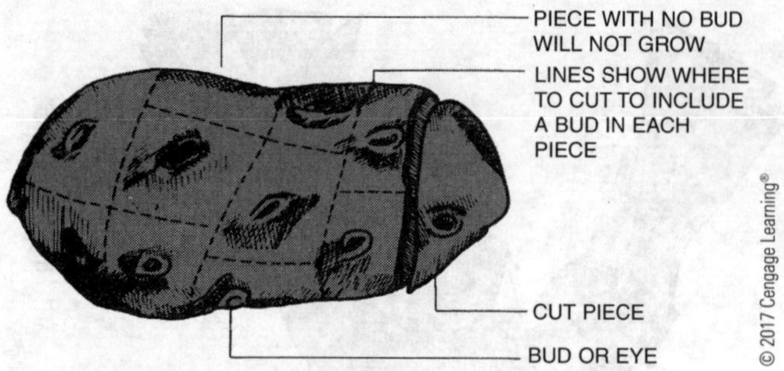

▲ **Figure 9–8** Division of the Irish potato (propagation by production of tubers).

PROPAGATION PROCEDURE Propagation of plants by tubers is done by cutting the tubers into small pieces. Each piece must contain at least one bud or eye (Figure 9–8). The pieces are planted in the same manner as larger seeds. No storage is required; the cut surfaces are allowed to dry, and the pieces are planted immediately.

Tuberous Roots

Tuberous roots are thickened roots that contain large amounts of stored food. They differ from tubers in that they are roots and have buds only at the stem end. Roots are produced at the opposite end (Figure 9–9).

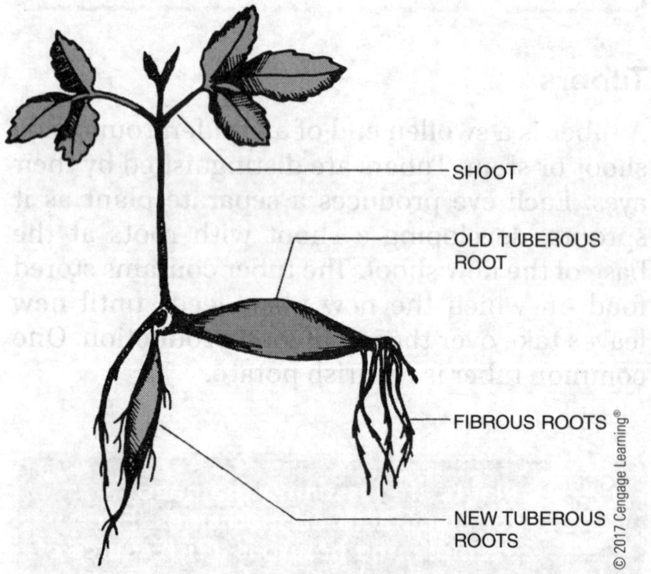

▲ **Figure 9–9** Tuberous root (dahlia).

PROPAGATION PROCEDURE Tuberous root crops are propagated by dividing the crown, or cluster of roots, when the plant is dormant. The plant is dug in the fall after frost has killed the top, or it has died back for other reasons. The soil is washed away from the roots, and they are allowed to dry. The roots are stored in dry sawdust, peat, or other materials at a temperature of 40 to 50°F (4 to 10°C) to prevent shriveling or complete drying out. Just before planting in early spring, the clumps or crowns are cut apart so that each piece has a bud. If sprouts are present, it is easier to see if a bud is part of the cut piece. The new pieces are planted in the garden or nursery row after the danger of frost has passed.

The sweet potato, a tuberous root, sends out many shoots from adventitious buds. (*Adventitious* buds are those that occur at sporadic and unexpected places on the vegetative structure.) The shoots that grow from these buds root very quickly if the tuber is placed in a moist rooting medium. The shoots are pulled off and planted as rooted shoots without any part of the old tuberous root attached (Figure 9–10).

GREEN TIP When making propagated cuttings, use a knife or Felco pruners to make the proper cuts to enhance the rooting ability of the cutting.

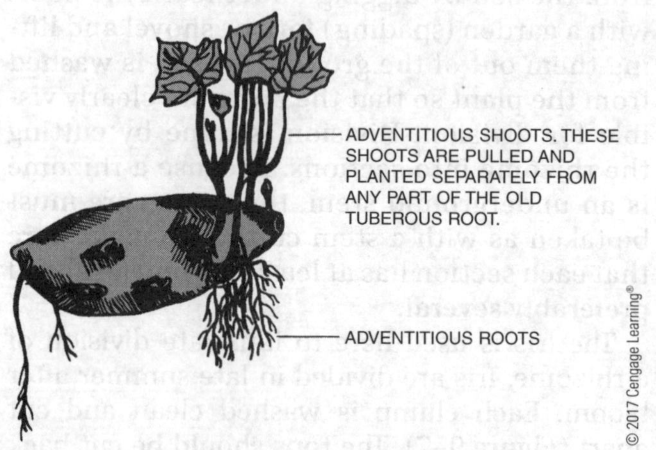

▲ **Figure 9–10** Tuberous root (sweet potato).

SAFETY FIRST

Use caution when collecting the cuttings; the knife and/or Felco pruners are sharp! For more information on occupational safety and health in the workplace, please visit the OSHA site at www.osha.gov.

Summary

Separation and division are asexual methods of propagating some plants, which are easily duplicated by these two methods. Separation is a natural method developed in nature. Division requires that plants be cut apart to separate clumps of roots or modified stems. Both methods result in exact duplicates of the parent. These methods are widely used by the nursery industry and gardeners to propagate many perennial plants.

Review Questions

1. How are the processes of separation and division similar? How are they different?
2. What are the five specialized natural plant structures that are used in propagation? How is each structure used for propagation?
3. For each structure discussed in question 2, name a plant that uses this method of propagation.

Student Activities

1. Propagate at least one plant by separation and one plant by division. Plant on the school grounds or at home, if possible.
2. Start a sweet potato in a glass of water or in moist sand. Observe the roots growing from one end of the tuberous root and the shoots from the other. This is known as polarity. The commercial grower increases the number of shoots for propagation by giving the sweet potato root a hot-water treatment to eliminate the polarity.
3. Cut a sweet potato root in half, and heat the pieces at 110°F for 26 hours. Place in moist sand or water. Observe the greater number of adventitious shoots that grow when polarity has been eliminated. Cover the base of the shoots with moist sand as they grow and roots will develop.
4. If time permits, choose other plants that are propagated by separation and division and experiment with them. Some other plants that propagate by separation or division are day lilies, hostas, orchids, and caladiums.
5. Search online for college courses in the following areas: plant division; botanical division.

Suggested Supervised Agricultural Experience (SAE) Activity

1. Work for a local nursery where bulb-producing plants are raised. Record this experience and use it for a placement SAE.

Additional Resources

http://pubs.ext.vt.edu/426/426-002/426-002.html

www.clemson.edu/extension/hgic/plants/landscape/flowers/hgic1150.html

UNIT 10

Grafting

Objective

To perform three methods of grafting so that success in growing 60 percent of the grafted plants is achieved.

Key Terms

cambium
compatible
grafting
high tunnels
rootstock
scion
stenting

Competencies to Develop

After studying this unit, you should be able to:

- identify three types of grafts from a sketch.
- list at least one plant that is propagated by each of the three grafting methods.
- list five reasons for grafting.
- Graft herbaceous stem tomatoes to increase production and disease resistance.

Materials

- rootstock and scion wood
- grafting knife
- tying material (plastic propagation tape or rubber ties)
- label and label marking pen or pencil
- mulch
- polyethylene tent or other wilt-proofing material
- aluminum foil
- light hammer for driving knife
- cleft grafting tool or other wedge
- grafting wax
- tomato grafting knife and grafting clip

Grafting of plants is an ancient technique practiced as early as 1000 B.C. Centuries later, the method was employed widely by the Romans. Grafting is a process by which two different plants are united so that they grow as one. The scion is the newly installed shoot or top of the plant. The rootstock (or *understock*) is the seedling or plant used as the bottom half of the graft. The rootstock becomes the root system of the newly grafted plant. The union between scion and rootstock is a physical union (growing together of the tissue) that allows free movement of plant sap across the graft and from the rootstock to the new top and back again.

Many times, grafting is used as a method of rapidly increasing the number of a desirable plant. Grafting is also used to give plants stronger, more disease-resistant roots. In grafting, it is important that both the rootstock and scion be disease free. The two must also be **compatible** so that they can grow together to form a strong union.

If rootstock is carefully selected, the stock will have little effect on the newly attached top of the plant. The top will grow much the same as it grew on the parent plant before the graft and will generally duplicate all the characteristics of the parent plant. In some cases, rootstock reduces vigor and the overall size to which the new plant will grow. Dwarf fruit trees are a good example of this.

Grafting is not as widely used as budding in the commercial propagation of plants but is often used for the following:

- Topwork a large tree by grafting a different, usually improved, variety or varieties to many of the limbs of the tree.
- Insert a different variety on part of the limbs of a tree for cross-pollination.
- Propagate plants that may be difficult to bud.
- Produce vegetables such as tomatoes with improved production and pest resistance.

GREEN TIP Grafting offers the opportunity to create a palette of unique plants that accent with special qualities to enhance the interest in the environment. Grafting also produces high-quality food for consumption.

TRY THIS Visit a local apple orchard or nursery and have the operator demonstrate topworking a tree. Ask the operator questions about grafting and its usefulness in orchard production.

Requirements for Successful Grafting

The following conditions must be met to ensure a successful graft.

COMPATIBILITY The two plants must be related to one another closely enough so that the stock and scion are able to grow together. As a general rule, it is best to graft a scion to a rootstock of the same type of plant; for example, apple is grafted to apple, and pear is grafted to pear. Certain varieties are exceptions, however. Some Japanese-type plums, almonds, and apricots, for example, experience best growth on peach roots. The almond and apricot are both able to grow on peach roots and cannot be grafted to one another.

Knowing which families and varieties of plants grow together best when grafted is extremely important. When selecting rootstock for commercial production, also consider the climate and conditions of the local growing area.

SCION WOOD The scion wood should be 1 year old and of vigorous growth.

TIMING Grafting is generally done when the stock and scion are dormant or have no leaves. The rootstock may be actively growing, but the scion should not be. Of course, when evergreens are grafted, needles or leaves are present on the shoot.

MATCHING OF TISSUE The **cambium** layer of the two matched plant parts, the **scion** and the **rootstock**, must come in close contact and be held tightly together. The cambium must not be allowed to dry out.

WATERPROOFING Immediately after the graft is made, all cut surfaces must be covered with a waterproofing material such as grafting wax, plastic ties, or rubber ties (Figure 10–1 A, B, C).

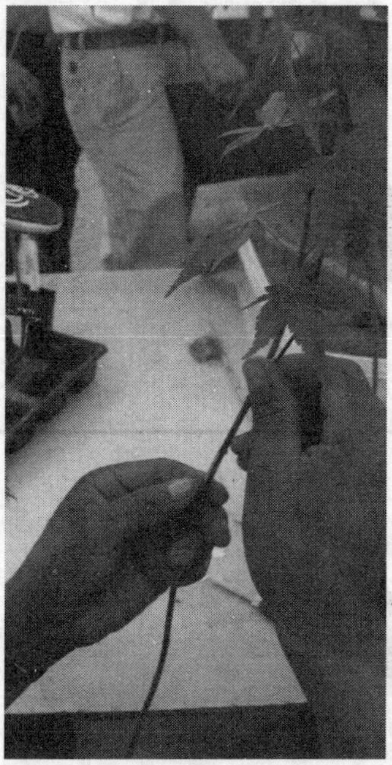

▲ **Figure 10–1A** Insert the scion into the understock cut, and line up the cambium layers of the scion and understock. Hold this union into place.

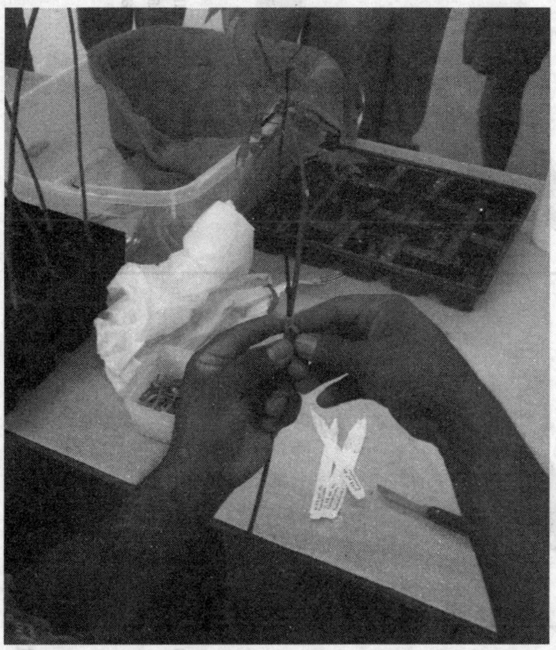

▲ **Figure 10–1B** Top graft is completed showing the proper wrap. Place this grafted plant in a shade house and allow it to form the graft union for 5–8 weeks.

A discussion of three methods of grafting follows: the whip (tongue) graft, the side veneer graft, and the cleft graft. A brief discussion of **stenting** (a modification of whip or cleft grafting) also follows.

▲ **Figure 10–1C** Now wrap the scion and understock with a grafting rubber to hold them together; as you near the end of the grafting rubber, slip one end under the wrapped rubber to tie it off. Pull the rubber firm.

Whip or Tongue Graft

The whip or tongue graft, accomplished during winter months, is used when small (1/4 to 1/2 inch) material is being grafted. Fruit trees are examples of plants propagated by the whip graft. If possible, the scion and root should be of the same diameter. The scion should contain three buds. The root piece should be at least 4 to 8 inches long and contain small fibrous roots. Figure 10–2 illustrates the steps in the whip graft.

The grafting cut is made below a bud on the stock. It should slant at an angle such that a smooth surface about 1 to 1 1/2 inches long is produced. A very sharp knife allows the cut to be made with a single pass and usually gives a straight, smooth surface. This smooth surface is essential so that close contact between the root and scion is made all along the cut surface. The cut made on the root piece should be exactly the same length and slope as that on the scion. This allows the two parts to fit together evenly. The cambium, the thin tissue just under the bark of the scion or stem, must

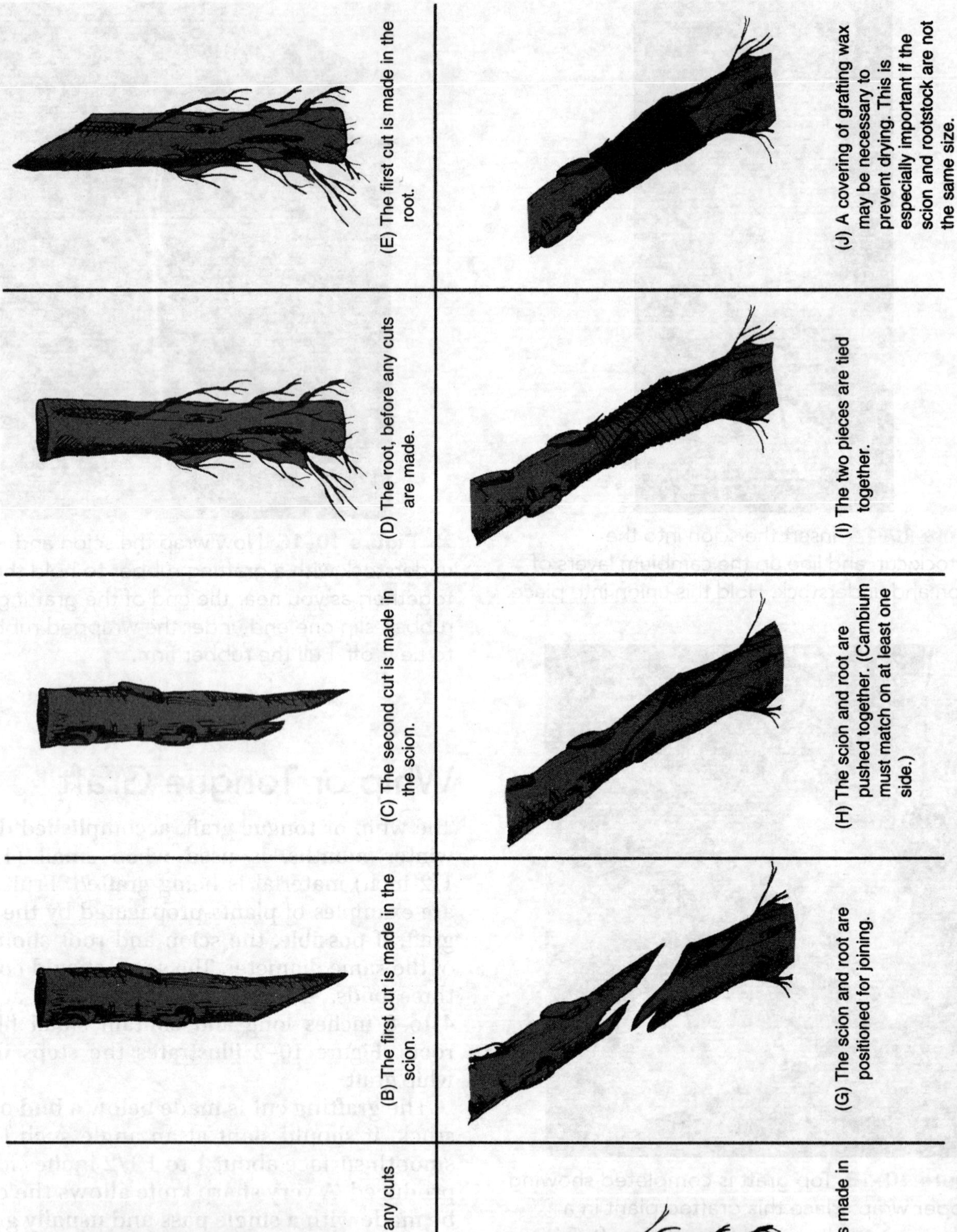

▲ **Figure 10–2** Steps in the whip graft.

match the growing area at the edge of the root piece on at least one side. If the two pieces are the same diameter, it will match on both sides.

Before the two pieces are placed together, a second cut is made on the first cut surface in the reverse direction, starting about one-third of the distance from the tip and cutting nearly parallel to the first cut. This cut is made about half as long as the first cut. The two pieces are then slipped together with the tongues interlocking. Great care must be taken to match the cambium on at least one side.

The two pieces are then tied tightly together with plastic propagation tape or rubber bud ties and stored in moist sand or peat moss to heal. Storage for the first 3 to 4 weeks is at 50 to 55°F (10 to 13°C). After that, storage is at 32 to 40°F (0 to 4°C) to prevent top growth. The grafted plant may also be planted directly in the nursery row. In these cases, the graft union must be below ground level.

The new plant should send out shoots and begin to grow with warm spring weather.

▲ **Figure 10–3** Growing grafted nursery stock material in containers.

Side Veneer Graft

The side veneer grafting method is a very effective way to graft evergreens such as the blue Colorado spruce (Figure 10–3). To obtain precisely the shade of blue these trees possess, a small piece of stem from a selected blue spruce must be grafted to a small spruce seedling used as rootstock. The selection of the stem or scion is very important because these seedlings vary in color, and very few have the particular shade of blue that is so desirable.

Grafting Procedure

The procedure for this graft is a simple one. Small seedlings may be grown in pots or in a nursery row. They graft more easily if they are about pencil size or a little larger. The grafts are made in early spring just before growth begins.

A shallow cut about 1 1/2 inches long is made into one side of the seedling rootstock just above the soil level (Figure 10–4A). The cut is not complete at the bottom but stops so that a second short downward cut leaves a ledge when the cut piece is removed (Figure 10–4B). The 4-inch scion, or new top piece, is cut on the opposite side at the bottom, making an end that fits on the ledge left on the seedling rootstock (Figure 10–4C). All cuts should be smooth and even so that the scion and rootstock surfaces fit together along the entire cut surface (Figure 10–4D).

The cambium layer of the scion and seedling rootstock must match on at least one side. The two are placed with cut surfaces matching and are tied securely together with rubber bud ties or similar material (Figure 10–4E). Placing mulching peat moss or other moist material over the graft union is very helpful in promoting healing and helps to ensure a higher percentage of successful unions.

When grafts are exposed to sun and drying winds, a polyethylene tent may be tied to a stake and drawn over the seedling and the graft until the graft has healed and the new growth has started on the top. Aluminum foil is secured on the south side of the tent to keep down the

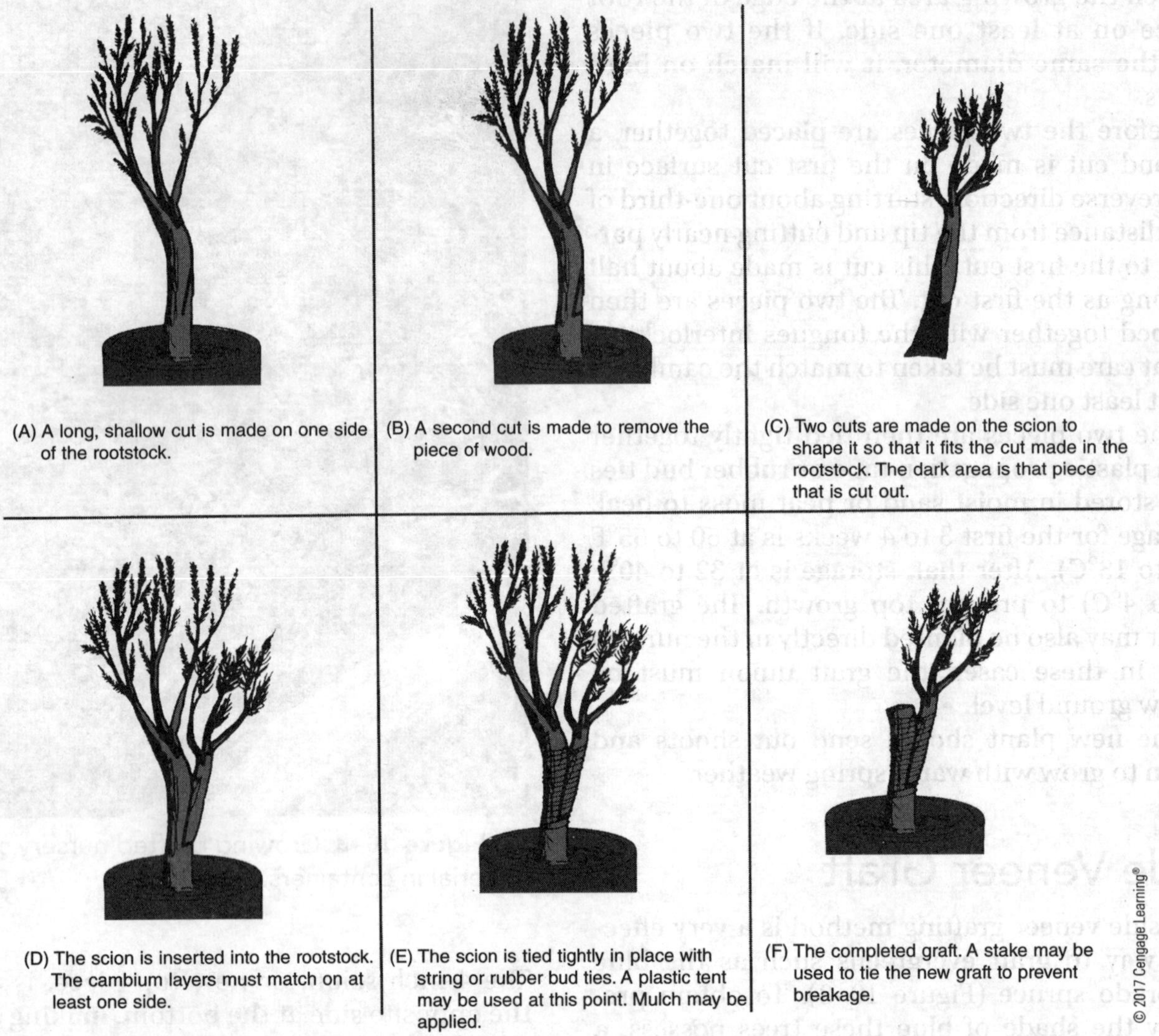

(A) A long, shallow cut is made on one side of the rootstock.

(B) A second cut is made to remove the piece of wood.

(C) Two cuts are made on the scion to shape it so that it fits the cut made in the rootstock. The dark area is that piece that is cut out.

(D) The scion is inserted into the rootstock. The cambium layers must match on at least one side.

(E) The scion is tied tightly in place with string or rubber bud ties. A plastic tent may be used at this point. Mulch may be applied.

(F) The completed graft. A stake may be used to tie the new graft to prevent breakage.

▲ **Figure 10–4** Steps in the side veneer graft.

temperature inside the tent. When the tent is removed, it should be opened gradually over a period of a week to harden off the new graft.

Spraying the stem or shoot grafted on the rootstock with one of the new plastic materials such as Wilt-Pruf is also helpful in holding down moisture loss and drying the new graft until the union heals.

As new growth starts, the top of the seedling rootstock is gradually cut away to force buds on the newly attached scion into active growth. The entire top of the seedling should be removed as soon as new growth is well established on the scion. The grafting process is now complete, but care must be taken in handling the new grafts so that the top is not broken loose from the rootstock (Figure 10–4F).

Cleft Graft

The cleft graft is used most often in topworking trees or grafting to a rootstock that is considerably larger than the scion. It is done in early spring just before the buds start to swell. The scion is cut from 1-year-old wood and should include about three buds.

Grafting Procedure

The selected rootstock or tree to be topworked must first be cut back to the point at which the graft is to occur. The limb or trunk is sawed off at a right angle (Figure 10–5A). A heavy knife is used to split the rootstock or understock through the center until a crack in which to place the scion opens wide enough (Figure 10–5B). A metal or wooden wedge is then placed in the center of the stem to hold the split open (Figure 10–5C). A piece of scion wood is selected, and two scions are cut, each containing three buds. The scion is cut with a long, smooth tapered cut on each side, making a wedge-shaped stick. The cut should be about 1 inch long and such that the wedge is slightly thicker on one side than the other (Figure 10–5D).

The thicker side is cut so that a bud is directly above the top of the wedge. This thicker side is then placed against the outside of the rootstock where it is pressed firmly in contact with the cambium of the understock (Figure 10–5E). The scion is pushed down into the split in the rootstock until the entire cut area is inside the split. Do not push the scion hard enough to cause tearing of the tissue. If necessary, the wedge is driven in farther to widen the split in the rootstock—this allows the scion to slip in more easily. Because the understock is much larger and has thicker bark than the scion, care must be taken to match the cambium layer of the scion with the cambium layer of the rootstock, not the bark of the rootstock. If properly placed, the scion will not be smooth with the outside of the understock, but will be set in (Figure 10–5F). The cambium layers must match for a successful graft.

After the two scions are in place, the wedge used to hold the split open is carefully removed. The grafter should check to be sure that the cambium layers are in contact for the full length of the cut area on the scion. The pressure of the split closing is usually sufficient to hold the scions in place. No nailing or tying is necessary. The entire cut surface is then covered with grafting wax to prevent drying.

If both scions grow, one should be removed after the first season so that there is no sharp crotch on the tree or limb at the graft union.

If cleft grafting is used to topwork trees, some type of support may need to be tied to the tree branch and then to the scion to support the scion in extremely windy conditions. Wind could tear the newly attached scion loose during the first year of growth.

Stenting

A relatively new type of grafting called stenting is used on roses as well as other plants. With roses, a marketable plant can be produced 1 year sooner than with normally used budding. This method

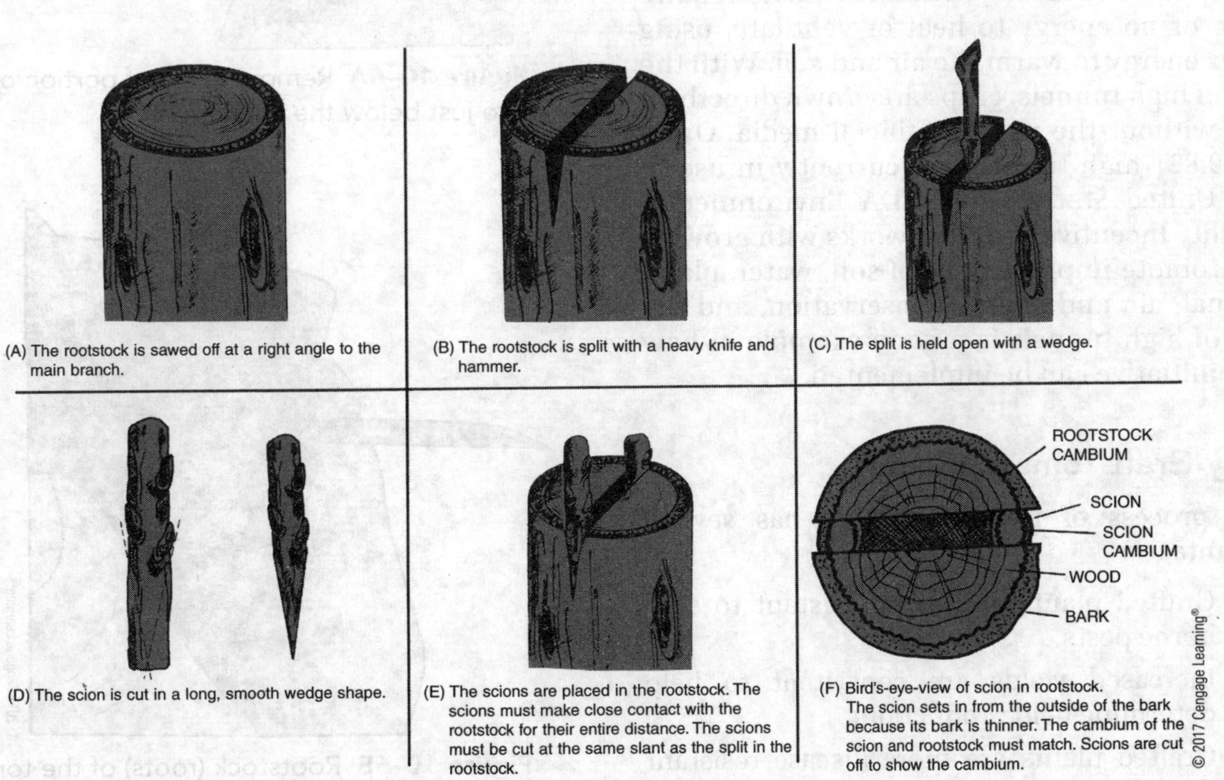

▲ **Figure 10–5** Steps in the cleft graft.

involves grafting a selected scion (top part of graft) of the desired plant onto a piece of *stem* that produces good roots. The grafted stems are then placed in a rooting chamber (the same as for semihardwood cuttings) for the graft to heal and roots to form on the bottom stem. Whip (or tongue) grafting or cleft (or saddle) grafting may be used to perform this grafting process. The graft is then treated exactly as a semihardwood cutting (see Unit 7).

Grafting Tomatoes

The technique of grafting tomatoes has increased their yields, improved their resistance to soil-borne pests, and improved environmental tolerance. This phenomenon has developed in situations where tomatoes are grown in the same soils year after year. In addition to the use of grafting to increase tomato yields, horticulturalists are addressing the increasing demand for more fresh vegetables by using **high tunnels** as a way to extend the growing season for the production of vegetables. High tunnels are a low-cost alternative to the standard greenhouse. They are freestanding plastic-covered structures that require little or no energy to heat or ventilate, using solar energy to warm the air and soil. With the use of high tunnels, crops are grown directly in soil without the use of artificial media. Overall, 9,821 high tunnels are currently in use in the United States. The USDA Environmental Quality Incentive program works with growers to promote improvement of soil, water, plant, animal, air, and energy conservation, and the use of high tunnels is a good example of how this initiative can be implemented.

Why Graft Tomato Plants?

The process of tomato grafting has several advantages:

- Grafted plants are more resistant to soil-borne pests.
- Increased yields are consistent to help determine field production.
- Grafted plants are more disease resistant than heirloom cultivars that lack disease resistance and are more susceptible to situations in the field, high tunnels, and/or greenhouses.
- Grafted plants have increased stress tolerance.
- Grafted plants increase water and nutrient uptake.

Grafting Procedure

The grafting line is cut on a 45-degree angle below the cotyledons (first leaves of the seedling) to prevent suckers on the tomato plant (Figure 10-6A). Figure 10-6B illustrates the rootstock from the grafting procedure. Insert

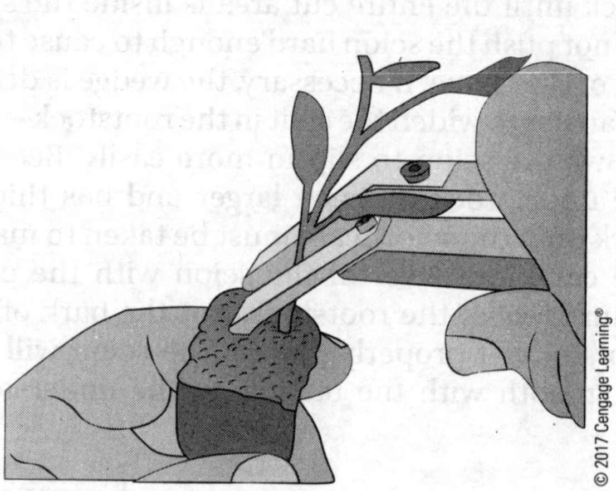

▲ **Figure 10–6A** Remove the leaf portion of the tomato just below the cotyledons.

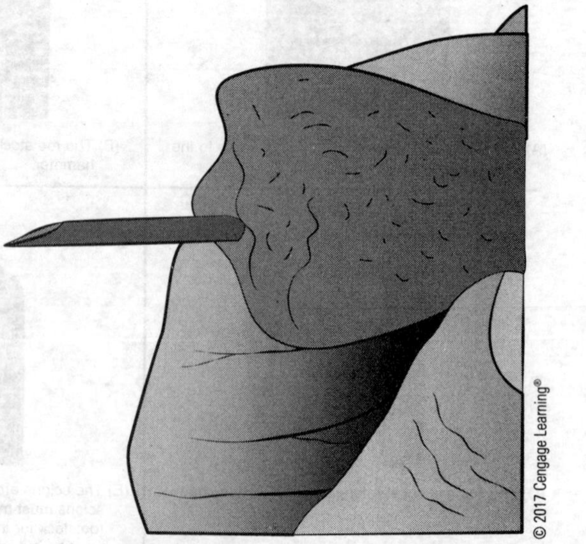

▲ **Figure 10–6B** Rootstock (roots) of the tomato grafting procedure.

the tomato grafting clip onto the rootstock (Figure 10-6C), and then complete the graft by joining the scion and the rootstock of the tomato (Figure 10-6D).

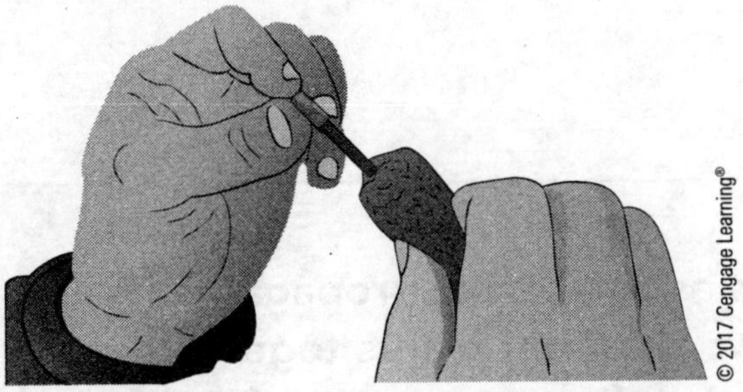

▲ **Figure 10–6C** Insert the tomato graft clip.

▲ **Figure 10–6D** Complete the graft by joining the scion and the rootstock of the tomato.

Grafting Factors That Must be Considered When Grafting Tomatoes

When preparing to graft tomatoes, it is necessary to consider several different factors. First, tomatoes should be grafted when they are in the 2–4 leaf stage and when the stems are 2–2.5 millimeters in diameter. Plan to graft in the early morning, when the plant is turgid, or full of water. About 2-4 hours before grafting, move the seedlings into a shaded area to reduce water stress. Following the grafting process, the seedlings will require about 7 days at 100% relative humidity at 85°F to heal completely. During this healing time, maintain complete darkness for 3–4 days, and then slowly increase the light before returning the plants to the greenhouse until they are ready to be transplanted. Note that grafting adds about two weeks of additional time to transplant production. Despite this additional time, grafting will increase production of tomatoes and create resistance to bacterial wilt and other soil-borne diseases. Figure 10–7 illustrates a time line for grafting tomatoes.

> **GREEN TIP** When grafting, use a grafting knife to make the proper cuts at a 45-degree angle to enhance the ability of the graft union to heal.

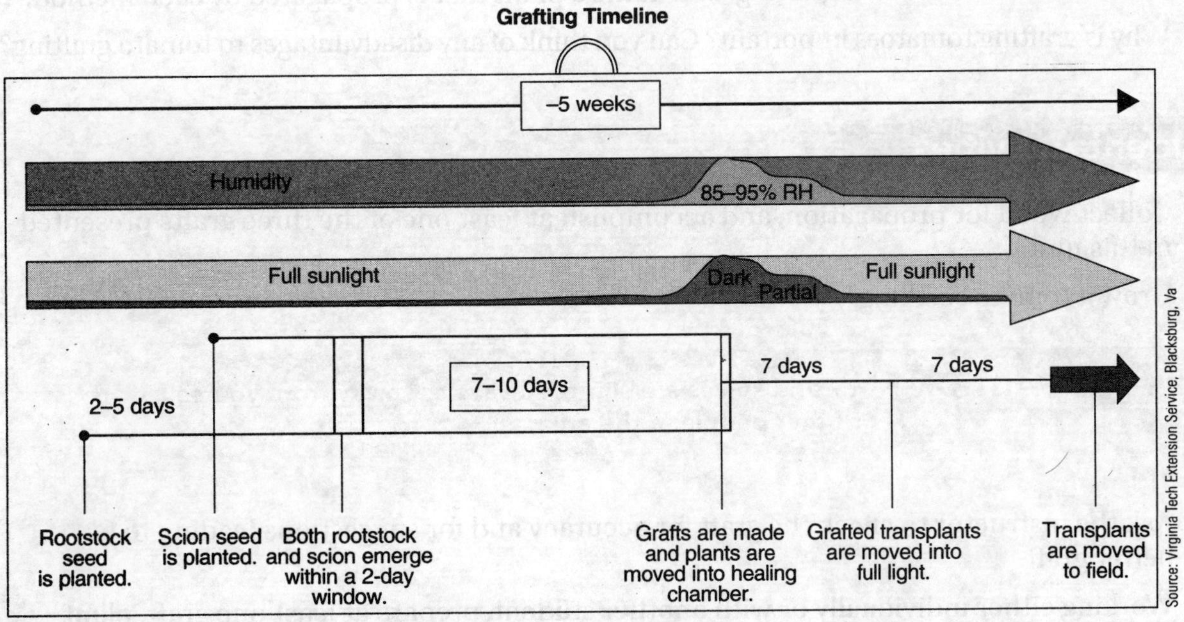

▲ **Figure 10–7** Time line for grafting tomatoes.

SAFETY FIRST

Use caution when collecting the cuttings; the knife and/or Felco pruners are sharp! For more information on occupational safety and health in the workplace, please visit the OSHA site at www.osha.gov.

Summary

Grafting is used primarily on trees as an asexual propagation process. The reason for placing two different plants together is to use one plant with a strong or dwarfing root system for the bottom of the plant and a desirable variety for the top. Grafting is also used to rework older trees and graft a new superior variety onto the branches to improve the quality of fruit or flower. Grafting tomatoes has met a significant demand in tomato production by increasing disease resistance and productivity in heirlooms, improving water and nutrient uptake, and increasing stress tolerance.

Review Questions

1. Discuss the process of grafting. What are the advantages of its uses?
2. Describe the different conditions that are necessary to achieve a successful graft.
3. Discuss three methods of grafting, and name a plant that is propagated by each method.
4. Why is grafting tomatoes important? Can you think of any disadvantages to tomato grafting?

Student Activities

1. Collect wood for propagation, and accomplish at least one of the three grafts presented in this unit.
2. Grow a tomato seedling.

CAUTION Grafting knives are sharp. Always cut away from yourself or other people in the area.

Ask the instructor to check the graft for accuracy and for suggestions leading to a better graft.

3. Working either individually or with another student, prepare at least one graft, plant the resulting grafted plant, and allow it to grow. (Needed supplies are the scion wood,

rootstock, a sharp knife, and waterproof material to tie the graft union.) After the plant is grafted, label it according to variety. If possible, invite an experienced propagator to assist in the actual grafting activities.

4. Research various methods of grafting plants online. Create a visual presentation that describes two of the grafting methods you researched.
5. Establish a grafting tomato project by grafting heirloom tomatoes to newer hybrid tomato understock to increase disease resistance and increase tomato production.

Suggested Supervised Agricultural Experience (SAE) Activities

1. Conduct a placement SAE at a local apple orchard or nursery. This could also become a research SAE if the manager of the orchard or nursery will allow some testing and experimenting.
2. Develop an entrepreneurship SAE revitalizing an old orchard by using grafting as the main propagation method.

Additional Resources

www.extension.umn.edu/garden/yard-garden/fruit/grafting-and-budding-fruit-trees/

www.hort.cornell.edu/grafting/reasons/ReasonsGBLeft.html

www.ladybug.uconn.edu/bridgegraftingandinarching.htm

UNIT 11 Budding

Objective

To propagate at least one plant by the budding process.

Key Terms

budding
bud shield
budwood
vegetative bud

Competencies to Develop

After studying this unit, you should be able to:

- name at least three plants commercially propagated by budding.
- list the seven steps in the budding process.
- identify on a diagram a piece of budwood, rootstock, a budding knife, and rubber bud ties.
- differentiate between a fruit bud and a leaf or vegetative bud by cutting each from a bud stick and labeling.
- propagate a plant using the T-budding process.
- propagate a plant using the chip budding process.

Materials

- budding knife
- rubber bud ties
- plastic bag to collect budwood
- label and waterproof pen or pencil
- rootstock
- budwood
- plant for rootstock
- a sharp budding knife

Methods of Budding

Methods of **budding** include T-budding, patch budding, inverted T-budding, flute budding, I-budding, and chip budding, among others. The most commonly used methods in the United States are T-budding and chip budding. The T-budding operation is especially popular in the propagation of tree fruits and roses and is a much quicker process than grafting. In fact, some rose budders insert up to several thousand buds a day if other people present are to do the tying. Chip budding is used on grapes and is being used more on other plants because of the longer time period during which it can be done.

The process of budding is actually a form of grafting; in fact, budding is sometimes called *bud grafting*. Budding differs from grafting in several ways, however. In budding, a single bud is used instead of a scion so many more plants can be reproduced from the same amount of parent wood. Budding is also accomplished more quickly. Another difference is the time at which the operation takes place—T-budding is accomplished when the rootstock is in active growth rather than in the dormant (resting) stage, either in spring (March or April), in June, or in late summer or fall (July through September).

T-Budding

T-budding is generally done on small, 1- or 2-year-old seedlings. Small branches in the tops of larger trees are also budded. Whichever type of rootstock is chosen, it is important that it be actively growing, disease resistant, and able to give the desired growth. If there is no sign of disease, and the bark separates easily from the wood, the operation should be successful. It is also important that mature vegetative or leaf buds of the desired variety are available. As in other forms of grafting, the scion and rootstock must be compatible.

Some plants commonly propagated by budding include apples, pears, peaches, plums, cherries, roses, and the citrus group. Late summer is the best time to propagate fruit trees by budding. The rootstock generally grows from seed planted in fall to the size of a lead pencil by July or August of the following year. Leaf buds on the stock or parent plant are generally mature enough for use at that time.

GREEN TIP One apple tree, which can be planted on the smallest city lot, can yield up to 15 to 20 bushels of fruit per year. One apple tree not only can provide fruit for people but also food for birds and other animals. Plant an apple tree today!

Collecting Budwood

Bud sticks, small shoots of the current season's growth, are collected on the same day the buds are to be inserted (Figure 11–1A). The **budwood** is kept wrapped in waterproof paper so that it does not dry out. The bundle of bud sticks is then labeled according to variety and date cut. Shoots that have a vigorous growth pattern usually have the best and most usable buds on them.

Remember that **vegetative buds** are necessary for propagation—fruit buds will not grow into new plants. Vegetative and fruit buds can be easily distinguished. The vegetative or leaf bud is slimmer and more pointed than the fruit bud, and fruit buds generally appear on the base or bottom end of the shoot.

Collect budwood of the proper cultivar. Carefully cut all the leaves from the stick, leaving a short piece of the leaf or petiole to act as a handle

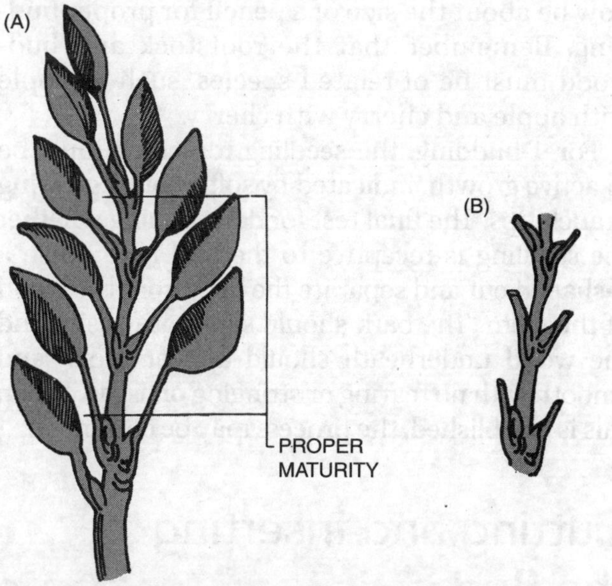

(A) Budwood with leaves still attached. Buds in the bracketed area are probably best for budding because of maturity.
(B) Budwood with top and bottom trimmed and leaves removed. The leaf petiole remains as a handle for holding the buds.

▲ **Figure 11–1** Preparing the budwood.

to hold the bud after it is cut from the stem (Figure 11–1B). Wrap the budwood in waterproof paper or place in a plastic bag.

PROCEDURE
Steps in the Budding Process

1. Plant seeds for seedling rootstock (the previous fall).
2. Select the variety of budwood to be propagated.
3. Determine the correct date to bud, as indicated by both the bud maturity of the desired plant and the active growth of the rootstock.
4. Cut the budwood, label it, and protect it to prevent drying out.
5. Perform the budding process.
6. Check to see if buds have taken.
7. Cut off the rootstock above the bud (the following spring).

Rootstock

The rootstock, which at this point has been developing from seed for a year or more, should now be about the size of a pencil for proper budding. Remember that the rootstock and budwood must be of related species, such as apple with apple and cherry with cherry.

For T-budding, the seedling rootstock must be in active growth, indicated by soft, rapidly growing branch tips. The final test for determining whether the seedling is receptive to the bud is to make a T-shaped cut and separate the bark from the wood of the stem. The bark should slip loose easily, and the wood underneath should appear moist and smooth with no tearing or stringing of tissue. When this is established, the process may be continued.

Cutting and Inserting the Bud

The next step is to prepare the seedlings being used for rootstock. The budding process is easier if a team of two people work together. One person makes the cut with a sharp budding knife and inserts the bud while the second person ties the bud securely in place.

CAUTION
The person making the cut should exercise care when handling the knife.

The person making the cut kneels beside the row of seedlings, bending them toward himself or herself. At a spot about 1 to 2 inches above ground level and where the stem is smooth, a 1-inch vertical cut is made through the bark. The cut should be made on the north side of the stem if possible to protect the bud from the sun both in summer and winter. A horizontal cut is then made across the tip of the vertical cut to form the T shape for which this process is named. The knife is given a twist so that the bark and wood separate (Figure 11–2A). If necessary the two sides of the vertical cut should be lifted at the top with the edge of the knife blade, opening the slit in the bark (Figure 11–2B). As mentioned, if the bark separates easily, the rootstock is growing rapidly enough to bud.

The next step is to cut the leaf bud from the bud stick. Generally, buds in the middle portion of the

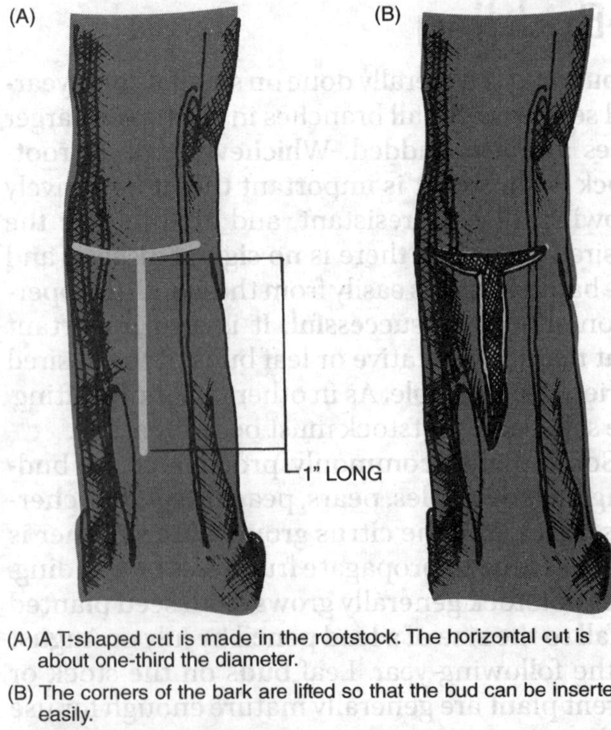

(A) A T-shaped cut is made in the rootstock. The horizontal cut is about one-third the diameter.
(B) The corners of the bark are lifted so that the bud can be inserted easily.

▲ **Figure 11–2** Preparing the rootstock.

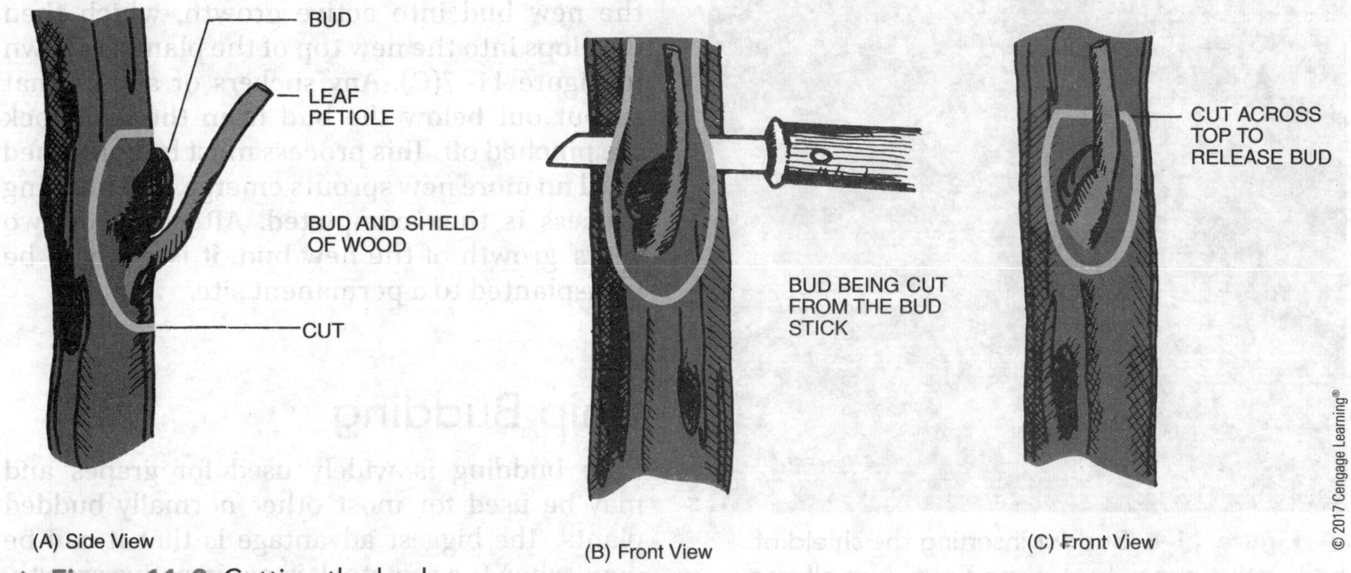

▲ **Figure 11–3** Cutting the bud.

bud stick are used for budding because buds on the terminal end may not be mature enough, and buds on the basal end may be flower buds. The buds that are used may vary with the maturity of the bud stick. Whichever is used, the bud in the axil of the leaf stem must be visible and well developed.

The bud is cut with a small shield of bark and a sliver of wood attached (Figure 11–3). The cut is made starting about 1/2 inch below the bud and cutting just deeply enough to include a small sliver of wood. The cut is continued under the bud and past it to about 1/4 inch above the bud. A second cut made at a right angle to the bud stick at the end of the first cut releases the bud from the bud stick. The **bud shield** should appear like the one shown in Figure 11–4.

The bud is immediately inserted into the T-shaped cut made in the rootstock and pushed down into the open slit in the bark until the top of the bud shield is even with the top of the T cut in the rootstock. This allows the bud shield to be completely inserted underneath the bark of the rootstock and to fit snugly against the wood (Figures 11–5 and 11–6).

To hold the new bud firmly in place, the cut area is now tied with a rubber bud tie or a one-piece bud tie tape, which fits over the bud. The entire cut area is covered with the bud tie; only the bud itself is exposed. The bud tie is tucked under itself at each end to hold it in place. These ties, made of a special

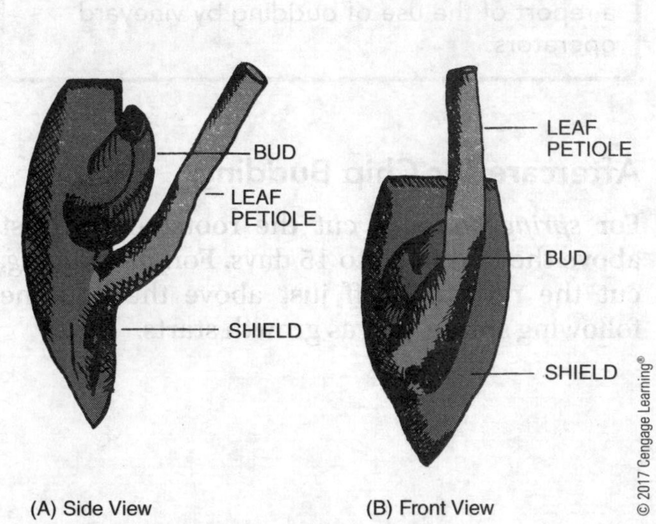

▲ **Figure 11–4** The leaf bud shield.

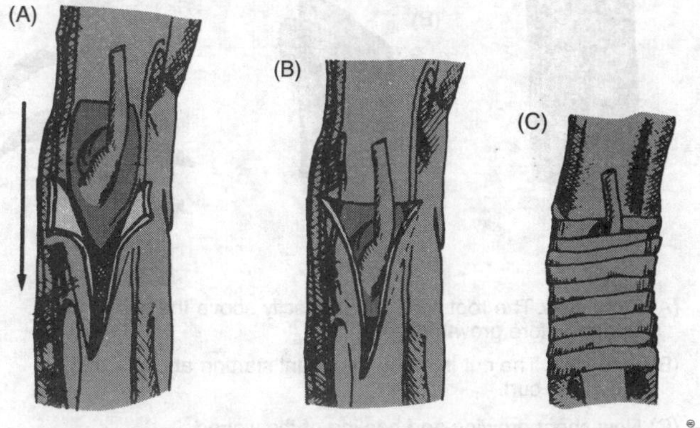

(A) The bud is inserted in the T cut on the rootstock by pushing the bud shield down and underneath the flaps of bark.

(B) The bud is now in place underneath the bark. The top of the bud shield is even with the horizontal cut on the rootstock.

(C) The bud is tied firmly in place with a rubber band tie. Nothing more is done until the following spring.

▲ **Figure 11–5** Insertion and tying of a bud.

Courtesy of Stark Brothers Nurseries and Orchards Co.

▲ **Figure 11–6** Budder inserting the shield of bark with a peach bud in the T cut. A small tree is being used as the rootstock.

material, are designed to disintegrate or break in about 3 weeks, so they do not require removal.

In about 3 weeks, an inspection is made to see if the bud has *taken* (has attached to the stem and is still alive). If the leaf petiole has dropped off and the bud appears to be plump and of normal color, the bud has probably grown to the rootstock. Nothing else is done to the bud or the rootstock until early the following spring. At that time, the top of the seedling rootstock is cut off directly above the bud as shown in Figure 11–7(A) and (B). This forces

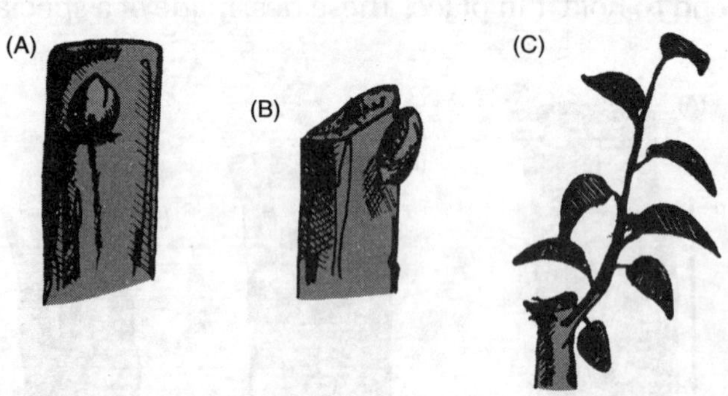

(A) Front view. The rootstock is cut directly above the bud in the spring before growth begins.

(B) Side view. The cut is made at a slant starting about ¼ inch above the bud.

(C) New shoot growing and healing of the wound.

▲ **Figure 11–7** Continuation of the budding process the following spring.

the new bud into active growth, which then develops into the new top of the plant as shown in Figure 11–7(C). Any suckers or shoots that sprout out below the bud from the rootstock are pinched off. This process must be continued until no more new sprouts emerge. The budding process is then completed. After one or two years' growth of the new bud, it is ready to be transplanted to a permanent site.

Chip Budding

Chip budding is widely used for grapes and may be used for most other normally budded plants. The biggest advantage is that it can be done when the rootstock is not in active growth. You will recall that in T-budding, the rootstock must be actively growing so that the bark can slip loose without tearing, allowing the bud to be inserted beneath it. This is not necessary for chip budding because the bark is not lifted from the wood. This means that chip budding can be done in summer or fall.

In chip budding, selection of rootstock and budwood is the same as for T-budding. The chip budding procedure is shown in Figure 11–8.

> ***TRY THIS*** Research the chip budding method used in viticulture. If possible, visit a local vineyard to receive a demonstration of the use of budding in the grape industry. Then prepare a report of the use of budding by vineyard operators.

Aftercare for Chip Budding

For *spring budding,* cut the rootstock off just above the bud in 10 to 15 days. For *fall budding,* cut the rootstock off just above the bud the following spring, just as growth starts.

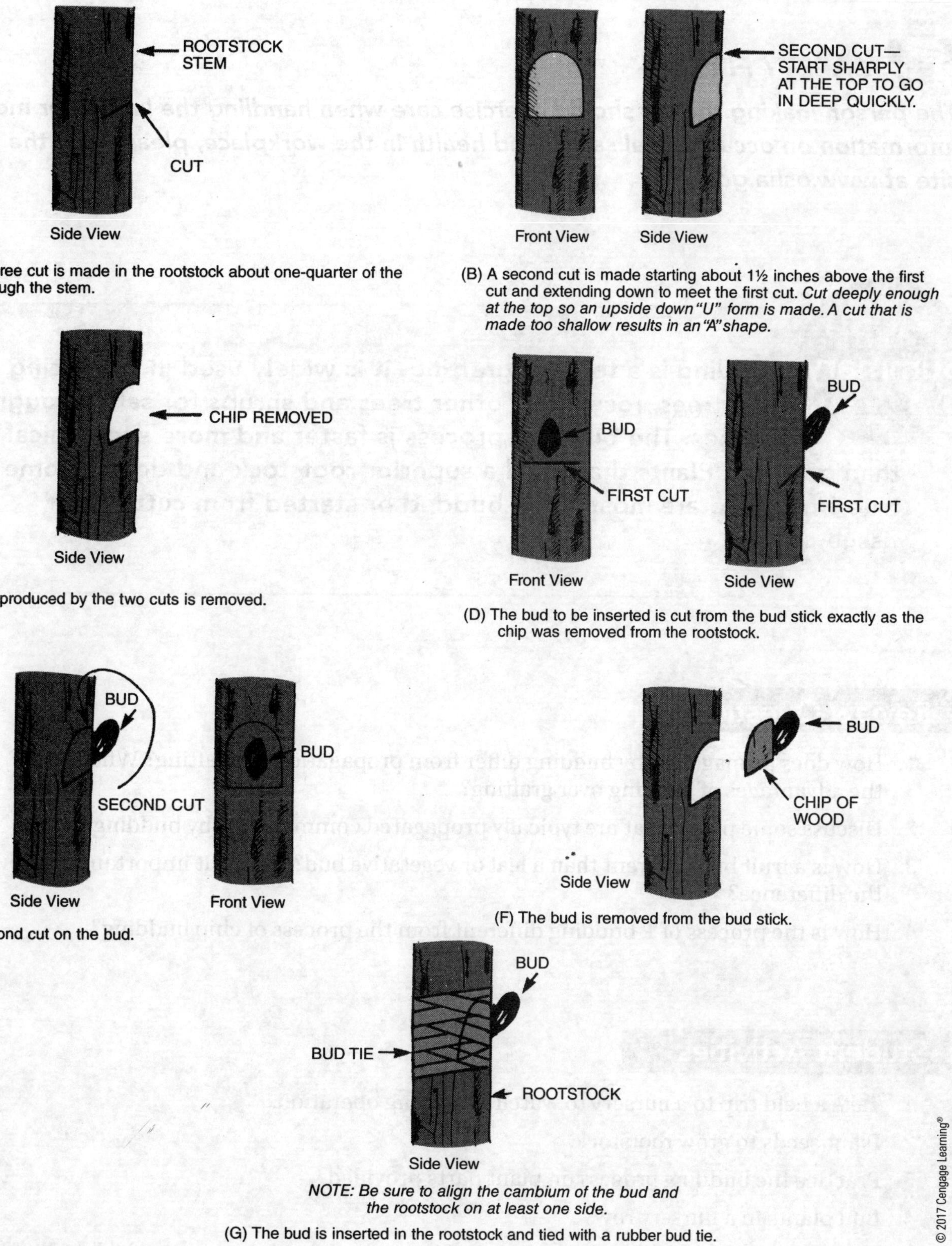

(A) A 45-degree cut is made in the rootstock about one-quarter of the way through the stem.

(B) A second cut is made starting about 1½ inches above the first cut and extending down to meet the first cut. *Cut deeply enough at the top so an upside down "U" form is made. A cut that is made too shallow results in an "A" shape.*

(C) The chip produced by the two cuts is removed.

(D) The bud to be inserted is cut from the bud stick exactly as the chip was removed from the rootstock.

(E) Second cut on the bud.

(F) The bud is removed from the bud stick.

NOTE: Be sure to align the cambium of the bud and the rootstock on at least one side.

(G) The bud is inserted in the rootstock and tied with a rubber bud tie.

▲ **Figure 11–8** Chip budding procedure.

SAFETY FIRST

The person making the cut should exercise care when handling the knife. For more information on occupational safety and health in the workplace, please visit the OSHA site at www.osha.gov.

Summary

Budding is a form of grafting. It is widely used in producing fruit trees, roses, and other trees and shrubs for sale through nurseries. The budding process is faster and more economical than grafting. Plants that need a superior rootstock and do not come true from seed are most often budded or started from cuttings or tissue culture.

Review Questions

1. How does propagation by budding differ from propagation by grafting? What are the advantages of budding over grafting?
2. Discuss some plants that are typically propagated commercially by budding.
3. How is a fruit bud different than a leaf or vegetative bud? Why is it important to know the difference?
4. How is the process of T-budding different from the process of chip budding?

Student Activities

1. Take a field trip to a nursery to watch a budding operation.
2. Plant seeds to grow rootstock.
3. Practice the budding process on plant parts provided.
4. Bud plants in a nursery row.
5. Write a description of the physical condition of a bud that has grown quickly and of one that is dead. Pinpoint the differences in appearance.
6. Search online for journal articles that discuss the following topics: asexual reproduction; grafting and budding.

Suggested Supervised Agricultural Experience (SAE) Activity

1. Develop a test plot for various types of budded crops. Research and record data for a research SAE.

Additional Resources

www.hort.cornell.edu/grafting/methods.alpha/TBudMeth.html

http://ohioline.osu.edu/hyg-fact/1000/1082.html

UNIT 12 Layering

Objective

To propagate one plant by one of the layering processes described in the unit.

Key Terms

air layering
compound layering
girdle
simple layering
stool layering
trench layering

Competencies to Develop

After studying this unit, you should be able to:

- describe two situations in which layering is used to propagate plants.
- list five plants commonly propagated by layering and two plants that propagate naturally by layering.
- list the eight steps in the layering process.
- perform simple layering as described in the text.

Materials

- parent plant
- shovel
- sharp knife or pruning shears
- peat moss or sawdust
- two stakes, one with hooks

Layering is a method of asexual propagation in which roots are formed on a stem or root while it is still attached to the parent plant. The stem or root that is rooted is called a *layer*. The layer is cut free from the parent only after rooting has taken place. This is the major difference between layering and other methods of asexual reproduction.

As a method of propagation, layering has advantages and disadvantages. It is a relatively simple process but requires more time than many other methods of asexual propagation because it requires a great deal of work by hand. Also, fewer plants can be started from each parent plant than with the use of cuttings. The major advantage of layering is the degree of success with which some plants will root when layered as compared with results achieved with other methods. (Some plants that are difficult to root by other methods may be rooted by layering.) In fact, many plants root naturally by layering (Figure 12–1).

Cane fruits such as the black raspberry and the boysenberry root easily and economically by *tip layering*, one form of natural layering. In this method of natural layering, a shoot with the current season's growth bends to the ground and then turns sharply upward. Roots develop from the point at which the shoot meets with the soil (Figure 12–2). For another example, refer to Figure 44–2 in Unit 44. Other plants that root naturally by layering are strawberries (by runners), red raspberries and spirea (by root suckers), and African violets (by crown division).

> **TRY THIS** A sample of a raspberry bush with the roots will be provided by the instructor. Sketch the root system and describe, in detail, how the plant functions, including the responsibilities of each part.

(A)

(B)

▲ **Figure 12–1** Natural layering of rhododendron. (A) The stem is cut free from the parent plant and separated where roots have developed. (B) Separated new plant showing root development. The new plant is transplanted to a nursery bed to develop further.

Some plants do not reproduce naturally by layering but can be propagated in this manner with assistance from the horticulturist. Two methods used for this purpose are simple layering and air layering.

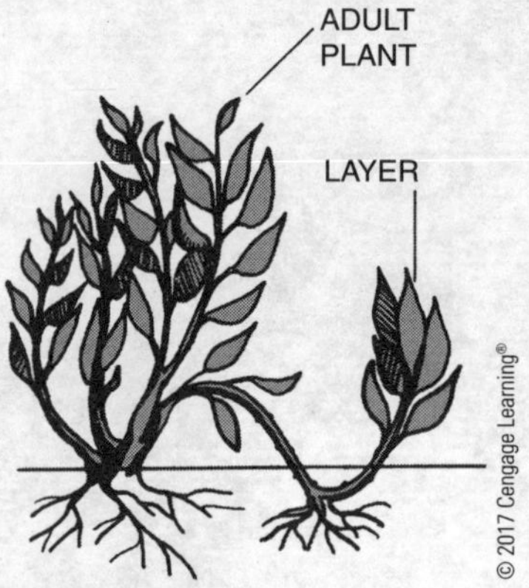

▲ **Figure 12–2** Tip layering.

Simple Layering

In **simple layering**, a branch from a parent plant is bent to the ground where it is partially covered at one point with soil. The terminal end remains exposed (Figure 12–3). This process is usually accomplished in early spring.

Because the root system of the new plant is so limited, it is necessary to water it liberally. Some shade may also be needed until the root system is well established.

▲ **Figure 12–3** The stem is cut halfway through where it is to be inserted in the soil (A) and then twisted 90 degrees (B). If plant stems tend to break when twisted, it is best to girdle the stem, removing all the bark and cambium for a 1-inch length on the stem, as shown in Figure 12–6.

GREEN TIP The more trees you plant, the better the air you breathe. Trees absorb pollutant gases, nitrogen oxides, ammonia, sulfur dioxide, and ozone, and then filter particulates out of the air by trapping them on their leaves and bark.

PROCEDURE
Steps in Simple Layering

1. Select a stem with 1-year-old wood that can be bent to soil level.
2. Make a cut about halfway through the top of the stem at the point at which the stem is to be inserted in the ground (Figure 12–3). Use a sharp knife for the best cut, or girdle the stem completely of all bark and cambium.

CAUTION Do not cut toward yourself or another person.

3. Twist the stem 90 degrees, and raise the stem to an upright position. (It is not necessary to twist girdled stems.) Some stems may break if twisted, as stated earlier. *For stems that break easily, do not cut and twist, but rather girdle the stem as shown in air layering* (see discussion in this unit).
4. Dig a hole or trench 3 to 6 inches deep at the point at which the twist in the stem touches the ground.
5. Dust the stem wound with rooting hormone, and push the stem into the opening in the ground.
6. Drive a peg with hooks on it next to the stem to hold it upright (Figure 12–4). Another peg holds the layer in the ground.
7. Fill the trench with soil, and mulch the top to help hold in moisture. A brick or stone may be placed on top to keep the layer in position. Keep the soil moist (by watering, if necessary). The layer should be rooted by the next spring (about 1 year).
8. Cut the layer free of the parent plant when it is well rooted, and transplant to a new location (Figure 12–5).

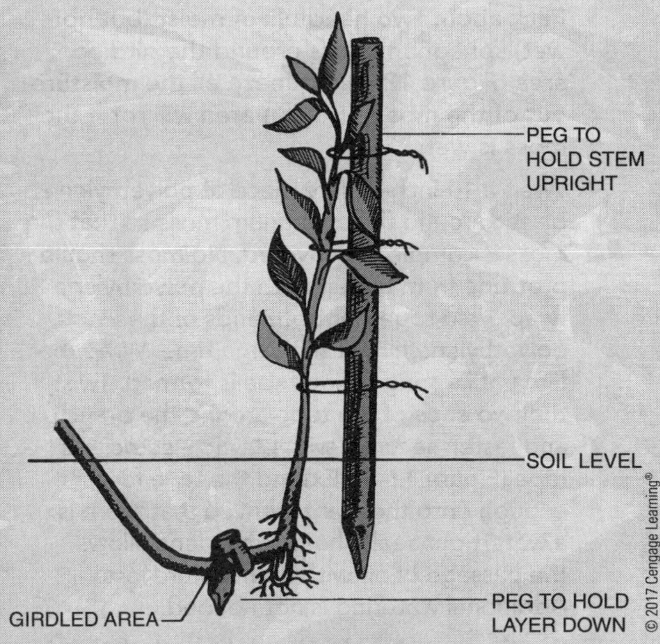

▲ **Figure 12–4** Pegs hold the layer upright and in the ground.

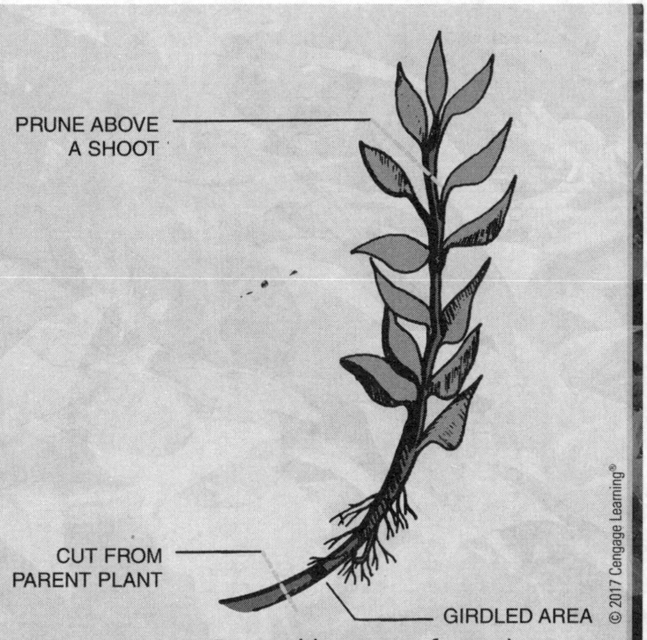

▲ **Figure 12–5** Rooted layer cut from the parent plant.

Air Layering

The ancient Chinese practiced a technique called **air layering**, a process that eliminates burying part of the parent plant in the soil. Instead, a part of the plant stem is slit or girdled (to **girdle** is to completely remove the bark and cambium around a plant) and then surrounded by a moist growing medium in some type of enclosure. Roots form where the plant has been wounded. The materials available in earlier years made the task much more time-consuming and difficult for the Chinese than it is today. The major problem in the past was keeping the layered area moist. The development of polyethylene film has made air layering a highly successful procedure. It is used commercially to propagate certain tropical and subtropical trees and shrubs that are difficult to root by other procedures. The Persian lime (*Citrus aurantifolia*) is commercially propagated in this way. The magnolia and the rubber plant (*Ficus elastica*) may also be propagated by air layering. In fact, most plants may be propagated using this method, but usually it is not economically feasible.

Air layers are generally made in the spring on wood of the previous year's growth or in late summer on partially hardened growth of

PROCEDURE
Steps in Air Layering

1. Girdle the selected stem in a 1/2- to 1-inch wide band. Girdle at a spot that is 6 to 12 inches from the top of the branch so that 6 to 10 leaves are left between the girdled spot and the branch tip. These leaves are needed to produce carbohydrates and hormones needed for good root development. Because the stem is girdled, these materials, produced in the leaf, accumulate at the girdled point on the stem of the plant, greatly stimulating root formation. Scrape the girdled area to ensure that all bark and cambium are removed, thus slowing the healing process (Figure 12–6).
2. Treat the girdled area with rooting hormone (Hormodin #3, Rootone #10, or a concentration of 0.5 percent to 0.8 percent indolebutyric acid in water). Treat the bark edges on the branch tip side of the girdled area especially well.

▲ **Figure 12–6** The stem is girdled to stimulate root formation.

3. Pack about two handfuls of moist (but not wet) sphagnum moss around the girdled area (Figure 12–7). Squeeze all the moisture out of the moss—the cut area will rot if the moss is wet.
4. Wrap a 10-inch-square piece of polyethylene plastic around the sphagnum moss so that the moss is completely covered. No moss should protrude from underneath the polyethylene wrap. Fold together both ends of the polyethylene film at the same time. Wrap the film until a snug-fitting tube is formed. Twist the two ends of the tube around the branch and fasten securely with plastic electrician's tape (Figure 12–8). Extend the tape far enough onto the plant stem so that there is a watertight seal. The polyethylene allows the passage of air with no moisture loss. Additional watering is not needed.
5. Rooting may be observed through the polyethylene plastic. When strong roots have developed, the layer may be removed for transplanting. If the plant being layered is deciduous, wait until the leaves fall from the plant and it is dormant. If the plant is an evergreen, remove the layer when no new active growth is occurring. Simply cut the layer free below the rooted area.

If the top of the new plant has a great deal of growth in relation to the root system, plant it in a pot or flat and place it in an area of high humidity for a short time. After additional root development has occurred, the plant may be gradually hardened off by exposure to a drier atmosphere in preparation for the permanent planting site.

▲ **Figure 12–7** Sphagnum moss is packed around the girdled area.

▲ **Figure 12–8** The two ends of the polyethylene tube are twisted around the branch and fastened securely with electrician's tape.

the current season. Older wood is much more difficult to root.

Other Methods of Layering

Three other methods horticulturalists use for layering plants are trench layering, stool (or mound) layering, and compound layering.

Trench Layering

In **trench layering**, the mother plant is bent to the ground and buried in a trench (Figure 12–9A). As shoots arise from the buried buds, roots form on the covered portion of the plant. The shoots may then be separated from the mother plant (Figure 12–9B).

Stool Layering

Stool layering begins with the planting of a rooted layer in soil. After one season's growth, the parent plant is cut back to soil level (Figure 12–10A). The stem is then covered with a mound of soil (12–10B). Soil is added to the mound periodically as the shoots grow (Figure 12–10C). At the end of the season, the new shoots are rooted and dormant (Figure 12–10D). The shoots are cut free and planted in early spring. One year has passed from the beginning of the layering process to the planting of the new shoots.

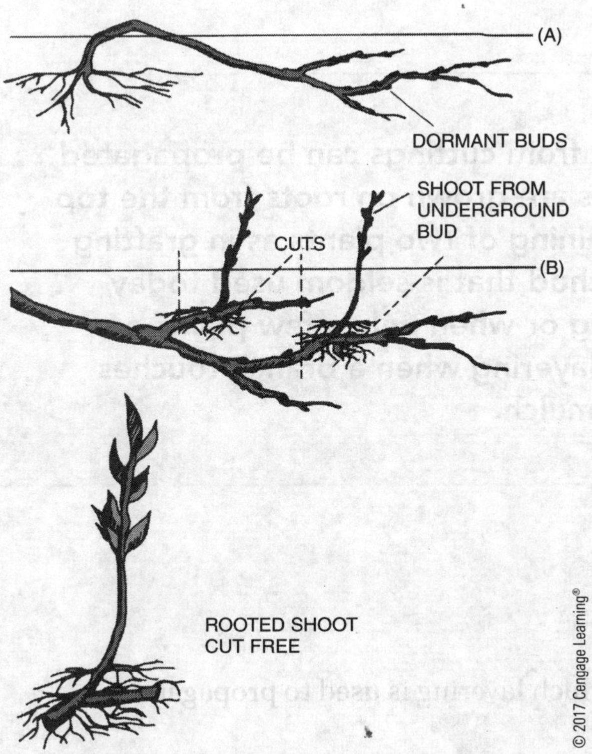

▲ Figure 12–9 Trench layering. (A) Shows the mother plant in the trench exhibiting dormant buds. Soil is added to the trench periodically to cover the developing shoots. At the end of the growing season, the layers are separated from the mother plant, as shown in (B).

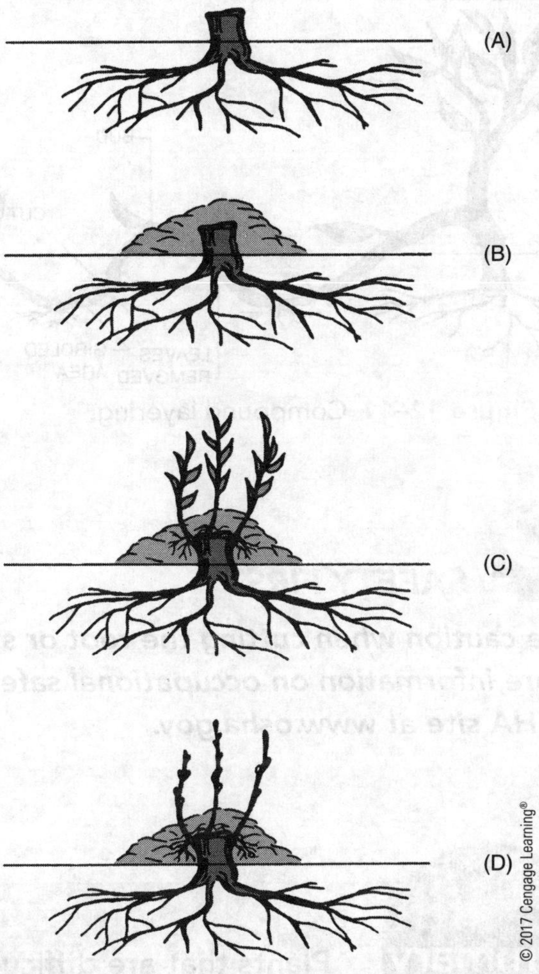

▲ Figure 12–10 Stool (or mound) layering.

Compound Layering

Compound layering, accomplished in springtime, is very similar to simple layering except that a stem is covered by soil at two or more points along its length. Long-stemmed, vinelike plants, such as wisteria, clematis, and grape, work well with this system (Figure 12–11).

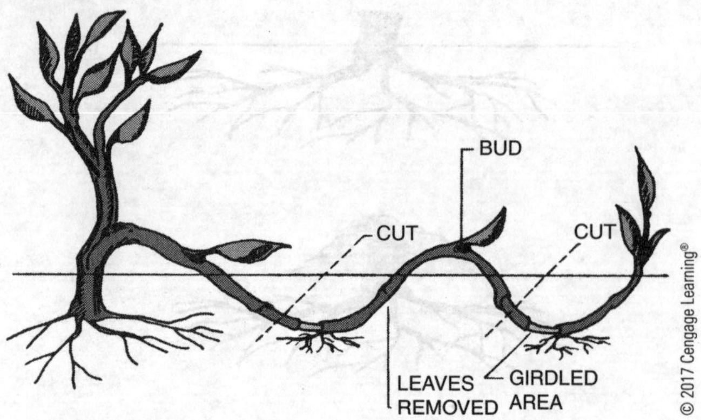

▲ **Figure 12–11** Compound layering.

The stem is girdled at a point below ground near where new roots are expected to form. The portion of the stem above soil level must have at least one bud to form a new shoot that will grow into the top of the new plant.

After rooting takes place, usually by the end of one growing season, the stem is cut to include roots and a new shoot on each new plant. Several plants may be produced from a single stem by compound layering.

SAFETY FIRST

Use caution when cutting the root or stem to make the layering cut on the plant. For more information on occupational safety and health in the workplace, please visit the OSHA site at www.osha.gov.

Summary

Plants that are difficult to root from cuttings can be propagated by layering. Because these plants are grown on roots from the top part of the plant, there is no combining of two plants as in grafting and budding. It is an old propagation method that is seldom used today, except for plants that root best by layering or when only a few plants are needed. Many plants naturally spread by layering when a branch touches the ground and is covered with leaves or mulch.

Review Questions

1. What is layering? What are some situations in which layering is used to propagate plants?
2. What are the advantages and disadvantages of layering?
3. How has modern technology improved the process of air layering?
4. Discuss at least five plants that are commonly propagated by layering.
5. Discuss two plants that propagate naturally by layering.

Student Activities

1. Visit a horticulturist and observe a layering process. Record the steps in an electronic device so that you can practice the steps back in the classroom lab.
2. With a group of students, perform a layering process on a parent plant. Keep records on the experiment, including the date on which the plant is layered and the date on which it roots. Plants suggested for layering are azalea, rhododendron, spirea, forsythia, wisteria, and grape.
3. Research online the black raspberry and small fruit industries in your state. Write a report that includes the data from your research.

Suggested Supervised Agricultural Experience (SAE) Activity

1. Use the budding techniques while developing your own raspberry bushes and business to satisfy the entrepreneur SAE.

Additional Resources

http://content.ces.ncsu.edu/plant-propagation-by-layering-instructions-for-the-home-gardener/

www.hort.purdue.edu/ext/ho-1.pdf

SECTION 4

Greenhouse Management and Crops

UNIT 13

Poinsettias

Objective

To produce a saleable poinsettia crop.

Key Terms

acidic
alkaline
bracts
breaks
cultivars
drench
growth regulator
pinching
relative humidity
sleeves
soluble fertilizer

Competencies to Develop

After studying this unit, you should be able to:

- prepare a soil mixture for growing poinsettias.
- transplant 10 rooted poinsettia cuttings.
- identify one growth regulator and use it correctly.
- outline a growing schedule for poinsettias.
- identify four insects that affect poinsettias and the control measures for each.
- list the common names of three diseases that affect poinsettias and the control measures for each.

Materials

- 2 1/4-inch rooted poinsettias
- loamy soil
- sand
- horticultural-grade perlite
- Pro-Mix B
- 4-, 5-, and 6-inch pots
- smartphone app for Plant Growth Regulators
- smartphone app for new varieties (see gro getter)
- for insect and disease control, contact your greenhouse products supplier

The poinsettia has long been a traditional flower used during the winter holiday season. Poinsettias were first introduced into the United States from Mexico by J. R. Poinsett, the first ambassador to Mexico. Poinsettias were used as exotic plants in conservatories and botanical gardens until 1900. Through the work of Albert Ecke, poinsettias were then grown in southern California as cut flowers. Today, the growing of poinsettias is a big part of many horticultural operations (Figure 13–1).

▲ **Figure 13–1** Poinsettia varieties are grown for the winter holiday season. In warmer climates, poinsettias are grown outdoors.

Greenhouse Management

The greenhouse structure used to start and grow plants throughout the year must be located to receive the maximum available sunlight. Other factors to consider in greenhouse production are temperature, moisture, ventilation, and climate.

Temperature in the greenhouse is an important key to growing a successful crop. To obtain optimum growth of plants, it is critical that the correct temperature be maintained throughout the greenhouse. Great variation in the greenhouse temperature can cause fast vegetative growth or the lack of plant growth. The general rule of thumb for greenhouses is the day temperature should be 5 to 10 degrees higher than the night temperature. It is important to monitor the greenhouse temperature with thermometers at the plant level to ensure optimum growth. Thermometers should be kept out of direct sunlight, facing a northern exposure with a wooden hood on each side. A maximum/minimum thermometer works well in a school greenhouse because it helps students to realize the high temperature and the low temperature of the day.

Plant growth depends on moisture. It helps to hold the cell shape, to nourish, and to carry dissolved nutrients present in media into the plant for optimum growth. The amount of moisture in the greenhouse will also influence the **relative humidity**. Watering plants in the greenhouse is an important task because the moisture supply controls the growth of plants. The amount of water needed depends on the type of plant and the outside conditions on a particular day (i.e., bright sunshine or clouds). The appearance or feel of the medium is an indication for watering plants. Good judgment must be used when watering.

Ventilation is the movement and exchange of air in the greenhouse, which is important for optimum plant growth. The individual greenhouse grower is able to feel the air movement in the greenhouse. This ensures the correct temperature and relative humidity and helps control plant diseases.

Climate directly influences the type of greenhouse structure and heating and cooling systems. Talk to a state land grant college, local extension service, or grower in the area when determining the greenhouse structure.

GREEN TIP Greenhouses provide many benefits. They make use of solar energy by converting it into heat. The heat comes from the sun; the greenhouse holds the heat inside, maintaining a warm atmosphere at night for better plant growth.

The Poinsettia Flower

The poinsettia flower is a small, yellow flower that grows at the terminal end of the plant. Just below the flower are the **bracts**, the leaves that give poinsettias their color. The most common bract color is red, but through years of research, other colors have been developed. Some of these colors are white, pink, yellow, marble, a light pink and white bicolor, white variegated, and red and white variegated.

Poinsettias require a short day for the production of the flower, which in turn is necessary to produce the color in the bracts. Poinsettias should be grown in full sunlight. During months with long days, the plants develop leaves and increase stem length. The plant forms the flower buds naturally during late September and early October. During November and December, color develops in the bracts.

Growing Poinsettias

Planting and Spacing

Wholesale growers can supply 2¼-inch rooted poinsettia cuttings for planting, or producers with a smaller greenhouse can buy prefinished poinsettias from wholesalers and grow them to finish just prior to the Christmas holiday. Before purchasing the cuttings or prefinished plants, check to be sure they are healthy.

When the cuttings are planted, several things must be considered. The size of the pot is important because it partially determines the size of the full-grown plant. Poinsettias are grown in a variety of pot sizes—4-inch, 5-inch, 6-inch, 7-inch, 8-inch, and 10-inch sizes—in either hanging baskets or azalea pots. The size of the plant also depends on the number of plants per pot and greenhouse space allowed for each potted plant. All of these factors also determine the price of the saleable plant.

			2 FLOWERS PER SQUARE FOOT	
POT SIZE	PLANTS PER POT	TREATMENT	SPACING INCHES	SQUARE FT/POT
4 inches	1	none	9 × 9	0.56
4 inches	1	pinched	9 × 9	0.56
5 inches	3	none	15 × 15	1.56
5 inches	1	pinched	15 × 15	1.56
6 inches	3	none	15 × 15	1.56
6 inches	1	pinched	15 × 15	1.56

Figure 13–2 Poinsettia planting guide.

From *Ball Red Book*, 15th ed., 1991. Courtesy of Ball Publishing.

Figure 13–2 gives the correct pot size and number of rooted cuttings per pot, treatment, and spacing for each pot in the greenhouse. Do not crowd poinsettias in the greenhouse. They need space to produce quality bracts and foliage.

Temperature

The greenhouse temperature is critical for growing top-quality poinsettias (Figure 13–3). The temperature should be 65°F (16°C) at night and 70 to 75°F (21 to 24°C) during the day. It is important to maintain these temperatures during September and October while flower bud initiation is taking place. After the bracts start to show development, the nighttime temperature can be reduced to 60 to 62°F (16 to 17°C). Newer **cultivars** are on the market today that require a cooler finishing temperature after the bracts have developed. It is best to consult the wholesaler of your geographic area for specific recommendations (Figure 13–4).

Finishing Tips

In the final stages of production, the poinsettia needs continued attention to the cultural practices and greenhouse conditions. It is vital to maintain good ventilation and air circulation. Maintain normal temperatures of 75 to 80°F (24 to 26°C) during the day and 64 to 66°F (18 to 19°C) at night. By following these procedures, poinsettia quality and longevity will increase.

DATE	TEMPERATURE (°F) NIGHT	DAY	GREENHOUSE PROCEDURES
September 10	64–66 (18–19°C)	70–75 (21–24°C)	Arrival of 2¼-inch rooted cuttings from a reliable greenhouse grower. Transplant three rooted cuttings into a 6-inch azalea pot. Treat with a fungicide to control disease. Mix with water: Dexon—35% WP, 1¼ teaspoons per 5 gallons of water.
September 11	64–66 (18–19°C)	70–75 (21–24°C)	Start slow-release fertilizer program. Apply 1 level tablespoon Osmocote 18-9-9 per 6-inch azalea pot. Start a constant liquid feeding program.
September 23	60 (16°C)	75 (24°C)	Space each 6-inch pot in an area of 15 inches by 15 inches. Spray with growth regulator such as Cycocel 3000 ppm (3½ ounces per 1 gallon of water). Lower temperature to 65°F. It is during this time that the flower buds form. Formation of flower buds is necessary for the production of color in the bracts of the poinsettia.
October 10	68 (18°C)	75 (24°C)	Increase night temperature to force color formation in bracts.
October 15	68 (18°C)	75 (24°C)	Drench soil with Dexon—35% WP, 1¼ teaspoons per 5 gallons of water.
November 15	68 (18°C)	75 (24°C)	Color should be showing in upper bracts. Use a fan to provide continuous air movement in the greenhouse. Drench soil with Dexon—35% WP, 1¼ teaspoons per 5 gallons of water.
December 5	62 (17°C)	75 (24°C)	Lower night temperature to develop a bright color in the bracts. By this time the bracts should have achieved maximum size.
December 10	64–66* (14–17°C)	75 (24°C)	Poinsettias are ready for sale.

WP—wettable powder
ppm—parts per million
*Lower nighttime temperature improves bract color, maintains plants at an optimal stage of development, and improves plant performance for the consumer.

Figure 13–3 Growing schedule for poinsettias.

▲ **Figure 13–4** Growing poinsettias under optimum conditions will produce quality bracts and foliage.

Pinching

Pinching poinsettias is a process in which the grower removes the terminal end of the plant. Pinching is accomplished by taking the top of the plant between the thumb and index finger and carefully breaking the top of the poinsettia completely off. Four or five nodes should remain above the soil level. Nodes occur at each point at which a leaf is attached to the stem. Therefore, four or five leaves should remain on the plant after it is pinched. After pinching, new shoots called **breaks** will develop from the bud in the axis of the node. As each shoot grows, it will produce a flower. After the flower develops, color will form on the bracts of the poinsettia. The major reason for pinching poinsettias is to increase the number of bracts per plant. It also causes the plant to be fuller.

Soil Mixes

Poinsettias may be grown in many different media. Whatever type of medium is used, it must be porous and well drained to allow for maximum root development. Poinsettias also require a generous supply of fertilizer and moisture.

One suitable soil mixture consists of one-quarter soil (loam), one-quarter construction-grade sand, one-quarter perlite (horticultural grade), and one-quarter peat moss. Another mix that works well is Pro-Mix BX (in bags containing 6 cubic feet) and horticultural-grade perlite (in bags containing 2 cubic feet). Perlite aids in drainage of the mixture. The Pro-Mix BX and perlite should be mixed to evenly distribute the perlite.

A water-powered Dosatron fertilizer injector (Figure 13-5) mixes the water with the fertilizer nutrients with precise PPM (parts per million) concentration to supply the correct quantity of plant nutrients for poinsettias.

▲ **Figure 13–5** A water-powered Dosatron fertilizer injector mixes the water with the fertilizer nutrients with precise PPM (parts per million) concentration to supply the correct quantity of plant nutrients for poinsettias.

It is important that all soil mixes be free of any insects and disease organisms and possess good nutrient-holding capabilities. If soil is used, it should be pasteurized at 180°F (80°C) for 30 minutes.

Fertility

Poinsettias require a regular fertility program. The fertilizer used should be a complete fertilizer, which means that it contains nitrogen, phosphorus, and potassium. Fertilizers used to produce quality poinsettias are in a liquid or slow-release form. The liquid fertilizer program may be constant with each watering or intermittent with every third watering. A prepared commercial water-soluble fertilizer formulation, such as 20-10-20 Peat-Lite at 200 to 250 ppm, is an excellent choice. It is a good general feed for poinsettias and at the same time contains secondary and micronutrient supplements. When using this water-soluble fertilizer, follow the manufacturer's recommendations for application. If the slow-release fertilizer is used as a part of the fertility program, it is applied in a dry granular form that has been prepared to dissolve slowly over a selected period of time (e.g., 3 months).

The pH of the poinsettia medium is important because it controls the resources of available nutrients needed by the plant. A soilless medium should have a pH of 5.8 to 6.2, whereas for a medium that contains soil, the pH should be 6.0 to 6.4. These pH values will ensure optimum availability of essential elements to the plant. Also, perform a water analysis test to check the pH level of the water being applied to the crop. If the pH is **alkaline**, it may be necessary to use an acid treatment in the water to lower the pH level. It is important to monitor media pH throughout the growing season. (pH means "potential hydrogen" and is used to determine the **acidity** or alkalinity of the medium.). The ultimate goal is improve productivity, quality, and efficiency.

Many greenhouse growers will grow a variety of poinsettia cultivars. Some cultivars require a higher fertility program than others. Therefore, it is advisable to fertilize at the lowest possible concentration and provide supplemental fertilizer to those cultivars that require the higher levels for normal growth.

Using Growth Regulators

Growth regulators were developed to produce a better-quality plant. Growth regulators have two main purposes: to control the height of the plant, therefore causing the plant to be saleable, and to improve the quality of color in the poinsettia bracts by darkening them.

Some poinsettia growers use a soil **drench** growth regulator, which is mixed with water, applied to the soil mixture, and is then absorbed by the poinsettia plant, controlling the height of the plant.

Another type of growth regulator is applied as a foliar spray. The poinsettia foliage is sprayed until droplets form on the leaves. The leaves absorb the regulator, thus controlling their height. The rate of application is determined by the manufacturer's directions. Using your PGR smartphone app determine the parts per million (ppm) application.

Height of greenhouse plants, including poinsettias, can be controlled to some extent by adjusting the light level and duration and by regulating day and night temperature differential as follows:

- Brighter light results in more compact plants.
- Poinsettias can be kept more compact by creating longer nights by shading from 5 P.M. to 8 A.M.
- Plant height can be shortened by maintaining lower daytime than nighttime temperature. Research has found that stem elongation occurs in the first 3 to 4 hours beginning at daybreak. Lowering the temperature for just 2 hours early in the morning is effective and may result in less need for chemical growth regulators.

CAUTION
Higher night temperature can result in plants discoloring and leaves turning down.

Before using these controls, contact your local commercial greenhouse supplier, either by phone or with a site visit. Many suppliers also provide advice online for growing poinsettia crops.

This information applies to chrysanthemums and lilies also. Check with a professional as to what works best for each crop.

TRY THIS Develop and manage the growth of poinsettias for a holiday FFA chapter fundraiser. FFA members should help grow, maintain, and sell the poinsettias for the fundraiser project.

Maintaining Poinsettias

Transporting the Plant

When poinsettias are purchased from the local florist and transported to a new location, they should be handled very carefully. The plants should be protected from cold temperatures, and special care should be taken that the branches are not broken. This can be accomplished by carrying the plant in a sleeve.

Sleeves are made for the purpose of protecting plants. They are constructed in several materials such as paper or plastic. Sleeves are tapered in a funnel shape. The base of the sleeve is the same size as the pot, allowing the plant to be completely covered by the sleeve. Poinsettias are very sensitive plants.

When removing the sleeve from the plant, cut it off with scissors. Lift the sleeve upward and off the plant. Be very careful not to break any of the plant's branches or bracts. Proper care during transportation will prevent any bruising or mechanical damage to the leaves and bracts.

The storage of poinsettias in the sleeve for longer than 24 hours will cause damage to the plant. During the sleeving process, the leaves and bract are pulled into an upward position, causing the plant to produce an ethylene gas. This will result in the poinsettia having a droopy appearance. The sleeve should extend 2 to 3 inches above the leaves and bracts of the plant to provide the best protection. Be especially aware of the environmental conditions during transportation. Chilling the plants will cause damage too, but it will not be noticeable for several days.

Figure 13-6 illustrates a sample care tag accompanying poinsettias that are purchased from florists. By following these instructions, poinsettias can be maintained throughout the entire holiday season.

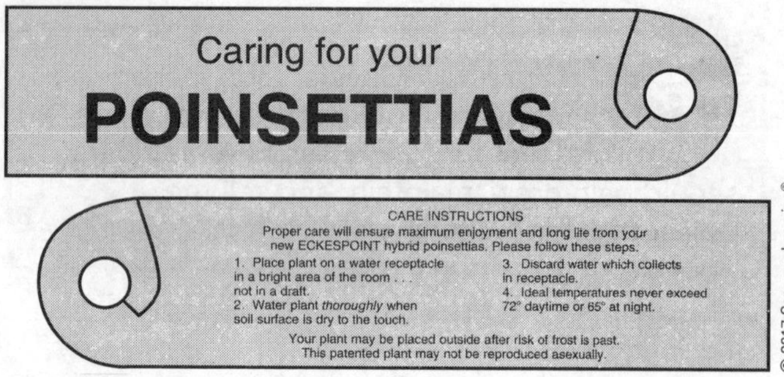

▲ **Figure 13–6** Instructions for care of poinsettias.

Controlling Insects

For optimum growth, poinsettias must be protected from insects. Some of the most common pests that affect poinsettias in the greenhouse are shore flies, whiteflies, fungus gnats, thrips, spider mites, and mealybugs (Figure 13-7). The use of chemical pesticides has been an effective management tool, but with pesticide law changes and increased restrictions and costs, other pest management avenues must also be explored. In Unit 16, integrated pest management (IPM) and biological control are discussed as alternative and effective pest management approaches to insect control. For the appropriate control of these pests, contact your greenhouse supply company technical advisor for recommendations. The

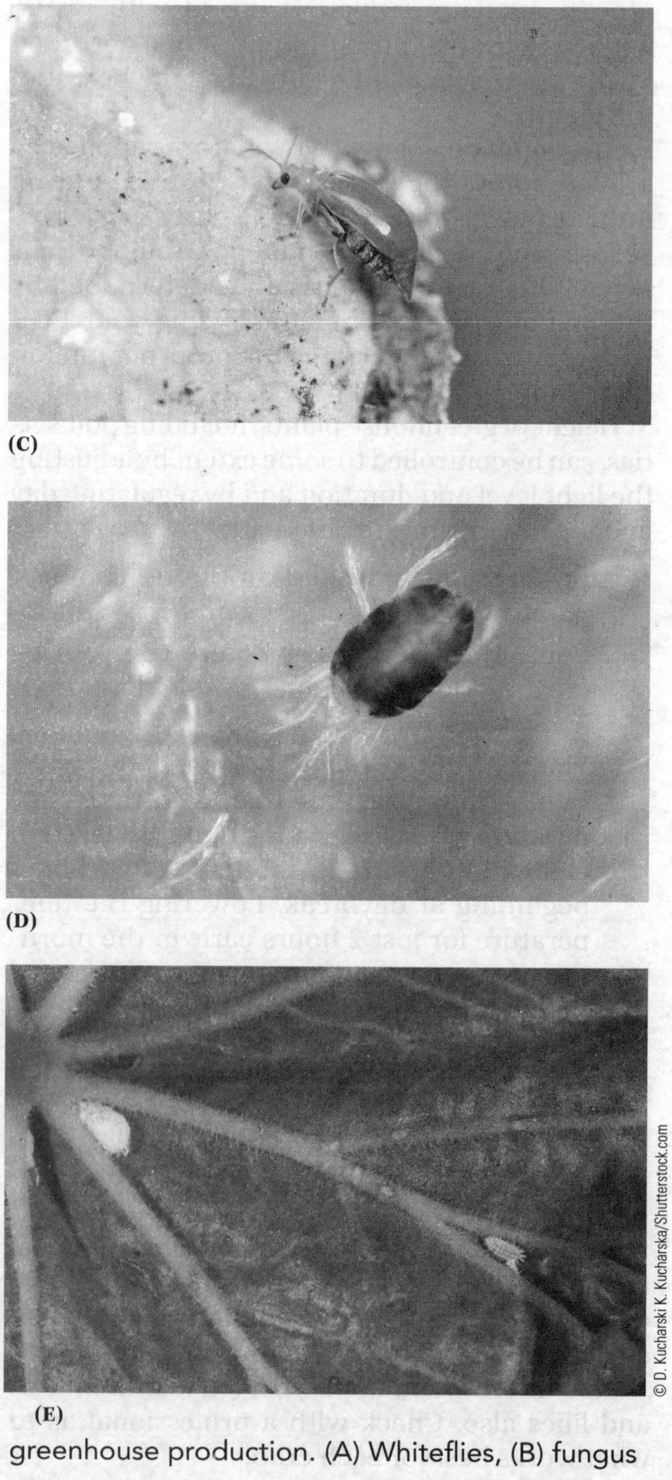

▲ **Figure 13–7** Common insect pests of poinsettias in greenhouse production. (A) Whiteflies, (B) fungus gnats, (C) thrips, (D) spider mites, (E) mealybugs.

other management approach for best production is being sure the greenhouse is kept clean.

Controlling Disease

Just as poinsettias must be protected from attack by insects, so must they also be protected from attacks by disease organisms. One of the more common problems of poinsettias growers is root rot. To help prevent this problem, purchase rooted poinsettias from wholesalers who supply disease-free stock. Also use clean growing media, have an excellent sanitation program, and have the proper growing conditions. When poinsettias are transplanted, treat them with a fungicide for further control.

Although it may seem that poinsettias have more than their share of attacks by disease organisms, keep in mind that fungicides provide good control.

SAFETY FIRST

In this unit we are using chemicals such as soluble fertilizer, insecticides, and fungicides. It is important to follow OSHA regulations to the law. Everyone involved must have the proper training to work with the specified materials to grow a poinsettia crop. Everyone should use the correct PPE to work with these products. For more information on occupational safety and health in the workplace, please visit the OSHA site at www.osha.gov.

Summary

The poinsettia is the traditional flowering plant used during the winter holiday season. It was introduced into the United States from Mexico by the first ambassador to Mexico and through the work of Albert Ecke; poinsettias were then grown in southern California. Poinsettias are grown in greenhouses with a day temperature of 75°F (24°C). The poinsettia is the small yellow flower at the terminal end of the plant. The color of the plant develops in the bracts, and poinsettias in a variety of colors are now available. Poinsettias are grown in different media with a fertility program of soluble fertilizers as well as slow-release fertilizers for normal growth. Plant growth regulators are used to control the height of the plant. It is important to control the diseases and insects on poinsettias. During transportation, sleeves must be used on all plants to provide optimum protection.

Review Questions

1. Discuss the five main factors that need to be considered by a greenhouse grower.
2. Discuss growing considerations related to the poinsettia.
3. Why is the pH of the poinsettia medium an important consideration?
4. What different methods can a greenhouse grower use to control the height of plants?

Student Activities

1. Demonstrate the proper techniques involved in mixing soil for growing poinsettias.
2. Propagate poinsettias from cuttings. Compare the full-grown plants with those propagated from 2 ¼-inch rooted cuttings.
3. At the proper stage of development, pinch the poinsettias grown in Activity 1 to increase the number of flower bracts per plant.
4. Contact a local university and inquire about field days on poinsettia production. If possible, become involved in some of the activities.
5. Tour a local greenhouse operation, and observe the greenhouse management techniques.
6. Show a video on greenhouse management techniques and procedures for optimum plant production.
7. Research online for current articles on pest management for poinsettias. Choose an appropriate article, and write a short summary.
8. Explore the www.greenhouse.cornell.edu/crops/nfmanagement.htm website.
9. Explore the www.ecke.com/poinsettias/website to make posters and signs to advertise poinsettias grown at school as fund-raisers.

Suggested Supervised Agricultural Experience (SAE) Activities

1. Work for a local flower shop doing design, plant care, deliveries, and so on to satisfy a placement SAE.
2. For an entrepreneurship SAE, grow and sell plants through the high school greenhouse.

Additional Resources

www.aces.uiuc.edu/vista/html_pubs/point/point.htm

http://urbanext.illinois.edu/poinsettia/facts.cfm

http://ohioline.osu.edu/hyg-fact/1000/1248.html

UNIT 14

Chrysanthemums

Objective

To produce a commercially acceptable potted chrysanthemum crop, beginning with rooted cuttings.

Competencies to Develop

After studying this unit, you should be able to:

- outline in writing a growing schedule for potted mums for any preselected market date.
- pot rooted cuttings.
- regulate formation of chrysanthemum flower buds by control of night length.
- determine the proper cultural requirements, such as fertilizing and watering, for production of a marketable chrysanthemum crop.

Key Terms

8-week variety
disbud
emulsifiable concentrate
Epsom salts
foot-candles
pH
pinching
pine bark
response time
soilless culture
pasteurize
vegetative period

Materials

- rooted chrysanthemum cuttings
- growing medium and 5-inch or 6-inch pots
- lights or shade cloth as needed to control length of night
- chemical growth retardants; contact your greenhouse supplier for the recommended PGR fertilizer: liquid and slow release
- insecticide and fungicide if needed

According to sales figures, the chrysanthemum (mum) is the most popular cut flower sold in the United States. It also competes very well in sales as a potted plant. Because it is now possible to artificially control the length of day and night in the greenhouse, the mum can be grown throughout the year. The most popular holidays for mums are Mother's Day, Easter, Memorial Day, and Thanksgiving.

The mum is considered an easy plant to grow. It offers a wide range of colors and keeps well in the home after being sold. In this unit, the mum is considered a winter crop, flowering for sale at Easter time. It will be prepared for sale as a potted plant.

FAST FACT

NASA did a study to find out which plants were best to filter the air of the space station. The findings noted that the chrysanthemum was one houseplant that could filter the air—not only producing oxygen from CO^2 but also absorbing benzene, formaldehyde, and trichloroethylene.

Determining Market and Variety

If sale of mums at Easter is desired, a schedule must be established that assumes the plants will be blooming before that date. This necessitates starting the plants a number of weeks before Easter.

The growing schedule is largely dependent on the variety grown. Some varieties bloom in 8 weeks after the start of short days; others bloom in 10, 12, and 14 weeks. This is known as the **response time** (response to short days). Varieties of chrysanthemums are grouped in terms of response time. For example, a catalog might specify an **8-week variety** or *10-week variety*.

Regulating Light

Mums are considered a short-season crop; that is, the plant sets flower buds and blooms only when nights are long and days are short. To grow mums satisfactorily, the length of day and night must be regulated at two times.

1. When rooted cuttings are first potted (for 1 to 3 weeks), days must be kept long and nights short. (Plants should be exposed to no more than 7 hours of continuous darkness.) This allows the plant to establish good vegetative growth before the formation of flower buds.
2. When summer flowering is desired, days must be kept short and nights long to encourage the proper timing of flower bud formation and flowering.

Mums naturally set flower buds in the fall. This means that unless lighting is regulated, the newly started cuttings will set blossom buds immediately after planting in the fall and winter, when nights are long. This causes the plants to blossom when they have not attained enough growth to be attractive.

To prevent flower buds from forming too soon, artificial lighting must be used to lengthen the day or, more accurately, to shorten the night. To delay flower bud formation, the plants must not be exposed to an uninterrupted period of darkness over 7 hours in length. The best way to accomplish this is to add a period of light in the middle of the night.

The intensity of light to which the plants should be exposed is about 7 **foot-candles**. Bulbs with reflectors are usually needed for best results. Two 4-foot beds may be lighted with one row of 100-watt bulbs. Space the bulbs 6 feet apart and 5 feet above the bed's soil level.

EXAMPLE If sunset occurs at 5:00 P.M., and the sun rises at 7:20 A.M., lights could be turned on for 4 hours from 10:00 P.M. until 2:00 A.M. This means that darkness is provided from 5:00 P.M. until 10:00 P.M. (5 hours) and from 2:00 A.M. until 7:20 A.M. (5 hours and 20 minutes). In this way, both periods of darkness are shorter than 7 hours. Under these conditions, the formation of flower buds is delayed, and the plants are allowed to increase in plant size and stem length during the **vegetative period.** If other plants are growing nearby that would be affected by the extra light, shade curtains should be installed to shield them from the light. In warm, sunny climates where plants naturally grow taller, lighting is not needed.

For summer flowering, shading is needed because days are long. Mums set flower buds only when exposed to short days (12 hours or less) and to night temperatures of 60°F (16°C) or lower. In summertime, black sateen shade cloths are used to shorten the days. Wires strung the length of the growing bed are used to support the cloth. The cloth is pulled over the bed in the evening and left there long enough to shorten the day to 12 hours or less. In very hot weather, the cloth should not be pulled over the plants until 6:00 or 7:00 P.M. If the cloth is pulled over the plants any earlier, heat buildup under the cloth could damage the plants.

▲ **Figure 14–1** Typical rooted chrysanthemum cutting ready for planting in the pot.

Obtaining Starter Plants

Mums are propagated by softwood cuttings. They are relatively easy to root. There are, however, a variety of diseases that may be passed along from parent plants to the rooted cuttings. For this reason, it is very important to purchase rooted cuttings from specialists who can guarantee that the cuttings are disease free. Contact a local greenhouse operator for a source. The operator may offer to order for you if you need only a few cuttings. Order at least several months in advance to be sure the selected varieties are available. Figure 14–1 shows a typical rooted cutting.

Establishing a Growing Schedule

Before ordering rooted cuttings, a growing schedule must be established. The schedule must be planned starting at the end of the schedule, the blooming date.

EXAMPLE Assume that Easter Sunday occurs on April 10. If the mums are being grown as a class project, and the school closes for the holiday on Friday, April 8, the mums must bloom prior to that date. To be sure that the blooms are available for sale the week of April 4 through 7, the blooming date should be set for April 1. (Mums hold their blossoms for a fairly long time.) The schedule would appear as shown in Figure 14–2.

According to variety, mums receive what is called *short treatment, medium treatment,* or *tall treatment.* These labels are necessary because some varieties naturally tend to grow short, some medium, and some tall in height. So that the plants grow to the same height and appear uniform, the length of the vegetative period (the long day in summer before shade is used, and the short night in fall and winter when artificial light is used) must be regulated differently for each of the three groups.

For example, the mum crop being prepared for the Easter market is grown in winter; therefore, the length of the vegetative period is controlled by using lights to shorten the number of hours of continuous darkness. After planting the rooted cuttings, the lights are left on for 2½ weeks for the short treatment varieties, 2 weeks for the medium treatment varieties, and 1 week for the tall treatment varieties. Notice that the planting date changes in Figure 14–2 according to variety. The schedule shown in Figure 14–2 is for 10-week response varieties; the schedule for 8-week response varieties is 2 weeks shorter. These varieties are simply planted 2 weeks later.

TREATMENT DATE	SCHEDULE					
	PLANT	SHADE	LIGHTS ON*	PINCH	B-NINE (2 WEEKS AFTER PINCHING)	BLOOM
Short (2 1/2 weeks of lights)	Jan. 4	not needed	Jan. 4–Jan. 21	Jan. 22	Feb. 5	April 1
Medium (2 weeks of lights)	Jan. 7	not needed	Jan. 7–Jan. 21	Jan. 25	Feb. 8	April 1
Tall (1 week of lights)	Jan. 14	not needed	Jan. 14–Jan. 21	Jan. 29	Feb. 12	April 1

*From end of "lights on" to bloom date is 10 weeks (when long nights start).

Figure 14–2 Growing schedule for Easter flowering (10-week response variety).

Planting Rooted Cuttings

Container and Medium (for Potted Mums)

A 6-inch plastic pot is a good choice for planting rooted mum cuttings. This size pot requires five cuttings per pot. Varieties that send out more side shoots may be planted with only four per pot. If a 5-inch pot is used, plant four cuttings to each pot. Some markets use a 4-inch pot and place two cuttings in each pot.

The medium should be one that is well drained and yet holds moisture well. If natural soil is used, it must be **pasteurized** first. Peat moss for moisture retention and sand or perlite for drainage should be added to soil in this ratio: one-third soil, one-third sand or perlite, and one-third sphagnum peat moss. There are other growing mixes on the market that contain no soil and are very effective. The medium should be such that water added to the surface of the pot drains quickly through the pot and out the bottom. If the soil mix has the proper texture, it will not crack when dry. Be especially careful that the potting medium does not dry out after the mums are potted. Mums should never be allowed to wilt from dryness. To prevent this, provide a constantly moist, but not wet, soil. Add water when the top of the potting medium first appears to be drying out.

In addition, mum media is available in **soilless culture**—Peat-Lite, **pine bark**, and other commercial mixes. The **pH** of the media should be 5.5 and afford the mum proper drainage. The water being applied should be tested for pH and corrected to reflect a pH of 5.5 to 5.8.

TRY THIS Use a basic water quality test kit to determine the pH of the water being used to grow your mums. Record the pH values; if the pH is not within appropriate range, explain what must be done to correct the water quality.

Caring for Potted Cuttings

Temperature Control

When first placed in the growing area, the greenhouse temperature should be kept at 63°F (17°C) at night (Figure 14–5). The temperature should be held to 63°F (17°C) on cloudy days and 70°F (21°C) on sunny days. This temperature range is maintained through the first 2 or 3 weeks of short days. The higher temperatures promote faster vegetative growth. If the temperature drops below 60°F (16°C) during this period, flower buds may not set evenly. Uneven bud set may be one of the first signs of improperly regulated temperature.

At the end of this period, the temperature should be dropped to 62°F (16°C) at night, to 62–65°F (1–18°C) on cloudy days, and to 70°F (21°C) on sunny days. For the last 3 weeks of growth, the temperatures should be dropped to 55°F (13°C) at night and on cloudy days, and to no higher than 65°F (18°C) on sunny days. This cooler temperature at the end of the growing period tends to harden off the plants.

PROCEDURE
Steps in Planting

As mentioned, mum cuttings should be planted in 6-inch shallow pots (azalea pots), with four or five rooted cuttings per pot. Select cuttings for each pot that are all approximately the same size to provide more uniformity to the finished product.

1. Fill the pot with planting medium to within 1 inch of the desired level after planting. Press the medium firmly in the pot so that it will not settle later.
2. Space the cuttings evenly in the pot. Slant the cuttings at approximately a 45-degree angle so that the top of the cuttings protrudes over the edge of the pot (Figure 14–3). This angle encourages the growth of more shoots and, thus, better, more compact plants.
3. Place just enough soil in the pot to cover the roots (about 1 inch or less). When the pot is planted and filled, the medium level should be about 1 inch below the top of the pot. This makes watering the cuttings an easier job.
4. Water the cuttings until the water runs through the bottom of the pot.
5. Place the pot in the growing area (Figure 14–4). Later, these pots will require more space. Pots should have at least 1 inch of space between them for the first 3 to 4 weeks and later about 8 inches between pot rims. This wider spacing gives the plant room to spread out, resulting in a better-shaped plant with more blossoms. Watch for signs of overcrowding within the pots. One of the first signs may be drying of lower leaves. (However, this may also signal underfeeding, disease, or poor drainage.)

▲ **Figure 14–3** Mum cuttings planted in a 6-inch pot. Notice that the cuttings are slanted toward the outside of the pot at an angle of about 45 degrees. The roots of the cuttings are barely covered.

▲ **Figure 14–4** Pots of mum cuttings placed in the growing area of the greenhouse. It will be necessary to provide more space between the pots in about 3 weeks.

TIME PERIOD	TEMPERATURE (°F) NIGHT	CLOUDY DAY	SUNNY DAY
From planting through the first 2 weeks of short days*	63	63	70
Up to last 3 weeks of growing cycle	62	62–65	70
Last 3 weeks of growing cycle	55	55	65

*If temperatures are cool in the early stages of short days, an uneven bud set may occur.

Figure 14–5 Temperature control schedule.

Fertilizing Cuttings

Mums may be fertilized using one method or a combination of several different methods.

FERTILIZING BY CONSTANT APPLICATION A very diluted fertilizer may be added to the irrigation water, supplying the mums with constant fertilizing. Wait a few days after planting to begin this program. If this form of fertilizing is chosen, a water-soluble fertilizer prepared specifically for mums should be used according to directions. The fertilizer should be high in nitrogen and potash. For example, if a 15-10-30 fertilizer is chosen, it should be mixed in a stock solution of 18 ounces per gallon of water and diluted again to 1 part solution to 100 parts water before application or as it is being applied through a siphoning device. Mums benefit from the addition of magnesium (Mg) to the irrigation water from **Epsom salts**. Follow the recommendations of the greenhouse supplies for the best rate.

CAUTION
To prevent accidentally damaging plants, always follow the directions on the fertilizer container label.

Note: During cloudy weather, less fertilizer and water are needed. This fertilizing system works well to reduce both at the same time when necessary.

FERTILIZING BY SLOW RELEASE A slow-release fertilizer may be mixed in the potting medium at the time of planting. Use 8 ounces of fertilizer for every bushel of potting medium. This fertilizer will furnish all or most of the fertilizer needs for a period of 3 to 4 months. If plants appear to need more nitrogen, additional fertilizer may be added in the irrigation water. Note: Do not pasteurize a soil mix after a slow-release fertilizer has been added.

FERTILIZING BY TIMED APPLICATIONS Fertilizer may be applied once every 2 weeks in the irrigation water or to each pot as needed. This method is not as effective as either of the other methods listed, and the crop produced may not be as good. To acquire the optimum application of soluble fertilizer with the irrigation system, use a fertilizer injector system to feed the correct PPM of soluble fertilizer (see Figure 13–5 in Unit 13 for the correct setup of a fertilizer injector).

Watering

How often mums are watered varies with the temperature and humidity in the room and the water-holding capacity of the medium. In hot, dry weather, the plants may require watering twice a day. In cool, cloudy weather, once a day or every other day may be sufficient. Check the soil mix frequently for dryness, and add water whenever the top of the medium is beginning to appear dry. Do not keep the soil saturated, however, as the plants' roots may rot. At the same time, allowing the plants to wilt will cause damage to the crop. Add enough water at each watering so that water runs through the pot and out the drainage holes.

CAUTION
Newly planted cuttings require special care and attention the first week. During this time, be especially watchful for signs of dryness.

Pinching and Disbudding

Pinching means simply to break or pinch off tips from plant stems. Hold the tip of the stem and pinch off 1/4 to 1/2 inch of growth (Figure 14–6). This causes side shoots to develop, thereby resulting in more branches to set flowers. The final result is a more compact, flowerful plant.

▲ **Figure 14–6** A plant tip is held directly above the mum from which it was pinched. The tip measures about 1/2 inch in height. Because of the pinching, many more side shoots will form, resulting in a bushier, more flowerful plant.

The ideal potted mum should have about 15 to 20 flowers per pot. If more buds are present, **disbud** (remove) some or all of the smaller side buds. This will result in larger, more attractive individual flowers (Figures 14–7A and B).

Growth Retardants

The chemicals B-Nine and A-Rest, growth retardants, tend to shorten stem growth, which causes the blossoms to grow closer together, resulting in a more compact plant. B-Nine is sprayed on the foliage about 2 weeks after pinching or when side shoots are about 2 inches long. Do not wet the plant foliage for 24 hours after applying B-Nine to prevent the chemical from being washed off. The treatment may be repeated if plants appear to be growing too tall. Follow label directions for the second treatment. Use PGR to determine the correct PPM per application of the growth regulator (refer to Figure 5–9 in Unit 5).

CAUTION
Growth retardants are powerful chemicals that may cause injury to plants, the environment, or humans, if used incorrectly. Always read label directions carefully before using these chemicals.

Controlling Insects and Disease

Preventing attacks on plants by insects and disease organisms is a far better method of plant protection than fighting a serious infestation.

▲ **Figure 14–7** (A) One stem of a mum plant showing many buds at tip and along the stem. (B) The same stem after all buds except the terminal bud are removed. The removal of the buds results in larger, single flowers on each stem.

Good sanitation and cultural practices are often all the control that is needed. If insects do become a problem, spray with an **emulsifiable concentrate** spray. Use a recommended fungicide, if necessary, for fungus disease.

Note: Wettable powders leave a residue on plants and should not be used within 6 weeks of sale date. Emulsifiable concentrates do not leave visible residues. (An emulsifiable concentrate is a chemical in liquid form that will mix with water.) Emulsifiable concentrates and wettable powders may not mix well if applied in the same solution.

CAUTION

Unless insect or disease problems arise, do not spray. When pest control is necessary, never use highly toxic substances. When any pesticide is used, read directions carefully, and follow all safety precautions.

▲ Figure 14–8 A potted mum ready for market.

▲ Figure 14–9 The yellow diamond, a relatively new mum variety, that has open, showy petals.

Space Requirements for Year-Round Production

To produce 50 pots of mums every 2 weeks year-round, about 380 square feet of bench space is needed. Provision must be made for lighting on at least part of the space (Figures 14–8 and 14–9).

Garden Mums

Many horticulturists prefer to cultivate plants that can be set outdoors after blooming in the pot. This may be done by using a special variety, the garden mum, for the spring crop. After planting outdoors in the spring, the garden mum can be expected to bloom again the following fall.

Plant cuttings of garden mums in 3- to 5-inch pots and grow according to the schedule for that particular variety. Most garden mums are the short-response variety and flower in 8 to 9 weeks of short days. This gives a quick return on a very popular crop.

SAFETY FIRST

In this unit, we are using chemicals such as soluble- and slow-release fertilizer, insecticides, and fungicides. It is important to follow OSHA regulation to the law. Everyone involved must have the proper training to work with the specified materials to grow a chrysanthemum crop, and everyone should use the correct PPE to work with these products. For more information on occupational safety and health in the workplace, please visit the OSHA site at www.osha.gov.

Summary

Chrysanthemums (mums) are the most popular cut flower sold in the United States. Mums are very popular at Mother's Day, Easter, Thanksgiving, and Memorial Day. Mums are easy to grow with a wide range of colors. When growing mums, one must determine the market to target. If Easter is a desired schedule, then it is necessary to plan with the varieties to grow. Mums are short-day plants; that is, for the plants to bloom, they need short days. Mums are started with rooted cuttings; for example, five rooted cuttings are planted in a 6-inch pot. Mums need plant food from fertilizer in soluble or slow-release forms to produce the desired flowers. The plants need to be pinched and disbudded, and height is controlled with growth regulators. Each grower must continue to monitor mums for insects and diseases. Garden mums are cultivated outdoors.

Review Questions

1. What makes the chrysanthemum a popular plant to grow among greenhouse growers?
2. Describe a growing schedule for chrysanthemums that need to be ready for Mother's Day (early to mid-May).
3. What are the purposes of pinching and disbudding chrysanthemums?
4. Discuss different ways to control pests when growing chrysanthemums.

Student Activities

1. Develop a potting soil for chrysanthemum cuttings. Determine by feel and appearance whether the potting medium has the proper characteristics.
2. Plant at least one pot of four cuttings of mums and care for it throughout the growing period.
3. Develop a complete growing schedule for an Easter mum crop. Compare it with your instructor's schedule for accuracy.
4. Visit a commercial greenhouse operator who produces mums. Observe the operation and discuss a prepared growing schedule with the grower.
5. Grow rooted mum cuttings at home during the summer for fall production and sale as a supervised agricultural experience (SAE). (This idea courtesy of Keith Staley, Middletown High School, Maryland.)
6. Search online using the following terms: mum production; garden mums. Create a bullet list of the tips to effective greenhouse mum production.

Suggested Supervised Agricultural Experience (SAE) Activities

1. For an entrepreneurship SAE, grow and sell chrysanthemums in a private greenhouse.
2. Participate in growing chrysanthemums for an annual FFA fundraiser. Record your hours and income from the fundraiser for an entrepreneurship SAE.

Additional Resources

www.mums.org/

http://extension.psu.edu/pests/ipm/pestproblemsolver/perennials-annuals/point-of-purchase/chrysanthemums.pdf

UNIT 15

Easter Lilies

Objective

To produce a marketable crop of Easter lilies.

Key Terms

dolomitic limestone Mg_2CO
fungicide
Lilium longiflorum
Pythium
Rhizoctonia
systemic
vernalized

Competencies to Develop

After studying this unit, you should be able to:

- select the proper container, medium and planting area and demonstrate by doing the planting or write up the procedure.
- prepare or select the proper medium for growing lilies, and explain the percentage of each material in it. If soil is used in the planting media, it should be pasteurized to control soil born disease organisms.
- pot plant Easter lily bulbs to the proper 2-inch depth, with properly filled pots.
- calculate the planting date for lilies to flower by Easter Sunday.
- outline in writing a fertilizer schedule for the crop.
- describe in writing which growth regulators are used and why.
- describe in writing the disease control methods needed to produce a clean crop.

Materials

- chemicals or steam to sterilize the working area
- 6-inch, three-quarter-size pots
- well-drained planting medium
- precooled (vernalized) lily bulbs
- ground limestone
- dolomitic limestone (Mg_2CO)
- greenhouse growing area with good temperature control
- the fertilizers potassium nitrate and calcium nitrate
- superphosphate
- contact your greenhouse supplier for the best plant growth regulator for height control, if necessary

The Easter lily, **Lilium longiflorum**, is a native of the Ryukyu Islands of Southern Japan with a white trumpet flower. Easter lilies are major holiday potted plants as well as popular cut flowers grown in the United States with a wholesale value of $37.4 million. We will deal with the Easter lily here as a potted plant being grown for the Easter holiday.

Varieties or Cultivars

The two most popular varieties of lilies on the market are called 'Ace' and 'Nellie White.' 'Ace' produces more flowers and grows slightly taller than 'Nellie White.' If you prefer not to use or to limit the use of growth regulators to dwarf the plants, 'Nellie White' is the best choice. 'Nellie White' is an excellent plant for potted culture and may be a slight favorite for this purpose.

Bulbs

According to the US Department of Agriculture, the leading producers of lily bulbs are (1) Michigan, (2) California, (3) Pennsylvania, and (4) Ohio. They are shipped in crates packed in slightly moist peat moss (Figure 15–1). The lily bulb is a soft structure with no heavy protective coating and loses moisture rapidly; bulbs should never be allowed to dry out. To simplify the growing procedure, it is wise to purchase precooled bulbs. Bulbs from 7 to 9 inches in circumference are large enough. Large bulbs may produce more flowers, but a controlled lower temperature can offset this difference.

▲ **Figure 15–1** A crate of Easter lily bulbs as shipped from California to an Eastern grower: Note that the bulbs are packed in peat moss to prevent drying out.

GREEN TIP Use the ashes from a fireplace in your garden to add phosphate, potassium, and other trace elements. Just thinly spread the ashes on your garden and dig in where you can. This application raises the pH of the soil and is most beneficial in soils that are acidic.

Establishing a Growing Schedule

It is very important to plant Easter lily bulbs on the proper date so that the plants will flower just before the Easter holiday (Figure 15–2). The demand for potted Easter lilies drops off after Easter. How soon the bulbs will flower after planting is determined by two things: the length of the cold treatment given and, after bud set, the temperature at which they are grown.

According to work done at Cornell University, the number of days to flowering time is influenced by how much cold treatment the bulbs have had. The approximate days to flowering may be predicted using the table in Figure 15–3.

2016—March 27	2023—April 9
2017—April 16	2024—March 31
2018—April 1	2025—April 20
2019—April 21	2026—April 5
2020—April 12	2027—March 28
2021—April 4	2028—April 16
2022—April 17	

Figure 15–2 Dates for Easter Sunday, 2016–2028.

DATE BULBS ARE RECEIVED	PROBABLE WEEKS OF PRECOOLING	APPROXIMATE DAYS TO BLOOM (60°F [16°C] AT NIGHT, 70°F [21°C] DURING THE DAY)
November 25	5	125
December 2*	6	120
December 9	7	114
December 16	8	108
December 23	9	103
December 30	10	100

Figure 15–3 Flowering times.

*Based on the information in Figure 15–3, if Easter Sunday falls on April 19, and precooled (**vernalized**) bulbs were received on December 2, the flowering date would be in approximately 120 days. The potting date would then be calculated to be December 21. It may be a good idea to plant 1 week earlier than this date to be sure. Plant growth can be slowed down later, if necessary, by lowering the greenhouse temperature.

If the bulbs need to be held 1 or 2 weeks before planting, store them in the shipping crate at 40°F. Do not allow the bulbs to dry out, but do not soak them either; misting the crate may help.

Planting Bulbs

Growing Medium

Lilies should be planted in a very well-drained potting mix. The mix should be high in perlite, sand, or coarse vermiculite to improve drainage. Many commercial mixes are available, but if you mix your own, use at least one-third perlite, sand, or vermiculite. If any soil is used in the mix, it should be pasteurized at 180°F for at least 30 minutes. The mix should be one part loam, one part peat moss, and one part perlite. The best policy is to use a soilless mix of peat and perlite or sand. A good soilless mix is one-half peat moss, one-quarter perlite, and one-quarter vermiculite.

The pH of the medium should be adjusted to 6.8 to 7.2 with limestone. To adjust the pH of most mixes requires about 7 or 8 pounds of limestone per cubic yard of mix. About 3 pounds of this should be **dolomitic limestone**, and the other 4 or 5 pounds should be regular ground limestone. About 12 to 16 ounces of regular superphosphate is also helpful.

Potting

Bulbs are planted in 6-inch, three-quarter-size pots. Each bulb should be placed so the tip is covered with 2 inches of potting medium (Figure 15–4). Remember that the lily bulb is very tender and must be handled carefully. Do not push the medium around the bulb hard enough to bruise it, but do pack it firmly. *After the medium is settled,* the pot should be filled to within 1 inch of the top, and the bulb tip should be 2 inches below the surface of the medium.

The pot should be watered well to thoroughly soak the medium. Watering after planting should be done often enough to keep the medium *moist, not wet.*

Immediately after potting, drench the medium in the pot with a combination of Lesan and Terraclor to help control root rot. Mix according to directions on the label, and apply one-half pint per pot.

CAUTION
Wear chemical-resistant gloves to protect your hands when working with chemicals. Nitrile or neoprene gloves are examples of lower-cost PPE frequently worn for chemical protection.

▲ **Figure 15–4** A bulb is placed in the pot so that it will be covered with about 2 inches of potting medium when the pot is filled.

Caring for Potted Lilies

Fertilizing

When shoots emerge from the soil, the pots should be watered with a mixed solution of 3/4 pound potassium nitrate per 100 gallons of water (0.12 ounces per gallon) and 1-1/2 pounds calcium nitrate per 100 gallons of water (0.24 ounces per gallon) every 2 weeks (Figure 15–5).

When flower buds are 1/2 inch long, use only calcium nitrate at 2 pounds per 100 gallons of water (0.32 ounce per gallon) weekly until the plants are sold (Figure 15–6). Do not use an acid-forming nitrogen source such as ammonium sulfate. The addition of a slow-release nitrogen fertilizer at 1/2 teaspoon per 6-inch pot will help to prevent nitrogen deficiency and enhance the shiny, healthy appearance of the leaves of the lily. It is important to keep the plants well fertilized during the growing season.

▲ **Figure 15–5** When plants reach the stage of growth shown in this picture, it is time for the first fertilizer application.

▲ **Figure 15–6** A plant in full flower, ready for market.

Watering

Soak the containers when the bulbs are first planted. After that, keep the medium moist but never wet. In other words, wait to water until the medium is barely moist on top. If the plants are allowed to dry out, plant quality and the number of flowers are decreased. If they are overwatered, root rot is a strong possibility. Good air circulation around the plants will help to prevent very wet conditions by drying the soil quicker after each watering. Overwatering can cause root rot problems in the crop.

Light

The potted lilies should be placed in full light. If the greenhouse glass has shade fabric, dirt, or shade spray on it, these should be removed.

Growing Temperatures

Day and night temperatures for Easter lilies should be closely controlled. A nighttime temperature of 60°F (16°C) and a daytime temperature of 70°F (21°C) should be carefully maintained. This is especially important up to the time flower buds form, usually about the middle of February. If temperatures go above 70°F at night, flower bud formation may be delayed.

After flower buds have formed, temperature control is not as important. Temperature may now be used to control growth and the time of flowering. Warmer night temperatures especially will speed up crop progress and cause earlier flowering. Cooler temperatures during both day and night will slow development and result in a later flowering date. Plants will need to be watered more often under warmer temperatures and less often under cooler temperatures.

TRY THIS Create a list of methods that greenhouse operators use to control the temperature for the delicate range within which Easter lilies must be grown. Use the Internet and other alternative resources for your research.

Growth Regulators

Extending the day length to 18 hours will cause lilies to grow taller. Long days are started when plants are 4 to 6 inches tall and continued for 5 to 6 weeks. Tall lilies are needed for the cut flower trade, but they are not desirable in plants sold in pots. Shorter-than-normal day length will shorten stem growth, but chemical growth retardants are more effective.

Contact your greenhouse supplier for the most economical growth retardant to use. For lilies that normally are short growers, such as 'Nellie White' and 'Ace,' a small amount of chemical is used. *Note:* Read the label for proper use and cautions. Plants are treated when they are from 4 to 7 inches tall. Chemical retardants do not affect the number of leaves or flowers; their purpose is to shorten the stem internodes or distance between leaves.

Keep a record of the chemical treatment used and its effect on the plants. Adjustments can be made in the amount of chemical used for the following crop based on this record. More chemical use results in shorter plants. *Note:* Follow label directions.

Cultural practices also affect plant height. A proper fertilizer program results in shorter plants than would otherwise develop without proper fertilization.

Controlling Insects and Diseases

ROOT ROT The best method of disease prevention is to be sure the growing medium is pasteurized. This includes the pots, medium, potting table, greenhouse bench, workers' hands, and so forth. The bulbs should not come into contact with anything that might be contaminated with the fungi ***Rhizoctonia*** or ***Pythium***. These are the two fungus disease organisms responsible for root rot in lilies. Growing benches may be drenched with Clorox in a 1:9 dilution with water. Steam is also effective to pasteurize the growing environment.

If proper sanitation is provided and good cultural practices are maintained, such as starting with a well-drained medium and *not* overwatering, root rot will probably not be a problem.

Immediately after planting, water the pot with a good preventive **fungicide**. Contact the local extension service for a recommended fungicide to use.

INSECTS The most common insect pests are the aphid, mite, and fungus gnat. If these pests are present, use any good **systemic** insecticide labeled for lilies and indoor use.

CAUTION
Use extreme caution with all chemicals.

Storage

According to researchers at Cornell University, if plants flower too early for Easter, they can be stored for up to 2 weeks in the dark at 31°F (21°C). Storage should start when the first bud begins to open. Plants should be sprayed with zineb (1/2 pound per 100 gallons of water) before storage to protect against botrytis. Any open flowers should be removed before storage because they are more susceptible to botrytis. Plants should be sprayed with a fungicide before storage to protect against botrytis.

Removal from storage should be done one full day before full bloom is desired. Move plants to a shaded area to warm up before placing them in full sunlight in the greenhouse.

SAFETY FIRST

In this unit, we are using chemicals such as soluble- and slow-release fertilizer, insecticides, and fungicides. It is important to follow OSHA regulation to the law. Everyone involved must have the proper training to work with the specified materials to grow an Easter lily crop, and everyone should use the correct PPE to work with these products. For more information on occupational safety and health in the workplace, please visit the OSHA site at www.osha.gov.

Summary

The Easter lily *(Lilium longiflorum)*, a native of the Ryukyu Islands of Southern Japan, has a white trumpet flower. In the United States, Easter lilies are major holiday potted plants and popular cut flowers with a large wholesale value. Michigan, California, Pennsylvania, and Ohio are the leading producers of lily bulbs. Bulbs need a well-drained growing medium with a neutral pH level. Lilies are potted with one bulb per 6-inch pot. Lily bulbs need continued care with fertilizer, watering, and light and temperature control. The lily needs insect and disease control for a productive flower to be produced.

Review Questions

1. Why is it necessary to properly sterilize containers, and pasteurize medium and planting area before planting Easter lilies or other plants?
2. What disease-control methods are necessary to produce a clean crop of Easter lilies?
3. If Easter falls on April 12, and the vernalized bulbs are received on November 20, when should the bulbs be potted?
4. Discuss the care needed for potted Easter lilies.
5. Your Easter lilies have flowered a week ahead of schedule. What do you do to keep them flowering for Easter?

Student Activities

1. Visit a greenhouse where Easter lilies are grown, and interview the manager. Prepare questions ahead of time, and ask if you can record the interview with a smartphone camera. Share the video with your class, so that they can experience the interview.

2. Have your local greenhouse grower come to class and speak with you about the crop and demonstrate his or her methods.
3. Plant and grow a crop of Easter lilies for sale. Use this experience to apply for an FFA Proficiency Award.
4. Using the Internet, research the history of Easter lilies and lily production. Write a short report describing the origins of the Easter lily industry.

Suggested Supervised Agricultural Experience (SAE) Activities

1. Develop and conduct an entrepreneurship SAE by growing, selling, and delivering Easter lilies to customers through direct marketing.
2. Work at the sales counter of a local greenhouse/nursery as a placement SAE.

Additional Resources

https://lancaster.unl.edu/hort/nebline/easterlily.shtml

http://pss.uvm.edu/ppp/articles/eastlily.html

www.clemson.edu/extension/hgic/plants/indoor/flowering/hgic1555.html

SECTION 5

Integrated Pest Management

UNIT 16

Integrated Pest Management and the Biological Control of Pests and Diseases

Objective

To understand the control of plant pests through natural, biological means.

Key Terms

bacteria
biological control
competitors
endophytes
entomopathogenic (beneficial) nematodes
fungus
nematodes
parasites
parasitoids
plant stimulants
predators
sticky cards

Competencies to Develop

After studying this unit, you should be able to:

- explain orally or in writing what biological control means.
- explain orally or in writing what integrated pest management means.
- outline a pest control program, explaining when biological control should be used and at what point chemicals must be used.
- list at least three insects that have been effectively controlled without man-made chemicals.
- list at least one plant disease controlled by biological means.
- list at least one weed and the biological method used to control it.

Materials

- collected insect predators for identification
- smartphone apps:
  - Pest Portal
  - Spectracide
  - Calibrate My Sprayer
  - GardenCompass
  - Ortho Problem Solver
  - AG PhD
  - Greenhouse Management
  - IPMPro

Integrated Pest Management

The new integrated pest management (IPM) approach to pest control of plants has been tested and found to be very effective in not only controlling pests but also in greatly reducing the need for applying chemicals. The key to an effective IPM program is regular monitoring to detect the pest when it first becomes active. This is done through regular inspection of plants by specialists and by using traps, sticky cards (cards that trap the adult stages of flying insects), or other collection devices to spot the first few pests on the scene. **Sticky cards** have a two-sided (yellow/blue) design that allows you to trap a wider variety of the most common plant-damaging insects. For example, the yellow card traps aphids, whitefly, and fungus gnats, while the blue card traps thrips (Figure 16–1). Action should be taken immediately, before the number of pests has a chance to increase.

IPM programs work to keep pests and their damage at a marketable accepted level by employing the following techniques: scouting, trapping, monitoring, pest identifying, threshold maintaining, and combining mechanical, biological, and chemical applications controls.

The first control used should be nonchemical such as hand collecting of pests and pruning out infected or diseased limbs or plants. Next, a biological control system is considered and used if deemed adequate. Lastly, chemicals are used. Treatment should be early when only a small area is affected, spraying only infected plants, thus reducing the need to spray entire fields or other large areas. Often, one well-timed spray application on a small area is all that is needed for control during the entire season. The least-toxic chemicals available should be used first. Spray oils, which are now highly refined and effective, are very safe for the plant. If used early in the season, spray oils may eliminate the need to spray with more toxic chemicals later, thus protecting the environment. IPM is working well and, for many pests, has eliminated the need for costly regular applications of chemicals once thought necessary.

Plant diseases are not quite as easy to control with IPM as insects. Good sanitation to prevent carryover of fungus spores, early identification, and prompt control can greatly reduce the need for regular preventive sprays. Disease-resistant varieties can also reduce or eliminate the need for chemical control of some diseases.

IPM can work. A report in Maryland stated that nurseries employing the IPM program used 98 percent fewer pesticides than reported for other nurseries. This reflects tremendous savings in money spent for chemicals and also reduces pollution of the environment.

▲ **Figure 16–1** Yellow sticky tag to monitor adult flying insects.

GREEN TIP IPM helps to control pests and diseases through natural and biological methods to reduce the amount of chemical controls used in the environment.

TRY THIS Visit a local orchard or a commercial or school greenhouse with a group of students to help with the IPM program. Practice hand collection of pests and pruning out infected or diseased limbs or plants to help the business with its pest management program.

A History of Biological Control

The **biological control** of plant pests is the oldest control method available. All creatures in nature have natural enemies, such as diseases, **predators**, or **parasites** that repel, kill, and consume them. Plant pests are controlled by naturally occurring **bacteria**, fungi, insects, **nematodes**, spiders, birds, toads and frogs, lizards, snakes, voles (micelike creatures), and others (Figure 16–2). If these natural enemies were not here, the world would be overrun with insects. If natural enemies did not exist, a single aphid could be responsible for the production of more than 318 million aphids in 1 month. The fact that aphids do not increase in numbers to this extent in nature (where no spraying is done) is proof that biological control works (Figures 16–3 and 16–4).

In the first half of the twentieth century and in earlier years, when farms were smaller, and crops were grown on smaller acreage and in rotation, the natural predators did a good job of controlling, but not eliminating, many of these pests. Then came the large, specialized grower with hundreds or thousands of acres of one crop, which was a paradise for plant pests. Whichever direction a fungus disease spore blew in the wind, it hit a plant it could grow on. Soil-borne diseases multiplied as the same crop was planted in the same field year after year. Pests survived in large numbers under the same conditions. Why didn't the plant pest-control organisms also build up and control these pests? There are three main reasons the pests' natural enemies did not keep up with and do their control job.

1. Many of these organisms needed a different environment as adults to survive. They may have needed nectar from wildflowers or some cover in which to lay their eggs. Large acres of the same crop destroyed this natural habitat and the adult pest-control organism died out.
2. The American consumer wanted a perfect food product with no blemishes from insect bites. To do this, intense chemical

▲ Figure 16–3 A braconid wasp (1 ½ inches long) laying an egg in an aphid. The wasp larva will kill the aphid.

▲ Figure 16–4 A ladybug larva preparing to eat an aphid.

▲ Figure 16–2 The green lacewing, one of the most effective biological controls of aphids and many other soft-bodied insects.

control was used, killing good organisms as well as pests.

3. The farmer felt that the only good insect was a dead insect and started using chemicals to kill them all. For biological control organisms to survive, there must be some pests on which they can feed.

For biological control to work, humans must rethink their strategy on pest control, and the *integrated* pest management methods may provide the answer. We must be willing to live with a few pests when no real economic damage is being done, and we must learn what environment the natural biological control agents need to survive.

It is now known that many adult parasitic pest-control insects need wild flowering plants such as mustard, wild carrot, yarrow, dill, anise, black-eyed Susan, catnip, and milkweed, to name a few, as food sources of nectar and pollen. Some areas set aside for these wildflowers would help to ensure survival of the adult insects and guarantee a new generation of natural pest-control insects.

Examples of Success

Brief mention will be made here of the successful biological control of some insects that cause serious damage. More detail is given later in this unit.

In the 1940s and 1950s, the Japanese beetle was a serious pest in the eastern United States. Millions of dollars were spent on chemical controls, which had to be repeated year after year. When two natural enemies of the beetle were brought to the United States and placed in the environment, the beetle was controlled to the point where it is not as damaging, and chemical sprays are less often needed. The biological controls live from year to year and need not be reapplied.

Researchers at Cornell University released and established the European seven-spotted lady beetle (*Coccinella septempunctata*) as a predator of aphid pests. This species is highly adaptable to many environments and should survive to provide continued control of aphids and greatly reduce the need for chemical control (Figure 16–5).

The US Department of Agriculture has had great success against isolated pockets of gypsy moth infestations. Through the use of biological agents and traps baited with sex attractants, the department has had consistent success in eradicating the moth. In 30 of 35 infested sites, the moths were eliminated with no chemical pesticides.

Source: US Department of Agriculture

▲ **Figure 16–5** A seven-spotted ladybug adult.

Other exciting success stories using biological control are being reported for the Colorado potato beetle, which is resistant to nearly all chemical pesticides, the alfalfa weevil, the Mexican bean beetle, the musk thistle, and many others.

The Mexican bean beetle can be controlled on a crop if a special management system is set up. Some beans planted early are used to initiate the bean beetle population and start the parasitic control agent on these beans. When the commercial crop is planted later, enough parasites have built up to maintain control on the larger planting. It is necessary to build up the pest-control agents in this way or grow them in a laboratory for release in the field. Otherwise, the effectiveness is diminished until the predators have time to increase in numbers large enough for satisfactory control.

Some predators remain in the environment and give control in the future; others must be released from the laboratory each year. Through selection and genetic changes, researchers are developing control agents that can live longer in the environment.

GREEN TIP Aromatic herbs such as lavender (*Lavendula*) repel rabbits, thus eliminating the need for chemical intervention and offering a more environmentally friendly approach to protect plants against this pest.

The Need for Biological Control

The most exciting thing about this success in biological pest control is that we are finally looking to nature for control of plant pests rather than using highly poisonous chemical materials. Many of these chemicals are no longer effective anyway because insects become resistant to them. The Chinese use biological control to a very large extent, and we can learn much from them in this respect.

Despite close monitoring, new plant pests and weeds are still being introduced into the United States. It is now estimated that 11 new immigrant species make the United States their home each year and that about 7 of these will become "pests of some importance." Bringing in the natural enemies of these new pests will be important in controlling them. Nature normally has a system of balances in which each creature, large and small, has some natural enemy to help control the population. It is easy to understand how a new insect, weed, or disease, introduced in an area where there are no natural enemies, would multiply rapidly.

Chemical-Free Control

Insects are controlled to a great extent by natural forces. Their natural enemies—birds, other insects such as the ladybug and lacewing, and diseases—help to keep the pest population under control. For example, tiny wasps lay eggs on the tomato hornworm (Figure 16–6); the eggs hatch into worms that eat the tomato worm from the inside out. Other parasitic wasps kill insects such as the alfalfa weevil (Figure 16–7) and the Mexican bean beetle (Figure 16–8). If natural enemies are not present in a certain area, they may be introduced to provide the necessary control. Figure 16–9 shows a small, wormlike larva of a wasp that was introduced to the United States to aid in the control of the Japanese beetle.

▲ **Figure 16–6** A parasitized tomato hornworm with tiny, white eggs of the braconid wasp attached to its back. The eggs will hatch, and the tiny larvae will enter the tomato worm and kill it.

Source: US Department of Agriculture

▲ **Figure 16–7** The tiny stingless wasp, a parasite of adult and larval stage alfalfa weevils, depositing an egg in a weevil larva. When the egg hatches, the wasp larva eats its way out, killing the weevil.

Source: Dr. Gary Cunningham, Animal and Plant Health Inspection Service, Hyattsville, Maryland

▲ **Figure 16–8** A tiny stingless wasp (*Pediobius foveolatus*) lays eggs inside a Mexican bean beetle larva.

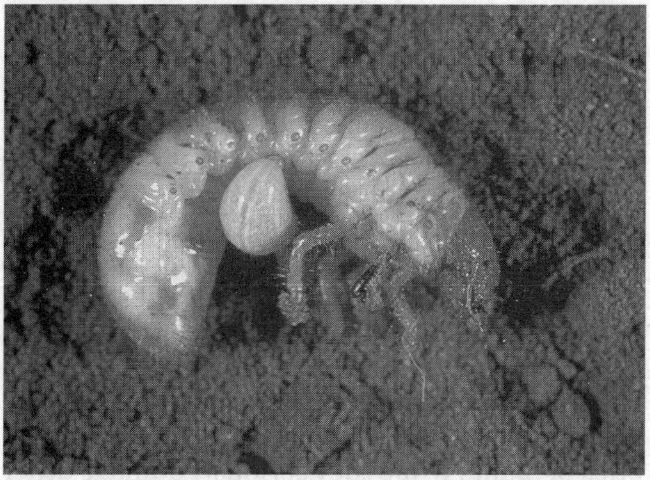

▲ **Figure 16–9** A digger wasp parasite of the Japanese beetle in its larval form feeding on the larva of a beetle. The US Department of Agriculture introduced the parasite from Japan to the United States, where it prevents the Japanese beetle from becoming a serious economic threat.

Commercial preparations containing pest-control organisms that attack insects can also be used for chemical-free control. Naturally occurring disease organisms that attack insects, such as fungi, viruses, and bacteria, are grown on host insects in the laboratory and then prepared for use as pesticides. In this way, the natural diseases of the insects are multiplied to such an extent that the insect population is greatly decreased or eliminated.

Before chemical pesticides are used, the insect or disease buildup is carefully monitored by university extension specialists and growers. If a problem is found to exist, natural enemies may be used such as parasitic wasps for the bean beetle (Figure 16–8). Thuricide (*Bacillus thuringiensis* [Bt]) is a bacterium that kills many worms or caterpillars. The bacillus kills by causing high gut pH, thus upsetting the digestive tract and causing fatal starvation. Caterpillars such as the spruce budworm, cabbage looper, imported cabbage worm, and gypsy moth are killed.

Another example of new experimental biological control is the use of the insects' own uric acid as poison. There is also a control for *Verticillium* wilt: a **fungus** *Talaromyces flavus*, which fights other harmful fungi. If established naturally, these new controls could save millions of dollars each year in crop damage and chemical pesticide cost. These biological controls are not harmful to other living things and do not harm the environment.

GREEN TIP Biological control agents are excellent sustainable green alternatives because they reduce or eliminate the dependency on chemical control of pests, thereby reducing exposure of humans, plants, and the environment to hazardous substances.

Biological Control Systems

As mentioned, biological control or natural enemies of plant pests should be considered first in any pest-control program. After being mass produced, these biological organisms are added to the soil or sprayed on plants. In most cases, these controls are very effective and do not poison the environment.

When using biological control effectively, users must understand that they cannot kill every pest in 1 or 2 days. This is not a problem if the pest does not enter the part of a plant to be marketed or eaten by humans. In other words, a potato beetle can eat some plant leaves without damaging the crop, which grows underground. If large numbers of beetles eat lots of leaves and the crop is reduced, this is a problem. Biological control may keep the number of beetles down to a number that does not reduce crop production. This is successful control, even if a few beetles do survive.

As discussed, biological control organisms are living things, such as viruses, bacteria, fungi, or insects (Figure 16–10). They may also be naturally occurring chemicals that repel insects, sterilize them so eggs do not hatch, or stop them from maturing to adulthood. Most of these living organisms or chemicals are specific; in other words, they affect only one insect or disease.

Specific Examples of Biological Control

Biological control agents work in a number of ways.

- The plant itself is resistant to the disease or insect.
- Chemicals in the plant are toxic or repellent, or sticky hairs on the leaves and stems repel insects.

Source: Photographer Unknown, Agricultural Research Society, US Department of Agriculture

▲ **Figure 16–10** A spined soldier bug attacking a caterpillar larva.

- Biological control agents produce chemicals that injure or kill the plant pest.
- The agents grow on or in the pest at different life stages (egg, larva, nymph, and adult) and kill it (parasitizing).
- The agents eat the pest or suck out its body fluids.
- The agents compete for food and space with plant pests in the soil or eliminate the food supply.

Natural enemies available commercially include **parasitoids**, predators, and **entomopathogenic (beneficial) nematodes.**

Disease- and Insect-Resistant Plants

Development of insect- and disease-resistant varieties of plants is a most important step in biological control, and many such plants are available that should be selected and used (Figure 16–11). Some of these varieties are so resistant to pests that no chemical sprays are needed. For example, the tomato variety Better Boy is listed as VFN resistant. The V stands for *Verticillium* wilt, F for *Fusarium* wilt, and N for nematodes. Therefore, this tomato variety needs no help in controlling all three of these pests.

Insect pests never bother the neem tree or the Ginkgo tree. We know that the neem tree has a chemical that deters or kills insects. But there are other reasons insects do not attack certain varieties of plants while they seriously damage other varieties of the same type of plant. For example:

- *Rhododendron 'Mist Maiden'* never has lace bug damage because of a thick layer of hairs (indumentum) on the underside of its leaves where the lace bug feeds.
- *Tomato variety 'Brimmer'* deters the tomato fruit worm because of its thick skin.
- *Bush lima bean 'Henderson'* deters all small insects because of leaf hair.

Endophytes

Microorganisms called **endophytes** are fungi and bacteria that live in other plants without harming them. The endophytes protect plants against insect attack. Some lawn grasses naturally infected with fungal endophytes have excellent resistance to lawn insects. Researchers are now inserting genes into the endophytes from other bacteria such as *Bacillus thuringiensis* to provide specific insect control in other plants. One example is control of the corn borer in corn. The treated endophytes are forced under pressure into the corn seed through tiny naturally occurring cracks in the corn seed. When the plant grows, the endophyte grows with it and controls the corn borer without harming the plant.

Natural Pesticides

For years, we have used rotenone and pyrethrums as natural insecticides. Some of the newest and most effective insecticides are the synthetic pyrethroids made by copying the chemical

VEGETABLE	PLANT VARIETIES	PLANT PEST RESISTED
Asparagus	Jersey Giant, Jersey Knight	Asparagus rust, Fusarium wilt, crown and root rot
Bean, bush or snap	Tendercrop	Bean mosaic virus and pod mottle virus
	Top Crop	Bean mosaic virus and pod mottle virus
Bean, pole	Kentucky Wonder	Rust
Bean, lima	Eastland	Downy mildew
Broccoli	Emperor Hybrid	Black rot, Downy mildew, Hollow stem
	Green Dwarf	Downy mildew
Corn	Burpee's Honeycross, Ambrosia	Wilt
Cucumber	Saladin Hybrid	Cucumber mosaic virus, Powdery mildew, Bacterial wilt
	Sweet Success	Cucumber mosaic virus, Scab, Target leaf spot
	Amira Hybrid	Cucumber mosaic virus, Downy mildew, Powdery mildew
Melon	Bush Charlston Gray	Fusarium wilt, Anthracnose
	Ediato Muskmelon	Downy mildew, Powdery mildew
	Dixy Queen	Wilt
	Sweet Favorite, Crimson Street	Anthracnose, Fusarium wilt
Pea	Sugar Bon snap pea	Powdery mildew
	Green Arrow garden pea	Downy mildew, Fusarium wilt
Pepper	Golden Summer hybrid	Tobacco mosaic virus
	Bell Boy	Tobacco mosaic virus
Potato	Kennebec	Late blight
Tomato	Super Beefsteak	VFN*
	Roma	VF*
	Better Boy	VFN*
	Park's OG Whopper Improved	VFFNT*
	Better Bush	VFN*
	Celebrity	VFFNT*
	Big Pick	VFFNTA*
Trees		
Chinese Chestnut		Chestnut blight
American Chestnut	Dunstan Hybrids	Chestnut blight
Ginkgo		All
Apple	Liberty	Scab, Cedar rust, Fire blight, Mildew
	Priscilla	Scab, Fire blight, Cedar apple rust
	Prima	Scab, Fire blight, Mildew

*V—Verticillium wilt
F—Fusarium wilt
N—Nematodes
T—Tobacco mosaic
FF—Race 1 and 2 Fusarium wilt
A—Crown wilt

Figure 16–11 Some plants that resist insects and diseases.

found in the chrysanthemum. In addition, the following six plants show real promise as natural pesticides:

- Basil. An effective repellent and larvacide for mites, aphids, and mosquitoes.
- Big sagebrush. The extract of its branch ends is very effective in preventing the Colorado potato beetle from feeding.
- Chilcuan. Root extracts are effective against many agricultural pests.
- Mamey. Flowers, fruit, and leaves are effective against a large number of insect pests.
- Calamuse. An insect repellent that is especially effective on stored grain insects.
- Neem tree. By far the most promising of these six, its seed extracts are outstanding repellents to insects.

THE NEEM TREE The chemical azadirachtina is a natural pesticide taken from the neem tree. The chemical is most concentrated in the seeds. Azadirachtin acts very effectively as an insect repellent and as a growth regulator, interrupting the growth cycle of insects. It does not kill adult insects, but it prevents them from eating. It does kill the larval stage as it stops the growth cycle of the insect.

The neem extract is not toxic to warm-blooded animals and does not harm beneficial insects because they do not eat the plant and swallow the azadirachtin. It provides effective control for more than 170 different plant insects. As a systematic chemical, it gets into the plant sap and offers longer protection because rain cannot wash it off.

Azadirachtin is effective as an insect feeding deterrent at concentrations as low as 1/10 ppm of water. It also breaks down rapidly in the environment.

The naturally occurring chemical from the neem tree is now commercially available under trade names such as BioNEEM and Margosan-O, which control insects only. It shows great promise as a naturally occurring, safe pesticide for the future. Neem oil is effective as both a fungicide *and* an insecticide.

Pest Life Cycle

To best control a pest, its life cycle must be known. This is true for chemical control as well. There are times in the life cycle of a pest when it is easier to control and times when control is not possible. Also, a pest that gets inside a plant part, such as a fruit or vegetable meant for consumption, should be controlled in the egg or first hatch stage, if possible, before the plant is damaged, or the adult must be prevented from laying eggs in the first place. An excellent nonpolluting control for eggs is spray oil. New highly refined oils can be used year round instead of just when plants are dormant, as with the older types of oils. Many insects, if hit by oil spray during egg or early hatch stage, are effectively controlled. The oils are not toxic and do not harm the environment.

Crop Rotation

Crop rotation is a method of controlling some types of insects that do not move rapidly from place to place. Crop rotation consists of planting a different crop in a field from year to year, rather than growing the same crop on the same land for several years. Many soil-borne diseases, such as tomato wilt, can live in the soil for only 2 or 3 years if the plants on which they feed (host plants) are not present. Thus, if crops are rotated so that the host plant or related plants are not grown in a field more than once in 3 years, the disease will die out. In the case of tomatoes, this means that tomatoes, eggplant, or potatoes (plants that also support tomato wilt) should be planted in a field only once every 3 or 4 years. A typical rotation might be: tomatoes the first year, beans the second year, corn the third year, and then tomatoes again the fourth year.

Rotations also help in the control of many insects that do not travel for long distances, such as weevils and nematodes.

Hand Picking

Where only a few insects or diseased plants are present, it is effective to pick them off by hand and destroy them.

Biological Control of Insects

Figure 16–12 shows specific types of biological control agents and the various plant pests that these agents control.

BIOLOGICAL CONTROL AGENT	PLANT PESTS CONTROLLED	ADDITIONAL INFORMATION
Parasitic wasp, Trichogramma wasp	Colorado potato beetle	The wasp lays its eggs on the beetle's eggs. Use of this wasp reduced need of chemical spray from 16 to 1 spray of Vydate in the early season. Eight thousand wasps per acre give good control.
Parasitic wasp, Trichogramma wasp	Tomato hornworm	The wasp lays its eggs on the hornworm. Wasp larvae enter the worm.
Parastic wasp, Trichogramma wasp	Cabbage looper and other worms	The wasp lays its eggs on the eggs of the moths and butterflies.
Bacillus thuringiensis (can be bought under the names Dipel, Biotrol, Thuricide, and Javelin)	Colorado potato beetle; all caterpillars, including gypsy moth; cabbage looper and imported cabbage worm	A bacterium that is extremely effective on caterpillars, upsetting their digestive tract so they stop eating in about 2 hours and die in 2 days.
Predatory mite	Spider mite and whitefly	Some of these mites are resistant to chemicals.
Beauveria bassiana	Colorado potato beetle	This fungus gives 77% control by attacking the pupal stage in the soil and also adults. Small fungus spores only need to touch the beetle to cause infection.
Orange peel	Many insects controlled, such as ants, fleas	Still experimental, but looks promising.
Ladybug	Aphids	Eats aphids at all stages of growth.
Nedalia beetle, Parasitic fly	Cotton cushion scale on citrus	Imported in 1988 to California and gave complete control of the scale insect.
Green lacewing	Aphids	Both nymphs and adults eat aphids and lace bugs
Green lacewing biologicals	Mealybugs and lace bugs	Operates in cooler temperatures than many.
Bacillus popilliae	Japanese beetle	Known as milky spore disease; kills grubs in the soil.
Digger wasp	Japanese beetle	A tiny parasite kills the grubs in the soil. See Figure 16–4.
Pediobius foveolatus	Mexican bean beetle	Eggs are laid on the bean beetle larvae. Wasp larvae destroy the bean beetle larvae.
Marigold and asparagus roots	Nematodes	Planting marigolds in an area greatly reduces the number of nematodes in the soil. Asparagus roots produce a chemical that causes a decline in the number of stubby root nematodes.
Spined soldier bug, *Podisus* spp.	Eats 100 kinds of harmful insects	One of the most effective insect predators.
Diatomaceous earth	Slugs, snails, mites, aphids	Finely ground sharp silica material kills soft-bodied insects, slugs, and snails.
Predatory nematode	Grubs of weevils, cutworms, fungus gnats, cabbage root weevil, wireworms	Controls many insects that spend some of their life cycle in the soil.

Figure 16–12 Biological insect control.

Biological Control of Plant Diseases

Ninety-nine percent of soil organisms are beneficial; less than 1 percent cause diseases. Yet soil fungi and bacteria cause many plant root diseases. Seed decay, blights, root rots, wilts, and damping-off of seedlings are some of these diseases, which result in costs of an estimated $4 billion each year in the United States. The good soil microbes are now being genetically improved to live longer in the soil and to be resistant to chemicals used to fight diseases. If good soil microbes are resistant to fungicides or other chemicals used to fight disease, they will not be destroyed when these materials are used and thus will remain to aid in the fight against diseases. Figure 16–13 lists specific types of biological agents and the various plant diseases that these agents control.

New Strains of Biocontrol Microbes for Plant Diseases

Over the past 10 years, scientists at plant research laboratories in Beltsville, Maryland, have been able to use chemicals and ultraviolet light to produce improved biological agents to fight soil-borne diseases. Some of the microbes were not killed by benomyl, a commonly used fungicide. Some of the new strains were stronger in their action against disease, and some had longer shelf life in the laboratory and longer life in the soil. The fungi *Trichoderma* spp. and *Talaromyces flavus* attack soil-borne plant diseases. The chemical/UV light treatment has made them more effective control agents.

In addition, a new strain of mite *(Melaseiulus occidentalis)* is resistant to several pesticides. This beneficial mite is used to control spider mites feeding on plants.

Newer methods are being developed to mass-produce these good microbes so they are available in large quantities and at reasonable prices.

Biological Control of Weeds

Plant diseases that attack only a specific weed are being developed. An example is a type of rust that affects only the skeleton weed (Figure 16–14). Another fungus gives good control of the weed joint vetch.

Insects also can be used in the same way. An imported fly has been released in Maryland recently to control a noxious weed, the musk thistle. The weevil *Rhinocyllus conicus* also controls musk thistle.

BIOLOGICAL CONTROL AGENT	PLANT DISEASE CONTROLLED	ADDITIONAL INFORMATION
Tilleteopars	Powdery mildew	A yeast that effectively controls powdery mildew in the greenhouse.
Talaromyces flavus	Potato wilt	A fungus that is adapting well to the soils where potatoes are grown and giving reduction of the disease.
Talaromyces flavus	Verticullium wilt	Reduced wilt on eggplant by 75% when added to pots containing small plants for later field growing.
Crop rotation and raised beds	Phytophthora, root rot, many wilts	Many soil-borne diseases cannot live in the soil for more than 2 or 3 years if a host plant is not available.
Plant disease resistance	Wilts, bacterial diseases, rust diseases	As new disease-resistant plants are produced, less and less spraying with chemicals will be needed. The Ginkgo tree is immune to all insects and diseases. There are hundreds of disease-resistant fruits, vegetables, grasses, and flowers, and they should be used whenever possible. Many more new ones are on the way as disease resistance is found in wild plants and bred into commercial varieties.

Figure 16–13 Biological plant disease control.

BIOLOGICAL CONTROL AGENT	WEED CONTROLLED	ADDITIONAL INFORMATION
Chrysolina quadrigemina, Apolocera Plagiata (California moth)	St. Johnswort weed	St. Johnswort was a serious weed on much grazing land in California. This shiny blue beetle and moth are on the way to controlling the weed. They feed only on this weed and its close relatives.
Fungus rust, *Puccinia chondrillina*	Skeleton weed	A rust disease destroys the seed pod of the plant, preventing it from reproducing and stunting its growth. This is still experimental.

Figure 16–14 Biological weed control.

Allelopathy

Some plants secrete chemicals that inhibit root growth or seed germination of other plants. For example, sweet potatoes reduce growth of the weed yellow nutsedge. As little as 10 ppm of sweet potato root extract inhibits nutsedge, velvetleaf, and pigweed seed germination.

Why Use Biological Pest Control?

The following are some reasons to use biological pest control:

- In many cases, the dollar cost is much less in the long run.
- These controls do not poison the environment. Most poisonous chemicals from plants are more easily broken down and made harmless in the environment.
- Insects are becoming resistant to many chemical pesticides.
- Many man-made chemicals are suspected of or have been proven to cause cancer or pose other health hazards to humans and other mammals.
- Some chemical pesticides are getting into our drinking water.
- The Environmental Protection Agency (EPA) is restricting the use of many chemical pesticides, so fewer chemicals will be available.

The EPA has determined or preliminarily determined 31 pesticides as having some tumor-causing effects on animals. Although this does not mean all these chemicals will be proved dangerous to humans when used for pest control, it does mean that they are suspect and will be monitored and tested more. It just makes sense to use biological agents that are known not to be harmful to control plant pests whenever possible.

Much is yet to be done before we can put away all the chemicals, but progress is being made to reduce the need for them. Through the use of insect- and disease-resistant plants and other forms of biological control, fewer chemicals may be needed in the future.

Microorganisms are also being used to clean up chemical spills in the soil and water by literally eating and breaking down the chemicals into harmless materials. One such organism can do so to large vats of pesticide in 48 hours.

Through genetic engineering (changing the heredity of an organism), selection, and understanding of habit, there are limitless possibilities for the future in biological control.

New Biological Control Methods under Study

The following biological control methods are currently being studied:

- *ClandoSan.* A by-product of crab and shellfish industries that controls nematodes by stimulating the growth of the nematodes' natural enemies that then destroying the nematodes.
- *Vectobac.* A biological insecticide for fungus gnats applied as a soil drench. Viable endospores and crystals kill in 24 to 48 hours.

- *Potato with a built-in insect repellent.* The new hybrid potato produces leptine, a chemical that repels the Colorado potato beetle.
- *Corn vaccination.* A single gene from the bacterium *Bacillus thuringiensis* is forced into the corn seed and becomes a part of the seed passed on to the plant. The gene produces a toxin that is deadly to the corn borer.
- *Japanese beetle control.* Nematodes *Heterohabditis* HP88 achieved 90 percent control of beetle larvae. This is as good as or better than any chemical compound. The nematodes overwinter.

Newest Research on Biological Pest Control and the Results

The following are a few of the most exciting new biological controls being studied that have promise for successful use in the agricultural industry. Man-made chemicals are not gone yet, but many are being replaced with biologicals.

- *Fungus as a fumigant.* Professor Gary Strobel of Montana State University discovered the fungus *Muscador albus*, which lives within a plant and secretes a gas that stops the growth of plant pests. This may be the first biological fumigant. It is being tested as a possible substitute for methyl bromide and may also have many nonagricultural uses.
- *Three new neem products.* Neem seed oil can kill pests such as whiteflies, aphids, mealybugs, mites, and lace bugs, as well as protect against rusts and powdery mildew. New products available are Trilogy, Triact, and Rose Defense. Researcher James Locke reported that in numerous tests, a spray of 1 percent neem oil in water stopped 95 percent to 100 percent of the powdery mildew on hydrangeas, lilacs, and phlox. The neem oil also stopped infections that had already occurred. Neem oil covers plant leaves and stops fungus spores from penetrating the leaf.
- *Substitute for malathion.* A compound produced by *Streptomyces avermitilis* could be used as a substitute for malathion (an organophosphate often used to control the fruit fly and many other insects). In tests against the Caribbean fruit fly, abamectin was 100 percent effective against the fruit fly at a concentration of 0.03 percent of the amount of malathion used to do the same job. The compound paralyzes the fly's nervous system.
- *Beneficial fungus.* Spores of a beneficial fungus, *Gliocladium vireus GL-21*, applied in granular form to soils (EPA registered, Soil Gard) inhibits damping-off diseases. It is being considered as an alternative to methyl bromide.
- *Control of soil fungus.* Two bacteria, *Enterobacter cloacae* and *Burkholderia cepacia*, compete with and control many soil fungus diseases. They are applied by coating on seeds before planting.
- *Weed killer.* Could *Ailanthus altissima* (tree of heaven) be the next top weed killer? Dr. Rod M. Hessey, associate professor of biology at Penn State's Schuylkill campus, found that a chemical compound extracted from the bark of the tree of heaven can be used as a weed killer (herbicide). The chemical kills all plants it comes in contact with. If the chemical can be produced commercially, it may be the next number one herbicide.
- *Armicarb (potassium bicarbonate).* This biological fungicide controls a wide range of foliar diseases. It is a natural and safe compound.
- *Replacement for methyl bromide.* Benzaldehyde from peach seeds is being studied to replace methyl bromide.
- *Kaolin.* Kaolin is a clay product sprayed on plants to control leafhoppers, leaf miners, aphids, thrips, and mites. Late-season apple diseases such as sooty blotch and flyspeck are also controlled. Kaolin suppresses the plum curculio and codling moth and gives full control of pear psylla and rust mite.
- *Bacterium kills pests.* Five plant pests—the Colorado potato beetle, corn rootworm, diamondback moth, green stinkbug, and silverleaf whitefly—cost farmers nearly $3 billion yearly in crop losses and control.

Laboratory tests have shown that the bacterium *Chromobacterium suttsuga* produces toxins that kill these pests.

- *Pro-Mix.* A growing medium from Premier Horticulture now contains a biofungicide that prevents root diseases and promotes root system development.
- *Rosemary oil.* A new broad-spectrum fungicide and a postemergence herbicide for use on vegetables are made from rosemary oil.

How Natural Biopesticides Work

Natural biopesticides work in the following ways:

- *As parasites.* They are host specific and search out specific insects to kill. Many small wasps are parasites.
- *As predators.* Predators search for organisms to eat. They generally eat more than one type of insect. Examples of predators are green lacewings, ladybugs, assassin bugs, spined spider bugs, and spiders.
- *As* ***competitors.*** Beneficial microorganisms, especially in the soil, colonize all the space in the soil and on plant roots. The disease-producing organisms then have no food or space to grow.
- *As* ***plant stimulants*** *or vaccines.* They cause a plant to turn on or increase its natural immunity to insects and diseases.

The Soil Is Alive with Natural Pesticides

Researchers have found the soil to be an excellent source of plant insect and disease control agents. A soil high in organic matter, added by compost, and mulched with organic material is alive with microorganisms that control plant diseases and some insects. At least 85 natural compounds from soil fungi have been found that are toxic to insects.

By adding compost and good microorganisms to the soil, many root diseases can be controlled. The good microorganisms grow on roots and crowd out the disease-causing microorganisms. Healthy roots produce more vigorous plants, which are also better able to fight insects and diseases of the leaf and stem.

For example, strains of the bacteria *Enterobacter cloacae* and *Burkholderia cepacia* protect roots against *Pythium ultimum* and *Fusarium.* Of seeds treated with the bacteria, 90 percent grew when planted in infected soil, whereas most untreated seedlings died.

There are many other examples. In some cases, the good fungi jump-start the plant's immune system against a disease as a vaccine does for humans. Some of these natural control agents work so well they are being considered as substitutes for methyl bromide, a chemical used to pastuerize the soil. Methyl bromide destroys the earth's ozone layer and was banned in the United States in 2005.

Insect Growth Regulators

Insect growth regulators (IGRs) are the newest class of biological insecticides. They work against the insects' own internal chemical system and change the insects' normal growth cycle.

Young immature insects go through a series of molts during which they shed their outer skin or exoskeleton (outside skeleton). As they grow larger they must shed the old skin and grow a new larger one. One type of growth regulator called *chitin inhibitors* prevents the proper formation of the exoskeleton or the shedding of the old skin. Another form of growth regulator called a *hormone analog* prevents the insect from maturing into a normal adult. IGRs have advantages and disadvantages.

Some advantages include the following:

- Effective at very low concentrations, as low as 0.001 percent
- Provide long-term control
- Kill pesticide-resistant insects

Some disadvantages include the following:

- Take longer to work than chemical pesticides
- Unstable in the environment and quickly break down, becoming ineffective
- Must be carefully applied to be effective

SAFETY FIRST

This unit advocates biological control, or integrated pest management, to create biodiversity that enhances health and the biological system. Although biological control agents are generally safer to use than toxic, man-made pesticides, always read the warning labels and follow directions for their use carefully. For more information on occupational safety and health in the workplace, please visit the OSHA site at www.osha.gov.

Summary

Biological control of plant pests is becoming more widespread as new and more effective control agents are developed. Current research is resulting in the production of environmentally safe, natural products. Many of these biological pesticides are as effective or more effective than the existing man-made chemicals. As toxic man-made pesticides are pulled from the market, they are being replaced by biological control agents that are sustainable green alternatives. These new agents are safer and often cheaper, and insects and diseases have shown little, if any, resistance to them. This back-to-nature movement is proving to be successful and environmentally safer. Horticulturists today are employing sustainable pest management strategies in their priorities to use products that are more environmentally safe and friendly. This in turn results in products, such as fruits and vegetables, that are safer for human use and consumption.

Review Questions

1. What does biological pest control mean?
2. Why is it important to know the life cycle of a pest to effectively control it?
3. Discuss some crops that might be grown in a 3- or 4-year rotation for soil disease control.
4. Why is it so important to be looking to natural pest control at this time?
5. How does the biological control of insects differ from the biological control of plant diseases?

Student Activities

1. Access your state agricultural college's website, and review research on biological control being done in your area.
2. While on a walking field trip, see how many beneficial (biological control) insects you can find. Create a list of the beneficial insects that you found, and then draw a sketch of the insect next to its name on the list.
3. Search online to find examples of an integrated pest management (IPM) program. Describe how it works.
4. Explore the Cornell University Horticulture website for current research on using biological controls.

Suggested Supervised Agricultural Experience (SAE) Activities

1. Participate in a placement SAE at a local orchard, nursery, or greenhouse landscape management program. Develop specific skills and experience working with the IPM program at the orchard.
2. Conduct a research SAE studying the effectiveness of several different biological controls.

Additional Resources

http://entomology.cals.cornell.edu/

www.ipm.ucdavis.edu/IPMPROJECT/fundedprojects.html

www.ars.usda.gov/main/site_main.htm?modecode=80-62-05-10

www.epa.gov/agriculture/tipm.html

UNIT 17

The Safe Use of Pesticides/Biopesticides

Objective

To use pesticides safely and competently.

Key Terms

acute toxicity
biological agent
biopesticides
chronic toxicity
dermal contact
inhalation
oral contact
pesticide
personal protective equipment (PPE)
Safety Data Sheet (SDS)
toxicity

Competencies to Develop

After studying this unit, you should be able to:

- identify the three main routes by which pesticides enter the body.
- examine five pesticide labels and
  - identify the type of each and its degree of toxicity.
  - demonstrate the recommended precautions in the mixing and handling of each.
- list first aid steps to be taken in case of poisoning by one pesticide from each of the three families of pesticides.

Materials

- pesticide labels or containers with labels
- sprayer and duster for hand application of pesticide
- all articles of protective clothing specified on pesticide labels, including rubber gloves, respirator, rubber boots, waterproof coveralls, goggles, and rubber hat
- local pest control schedules and calendars from a local university or extension service
- pesticides for application
- smartphone apps:
  - Pest Portal
  - Spectracide
  - Calibrate My Sprayer
  - GardenCompass
  - Ortho Problem Solver
- Material Safety Data Sheets (MSDS)/Safety Data Sheets (SDS)

The Worker Protection Standard (WPS)

Concern for worker protection led to the passage of the original Worker Protection Standard (WPS) in 1992. In 2014, the EPA proposed changes to this law that include a new set of rules to reduce pesticide-related illnesses for all who come in contact with agricultural chemicals. The WPS does not regulate or ban any chemical. However, it sets guidelines to follow during and after application, such as how soon workers can reenter a treated area, warning signs put up to point out treated areas, information so applicators are aware of the toxic effect of chemicals, and what protective clothing to wear. Facilities to wash up and change clothes after chemical use must also be provided within one-quarter mile of the area being treated with chemicals.

All pesticide labels must meet the new standard. Dealers will be allowed to sell pesticides with old labels until the new law takes effect. Exact dates for compliance can be found by contacting the Environmental Protection Agency (EPA) or your state pesticide regulation agency.

Note: The WPS does not cover homeowner use of pesticides.

Types of Pesticides

A **pesticide** is any natural or man-made material used to control *pests* (plants or animals that are harmful to people or crops they cultivate). Biological controls may be living organisms. Seven types of pesticides are of concern to horticulturists, each classified according to the type of pest it destroys:

- *Insecticides* are used to control insects. The insects are killed by body contact with the chemical, by swallowing the insecticide, or by a **biological agent**.
- *Miticides* are used to control *mites* (tiny spiderlike animals) and ticks. Mites and ticks are usually killed by coming in contact with the chemical or by being attacked by a biological organism.
- *Fungicides* are used to control fungus disease. To be effective, they must come in contact with the fungus. Fungicides are usually used to prevent a plant from becoming diseased and are applied before the disease is present.
- *Herbicides* or *weed killers* are used to kill unwanted plants. *Nonselective* herbicides kill all plants, whereas *selective* herbicides kill only certain plants.
- *Rodenticides* kill rodents, such as rats and mice. These chemicals are usually applied as bait; the rodents are poisoned by eating the chemical.
- *Nematocides* kill nematodes (tiny hairlike worms that feed on the roots of plants). Nematocides are usually applied in the form of *fumigants*—substances that produce a smoke, vapor, or gas when applied—or as biological agents.
- *Molluscicides* are used to kill slugs and snails (types of *mollusks*). These chemicals are usually applied as bait, which attracts the slugs and snails and poisons them.

Selecting the Proper Pesticide

The first step in selecting a pesticide is to identify the pest to be controlled and determine whether control is needed. If the pest is doing very little damage or if the crop is ready for harvest, the control may not be needed.

If control is needed, methods that do not employ chemicals should be considered first. The spread of disease might be halted simply by proper *sanitation*, removing and destroying any diseased plant material, and disinfecting all tools and work areas. (This procedure should be followed while making cuttings even if there is no sign of disease.)

If the pest to be controlled is an insect, insects that eat other insects (such as the ladybug) may be effective and less expensive than chemicals. Such insects may be purchased and released on the infected plants. It may be unnecessary to introduce the insect to an area: natural parasites (such as the tiny wasp that kills tomato worms) may be present in large enough numbers to minimize attack by the pest. In the same way, mulches, rather than herbicides, may be used for weed control to minimize the use of harsh chemicals.

It is important to select the best pesticides for a particular problem. If correctly chosen, the pesticide will meet the following requirements:

- It kills or controls the pest.
- It does not injure the plant on which it is used.
- Of all the materials that could be used for the problem at hand, it is the least harmful to the environment.
- It is suitable for use with the available equipment.
- The label recommends its use for the plant and pest for which it has been chosen.

Toxicity

Because many pesticides can be poisonous to people and the environment, the United States has established standards for their handling and use through the EPA. Other countries have similar regulations. Due to their poisonous nature, pesticides must be used with a degree of caution, depending on their **toxicity** level and how poisonous they are.

To poison or injure people or animals, pesticides must be present in or on the victim's body. There are three main routes by which poisons enter the body:

- ***oral contact.*** By swallowing.
- ***dermal contact.*** By contact with the skin.
- ***inhalation.*** By breathing.

When dealing with the danger of children coming in contact with poisons, the major concern is preventing the child from taking the poisons orally. On the other hand, the individual who applies pesticides is more likely to be poisoned by contact dermally or through inhalation. Chemicals such as organophosphates are absorbed rapidly through the skin. Any chemical that vaporizes, has a strong odor, or is a fine dust or mist can be inhaled and absorbed through the lungs. Pesticides pass very rapidly through skin on the back of the hands, wrists, armpits, back of the neck, groin, and feet. Cuts or scrapes also allow pesticides to enter the body more easily.

> **CAUTION**
> When handling pesticides, wear **PPE**. Never allow a pesticide to come in contact with the skin. Skin contact with a pesticide may cause blistering, itching, cracking, or color changes. Some pesticides may also pass through the skin and cause internal problems. If accidental skin contact with a pesticide does occur, wash the affected area immediately with soap and water.

Types of Toxicity

Acute toxicity is a measure of how poisonous a pesticide is after a single exposure. Pesticides are generally rated according to acute toxicity. **Chronic toxicity** is a measure of how poisonous a pesticide is over a period of time and after repeated exposure. Chronic toxicity is a danger of chemicals that accumulate in the body, such as chlorinated hydrocarbons.

Measuring Acute Toxicity

Although there is no standard measure for chronic toxicity, acute oral and dermal toxicity and acute inhalation toxicity can be measured as described next.

ORAL AND DERMAL TOXICITY The method used to measure acute oral and dermal toxicity is LD_{50}. The *LD* stands for *lethal dose* (amount necessary to cause death), and the 50 means that 50 percent of test animals (generally rats and rabbits) are killed by this dose. The lower the LD_{50} number of a pesticide, the more poisonous it is. LD_{50} values are given in milligrams of substance per kilogram of a test animal's body weight. This is a metric measurement that is the same as parts per million.

INHALATION TOXICITY Acute inhalation toxicity is measured by LC_{50} values. *LC* stands for *lethal* (deadly) *concentration.* LC_{50} values are measured in milligrams per liter. A *liter* is a volume measurement (in the metric system of measurement) equal to about 1 quart. Again, the lower the LC_{50} number, the more poisonous the pesticide is.

Standard Label Information

Labels on pesticide containers always include mixing instructions, which must be followed carefully. Other standard points of information that appear on pesticide containers include the following:

- name and address of manufacturer
- trade name (may or may not be the same as the chemical name)
- active ingredients, including the official common name of each ingredient; when an accepted common name is not available, the chemical name appears
- type of pesticide such as insecticide or fungicide
- form of substance such as dust, wettable powder, or emulsion
- EPA registration number
- storage and disposal precautions
- hazard statement (read carefully)
- directions for use (read carefully and follow)
- net contents
- worker protection procedures under the new WPS regulations

CAUTION

Pesticide product label changes are in effect frequently, so the labels represented in this text are for illustrative purposes only and should not be used as instructions for field application. Manufacturers will not be held accountable for the accuracy of the recommendations and precautions stated in the text. Whenever a pesticide is purchased or used, strictly follow the recommendations and precautionary information as stated on the product package label.

Signal Words

To alert the user to the toxicity of a pesticide, certain *signal words* appear on the container label, such as "Danger," "Warning," and "Caution." There are four categories of acute toxicity and three signal words (Figure 17–1).

It is essential to read labels to determine the toxicity of pesticides and the necessary precautions in their handling. The label should be examined before the product is purchased to determine whether the chemical is effective against the particular pest that is causing the problem and whether, perhaps, it is too toxic to work with.

		LETHAL DOSE			
CATEGORIES	SIGNAL WORD REQUIRED ON LABEL	LD_{50} mg/kg Oral	LD_{50} mg/kg Dermal	LC_{50} mg/l inhalation 24-hr Exposure	PROBABLE ORAL LETHAL DOSE FOR A PERSON WEIGHING 150 POUNDS
I. Highly Toxic	DANGER skull and crossbones Poison	0–50	0–200	0–2	a few drops to a teaspoonful
II. Moderately Toxic	WARNING	over 50–500	over 200–2,000	over 2–20	over 1 teaspoonful to 1 ounce
III. Slightly Toxic	CAUTION	over 500	over 2,000–20,000	—	over 1 ounce to 1 pint or 1 pound
IV. Relatively Nontoxic	CAUTION (or none)	over 5,000	over 20,000	—	over 1 pint or pound

Adapted from US Environmental Protection Agency

Figure 17–1 Categories of acute toxicity.

▲ **Figure 17–2** A lawn and garden insecticide for home gardener use.

An example of a pesticide for home gardener use is shown in Figure 17–2. Further detail and instructions for use can be found on the label on the back of the product (Figure 17–3).

> **CAUTION**
> Before purchasing or using a pesticide, read and understand the label. Watch for signal words such as DANGER, WARNING, and CAUTION (see Figure 17-1) indicating toxicity of the chemical. Be prepared to take any precautions listed. If the pesticide is highly toxic, try to purchase a material that is as effective yet less toxic.

> **CAUTION**
> When considering use of any pesticide, be sure that the intended use appears on the label. It is a violation of federal law to use any pesticide in a manner that is inconsistent with the labeling. For example, it is unlawful to use a pesticide on any pest or on any site that is not specifically listed on that label.

Safety Precautions

Safety precautions such as the following must be considered before mixing or applying a pesticide.

Read the label carefully. Follow all directions and precautions exactly.

Check the recommended uses. Be sure it is the correct substance for the intended use.

Have clean water and detergents available. These are needed to wash up spills.

Wear protective clothing. Be sure you have proper gloves, a respirator, and any other protective gear called for on the label. Some pesticides require little protective clothing (Figure 17–4). Other highly toxic materials require complete coverage of the applicator's entire body with waterproof material (Figure 17–5).

> **CAUTION**
> The proper protective clothing can be determined only by reading the label on the pesticide container.

Use extra caution with concentrated chemicals. (Chemicals are considered concentrated prior to mixing or diluting with water or dust.) Protective covering is especially important when handling the concentrated chemical in mixing and filling the spray tank. Always wear protective gloves and goggles. If a concentrate is spilled or splashed on clothing, wash and change clothes immediately. Always protect the eyes with goggles when handling concentrates; the eyes absorb pesticides rapidly. Any pesticide spilled on the ground or floor should be removed immediately.

If you have a medical emergency, need consultation on diagnosis, therapy and analysis, please call 800-420-9347.
If you have questions relating to product use or application, please call 866-945-5033. If you have a product spill or transportation emergency, please call 800-424-9300. For Material Safety Data Sheets, please call 866-945-5033.
Product of China; Contains U.S. and Imported Ingredients.

EPA Reg. No. 432-1209-71004
EPA Est. 67572-GA-1a, 67572-GA-2b
Subscript used is first letter of run code on container.

Sevin is a registered trademark of Tessenderlo Kerley, Inc.
GardenTech is a registered trademark of Gulfstream Home and Garden Inc.

Distributed by: TechPac, LLC.,
1600 Parkwood Circle, Suite 700
Atlanta, GA 30339

1202OLD 2

FOR OUTDOOR RESIDENTIAL USE ONLY

DIRECTIONS FOR USE

It is a violation of Federal law to use this product in a manner inconsistant with its labeling.

STORAGE AND DISPOSAL

STORAGE: Store pesticide in original container and place in a locked storage area.
DISPOSAL: If empty: Non-refillable container. Do not reuse or refill this container. Offer for recycling if available. Place in trash or offer for recycling if available.
IF PARTLY FILLED: Call your local solid waste agency or 866-945-5033 for disposal instructions. Never place unused product down any indoor or outdoor drain.
RESTRICTIONS: Do not apply this product in a way that will contact any person or pet, either directly or through drift. Keep people and pets out of the area during application. Do not allow people or pets to enter the treated area until dusts have settled.
NON-REFILLABLE CONTAINER. DO NOT CUT PACKAGE TO EXPOSE CONTENTS. Discard package when empty. Not for use on plants grown for sale or commercial use, research, or commercial seed production.
PRECAUTIONS: Some phytotoxicity may occur on tender foliage in the presence of rain or high humidity of several days duration following application. Begin applications when insect or damage is first observed.
DIRECTIONS: Wearing household latex or rubber gloves, use this ready-to-use shaker container to apply product. Dust lightly to cover leaf surfaces with a thin, even film of dust. Apply when air is calm. Start applying at the farthest corner of the treatment area and work backward. For plants above chest height, consider using a liquid carbaryl product.

HOME GARDEN VEGETABLES AND SMALL FRUITS

Do not repeat applications more than once a week (7 days)

PREHARVEST USE RESTRICTIONS

Observe the following preharvest intervals (number of days between last application and harvest)

PREHARVEST INTERVAL

GARDEN CROP	PREHARVEST INTERVAL (DAYS)
Garden beets (roots), radishes, rutabagas, turnips (roots), carrots, potato, blackberries, boysenberries, dewberries, loganberries, raspberries, blueberries, grapes and strawberries	7
Collards, garden beets (tops), kale, lettuce (head and leaf), mustard greens, spinach and turnip (tops)	14
Dried shelled legume beans and peas including: field bean, kidney bean, dry lima bean, navy bean, pinto bean, blackeyed pea, cowpea, Crowder pea, mung bean, southern pea, lentil (dry)	21
Edible-podded legume vegetables including: snap bean, wax bean, yardlong bean, jack bean, edible-pod pea, snow pea, sugar snap pea, broccoli, brussel sprouts, cabbage, cauliflower, kohlrabi, cantaloupe, cucumber, melons, pumpkins, squash, eggplant, pepper and tomatoes	3

VEGETABLES

CROP	INSECT PESTS	COMMENTS
Edible-podded legume vegetables including: Snap bean, wax bean, yardlong bean, jack bean, ediblepod pea, snow pea, sugar snap pea. Dried shelled legume vegetables including: field bean, kidney bean, lima bean (dry), navy bean, pinto bean, adzuki bean, blackeyed pea, cowpea, mung bean, southern pea, lentil(dry). (do not use on lentils in CA.)*	Bean leaf beetle, cucumber beetles, corn earworm, flea beetles, Japanese beetles, leafhoppers, Lima bean pod borer, Lygus bug, Mexican bean beetle, stink bugs	Maximum number of applications per year: 4
Broccoli, brussel sprouts, cabbage, cauliflower, kohlrabi	Armyworms, corn earworm, diamond-back moth larvae, flea beetles, harlequin bugs, imported cabbageworm	Application is only permitted within thirty days from the date of crop emergence or the date of transplanting. Will not control cabbage loopers. Maximum number of applications per year: 4
Collards, garden beets, kale, lettuce, mustard greens, radishes, rutabagas, spinach, turnips	Armyworms, corn earworm, diamond-back moth larvae, flea beetles, harlequin bug, imported cabbageworm, leaf-hoppers, stink bugs, tarnished plant bug	Application is only permitted within thirty days from the date of crop emergence or the date of transplanting. Will not control cabbage loopers. Maximum number of applications per year: 4
Cantaloupe, cucumbers, melons, pumpkins, squash	Cucumber beetles, flea beetles, melonworms, pickleworm, squash bugs	Maximum number of applications per year: 6
Carrots	Flea beetles, leafhoppers, Lygus bugs, spittlebugs, stink bugs	Maximum number of applications per year: 6
Eggplant, peppers, tomato	European corn borer, Fall armyworm, flea beetles, leafhoppers, tomato fruitworm, tomato hornworms, stink bugs	Maximum number of applications per year: 7
Potato	Colorado potato beetle, flea beetles, leafhoppers, Lygus bugs	Maximum number of applications per year: 6

* Not for use on fresh/succulent peas and beans

SMALL FRUIT

PLANT	INSECT PESTS	COMMENTS
Blackberries, boysenberries, strawberries, loganberries, dewberries, raspberries	Japanese beetles, leafhoppers, leafrollers, raspberry aphid	Maximum number of applications per year: 5
Blueberries	Blueberry maggot, cherry fruitworms, cranberry fruitworms, Japanese beetles	Apply 3 weeks before harvest and repeat 10 days later if necessary. Maximum number of applications per year: 5
Grapes	Grape berry moths, grape leaffolder, grape leafhopper, grape leaf skeletonizer, Japanese beetle, June beetle, red-banded leafroller	Apply just before first brood leaffolder larvae emerge from rolls and as needed for leafhoppers. Maximum number of applications per year: 5
Strawberries	Meadow spittlebug, strawberry leafroller, strawberry weevil	Maximum number of applications per year: 5

ORNAMENTAL SHRUBS AND FLOWERS

For outdoor use only. To kill bagworm, blister beetles, boxelder bug, boxwood leafminer, flea beetles, Japaneese beetls, lacebugs, leafhoppers, leafrollers, periodical cicada, plant bugs, psyllids, rose aphid, rose slug, scale insects in the crawler stage, tent caterpillars and exposed thrips, apply thoroughly to the infested plants at the first sign of damage. Do not repeat application more than once weekly or more than 6 times per year. For best results on scale, apply in spring and early summer when scale crawlers are present.

HOME LAWNS

To kill earwigs, fall armyworm, leafhoppers, millipedes, sod webworm, and sowbugs, sprinkle lightly on grass. To kill ants (except pharaoh and carpenter ants), chiggers, fleas, spiders (except Brown recluse and Black widow) and ticks, sprinkle lightly on grass. Will only kill pests present at the time of application. Pest must be directly contacted by product to be effective. For best results, apply after rain or watering and do not water for at least 2 days after application. Repeat treatment as needed but not more than every 2 weeks and not more than four times per year.

EARWIGS

This material is effective in killing earwigs and when properly used will reduce the number gaining entrance to homes. Using this ready-to-use shaker container, lightly dust a band 3 to 4 inches wide around the outside foundation wall of the home. Earwigs breed over wide areas around yards and fields and are driven by cold weather to seek shelter in warm buildings. As they move along the foundation wall seeking entrance, they come into contact with the treated barrier and are killed. Be sure to treat in front of all steps and other entrance ways as well as along the foundation wall. Apply also around refuse piles, lumber, mulch and other areas in the yard offering areas of concealment to earwigs. Repeat application in 10 days or when earwigs become troublesome. If the dust barrier is washed away by rain, it should be replaced with a new application if earwigs are still a problem.

This container is sold by weight, NOT by volume. You can be assured of the proper weight even though some settling of contents normally occurs during transportation and handling.

Important: Read the entire Directions before using this product. Disclaimer: Manufacturer warrants that this product conforms to the chemical description on the label and is reasonably fit for the specific purposes referred to in the Directions for Use. To the extent consistent with applicable law, manufacturer makes no other express or implied warranty of fitness or merchantability nor any other express or implied warranty. To the extent consistent with applicable law, in no case shall Manufacturer be liable for consequential, special or indirect damages resulting from the use, handling or application of this product.

PRECAUTIONARY STATEMENTS
Hazards to Humans & Domestic Animals
CAUTION: Harmful if swallowed. Harmful if absorbed through skin. Avoid contact with eyes, skin or clothing. Wash thoroughly with soap and water after handling and before eating, drinking, chewing gum, or using tobacco. Remove contaminated clothing before reuse.

FIRST AID THIS PRODUCT IS A N-METHYL CARBAMATE AND IS A CHOLINESTERASE INHIBITOR

If Swallowed
- Call poison control center or doctor immediately for treatment advice.
- Have person sip a glass of water if able to swallow.
- Do not induce vomiting unless told to do so by the poison control center or doctor.
- Do not give anything by mouth to an unconscious person.

If on Skin or Clothing
- Take off contaminated clothing.
- Rinse skin immediately with plenty of water for 15-20 minutes.
- Call a poison control center or doctor for treatment advice.

Have the product container or label with you when calling a poison control center or doctor, or going for treatment. You may also contact 1-800-420-9347 for emergency treatment information.

6 13499 07017 3

NOTE TO PHYSICIAN CARBARYL is an N-methyl carbamate insecticide which is a cholinesterase inhibitor. Specific antidote consists of atropine. Narcotics and other sedatives should not be used. Drugs like 2-PAM (pyridine-2-aldoxime methiodide) are NOT recommended.

ENVIRONMENTAL HAZARDS
This pesticide is extremely toxic to aquatic invertebrates. To protect the environment, do not allow pesticide to enter or run off into storm drains, drainage ditches, gutters or surface waters. Applying this product in calm weather when rain is not predicted for the next 24 hours will help to ensure that wind or rain does not blow or wash pesticide off the treatment area. BEE CAUTION: MAY KILL HONEY BEES AND OTHER BEES. This product is highly toxic to honey bees and other bees. Do not apply to plants in bloom.

WARNING: This product contains carbaryl, a chemical known to the State of California to cause cancer and reproductive harm in laboratory animals.

Sevin is a registered trademark of Tessenderlo Kerley, Inc. GardenTech is a registered trademark of Gulfstream Home and Garden, Inc.

▲ **Figure 17–3** Detail on how to use the pesticide in Figure 17–2.

Apply the chemical with care. When applying pesticides, be sure that the chemical is directed only at the target area and that as little as possible is allowed to spread anywhere else. Apply only the amount needed to do the job.

Always mix just enough for the job at hand. The label lists the proper amount needed for various crops and pests. The sprayer may need *calibration* (adjustment of the nozzle to a certain size) to ensure that the correct

▲ Figure 17–4 The chemical being applied here is relatively nontoxic. Note the lack of protective clothing.

▲ Figure 17–5 Personal Protection Equipment (PPE) must be used to apply toxic chemicals.

volume of chemical is applied. Applying at the proper rate is especially important when working with herbicides (weed killers).

Guard against inhalation or ingestion (swallowing) of the chemical; do not inhale its fumes, dust, or mist. Never eat or smoke while handling any pesticide.

Consider weather conditions. Pesticides should be applied when weather conditions are favorable. Be aware of forecasted weather conditions before application of the chemical. If used outdoors, chemicals applied during heavy rain will wash away. Strong winds prevent applications of sprays and dust from hitting and adhering to the target crop. Some chemicals must be applied only within certain temperature ranges. Read the label and use good judgment. When applying pesticides, avoid contact with the mist or dust that drifts through the air. Do not spray where the pesticide can drift into streams, lakes, or apiaries (bee yards) or near sensitive plants or animals that might be injured.

Store and dispose of chemicals properly. Store all concentrated pesticides in the original container. Keep in a locked area, away from children and animals. Do not store near foods, animal feed, fertilizer, or in a building that houses people or animals. Sprayers and dusters should be treated in the same manner for storage.

Dispose of all diluted chemical that is in the spray tank. If any is left after spraying or dusting, try to use it on another area or crop where it will not contaminate or damage the environment, livestock, or people. However, the best way to eliminate problems associated with leftover chemicals is to mix only the amount needed for the job at hand. Empty containers must be disposed of out of the reach of all persons and animals and in such a way that they will not contaminate the environment. Never reuse a pesticide container. Never set empty or partly empty containers out to be collected as trash unless they are clearly labeled.

> **CAUTION**
> Before disposing of any pesticides or empty containers, check local regulations because the procedures for disposal may vary greatly.

Clean up thoroughly. The sprayer, all measuring equipment, gloves, and any other protective clothing should be washed thoroughly. The sprayer tank and pump should be rinsed three times with large amounts of water. Some of the water should be run through the sprayer each time to flush the hose and spray the pump. The rinse water is then dumped into the mixing tank. Protective gloves should be washed thoroughly with soap and water while still on the hands and hung up to dry. When the pesticide application is finished and the sprayer or other application equipment is cleaned and safely stored, the applicator should shower and change clothes immediately. Contaminated clothing should be handled carefully and washed separately.

TRY THIS Create a computer-generated poster highlighting the proper safety procedures to use while handling pesticides or **biopesticides**. Hang the posters up in an area where such substances are stored and handled.

Pesticide Poisoning

If the warnings mentioned are not heeded, or if accidental spillage results in poisoning, it is important that any symptoms of poisoning be recognized quickly. In cases of spillage or other types of exposure to a toxic chemical, the most important consideration is to remove the contaminated clothing and wash yourself thoroughly with a good soap or detergent and plenty of water. A shower should be used rather than a tub bath because fresh, clean water is continually supplied.

FAST FACT
Dilution of the poison is the most important first aid practice.

First Aid for Pesticide Poisoning

First aid is just what the term indicates—the first help given to a victim until medical professionals arrive. If poisoning has occurred, telephone a doctor immediately and read the pesticide label to him or her. Before assistance arrives, you can do the following.

- For any pesticide spilled on the skin, wash with plenty of soap and water to dilute the chemical.
- Remove contaminated clothing. Shower, dry, and wrap or dress the victim in warm blankets or clothing. Cover any chemical burns with loose, clean, soft cloth.

CAUTION
Be careful that you do not have contact with the pesticide while coming to someone's aid.

- For eye exposure, hold the eye open and flush with clean water for at least 15 minutes.
- For inhaled poisons, carry the victim immediately to fresh air. Do not allow the victim to walk. Loosen clothing and apply artificial respiration if breathing has stopped. Keep the patient quiet and warm. Do not give the victim alcohol.

The label on the pesticide container may give additional first aid information. Use any information available. After notifying a doctor, follow his or her advice. Provide expert medical attention as soon as possible. Do not panic and frighten the victim; instead, remain calm and reassure the victim that help is on the way.

CAUTION
Always choose the least toxic chemical that will do the job.

EPA Rules for the Use of Chemicals

The EPA published guidelines for the handling of chemical pesticides in an additional attempt to reduce the health risk of using pesticides. The agency established four toxicity categories and regulations stipulating how soon it is safe for a person to reenter an area after it has been treated with chemical pesticides. The categories are as follows:

- *Toxicity I chemicals.* Areas treated with chemicals carrying a "Danger—Poison" signal word. Warning signs must be posted, and no one may reenter the treated area for 48 hours after application.
- *Toxicity II chemicals.* Areas treated with chemicals carrying a "Warning" label. Restriction from the treated area is for 24 hours.

- *Toxicity III chemicals.* Areas treated with chemicals carrying a "Caution" label. People may reenter the area as soon as the dust or spray mist settles.
- *Toxicity IV chemicals.* Areas treated with pesticides carrying no warning on the label. People may reenter the area immediately.

Safe Pesticides

A new approach in the development of pesticides has the goal to produce chemicals that are toxic to insects but not to warm-blooded organisms such as people. Some pesticides now in use affect the nervous systems of insects. The problem is that they also affect any other organism with a nervous system, including people.

Some progress has been made in the development and use of pesticides that affect only insects. A good example is chemicals that prevent the molting or shedding and hardening of the insects' skin. If insects cannot shed and then harden a new skin as they grow, they die. Some new biologicals act as repellents or interrupt food digestion.

Future research will focus on more man-made and natural pesticides that kill only insects and, if possible, only the bad insects.

GREEN TIP Researchers are rapidly developing natural control measures to replace the more toxic chemicals, such as garlic spray and hot pepper wax as barriers.

SAFETY FIRST

A Safety Data Sheet (SDS) (formerly known as the Material Safety Data Sheet [MSDS]) is a document that gives detailed information about a pesticide product and company identification, composition/information on the ingredients, hazard identification, first aid measures, firefighting measures, accidental release measures, handling and storage, exposure control/personal protection, physical and chemical properties, stability and reactivity, toxicological information, ecological information, disposal considerations, transport information, and other pertinent safety information. For more information on occupational health and safety in the workplace, please visit the OSHA site at www.OSHA.gov.

Summary

Safety in the use of pesticides cannot be overstressed. Through elimination of the availability of highly toxic chemicals and control of others so that only professionals use the most dangerous ones, it is now safer for the average gardener to use pesticides. However, this is no excuse to be careless. The label must be read carefully and followed exactly because there is still considerable danger in the use of some chemical pesticides. Researchers are rapidly developing natural control measures, which will continue to replace the more toxic chemical pesticides.

Review Questions

1. What are the seven categories of pesticides?
2. What is the first step in selecting a pesticide?
3. What is the purpose of signal words that are found on pesticide labels? What are four signal words, and what category of pesticide does each word signal?
4. What is a toxic chemical? What is the difference between acute and chronic toxicity?
5. What is a SDS? What is included in it?

Student Activities

1. Obtain labels from containers of commonly used pesticides. Draw up a list of pesticides, including the following:
   a) name of pesticide
   b) crops on which pesticide is commonly used
   c) pests controlled by the pesticide
   d) toxicity level of the pesticide and necessary precautionary measures, including protective clothing
2. Demonstrate the proper procedure for mixing and applying a pesticide.
3. Try on and properly adjust a respirator, rubber gloves, waterproof hat, and goggles.
4. Search online for educational videos that highlight pesticide safety. Share the best one with the class.

Suggested Supervised Agricultural Experience (SAE) Activities

1. Conduct a research SAE studying the effect of pesticides on the water sources surrounding the application location. Record all hours worked, experience, and findings from your SAE.
2. Develop a pesticide application schedule for an entrepreneurship SAE growing fruit trees, greenhouse production, landscape maintenance, or agronomy.

Additional Resources

www.pest.umd.edu; http://www.epa.gov/pesticides/health/safely.htm

www.epa.gov/pesticides/health/safely.htm

http://pods.dasnr.okstate.edu/docushare/dsweb/Get/Document-2282/F-7450web.pdf

UNIT 18 Insecticides

Objective

To identify insect pests and select and apply the appropriate synthetic (man-made), biological, or biorational (derived from items in nature) insecticides.

Key Terms

aerosols
attractants
biological
biorational
insect
insecticides
pheromones
repellents
synthetic compounds

Competencies to Develop

After studying this unit, you should be able to:

- identify common insect pests and select an effective control method for each.
- describe the seven ways in which insects are killed by insecticides and the type of insect against which each is most effective.
- list the names and characteristics of the three major groups of insecticides (according to their chemical makeup).
- compare the six ways in which insecticides are applied.
- explain the relationship between the life cycle of insects and timing of insecticide application.
- identify the common biological controls of insects.

Materials

- illustrations of insect pests
- live insects
- recommendations for spraying from the local extension service
- insecticides with labels (none should be highly toxic)
- spraying and dusting equipment
- safety gear, including elbow-length protective gloves and respirator
- smartphone apps:
  - Pest Portal
  - Spectracide
  - Calibrate My Sprayer
  - GardenCompass
  - Ortho Problem Solver
- safety data sheet (SDS) (Refer to Unit 17)

Insecticides, one of the seven types of pesticides mentioned in the previous unit, are used by most people at one time or another in the control of flies, mosquitoes, and other insect pests. They are used widely by horticulturists to protect plants from insect damage. Although insecticides can effectively and safely control insects, they can be dangerous and even deadly to people when used improperly. For this reason, it is extremely important that they be handled with care and only according to directions on the label.

This unit deals with chemical insecticides, not the biological insecticides discussed in Unit 16. Because changes in available insecticides are occurring rapidly, a list of suggested insecticides is no longer included in this unit.

What Is an Insect?

An **insect** is a small animal with three clearly defined body regions and three pairs of legs. The three body regions are the head, thorax, and abdomen (Figure 18–1). The proper identification of insects is an important step in their control. Only after the insect is identified can the best method of control be selected. Guides to insect identification are available from the US Department of Agriculture.

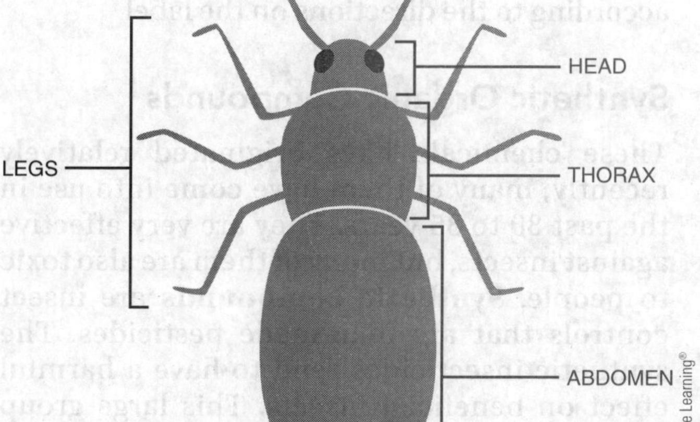

▲ **Figure 18–1** Parts of an insect. Notice the three definite body regions (head, thorax, and abdomen), with the legs attached to the thorax.

How Insecticides Kill

To be considered good insecticides, chemicals must kill a specific type of insect while doing little or no damage to the plant. They must also be as safe as possible for the handler and environment.

STOMACH POISONS These poisons work against insects that actually eat a part of the plant. The chemical is sprayed or dusted on the plant; as the insect eats the plant, it is poisoned through the stomach. Insects that chew their food, such as caterpillars, grasshoppers, and beetles, are controlled by stomach poisons. Rotenone is considered to be an effective stomach poison.

CONTACT POISONS These insecticides kill insects when they are hit by or come in contact with the poison; the insect does not have to consume it. Any type of insect can be controlled by contact poisons, including insects that suck plants, such as aphids and leafhoppers. Contact poisons kill by upsetting the insect's nervous system or breathing system. An example of a contact insecticide is malathion.

New, highly refined oil sprays are excellent contact poisons against insect eggs and soft-bodied insects, such as aphids, mites, scale, and whiteflies. Oils do not contaminate the environment and are safe to use.

SYSTEMIC POISONS These chemicals enter the plant sap and move throughout the entire plant. When insects eat parts of plants or suck juice from plants, the chemical is swallowed. Systemic poisons are effective in the control of insects with either chewing or sucking mouthparts. They are especially effective in controlling insects that hide under leaves or underground and that therefore are not affected by contact sprays. If used on plants that are consumed as food, the chemical must be given time to break down within the plant and become nonpoisonous before the harvest date for the crop. These poisons are sprayed on the plant foliage or mixed with water and applied to the roots. Both methods result in the chemical spreading throughout the entire plant. One systemic insecticide is Orthene (acephate).

FUMIGANTS Fumigants are actually contact poisons applied in gaseous form. The gases or fumes kill the insect after entering its system through breathing pores. The insect must actually absorb the poison for this insecticide to be effective. Fumigants are used to control soil-borne insects that damage roots, such as weevils and June beetle grubs. One example of a fumigant is methyl bromide. In the United States, most uses of methyl bromide were phased out on January 1, 2005.

Fumigants are also used in greenhouses. When applied in the greenhouse, the fumigant, consisting of tiny particles of insecticide, settles on the insect and is breathed into its system.

REPELLENTS **Repellents** generally do not kill insects but, instead, drive them away before they attack the plant. One example of a repellent is aluminum foil, which, when placed around plants, repels flying insects (Figure 18–2).

ATTRACTANTS AND PHEROMONES **Attractants** and **pheromones** work in the opposite way from repellents; they lure insects to their deaths. One example is the Japanese beetle bait used in traps to catch the beetle. Another example is the sex lure used to trap the gypsy moth. The sex lure used in these traps is made from naturally occurring or synthetically produced hormones that attract the adult insects. The male gypsy moth is irresistibly attracted to the sex lure; this attraction is stronger than the normal attraction he has to the female moth. Because most of the male moths are caught in the trap, the females remain unfertilized and lay sterile, unfertilized eggs that do not hatch. The sex lure acts in the same manner on the Japanese beetle.

▲ **Figure 18–2** Aluminum foil strips around cucumber plants protect the plants from flying aphids by reflecting ultraviolet rays from the sky, making the area undesirable to the aphids.

The Chemical Makeup of Insecticides

Insecticides are classified in three major groups according to their chemical makeup.

Inorganic Compounds

Inorganic compounds are of mineral origin, which means a mineral is used as the basis for the poison. They usually work in the form of stomach poisons. Sulfur is one type of inorganic insecticide that is also widely used as a fungicide.

Organic Compounds

Organic compounds are those derived from plants. They usually work in the form of stomach poisons or contact poisons. Two examples of organic compound poisons are rotenone and pyrethrum, which are very safe when used according to the directions on the label.

Synthetic Organic Compounds

These chemicals have originated relatively recently; many of them have come into use in the past 30 to 35 years. They are very effective against insects, but many of them are also toxic to people. **Synthetic compounds** are insect controls that are man-made pesticides. The synthetic insecticides tend to have a harmful effect on beneficial insects. This large group of insecticides is subdivided into three smaller groups: chlorinated hydrocarbons, organophosphates, and carbamates. They are not found naturally as are the organic compounds listed previously; instead, they are produced in the laboratory.

CHLORINATED HYDROCARBONS These insecticides contain chemicals that have long *residual* control; that is, they continue to kill long after the initial application. DDT is one of the best-known chemicals in this group. It is no longer sold in the United States, however, due to concern regarding the buildup of the poison in the environment and the bodies of warm-blooded animals. Chemicals from this group are being phased out.

ORGANOPHOSPHATES This group of insecticides is very effective in controlling insects, but it also contains some of the chemicals most toxic to warm-blooded animals. Parathion is so toxic that one drop of the concentrated chemical in the eye can kill a person. These chemicals are also absorbed rapidly through the skin. Malathion, another chemical in this group, is a relatively safe chemical that can be used to control many insects.

CAUTION
Carefully read the insecticide label for instructions in use of these chemicals.

Organophosphates break down quickly in the environment (within 15 to 30 days) and do not build up in the bodies of warm-blooded animals. This group may also soon be phased out.

CARBAMATES This group of synthetic organics contains some of the safest insecticides on the market. Carbamates such as carbaryl (Sevin) are very effective in killing some insects although not effective in killing others. They are slightly toxic to warm-blooded animals. Carbamates break down rapidly (in 2 to 7 days) and leave no residue to contaminate the environment. These chemicals do not build up in the bodies of warm-blooded animals.

Chemical Sterilants

The most successful use of sterilants involves gamma radiation treatments to sterilize insects so they cannot produce offspring. One method is to sterilize male insects and release them to mate with females. Because the eggs are not fertilized, they will not hatch when deposited by the female. This is a very effective method in the control of certain insects such as the screwworm, which infests cattle. This process is also being tested for use on plant insect pests such as the gypsy moth.

As insects become resistant to existing chemical pesticides, new types of controls are developed. These new chemicals are totally different in their action on insects and are more effective for now and into the future. It is expected that insects will eventually become resistant to these also, and the search for more biological controls will continue. Biological controls seem to be more effective over a longer period of time.

Application of Insecticides

Dusts

Insecticides are applied with dusters if purchased as dusts. The chemical is already diluted when bought; it requires no mixing before application. Dusts are easy to apply with inexpensive equipment but tend to blow or drift from the target (Figure 18–3). This drifting can damage other crops or contaminate the environment.

Wettable Powders

Wettable powders resemble dusts in appearance but are concentrated and must be diluted with

▲ **Figure 18–3** Nontoxic dust being applied. Note how the dust drifts away from the target plants.

▲ **Figure 18–4** Sprayer being used to spray potatoes. There is very little drift with spray, a water solution.

water before application. Read the directions for the amount of insecticide to mix per gallon of water. Wettable powders tend to settle while in solution and must be stirred while being sprayed. Some powders are pressed into granules that release less dust into the air during mixing.

With a water-mixed spray, better coverage of plants is usually possible and less drift occurs (Figure 18–4).

Emulsifiable Concentrates

Emulsifiable concentrates are liquids that are mixed with water in the same manner as wettable powders. The concentrate is safer to handle because there is no powder or dust to blow during mixing. Emulsifiable concentrates do not settle and separate from the solution, as wettable powders do, and they give the same good coverage of plants. Shelf life of these chemicals is generally longer than that of dusts or wettable powders if the container is kept closed and stored in a cool, dark place.

Granules

Granules are insecticides in the form of small pellets that are spread on the soil surface where they either penetrate the soil after the application of water or turn into gases that fumigate the insects. Some of the systemic insecticides mentioned in this unit are sold in the form of granules. They may be applied to the soil in granulated form and watered in, after which the chemical is absorbed by the roots of the plants. One of the newest systemic insecticides available in granular form is imidacloprid, sold under the trade names Merit, Marathon, and others.

Baits

Baits are poisonous materials that attract insects. The pest eats the bait and is killed by the poison. Slugs, snails, cutworms, grasshoppers, and weevils are examples of pests that are controlled in this manner.

Aerosols

Aerosol insecticides are contained in pressurized cans. They are generally used for small insect control jobs around the house or in the greenhouse. The chemical is already diluted and ready for use when purchased (Figure 18–5). Many **aerosols** in the form of bombs are available for use in the greenhouse. The bombs are punctured and the chemical is slowly released as a fine mist.

▲ **Figure 18–5** Sprays commonly used by the homeowner. The can on the left is an aerosol. The others use manual power to spray or dispense the chemical.

Spikes

Spikes of various sizes and shapes containing systemic pesticides are available. Small spikes are pushed into the soil in pots and larger ones driven in around the plants in the landscape. Plant roots pick up the chemical, and it spreads throughout the plant. The pesticide is released over a long period of time, giving extended protection.

Equipment

Hand-Operated Dusters and Sprayers

Application of insecticides on school grounds or in greenhouses can be done safely using inexpensive hand-operated sprayers and dusters. An explanation of the safe use of each of these pieces of equipment will be given by the instructor.

Larger Dusters, Sprayers, and Granular Application

Large-scale commercial application of sprays, dusts, or granules is accomplished by the use of airplanes, helicopters, tractor sprayers, or trucks with special granular spreaders or sprayers. Figure 18–6 illustrates the use of a large boom orchard sprayer.

Airplanes and helicopters are used to spray and dust large acreage of crops, and crops that cannot be treated by ground-operated equipment (Figure 18–7). Commercial applicators apply the specified chemical and charge the grower at a cost per acre. This is an economical way to apply insecticides when large areas must be covered quickly.

Courtesy of AgTec Division, Ag-Chem Equipment Co. Inc.

▲ **Figure 18–6** A large commercial sprayer applying pesticide to field-grown rhododendrons.

Source: U.S. Department of Agriculture

▲ **Figure 18–7** Airplanes are used to apply pesticides on a large scale. Many acres can be covered rapidly via this method.

Applying the Proper Amount

Whatever method of application is used, it is important that the proper amount of the chemical is applied to a given area. This requires calibration of sprayers and dusters. Calibration involves measuring the amount of chemical applied during a specified period of time or over a measured area and then adjusting the applicator or speed of application to obtain the proper amount of chemical required.

Dusts and sprays must be applied so that all leaf surfaces are completely covered. When dusts are applied to plants, they can be seen settling on the surface of the plants, thereby making it easy to see the extent of the coverage. When sprays are used, leaves of the treated plants should be wet just to the point of dripping.

CAUTION

Safe, efficient control of insects requires that the proper concentration of the chemical be mixed in sprays. Read the label for proper mixing instructions.

TRY THIS Create a display of insecticide equipment. Label each piece of equipment with a description of the function of each piece of equipment.

Identifying Insect Damage to Plants

To identify the insect causing damage to plants, the damage itself must be examined. Chewing insects actually eat away part of the plant. This results in holes in the leaves or missing pieces of bark. Chewing insects are controlled by a stomach poison or a contact spray (Figure 18–8).

Sucking insects suck the juices from the leaves of plants. The injury caused by sucking insects is not as easily detected as that caused by chewing insects. Some signals of damage by sucking insects include twisted plant tips and rolled leaves. Plants that are infested by sucking insects may show no signs of infestation at all, but simply appear to be less healthy. Figure 18–9 shows damage to a plant by a sucking insect, the aphid. Sucking insects must be controlled with a contact chemical, a fumigant, or a systemic.

Source: U.S. Department of Agriculture

▲ **Figure 18–8** Damage to leaves by Japanese beetles. These chewing insects are actually eating parts of the leaves. Notice that the tough leaf veins are left uneaten.

▲ **Figure 18–9** Aphids attacking a plant. These insects suck juices from the plant, thereby greatly reducing plant vigor and yield. Notice the ants that are also present. Ants tend aphids like farmers tend cows. They collect honeydew (a sweet liquid) from the aphids as food and protect the aphids.

Timing in Insect Control

Insects must be killed when they are actively feeding or moving about on the plant. When in the resting *(pupa stage)* or egg stage, very few chemicals can affect insects; however, the new spray oils give control in the egg stage. Before spraying, it is important to determine what stage of growth the insect is in by examining the insect on the plant.

Insects go through one of the following basic life cycles. The stages during which the insects are controlled are shown in shaded boxes.

- Complete metamorphosis:

  egg — larva — pupa — adult

- Incomplete metamorphosis:

  egg — nymph — adult

If the entire insect population of a plant is in an inactive stage, insecticides, other than oil, must not be applied until the eggs hatch or the pupa emerges as an adult. If insects in both active and inactive stages are present, insecticide should be applied immediately, with a second application made in 7 to 10 days when the eggs have hatched and before the insect has a chance to mature and lay more eggs. Figure 18–10 illustrates the life cycle of the Japanese beetle.

▲ **Figure 18–10** The adult Japanese beetle is controlled with an insecticide applied as a spray only during June, July, August, and September. The rest of its life cycle is spent underground where it does a great deal of damage to grass roots. Controlling Japanese beetle can include biological, cultural, and chemical approaches. Employing IPM, insecticides, in the form of soil chemicals, or introducing natural enemies, such as "milky sport," are used during the larval stage, which is October or early spring.

The integrated pest management (IPM) system relies on exact timing and less use of chemicals to control pests.

Plants as Insect Repellents

According to organic gardening experts, certain plants may be used to prevent attacks by insects on other plants. Figure 18–11 lists some of those plants, along with the insects they repel.

PLANT	REPELS
mint	flea beetles, cabbage butterflies
onions, garlic,* chives, leeks	aphids
marigolds	root nematodes
nasturtiums	aphids, cucumber beetles
sage	carrot flies, cabbage pests
horseradish	potato bugs, other flying insects

Figure 18–11 Plants reputed to repel insects.

*Garlic also inhibits the growth of asparagus.

The Organic Gardener

Organic is defined as relating to, produced with, or based on the use of fertilizer that originates from plant or animal without the use of chemically formulated fertilizers or pesticides. These gardens rely on insect repellents that do not include chemicals in any form—natural enemies, plants, extracts of ground insects used as sprays, natural insecticides, crop rotation, sanitation, and resistant varieties.

GREEN TIP The green industry is using more natural materials and insect predators to clean up the environment and make our food supply safer.

The incorporation of organic matter into the soil and the use of mulches result in better plant growth due to better moisture retention in the soil, more microorganisms that inhibit or kill disease-producing organisms, and increased availability of minor plant food elements and other plant growth stimulants. This leads to the production of healthier, faster-growing plants that are better able to resist attack by insects and disease.

Natural control of insects can be accomplished by using **biological** and **biorational** controls that are derived from resources of nature. Biological controls are living things that can attack the pests. Some examples of biological controls are milky spore, nematodes, and *Bactius themoginus* (Bt).

Biorational controls are nonliving materials that are not man-made synthetic material chemical pesticides. Biorationals can impact pests in different ways, such as giving the plant a bad taste that deters pests from feeding on the plant, disrupting the insect's enstar stage of the life cycle, or having the pest ingest the product to outright kill the pests. Some examples of biorational controls to use are neem oil (oil extract), pyrethrin (extract from chrysanthemums), and insecticidal soap (derived from fatty acids). Bioproducts are environmentally friendly, do not last as long, and break down into the ecosystem; however, they are also more expensive than **synthetic pesticides** and more challenging to incorporate into conventional systems.

SAFETY FIRST

In this unit, we are using chemicals such as insecticides. It is important to follow OSHA regulations to the law. Everyone involved must have proper training to work with the specified materials to use insect control measures. Everyone should also wear the correct PPE to work with these products. For more information on occupational health and safety in the workplace, please visit the OSHA site at www.OSHA.gov.

Summary

Insecticides are gradually making a change from man-made chemicals to natural materials and insect predators. Many of these natural materials are as effective or more effective than existing chemicals. Some new man-made materials are copies of materials found in soil fungus and bacteria and plants, and others are insects. We are cleaning up the environment and making our food supply safer. Many insecticides found on the market just a few years ago are no longer approved for use.

Review Questions

1. Discuss six different ways in which insecticides work to control insects.
2. Discuss common insect pests that are found in your local area. Based on what you learned in this unit, name different insecticides that may be used to control each one.
3. List the three major groups or families of insecticides and one chemical belonging to each.
4. What safety precautions should you follow when using insecticides?
5. Discuss the pros and cons of controlling insects through biological controls versus insecticides. What factors should you consider when choosing one method over the other?

Student Activities

1. Observe insect damage to plants. Identify the insect involved and select the proper pesticide for effective control.
2. Use one of the many insect ID smartphone apps to identify insects that need to be controlled.

3. With guidance from your instructor, demonstrate the proper mixing and application procedures for an insecticide for a live pest discussed in this unit and then evaluate the results.
4. Use each piece of equipment described in this unit to properly apply an insecticide. The instructor will demonstrate the proper filling and spraying technique.
5. Visit a local greenhouse operation, arboretum, golf course, nursery, or turf to observe insect control techniques. Record the mixture rations and application rates.
6. Search online for collegiate programs focusing on entomology. Compile a list of possible postsecondary programs that are of interest to you.

Suggested Supervised Agricultural Experience (SAE) Activities

1. Participate in a placement SAE working in product distribution for an agribusiness that sells and distributes insecticides.
2. Participate in a placement SAE as an intern for a horticultural equipment purveyor.
3. Participate in a placement SAE as an intern with the National Park Service.

Additional Resources

www.epa.gov/caddis/ssr_ins_int.html

http://ipmworld.umn.edu/chapters/ware.htm

www.ers.usda.gov/topics/farm-practices-management/chemical-inputs/pesticide-use-markets.aspx

UNIT 19 Fungicides, Rodenticides, Molluscicides, and Nematocides

Objective

To identify fungus diseases, rodents, mollusks, and nematodes, and prescribe a chemical control for each.

Key Terms

fungicides
molluscicides
nematocides
rodenticides

Competencies to Develop

After studying this unit, you should be able to:

- identify fungus diseases on plants and set up a spray or dusting schedule for effective control.
- recognize at least two types of rodents and apply a rodenticide in a bait station.
- recognize slugs and snails on sight and develop a slug control program, including chemicals used in the bait station or as contact poisons.
- identify two nematocides and explain in writing their application to specific situations.

Materials

- plants infected with fungus disease
- one live snail and one live slug
- one plant with nematode damage to roots and one healthy plant of the same variety
- sample fungicides, rodenticides, molluscicides, and nematocides
- smartphone apps:
  - Pest Portal
  - Spectracide
  - Calibrate My Sprayer
  - GardenCompass
  - Ortho Problem Solver
- safety data sheets (SDS) (Refer to Unit 17)

Fungicides

Fungicides are pesticides used to control plant diseases caused by fungi. *Fungi* are tiny nongreen plants such as rusts, mildews, molds, and smuts that lack chlorophyll and live as parasites on green plants. Fungus diseases are spread by *spores* that resemble tiny seeds floating through the air. As fungi grow on the green plants, the plant tissue is killed, or the plant is weakened. These diseases must be controlled if attractive, highly productive crops are to be grown.

> **GREEN TIP** To make a natural, chemical-free fungicide, all you need is milk and water. Pour an equal amount of each into a bottle and shake. Spray your plants twice a week with this fungicide for a healthy garden.

How Fungicides Work

Fungicides are most effective when used to prevent germination and/or growth of the fungus spore, thus preventing the attack on the plant. Some new fungicides, however, can control a fungus disease even after the disease is present in the plant. Systemics work by entering the plant and circulating in the sap. They work primarily as other fungicides do, that is, to prevent the disease rather than cure it after it is established in the plant.

Fungicides are generally not as toxic to humans and warm-blooded animals as are insecticides. Beneficial fungi in the soil kill disease-producing fungi. These natural fungicides are some of the most effective. Figure 19–1 illustrates a typical fungicide label.

> **GREEN TIP** New sustainable biological and biorational controls in the green industry are available to control pests and enhance a safer environment.

Effects of Fungus

Fungus diseases attack all parts of the plant: the leaves, stem, fruit, seed pods, and roots.

LEAVES Plant leaves are attacked by many so-called *leaf spot diseases.* The term *leaf spot*

Courtesy of Southern Agricultural Insecticides Inc.

▲ **Figure 19–1** Typical fungicide label. Notice the different types of information that appear on the label.

describes the appearance of the diseased leaf. The leaf is marked by round or irregular spots that are beginning to turn a color different from the rest of the plant (Figure 19–2). The spot or spots on the leaf grow larger and larger as the fungus disease spreads. The entire leaf may be killed, but this is unusual.

Leaf spot diseases damage the plant by destroying the cells that manufacture food. As more cells are killed, the plant is less able to make food and becomes weaker. The appearance of the plant is also affected. A plant sold for its beauty may be damaged so much that it is unsaleable. A plant that produces a crop of fruit or seeds does not produce as large a crop when affected by leaf spot disease because the leaves cannot manufacture enough food for best growth of the fruit or seeds.

Leaves are also attacked by another fungus disease, known as *rust disease* (Figure 19–3). Again, the name describes the appearance of the disease; the leaf, which takes on a rusty color in the beginning, is gradually killed by the disease. Mildews and molds are other fungus diseases. Their names describe their appearance on the plant.

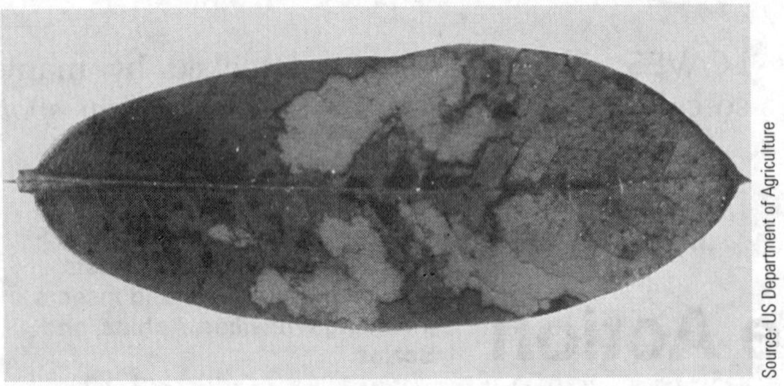

▲ **Figure 19–2** The effects of leaf spot disease on a rhododendron leaf.

▲ **Figure 19–3** Rust disease on rhododendron leaves.

Fungicides are sprayed on leaves to prevent the entry of the fungus disease into the plant. When the fungus spore sprouts on the plant leaf, the fungicide kills it. A spray or dusting calendar complete with dates must be established so that the fungicide can be applied on a regular schedule. Systemic fungicides such as Subdue, which enter the plant sap, are more effective than nonsystemic ones in preventing leaf diseases. Neem oil, a biological, is also an excellent fungicide.

FAST FACT

Most fungicides are used to prevent disease and must therefore be sprayed or dusted on the plant before the disease is present to prevent spread of disease. Some fungicides do cure or eliminate disease already present on the plant, but even these chemicals work best as preventers of disease and not as a cure.

STEMS, FRUIT, AND SEED PODS These parts of plants are also attacked by fungus diseases. The effect and control are the same as those listed for leaves. Anthracnose, for example, is a fungus disease that attacks the stems of many plants. High humidity and rain help to spread fungus diseases of stems and leaves. Most fungus diseases grow best in high-moisture conditions.

ROOTS Plant roots are attacked by many types of root rot or root mold. *Root rot* and *root mold* describe the appearance of the root after attack by the fungus disease. Root diseases are controlled by drenching the soil with fungicides or by spraying.

Fungus root diseases are most often found in soils that do not drain well and are wet too much of the time. Improved soil drainage generally helps to control fungus diseases of roots.

Small seedlings are often attacked by fungus diseases such as *damping-off*. The use of sphagnum moss to grow seedlings often prevents the start of the disease. Adding composted organic matter and beneficial soil bacteria also helps control root rot diseases. Dusting or drenching the soil mix with Captan often gives excellent control of seedling root and stem rot.

One biological fungicide is the commercially available Actinovate. The active ingredient is a soil bacterium, *Streptomyces lydicus* strain WYEC 108. When applied to the soil, the bacterium grows around the root system, protecting roots from harmful soil diseases. It also produces an active chemical that kills certain fungi.

Fungus diseases can be reduced greatly by following these good cultural practices:

1. Select well-drained soils to protect against root diseases.
2. Maintain soil aeration and help control diseases by adding compost.
3. Practice crop rotation.
4. Use compost as a source of plant food.
5. Space plants wide for good air movement.
6. Test the soil and adjust pH.
7. Use organic preventive sprays whenever possible.
8. Select disease-resistant plants.
9. Clean up the garden at season's end to remove all plants.

TRY THIS Test the effects of fungus on a plant. In the classroom laboratory, introduce fungal spore suspensions to the plant in the form of a foliar spray, stem injections, or soil drenches. For example, the entomopathogenic fungus *Beauveria bassiana* can be introduced to coffee seedlings using any of these three application methods. After the fungus has been applied to the plant, record the changes that occur in the plant over a one-week time period.

Application

Fungicides are sprayed or dusted on plants or poured around the roots as drenches. The methods of application are the same as those for insecticides with one difference regarding the timing. Insecticides are usually applied after it is determined that an insect is on the plant and is causing damage, whereas fungicides are applied to prevent attack. The chemical acts as a protective film to kill the fungus spore as it sprouts and before it can penetrate the surface of the plant.

Rodenticides

Rodenticides are chemicals used to control rats, mice, and other rodents. Most commonly, rodenticides kill as stomach poisons. They are applied as bait that is attractive to the rodent and is eaten. In limited cases, rodenticides are spread or sprayed on the rodent or in an area through which the rodent travels. This technique is usually avoided, however, because other wildlife or domesticated animals could be poisoned accidentally.

A common chemical used for rats and mice is warfarin. This chemical is applied to grain in *bait stations* where the rodent feeds. Field mice in orchards are often poisoned by a coating of deadly poison on pieces of apple. The apple is placed in bait stations for the mice to eat.

CAUTION
Always place poisoned baits in bait stations where other animals cannot reach the poison. Place the bait under a wooden or metal box with a small hole cut in it that is just large enough for mice to enter.

Molluscicides

Molluscicides are chemicals used to control snails and slugs. They are applied as poison baits or contact poisons. These chemicals kill as stomach poisons when eaten or as contact poisons when the slug or snail is covered by the poison or crawls through an area to which the poison has been applied. Slugs and snails usually hide during the daytime; therefore, it is difficult to hit them while spraying or dusting.

Snails and slugs look much alike, except that slugs do not have a shell and snails do (Figure 19–4). Both travel on the soil surface, leaving a slimy trail behind, and both feed on plant leaves and tender stems at night. Because they feed at night, locating slugs and snails actually doing damage to plants may be difficult, thereby causing other insects to be blamed for the

Courtesy of NYS Experiment Station

▲ **Figure 19–4** (A) Snail, (B) slug.

damage. To pinpoint snails and slugs as the pest, look for the slimy trail. To find slugs and snails, lift stones and mulch and look underneath. Any cool, moist place is a good hiding place for them.

Control

Rough sand, ashes, gravel, or cinders act as barriers to slugs and snails; the rough surface provided by these materials is difficult for them to crawl on. Another chemical-free control method is to set a shallow pan filled with beer in the soil, with the top of the pan at the soil level. The slugs follow the smell of the beer into the pan and drown. Yeast mixed with water is also effective.

Various slug killers are sold by pesticide dealers. Some effective chemicals include Slugit, which is a stomach and contact killer; Slug-kill (15 percent methaldehyde dust), which is a contact poison; and Zectran. Escar-Go is a new slug control that is not toxic to other organisms.

CAUTION
Keep these and all pesticides out of the reach of children.

Nematocides

Nematocides are chemicals used to control nematodes. Nematodes are small hairlike worms that feed on plant roots. These tiny worms bore into the root, usually causing knots or bumps to appear on the attacked root (Figure 19–5). Nematodes live in the soil and swim in soil water to move from plant to plant. Some nematodes also live inside plant leaves.

Control

Soil fumigants are the most effective treatment for nematodes. The nematocide must contact the nematodes as a gas or liquid passing through the soil. Two chemicals meant specifically for nematodes are Dasanit and Nemagon. Systemic insecticides are effective on leaf nematodes.

Source: US Department of Agriculture

▲ **Figure 19–5** Nematode damage to plant roots. The plant on the right is badly infested with nematodes, and the root system is severely damaged. The plant on the left has a normal root system.

SAFETY FIRST

In this unit, we are using chemicals such as fungicides, nematocides, molluscicides, and rodenticides. It is important to follow OSHA regulations to the law. Everyone involved must have proper training and wear the correct PPE to work with the specified materials for insect control measures. For more information on occupational health and safety in the workplace, please visit the OSHA site at www.OSHA.gov.

Summary **Fungus disease, rodents, snails, slugs, and nematodes can be serious plant problems. They can be economically controlled, and new biological controls are being developed. A vigorous plant with strong roots in an organic soil can survive these attacks much better, often without chemical treatment.**

Review Questions

1. Discuss factors that might lead to the development of fungus diseases and how each might be prevented through good cultural practices before the application of a fungicide is needed.
2. Discuss ways to prevent or control snails and slugs that do not require the application of molluscicides.
3. What are cautions to keep in mind when considering the use of rodenticides?

Student Activities

1. Under the instructor's supervision, apply a fungicide to a plant on the school grounds to protect it against fungus disease. Use a small aerosol. As the spraying progresses, check for disease control and record the results. If two plants are available, spray one and keep one as a control.
2. Put together two bait stations, one for rodents and one for slugs and snails. Actually use the stations under the supervision of your instructor if circumstances permit.

CAUTION These activities should be accomplished only under the supervision of the instructor. When applying any pesticide, carefully read and follow all instructions and precautions.

3. Search online for residential plant disease control, rodenticides, fungicides, and molluscicides. Create a flyer for homeowners with tips on plant disease control using rodenticides, fungicides, and molluscicides based on your research.

Suggested Supervised Agricultural Experience (SAE) Activity

1. Conduct a diversified crop production SAE. Record all investments and labor as part of an entrepreneurship SAE.

Additional Resources

www.pest.umd.edu

http://edis.ifas.ufl.edu/pp154

www.clemson.edu/extension/horticulture/fruit_vegetable/peach/diseases/br_fr_issues.html

http://msue.anr.msu.edu/news/fungicide_disease_management_tactics_for_vegetables_webinar

UNIT 20

Herbicides

Objective

To select and apply herbicides so that weed control is accomplished without damaging the crop being cultivated or contaminating the environment.

Key Terms

biocontrol
calibrated
drift
herbicide
nonselective herbicide
postemergence
preemergence
preplanting
selective herbicides

Competencies to Develop

After studying this unit, you should be able to:

- identify a weed problem and select a herbicide to control the problem.
- define and differentiate between selective and nonselective herbicides.
- describe three ways in which herbicides destroy weeds.
- outline in writing how a sprayer is calibrated.
- list three possible reasons for failure of a herbicide to work properly.
- properly apply a herbicide.

Materials

- weed killers complete with label and recommendations
- weed charts for weed identification
- sprayer for use in calibration and application of herbicides
- an area of lawn or crop land for use in locating and identifying weeds and applying herbicides
- smartphone apps:
  - Pest Portal
  - Spectracide
  - Calibrate My Sprayer
  - GardenCompass
  - Ortho Problem Solver
  - AG PhD
- safety data sheets (SDS) (Refer to Unit 17)

An **herbicide** is a chemical that kills weeds. Herbicides comprise a large percentage of the total pesticides used each year. In the past 20 years, many new weed killers have been introduced on the market, and each year new uses are found for traditional weed killers. In the United States, estimates of damage and loss of production to crops caused by weeds is in excess of $4.5 billion each year. Herbicides are used to kill weeds in crops, on lawns, along roadways, and in many other places.

Weeds come in many varieties and vary from region to region, but each one is considered to be a plant out of place. Weeds compete with crops for sunlight, moisture, and nutrients, and they can have a tremendous effect on crop yields. Grass that is growing in a lawn is not a weed but is considered a weed if it is growing in the garden.

Weed control can be accomplished without the use of chemicals. In some cases, mulching and the cultivation of crops may be more effective and just as economical to control weed growth. For example, it would be very difficult to spray weed killers in a small garden without damaging other plants. However, in large acreages of single crops, selective weed killers can be used to greatest advantage.

Many different types of chemicals are used as weed killers. They are generally not as toxic dermally (on the skin), although some are very poisonous if swallowed. Chemicals used as herbicides may be changed or withdrawn from the market.

GREEN TIP Research and development of biological herbicides are presently being expanded. They provide effective control of broadleaf weeds and are environmentally friendly.

Types of Herbicides

Nonselective Herbicides

Nonselective herbicides kill all plants to which they are applied. These herbicides are used in places where no plant growth is wanted. Lumber yards, railroad tracks, fence lines, driveways, and parking areas are a few examples of areas where these chemicals are used. An example of a weed killer in this category is Roundup.

Nonselective weed killers sometimes may also be used as selective weed killers. Examples of this type of substance include Atratol and AAtrex (atrazine). Used at a rate of 2 to 4 pounds of active chemical per acre, it kills only small weeds or selected weeds; if application is increased to from 5 to 10 pounds per acre, it kills all plants.

Selective Herbicides

Selective herbicides are just what the name implies—chemicals that kill some plants but not others. This is by far the largest group of weed killers. Weed control in a particular crop can be accomplished safely and effectively with selective weed killers. These chemicals are designed to kill the weed but not the crop.

In some cases, a selective weed killer that kills broadleaf plants will not kill grasses, which have narrow, bladelike leaves. This makes it possible to spray a lawn to kill dandelions, which have a broad leaf, and not kill the grasses, which have narrow leaves. Because corn is a grass, broadleaf weeds in a corn field can be killed with a selective herbicide without damaging the corn.

Herbicides that have the potential to kill a particular crop at high concentrations can be used at low concentrations to kill young weeds that are growing in the crop and not damage the crop. By applying the weed killer at a low concentration, young, tender weeds are killed, and the older, more mature crop plant is not harmed. Strawberries are protected from weeds in this way.

The Effect of Herbicides

Herbicides work by upsetting the metabolism, or life functions, of the plant (Figure 20–1). The plant either starves to death or wears itself out because of the increased rate of activity caused within its system by the chemical. Herbicides are designed to affect the metabolism of some plants and not others. When using herbicides, follow the recommendations on the label on the container. It is the law.

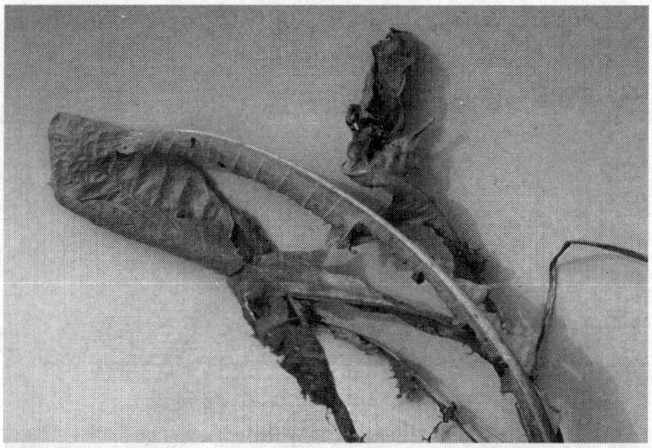

▲ Figure 20–1 A weed killer was used to kill the roots of this weed, resulting in the top drying out and turning brown.

The following are some specific examples of the effect of herbicides.

Dimension

Dimension is a herbicide that is used to kill both broadleaf weeds and grassy weeds in turfgrass sod as well as those in tree and shrubbery plantings.

APPLICATION Dimension may be applied **preplanting** (before planting has taken place), **preemergence** (before emergence of seedlings aboveground), or **postemergence** (after seedlings have grown aboveground but before the crop matures). However, it must be applied before the weeds germinate.

BEHAVIOR IN THE PLANT Dimension is absorbed through the roots. It moves through the plant in the xylem or wood, accumulates in the stem tips and leaves, and kills the plant by preventing photosynthesis from occurring. Because the plant cannot manufacture food, it starves to death.

2,4-D

2,4-D is a systemic herbicide that is still used to control broadleaf weeds but is being phased out. Most dicot (broadleaf) plants are susceptible to the chemical at normal application rates.

APPLICATION Application is usually by spray directed on the plant foliage. Both plant leaves and roots absorb 2,4-D.

BEHAVIOR IN THE PLANT The chemical moves through the plant phloem or bark. Movement is more rapid in fast-growing plants, so plants in active growth are killed more easily. The chemical 2,4-D causes abnormal growth in the weed and affects respiration, food reserves, and cell division.

Treflan

Treflan is a selective preemergence herbicide that is used for the control of annual grasses and certain broadleaf weeds in vegetable crops and ornamental turf. The chemical in Treflan is trifluralin.

APPLICATION Application is generally at planting time before weeds come up. The chemical is usually diluted with water and applied as a spray.

BEHAVIOR IN THE PLANT Treflan is not absorbed by leaves, and it does not move through the plant. Rather, it kills sprouting seeds. The exact way it acts is not yet known.

Herbacide Summary

The three herbicides just discussed are not generally used by the average gardener, but rather by professionals. They each kill weeds in three different ways.

- Dimension → prevents photosynthesis
- 2,4-D → upsets cell division, respiration, and food reserves
- Treflan → destroys sprouting seeds (specific action not yet known)

All weed killers do not fall into these three groups, but they are representative of a great percentage of herbicides

A New Approach to Weed Killing

In a new method of weed killing, a natural protein from a soil fungus causes weed cells to kill themselves within 24 hours, which is a different mechanism than that triggered by any other herbicide. It works because, in general, plants defend against diseases by killing their own cells surrounding the disease-causing organism. This isolates the killer and starves it to death. When this new fungal protein is applied to plants, the weed turns on its defensive response to such a high degree that it kills itself. Only broadleaf weeds are affected. This may be the first weed killer for organic farmers.

Recently, researchers at Penn State discovered a chemical in tree-of-heaven that kills all plants it comes in contact with. This chemical shows great promise as a new weed killer.

Biological Weed Control

The employment of biological weed control of noxious weeds uses a nonchemical form of weed control. In one approach, the environmentally friendly adult flea beetle is a beneficial insect and a natural predator that controls leafy spurge weeds, which are invasive plants, by eating the entire plant. This weed control process has proven to be cost effective and is saving our environment by avoiding placing dangerous chemicals in the soil. This is known in the field as **biocontrol** because a living organism is being used to control a pest (see Figure 20–2). This approach is cost effective, and suppliers have the natural predator readily available to ship to their clients.

Source: Agricultural Research Service, US Department of Agriculture

▲ **Figure 20–2** The *Aphthona flava* flea beetle is a good biological control for leafy spurge weeds.

> **GREEN TIP** Research and development of biological herbicides are presently being expanded. They are providing effective controls for micot and dicot weeds and are eco-friencly.

Application of Herbicides

Herbicides are either sprayed with ordinary sprayers or applied already mixed in fertilizers or in granules. The same type of equipment is used as with insecticides. However, a sprayer that is used to apply herbicides should not be used to apply insecticides or fungicides unless it is thoroughly cleaned. This does not mean simply rinsing out the sprayer; it must be washed with a substance that neutralizes the herbicides, such as ammonia. Read and follow the directions carefully; otherwise, enough herbicide may remain in the sprayer to damage desirable plants sprayed for insects or diseases. When small hand sprayers are used, a separate sprayer should be used for the application of herbicides.

Time of Application

Application of herbicides must be made at one of three times to avoid injuring desirable plants.

PREPLANTING Herbicides are mixed into or sprayed onto the soil or seedbed. Some herbicides are effective only when incorporated with or mixed into the soil; others need no mixing. Read the label before buying.

PREEMERGENCE Treatment is made before any plant growth appears or after the crop comes up but before the weeds appear. Read the label to see if the chemical should be applied before the crop emerges or after the crop emerges and before the weeds emerge. When used at the recommended concentration, these chemicals usually prevent seeds from germinating or kill only small, tender weeds without damaging desirable plants.

POSTEMERGENCE Applications must be made very selectively in terms of the chemical used. These chemicals are applied after the crop plant has come up.

CAUTION
The horticulturist must be very careful to choose a herbicide that kills the weeds but not the crop because at the time of application, both weeds and crop are aboveground.

Calibration of Sprayers

It is extremely important that the proper amount of actual chemical be applied to a particular area when herbicides are sprayed. Even if the right mixture or concentration of spray is in the spray tank, it is possible to apply the wrong amount of weed killer. To avoid this problem, the sprayer should be properly **calibrated** or adjusted before use. The correct amount of active chemical must be applied on a measured area for the most effective control.

LARGE SPRAYERS The first step in the calibration of large sprayers is to measure and mark off an acre. Spray the acre with water and refill the spray tank, measuring exactly how many gallons of water are required to refill the tank, the amount of water applied to the acre. This is the rate of application per acre. If 10 gallons of water are applied per acre, and 2 pints of chemical are to be applied per acre, then 2 pints of chemical are added to each 10 gallons of water. For example, for a 100-gallon tank, add 20 pints, or 2.5 gallons, of chemical.

CAUTION
The rate of travel over the area and the sprayer pressure must be the same as in the trial run.

SMALL HAND SPRAYERS Follow these steps to calibrate a small hand sprayer.

1. Lay out an area measuring 10 feet × 100 feet (1,000 square feet).
2. Fill the sprayer with water.
3. Spray the 1,000-square-foot area.
4. Measure the number of pints of water required to refill the sprayer to the same level as before. This is the number of pints applied per 1,000 square feet.

If 8 pints (1 gallon) are sprayed on the 1,000 square feet of land, and 2¼ teaspoons of chemical are required per 1,000 square feet, add 2¼ teaspoons of chemical to each 1 gallon of water. This is the same rate as 1 pint per acre. (One acre is 43,560 square feet.)

CAUTION
Spray with the same pressure and speed of application as in the trial run.

Another way to determine the amount of water that was applied to the 1,000-square-foot area is to measure the length of time required for the application. Spray into a container for the same length of time and measure the amount of water collected.

CAUTION
Use the same spray pressure as when spraying the 1,000-square-foot area.

Drift

Any substance that is sprayed can blow or **drift** in the wind. If a herbicide drifts onto a nontarget plant, there is a chance that this plant will also be damaged or killed. When spraying herbicides, use spray nozzles that apply large droplets of spray. These larger droplets do not blow or drift as much as a fine mist, which greatly reduces the danger of damage to other crops nearby. Tomatoes and tobacco are injured especially easily if 2,4-D drifts over them. Injury to other crops by drifting chemicals is a greater problem with herbicides than with other types of pesticides.

> **TRY THIS** Study for, prepare for, and participate in the FFA Agricultural Technology and Mechanics Career Development Event (CDE). You will need to form a team of students to participate in the event. Sprayer application calculations are often a component of this FFA event.

Breakdown of Herbicides

Most herbicides are broken down in the soil by soil microorganisms and eventually become harmless materials. The length of time required for this breakdown varies from several weeks to several years. Read the label for recommendations. If a herbicide persists or stays in the soil for 1 year or more, it may kill or damage the crop planted in that soil the following year or later in the same season. Crops that are not damaged by the chemical should be planted there until the soil is clean enough for crops that could be damaged by the herbicide.

Because preemergence herbicides are only effective in controlling germinating weed seeds, they must be applied on clean, cultivated soils before weed seeds germinate.

These recommendations should be used only as a guide in selecting the proper test herbicide. Under no circumstances should a herbicide be applied over the entire nursery without first spending one growing season testing it in a limited area under the same soil conditions and the same growing program. Herbicides are noted for not being equally dependable under all growing conditions.

Always read herbicide labels for instructions, including the species that have been cleared for use for that herbicide. Even when materials are used as recommended, the manufacturer cannot be held responsible for injury when they are applied to ornamentals not listed on the label.

Why Herbicides Sometimes Fail

Why herbicides are sometimes ineffective is a very complex question. The following are some of the more common reasons for failure.

Incorrect Selection of the Herbicide

Unsatisfactory weed control may result from use of the wrong herbicide. Many weed killers are selective; that is, they kill only certain weeds. Therefore, selecting the right chemical is critical to controlling a specific weed. This requires accurate identification of the weed to be controlled. To use herbicides effectively, the applicator must be able to accurately identify weeds.

Improper Application of the Herbicide

Herbicides are designed for application in a variety of ways and at various stages of the growing process. Poor application and poor timing of application are reasons for herbicide failure. For example, preemergence herbicides must be spread evenly on the surface of the soil. To make contact with sprouting weed seeds, the entire soil surface must be covered with the chemical; if not, the herbicide could be rendered ineffective.

Some chemicals also must be washed into the germination zone by rainfall or irrigation within 7 to 10 days after application. If no rain occurs or irrigation is not possible, poor weed control will result. Preplant weed killers such as Eptam must be mixed into the soil immediately after application; if not, they evaporate into the air. Much of the active chemical is thus lost, and control is poor. Postemergence weed killers such as 2,4-D must be sprayed onto the plant to be killed or over the roots of the plant so that the chemical can be absorbed into the plant. The chemical 2,4-D also works most effectively if the weeds are in active growth.

Sprayer calibration is also very important for proper application of herbicides. This may involve tests to determine the proper amount of chemicals to apply per acre of land. Rates that are too low result in poor weed control; rates that are too high could kill the crop as well as the weeds.

Environmental Variables

THE MATURITY OF THE WEEDS Weed maturity is an important variable; most weeds are killed more easily when they are young and in active growth. In fact, this in the only time many weed killers work effectively. An old, hardened plant requires much more chemical or may even require a different type of weed killer. An exception to this is the herbicide glyphosate (Roundup), which works better on mature plants.

RAIN Rain that falls just after application of the herbicide can be an advantage with some preemergence chemicals. However, excessive rain can leach or wash the chemical through the soil or off foliage and dilute it so much that weeds are not controlled.

SOIL TYPE AND ORGANIC MATTER CONTENT The type of soil and its organic content affect the amount of active chemical that must be applied. Some chemicals are absorbed or held on the surface by clay particles and organic matter. There are two dosages for some chemicals. For example, Dacthal is recommended in doses of 10 to 15 pounds of active ingredients per acre: 10 pounds is recommended for sandy soils, and 15 pounds for silt and clay soils. A soil high in humus absorbs and makes some of the weed killer inactive, requiring a higher rate of chemical application.

Not all chemicals are affected by organic matter in the soil to the same degree. For example, dalapon is not absorbed at all, whereas Dacthal is easily absorbed and would be greatly affected by the organic matter level of the soil.

Weed control with chemicals is not a simple job; many variables must be considered to achieve effective control. Herbicides have limitations and are not a magic cure-all for weed problems, but if used carefully and accurately, good results can be expected.

SAFETY FIRST

In this unit, we are using chemicals such as herbicides. It is important to follow OSHA regulations to the law. Everyone involved must have proper training and wear the correct PPE to work with the specified materials for weed control measures. For more information on occupational health and safety in the workplace, please visit the OSHA site at www.OSHA.gov.

Summary

Weed killers, called herbicides, work in many ways and on various plants. It is very important to select the proper chemical and apply it properly. A new weed control herbicide has been discovered that is very effective against mocot and dicot weeds. This good news should encourage more growers to use organic methods and produce more healthy food. Always read the label to determine the proper use of the herbicide.

Review Questions

1. Describe the differences between a selective and a nonselective herbicide. Under what circumstances would you use each one?
2. Discuss some options for controlling weeds other than the application of herbicides.
3. What are some ways to control drift?
4. Discuss three organic variables that must be considered prior to the application of a herbicide.
5. Discuss the reasons why herbicides sometimes fail.

Student Activities

CAUTION Complete these projects only under the supervision of your instructor.

1. Working in groups of three to five students, demonstrate the proper calculation and mixing of at least one herbicide.
2. Visit a commercial applicator for a demonstration on the calibration of a pesticide applicator.
3. If land is available, mark off plots of 100 square feet. Apply the herbicide mixed in Activity 1 to an assigned area. Work in the same groups.
   a) Identify the weeds present, and estimate the number of each.
   b) Select the proper weed killer.
   c) Mix and apply the chemical.
   d) Record the results.
4. a) Calculate how much chemical to use in 100 gallons of water if the weed killer is to be applied at the rate of 1 pint per acre, and the sprayer sprays 20 gallons of water per acre at the speed traveled.
   b) If the speed at which the sprayer is pulled changes from 2 miles per hour to 4 miles per hour, how will this affect the amount of spray applied per acre?
5. Use an online sprayer calibration calculator (e.g., www.omafra.gov.on.ca/english/crops /sprayer/ep75.htm).
6. Conduct research online to identify various nonchemical weed controls. Make a list of controls, and describe how each one works.

Suggested Supervised Agricultural Experience (SAE) Activities

1. Conduct an agriculture placement SAE in the area of agricultural mechanics. Examples include orchard equipment operator/mechanic and field applicator for an agribusiness.
2. Participate in a placement SAE at a local agricultural repair facility. Responsibilities should include custom sprayer calibration for farmers.

Additional Resources

https://agnr.umd.edu/

www.clemson.edu/extension/pest_ed/safety_ed_prog/application_equip.html

www.weedsoft.org/Apps/TMCalc/TMCAcceptance.aspx

http://store.msuextension.org/publications/AgandNaturalResources/MT200914AG.pdf

SECTION 6

Container-Grown Plants

UNIT 21

Dish Gardens, Miniature/Fairy Gardens, and Terrariums

Objective

To construct and care for dish gardens, miniature/fairy gardens, and terrariums.

Key Terms

activated charcoal
chloric
fertility
limestone
pathogens
pH
soilless culture
sphagnum moss
stomata
terrarium
transpiration

Competencies to Develop

After studying this unit, you should be able to:

- select a container, a medium, and plants for a dish garden, fairy and/or miniature garden, and terrarium, respectively.
- plant a dish garden, fairy and/or miniature garden, and terrarium.

Materials

- 2½-inch-tall potted foliage plants
- 6-inch wooden pot labels
- sphagnum moss or peat moss
- small pebbles or crushed rock
- dish garden container
- appropriate plants
- smartphone apps:
  - Weather
  - Greenhouse Management
  - Spectracide
  - Pest Portal
  - Calibrate My Sprayer
  - PGR Mix Master
  - Ortho Problem Solver
  - LikeThat Garden

Dish Gardens

Dish gardens provide a very attractive and appropriate living gift. They are often used as attractive presents to friends on special occasions or as gifts to hospital patients. Dish gardens should be arranged so that the plants and container come together to form a design pleasing to the eye. There should be a focal point with other plants complementing or reinforcing this point. A formal design has the focal point, the plant that is the largest or showiest, in the center with equal balance of plants on the sides and back. The focal point in an informal design is off-center, on an edge, or in a corner of a rectangular container.

GREEN TIP Dish gardens are an attractive living gift for the home or office that provide a natural, sustainable ecosystem in a closed ecological unit.

Container Selection for Dish Gardens

The choice of the container for a dish garden is influenced by the personal preference of the individual or the designer. The container should be of the proper size, shape, color, and texture for the occasion and for the plants that go into it. Containers may range from baskets to ceramic bowls in a myriad of shapes and depths. Cacti and other succulents look best in rustic pottery. Flowering plants look best in glazed, light-colored containers. Plastic containers set inside baskets are very popular now, especially for foliage plants. The container should not distract from the plants.

Containers should be at least 3 or 4 inches deep and deeper if larger plants are used. The plant's root system must fit down into the container. The container must also provide drainage so plant roots do not stand in water. Plant roots need to be able to breathe and grow, and too much water will suffocate the roots. This is accomplished by placing a 1- to 2-inch layer of coarse gravel in the bottom of the container (this requires a deeper container).

Plant Selection

It is important to select plants that thrive under the same growing conditions. Plants requiring dry conditions, such as cacti, should not be interplanted with plants that thrive in moist soils. Figure 21–1 shows plants used in a small dish garden.

▲ **Figure 21–1** The seven plants selected for this dish garden are (clockwise from left front) *Dracaena godseffiana*, *Nephthytis* 'White Butterfly' (*Syngonium*), *Dieffenbachia*, green ivy, *Croton* 'Gold Dust,' *Dracaena* 'Florida Beauty,' and *Peperomia* 'Rainbow.'

Nearly any indoor plant that will fit in the dish being constructed will work. Plants that will survive some neglect, especially improper watering, will probably grow best. Because a number of plants are crowded in a small place, they may dry out quicker, needing frequent watering. On the other hand, dish gardens do not have drainage holes, so they can be easily overwatered.

Selection of Planting Medium

Many greenhouse growers of dish garden plants are using **soilless culture** media:

- The mix must be uniform in its structure, **fertility**, acidity (**pH**), and texture.
- The media must be free of harmful **pathogens** (disease organisms, insects, and/or weed seeds).
- Soilless media are great sources of air that are lightweight and easier to handle and ship.
- Soilless cultures have good moisture retention and drainage because they are blended with foundation materials of perlite, peat moss, sphagnum moss, vermiculite, pine fines, and **limestone**.

Watering

Moisture is the most important plant element. As has been mentioned previously, it is the primary source of growth and plant functions. The proper watering of dish gardens is important. The medium should be moist to the touch and allowed to dry before watering again. If the planter is overwatered, the excess water will drain into the gravel and keep the roots from standing in the moisture. Too much water will cause the lower leaves of the plants to be chlorotic (yellowing over the entire leaf).

Placing Plants in the Container

The tallest or most striking plant is used as the focal point. It is placed in the center in formal designs and off to the side or in a corner in informal dish gardens. Move supporting plants behind, around, and in front of the focal point plant until the most attractive design is reached. Plants with different leaf textures or colors look best next to each other. Plants that hang down should be placed at the edge of the container. All plants should be visible to the viewer and blend together well with no single plant being too bold or conspicuous so as to overpower the rest of the planting. Figure 21–2 shows the sequence of starting from the back and filling the dish.

▲ Figure 21–2 Working from the back to front, the back three plants and the center plant are inserted first.

▲ Figure 21–3 The completed dish garden inserted into a decorative basket.

After the plants are properly placed in the dish, fill in between the root balls with a potting soil mix best suited for the plants used. Brush any potting medium off plant leaves with a soft brush. Water the plants to settle the medium around the plant roots. Fertilizer should be used sparingly, if at all. The plant will outgrow the dish soon enough without fertilizer.

If a plain plastic dish is used, it is placed inside a basket or other container (Figure 21–3). Artificial additions look out of place, but ribbons may be added to increase the appeal of the arrangement.

> **GREEN TIP** You can use human food waste as organic nutrient plant food. Did you know that egg shells are rich in the element of calcium, which is an excellent resource to help the plant grow and be healthy? The next time you cook eggs, crush the egg shell and incorporate them into the plant medium.

Dish gardens will create special accents. They are often given as gifts for special occasions. They look good in an instant, and are a good way to add a touch of the outdoors to interior spaces. Dish garden plantings add that extra appeal to décor.

▲ **Figure 21–4** A larger dish garden in a glazed white container.

Variations of dish gardens include plants planted in driftwood, fiber, or sphagnum peat moss in wire frames, lava rock, or other unique containers. Many of these are designed for long life, and dwarf slow-growing plants should be used.

Dish gardens make good class fund-raisers and can be sold throughout the school year. Figure 21–4 shows a larger dish garden in a white container.

Dish Garden Care

Many of the tropical dish garden plants have been grown in optimum conditions. The micro-environment for the dish garden is a key to the success of a productive garden.

LOCATION The first consideration for a dish garden is the location in the microenvironment. The location must have adequate lighting, whether natural filtered light or supplemental light provided by special grow lamps that provide the foot-candles of illumination for growth.

HUMIDITY Indoor plants thrive well in conditions of 30–70 percent relative humidity. This condition can be provided by setting the dish garden in a shallow tray with pebbles or marbles lining the base of the tray and enough water to cover them to ensure an increase in the relative humidity. Another resource is spraying the foliage with a fine mist of water daily to provide the needed moisture.

TEMPERATURE Dish gardens will do well if they are placed in temperatures of the home. Most of the plants prefer a warmer environment, but the key is that the relative humidity is maintained.

CARE OF THE FOLIAGE Wipe the leaves with a moist towel to remove any particles that settle on the leaf surfaces. These particles reduce the plant's ability to absorb the light rays and moisture in the atmosphere.

TRY THIS Using the information in this unit, create and maintain a dish garden.

Fairy Gardens/Miniature Gardening

Fairy gardens, also called miniature gardens, are small container gardens that include both living plants and tiny structures. They are trendy and surging with popularity in today's millennial world. These gardens provide a unique way to explore individual creativity and talents because they look at gardens in an entirely different fashion. This gardening opportunity can be implemented in areas of limited outdoor space. For example, miniature gardens may create a focal point in a small garden; they may be built in the top of a birdbath, inside a small wooden box, or on a table top in a high-rise atrium overlooking a city; or they may be used to create an indoor fairy garden with miniature garden structures, equipment, and character figures.

This type of gardening offers the opportunity to unleash your creativity and still apply basic design principles that will be discussed later in this text, such as unity, scale, harmony and balance, and repetition. Employing these basic components of design will create the curb appeal needed to ensure a focal point of interest for the garden viewer.

Location

When locating a spot for a fairy garden outdoors, choosing a slightly obscure location will help inspire the imagination and create more interest. For example, locating a fairy garden just as one turns a corner to accentuate the element of surprise can be an effective approach.

In the past few years, the market demand for fairy gardens has greatly increased. Many garden centers are designating special sections to plants and other products used to create fairy gardens. This effort is to appeal to those individuals who have the opportunity to explore gardening on a smaller scale but yet still want to experience the world of mother nature.

Container Selection and Design for Miniature or Fairy Gardens

When designing and building a miniature garden or fairy garden, choose a garden container that has a low profile. The container should be no more than 6 inches tall.

In addition to the container, your design might include the following features:

- Stones for pathways. Smaller scale stones work best. These stones can be found at aquarium supply stores. Pea gravel also works well and can be purchased at most garden centers.
- Miniature accessories. With the popularity of fairy/miniature gardens, many garden centers have a wide variety of miniature furniture, houses, gazing balls, and much more.

Plant Selection

Many garden centers offer miniature plants, grown specifically for fairy gardens, miniature gardens, and terrariums. Some examples of plants to use are thyme, sedums, cacti, Japanese maple seedlings, creeping jenny, miniature *aureus*, *Adiantum*, *Agave*, and *Adonis*. These plants are perfect in scale for fairy/miniature gardens.

Selection of Planting Medium

The media may consist of two-thirds good-quality garden soil mix and one-third pine bark fines, or a professional prepared potting soil mix. The professional potting mixtures will sometimes use other media drainage material such as perlite. In that case, it would be best to cover the media with soil to conceal the perlite. Soil, mulch, or peat moss work well.

Watering

When building an indoor fairy/miniature garden, make sure the plants you use require the same amount of water and sunlight. Examples of different miniature gardens are illustrated in Figures 21–5 through 21–8.

Terrariums

The original **terrarium**, a container (usually covered) in which plants are grown, was developed by Dr. Nathaniel Ward, a nineteenth century

Courtesy of Patti Kuhlmand, www.wholesalefairygardens.com

▲ **Figure 21–5** Miniature garden design.

▲ **Figure 21–6** Miniature/fairy garden – Military.

▲ **Figure 21–7** Miniature/fairy garden – Wizard of Oz.

▲ **Figure 21–8** Miniature/fairy garden in a cart.

English surgeon. Quite by accident, in an experiment unrelated to horticulture, Dr. Ward made his terrarium discovery. He had been attempting to grow bog ferns but found that the polluted air from the London factories near his home inhibited their growth. During the same period, Dr. Ward investigated many aspects of nature, including the life cycle of butterflies. As a part of this work, he placed the pupa (cocoon) of a butterfly in a glass jar to keep it alive, along with some soil from the same area. After several weeks, he noticed a seed growing inside. He found that when the seedling lost moisture from the leaves (transpiration), it collected on the walls of the terrarium. As the water formed and moved down the sides of the container, it was reabsorbed by the plant and the soil. This established a rain cycle for the plant inside the container, making it an ideal place to grow healthy plants with a minimum of attention. Despite the rain cycle that developed, Dr. Ward found it necessary to occasionally add a few drops of water to prevent the plant drying out. After several weeks, the plant continued to grow without any particular attention.

Dr. Ward went on to experiment with growing plants in containers of various sizes. As a result of this work, the *Wardian case* was developed. This enclosed case containing media and moisture in which plants could be started and grown evolved into the modern-day terrarium. It was originally used to transport sensitive tropical plants that needed protection from sea air and the rough weather of sea voyages. Over the past 150 years, the case has continued to be used for both growing and transporting plants. Today, many nurserymen, horticulturalists, and plant enthusiasts are using the case for starting and displaying plants.

Container Selection for Terrariums

Various containers are suitable for the construction of terrariums (Figure 21–9). When choosing a container, several things should be considered. The terrarium should harmonize with, as well as accent, the furnishings and the materials already present in the surroundings. For example, a modernistic terrarium would not be a good choice for a room furnished with antiques. Terrarium size should also be chosen in relation to the size of the area in which it is to be located. A terrarium that is too large or too small is less effective than one that is in proportion with the rest of the room.

COLOR OF GLASS The chosen container must allow sufficient light to reach the plants so that photosynthesis can occur. Dark tinted glass should be avoided because it filters out too much sunlight; plants are unable to grow in this environment.

SIZE OF CONTAINER OPENING The size of the opening in the container limits the type of plants that can be used. For example, if a 1-gallon bottle is selected, small plants that fit through the bottle's narrow neck must be chosen. A wide variety of plants is possible when a container has a large opening (Figure 21–10).

Check to be sure the container is free from cracks or scratches. It should be clean and dry. Water spots on the inside and outside should be removed so that the finished terrarium appears clean and attractive. If a closed container is chosen, it must be water tight. This helps maintain the necessary humidity in a terrarium to allow the best growth of plants.

Selection of Planting Medium

Terrarium containers require special preparation prior to planting. Several layers of material

▲ Figure 21–9 Various terrarium containers.

▲ Figure 21–10 A wide variety of plants are possible in the design of your terrarium when a container with a large opening is used.

are placed in the terrarium in preparation for the planting. These include a layer of drainage material, sphagnum moss, charcoal, and media mix.

DRAINAGE MATERIAL Proper drainage is important to allow the movement of excess water through the root zone area. It also helps to maintain the proper balance between air and water in the container. Excess water in the root zone of the terrarium makes it impossible for plants to absorb oxygen from the air in the soil mixture. The average soil mixture contains about 25 percent air, 25 percent water, 45 percent mineral matter, and 5 percent organic matter. If the plants in the terrarium are not provided with the best possible drainage in the root zone, they will turn yellow and eventually die from excess water and lack of oxygen.

Drainage in terrariums can be provided by crushed pea gravel, washed river gravel, or pieces of broken clay pots. These materials are available at florists, nurseries, and garden centers. The material selected is the first item to be placed in the container and spread directly on the bottom of the terrarium.

SPHAGNUM MOSS AND CHARCOAL **Sphagnum moss**, composed of sterile leaf and stem tissues from acid bog plants, is spread in a layer over the drainage material in the container. The main purpose of the moss layer is to prevent the soil mixture from seeping into the drainage layer. A layer of **activated charcoal** is spread over this mulch to absorb any odors in the terrarium. These odors usually develop from bacteria that results from the breakdown of the soil particles in the terrarium.

SOIL MIXTURE A good soil mixture is placed over the charcoal. This is important because a wide variety of plant materials are usually used in terrariums. A good general potting mixture consists of two parts loamy soil, one part organic matter (peat moss), and one part horticultural grade perlite. If this mixture is used, it must first be sterilized to kill all weed seeds and harmful disease organisms that might be present. To sterilize the soil, spread a layer of the mixture in a shallow pan, place it in an oven, and bake at 180°F for 30 minutes. Commercial media also work very well in terrariums. If only a small amount of media is required for a terrarium, it is most practical to purchase commercial terrarium soil from a florist, nursery, or garden center. Fertilizer should not be added to closed terrariums. When the fertilizer breaks down, it releases ammonia that causes the plants to burn and eventually die.

Selecting Plants for the Terrarium

The process of selecting plants for a terrarium can be a very interesting and enjoyable experience. Usually, the first consideration in plant selection is the appearance of the plant and how well it will fit into the overall design of the terrarium. One very important plant characteristic is leaf color. Some plants have solid foliage colors, while others are variegated (have more than one color). For example, a leaf may have a dark green center and a white outer edge. A combination of solid colored and variegated plants creates an attractive contrast in a terrarium.

The texture, shape, and finish of the leaves and the shape and size of the plant are also important factors. Plants selected for terrariums are usually small, relatively inexpensive, and easy to work with when planting. Small, newly rooted cuttings or small potted plants produce the best results. However, larger plants work well if a large container with a good-sized opening is selected.

Terrariums vary in their environmental requirements, depending on the type of plants that are used. Because the entire terrarium provides the same environment, it is wise to choose plants that have similar basic needs in terms of light, soil, and moisture. Terrariums are often constructed by selecting one of the groups described next and using a variety of plants from that group.

The following section describes the types of plants that are most commonly used in terrarium design.

CACTI, AIR PLANTS, AND SUCCULENTS Cacti, air plants, and succulents require natural desert conditions and full sun. They do not require a great deal of humidity, so they work well in open containers. During the spring and summer, cacti and succulents are watered to promote active growth. They enter a period of rest in winter, during which watering is reduced. The media previously recommended (two parts loamy soil, one part organic matter, and one part horticultural perlite) with the addition of one part coarse sand is recommended for these plants.

WOODLAND PLANTS Woodland plants are native to forests or woods. When grown away from their natural environment, they require similar conditions to those present in a typical woodland setting. The woodland terrarium requires dense shade, soil with a high organic content, and a good supply of moisture. The soil mix previously mentioned (one part loamy soil, one part organic matter, and one part horticultural grade perlite) usually works well in woodland terrariums.

Familiar woodland plants include ferns, mosses, yellow root, ground ivy, jack-in- the-pulpit, lichens, oxalis, partridge berry, trailing arbutus, and violets (Johnny jump-ups). These plants grow well in enclosed terrariums because of the humidity created by the accumulation of moisture in the closed container. Generally, woodland terrariums are easy to manage and keep clean.

TROPICAL PLANTS Tropical plants respond favorably to a more humid type of terrarium than either cacti or woodland plants because their natural environment is a moist, tropical region. Light requirements for this group vary with different plants. It is advisable to use those plants that require low light. Although there is a great variety of large tropical plants, the smaller varieties are best for terrariums because they are easier to work with and less expensive.

Planting the Terrarium

Planting a terrarium is not difficult, but it does require a few basic steps as described next.

PLANTING EQUIPMENT A few simple tools can be used to simplify planting terrariums. A funnel is used to work the soil mixture into the container so that the soil is evenly distributed in the terrarium. When performing this operation, it is best to avoid getting soil on the sides of the container because it is difficult to remove after planting. A wooden dowel is used to form planting holes in the soil mixture. One end is sharpened for digging while the other end is used to tamp the soil around the plant roots. When planting in a narrow-necked bottle, a bamboo stick slit about 2 inches on one end can be used to hold plants. The plant is slipped into the slit in the stick to hold it securely while planting. A bent spoon is helpful in adding soil around the plants. After all the plants have been placed inside the terrarium, use a mister to spray the foliage.

GREEN TIP

Plants give off moisture (**transpiration** process) from small openings on the underside of the leaves known as **stomata**. When plants are in a closed terrarium, the moisture collects on the sides of the container and drops back into the soil so that the plants may use it again. If the moisture is concentrated in the terrarium, the container may fog up. In this case, remove the top and allow the moisture to escape. This establishes the balance of air and moisture in the terrarium. After the fogging has cleared, replace and seal the container top.

If a terrarium lacks moisture, the leaves of the plants may droop, and the soil may appear lighter than normal. When this occurs, spray the terrarium with a mist of water until beads of water form on the leaves of the plants. Seal the terrarium and allow the rain cycle to become reestablished. The plants can then continue their natural growth process.

CARING FOR THE TERRARIUM After the terrarium has been planted, it should be checked periodically. It is especially important to see that it continues to have the proper light, moisture, and temperature.

Disorder Symptoms

MUSTY ODOR This can be corrected by the addition of activated charcoal to the terrarium. The charcoal absorbs the bacterial odor and aids in drainage.

CHLORIC (YELLOWISH) AND STUNTED GROWTH OF PLANTS The **chloric** condition could result from improper drainage, improper ventilation, or improper soil mixture.

DECAYING FOLIAGE This may be the result of several things. Improper ventilation, one possible cause, may be corrected by removing the sealed lid to ventilate the terrarium. High nitrogen fertilizers may be present in the soil mixture, giving off ammonia and thus burning the plant leaves. In this case, it is best to replace the soil mixture.

BROWN FOLIAGE This indicates that the terrarium has not received enough water.

Pruning

In some cases, the growth of plants in the terrarium is so great that they may become too big for the container. This may be due to the plant type or very favorable cultural conditions. Whatever the reason, the plants must be pruned to remove excess foliage. Pruning plants causes them to branch out and become fuller.

Use a single-edge razor blade for pruning. If the terrarium container is a long-necked bottle, insert a single-edge razor blade into a bamboo stake.

> **CAUTION**
> Remember to remove any material that is pruned from the terrarium to prevent it from decaying inside the container.

Insect Pests

Insect pests are controlled most efficiently by planting the terrarium with clean, insect-free plants. At times, however, pests may be present in the soil and not noticed when the terrarium is planted. Some insecticides may be used for insect control in terrariums; consult the local extension service for recommendations. Specific insects that plague terrarium plants include the whitefly, mite, aphid, mealybug, scale, and ant.

WHITEFLIES Whiteflies have very small, white, wedge-shaped wings. When the plant is disturbed, these insects fly around the plant's leaves where they can be seen. Whiteflies feed off the underside of the leaf.
Control: Spray with Organic Neem Oil

MITES These are very small insects that are difficult to see. If you suspect that a plant has mites, take a sheet of white paper, insert it under the leaf, and tap the leaf lightly. If mites are on the plant, they will fall onto the paper. Check the paper for small moving insects. Draw a finger across the insect and the paper; if mites are present, there will be a red mark on the paper.
Control: Apply a systemic insecticide to the soil. Systemic insecticides used to control sucking insects are absorbed into the plant's system through its leaves and roots. As the insect sucks the plant juices containing the insecticide, the body system of the insect is affected, thus causing death.

MEALYBUGS Mealybugs are small, flat, white insects that resemble a small spot of cotton on the plant. Mealybugs suck the plant juices until the plant dies.
Control: Swab the areas of the leaves where the bugs are located with alcohol.

SCALE INSECTS These pests harm plants by sucking their juices. Scale insects are small, gray to brown in color, and flat, with a soft or hard shell. They can be found sticking to the stems of plants. After scale attacks, the plant becomes yellow, then turns brown, and finally dies.
Control: Spray with Organic Neem Oil

ANTS Tiny black ants are drawn to some plants by the presence of honeydew on their leaves. Honeydew is sap that is secreted from the leaves of plants after an attack by aphids, scales, or sometimes a fungus. These ants usually attack the plant's root system.
Control: Spray with Organic Neem Oil

Watering

If moisture beads are present on the sides of the terrarium container, there is a good chance that the rain cycle has been established. Be cautious of overwatering. If the terrarium is overwatered, the plants may die from poor drainage and a lack of air circulation in the soil. In the case of overwatering, it is best to open the top of the terrarium and allow it to dry out somewhat. When enough water has evaporated, reseal the terrarium to reestablish the rain cycle.

Temperature

The ideal temperature of terrariums depends to some extent upon the type of plants they contain, but a temperature of 65°F to 75°F is suitable in most cases. Remember that the temperature is greatly influenced by the location of the terrarium.

For example, if it is located in or very close to a window, it will experience greater changes in temperature than a terrarium that receives indirect light.

Light

Different types of plants require different amounts of light. Remember that plants in terrariums located near windows grow toward the light source (phototropism).

ARTIFICIAL LIGHT Terrariums are excellent decorative additions to the home, office, or school. The sunlight in these areas may be very limited, however, making artificial light necessary. Special lights on the market today are especially good for this purpose. A fluorescent light is a better choice than an incandescent light (standard light bulb).

Various manufacturers of fluorescent lights publish information concerning the placement of the lights for best results in the terrarium. On average, it is recommended that they be located about 15 to 18 inches away from the terrarium. The light should be provided from 12 to 14 hours per day. Flowering plants require more light than others.

Fertilizers

If plants are grown in a good media, additional fertilizer is not necessary. Fertilizer tends to stimulate excessive growth of terrarium plants causing the terrarium to appear unkempt and cramped.

If the plants must be fertilized, use an organic fertilizer such as bone meal. Simply sprinkle a small amount of fertilizer in the terrarium. Organic fertilizers release slowly and do not burn plants, as inorganic fertilizers tend to do.

Preventive Disease Control

When the terrarium has been completed, preventive measures should be taken to control disease organisms that may be present in the terrarium. One chemical that is commonly used is Benlate. Follow the manufacturer's directions for its use. Use a spray bottle to ensure even distribution of the chemical on the foliage and soil.

PROCEDURE
Planting a Terrarium

The following steps comprise the planting process in the construction of terrariums:

1. Push the wooden dowel into the soil mixture to form holes in which to place the plants.
2. With the plant held by the split in the bamboo stake, transplant the plant into the bottle. Gently bending the leaves does not harm the plant.
3. Position the plant, and then carefully remove the plant from the bamboo stake with the dowel. Lightly tamp the soil around the plant roots with the dowel to help hold the plant straight and firm.
4. Add additional plants until the terrarium design is complete.
5. Water the completed terrarium with a mister. Spray the foliage until water droplets form on the leaves.
6. Finally, seal the terrarium and enjoy watching it grow.

SAFETY FIRST

Wear PPE to ensure no contamination from the media mix, chemicals, and the fertilizer being used in the construction of the terrariums, fairy gardens, and dish gardens. OSHA's regulations ensure that employers and employees know about the work hazards and how to protect themselves to prevent injuries in the workplace. For more information on occupational health and safety in the workplace, please visit the OSHA site at www.OSHA.gov.

Summary Dish gardens, fairy gardens, and terrariums are attractive living gifts of mostly small- to medium-size foliage plants, broadleaf evergreens, conifers, flowers, and colored leaves to accent your designs. The design of the dish garden, fairy garden, and/or terrarium is a combination of the selection of plant material and arrangement in an attractive container. It is important to select plant material for the dish gardens, fairy gardens, and/or terrariums with similar environmental requirements of light, water, temperature, and growth maturity. The longevity of dish gardens, fairy gardens, and/or terrariums is primarily related to the aftercare provided throughout the life span of the plants within the planter.

Review Questions

1. How are dish gardens, miniature gardens, and terrariums similar? How are they different?
2. What are the advantages of planting a miniature garden instead of a full-size garden?
3. What considerations are important when selecting plants for a terrarium?
4. Imagine you are assisting a client who is an avid gardener and must downsize from a large home with a full-size perennial garden to a small city apartment. How would you help your client transition to miniature gardening?

Student Activities

1. Ask a local florist to visit the class and demonstrate the various techniques used in dish garden, fairy garden, terrarium construction and planting.
2. Take a walking field trip to a wooded area, and collect native plant materials for dish gardens, fairy gardens, and terrarium plantings.
3. Propagate plants for the terrarium. For woodland plants, students might visit a nearby forest or woods and collect seeds, sow them, and select the terrarium plants from the results.
4. As a class project, design and plant a dish garden, fairy garden, or a terrarium. Maintain the plants at the proper temperature and moisture level.
5. As the plantings in the three gardens grow, observe and record the results. Care for the plants, employing the techniques found in the text. Prune when necessary.
6. Using a smartphone, search online to glean design ideas for a dish garden, fairy garden, and terrarium.
7. Visit your local garden center and ask the manager to demonstrate how to make a fairy garden or terrarium for the class. While at the garden center, purchase supplies to design a fairy garden and terrarium for home.

Suggested Supervised Agricultural Experience (SAE) Activities

1. Grow and sell plants through a private or school greenhouse. This could qualify as either a placement or entrepreneurship SAE.
2. Design and create dish gardens and retail them through a local business or farmers market to qualify for an entrepreneurship SAE.

Additional Resources

https://aggie-horticulture.tamu.edu/syllabi/302/new/topic/dish.htm

www.ext.colostate.edu/pubs/garden/07238.html

http://extension.oregonstate.edu/nep/edmaterials/Gardening/infosheet-containerplants.pdf

UNIT 22

The Art of Bonsai

Objective

To construct and maintain an attractive, healthy bonsai planting.

Key Terms

bonsai
pottery
potting medium
repotting
root-bound
unglazed
wiring

Competencies to Develop

After studying this unit, you should be able to:

- explain the purpose and general technique of bonsai.
- list the steps in the construction of a bonsai planting, orally or in writing, followed by a demonstration of the technique.
- describe the aftercare of bonsai plantings.
- explain how to repot bonsai plantings.

Materials

- bonsai containers
- 9- to 20-gauge copper or painted aluminum wire (for shaping)
- screen wire (to cover drain holes)
- potting medium and gravel
- chopsticks or similar sharp sticks
- pruners (some should be for fine work)
- potted plants (to be trained to bonsai)

The word **bonsai** can be divided into two parts, *bon* means "tray" in Chinese and *sai* means "little tree." According to historical documents, bonsai started in China 1,200 to 1,500 years ago. The first exhibition of a bonsai collection outside of China or Japan was in 1828 at the World's Fair in Paris and later in London in 1909. Bonsai is the art of dwarfing plants by growing them in shallow pots and trays and by judicious trimming of roots and tops. The end result is a plant that resembles a large tree in miniature. By applying the techniques of bonsai, the horticulturist is able to produce an 18-inch tree that resembles an 80-foot tree in all aspects except size. Part of bonsai tradition is to make these plants appear old and rugged or graceful and wispy, just as they appear in nature after the passing of many generations (Figure 22–1). Bonsai designs may be simple or complex, ranging from a simple one-tree planting to a miniature forest (Figure 22–2).

▲ **Figure 22–2** The Goshin bonsai is considered the most famous American Bonsai. Trained since 1953, Goshin is a forest planting and is considered the crown jewel of the Museums of the North American Collection. "Goshin" means "protector of spirit" and was donated by John Y. Naka to the National Arboretum, Washington, DC, in 1984.

GREEN TIP Bonsai are excellent for those who have small areas to grow plants; they take up little space but are reflective of a woodsy environment.

▲ **Figure 22–1** Japanese white pine (*Pinus parviflora*), 150 years old. From the collection of Late His Majesty Hassan II, King of Morocco. Now at the National Arboretum.

Planting the Bonsai

Choosing the Plants

Plants used in bonsai may be evergreen or deciduous. The plants that are selected should have small leaves to help them appear more as reduced-size models of larger trees or plants. Plants selected should also be able to grow in crowded conditions. Flowers, berries, and fruit can add to the attractive quality of bonsai specimens. The following species are some of the best choices for bonsai plantings.

Evergreens

Japanese white pine (*Pinus parviflora*)
Creeping juniper (*Juniperus procumbens* 'Nana')
Hinoki cypress (*Chamaecyparis obtusa*)

Azaleas (Satsuki and Kurume)
Japanese yew

Deciduous plants
Japanese maple (*Acer palmatum*)
American hornbeam (*Carpinus caroliniana*)
Japanese flowering apricot (*Prunus mume*)
Japanese flowering quince (*Chaenomeles lagenaria*)

Source of Plants

There are two sources of bonsai: naturally dwarfed and artificially dwarfed plants.

NATURALLY DWARFED PLANTS It is sometimes possible to find old plants growing in the wild that have been twisted and dwarfed by cold or dry conditions. Collecting such plants is discouraged due to depletion of wild plants. It is best to grow seedlings for bonsai use.

ARTIFICIALLY DWARFED PLANTS The best source of plants is a nursery, where potted plants that have the potential to be pruned to the desired shape may be purchased. Plants with small leaves or needles that have interesting trunk lines should be chosen. They should also be selected according to their ability to grow well in the area (Figure 22–3).

Plants for bonsai may also be grown from seed or cuttings. These plants are trained from the very beginning by restricting their growth and by pruning to obtain the desired shape. The first step is selection of the plant. Figure 22–4 shows a container-grown plant selected for a bonsai.

▲ **Figure 22–4** An evergreen plant (*Chamaecyparis obtusa* 'Nana Gracilis') chosen for the bonsai created in this unit. When selecting container-grown plants for bonsai, the desired shape must be visualized and one branch selected for the main trunk.

▲ **Figure 22–3** A 250-year-old Sargent juniper trained to resemble an old, twisted tree. The juniper is 29 inches tall.

Containers

The bonsai container should

- have one or more drain holes.
- be shallow.
- have a length two-thirds the height of the plant placed in it.
- not detract from the beauty of the plant placed in it. Subdued earthy colors are best; browns, greens, and grays are most natural.

▲ **Figure 22–5** A shallow, unglazed container with screen wire covering the drainage holes and aluminum wire up through the holes to tie down and support the root ball.

- have a simple form (round, square, or rectangular).
- be made of **pottery** and be **unglazed** on the inside.

Figure 22–5 illustrates one type of bonsai container.

Tools

Various tools used to cut and prune plants in preparation for bonsai are shown in Figure 22–6. All of these tools were used to prepare the bonsai shown in this unit.

TRY THIS Using the Internet, access the United States National Arboretum (USNA) website. Take the National Bonsai and Penjing Museum Virtual Tour to view some impressive examples of bonsai.

Dwarfing Plants

Determining Plant Shape

The first step is to visualize the bonsai design in its final form. Will it have one trunk or several? Will there be more than one plant in the container?

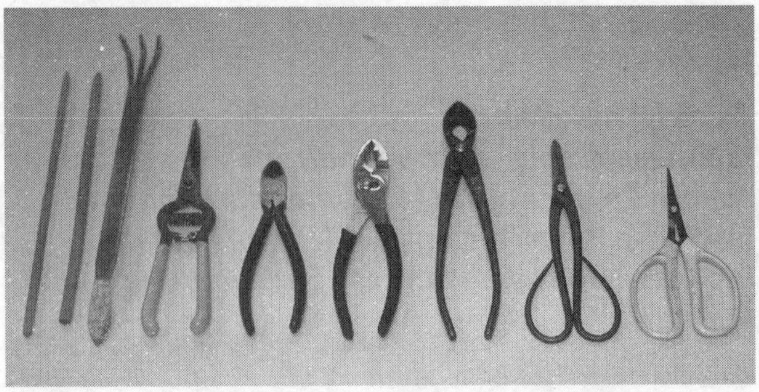

▲ **Figure 22–6** Tools used in preparation of plants for bonsai.

Is the plant to be set straight or slanted in the container? Will roots show above the soil level or be covered? These are some of the questions that must be answered before pruning. When pruning, remember that the trunk, the rootage, apex, and branches of the plant are the focal points of the design.

Pruning the top of the tree is essential in developing the rugged but miniature shape that is characteristic of bonsai. Any active shoots must be shortened and undesirable buds or shoots removed. There should be some space between limbs or side branches to give the effect of a tall tree (Figures 22–7 and 22–8).

Initial Pruning

Remove all of the unwanted branches. This step must be done very carefully to ensure that enough branches remain to form the desired

▲ **Figure 22–7** The first cut in starting the bonsai is to shorten the plant.

▲ **Figure 22–8** Other branches to be saved are selected, and all other branches are removed.

shape for the entire plant. Figure 22–7 shows the same plant as the one in Figure 22–4 after the first pruning is done.

> **GREEN TIP** Use recycled steel or copper wire manufactured from old cars, refrigerators, and washing machines.

Wiring

The remaining branches are wired with copper or aluminum wire (Figure 22–9). Branches are wired to force growth in a direction other than the direction in which they naturally grow. Copper or painted aluminum wire is desirable because it does not rust and is not highly visible. Heavy wire is used for the larger branches and trunk and smaller wire for small branches. The size of the wire used varies from 9 to 20 gauge. The wire must be pliable enough to bend and yet rigid enough to hold the plants in shape. Cut the wire one-third longer than the combined length of the trunk and branch to be wired. Stick one end of the wire into the soil mix to hold it firmly and wire the trunk. When **wiring** branches, use two wires starting at different levels on the trunk. The two wires overlap and stay in place better. Wrap the branches with spirals about 1 inch apart, being careful not to tear the bark on the twigs.

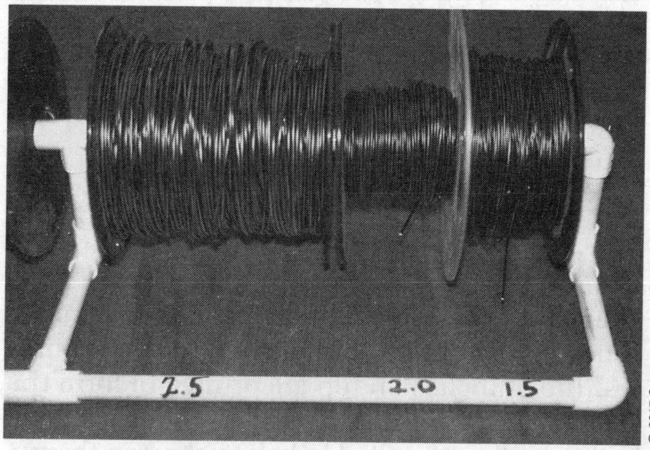

▲ **Figure 22–9** Painted aluminum wire to twist around and hold branches in place.

Avoid wrapping plant foliage under the wire because this foliage may die and disfigure the plant. When the wrapping is finished, cut the end of the wire off close to the branch so that the ends do not show.

When two branches are to be pushed apart, one end of the wire is used on each branch. The first branch is wired and pushed down, and then the second branch is wired (Figure 22–10). Notice that the two pieces of wire are twisted in opposite directions. After the wire is twisted on both branches, the branches are pushed in the desired direction. The wire holds them in place. The ends of the wire are cut off close to the branches so that pieces of wire do not protrude from the plant. The wire should not be left on long enough to girdle the branches as they grow.

▲ **Figure 22–10** Selected branches wired and twisted into place to form the shape of the new bonsai.

CAUTION
Be careful not to twist the wire too far and break the branches or to twist the branches from the trunk.

Additional Pruning

The top of the plant is now pruned to remove some of the small branches and shorten others. The branches on top are pruned to form the desired shape. The needles of the bonsai shown in this unit were pulled in half to shorten them.

Root Pruning

After the top is pruned, attention is focused on the roots. Dig the soil ball apart by gently pulling down with a chopstick or other tool through roots that make up the root ball (Figure 22–11). Continue until most of the soil is removed, and the roots are exposed. Expect many roots to be torn off in this process. The large cutters are used to cut off the center tap root (Figure 22–12). Many of the roots must be cut back so that the new plant fits in the bonsai container (Figure 22–13).

Placing the Plant in the Container

The plant is now ready to be placed in the bonsai container. The container is prepared by placing

▲ **Figure 22–11** The root ball is dug down from the top and in from the sides to make it shallower and smaller in diameter.

▲ **Figure 22–12** The bottom of the root ball is cut off.

▲ **Figure 22–13** Root ball cut to final size to fit the container. Note that many small fibrous roots remain.

a screen wire over the drainage holes in the bottom. A thin layer of gravel is then spread over the screen wire to ensure proper water drainage (Figure 22–14). The plant is set on top of the gravel and pushed down so that the root level is higher than the edge of the container. It may be necessary to cut off the bottom of the root ball, but save enough fibrous roots. Notice that the plant is placed in the center of the round container. Bonsai are generally not placed in the exact center of their containers except for hexagonal, round, or square pots. Before adding soil, tie the plant to the container by guiding a piece of copper or aluminum wire up through a drain hole and over part of the roots or main stem.

▲ **Figure 22–14** Small gravel in the bottom of the pot, ready to plant.

Adding Potting Medium

A **potting medium** with good drainage is now placed on top of the root system and gently worked through the roots with a pointed stick to fill any air pockets (Figure 22–15). The potting medium should be of such structure that it is well drained and yet holds moisture. The Japanese use fired silt clay called Akadama. The mix used by Dr. Daniel Chiplis, Assistant Curator at the U.S. National Arboretum's National Bonsai and Penjing Museum, to plant the bonsai shown in this unit is as follows:

- 1 part rock (which is 2 parts crushed granite and 1 part lava stone)
- 1 part organic (pine or fir bark and a nursery mix of decomposed peat moss, bark, sand, and vermiculite or perlite)
- 1 part fired clay particles (turface)

The mix is screened to take out all fine particles that might plug up pore spaces and reduce drainage.

The medium should cover all of the roots except for large roots near the trunk, which are sometimes left exposed for effect. Exposed roots cause the plant to look like an older tree whose large roots have gradually been exposed.

Fill the container to a point within 1/2 inch of the top of the edge and form a gradual mound toward the center of the container to cover the plant roots. Work the medium down among the roots so that there are no large air pockets.

▲ **Figure 22–15** The growing medium is worked down around the roots with a small stick. No large air spaces should remain around the roots.

Watering

Soak the newly potted plant in water until the roots and medium are wet. Figure 22–16 shows

▲ **Figure 22–16** The finished bonsai. Note the taper of the trunk and the spacing of branches causing the miniature plant, about 12 inches tall, to closely resemble an old tree. Moss placed on top of the soil breaks up the smooth surface, adding appeal to the finished planting.

the finished product. The new plants should be protected from wind and sun for at least 2 weeks until they recover from the severe pruning.

Care and Management of Bonsai

Aftercare of the plant itself consists of pruning the roots every 1 to 3 years so that they fit the container and shaping the top to obtain the desired effect (Figure 22–17).

Watering

Keep the soil moist; the bonsai has a small root system and must not be allowed to dry out. In summer, this means watering every day. Add water until it runs through the bottom of the container. Do not keep the roots wet. Indoor bonsai may need watering only two or three times a week.

Source: US National Arboretum (http://usna.usda.gov)

▲ **Figure 22–17** The *Bougainvillea glabra*, Bougainvillea. Donated by Harold M. Harvey to the National Arboretum in 1990, this bonsai has been in training since 1985.

Location

During summer months, keep the bonsai outdoors in light shade or sun, depending on the type of plant. Protect it in a cold frame or greenhouse in winter months to prevent the roots from freezing to temperatures below 23°F (−5°C) and winds from drying the plant out. Storage in a cool basement in winter is adequate. The plant must be watered and cared for while it is inside. A bonsai should not be kept inside except when protection from heat or cold is needed.

PROCEDURE
Repotting

1. Remove the plant from the container.
2. Straighten out the roots if curled in a ball.
3. Remove one-third of the soil from the outside of the root ball, being careful not to break up the rest of the root ball. This process is much easier if the soil is dry.
4. Trim the roots back to the soil left in the root ball with sharp scissors or shears. (One-third of the root length is cut off.)
5. Repot the tree, using the same procedure as was used when the bonsai was originally potted. The new root ball should have 2 inches of space between it and the outside edge of the pot. New soil is worked firmly into this area with a small stick. The 2 inches of new soil provide room for new active feeding roots to develop and keep the root system active. Soil is packed tightly against all roots so that no air space is left. Take the time to do this job thoroughly. The soil added should be dry so that it sifts down around the roots.
6. If support is needed, tie the plant to the container by guiding a piece of copper or aluminum wire, threaded through a small plastic tube where it contacts the root system. The wire will not show because it is covered with soil. Do not wrap the wire completely around the roots, or they will be girdled.

Fertilizing

Organic materials such as cottonseed meal, well-rotted animal or vegetable manure, or the newer liquid fertilizers work well. Sprinkle fertilizer on the surface (2 teaspoons per plant of organic fertilizer every 6 weeks from midspring to August 1). Liquid fertilizers, at one-quarter strength, may be applied with watering. Apply them about three times a year, beginning when active growth starts, and repeat every 4 weeks until August 1.

Repotting and Training Procedure

As the new plant grows, additional pruning is required. **Repotting** should be done whenever the plant becomes **root-bound** or roots have curled around the root ball. It is very important to repot bonsai to keep the soil fresh and the roots actively growing. Repotting should generally be done every 1 to 2 years for deciduous trees and every 2 to 3 years for evergreens. Fast growers need repotting sooner than slower growers. The best time to repot is in early spring, just before new growth begins. When bonsai plants are repotted, root pruning is the most important consideration. Roots must be pruned to bring about the growth of new roots. Many small feeding roots will emerge from the cut ends of large roots.

Bonsai can be both a rewarding hobby and a business. Anyone with the patience to work with the plant and wait for its beauty and character to develop can enjoy growing bonsai. It is not necessary to wait 50 to 100 years for an attractive bonsai. The small plant used for demonstration purposes in this unit will develop shape, character, and beauty in just a few years.

SAFETY FIRST

Use extra caution when using sharp tools to design a bonsai. For more information on occupational health and safety in the workplace, please visit the OSHA site at www.OSHA.gov.

Summary

Bonsai can be a good hobby for gardeners who have small growing areas. The plants are grown in trays or other containers and kept pruned. They require a small amount of space. In this small space, the gardener can enjoy what appears to be a large tree on a small scale. If properly cared for, these plants will last for generations.

Review Questions

1. Describe the eight steps that are required in the construction of a bonsai planting.
2. How does bonsai planting differ from miniature or dish gardening?
3. What characteristics make certain plants more appropriate for bonsai planting than others?

Student Activities

1. Select a container and suitable plant and prepare a bonsai planting following the steps in the text. This may be done in groups of two to four students or individually.
2. Search online for bonsai artist profiles. Select one that interests you, and write a short biography of their career.

Suggested Supervised Agricultural Experience (SAE) Activities

1. Apprentice with a bonsai artist to create a bonsai design that can be re-created and retailed as an entrepreneurship SAE.
2. Conduct your own experiment growing small trees in the bonsai technique for a research SAE.

Additional Resources

www.hav.com/bonsai

www.arboretum.harvard.edu/bonsai/care.html

https://pubs.ext.vt.edu/426/426-601/426-601.html

www.ndsu.edu/pubweb/chiwonlee/plsc211/student%20papers/articles02/avoorhees/bonsai.txt

UNIT 23

The Interior Landscape: Houseplants and Plantscaping

Objective

To grow and maintain healthy interior landscape plants.

Key Terms

atrium
interiorscapes
phalaenopsis
plantscaping
slow-release fertilizer
soluble fertilizer

Competencies to Develop

After studying this unit, you should be able to:

- list the four major concerns in caring for interior plants.
- select a medium to use for growing various interior plants.
- describe the best cultural conditions (water, light, and soil requirements) for 10 selected interior plants of your choice.
- list the four methods of watering interior plants.

Materials

- 15 interior plants from the plant list in the text
- one bag of soluble fertilizer
- one bag of slow-release fertilizer
- planting medium and pots
- smartphone apps:
  - gro getter
  - LikeThat Garden
  - Greenhouse Management

Foliage plants have been a special part of people's lives and environments throughout time. Of course, plants that bear food and materials for making articles have always been essential to life, but the plant that simply adds beauty to its surroundings serves a special purpose. Since the initiation of the environmental protection and "green survival" movements, the popularity of interior plants has grown rapidly. In 2000, according to the USDA Horticulture Crops Summary, $574 million wholesale worth of interior plants were sold.

© rodho/Shutterstock.com

▲ **Figure 23–2** Interior plantscaping creates attractive natural barriers to direct the flow of pedestrian traffic. It also provides an area where shoppers can relax in a peaceful environment.

This increased interest in live indoor plants is reflected in both public places and private homes. Shopping malls, office buildings, and motel complexes are now using live plants to landscape indoor areas. This method of interior plant use is called **plantscaping**. The ancient practice of building homes around an **atrium** (indoor garden accessible from all main rooms of a house) is receiving renewed attention. People who live in crowded urban areas often find the idea of a private garden of interior plants particularly appealing. Interior plants can be used solely for the interest they create or to serve a specific function. Plants act functionally when used as room dividers and screens and to accent pieces of furniture (Figures 23–1 and 23–2).

© JJ Studio/Shutterstock.com

▲ **Figure 23–1** Interior plantscaping has expanded throughout the country. It enhances shopping malls and office complexes by providing a natural garden atmosphere.

The interior plant business has developed new cultivars with improved leaf color, twisted plant stems, and improved flowering quality for **interiorscapes**. The demand for more high-quality, new interior plants is increasing. Therefore, the future of this segment of the green industry is expanding. The **phalaenopsis** orchid has increased in popularity to the point of being the leading interior plantscaping plant, valued for its flowers and its ability to adapt to be grown indoors. The wholesale value of the orchids in Florida had increased to $37 million in 2000, according to the USDA in 2001.

GREEN TIP Interiorscapes help to create an environment that ties people to nature in the home, office complex, and hotels and businesses.

Caring for Interior Plants

To care for interior plants properly, the following questions must be answered for each type of plant being grown:

- What is the proper way to water the plant?
- How much light does it need?
- What is the best temperature for optimum growth?
- What type of planting medium is needed?

Watering

To be sure that houseplants are watered properly, the plant's behavior must be observed carefully. There are several signs of improper watering. Drooping leaves or leaves that do not seem as full as they should may indicate a lack of water. Too much water can cause leaves to wilt or turn yellow and finally drop off the plant. Other moisture control factors besides the amount of water given a plant are important to consider as well, such as using the proper medium and providing good drainage.

There are several methods of watering interior plants:

- Drench/let dry. Enough water is added to moisten the medium, which is then permitted to dry before watering again. In the drenching method, the desired amount of water is added all at one time.
- Drench/let dry slightly. Enough water is added to moisten the medium. It is then permitted to dry slightly so that it is damp to the touch.
- Keep medium constantly moist. Enough water is added so that the medium is damp to the touch at all times. Remove any excess water in the base of the container.

Lighting

As we landscape interiors with tropical and semitropical plants, the interior environments cause stress because of low available light. Therefore, to create a long-term plantscape, it is necessary to create appropriate artificial environments to grow certain plant cultivars. Interior plants can be grown under natural light, filtered sunlight, and incandescent, fluorescent, and high-intensity discharge lights. All of these light sources allow the plants to receive enough energy for photosynthesis (process of manufacturing food).

Light meters (small machines that measure illumination or light) can be used to determine whether the location selected for the plant is suitable. Light meters are sold at florists, garden centers, and department stores with plant shops. Light intensity requirements, which vary from plant to plant, are classified into the following groups:

- Direct sun. Plant should receive the full intensity of natural sunlight.
- Partial sun or weak sun. Plant should receive less than 50 percent of natural sunlight.
- Indirect or filtered light. Plant should receive no direct sunlight at all.

In general, the more light that is available, the greater the variety of plants that may be grown. In most cases, foliage plants require less light than flowering plants.

Artificial Lighting

The advent of the fluorescent light greatly influenced the art of growing interior plants. The fluorescent light enables the horticulturist to

- root cuttings of interior plants more easily.
- start seeds of new and different interior plants.
- grow tropical plants, cacti, and succulents with greater ease.

Temperature

Many times, interior plants are exposed to a great range of temperatures within one 24-hour day. For this reason, plants must sometimes be given special protection. For example, plants located in windows during winter months may receive full sunlight during the day. During the night, however, the temperature next to the window will be much cooler. If the temperature is below 32°F (0°C), and the plants are very close to the window, they may freeze. The temperature recommended for interior plants (50 to 85°F) is the daytime temperature; the night temperature can generally drop 10 to 15° F (10 to 30°C) below this temperature without harming the plants.

The common use of air conditioners in homes and offices can be a source of trouble for plants. Plants can be protected, however, by locating them away from drafts and direct air flows.

Humidity

Humidity is the percentage of moisture in the air, which is an important factor to consider when growing interior plants. Humidity varies with seasons; it is lower than usual in the winter; therefore, additional moisture is needed in the atmosphere. During the summer, the humidity is usually higher than in the winter.

Humidity around interior plants can be increased by

- spraying a fine mist of water on the foliage of the plant with an inexpensive plastic spray bottle or any type of mister.
- using a watertight tray filled with small pea gravel to hold the potted plants. Water is added to the tray to come to the top of the gravel and plants are set on the gravel, making certain that the pots are not actually in the water.
- using a humidifier to provide moisture in the air.
- placing the plants in the part of the house that has the highest humidity. For example, areas of highest humidity are probably the bathroom or the kitchen just above the sink.

Plant Media

The plant media for indoor plants must be loose and airy and must hold moisture and nutrients well. Many good container plant mixes are available in the market today. Some are a mixture of fine composted pine bark and sphagnum peat moss as the bulk of the medium. The pine bark provides drainage and aeration, and the peat moss adds organic matter to the plant media for improved plant nutrients and holds moisture. Pine bark may also help in root disease control. Other mixes contain sphagnum peat moss, vermiculite, and perlite. These mixes also are very satisfactory. If the percentage of perlite and vermiculite is high, soil mixtures drain well. As more peat is added, mixtures hold more moisture. For cacti and other succulents, a mixture of sphagnum peat moss, reed sedge moss, perlite, and sand provides a faster-draining mix. Some mixes contain sphagnum peat moss and vermiculite only. Many of the mixes contain limestone to adjust the pH, especially for cacti and succulents. A wetting agent is present in many of the mixes to help wet the mix faster when it is watered. When sphagnum peat moss dries out, it becomes difficult to wet without a wetting agent.

It is obvious that sphagnum peat moss and bark are the main ingredients in soilless mixes, with perlite, vermiculite, and sand used in various amounts to provide drainage and aeration.

Fertilizers

Two types of interior plant fertilizers are available in the market today: slow release and soluble. **Slow-release fertilizer** is applied to the soil mixture in the form of small beads, which are coated with plastic and dissolve over a period of time. As it comes in contact with normal soil moisture, plant nutrients are steadily released. The plant absorbs the released nutrients as needed, thus eliminating the danger of overfeeding or fertilizer burn.

Soluble fertilizer is available in liquid or solid form. Solid forms of soluble fertilizer must be dissolved in water before being applied to interior houseplants. Applying this fertilizer dry may burn the roots of the plant. Soluble fertilizers in liquid form are concentrated, requiring that water be added to the mixture before application.

CAUTION

To avoid damaging plants, always follow the manufacturer's recommendations when using any fertilizer. Some fertilizers have specific requirements prior to their use that must be followed exactly.

Containers

Many beautiful ceramic, pottery, fabric, and natural tree bark plant containers are available today. Plants grow in almost every kind of container, provided that drainage is available. However, if there are no drainage holes in the container, the soil mixture will become waterlogged. Plants that take up the oxygen in the medium cannot survive under these conditions. In addition to drainage, there must be some air spaces in the mixture so that the plant roots breathe properly. Proper air circulation is an important ingredient of the soil mixture.

Clay has long been the standard type of container for growing plants, but most greenhouse growers have now changed to other types of containers such as plastic, glass, fiberglass, and metal. All plant containers should be set in some type of saucer to catch excess drainage water. Because clay is porous, water can escape through the clay saucers to table surfaces. This problem can be eliminated by coating the saucer with plastic or paraffin wax.

Potting Plants

After the proper container and soil mixture have been selected, potting or repotting interior plants is a simple process (Figure 23–3). First, some type of drainage must be provided in the bottom of the pot. For clay pots, a crock may be used. A *crock* is a broken piece of clay pot. Place the crock so that it is curved upward over the drainage hole in the bottom of the pot. If it is placed with the curve down, it may seal the drainage hole and cause problems by not

(A) Select a pole three to five times taller than the container in which the totem is to be used.

(B) Wrap the pole with a layer of sphagnum moss 2 to 3 inches deep, and bind the moss with string.

(C) Push the pole into the soil.

(D) Wind the vine around the totem.

(E) Use hairpins to fasten the vine to the totem.

(F) Keep the moss damp; roots will grow into the moss, and the leaves will form a solid mass.

▲ **Figure 23–3** Making a totem.

allowing excess water to drain out of the pot. Stones may also be used in containers. If they are used, be sure that the stones do not completely close the drainage holes in the bottom of the pot.

Add the selected potting mixture to the container so that it covers the drainage stone or crock. Place the plant in the container, making sure to check the depth of the plant. If the plant is below the rim of the pot, adjust it so it is 1/2 inch below the top of the pot. This allows for an area at the top of the pot to hold additional water while the plant is being watered. If no space is allowed, the water will run off over the edge of the pot. Not having received much of the water added, the plant will dry out much more quickly, thereby inhibiting growth.

Repotting

Interior plants should be repotted when they become pot-bound. To determine whether the plant is pot-bound, remove the plant from the original pot. If the roots of the plant are growing around the root ball of the plant and are firm to the touch, the plant should be repotted. If the plant is pot-bound, the roots must be separated; this allows the roots to develop and make new growth. Repotting should be done before new growth starts. If the plant is flowering, wait until after flowering is completed.

GREEN TIP Cinnamon is a natural fungicide. Use cinnamon when repotting plants to prevent root rot in broken or cut roots. Simply dust the broken or cut surfaces with cinnamon prior to potting.

Interior Plantscape Planters

As the interior plantscaping business continues to grow, more and more businesses are using interior plants for the aesthetic improvement of their offices. Various types of planters are used today. The moveable, decorative planter is one type that includes floor planters, hanging baskets, and file-top planters. Another type of planter is the fixed-bed planter. This one is constructed in place and has a drainage system that is linked to the office plumbing system.

When using either the moveable or fixed-bed planters, there are two planting procedures that may be employed: direct planting or double potting. In *direct planting,* the plant is removed from the nursery container and planted directly into the medium in the container bed. In *double potting,* the plant is not removed from the nursery container; both are set into the medium. This allows for easier seasonal changes and replacement of plants that have been damaged or experienced physical changes, such as yellowing leaves.

General Care

The leaves of interior plants should be cleaned occasionally with a fine spray of water at room temperature. Wipe the leaves on both the top and underside. This is a good way to remove dust and insects, help to maintain humidity, and keep the plant fresh-looking and attractive (Figure 23–4)

Controlling Indoor Plant Diseases and Pests

Interior landscape plants have few problems with diseases and insects. The secret to controlling these pests is to be able to recognize them and control them on the plant. Some of the most common insects affecting interior plants are aphids, whiteflies, mealybugs, mites, and scale.

Aphids are small green insects with piercing mouth parts. These insects are called plant lice because they suck the plant juices. Aphids work on the stem of plants and the underside of the leaves. The leaves then become sticky with honeydew, causing the growth of a black, sooty mold. Heavy infestation causes the leaves to curl, disfigure, and

▲ **Figure 23–4** Cleaning the leaves of a *Ficus decora* to remove dust.

turn yellow. Aphids can be controlled with Orthene, Safer Insect Soap, or spray oil.

The *whitefly* is a small, white, winged, sucking insect that works on the underside of the plant leaves. Whiteflies are heavy producers of eggs; therefore, infestation can occur fast. These egg masses are laid on the underside of the plant leaves. They cause the plant to turn yellow and have a mottled appearance. A whitefly infestation is easy to recognize because the adult fly will leave the plant with movement of the leaves. They can be controlled with Orthene.

Mealybugs are white, woolly masses like a ball of cotton. They will concentrate at the nodes of plants and the underside of leaves. They suck the plant juices to give the plant a faded yellow appearance. These bugs are controlled in an organic fashion by dipping Q-tips in alcohol and swabbing the insect, thus removing it from the plant. Orthene can also be used.

Mites are small spiderlike insects on the underside of the leaf. They suck the plant juices and cause the plant to turn a faded yellow. Mites are so small that it is impossible to see them with the naked eye. An easy way to determine mite infestation is to take a piece of white paper and hold it under the infested leaf; then tap the leaf, causing the insect to fall onto the paper. Look for spiderlike insects moving on the paper. If you suspect they are there, take your finger and strike it over the paper. If there are spider mites, there will be a streak of red across the paper. Mites can be controlled with Orthene, Safer Insect Soap, or oil spray.

Scale in its mature stage attacks the stems of indoor plants in a cluster. They look like small white, gray, or brown lumps. Scale is very noticeable on the infected plant as the leaves turn yellow and the plant itself loses vigor. Scale can be controlled by maintaining a clean environment. It can be removed by rubbing with a soap solution or spraying with Orthene or spray oil.

Botrytis and mildew are two common diseases of indoor plants. Botrytis causes indoor plant leaves to turn black and then develop a gray mold. They will rot off if the disease is allowed to advance. Environmental controls include providing good air circulation and sanitation. Mildew causes indoor plant leaves to get a heavy white coating. A good control is to provide the plants with proper sanitation, air circulation, and a fungicide as a last resort.

The Indoor Plant Chart

Following is an indoor plant chart that gives essential information for growing 51 indoor plants. Each plant is illustrated with a picture to aid in identification.

> **TRY THIS** Choose 10 of the plants from the indoor plant chart, and describe each one in writing. Be sure to include the following categories: watering, sunlight, temperature, soil mixtures, and ease of growth (explain the data).

The common name of the plant is given first, followed by the scientific name in italics. The rest of the information in the indoor plant chart is broken down as follows:

- *Watering* tells how to water each plant.
- *Sunshine* gives the amount of light exposure each plant needs, ranging from full sun to indirect light.
- *Temperature* indicates the optimum temperature at which to grow each individual plant.

Note: Soil mixture information has already been covered in the preceding section, so it is not included in this plant chart.

- *Ease of growth* is a rating given for each plant on a scale of 1 to 4:
  1. means that the plant is very easy to grow and requires little care.
  2. means that the plant is easy to grow but needs some special attention.
  3. means that the plant is very difficult to grow.
  4. means that the plant is very difficult to grow and must have very special care.

 indicates plant will grow in a closed terrarium.

 indicates plant will grow in a hanging basket.

 indicates plant will grow in a dish garden or pot.

 indicates plant will grow in an indoor planter.

Indoor Plant Chart

BOUGAINVILLEA

(Bougainvillea spectabilis 'Barbara Karst'*)*

WATERING:	Moist in summer/dry in winter
SUNSHINE:	Full sun
TEMPERATURE:	50°–60°F
EASE OF GROWTH:	2

GARDENIA

(Gardenia jasminoides)

WATERING:	Keep moist
SUNSHINE:	Direct sun
TEMPERATURE:	60–75°F
EASE OF GROWTH:	4

ASPARAGUS FERN

(Asparagus densiflorus 'Sprengeri'*)*

WATERING:	Drench/let dry slightly
SUNSHINE:	Weak (partial) sun
TEMPERATURE:	50–70°F
EASE OF GROWTH:	1

POINSETTIA

(Euphorbia pulcherrima)

WATERING:	Keep moist
SUNSHINE:	Direct sun
TEMPERATURE:	70–80°F
EASE OF GROWTH:	2

Indoor Plant Chart

GRAPE IVY

(Cissus rhombifolia 'Mandaiana'*)*

WATERING: Drench/let dry slightly
SUNSHINE: Weak (partial) sun to filtered light
TEMPERATURE: 50–70°F
EASE OF GROWTH: 1

SOUTHERN YEW

(Podocarpus macrophyllus)

WATERING: Evenly moist/not constantly wet
SUNSHINE: Bright to filtered
TEMPERATURE: 62–85°F
EASE OF GROWTH: 2

CHINESE FAN PALM

(Livistona chinensis)

WATERING: Moist
SUNSHINE: Filtered
TEMPERATURE: 65–80°F
EASE OF GROWTH: 2

KALANCHOE

(Kalanchoe pinnata)

WATERING: Drench/let dry slightly
SUNSHINE: Direct sun to partial sun
TEMPERATURE: 55–65°F
EASE OF GROWTH: 2

Indoor Plant Chart

CORN PLANT

(Dracaena fragrans massangeana)

WATERING:	Moist
SUNSHINE:	Weak (partial) sun
TEMPERATURE:	60–80°F
EASE OF GROWTH:	2

BOSTON FERN

(Nephrolepis exaltata bostoniensis)

WATERING:	Keep mixture moist, mist often
SUNSHINE:	Partial sun
TEMPERATURE:	50–70°F
EASE OF GROWTH:	1

AZALEA

(Rhododendron)

WATERING:	Drench/let dry slightly
SUNSHINE:	Weak (partial) sun
TEMPERATURE:	50°–65°F
EASE OF GROWTH:	3

NORFOLK ISLAND PINE

(Araucaria heterophylla)

WATERING:	Keep moist
SUNSHINE:	Weak (partial) sun
TEMPERATURE:	50–65°F
EASE OF GROWTH:	1

Indoor Plant Chart

DRAGON TREE

(Dracaena marginata)

WATERING: Moist
SUNSHINE: Weak (partial) sun
TEMPERATURE: 60–80°F
EASE OF GROWTH: 2

DUMBCANE

(Dieffenbachia amoena 'Tropic Snow'*)*

WATERING: Drench/let dry slightly
SUNSHINE: Filtered light
TEMPERATURE: 60–75°F
EASE OF GROWTH: 1

BROMELIAD

(Aechmea fastciata)

WATERING: Even moist (fill vase of plant with water)
SUNSHINE: Partial sun to filtered light
TEMPERATURE: 62–65°F night, 80–85°F day
EASE OF GROWTH: 2

FIDDLELEAF FIG

(Fiscus lyrata)

WATERING: Evenly moist, but not constantly wet
SUNSHINE: Filtered light
TEMPERATURE: 62–65°F night, 80–85°F day, keep out of draft
EASE OF GROWTH: 1

Indoor Plant Chart

CHRYSANTHEMUM

(Chrysanthemum sp.)

WATERING:	Moist to touch
SUNSHINE:	Direct sun
TEMPERATURE:	60–75°F
EASE OF GROWTH:	2

SPIDER PLANT

(Chlorophytum comosum)

WATERING:	Keep moist
SUNSHINE:	Weak (partial) sun
TEMPERATURE:	50–70°F
EASE OF GROWTH:	1

FICUS MIDNIGHT BRAID

(Ficus benjamina 'Midnight Braid')

WATERING:	Evenly moist, but not constantly wet
SUNSHINE:	Filtered light
TEMPERATURE:	62–65°F night, 80–85°F day
EASE OF GROWTH:	1

UMBRELLA TREE

(Schefflera digitata)

WATERING:	Drench/let dry slightly
SUNSHINE:	Weak (partial) sun
TEMPERATURE:	40–60°F
EASE OF GROWTH:	1

Indoor Plant Chart

PRAYER PLANT

(Maranta leuconeura)

WATERING: Keep moist
SUNSHINE: Weak (partial) sun
TEMPERATURE: 60–70°F
EASE OF GROWTH: 3

SNAKE PLANT

(Sansevieria trifasciata)

WATERING: Drench/let dry
SUNSHINE: Direct to indirect
TEMPERATURE: 50–75°F
EASE OF GROWTH: 1

JANET CRAIG DRACAENA

(Dracaena deremensis 'Janet Craig'*)*

WATERING: Moist
SUNSHINE: Weak (partial) sun
TEMPERATURE: 60–80°F
EASE OF GROWTH: 2

EXACUM

(Exacum affine)

WATERING: Quite moist
SUNSHINE: Filtered
TEMPERATURE: 62–65°F night, 80–85°F day
EASE OF GROWTH: 3

Indoor Plant Chart

DWARF DATE PALM

(Phoenix roebelenii)

WATERING: Quite moist
SUNSHINE: Filtered
TEMPERATURE: 62–65°F night, 80–85°F day
EASE OF GROWTH: 3

SPATHIPHYLLUM

(Spathiphyllum wallisii)

WATERING: Quite moist
SUNSHINE: Filtered to low light
TEMPERATURE: 62–65°F night, 80–85°F day
EASE OF GROWTH: 1

SPINELESS YUCCA

(Yucca elephantipis)

WATERING: Drench/let dry
SUNSHINE: Direct
TEMPERATURE: 55–70°F
EASE OF GROWTH: 1

SWEDISH IVY

(Plectranthus australis)

WATERING: Keep soil moist
SUNSHINE: Indirect sun
TEMPERATURE: 50–65°F
EASE OF GROWTH: 2

Indoor Plant Chart

ENGLISH IVY

(Hedera helix)

WATERING: Drench/let dry slightly
SUNSHINE: Partial sun
TEMPERATURE: 50–65°F
EASE OF GROWTH: 1

MING ARALIA

(Polyscias fruticosa)

WATERING: Evenly moist
SUNSHINE: Direct light
TEMPERATURE: 62–65°F night, 80–85°F day
EASE OF GROWTH: 1

CALADIUM

(Caladium spp.)

WATERING: Evenly moist
SUNSHINE: Filtered light
TEMPERATURE: 62–65°F night, 80–85°F day
EASE OF GROWTH: 1

NEPHTHYTIS OR ARROWHEAD PLANT

(Syngonium podophyllum 'White Butterfly')

WATERING: Quite moist
SUNSHINE: Filtered to low light
TEMPERATURE: 62–65°F night, 80–85°F day
EASE OF GROWTH: 1

Indoor Plant Chart

BANANA

(Musa maurelii)

WATERING:	Keep soil moist
SUNSHINE:	Partial sun
TEMPERATURE:	60–85°F
EASE OF GROWTH:	3

MEDICINE PLANT

(Aloe vera)

WATERING:	Drench/let dry
SUNSHINE:	Direct sun
TEMPERATURE:	55–70°F
EASE OF GROWTH:	1

BIRD OF PARADISE

(Strelitzia reginae)

WATERING:	Keep soil moist—summer; drench/let dry—winter
SUNSHINE:	Direct sun
TEMPERATURE:	50–75°F
EASE OF GROWTH:	3

HEART LEAF PHILODENDRON

(Philodendron)

WATERING:	Drench/let dry slightly, mist leaves often
SUNSHINE:	Indirect
TEMPERATURE:	50–75°F
EASE OF GROWTH:	1

Indoor Plant Chart

AMARYLLIS

(Hippeastrum spp.)

WATERING: Keep soil moist
SUNSHINE: Direct sun
TEMPERATURE: 55–65°F
EASE OF GROWTH: 3

POTHOS

(Scindapsus aureus or *Epipremnum aureus)*

WATERING: Drench/let dry
SUNSHINE: Indirect/filtered
TEMPERATURE: 50–75°F
EASE OF GROWTH: 1

JADE PLANT

(Crassula argentea)

WATERING: Drench/let dry
SUNSHINE: Partial sun
TEMPERATURE: 60–70°F
EASE OF GROWTH: 1

KING SAGO

(Cycas revoluta)

WATERING: Drench/let dry between waterings
SUNSHINE: Filtered light
TEMPERATURE: 62–65°F night, 80–85° day
EASE OF GROWTH: 1

Indoor Plant Chart

CROTON

(Codiaeum variegatum var. *pictum)*

WATERING:	Keep soil moist
SUNSHINE:	Partial sun
TEMPERATURE:	60–80°F
EASE OF GROWTH:	2

CHINESE EVERGREEN

(Aglaonema 'Emerald Bay'*)*

WATERING:	Saturate soil completely, then allow soil to dry before watering again
SUNSHINE:	Indirect/filtered
TEMPERATURE:	50–75°F
EASE OF GROWTH:	1

COLUMN CACTUS

(Cereus peruvianus)

WATERING:	Drench/let dry between waterings
SUNSHINE:	Direct
TEMPERATURE:	62–65°F night, 80–85°F day
EASE OF GROWTH:	1

STAGHORN FERN

(Platycerium bifurcatum)

WATERING:	Moist, mist daily
SUNSHINE:	Partial shade
TEMPERATURE:	60–80°F
SOIL MIXTURE:	Osmunda Fiber Mix
EASE OF GROWTH:	3

Indoor Plant Chart

PONYTAIL PALM

(Beaucarnea recurvata)

WATERING:	Drench/let dry
SUNSHINE:	Full sun/partial shade
TEMPERATURE:	60–80°F
EASE OF GROWTH:	2

AIR PLANTS

(Tillandsia sp.)

WATERING:	Mist lightly two to three times per week
SUNSHINE:	Bright filtered Light
TEMPERATURE:	50°F–80°F night, 50–90°F day
EASE OF GROWTH:	1

PHALAENOPSIS ORCHID

(Epidendrum amabile)

WATERING:	Keep moist
SUNSHINE:	Light to shade
TEMPERATURE:	62–65°F night, 70–80°F day
EASE OF GROWTH:	1

ANTHURIUM

(Anthurium sp.)

WATERING:	Water well; allow to dry down
SUNSHINE:	Bright indirect light
TEMPERATURE:	75°F–90°F
EASE OF GROWTH:	2

Indoor Plant Chart

CHENILLE PLANT

(Acalypha hispida)

WATERING: Keep moist
LIGHT: Full sun to part shade
TEMPERATURE: 75°F–85°F
EASE OF GROWTH: 2

PINEAPPLE FLOWERS

(Eucomis 'Katie'*)*

WATERING: Keep moist
SUNSHINE: Full sun
TEMPERATURE: 75°F–85°F
EASE OF GROWTH: 2

PERSIAN SHIELD

(Strobilanthes dyerianus)

WATERING: Keep moist
SUNSHINE: Full sun
TEMPERATURE: 65°F–80°F
EASE OF GROWTH: 1

SAFETY FIRST

Use the proper equipment to move large container plants, and use caution when pruning and cleaning up the display area to avoid injury. For more information on occupational health and safety in the workplace, please visit the OSHA site at www.OSHA.gov.

Summary

The interiorscape business has continued to grow, with a $574 million wholesale value of foliage and flowering plants in 2000. The interest in having an environment that ties people to nature is flourishing in homes, office complexes, and hotels. The demand for new cultivars has increased exploration for different plants with leaf variegation and texture, stem coloration, improved flowering, and improved dwarf species. People who live in crowded urban areas find private gardens of interior plants appealing.

Review Questions

1. Discuss the major concerns related to the care of indoor plants.
2. What are the different ways to water indoor plants?
3. How does slow-release fertilizer differ from soluble fertilizer? How do you determine which type of fertilizer to use on an indoor plant?
4. Discuss the most common diseases and pests that affect indoor plants and ways to control each one.

Student Activities

1. Visit a local retail florist to see the available indoor landscape plants.
2. Attend a home and garden show to observe the use of these plants.
3. Participate in the FFA Floriculture Career Development Event (CDE).
4. Review interior decorating magazines to see how interior plants are used in decorating.
5. Propagate a large variety of indoor plants, and use them for identification.
6. Read trade magazines, for example, *Florists' Review*.
7. Design and install an interior plantscape garden.
8. Have an interior plantscape professional speak to the class.
9. Search online for visual examples of interior landscapes. Choose the one that appeals to you the most, and describe which plants draw the viewer's attention.

Suggested Supervised Agricultural Experience (SAE) Activities

1. Conduct a placement SAE, working in a retail florist that provides indoor landscape plants.
2. Intern with an interior plantscape professional to learn the skills of the trade. Record the labor toward a placement SAE.

Additional Resources

www.safnow.org

http://architecture.unl.edu/degree-programs/interior-design

http://botanicus.com/Employment/InstallationAssistant.aspx

UNIT 24 Shrubs and Trees

Objective

To provide enough basic facts so that a container crop can be grown to marketable quality.

Key Terms

aeration
balling
durability
herbaceous
insulation value
pot-in-pot
QR code
site
windbreaks

Competencies to Develop

After studying this unit, you should be able to:

- list and explain at least four advantages and three disadvantages of container growing.
- list at least five types of plants that are commonly grown in containers.
- list or explain orally the characteristics of a good growing site.
- list the factors to consider in selecting a container.
- prepare a soil mix, fill a container, and successfully place a plant in a container.
- set up a growth control schedule.
- discuss marketing of the finished product.
- discuss marketing by using a QR code.
- outline at least one winter storage method where temperatures fall to below freezing.
- list at least one reason for shading container plants.

Materials

- containers
- plants to grow
- growing area
- pesticides, if needed
- weed killers, if needed
- growing medium
- fertilizers
- watering system
- hand pruner
- smartphone apps:
  - AG PhD
  - Pest Portal
  - Spectracide
  - LikeThat Garden
  - Leafsnap

Container growing of shrubs and trees is becoming more important and is gradually replacing field growing for many kinds of nursery plants. Container growing in the nursery trade means growing the plant to marketable size in a container. Many plants are grown in pots for short periods of time as seedlings, grafts, or rooted cuttings and later transplanted to a field or to a large container for container growing. This short-term potting culture is not considered container growing.

Advantages of Container Growing

Container growing has a number of advantages:

- Plants may be sold and transplanted year-round. Because the roots are in a container, they are not cut off and damaged as may be the case with field-dug plants. This also allows for more year-round use of labor. If plants are pot-bound, and roots circling the inside of the pot must be cut, this advantage is lost.
- More plants can be grown in the same area, thus reducing cost.
- Uniform growing medium (mix used instead of soil) is readily available, and plant nutrients and water may be more accurately controlled, which results in more uniform plant growth.
- Plants require no digging or **balling** (wrapping the root ball in burlap) of roots to hold the root ball together and prevent drying out, which saves labor.
- Growing medium is usually lighter than soil, making it easier to haul larger loads of plants.
- The container, if sturdy, protects the root ball from being crushed when plants are stacked on top of each other in transporting to the retailer or planting **site**.
- The growing medium is generally sterilized, thus reducing damage to plant roots from insects and disease organisms. This results in faster, healthier plant growth.
- The cost to produce a plant to marketable size is lower because many of the items listed previously result in less labor, fewer insect and disease problems, lower shipping costs, and, if managed properly, more uniform plant growth.

GREEN TIP Container growing of trees and shrubs has increased the available plant material to the general public, promoting a greener environment for the future.

Disadvantages of Container Growing

The major disadvantages of container growing are the following:

- The initial cost of getting established is higher. The growing area should be graded flat with a slight slope, treated for weed control, and covered with plastic or ground cover fabric.
- Personnel must be trained in potting technique, and watering becomes a more precise growth factor in container growing.
- More water is required because the plants cannot tap soil reserves. A good supply of irrigation water is a must.
- If plants are overwintered, some protection must be provided where temperatures fall much below freezing. Roots are damaged at higher temperatures than the top part of the plant. Because the roots are not down in the soil, they get colder in containers than in field-grown situations.
- Transplanting to the landscape is more difficult for some plants because the soilless mixes generally used are too different in structure from the soil in which the plant will be transplanted. (Adjustment may be made by adding soil to the container mix or by digging pine bark and sphagnum peat moss into the soil at the planting site to

Courtesy of Griffin Corporation

▲ **Figure 24–1** Notice the circling roots on the plant on the left. This plant is badly root-bound and will be difficult to transplant. The plant on the right was grown in a pot treated with a copper solution, SpinOut, and is not pot-bound.

lessen the difference.) Plant roots are thus encouraged to leave the container ball and grow out into the soil.

- If left in containers too long, plants develop a circular, pot-bound root system (Figure 24–1). This system may never grow well in the soil and may not provide adequate anchorage or spread out to absorb water and nutrients needed for normal growth.

Types of Plants Grown in Containers

No attempt will be made to list all the plants grown in containers in the nursery trade. It can be generally said, however, that nearly all shrubs and trees are container grown. Some plants such as yews are difficult and not grown in containers as often. The variety of plants ranges from garden flowers such as **herbaceous** perennials, ground cover, vines, broadleaf evergreen shrubs, and narrowleaf evergreen shrubs, to evergreen and deciduous trees. Indoor tender foliage plants are also container grown.

Deciduous trees and shrubs that are usually planted during the dormant season only (such as fruit trees) are generally still field grown and dug and shipped bare root (no soil on the roots). This is a cheaper method of handling and shipping large quantities of this type of plant. Flowering trees, however, often are grown in large containers.

Site

Site (the location where plants are to be grown) selection for container plant growing is very important. A flat, gently sloping area is necessary. Containers will fall over on too steep a slope, but enough slope should be provided to drain water off the surface so that no water collects around the containers (Figure 24–2). It is generally necessary to use a road grader to grade and slope the land.

Where plants must have individual attention, such as pruning, beds should be made no more than 5 feet wide with an 8-foot service road on at least one side of each bed for loading and unloading from a truck. Beds should be planned to make the best use of the irrigation (watering) system. If underground water lines are used, they should be installed at the time that the beds are graded and laid out.

The growing area should also be located convenient to the water supply and irrigation system. Because the plant roots must depend entirely on the medium within the pot for water and nutrients, a dependable supply of irrigation water is critical. On hot, windy days, plants may need to be watered daily or even twice daily.

▲ **Figure 24–2** Potted plants placed on gravel, under irrigation, to grow on. Notice the wide roadway between rows of pots for truck or wagon traffic to load and unload plants.

Windbreaks

Windbreaks are also very important in site selection. A strong wind can blow containers over. These containers do not get watered and soon dry out, becoming damaged to the point at which the plants are not marketable. Winds also interfere with sprinkler irrigation systems. A row of evergreen trees, a snow fence, or other windbreaks should be put in place. The height of the windbreaks will determine how far apart they should be placed. A good rule of thumb is that a windbreak is effective for a distance of about seven times its height. A 4-foot snow fence would thus be spread 28 feet apart, whereas 45-foot trees could be spread about 300 feet. Figure 24–3 shows a natural windbreak of evergreen trees surrounding a growing area.

Weed Control

Weed control is best done through the use of black plastic ground cover fabric or stones on the soil surface before the pots are put in place. A weed killer may be placed on the soil before the material is put down. Black plastic lasts only one season unless it is covered with 4 inches of gravel or crushed stone to shield it from sunlight. If the crop is to remain on the same site for more than 1 year, the gravel cover may be a good investment. The stones or gravel will also hold the plastic in place when plants are marketed, and new containers may be placed in the same beds without reworking them.

▲ Figure 24–3 An evergreen windbreak around a container growing area at a nursery.

Containers

Proper selection of the container that the plant will grow in is very important. The following factors should be considered:

Cost of the container must be low because container cost is one of the major expenses in the production of container-grown nursery stock.

Durability should be matched with the length of time the crop will grow in the container prior to marketing. An expensive container that will last 3 years would not be a good business investment if the crop is marketed at the end of 1 year. Containers are made of materials such as wood, plastic, metal, and pressed paper. Durability information should be obtained prior to purchase. Metal containers rust, and wooden containers rot out in the bottom if drainage is not good.

Appearance of the container may be almost as important as plant appearance to some retail buyers. A container that remains attractive through the marketing cycle is important. Painted metal and plastic containers last longer, and paper and peat containers have the shortest life.

Insulation value to the plant roots is also important. Where there are freezing temperatures during the growing cycle, the container must withstand freezing without becoming brittle and breaking. Plant roots grow close against the container wall and are affected by container temperature and insulating value. A metal container has poor insulating value, and roots would be damaged from heat or cold sooner in them than in a wooden or fiber container.

Shipping durability is another consideration. If plants are to be stacked one on the other, fiber pots or wooden baskets may crush and damage roots. Metal or hard plastic would protect root balls better under these conditions.

The *shape* of the container is another factor. Tapered containers tend to store in less space when empty because they can be stacked one in the other. Too much taper, however, could result in easy tipping or blowing over if the bottom is too narrow compared with the top. Tall, slim containers also blow over easier than short, broad ones. The container should be shaped to permit normal plant growth. This varies with different plants.

Drainage and soil **aeration** are important to all plants and especially for azaleas and rhododendrons. The potting medium must be coarse

▲ **Figure 24–4** Different-sized containers used in growing plants. Pot sizes are 1, 2, 3, 5, 10, and 15 gallons.

enough to allow good drainage and aeration. Wooden baskets also provide aeration to the roots. Without good drainage and aeration, roots will rot or at best be stunted in growth. See Figure 24–4 for some sample containers of various sizes.

The *size* of the container affects plant growth. A plant in a container that is too small will become root-bound and not grow as well as if it were in a proper-sized container. After the roots penetrate to the walls of the container, plant growth starts slowing. Plants should be moved to larger containers before they become root-bound, or they should be sold. Some standards set forth by the American Association of Nurserymen for minimum container size can be found on its website (www.anla.org).

The container must not contain any material that is toxic (poisonous) to the plant.

Selection of the proper container is a very important decision. The grower should get advice from an experienced grower or the state extension service. Purchasing the wrong container could be an expensive mistake.

New Container Designs

After considerable research, Dr. Carl E. Whitcomb (retired from Oklahoma State University) developed the field-grow fabric container for trees, which was designed to modify the root growth and prevent circling of roots of the plant grown in it.

The *field-grow fabric container* is used for growing trees in the field. A tree is planted in a fabric container using the same soil as in the field, and the container is placed in the soil with just the top inch or two of the container above ground (Figure 24–5). The pot restricts the tree's roots, and when the tree is dug up, none of the roots are cut off. The container is made of a new fabric with small holes that allows the root tip to grow through the pot for a short distance before it is stopped. As the root grows larger in diameter, the hole in the pot stops its growth, and many small side roots form on the inside of the container. The result is a root system with many more small roots in a smaller root ball. This smaller ball is easier to move, and it contains many more feeding roots than a field-grown tree without the pot (Figure 24–6).

The growth of trees in the pots is not different from that of trees grown in the open field. The big difference is in digging and transplanting. Trees grown in the fabric container are much easier to dig and survive much better when transplanted.

▲ **Figure 24–5** Evergreens growing in a field potted to fabric root control bags.

▲ **Figure 24–6** Compact fibrous root system of a tree grown in a root control bag. There is no circling of roots.

The major advantages are less labor cost in digging, far less loss from transplanting to the landscape, and a better growth rate after transplanting. The bags often are difficult to remove from the entwining root.

Another new and promising technique for reducing circling roots of pot-grown plants is to coat the inside of the pots with copper. The chemical compound copper hydroxide [$Cu(OH)_2$] is sprayed or rubbed onto the inside walls and bottom of the pot. One current Environmental Protection Agency (EPA)–registered copper compound labeled for root control is SpinOut, which is a copper paint made by Griffin Corporation.

Researchers at Texas A&M University found the copper spray to be very effective in reducing root circling in pots. It also resulted in a more compact root system with more fibrous roots in the interior of the root ball. In many cases, no root pruning at transplanting time to correct and stop circling roots was necessary.

When root ends touch the copper coating of the container wall, they stop growing. Instead of circling, the pot roots send out side roots (laterals) back inside the root ball. This produces a superior root ball, resulting in better plant growth and higher survival rate after transplanting. When removed from the pots, root ends again start extending out into the soil and grow normally.

A new container now on the market has the copper mixed into the plastic of the pot. This container is reputed to hold the copper chemical longer and therefore remain effective longer. It can be reused for more than one crop. Pot-bound root systems must have the circling roots cut or pulled out of the root ball to stop the circling growth pattern when transplanted to the garden. This is necessary to prevent girdling, to form a natural root system, and to get roots out into the soil where moisture is available.

A relatively new container-growing method is the **pot-in-pot** system, Figure 24–7. One pot is dug into the ground with only the top 1 or 2 inches above soil level. A second pot, holding the plant, is set down into the first pot.

Drainage is an important factor with this system. No water can stand around the bottom of the pots, so well-drained soil is very important.

▲ **Figure 24–7** Pot-in-pot container growing.

Some advantages of this system are the following:

- Root temperatures remain more uniform. Being below soil level protects from both heat and cold. The result is a healthier root system.
- Pots do not blow over.
- Top growth is better on many plants.
- Roots usually do not need any additional winter protection.

Some disadvantages include the following:

- The initial cost is higher.
- Some plants send roots out the bottom through drainage holes in both pots, making it difficult to remove them from the pots. Deep-rooting trees, especially those with tap roots, are worse in this regard.

The pot-in-pot system appears to have much merit and, where well-drained soils exist, could become widely used for container-grown plants.

Growing Mixes

After selecting the container for a particular crop, the next important decision is the medium or soil mix the plant will grow in. Again, some plants require a more porous or open mix than others to allow for better aeration or oxygen supply to roots. For example, rhododendrons and azaleas require a more porous mix than does a maple tree. A tight container may also require a more open mix than does a basket. Container mixes must be more porous than mixes for the same plant grown in the field or open bench. Containers do not allow for as good drainage or air movement as do good field conditions.

A good growing medium must

- provide good drainage (sand, perlite, and bark are often used).
- be porous for good aeration (sand, perlite, and bark are often used).
- hold moisture well (peat moss is most often used).
- hold plant nutrients (some soil may be added).
- have the proper pH range (different plants require different pH).
- be sterile (free of insects, disease organisms, and weed seeds).
- be economical and easily obtained.
- be consistent in quality (always the same contents in the same percentages).
- encourage good root development (be compatible to the plant).
- transplant well to the soil in the plant's permanent location (a plant grown in bark and peat may not transplant well to a clay soil).

Growers prefer to stay with a mix that works best for them. This is probably a good idea because fertilizers and watering can be scheduled better if the mix remains the same. When many different kinds of plants are grown, at least two mixes will probably be required: one mix for plants requiring a pH of 6.0 to 7.0 and another for acid-loving plants with a pH of 5.0 to 5.5. Fertilizers used as additives to the growing medium may differ for different plants. Many commercial mixes are available already mixed for use.

In large businesses where growers mix their own planting media, pasteurization may be done at the nursery. The mix should be steam heated to 180°F (83°C) for 30 minutes. Most bagged commercial mixes are sold as sterilized and should be ready to use directly from the bag.

Many soilless mixes containing sphagnum peat moss and perlite or vermiculite are very light in weight. If wind tends to blow containers over, it may be necessary to add sand or gravel in the bottom of the containers to hold them down. Be sure this material is pasteurized before placing it in the container. Fine pine bark (1/4 inch and smaller) is being used extensively as a container growing mix now. It provides excellent drainage but has poor moisture-holding capacity. A mix of one part sphagnum moss and three parts pine bark holds moisture better.

GREEN TIP If soil is used in the mix, pasteurization is necessary. After the mix is prepared, it is either fumigated or steam pasteurized at 180°F (83°C) for 30 minutes. Keep the pile covered with plastic until used. (Steam pasteurize before fertilizers are mixed in.)

When soilless mixes are used, all the plant nutrients needed for growth, including minor elements, must be added. To adjust the pH upward and to add calcium and magnesium, use dolomitic limestone. The amount of lime to add will depend on the pH of the mix and the desired final pH. Sulfur or iron sulfate will lower pH. A test for pH should be made to ensure accuracy. Superphosphate should be added at the rate of 6 pounds per cubic yard if tests indicate a need. Potash may also be added if tests indicate a need.

Because soilless mixes contain few trace elements such as manganese, iron, copper, or zinc, it is essential to add 1 or 2 ounces of slow-release trace elements per cubic yard.

Commercial slow-release fertilizers such as MagAmp and Osmocote are often mixed into the medium to provide the first supply of plant food for the newly potted plants. (Add after steam pasteurization.)

CAUTION

It is essential that all of the above materials are mixed thoroughly to avoid poor results.

Some of the materials in soilless mixes such as peat moss are difficult to wet after they dry out. A wetting agent may be added if this is a problem. Wetting agents improve uptake and uniformity of wetting. They also shorten the time needed to wet the mix.

Planting

Placing plants in the containers is generally done by hand. Some large nurseries have machines (Figure 24–8) that fill the containers with the growing medium, and some plants are planted by machine. Even with the use of machines, actual placement of the plant in the container is a hand operation. Plants are usually seedlings, rooted cuttings, dormant grafts, or plants being transferred from smaller pots or flats to containers. The plants may have been purchased from another nursery or grown by the same nursery.

Planting is usually done in the spring. If storage space is available to protect plants from freezing, planting may be done all winter to make better use of labor. The plant is placed in the center of the container (if only one plant is used) at the same depth it was growing before. The growing medium is firmed around it and watered as soon as possible after planting until water runs through the bottom. After planting, the growing medium should be about 3/4 inch down from the top of the container to allow space for water to collect when plants are watered. If more than one plant is placed in the container, the plants should be evenly spaced within the container.

© Gertan/Shutterstock.com

▲ **Figure 24–8** Potting machine.

Planting may be done in the field from wagons loaded with containers, pots, and plants, with the plants placed in the growing beds, or it may be done at a permanent spot. In permanent locations or on flatbed wagons or trucks, an assembly line is usually set up using a five-person work team. Each person has one job to do repeatedly, such as filling the container with growing medium, making planting holes, planting the plant, watering, and placing plants in the growing bed. A very efficient team can plant up to 8,000 plants per day. Different-sized holes for plants can be made using various machine attachments. As the hole is drilled, the medium is firmed in the container. The plant is placed in the hole by hand and firmed in. The next person levels the medium to the proper depth and moves the plant toward the wagon. Plants are hauled to the growing area, put in place, and watered immediately. If the plants being transplanted are in peat pots, the top of the pot must be torn off or covered by the growing medium in the container. Peat pot tops, if allowed to stick out above the medium, tend to dry out and pull moisture away from the roots of the young plant being transplanted.

The depth of planting is also important. Some plants prefer shallow planting and if planted too deeply will get stem rot, or roots will rot for lack of oxygen. For most plants, shallow planting with the roots just covered is better.

Container Spacing

Spacing of plants in the bed is very important. When plants are small, the containers may be placed so that they touch one another. As the plants grow, it may become necessary to remove every other container to allow space for the plant's branches to spread and grow normally. Ideal spacing is when the ends of the branches of plants almost touch the plant next to them (Figure 24–10). This allows maximum use of space and results in compact plants. Too close spacing results in shading out and browning or killing of lower branches or foliage. Too wide spacing wastes space and irrigation water if sprinklers are used. Plants are often pruned to shape them and cause compact growth, which also helps to prevent crowding and may allow closer spacing. Chemicals can be used in place of hand pruning to cause plants to grow more compactly. An example is the use of a chemical that kills all the tender tips on azaleas to cause branching and compact growth. Without the chemical, each tip would need to be pinched off by hand.

TRY THIS Recent technology has allowed robotics to advance efficiency in the container plant industry. Robots are being used more often to sort and accurately space container plants for optimum growth (as shown in Figure 24–9). Using the Internet, research the utilization of robots in the nursery and container plant industry. Select a specific robot used to space and place container plants and write a description of its uses.

© Boston Globe/Dina Rudick/Getty Images

▲ **Figure 24–9** Robotic plant mover used to increase labor efficiency to organize plants in a garden center lot.

© 2017 Cengage Learning®

▲ **Figure 24–10** Plants spaced with room to grow for the current season.

A definite growing schedule should be drawn up for each plant, including the space needed for the first year and any extra years needed for the plant to reach marketable size. Contact your local commercial grower to obtain a growing schedule for a container crop grown in your area.

Watering

Watering becomes the most important consideration in growing container plants after they are placed in the growing beds. The plants have a smaller root growing area than field-grown plants, and the container is subject to more heat and drying. A 1-gallon container requires about 1 pint of water at each watering. In hot, dry summer weather, an acre of container plants may require from 35,000 to 45,000 gallons of water a day. Water volume may exceed a flow rate of 300 gallons per minute. This volume of water requires a large, dependable water supply such as a pond, stream, or high-volume deep wells. Water must be analyzed for any pollutants that would be harmful to the plants, such as chemicals or salt.

GREEN TIP Collect rainwater for irrigating your yard, flower beds, or garden. Collecting rainwater could offset our freshwater needs by millions of gallons a year.

Plants need watering when the surface of the medium in the container feels dry to the hand. At this stage they should be watered immediately. Some plants may be best watered before this stage, and a knowledge of the plant and its water requirements is needed.

How often to water (how many days between waterings) depends on a number of factors:

- Weather (wet or dry, hot or cold, windy). Plants require more water in hot, dry, windy weather.
- Plant growth. If the plants are in active growth, more water is needed. Plants also need more water when in bloom.
- Planting medium. A mix with a higher percentage of peat moss will hold water better than a sandy or high percent perlite or pine bark mix.

- Kind of plant. Plants such as broadleaf evergreens require more water than narrowleaf evergreens or succulents such as cacti.
- Size of container. The smaller the container, the faster it will dry out.
- Type of container. Plastic containers hold moisture longer than fiber or pressed paper containers.
- Container surface mulch. A mulch on the surface of containers or shade to container walls will keep the root area cooler and slow the loss of water.

A definite watering schedule should be set up. Only by examining containers by hand to determine dryness can the exact watering needs be determined. It is important that the growing medium drains well (has large pore spaces) so that the oxygen supply to the plant roots is good.

There are three methods for watering container plants in outdoor growing areas: by hand, by trickle tube, or by overhead irrigation (sprinklers). Bench growing allows use of a capillary mat under the containers. With all of these systems, containers must be watered clear through from top to bottom.

Hand Watering

This method is generally too costly for use as a regular watering system except in very small operations. It is used when plants are first planted in the container and to spot water areas that an automated system may have missed. Containers should always be watered until water runs through the bottom of the container.

Trickle Tube Systems

These systems are the most efficient in saving water, although they are expensive. Water is delivered directly to each container by a small tube. Each container gets the same amount of water. Tubes are weighted on the end so they remain in place in the container top.

A typical layout for this system has a long plastic water main with secondary lines between every other row with trickle tubes coming out all along it to each pot (Figure 24–11). Strainers must be placed in the main lines to prevent dirt from clogging the small tubes. Liquid fertilizer may be fed through these tubes to fertilize the plants as watering is being done.

The length of time the system is left on at each watering must be determined by watching the containers. Watering is stopped when water first starts to run through the drain holes in the bottom of the container.

One thing to watch for in trickle tube systems is that rodents, such as mice and rabbits, do not chew the small tubes. Treatment with a spray containing thiram (Arasan) and a sticker, which helps keep the material on the tubing longer, helps to prevent rodent damage. Tubes should be inspected each day for damage or clogging so that all plants are watered.

Overhead Irrigation (Rotary Heads)

Overhead irrigation is perhaps the most often used system with container nursery crops. It is cheaper to install and operate. The biggest disadvantage is that it requires more water to do the

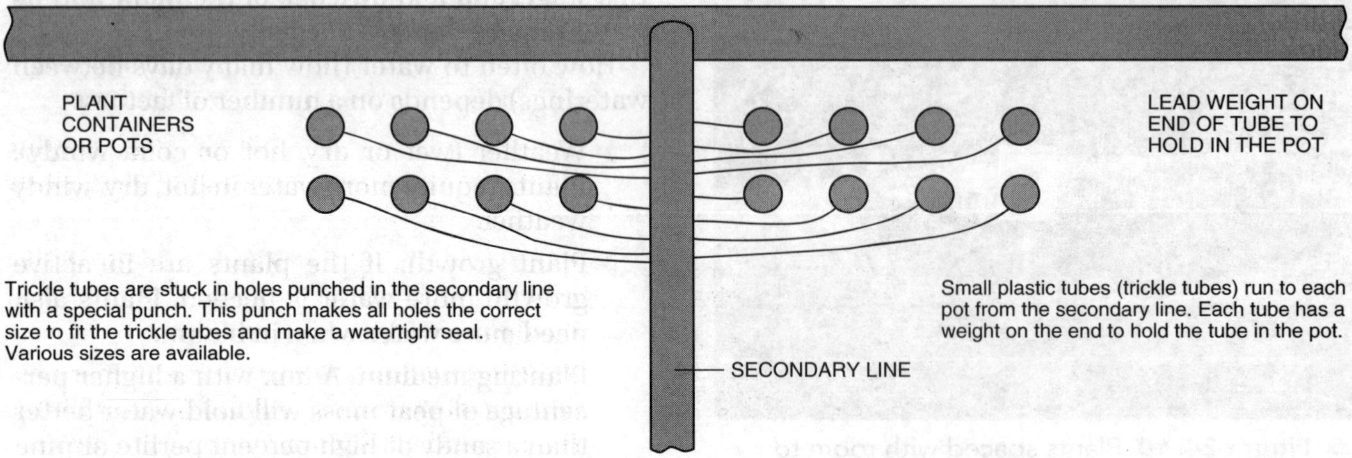

▲ **Figure 24–11** Trickle tube system.

▲ **Figure 24–12** Overhead irrigation used on container plants. Container growing requires a lot of water.

same job. Some water is lost between containers, and also to evaporation as it is sprayed out through the air above the containers.

There are two different styles of sprinklers: overhead and oscillator. The overhead sprinkler distributes water in a circular pattern (Figure 24–12). The oscillator spreads water in a rectangular pattern.

The procedure for establishing watering time is the same as that for the trickle tube system. Sprinkle until the containers are wet through to the bottom. The time between waterings can be determined only by examining containers to see when water is needed. If this is not possible, good drainage of containers must be assured and watering be done often enough to prevent drying.

Each sprinkler head or oscillator should have a cutoff valve so sections of the system can be cut on and off as needed. Some areas of the system may be empty of plants and not need watering, or workers may need to enter the area.

Fertilizers

Fertilizers may be applied to container-grown plants through several methods. The planting medium probably had some fertilizer mixed in it prior to planting. This was explained in the section on growing mixes.

Additional fertilizer will be needed as the plants grow.

Liquid Fertilizer

Fertilizer may be applied through siphoning devices into the irrigation water if plants are watered by hand or with trickle tubes. Fertilizers should not be applied with the irrigation water when overhead sprinklers or oscillator systems are used because much of it would fall between the containers and not get to the plants. This is expensive and wasteful and could pollute streams or ponds nearby.

If liquid fertilizer is used at each watering, it should be applied at a rate of 100 to 150 ppm of nitrogen. The fertilizer ratio of nitrogen, phosphorus, and potassium will vary with the plant being grown.

Periodic feeding every 2 or 3 weeks with liquid fertilizer at 300 to 500 ppm is not as effective but is often used. A definite fertilizer schedule should be set up for each plant grown.

Slow-Release Fertilizers

New slow-release fertilizers, such as MagAmp or Osmocote, are used with container plants in two ways. With Osmocote (18-6-12), for example, the material is mixed in the planting medium at a rate of 9 to 10 pounds per cubic yard for completely soilless mixes. (Do not steam the mix after the fertilizer is added.) The mix should be used within 2 weeks after the Osmocote is added. The MagAmp (7-40-6) slow-release fertilizer may be mixed into the growing medium and feeds for one full growing season. Supplemental nitrogen may be needed.

Osmocote can also be used as a top dressing fertilizer on top of the medium in the container. Because the release of plant food (available to the plant) from Osmocote is related to soil temperature, it can be added at any time. As the temperature rises, more nutrients are released, and as soil or medium temperature cools, nutrient release slows. One application of Osmocote on top of the container medium will feed the plant for 1 full year. The plant food is washed down each time the container is watered.

Plants that are kept in containers for more than 1 year need to have more fertilizer added. For liquid fertilizer, granular fertilizers, or Osmocote, 1 teaspoon of fertilizer (18 percent nitrogen) should be added for each 1 gallon of container capacity. Liquid or granular fertilizers should be applied in the spring, and Osmocote should be applied in the fall. Plants should be watered thoroughly immediately after fertilizing, if a granular or solid form is added.

The liquid fertilizer program is continued through the second or third season the same as for the first year. It should be started in early spring before growth starts and continued throughout the growing season. At the end of the growing season, at least one watering should be done with no fertilizer.

Granular fertilizers should be started at the beginning of the growing season and repeated every 8 weeks using 1 teaspoon per gallon of growing medium in the container. A maximum of three applications should be made per season.

Agriform 14-4-6 tablets for container use should be applied at the beginning of the growing season with repeat applications every 10 weeks. One 12-gram tablet per gallon of growing medium in the container is the proper rate.

Growth of plants should be noted and fertilizer rates changed as plant foliage and growth rate show a need for more or less fertilizer. As the growing season comes to an end (August and September in the mid-Atlantic states), nitrogen application should be reduced to slow growth and start the hardening-off process in preparation for winter. Additional phosphorus fertilizer also helps plants to set flower buds and harden off for winter. Watering less also tends to harden plants off for winter.

One advantage of Osmocote is that as temperatures lower at the end of the summer, the plant nutrients are released more slowly. This assists in slowing growth and hardening off plants for winter.

Fertilizer programs vary for each type of plant grown, so the exact fertilizer schedule should be set up before plants are potted. Also, some growing media hold fertilizer better than others. Bark and sand do not hold nutrients well.

Disease and Insect Control

Disease control is best started with a sterile planting medium, a sanitary growing site, and sanitary containers. If the site is clean and drains well, if the planting mix is pasteurized, and if new or pasteurized containers are used, the next source of disease or insects to check for is from the plants that are going into the containers. These plants should be examined prior to planting. Diseased plants should be destroyed and plants with insects should be sprayed immediately after placement in the growing area. If these precautions are taken, insect and disease problems should be minimal. Close inspection during the growing season will uncover problems, and spray schedules or other control methods may be set up. Because the chemicals used vary, and new materials are constantly coming on the market, no specific recommendations will be made here. Consult your local extension office or state agriculture college for details.

Pruning

Container-grown plants are pruned mainly to shape the plant or make it more compact. Plants for which this is not a problem, such as ground cover plants, are usually not pruned. Plants should be pruned to help them develop the desired shape for that particular plant. The pruner must know if a plant should have one central leader stem or if it grows naturally as a many-stemmed plant. The pruner then prunes to assist the plant in that development.

Pruning is best done immediately after planting and during the growing season to remove long, unattractive shoots. After this first season pruning, plants with single main trunks usually have side branches shortened each spring to promote compact growth. Pines usually have the new shoots (candles) cut to one-half their length in late spring before the needles have expanded. If this is done on all shoots, a compact tree will result. Plants such as boxwood or hedge-type plants may be sheared to the desired shape.

Pruning may be done in winter instead of spring to spread labor demands if cold injury to the pruned branches is not a problem in the growing area. Chemical pruning of azaleas was mentioned earlier under the "Container Spacing" section.

Winter Storage

In growing areas where winter temperatures drop much below freezing, some sort of protection must be provided to container-grown plants. The most serious winter injury is due to lack of root hardiness. The roots of most plants are not as hardy as the top part of the plant. Because the containers are up on top of the soil, the temperature around the roots is nearly the same as the air temperature. Roots of many woody plants die when the soil or growing medium temperature drops below 20°F (27°C). Plants with extensive root damage from cold usually die the following spring.

If plants are to survive temperatures at or below those listed in Figure 24–13, some sort of protection must be provided. Some growers place plants in plastic-covered greenhouses where temperatures can be controlled with heaters to just above killing temperatures (Figure 24–14). Other growers lay plants on their sides or stand them up under a low structure on the soil and stack containers as close together as possible without damaging the foliage. The plants are then covered with a single layer of clear, 4 mil polyethylene. Straw or other insulating material 4 to 6 inches thick is placed on top of this plastic and another layer of clear, 4 mil polyethylene is spread over the straw. Edges are sealed with soil and the plastic held down with old tires or other weight to keep the wind from blowing or tearing it.

Mice may get into the storage area and cause damage to plants over the winter. It is important to use some sort of poison to control these rodents.

A thermo-blanket system may also be used to protect container plants from winter root damage (Figure 24–15). The blanket is a microfoam made of white Styrofoam. It has high insulating value and transmits about 50 percent light. Because it reflects 50 percent light, temperatures under it stay uniformly cool. Contact a local grower or your agricultural college for more information on this new material.

TAKING PLANTS FROM WINTER STORAGE In spring, when temperatures get above the danger level, uncover the plants. Set them upright, and water well.

As mentioned earlier, some containers insulate better than others. This gives some root protection if the temperatures dropped below the root-killing level for short periods of time such as a few hours or one night only. The main advantage of the container insulating value, however, is to help keep the roots cooler in summer. This could help to control some root rot problems in plants such as rhododendrons. Many root-rotting organisms cause infection only at high soil temperatures.

If high summer temperatures are a problem with container crops, shade should be supplied to the growing area. Lath houses using wooden slats or a snow fence overhead give about 40 to 50 percent shade. Shade fabric, either plastic or cloth, may be used to provide varying degrees of shade. These materials require the support of posts and two-by-fours or other material to hold them in place above the plants.

Shade not only lowers temperature to the plant's top and to the roots in containers but also reduces the amount of water needed for irrigation as temperatures are lowered. Foliage burn is greatly reduced.

Shade in winter is also helpful in reducing foliage burn caused by the sun heating foliage and depleting moisture. This reduces leaf browning and results in a more attractive plant that is ready for sale the following spring.

Transplanting from the Container

One of the biggest problems when transplanting container-grown plants is drying out of the root ball. This results in root damage and, if severe enough, death of the plant. This problem is caused by the difference in texture of potting medium and the soil the plant is transplanted into. Soilless potting mixes are coarse and dry out rapidly. Most soils are finer in texture and hold more moisture. The soil may be moist, while the container plant's root ball is dry. The plant can die from lack of moisture in a moist soil because the plant's roots are confined to the original dry root ball. To solve this problem, plant roots must be pulled out of the ball or the medium must be washed or pulled away from them so roots can

SCIENTIFIC NAME	COMMON NAME[1]	°F	°C
Buxus sempervirens	Common Box	27	−2.7
Cotoneaster congesta	Cotoneaster	25	−3.8
Ilex crenata 'Dazzler'	Dazzler Holly	25	−3.8
Cotoneaster dammeri	Lowfast Cotoneaster	23	−5
Daphne cneorum	Rose Daphne	23	−5
Euonymus fortunei vegetus	WinterCreeper Euonymus	23	−5
Hypericum spp.	St. Johnswort	23	−5
Ilex cornuta 'Nellie R. Stevens'	Holly	23	−5
Ilex crenata 'Convexa'	Convexleaf Japanese Holly	23	−5
Ilex crenata 'Helleri'	Dwarf Holly	23	−5
Ilex crenata 'Hetzi'	Hetz Holly	23	−5
Ilex crenata 'Stokes'	Stokes Holly	23	−5
Ilex meserveae	Holly—Blue Girl and Blue Boy	23	−5
Ilex opaca	American Holly	23	−5
Magnolia x soulangeana	Saucer Magnolia	23	−5
Mahonia beali	Leatherleaf Mahonia	23	−5
Pyracantha coccinea 'Lalandi'	Firethorn	23	−5
Cornus florida	Dogwood	22	−5.6
Euonymus patens	Spindle Tree	22	−5.6
Ilex crenata 'San Jose'	San Jose Japanese Holly	22	−5.6
Magnolia stellata	Star Magnolia	22	−5.6
Cotoneaster dammeri 'Skogholmen'	Skogholmen Cotoneaster	20	−6.7
Leucothoe catesbaei	Leucothoe	20	−6.7
Rhododendron prunifolium	Deciduous Azalea Prunifolium	20	−6.7
Viburnum plicatum tomentosum	Double-File Viburnum	20	−6.7
Euonymus alatus	Winged Euonymus	19	−7.2
Rhododendron 'Hino Crimson'	Azalea Hino-Crimson	19	−7.2
Stephanandra incisa	Cutleaf Stephanandra	18	−7.8
Cotoneaster horizontalis	Rock Spray Cotoneaster	17	−8.3
Cryptomeria japonica	Japanese Cryptomeria	17	−8.3
Rhodoendron 'Exbury Hybrid'	Deciduous Azalea Group	17	−8.3
Taxus media 'Hicksi'	Hicks Yew	17	−8.3
Cytisus x praecox	Warminster Broom	16	−8.9
Ilex glabra	Inkberry Holly	16	−8.9
Koelreuteria paniculata	Golden Rain Tree	16	−8.9
Acer palmatum 'Atropurpureum'	Japanese Maple	15	−9.4
Euonymus fortunei 'Argenteo-marginatus'	WinterCreeper Euonymus	15	−9.4
Euonymus fortunei 'Carrieri'	WinterCreeper Euonymus	15	−9.4
Hedera helix 'Baltica'	Baltic English Ivy	15	−9.4
Kalmia latifolia	Mt. Laurel	15	−9.4
Pachysandra terminalis	Japanese Spruce	15	−9.4
Pieris japonica	Japanese Andromeda	15	−9.4
Rhododendron 'Purple Gem'	Rhododendron Purple Gem	15	−9.4
Rhododendron schlippenbachi	Deciduous Azalea Species	15	−9.4
Viburnum carlesii	Korean Spicebush	15	−9.4
Vinca minor	Periwinkle	15	−9.4

Source: Dr. France R. Gouin, University of Maryland Horticulture Department

Figure 24–13 Root-killing temperatures of container-grown ornamentals in winter storage.

Cotoneaster adpressa praecox	Cotoneaster	12	−11.1
Juniperus conferta	Blue Pacific Juniper	12	−11.1
Juniperus horizontalis 'Plumosa'	Andorra Juniper	12	−11.1
Juniperus squamata	Meyeri Juniper	12	−11.1
Taxus x media 'Nigra'	Media Yew	12	−11.1
Mahonia aquifolium	Oregon Grape, Holly	10	−12.2
Rhododendron x 'Gibraltar'	Deciduous Azalea Gibraltar	10	−12.2
Rhododendron x 'Hinodegiri'	Azalea Hinodegiri	10	−12.2
Thuja occidentalis	American Arborvitae	10	−12.2
Euonymus fortunei 'Colorata'	Purpleleaf Winter Creeper	5	−15
Leucothoe fontanesiana	Leucothoe	5	−15
Pieris floribunda	Floribunda Andromeda	5	−15
Juniperus horizontalis 'Douglasii'	Waukegan Juniper	0	−17.8
Rhododendron carolinianum	Carolina Rhododendron	0	−17.8
Rhododendron catawbiense	Catawba Rhododendron	0	−17.8
Picea glauca	Colorado Spruce	−9	−23.3
Picea omorika	Serbian Spruce	−9	−23.3
Potentilla fruticosa	Cinquefoil	−9	−23.3
Rhododendron P.J.M. Hybrids	Rhododendron P.J.M.	−9	−23.3

Temperatures that will injure primary and possibly secondary roots, but will not result in 100% kill of the root systems in moist soils.
Temperature data pooled from: John R. Havis. Root hardiness of woody ornamentals. HortScience 11(4), 1976, pp. 385–386, and Peter L. Steponkus, George L. Good, and Steven C. Wiest. Root hardiness of woody plants. American Nurseryman CXLIV(6), 1976, p. 16.
[1]Common names added by author.

Figure 24–13 Root-killing temperatures of container-grown ornamentals in winter storage. *(continued)*

▲ **Figure 24–14** A plastic-covered poly house used to overwinter container plants. Temperature is kept at or above 28°F (−2°C) at night and cooled during the daytime by opening vents and doors.

▲ **Figure 24–15** Container-grown plants are placed on the ground on their sides and covered with microfoam and white plastic to protect them from low-temperature damage to roots.

be extended at least 2 or 3 inches out into the surrounding soil. Extended roots must be held up away from the ball and soil worked in under and around them firmly *with the hands.* When planting fibrous rooted plants, do not tamp with your feet as this will break tender roots. Completely fill the hole, firming soil around the roots as you go. Mulch and water thoroughly to settle soil around the roots. Keep the roots moist but not wet, especially for the first year after transplanting.

The soil for planting should be improved with organic material such as a 50/50 mix of fine pine bark and sphagnum peat moss. The planting area should be improved in a 3- to 4-foot-wide area or at least three times the diameter of the root ball, but no deeper than the plant root ball. The plant should sit on firm undug soil in the bottom of the hole so it does not sink. Shrubs and trees should be planted no deeper than the top of the root ball. (No soil on top of the root ball.) Use of organic matter is more important with plants that have a fine, fibrous root system such as azaleas and rhododendrons, but most plants grow better in a soil high in organic matter.

▲ **Figure 24–16** A container-grown rhododendron.

Marketing

Container-grown plants can be marketed and transplanted any time the ground is not frozen. Most plants are planted in spring and fall as these are the seasons for best survival. Rainfall and temperatures are usually better for transplanting at this time of year. Figure 24–16 shows a rhododendron ready for market.

When a container plant reaches the size specified in the standards chart, it must either be marketed or transplanted to the next larger container. If plants are allowed to stay in the same container for another year, the roots will grow into a tight spiral ball. This root-bound condition slows or even stops root extension out into the surrounding soil when the plant is transplanted. This results in stunted growth or even death of the plant.

When container plants are retailed through stores without trained personnel to tend them, much damage and loss of quality can result. The biggest problem is improper watering. Top-quality plants can become almost worthless in a period of 1 to 2 weeks if roots are allowed to dry out and die. The plant must be properly cared for until planting in the permanent location is completed.

Today, the **QR code** (Quick Response code) has become an additional marketing tool in the horticultural business (see Figure 24–17). Major world horticultural producers are using QR codes on their plant tags and other marketing materials to more efficiently provide customer support to the general public. QR codes can be read by a smartphone by selecting a reader app for QR codes. When the QR code is scanned, the app will automatically direct the individual to the horticultural company website that provides vital information about the specific plant. This process gives the horticultural business an additional resource to help sell its plants. The QR code is a direct to consumer technique to accommodate the fast-paced world of today.

FIGURE 24–17A AND B Plant pot tags are super marketing tools to help sell plants. Plant tags have a picture of the flower, plant care, and QR code. Information on companion plants that can be chosen for the beautification of the landscape may also be included.

SAFETY FIRST

When working with pesticides and fertilizers, follow the label and recommendations of the manufacturer and soil analysis test. It is also important to exercise caution when using sharp hand tools and when lifting heavy tree containers to avoid personal injury. For more information on occupational health and safety in the workplace, please visit the OSHA site at www.OSHA.gov.

Summary

Container growing of trees and shrubs has developed into a sizable market share of production in the green industry. Many plants are grown for short periods as seedlings, grafts, rooted cuttings, and whips. Other plants are produced for sale to the general public directly in the container. With container growing, more plants can be produced in the same area, but it takes more labor to grow the plants. Container growing requires selecting containers, choosing growing mix, and performing weed control for optimum plant growth. Management skills are needed to implement appropriate container spacing, watering, fertility programs, disease and insect control, pruning, and winter storage to ensure top-quality and economical plants for the market.

Review Questions

1. Discuss four advantages and four disadvantages of growing plants in containers.
2. Name the different types of plants that are typically grown in containers. What types of plants are not as well-suited to container planting? Why?
3. What are some important considerations to keep in mind when choosing a container for a plant?
4. Discuss 10 features a good growing medium provides to container plants.
5. What are some different methods of watering container plants? What are the advantages and disadvantages of each?

Student Activities

1. Transplant a small plant into a commercially used container, and grow the plant to saleable size. Set up a growing schedule for things such as pruning, watering, fertilizing, pest control, and marketing date. Contact a local grower for help if needed.
2. Make an appointment to visit a local container grower and tour the operation, and note the growing system used. Discuss the tour in class. If possible, obtain a few plants from the grower to be grown as a class project or individual projects.
3. Have a professional grower come in to speak to the class and discuss/demonstrate growing and sales techniques.
4. Compete in the FFA Nursery/Landscape Career Development Event (CDE).
5. Using the Internet for research, study landscaping trees. Make a slideshow describing five species of landscape trees with a visual picture of each tree.

Suggested Supervised Agricultural Experience (SAE) Activities

1. Conduct a research SAE that compares the growth of container plants when placed in different spacing arrangements. Record the information and reach a conclusion as to which spacing is optimum.
2. Purchase and install container plants as part of your own landscaping business. Keep accurate records to qualify for an entrepreneurship SAE.

Additional Resources

http://urbanext.illinois.edu/containergardening/combining.cfm

http://extension.oregonstate.edu/nep/edmaterials/Gardening/infosheet-containerplants.pdf

SECTION 7

Using Plants in the Landscape

UNIT 25

Annual Bedding Plants

Objective

To plant and maintain annual flowers in the landscape.

Key Terms

accent
aesthetics
annuals
commercial
direct seeding
exposure
multifaceted
residential

Competencies to Develop

After studying this unit, you should be able to:

- identify four uses of annual flowers.
- design a bed layout using annual flowers.
- explain the steps in preparing the soil for annual flowers.
- list the seven steps in the aftercare of annual flowers.
- demonstrate proper transplanting techniques for annual flowers.

Materials

- sources of color pictures of annuals, such as nursery catalogs
- seed flats and growing medium in which to start plants
- seed and nursery catalogs (one catalog per two students)
- planting area in which to design a flower bed, using annuals
- all necessary ingredients for preparation of flower beds (peat moss, fertilizer, etc.)
- smartphone apps:
  - Pest Portal
  - Spectracide
  - Calibrate My Sprayer
  - GardenCompass
  - Ortho Problem Solver
  - LikeThat Garden

Annuals, as evidenced by the name, are plants that complete their life cycle in 1 year. The plant starts from seed, grows, blooms, sets seed, and dies in one season. Both vegetable plants and flower plants can be classified as annuals. Although both types bear flowers, the flowering plant is usually used for landscaping purposes. Professional landscapers are making more use of annuals in **residential**, public, and **commercial** landscapes.

The market for bedding plants (annuals) has grown in the past 10 years. Today, a wide variety of retail outlets such as roadside markets, farmers chain stores, garden centers, grocery stores, and shopping malls are selling annuals.

When purchasing bedding plants, it is important to make sure that the plants carry labels that give the price, required light **exposure** (shade or sun), and planting and care instructions.

Flowering Annuals

Flowering annuals are grown to **accent** and lend color to landscapes a short time after they are planted. They are usually started from seed indoors early in the spring and transplanted to the garden. The most popular annual flowering plants include marigolds, petunias, zinnias, ageratums, celosias, coleus, portulaca, pansies, and snapdragons. Figure 25–1 shows a mass planting of petunias.

Courtesy of Ball Horticultural Company

▲ **Figure 25–1** A planting of petunias adds a solid mass of color to the landscape.

Courtesy of Ball Horticultural Company

▲ **Figure 25–2** Impatiens used as a border planting to accent a garden walk in a park.

> **GREEN TIP** Annual bedding plants add a splash of color to sustainable environments that impact the attitude and well-being of each individual who experiences them.

Flowering annuals are generally used to

- provide a mass of color around a house foundation, in flower beds, or in front of evergreens (Figure 25–2) to enhance **aesthetics** of the residential home and/or park area.
- fill spaces between shrub plantings or other perennials and give color when these plants are not blooming.
- provide color in bulb beds after the bulbs have bloomed.
- supply cut flowers.
- plant along fences or walks.
- cover bare spots between or in front of larger shrubs.
- create seasonal color (Figure 25–3).

Courtesy of Ball Horticultural Company

▲ **Figure 25–3** Annuals are valued and used for seasonal color in the landscape.

- create special emphases of landscape features (i.e., statuary, fountains, and other artistic forms).
- direct traffic patterns in a garden.

© 2017 Cengage Learning®

▲ **Figure 25–4A** The integration of *Celosia sp.*, *Impatiens sp.*, and *Caladiums sp.* creates a display to entice customers to buy plants to adorn their gardens.

Selecting Flowering Annuals

Great improvements have been made in the quality and vigor of annual flowering plants. One example is the triploid hybrid marigold, which blooms within 6 weeks from seed. It sets no seed and blooms longer than the traditional marigold. If seed is allowed to set on annual plants, the seed production causes strength to be sapped from the plants and reduces blooming. For this reason, faded blooms should be cut from all annual plants. In the horticultural industry, this is referred to as "deadheading."

When deciding upon which annuals to grow, first consider the purpose of the plant and where it is to be planted. Be sure to consider height, keeping shorter plants in the front of beds. Select plants with colors that will blend in well with one another (Figure 25–4). Figure 25–5 provides other information concerning various annuals.

Courtesy of Ball Horticultural Company

▲ **Figure 25–4B** Annuals such as gazania and dusty miller create an aesthetic, pleasing flower bed.

NAME	HEIGHT	SPACING	COLOR	USES	HOW TO START	REMARKS
Ageratum (floss flower)	6″–12″	6″–9″	blue, pink, white	edging	seed and transplant**	Compact, excellent bloom. Needs full sun.
Alternanthera	12″	12″	a variety of colors	accent	seed	Creates background for flowering plants.
Antirrhinum (snapdragon)	12″–36″	6″–8″	yellow, pink, white, blue, red, salmon, rose, bronze, orange	cut lowers, flower bed	seed and transplant	Must be staked for straight flower spikes. Needs full sun. Excellent background plants.
Aster (aster)	8″–36″	10″–12″	blue, red, salmon, pink, white, yellow	flower beds	seed and transplant	Excellent cut flowers
Begonia (begonia)	6″–12″	6″–10″	pink, red, white	edging, potted plant, window box, hanging basket	cuttings or seed	May be used as a houseplant. Needs direct light.
Browallia (browallia)	12″–18″	8″–10″	blue, white	hanging basket, window box	cuttings or seed	Makes an attractive houseplant. Needs full sun.
Calendula (potted marigold)	12″–24″	10″–12″	yellow, orange	flower bed, hanging basket	seed and transplant	Flower petal used in cooking stews to add color. Needs full sun.
Callistephus (China aster)	12″–24″	10″–12″	blue, purple, white, yellow	flower bed, cut flowers	seed and transplant	Gives excellent summer color. Needs full sun.
Celosia (cockscomb)	12″–30″	10″–12″	red, orange, yellow pink	flower bed, cut flowers	seed and transplant	Excellent dried flower. Needs full sun.
Cleome speciosa (spider plant or spiderflower)	24″–36″	18″–36″	pink, white	flower bed	seed directly*	Makes an attractive houseplant.
Coleus (coleus)	12″–24″	10″–12″	red, bronze, yellow, pink foliage	flower bed, hanging basket	seed and transplant	Needs full to partial sun. Beautiful foliage plant.
Coreopsis (coreopsis)	12″–24″	6″–10″	yellow, red	flower bed	seed directly	Good for cut flowers.
Cosmos (cosmos)	48″–72″	12″–18″	red, pink, yellow	flower bed	seed and transplant	Keep near back of beds.
Dahlia variabilis (dahlia)	12″–24″	12″–15″	red, pink, yellow rose	flower bed, cut flowers	seed and transplant	Profuse bloomer. For maximum bloom, sow several weeks before other annuals.

* Seed directly—sow the seed directly into the soil where the plants are desired.
** Seed and transplant—start the seed indoors (in greenhouse, hotbed, or portable seed germination case). As the seedlings develop, they are transplanted from the seed flat to other containers where they grow until ready to be set out.

Figure 25–5 Bedding plant chart for annual flowers. *(continued)*

NAME	HEIGHT	SPACING	COLOR	USES	HOW TO START	REMARKS
Delphinium (larkspur)	18″–24″	8″–10″	blue, pink, white	flower bed, cut flowers	seed and transplant	Grows best in peat pots. Difficult to transplant.
Dianthus (pink)	6″–18″	6″–8″	white, pink	edging, cut flowers	cuttings or seed	Very fragrant flowers.
Dimorphotheca (cape marigold)	12″–18″	8″–18″	yellow, orange, white	flower bed	seed and transplant	Grows well in dry areas.
Dolichos lablab (hyacinth bean)			purple	accent	seed	Vine texture
Gaillardia (gaillardia)	12″–24″	8″–10″	red, orange	cut flowers, flower bed	cuttings	Loves seashore conditions. Does well in dry areas.
Gazania (treasure flower)	6″–10″	8″–10″	orange, yellow, white, tangerine, bronze	flower bed	seed and transplant	Bright blooms.
Gomphrena (globe amaranth)	9″–24″	8″–10″	blue, pink, whte	cut flowers, flower bed	seed and transplant	Makes good dried flower. Collect and hang in dry, dark place.
Gypsophila (baby's breath)	12″–18″	10″–12″	pink, white	flower bed, cut flowers	seed directly	This is the annual form; there is also a perennial plant.
Helianthus (sunflower)	18″–60″	12″–36″	yellow, brown	flower bed	seed directly	Seeds are edible.
Helichrysum (strawflower)	24″–36″	8″–10″	yellow, red, white	flower bed	seed and transplant	Makes good dried flower. Cut; hang in dry, dark place.
Hypoestes (polka-dot plant)	8″–24″	10″–12″	burgundy, red, rose, white, pink	hanging baskets, window boxes, flower bed	seed and transplant	Foliage accented by splashed colored leaves.
Iberis (candytuft)	10″–18″	6″–8″	white, pink, red	edging	seed and transplant	Makes attractive ground cover.
Impatiens (impatiens)	6″–18″	12″–18″	pink, red, white, multicolor	edging, hanging basket	seed and transplant	Does best in shaded area.
Ipomoea (morning glory)	60″–120″	10″–12″	blue	trellis	seed directly	Vine.
Lantana (lantana)	12″–36″	2″–15″	yellow, blue, red	hanging basket, flower bed, potted plant	cuttings	Makes excellent topiary plant for patio.
Lathyrus (sweet pea)	36″–60″	6″–10″	pink, blue, white	cut flowers	seed directly	Grows best in cool conditions.
Limonium (statice or sea lavender)	18″–30″	10″–12″	blue, pink, white	flower bed, cut flowers	seed and transplant	Makes good dried flower. Filler flower for bouquets.

Figure 25–5 Bedding plant chart for annual flowers. *(continued)*

NAME	HEIGHT	SPACING	COLOR	USES	HOW TO START	REMARKS
Lisanthius	6″–24″	12″–14″	blue, rose, orchid, white, pink, yellow	flower bed	seed and translpant	Large blooms, drought tolerant.
Lobelia (lobelia)	3″–6″	4″–6″	blue, pink, white	edging, hanging basket, potted plant	seed directly	Very attractive in hanging basket.
Lobularia (sweet alyssum)	4″–6″	4″–6″	white, pink, blue	window box, edging, hanging basket	seed and transplant	Blooms all summer.
Matthiola (stock)	12″–36″	8″–10″	white, pink, yellow, red	cut flowers, flower bed	seed and transplant	Fragrant flowers.
Mirabilis (four-o'clock)	24″–36″	10″–12″	yellow, red, white	flower bed	seed and transplant	Withstands city conditions.
Myosotis (forget-me-not)	6″–12″	6″–8″	blue, pink	edging, cut flowers	seed directly	Makes attractive cut flowers.
Nicotiana (flowering tobacco)	18″–24″	10″–12″	rose, white	cut flowers, flower bed	seed and transplant	Fragrant flowers after dark.
Osteospermum (African daisy)	15″	12″	white, pastels orange, red	accent, urn, flower bed	seed	Excellent flower quality.
Papaver (poppy)	18″–36″	10″–12″	red, yellow, white	cut flowers, flower bed	seed directly	Grows in masses. Bright colors.
Pelargonium (geranium)	12″–18″	10″–12″	red, white, pink, lavender	flower bed, window box, hanging basket	cuttings and seeds	Excellent flowering plant.
Pennisetum glaucum (millet)	24″	24″	bold metallic leaves, purple grassy leaves	urn, accent, flower bed	seed	12″ cattail spikes.
Pentas (star flower)	18″	18″–24″	pink, white, red, pastel	flower bed, hanging basket, planter	seed	Flower color full season.
Petunia (petunia)	6″–18″	10″–12″	pink, red, white, blue	edging, flower bed, hanging basket	seed and transplant	Most widely used annual. Use 'Wave,' 'Magic Carpet,' 'Dazzle,' 'Cascade.'
Phlox (phlox)	6″–18″	6″–8″	pink, blue, white, red	cut flowers, flower bed, window box, hanging basket	seed directly or transplant	Very intense colors.
Portulaca (rose moss)	4″–6″	3″–6″	red, orange, yellow	edging, hanging basket	seed directly	Reseeds by itself.
Rudbeckia (black-eyed Susan)	24″	24″	yellow daisy, dark center	flower bed, accent	seed	'Indian Summer,' 'Sonora Marmalade' are cultivars.

Figure 25–5 Bedding plant chart for annual flowers. *(continued)*

NAME	HEIGHT	SPACING	COLOR	USES	HOW TO START	REMARKS
Salpiglossis (giant velvet flower)	18″–36″	10″–12″	gold, scarlet, rose, blue	flower bed, cut flowers	seed and transplant	Very small seed.
Salvia (scarlet sage)	10″–36″	6″–12″	blue, red, white	cut flowers, flower bed	seed and transplant	Very showy color.
Scabiosa (pincushion flower)	30″–36″	12″–14″	multicolor	cut flowers	seed and transplant	Variety of plants from which to select.
Spilanthes (peek-a-boo)	18″	24″	yellow globe	flower bed, cut flowers, accent	seed	Companion crop to marigolds.
Tagetes (marigold)	6″–48″	6″–12″	orange, yellow	flower bed, edging, potted plant, hanging basket	seed and transplant or seed directly	Wide range of varieties. No insect problems.
Thunbergia (black–eyed Susan vine)	24″–60″	10″–12″	orange	climbing vine, hanging basket	seed and transplant	Nice addition to a summer/fall bouquet.
Tithonia (Mexican sunflower)	24″–30″	24″	yellow, orange, red	flower bed, cut flowers	seed	Drought-resistant. Attracts butterflies. Mass flower for bouquets.
Tropaeolum (nasturtium)	12″–18″	10″–12″	orange, yellow	flower bed, window box, hanging basket	seed and transplant or seed directly	Flowers and leaves are edible.
Verbena (verbena)	6″–12″	10″–12″	purple, red, white	flower bed, edging	seed and transplant	Good rock garden plant.
Vinca (common periwinkle)	8″–12″	12″	pink, pastel, white, blue, red, varie-gated colors	flower bed, urn, planter, edging and mounds	seed	Drought tolerant. Loves heat.
Viola (pansy)	6″–8″	6″–8″	multicolor	edging, flower bed, cut flowers, hanging basket	seed and transplant	Gives excellent color in summer.
Zinnia (zinnia)	8″–36″	6″–12″	multicolor	edging, flower bed, cut flowers	seed and transplant or seed directly	Gives excellent color in summer.

Figure 25–5 Bedding plant chart for annual flowers.

Hanging baskets of annual flowers are very popular. These add new dimension to the display of plants and their use in the landscape. Baskets are often used to create an attractive floral display or focal point on the patio or at the main entrance of the home.

Planning and Designing with Seasonal Annuals to Create an Effective Display

The seasonal color of a planting of annuals enhances entryways, eating areas, and outdoor settings, welcoming people to enjoy the aesthetics while relaxing and visiting. Annuals also can entice many species of unique wildlife (Figure 25–6). Factors to consider are clients' individual needs and preferences; flower color, height, and texture; bed size and design; and surrounding landscapes.

- Clients' individual needs, desires, and preferences play a major role in the selection of plant materials. Each designer will make recommendations to ensure that the seasonal exhibit is acceptable to the client.
- Flower color is important for a superb seasonal show. The most eye-catching seasonal designs are those gardens with a monochromatic color (using one color in the entire flower bed). Mixed colors in seasonal designs amplify the beauty of landscapes with interest and purpose. They instill a graphic flow of movement to the overall annual garden. When using mixed colors, one must think about the color wheel and implementing contrasting shades of complementary colors that blend well together to give an outstanding display with the seasonal palate.
- Height of plants in the garden is a vital consideration because it creates interest and depth in the garden. In the design of annual gardens, height has to be integrated with other elements of color, texture, exposure, spacing, and direction. The designer must decide where the height is going to be implemented (i.e., to the back, center, or either end of the garden). This will be determined from how the garden is viewed, that is, from all sides or only from one side. After the location for height has been selected, then the designer determines how much height and the types of taller plants to use to provide the desired accent in the garden.
- Texture is another key building block for accents in the seasonal garden. Texture refers to the size of the plant leaf. Plants with larger leaves (e.g., banana plant or canna lilies) stir interest in landscapes. Fine texture plants have small leaves (i.e., dusty miller or cosmos), which help to accent and blend with the entire flow of the garden design ingredient of aesthetics.
- Surrounding landscapes, for example, shrubs, ground covers, deciduous and evergreen trees, perennials, turfgrass, and hardscapes, necessitate the use of different textures.
- Bed size and design will be influenced by the color, texture, height, availability, and area that needs this seasonal treatment. Beds may be designed in various geometric shapes, for example, circle, crescent, square, rectangle, or serpentine.

▲ **Figure 25–6** Zinnias add beautiful summer color.

TRY THIS Attend a local flower show to view landscape design displays that use annuals. Choose two of the designs, and compare the following categories against each other: color, height, texture, surroundings, and the size of the design area. Write a short report describing your comparisons.

Planting Hanging Baskets

Hanging baskets come in various sizes and shapes. The most commonly used hanging baskets range from 4 to 14 inches in size. All baskets used must provide proper drainage; therefore, they should have drainage holes to promote proper plant growth. The standard medium used to start annuals is also a very good source of medium for hanging baskets.

In planting hanging baskets, the standard rule of thumb is to plant one less plant than the size of the pot. For example, if an 8-inch pot is used, select seven plants and place six of them around the outside perimeter of the pot and the other plant in the center of the basket. Proper care in watering, fertilizing, and deadheading (removal of dead and faded flowers) can help to produce an attractive hanging basket of annuals.

Planning and Designing the Flower Bed

Flower beds for annuals range in size and design according to individual needs and preferences. Figure 25–7 shows the design for a flower bed to be planted between a walkway and a home.

Preparing the Soil

It is best to prepare the soil in the fall before planting the next spring. The soil should be spaded or tilled to loosen it. Organic matter may be added in the soil by applying peat moss, pine bark fines, composted sewage sludge, composted leaves and grass clippings (Leafgro), and composted animal manure. The addition of sphagnum peat moss and pine fines mixed 50/50, applied 2 to 4 inches deep, and dug into the soil, improves the aeration, soil drainage, and moisture-holding capacity of the soil.

PROCEDURE
Growing Flowering Annuals

For a summer of color with annuals:

1. Decide which annuals to grow. Outline the plan for the flower bed in writing. Probably the easiest plants for the beginner to grow are marigolds and zinnias.
2. Seed 6 to 8 weeks prior to outside planting date (frost-free date). Figure 25–8 provides a schedule for when to start bedding plants in the greenhouse.
3. Prepare the soil well ahead of the planting date. Add organic amendments, fertilizer, or lime as needed.
4. Plant according to the landscape plan after the frost-free date.
5. Water, weed, and mulch as needed.
6. Pinch off all faded blossoms.
7. Provide for insect control.
8. Enjoy cut flowers.

When planting in beds that contain bulbs, do not dig deeply enough to damage the bulbs. Be sure that the bed does not dry out completely when dug. This drying could also damage the bulbs.

Before preparing new beds, test the soil for drainage. To do this, dig a 10-inch hole and fill it with water. Fill it with water again the following day. If all the water has not seeped into the soil within 10 hours on the second day, drainage is a problem. To improve drainage, raise the bed level above the normal soil level, dig a shallow trench around it, and add pine bark to the soil.

Fertilizing

To fertilize the bed, add about 2 pounds of a 5-10-10 fertilizer per 100 square feet in the spring as the bed is being prepared. If the soil pH is below 6.0, add ground limestone at the rate of 5 pounds per 100 square feet and rake it into the soil surface. If the plants do not have a healthy, dark green color, sprinkle a 10-10-10 fertilizer lightly around each plant on the soil surface. Do not place fertilizer on or against the plant.

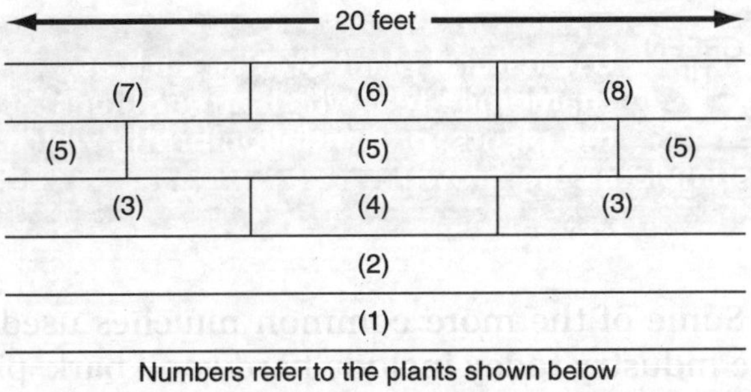

(2) Alyssum (snow crystal)

(1) Dianthus (coral lace mix)

(3) Salvia (red hot sally)

(4) Ageratum (blue puffs)

(5) Lisianthus

(6) Celosia (century)

(7) Aster (master)

(8) Snapdragon (Maryland)

▲ **Figure 25–7** One way to beautify a landscape is to plant a flower bed between a walkway and a home. Such a planting could be designed according to this plan: (1) Dianthus (coral lace mix), (2) Alyssum (snow crystal), (3) Salvia (red hot sally), (4) Ageratum (blue puffs), (5) Lisianthus, (6) Celosia (century), (7) Aster (master), and (8) Snapdragon (Maryland).

PLANT SCHEDULE FOR STARTING BEDDING PLANTS IN THE GREENHOUSE

NAME	WEEKS TO A SALEABLE PLANT	SOWING DATE
FLOWERS		
Ageratum	12–14	Jan. 22
Begonia	13–14	Dec. 19
Celosia	5–6	Feb. 14
Coleus	6–8	Feb. 14
Dusty Miller	12–14	Jan. 22
Geranium	12	Jan. 22
Impatiens	14	Dec. 19
Marigold	6–8	March 1
Petunia	12–14	Jan. 22
Salvia	6–8	March 1
Zinnia	6–8	March 10
VEGETABLES		
Broccoli	4–6	Feb. 1
Cabbage	4–6	Feb. 1
Cauliflower	4–6	Feb. 1
Cantaloupe	3–5	March 20
Cucumber	4	March 20
Tomato	6–8	March 1

Figure 25–8 Plant schedule for starting bedding plants in the greenhouse.

Many people are now growing beautiful annuals with soluble fertilizers. To make a soluble fertilizer, mix a complete fertilizer (20-20-20) with water at the rate of 1 tablespoon per gallon of water or according to directions. Fertilizer applied to plants in a solution is more readily available to the plants.

Mulching an Annual Flower Bed

The use of mulch for an annual bed is **multifaceted** in that it creates the aesthetics that enhance the annuals planted, provides a source of organic matter to the plants, helps to control weeds, conserves moisture, and promotes even ground temperatures for improved plant growth. When planting an annual flower bed, apply a 2- to 3-inch layer of mulch before the bed is planted. This procedure will allow for more even distribution of mulch as well as reducing mulching time for the planted bed.

GREEN TIP Use natural plant broths, such as chamomile tea, which strengthens the plants, or dandelion broth, which stimulates the growth.

Some of the more common mulches used in the industry today include hardwood bark, pine bark, cocoa bean hulls, buckwheat hulls, pine needles, well-rotted sawdust, and leaf mold. Note that some of these mulches will influence the pH level of the soil in the annual flower bed, so it is important that the soil pH level is followed closely to ensure that the optimum growth medium is provided for the plants.

Seeding

As mentioned, most annuals are seeded indoors and set outside as plants. Figures 25–9, 25–10, and 25–11 detail this process. The plants are set

▲ **Figure 25–9** To start seed indoors, fill the flat with properly prepared seeding medium. Small, square peat pots have been set in this flat, with one or two seeds placed in each pot. The peat pots are separated and transplanted individually to the flower bed after the danger of frost has passed. Very small seeds are planted in flats without peat pots and are transplanted to the pots at a later date.

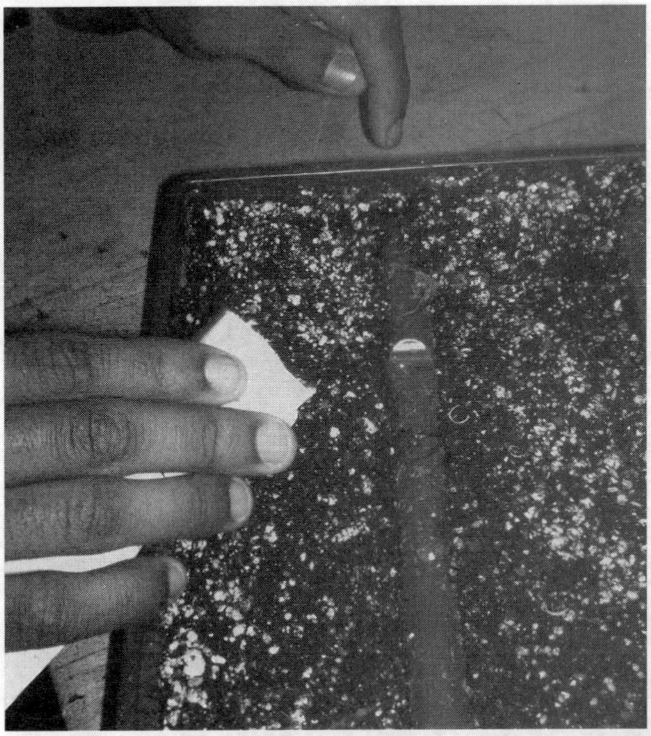

▲ **Figure 25–10A** Annual seeds should be sown carefully to obtain an even distribution of the seeds in rows.

▲ **Figure 25–10B** Most annual seeds should be covered for proper germination.

▲ **Figure 25–11** There are two methods for watering: (1) sprinkle the flat as shown, or (2) set the entire flat in water and allow it to soak up from the bottom. Both methods allow even distribution of water on the seeds. Very small seeds, such as petunia, must not be watered from the top with any force, or they will be washed down into the planting medium too deeply.

▲ **Figure 25–12** Direct seeded sunflowers.

outside in the flower bed as soon as the danger of frost has passed. Some annuals may be seeded directly outdoors; this is called **direct seeding**. Direct seeding is the process of planting seeds in the soil and allowing them to germinate. Some examples of these are baby's breath, cornflower, globe amaranth, poppy, cleome, strawflower, sweet alyssum, sweet pea, and sunflowers (Figure 25–12). Seeds of these plants should be sown as early in spring as possible.

Wait until the soil warms to about 60°F before setting the started plants outdoors. Seedlings in peat pots are set directly into the soil. Very small seeds that are started in flats will probably require transplanting to a larger container before setting outside. These may be replanted to peat pots so that they can establish better growth prior to being planted outside (Figure 25–13).

Plant seedlings at about the same depth as the depth at which they grew in the starting container. Firm the soil around the roots and water well. It is helpful to add a starter solution of 1 tablespoon of a water-soluble, high-phosphate fertilizer in 1 gallon of water after transplanting. Space the plants according to the bedding plant chart (refer to Figure 25–5).

▲ **Figure 25–13** To transplant the small seeded annual to a larger pot, first grasp the seedling by its true leaves. Gently dislodge it from its original pot and place it in the new pot. Fill the pot with soil mixture to the soil level of the original pot. Firm the soil by pressing down lightly with the thumb and index finger.

PROCEDURE
Caring for the Plants

Care for the rest of the season is relatively simple (Figure 25–14).

1. Water if the soil becomes too dry.
2. Pull weeds, apply preemergence herbicide, and mulch for weed control.
3. Prune all faded flower blossoms (deadheading).
4. Control insects and diseases (contact the local extension service for controls to use in your area).
5. Fertilize with a time-release fertilizer (14-14-14) according to application directions.
6. Cut flowers as desired for use in the home.
7. Prune or pinch when necessary for better growth. When the plants are herbaceous (soft stemmed), the stem may be shortened by pinching with the fingers. The pinching will result in a thicker, more attractive plant.

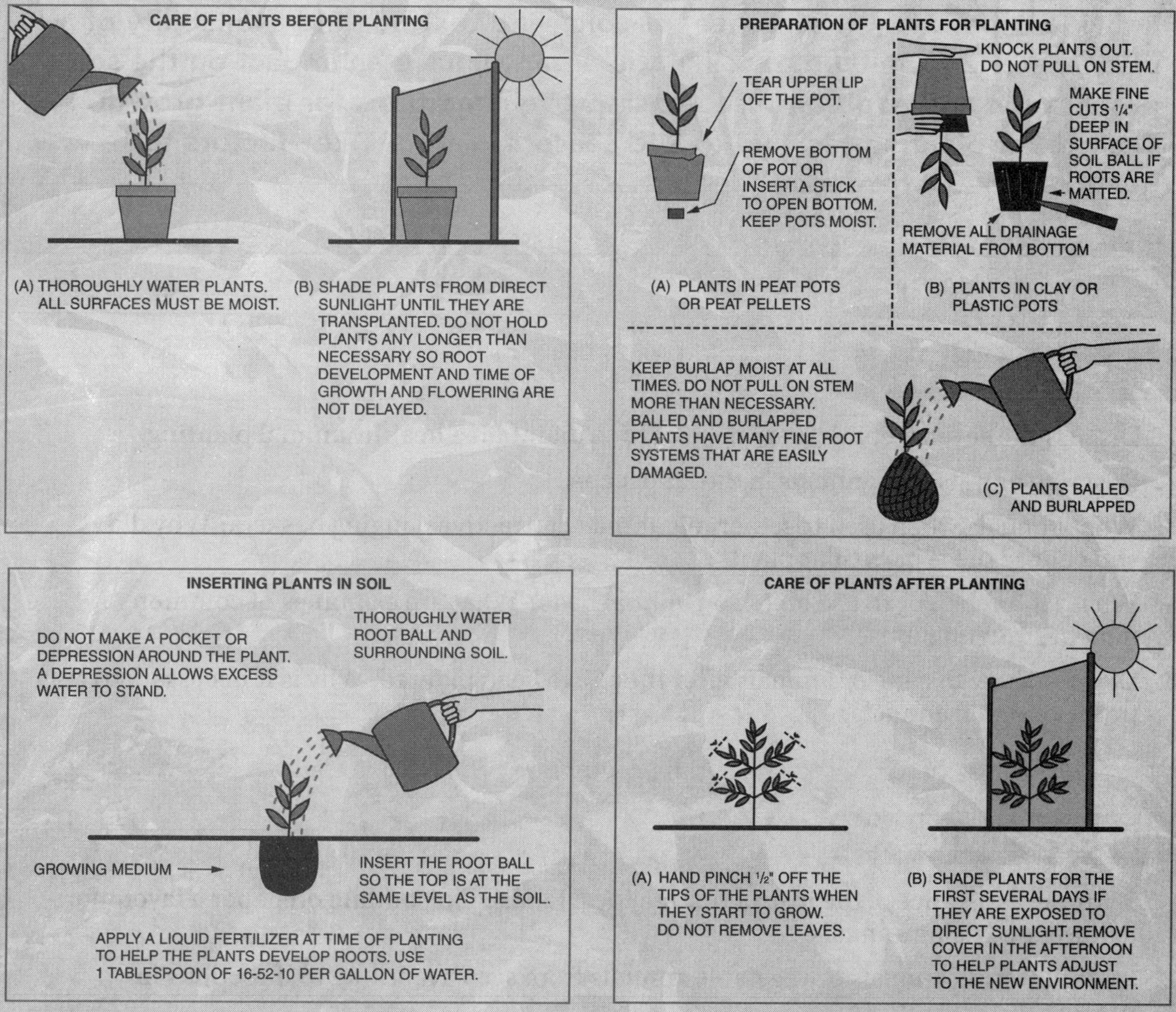

Adapted from the US Department of Agriculture

▲ **Figure 25–14** Care of plants before and after planting.

SAFETY FIRST

OSHA's regulations ensure that employers and employees know about the work hazards and how to protect themselves to prevent injuries in the workplace. For more information on occupational health and safety in the workplace, please visit the OSHA website at www.OSHA.gov.

Summary

Annual bedding plants create a splash of color in the landscape. This unit discussed four uses of annuals to enhance the landscape for a growing season. The design of a bed layout will vary with individual creativity, preference for color and texture, and availability of plant material. The type of soil in the area will have an impact on the soil preparation and additions used to improve conditions for plant growth. The proper planting techniques and seasonal care are key factors to beautiful annual gardens.

Review Questions

1. Discuss the seven steps that are necessary to maintain a healthy annual planting.
2. Describe four uses of annuals in the landscape.
3. What annuals when used together might make attractive hanging baskets? Why did you choose these particular plants?
4. Why is it important to use mulch on annual beds? What are examples of common mulches you might use?
5. Describe how to care for annuals after they have been planted. Why is it necessary to prune or pinch annuals?

Student Activities

1. With the assistance of the instructor or a local landscaper, outline on paper a layout for a flower bed using annuals.
2. Start flowering annuals or vegetable annuals from seed according to directions in the text.
3. Plant the flowering annuals started on the school grounds or at home. Plant the annuals according to a plan approved by the instructor.

4. Conduct an FFA fundraiser, raising and selling hanging baskets in the spring.
5. Compete in the FFA Nursery/Landscape Career Development Event (CDE).
6. Search the Internet to identify the closest retail locations that sell annual flowers. If possible, visit the locations, and make a list of the annuals available.
7. Write a description and function of each of the following smartphone app: Spectracide, GardenCompass, Ortho Problem Solver, and LikeThat Garden. In your opinion, which app seems most useful for designing an annual garden?

Suggested Supervised Agricultural Experience (SAE) Activities

1. Work installing annuals for a professional landscaping contractor to qualify for a placement SAE.
2. Raise hanging baskets and retail them as an entrepreneurship SAE.

Additional Resources

www.bhg.com/gardening/plant-dictionary/annual/

www.uky.edu/Ag/CCD/introsheets/beddingplantsintro.pdf

www.greenhouse.cornell.edu/crops/annualbed.html

www.prolandscape.com

UNIT 26 Perennials, Ornamental Grasses, Vines, Bamboos, and Prairie Gardens

Objective

To design, plant, and maintain a perennial garden.

Key Terms

accent
fragrances
hardiness
hybridized
mass
perennials

Competencies to Develop

After studying this unit, you should be able to:

- list five perennial flowers.
- design a layout for a perennial border.
- list the environmental factors that affect perennial gardens.
- list three ornamental grasses, vines, and bamboos.

Materials

- sources of color pictures of annuals, such as nursery catalogs
- seed flats and growing medium in which to start plants
- seed and nursery catalogs (one catalog per two students)
- planting area in which to design a flower bed, using annuals
- all necessary ingredients for preparation of flower beds (peat moss, fertilizer, etc.)
- smartphone apps:
  - LikeThat Garden
  - Spectracide
  - GardenCompass
  - Homegrown (Bonnie Plants)

Perennials

Perennials are plants that live from year to year and therefore do not require replanting. The tops may or may not die back in the winter or dry season. Some perennials bloom their first year, but most produce larger, more attractive flowers and develop stronger root systems in successive seasons. Perennials may be small flowering plants, shrubs, or flowering trees. Flowering trees are called *hardy perennials*. A *herbaceous perennial* is a nonwoody plant that lives from year to year but dies back to the ground each winter and grows back the following spring (e.g., hosta, hollyhocks, and daylilies).

Perennials have made a resurgence in the United States in the past several years, and many hardy perennials are native to this country. The demand for perennials has expanded with the growing interest of the general public in gardening and landscaping and the desire to **accent** with plants that will come back each year. Perennials adorn gardens with color and foliage, and they result in a long-term garden meant to be enjoyed. This interest in perennials has even resulted in the formation of the Perennial Plant Association, which each year selects a "Perennial Plant of the Year."

As mentioned previously, the marketing of perennials has expanded to include the use of smartphone apps with QR codes. This tool assists customers, especially novices, by providing instant feedback about a particular perennial, such as light and care needs, before they purchase it.

Many of the early perennial gardens contained the old favorites such as chrysanthemums, daylilies, irises, peonies, and delphiniums. Many of these plants have been in borders for more than 100 years. With the introduction of new cultivars, perennials have been **hybridized** to give a variety of colors, forms, textures, **fragrances**, and **hardiness** levels. Europeans have always been strong in the use of perennials in their gardens, and they have introduced many of the new cultivars. As we learn more about these perennials and their uses, demand for them will grow.

As densely populated areas continue to grow, and townhouses become more popular, modern perennial gardens are becoming smaller. With this scaled-down approach to landscaping, perennial plants have become an outdoor extension of indoor living, featuring unique color designs that create a special relationship with nature and the environment.

GREEN TIP Perennials, ornamental grasses, vines, bamboos, and prairie gardens have a sustainable impact on the environment. They regenerate the biomass of plant material resources to improve our biological system, conserve water, and reduce pollution of our environment. Ornamental grasses, such as switchgrass, for example, are being sought after to produce alternative fuels. Vines are also employed to ecologically conserve energy through heating and cooling buildings with the use of vertical gardens and rooftop gardens.

Perennial Beds and Location

The perennial bed (see Figure 26–1), should be designed first on paper to fit the garden area for a particular bed. Some basic guidelines to follow are the following:

- Use an island bed with an informal-shaped border, surrounded with a lawn area. The taller plants are planted in the center, with the design using a sequence of tall-to-short plants, with the shortest plants on the outer edge. The bed should be 5 feet wide to allow the individual to work the garden from all sides without walking over the bed. With the resulting reduction in soil compaction, damage to the plant

▲ **Figure 26–1** Perennial garden in full bloom with flowering perennials, multitude of texture, and movement with the grasses to create fantastic curb appeal.

roots will be minimal, thus lowering maintenance requirements. If a walkway through the bed is part of the design, then the size of the garden may be from 9 to 15 feet wide.

- Use a perennial border planted along a fence or property line as a divider. It should be up to 5 feet wide and as long as desired by the designer and the gardener.

SOIL, LIGHT, AND VIEW When planting a perennial bed, other factors must be considered such as environment, soil type, drainage, depth of the topsoil, organic matter content, and light exposure (full sun, partial shade, or full shade). The view from the house, patio, or lawn will also have a bearing on the location of the perennial bed or border.

Organic matter resources are peat moss, pine bark fines, composted leaves and grass (Leafgro), composted sewage sludge, composted animal manure, and other horticultural biosolids. Using additional organic matter mixed with the soil will improve the chemical reaction of the soil media known as the cation exchange capacity.

Selection

Perennials should be selected by personal preference, color scheme, texture, shape, growth habit, and the microclimate of the local area. Examine plants grown in the area, or ask for advice from a nursery or garden center, extension service, or horticultural instructor. The color pictures in seed and nursery catalogs are another source that can be helpful in plant selections. Check Figure 26–2 to determine whether the plants selected will grow under the light and soil conditions in the area in which they are to be planted and to be sure they will bloom during the season when added color is needed in the landscape (mid-spring, late summer, and fall). Plant flowers that bloom at the same time together for the best display of color.

When choosing plants, it is important to **mass** varieties in groups of three, five, nine, or more. A single variety of perennial may be used as long as the growth is full. For example, the use of daylilies as a single variety is good because they tend to grow in a clump and have several flowers blooming at one time. Plants with small growth habits should not be used individually as they give a sparse appearance in the garden.

To concentrate on early color, one must consider the use of spring-flowering bulbs. No perennial garden should be without some small, early flowering bulbs that multiply freely over the years. Some bulbs that can be used are the grape hyacinth, snowdrop, and crocus. Another variety to use is the lovely miniature iris *(Iris reticulata)* with fragrant purple flowers that appear in March. Daffodils are also important to color an early spring perennial garden.

Designing a Perennial Garden

Landscape design is often achieved by some designers utilizing computer-aided design (CAD) in a 3D image to showcase perennial gardens. This process of 3D imaging allows the designer to create picturesque gardens for the client. The client can then see what the garden will look like when it starts to grow and flower.

To design the perennial garden, you must consider the soil, time of blooming, and space, and have a plant list, including the colors and varieties that are best suited for your geographic area. Using a CAD software program or design equipment such as a T-square, a triangle, architect's scales, a #2 pencil, and a drafting board, lay out the desired perennial garden for the area. Keep in mind the height, color, and spread of the plants. A recommended rule to follow for planting distance is to have perennials planted at one-half of the mature height of the plant. For example, if a plant grows to 3 feet, then the distance between plants should be 18 inches O.C. (on center) (Figure 26–3).

In designing the perennial garden, the use of some deciduous shrubs as well as evergreens creates interest. The shrubs create a background for, and contrast with, the perennials.

You could also implement the border design that Gertrude Jeckyll (an artist and gardener of the 1800s) used. She was interested in perennials and in using them in borders. She had the unique ability to blend and contrast perennials having different sizes, shapes, and foliage. She did not limit her gardens to the use of perennials only, but worked with biennials and annuals as well. In designing the garden, she would create a sequence of blooming flowers from spring to fall by arranging the garden on a north-south axis, using blue flowers (cool colors) on the north and, moving south, yellow and then red flowers (warm colors). About

NAME	WHEN TO PLANT SEED	EXPOSURE	GERMINATION TIME (DAYS)	SPACING	HEIGHT	BEST USE	COLOR	REMARKS
Achillea millefolium (yarrow)	early spring or late fall	sun	7–14	36″	24″	borders, cut flowers	yellow, white, red, pink	Seed is small. Water with a mist. Easy to grow.
Alyssum saxatile (golddust)	early spring	sun	21–28	24″	9″–12″	rock garden, edging, cut flowers	yellow	Blooms early spring. Good in dry and sandy soils.
Anchusa italica (alkanet)	spring to September	partial shade	21–28	24″	48″–60″	borders, background, cut flowers	blue	Blooms June or July. Refrigerate seed 72 hours before sowing.
Anemone pulsatilla (windflower)	early spring or late fall for tuberous	sun	4	35″–42″	12″	borders, rock garden, potted plant, cut flowers	blue, rose, scarlet	Blooms May and June. Is not hardy north of Washington, DC.
Anthemis tinctoria (golden daisy)	late spring outdoors	sun	21–28	24″	24″	borders, cut flowers	yellow	Blooms midsummer to frost. Prefers dry or sandy soil.
Arabis alpina (rock cress)	spring to September	light shade	5	12″	8″–12″	edging, rock garden	white	Blooms early spring.
Armeria maritima alpina (sea pink)	spring to September	sun	10	12″	18″–24″	rock garden, edging, borders, cut flowers	pink	Blooms May and June. Plant in dry sandy soil. Shade until plants are well established.
Aster alpinus (hardy aster)	early spring	sun	14–21	36″	12″–60″	rock garden, borders, cut flowers	white	Blooms June.
Astilbe arendsii (false spirea) 'Europa' 'Fanal' 'Deutschland' 'Superba'	early spring	sun	14–21	24″	12″–36″	borders	pink, red, white	Blooms July and August. Gives masses of color.

Figure 26–2 Various flowering perennials. (*continued*)

Source: US Department of Agriculture Bulletin 114

NAME	WHEN TO PLANT SEED	EXPOSURE	GERMINATION TIME (DAYS)	SPACING	HEIGHT	BEST USE	COLOR	REMARKS
Begonia evansiana (hardy begonia)	summer in shady, moist spot	shade	12	9″–12″	12″	flower bed	yellow, pink, white	Blooms late in summer. Can be propagated from bulblets in leaf axils.
Bergenia purpurascens (bergenia)	late winter	light shade	10	18″	2′–3′	herb	pink, red	Hummingbirds love it.
Candytuft *(Iberis)*	early spring or late fall	sun	20	12″	10″	rock garden, edging, ground cover	white	Blooms in late spring. Prefers dry places. Cut faded flowers to promote branching.
Canterbury bells *(Campanula medium)*	spring to September (Do not cover.)	partial shade	20	15″	24″–30″	borders, cut flowers	white, pink, blue	Divide mature plants every other year. Best grown as a biennial.
Carnation *(Dianthus caryophyllus)*	late spring	sun	20	12″	18″–24″	flower bed, borders, edging, rock garden	pink, red, white, yellow	Blooms in late summer. Cut plants back in late fall and hold in cold frame.
Cerastium tomentosum (snow-in-summer)	early spring	sun	14–28	18″	6″	rock garden, ground cover	white	Blooms in May and June. Forms a creeping mat and is a fast grower. Prefers a dry spot.
Chinese lantern *(Physalis alkekengi)*	late fall	sun	15	36″	24″	borders, specimen plant	orange	Lanterns are borne the second year in the fall.
Columbine *(Aquilegia)*	spring to September	sun or partial shade	30	12″–18″	30″–36″	borders, cut flowers	wide color range	Blooms in late spring. Best grown as a biennial.
Coreopsis lanceolata	early spring	sun	5	30″	24″	borders	yellow	Blooms from June to fall if old flowers are removed.

Figure 26–2 Various flowering perennials.

NAME	WHEN TO PLANT SEED	EXPOSURE	GERMINATION TIME (DAYS)	SPACING	HEIGHT	BEST USE	COLOR	REMARKS
Daisy, Shasta (*Chrysanthemum maximum*)	early spring to September	sun	10	30″	24″–30″	borders, cut flowers	white	Blooms June and July. Best grown as a biennial in well-drained location.
Delphinium elatum (delphinium)	spring to September	sun	20	24″	48″–60″	borders, background, cut flowers	blue, lavender, white, pink	Blooms in June. Best grown as a biennial. Needs dry location.
Dianthus deltoides (pink)	spring to September	sun	5	12″	12″	borders, rock garden, edging, cut flowers	pink	Blooms in May and June. Best grown as a biennial. Needs dry location.
Echinacea purpurea (coneflower)	spring to September	sun	20	30″	30″–36″	borders, flower bed, cut flowers	white, pink, red, rose, purple	Blooms midsummer to fall. Shade summer plantings. Attracts butterflies. Deer resistant. Accent specimen massing.
Foxglove (*Digitalis purpurea*)	spring to September	sun or partial shade	20	12″	48″–60″	borders, cut flowers	pink, white, purple, rose	Blooms in June and July. Grown as a biennial. Shade summer plantings.
Gaillardia grandiflora (gaillardia)	early spring or late summer	sun	20	24″	12″–30″	borders, cut flowers	scarlet, yellow	Blooms from July until frost.
Gypsophila paniculata (baby's breath)	early spring to September	sun	10	48″	24″–36″	borders, cut flowers, drying	white, pink	Blooms early summer until early fall. Needs lots of lime.
Hemerocallis (daylily)	late fall	sun or partial shade	15	24″–30″	12″–48″	borders, among shrubbery	pink, red	Plant several varieties for longer blooming season.
Hibiscus moscheutos (mallow rose)	spring or summer	sun or partial shade	15 or longer	24″	36″–96″	background, flower bed	white, pink, red, rose	Blooms July to September.

Figure 26–2 Various flowering perennials. (*continued*)

NAME	WHEN TO PLANT SEED	EXPOSURE	GERMINATION TIME (DAYS)	SPACING	HEIGHT	BEST USE	COLOR	REMARKS
Iris	bulbs or rhizomes in fall	sun or partial shade	next spring	18″–24″	3″–30″	borders, cut flowers	blue, red, yellow, pink, bronze, wine	Blooms spring and summer if different varieties are used.
Liatris pycnostachya (gayfeather)	early spring or late fall	sun	20	18″	24″–60″	borders, cut flowers	rose-purple	Blooms summer to early fall. Easily started from seed.
Lupinus polyphyllus (lupine)	early spring or late fall (Soak before planting.)	sun	20	36″	36″	borders, cut flowers	white, yellow, pink, rose, red, blue, purple	Blooms most of summer. Needs excellent drainage. Does not transplant easily.
Mentha ssp. (mint)	late winter	sun	8	15″	6″–3′	culinary	white	Mint flavors.
Peony (*Paeonia*)	Plant tubers in late fall 2″–3″ deep.	sun	variable	36″	24″–48″	borders, cut flowers, flower bed	pink, red, white, rose	Blooms late spring. Difficult to grow from seed.
Phlox paniculata (summer phlox)	late fall or early winter	sun	25 irregular	24″	36″	borders, cut flowers	red, pink, blue, white	Blooms early summer. Color of flower varies.
Phlox subulata (moss phlox)	grown from stolons	sun		8″	4″–5″	borders	blue, red, white, pink	Blooms in spring. Drought resistant.
Poppy, Iceland (*Papaver nudicaule*)	early spring	sun	10	24″	15″–18″	borders, cut flowers	white, pink, red	Blooms early summer. Does not transplant easily.
Poppy, Oriental (*Papaver orientale*)	early spring	sun	10	24″	36″	borders, cut flowers	pink, red, rose, orange, white, salmon	Blooms early summer. Does not transplant easily.
Primrose (*Primula polyantha* and *Primula veris*)	January, in a flat on surface. Allow to freeze; then bring in to germinate.	partial shade	25 irregular	12″	6″–9″	rock garden	white, yellow, pink, red, blue	Blooms April and May. May be seeded in fall.

Figure 26–2 Various flowering perennials.

NAME	WHEN TO PLANT SEED	EXPOSURE	GERMINATION TIME (DAYS)	SPACING	HEIGHT	BEST USE	COLOR	REMARKS
Pyrethrum roseum (painted daisy)	spring to September	sun	20	18″	24″	borders, cut flowers, accent	various, including gold, pink, and lavender	Blooms May and June. Prefers well-drained soil.
Rudbeckia fulgida (pot of gold)	midsummer to September	sun, partial shade	20	30″	30″–36″	borders, flower bed, cut flowers, accent		Specimen attracts butterflies. Easy to grow. Deer and rabbit resistant. Wild flower.
Salvia (*Salvia azurea grandiflora* and *Salvia farinacea*)	spring	sun	15	18″–24″	36″–48″	borders	red	Blooms August until frost.
Sea lavender (*Limonium latifolia*)	early spring	sun	15	30″	24″–36″	flower bed, cut flowers, drying	pink, yellow, mauve	Blooms in July and August.
Sedum spectabile (sedum)	late winter	sun	10	10″	4″–15″	ground cover	pink, white	Fall foliage.
Stokesia cyanea (Stokes' aster)	early spring to September	sun	20	18″	15″	borders, cut flowers	white, blue	Blooms in September if started early.
Sweet pea (*Lathyrus latifolius*)	early spring	sun	20	24″	60″–72″	background	pink, white, purple, red	Blooms June to September. Easily grown as a vine on fence or trellis.
Sweet William (*Dianthus barbatus*)	spring to September	sun	5	12″	12″–18″ (Dwarf form also available.)	borders, edging, cut flowers	red, pink, white	Blooms May and June. Very hardy. Needs well-drained soil.
Veronica spicata (speedwell)	spring to September	sun	15	18″	18″	borders, rock garden, cut flowers	purple	Blooms June and July. Easy to grow.

Figure 26–2 Various flowering perennials.

(D) *Coreopsis* 'Early Sunrise'

(G) *Pennisetum* ornamental grass.

(A) *Sedum* stonecrop

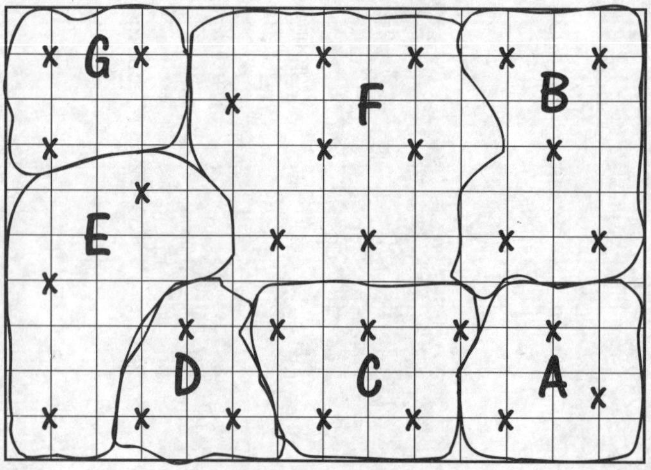

SCALE: 2 BLOCKS = 1'

(F) *Rudbeckia* 'Computer's Gold'

(B) Hollyhock

(C) Veronica

(E) *Liatris floristan*

Courtesy of Ball Horticultural Company

▲ **Figure 26–3** A perennial garden plan.

halfway down the garden the colors reversed to range back to the blues (cool colors) at the opposite (south) end of the garden.

Soil Preparation and Fertilizing

Remember that perennials live in the same soil for more than 1 year. It is necessary, therefore, to prepare the soil well. It will then provide medium conducive to root development. Organic matter should be dug in and drainage improved if necessary. Adequate fertilizer should be dug into the soil along with the organic matter to a depth of at least 1 foot. When the bed is being prepared, use 2 pounds of 5-10-10 (commercial fertilizer), 3 pounds of bone meal (organic fertilizer), or 2 pounds of 14-14-14 (time-release fertilizer) per 100 square feet of bed. The use of time-release fertilizer will ensure the availability of nutrients for the entire growing season. These fertilizers are readily available through green industry suppliers. A soil test analysis should be performed on each bed every year. Each applicator should follow the soil test recommendations for the correct amount of fertilizer each year. Careful preparation prior to planting results in a much more attractive floral display.

▲ **Figure 26–4** Putting the finishing touches on a mulch bed.

Mulching a Perennial Flower Bed

As we mentioned in the previous unit on annuals, the use of mulch for a perennial bed is also multifaceted. Just as for annuals, mulch creates the aesthetics that enhance the perennial planted, serves as a source of organic matter for the plants, helps to control weeds, conserves moisture, and promotes even ground temperatures for improved plant growth.

When planting a perennial flower bed, it is best to apply a 2- to 3-inch layer of mulch before the bed is planted. This will allow for more even distribution of mulch as well as reducing mulching time for the planted bed.

Some of the more common mulches used in the industry today include hardwood bark, pine bark, cocoa bean hulls, buckwheat hulls, pine needles, well-rotted sawdust, and leaf mold. Figure 26–4 illustrates an example of bark mulch being applied to a landscape. A precaution when using some of these mulches is that they will influence the pH level of the soil. Therefore, it is important that the soil pH level is followed closely to ensure that the optimum growth medium is provided for the plants.

Transplanting

Perennials are transplanted in the spring or the fall. When they are transplanted, the crown of the plant must be at the correct level. (The *crown* is the point where the top is connected to the roots, and where the new growth comes from.) Be sure the crown is set at the original depth; as the plant is set, firm in around the roots with your hands. Water to settle the soil in place and to prevent drying out of the roots. If the plants flower in the spring, they should be divided and replanted in the fall. If they flower in the fall, they should be divided and transplanted in the spring.

It is usually best to start perennial seeds indoors 6 to 8 weeks prior to the transplanting date. Plants that do not transplant well must be started in peat pots or direct seeded in the permanent location. The procedures for starting seeds and transplanting perennials are the same as those used for annuals.

Ornamental Grasses

Ornamental grasses are excellent for use as **accents** and, therefore, make attractive edging. They are also good ground covers. They have unique characteristics with their various colors, shapes, and textures of plumage. Also, ornamental grasses create a sense of movement in the landscape. As air circulates in the garden, the grasses wisp with movement to accent the garden. The leaves of ornamental grasses are excellent in dried arrangements as the foliage will vary from dark to variegated with seasonal changes in the plants. Most of these grasses prefer a hot, sunny, and well-drained location. Figure 26–5 lists some ornamental grasses and their characteristics.

Vines

Vines are important in today's landscape. Many types of vines are used to mask various landscape features and create a flowing effect throughout the garden. They are used to soften the appearance of walls, fences, porches, trellises, and pergolas. They are also very effective in small gardens to provide flowers and foliage where space is limited. Vines on a wall have a microclimatic effect. For example, the use of a deciduous vine on a wall creates a cooling effect in the summer, whereas in the winter it allows the wall to be exposed to the sun for a warming effect.

Vines climb in three different ways. English ivy forms small, rootlike holdfasts to a wall as a means of support. Sometimes these are modified tendrils with small circular discs at the tips. Other vines, such as clematis and grape, climb by winding tendrils or leaflike appendages that act as tendrils around the object on which they are growing. The third group, such as bittersweet and wisteria, climb by twining. Knowing in advance how each vine climbs allows the grower to provide the proper means of support. Figure 26–6 lists some typical vines and their characteristics.

NAME	EXPOSURE	SPACING	HEIGHT	BEST USE	REMARKS
Carex morrowii ('Variegata') (sedge)	semishade	9″–12″	12″–15″	border, bank	Leaves narrow, white margin.
Calamagrostis	sun	4′	3′–5′	accent	Looks great with black-eyed Susan.
Chasmanthium latifolium (northern sea oats)	sun	18″–24″	3′	accent	Winter interest.
Cortaderia (pampas grass)	sun	9′–12′	6′–9′	specimen	Silvery flower stalks.
Festuca glauca (blue fescue)	sun	4″–6″	10″–12″	border, bank	Blue-gray foliage.
Imperata (Japanese blood grass)	sun/shade	3′–4′	1′–2′	massing	Reddish foliage.
Leymus (blue wild rye)	sun	2′–3′	2′–3′	massing	Silvery blue foliage.
Miscanthus floridulus (giant miscanthus)	sun	9′–12′	12′–15′	accent, screen windbreak	Effective fall and winter.
Panicum virgatum (switchgrass)	sun	2′–3′	3′–4′	accent	Fall interest. Open panicles.
Pennisetum (dwarf fountain grass)	sun	4′	3′–8′	accent	Excellent fall color.
Phalaris arundinaccea picta (ribbon grass)	sun	12″–15″	18″	accent	White, variegate leaves. Spread by rhizomes.

Figure 26–5 Ornamental grasses.

NAME	REMARKS
Campsis radicans (trumpet vine)	This vine has a rich-looking orange flower. It is a very fast grower and will cling to most surfaces. It will grow in most soils in full sun.
Celastrus scandens (American bittersweet)	Bittersweet is fast growing, with orange and red berries on the female plants. It is very effective in the fall and grows well in most soils in full sun. With such bright-colored berries it is excellent for holiday decorations.
Clematis hybrida (clematis)	Clematis is best known for its excellent flowers, but must be trained on some support trellis, fence, or even a porch post. It needs well-drained soil that is slightly alkaline. The flowers like a sunny location, whereas the roots prefer a cool, shaded location. There are many cultivars on the market featuring striking colors of pink, purple, and white, just to name a few. It is best to prune in late winter or very early spring.
Parthenocissus tricuspidata (Boston ivy)	This vine is a very rapid grower. The foliage opens red in the spring, then turns a rich green in the summer. It does very well on walls.
Polygonum auberti (silverlace vine)	A very rapid grower, with white flowers in the summer. It needs a well-drained soil in full sun or in shade.
Wisteria sinensis (wisteria)	This vine is the old-timer, with fragrant blue-violet pendulous racemes (looks like a bunch of grapes) in late May. It was introduced into this country in 1816, and it requires deep, moist, well-drained soil.

Figure 26–6 Vines used in today's landscape.

Bamboo

Bamboo is being used more extensively now in landscaping, especially to create an oriental touch. Bamboo can range in height from 6 inches to 120 feet. Although it is not really winter hardy, when it is killed back by cold winter temperatures, it will often sprout back from the root when spring comes.

Bamboo has a rhizome root system. In some cases, when the landscape area calls for contained bamboo, it may be necessary to confine the roots to a given area. This can be done by planting the bamboo in a container, thus keeping the plant from spreading into other parts of the landscape. Bamboo is very difficult to contain in the open soil. Metal inserted several feet into the soil will slow the spreading of bamboo until the metal rusts out.

The following are types of bamboo used in the landscape industry.

- *Arundinaria variegata.* The leaves of this variety are grasslike with variegated foliage. It grows to 18 inches high in full sun to semishade.
- *Phyllostachys.* This variety grows tall, in a range of 30 to 70 feet, with feathering foliage. It needs full sun exposure and must be forced to grow in clumps.
- *Arundinaria pygmaea* (pygmy bamboo). This bamboo grows low with grassy green foliage. It needs full sun and spreads very rapidly by rhizomes. It should be contained to the designated area.

TRY THIS Research the three types of bamboo described in the preceding list. Create an educational poster using a computer design program. Include visual images of each species of bamboo, and then describe the different characteristics and uses of each.

Prairie Gardening

Many of the plants used in prairie gardens are native annuals, biennials, and perennials. These plants offer a unique way of naturalizing a garden. Prairie plants are grown informally in a meadow environment where masses and drifts of plants add beauty to the landscape.

Prairie plants attract butterflies, songbirds, and other wildlife, and, in the right combination, *Liatris* and *Asters*, for example, provide compatible colors and texture variation in the landscape.

Some nurseries have special seed mixes that grow best in different areas of the United States. The garden should be planned to include plants that grow well under local soil and climatic conditions and that bloom over a long period of time. Tall-growing plants should be placed in the center or back of the bed while lower growers should be placed at the edges of the front.

Most prairie plants grow best in full sun and are very drought tolerant. They should, however, be watered during the first season after planting and during extreme dry periods.

Soil

As for most other plants, rich soils, high in organic matter, result in stronger, healthier plants. Some plants will grow in rather poor, dry soils, but all do better in rich loam. The soil should be prepared the same for planting annuals or perennials as for seeding a lawn.

The most economical way to plant is by seeding just as if seeding a lawn. Follow instructions on the seed packets as to depth and thickness to seed. Small seeds may need to be mixed in sand to space properly. Plants can also be transplanted if a more immediate effect is desired. Prairie gardens are maintained by hand pulling weeds during summer and mowing in fall to remove old tops.

Annuals reseed themselves and perennials regrow from the roots to provide years of enjoyment with minimum effort. Additional seeding and planting or complete renovation may be necessary if undesirable weeds overwhelm the garden or desirable plants die out. Planting the proper mix of plants for the geographic area results in a more persistent garden.

Figure 26–7 lists prairie garden plants available for purchase and gives plant height, color, bloom time, moisture needs, best soil type, and light requirements. Use this list to plan your garden (Figures 26–8 and 26–9).

NAME	BOTANICAL NAME	HEIGHT	COLOR	BLOOM TIME	MOISTURE	SOIL	SUN
Lavender Hyssop	*Agastache foeniculum*	1′–3′	purple	Jul–Sep	D,M	S,L	○◑
Nodding Pink Onion	*Allium cernuum*	1′–2′	white/pink	Jul–Aug	M,W	S,L	○
Leadplant	*Amorpha canescens*	2′–3′	purple	Jun–Jul	D,M	S,L	○
Canada Anemone	*Amemone canadenis*	1′–2′	white	May–Jun	M,W	S,L,C	○◑
Pasque Flower	*Anemone patens*	<1′	white	Apr–May	D	S	○
Virginia Anemone	*Anemone virginiana*	2′–4′	white	Jun–Jul	M,W	S,L	◑
Angelica	*Angelica atropurpurea*	5′–10′	white	June	W	S,L,C	○◑
Columbine	Aquilegia canadensis	1′–3′	red-yellow	May–Jun	D,M	S,L	○◑●
Rock Sandwort	*Arenaria stricta*	<1′	white	June	D	S	○
Jack-in-the-Pulpit	*Arisaema triphyllum*	1′–2′	green	Apr–May	M,W	S,L,C	◑●

MOISTURE (Soil Moisture): D=Dry, M=Medium, W=Wet
SOIL (Soil Type): S=Sand, L=Loam, C=Clay
SUN: ○ =Full sun, ◑ =Partial sun, ● =Shade

Figure 26–7 Wildflower selection guide. Species are alphabetically listed according to their botanical names. *(continued)*

NAME	BOTANICAL NAME	HEIGHT	COLOR	BLOOM TIME	MOISTURE	SOIL	SUN
White Woodland Milkweed	*Asclepias exaltata*	4′–5′	white	Jul–Aug	D,M	S,L	○◑●
Red Milkweed	*Asclepias incarnata*	3′–5′	pink-red	Jun–Jul	W	S,LC	○
Prairie Milkweed	*Asclepias sullivantii*	3′–5′	pink	Jul–Aug	M	S,L,C	○
Common Milkweed	*Asclepias syriaca*	2′–4′	lavender	Jun–Aug	D,M	S,L,C	○
Butterflyweed	*Aslepias tuberosa*	2′–3′	orange	Jun–Aug	D	S,L	○
Butterflyweed for Clay	*Asclepias tuberosa, var. clay*	2′–3′	orange	Jun–Aug	M	L,C	○
Sky Blue Aster	*Aster azureus*	2′–3′	blue	Aug–Oct	D,M	S,L	○◑
Heath Aster	*Aster ericoides*	1′–3′	white	Aug–Oct	D,M	S,L,C	○
Smooth Aster	*Aster laevis*	2′–4′	blue	Aug–Oct	D,M	S,L	○
New England Aster	*Aster novae-angliae*	3′–6′	pink-purple-blue	Aug–Oct	M,W	S,L,C	○◑
White Aster	*Aster ptarmicoides*	1′–2′	white	Aug–Sep	D	S	○
Silky Aster	*Aster sericeus*	1′–2′	purple	Sep–Oct	D	S	○
Canada Milk Vetch	*Astragalus canadensis*	2′–3′	yellow	Jul–Aug	D,M	S,L	○
Blue False Indigo	*Baptisia australis*	2′–5′	blue	Jun–Jul	M	S,L,C	○◑
White False Indigo	*Baptisia leucantha*	3′–5′	white	May–Jun	D,M	S,L,C	○
Cream False Indigo	*Baptisia leuco-ophaea*	1′–2′	cream	May–Jun	D,M	S,L	○
Pale Indian Plantain	*Cacalia atriplicifolia*	4′–8′	white	Jul–Sep	W	S,L	○◑
Sweet Indian Plantain	*Cacalia suaveo-lens*	3′–6′	white	Jul–Sep	M,W	S,L	○
Poppy Mallow	*Callirhoe triangu-lata*	1′–2′	magenta	Jul–Aug	D	S	○
Marsh Marigold	*Caltha palustris*	1′–2′	yellow	Jul–Aug	W	S,L	○◑
Harebell	*Campanula rotun-difolia*	1′–2′	blue	Jun–Sep	D,M	S	○◑
Patridge Pea	*Cassia fasciculata*	1′–2′	yellow	Jun–Aug	D	S,L	○◑
Wild Senna	*Cassia hebecarpa*	4′–6′	yellow	Jul–Aug	M,W	S,L,C	○

Figure 26–7 Wildflower selection guide. Species are alphabetically listed according to their botanical names. (*continued*)

NAME	BOTANICAL NAME	HEIGHT	COLOR	BLOOM TIME	MOISTURE	SOIL	SUN
New Jersey Tea	*Ceanothus americanus*	2′–3′	white	Jul–Aug	D,M	S,L	○◑
White Turtlehead	*Chelone glabra*	2′–4′	white	Aug–Sep	W	S,L	○◑
Lanceleaf Coreopsis	*Coreopsis lanceolata*	1′–2′	yellow	Jun–Aug	D,M	S,L	○
Stiff Coreopsis	*Coreopsis palmata*	2′–3′	yellow	Jun–Aug	D,M	S,L	○
Canada Tick-Trefoil	*Desmodium canadense*	2′–5′	purple	Jul–Aug	M	S,L,C	○
Shootingstar	*Dodecatheon meadia*	1′–2′	white-pink	May–Jun	M	S,L	○◑
Narrow Leaf Purple Coneflower	*Echinacea angustifolia*	1′–3′	purple	Jun–Jul	D,M	S,L	○
Pale Purple Cone-flower	*Echinacea pallida*	2′–5′	purple	Jun–Jul	D,M	S,L,C	○
Purple Coneflower	*Echinacea purpurea*	3′–4′	purple	Jul–Sep	D,M	S,L,C	○◑
Fireweed	*Epilobium angustifolium*	2′–6′	pink	Jul–Aug	M,W	S,L	○
Rattlesnake Master	*Eryngium yuccifolium*	3′–5′	white	Jun–Aug	D,M	S,L,C	○
Joe Pye Weed	*Eupatorium maculatum*	4′–6′	pink	Jul–Aug	W	S,L,C	○
Boneset	*Eupatorium perfoliatum*	3′–4′	white	Aug–Sep	M,W	S,L,C	○
Sweet Joe Pye Weed	*Eupatorium purpureum*	3′–6′	purple	Jul–Aug	M	S,L,C	○◑
Flowering Spurge	*Euphorbia corollata*	2′–4′	white	Jul–Aug	D	S,L	○
Queen of the Prairie	*Filipendula ruba*	4′–5′	pink	Jun–Jul	M,W	S,L,C	○
Bottle Gentian	*Gentiania andrewsii*	1′–2′	blue	Aug–Sep	W	S,L,C	○◑
Wild Geranium	*Geranium maculatum*	1′–2′	lavender	Apr–Jul	M	S,L	○◑●
Prairie Smoke	*Geum triflorum*	1′	red	Jul–Aug	D,M	S,L	○
Sawtooth Sunflower	*Helianthus grosseserratus*	6′–12′	yellow	Aug–Sep	M,W	L,C	○
Showy Sunflower	*Helianthus laetiflorus*	3′–5′	yellow	Aug–Sep	D,M	S,L,C	○

Figure 26–7 Wildflower selection guide. Species are alphabetically listed according to their botanical names. (*continued*)

NAME	BOTANICAL NAME	HEIGHT	COLOR	BLOOM TIME	MOISTURE	SOIL	SUN
Downy Sunflower	*Helianthus mollis*	4′–6′	yellow	Aug–Sep	D	S	○
Western Sunflower	*Helianthus occi-dentalis*	2′–5′	yellow	Aug–Oct	D,M	S,L	○◑
Woodland Sunflower	*Helianthus stru-mosus*	2′–5′	yellow	Aug–Oct	D,M	S,L	○◑
Ox Eye Sunflower	*Heliopsis helian-thoides*	2′–5′	yellow	Jun–Sep	M,W	S,L,C	○◑
Alum Root	*Heuchera richard-sonii*	2′–3′	green/ orange	May–Jun	D,M	S,L	○◑
Wild Iris	*Iris shrevei*	2′–3′	blue	May–Jun	W	S,L,C	○◑
Blue Flag Iris	*Iris versicolor*	2′–3′	blue	Jun–Jul	W	S,L,C	○◑
False Boneset	*Kuhnia eupatori-oides*	1′–3′	white	Aug–Sep	D,M	S,L	○
Roundheaded Bushclover	*Lespedeza capitata*	2′–4′	white	Aug–Sep	D,M	S,L	○
Rough Blazingstar	*Liatris aspera*	2′–3′	purple-pink	Aug–Sep	D,M	S,L	○
Meadow Blazingstar	*Liatris ligulistylus*	3′–5′	purple	Aug–Sep	M	L	○
Dotted Blazingstar	*Liatris punctata*	1′–2′	rose	Jul–Sep	D	S,L	○
Prairie Blazingstar	*Liatris pyc-nostachya*	2′–4′	purple-pink	Jul–Aug	M,W	S,L,C	○
Dense Blazingstar	*Liatris spicata*	3′–6′	pink-purple	Aug–Sep	M,W	S,L,C	○
Turks Cap Lily	*Lilium superbum*	3′–5′	orange	Jul–Aug	W	S,L,C	○◑
Cardinal Flower	*Lobelia cardinalis*	2′–5′	red	Jul–Sep	W	S,L	○◑
Great Blue Lobelia	*Lobelia siphilitica*	1′–4′	blue	Jul–Sep	M,W	S,L,C	○◑
Lupine	*Lupinus perennis*	1′–2′	blue	May–Jun	D	S	○◑
Bergamot	*Monarda fistulosa*	2′–5′	lavender	Jul–Sep	D,M	S,L,C	○◑
Dotted Mint	*Monarda punctata*	1′–2′	lavender	Jul–Sep	D	S	○
Evening Primrose	*Oenothera biennis*	3′–6′	yellow	Jul–Aug	D,M	S,L	○
Wild Quinine	*Parthenium integ-rifolium*	2′–5′	white	Jun–Sep	D,M	S,L,C	○
Smooth Penstemon	*Pentemon digitalis*	2′–3′	white	Jun–Jul	M	S,L,C	◑●
Slender Penstemon	*Penstemon gracilis*	1′–2′	lavender	May–Jun	D	S	○◑

Figure 26–7 Wildflower selection guide. Species are alphabetically listed according to their botanical names. *(continued)*

NAME	BOTANICAL NAME	HEIGHT	COLOR	BLOOM TIME	MOISTURE	SOIL	SUN
Beardtongue	*Penstemon grandiflorus*	2′–4′	lavender	May–Jun	D	S	○
Purple Prairie Clover	*Petalostemum purpureum*	1′–3′	purple	Jul–Aug	D,M	S,L,C	○
False Dragonhead	*Physostegia virginiana*	1′–2′	pink	Aug–Sep	M,W	S,L	○
Jacob's Ladder	*Polemonium reptans*	1/2′–2′	blue	Apr–Jun	M	S,L	◑ ●
Great Solomon's Seal	*Polygonatum canaliculatum*	1′–4′	green	May–Jun	D,M	S,L,C	○ ◑
Prairie Buttercup	*Ranunculus rhomboideus*	1/2′	yellow	Jul–Aug	D	S	○
Yellow Coneflower	*Ratibida pinnata*	3′–6′	yellow	Jul–Sep	D,M,W	S,L,C	○
Meadow Rose	*Rosa blanda*	3′–4′	pink-white	Jun–Jul	M	S,L,C	○
Pasture Rose	*Rosa carolina*	1′	pink	Jun–Jul	D,M	S,L	○
Black-Eyed Susan	*Rudbeckia hirta*	1′–3′	yellow	Jul–Sep	D,M	S,L,C	○ ◑
Green Headed Coneflower	*Rudbeckia laciniata*	3′–6′	yellow	Aug–Sep	W	S,L,C	○ ◑
Sweet Black-eyed Susan	*Rudbeckia subtomentosa*	3′–6′	yellow	Aug–Oct	M,W	S,L,C	○ ◑
Branched Coneflower	*Rudbeckia triloba*	2′–5′	yellow	Jul–Oct	M,W	S,L	○
Wild Petunia	*Ruellia humilis*	1′–2′	violet	Jun–Aug	D,M	S,L	○
Royal Catchfly	*Silene regia*	2′–4′	red	Jul–Aug	M	L	○
Rosinweed	*Silphium integrifolium*	2′–6′	yellow	Jul–Sep	M,W	S,L,C	○
Compassplant	*Silphium laciniatrum*	3′–10′	yellow	Jun–Sep	D,M	S,L,C	○
Cupplant	*Silphium perfoliatum*	3′–8′	yellow	Jul–Sep	M,W	S,L,C	○ ◑
Prairie Dock	*Silphium terebinthinaceum*	3′–10′	yellow	Jul–Sep	M,W	S,L,C	○
Blue Eyed Grass	*Sisyrinchium campestre*	<1′	blue-white	Jun	D,M	S,L	○ ◑
Mountain Blue Eyed Grass	*Sisyinchium montanum*	<1′	blue	Jun	M	L,C	○
Sweet-Scented Goldenrod	*Solidago jejunifolia*	1′–2′	yellow	Aug–Sep	D	S	○
Grey Goldenrod	*Solidago nemoralis*	1′–2′	yellow	Aug–Oct	D	S,C	○

Figure 26–7 Wildflower selection guide. Species are alphabetically listed according to their botanical names. (*continued*)

NAME	BOTANICAL NAME	HEIGHT	COLOR	BLOOM TIME	MOISTURE	SOIL	SUN
Ohio Goldenrod	*Solidago ohioensis*	3′	yellow	Jul–Sep	M,W	S,L,C	○
Stiff Goldenrod	*Solidago rigida*	2′–5′	yellow	Aug–Oct	D,M	S,L,C	○
Showy Goldenrod	*Solidago speciosa*	1′–3′	yellow	Jul–Oct	D,M	S,L	○
Fame Flower	*Talinum rugospermum*	1/2′	pink	Jun–Aug	D	S	○
Meadowrue	*Thalictrum dasycarpum*	3′–6′	white	Jun–Jul	W	S,L,C	○◑
Spiderwort	*Tradescantia ohiensis*	2′–4′	blue	May–Jun	D,M	S,L	○◑
White Ohio Spiderwort	*Tradescantia ohiensis alba*	2′–4′	white	May–Jun	D,M	S,L	○◑
Blue Vervain	*Verbena hastata*	2′–5′	blue	Jul–Sep	M,W	S,L,C	○
Hoary Vervain	*Verbena stricta*	2′–4′	blue	Jun–Sep	D,M	S,L	○
Ironweed	*Vernonia fasciculata*	4′–6′	purple	Jul–Sep	W	S,L,C	○
Culver's Root	*Veronicastrum virginicum*	3′–6′	white	Jul–Aug	M,W	S,L,C	○◑
Birdsfoot Violet	*Viola pedata*	<1′	blue-purple	Apr–Jun	D,M	S,L	○◑
Heartleaf Golden Alexanders	*Zizia aptera*	1′–2′	yellow	May–Jun	M	S,L	○
Golden Alexanders	*Zizia aurea*	1′–2′	yellow	May–Jun	M,W	S,L,C	○

Figure 26–7 Wildflower selection guide. Species are alphabetically listed according to their botanical names.

▲ **Figure 26–8** Diane Ogg reviews her national winning prairie garden design with Carroll Shry at the National FFA Center, Indianapolis, Indiana.

▲ **Figure 26–9** A prairie garden in flower. This garden is planted with *Rudbeckia* (black-eyed susan) and *Echinacea* (purple coneflower).

SAFETY FIRST

OSHA's regulations ensure that employers and employees know about the work hazards and how to protect themselves to prevent injuries in the workplace. For more information on occupational health and safety in the workplace, please visit the OSHA site at www.OSHA.gov.

Summary Perennials are plants that regrow every year. Hardy and herbaceous varieties are available. Flowering perennial garden plants will provide an array of color during all seasons. Other plants valued in the landscape are ornamental grasses, bamboos, and vines. Plants used in prairie gardens are native plants and lend an aura of interest for naturalists.

Review Questions

1. What are the advantages of using perennials instead of annuals? What are the disadvantages?
2. Discuss the importance of color when planning a perennial garden. Name some perennials that have blue flowers, yellow flowers, red flowers, and pink flowers, respectively.
3. What are some perennials that grow well in full sun? What are some perennials that grow better in partial shade or full shade? Why is it important to be aware of light requirements when planning a perennial garden?
4. What are three varieties each of ornamental grasses, vines, and bamboos that are used in the landscaping industry? What are some ways grasses are integrated into a perennial garden?
5. What are the advantages of using a perennial plant tag with a QR code?

Student Activities

1. Draw a layout for a perennial flower bed to be planted on the school grounds or at home. If possible, plant the flower bed. Observe and record changes in the perennials as they enter the winter season.
2. Visit a local garden center that specializes in perennials.
3. If possible, visit Longwood Gardens in Pennsylvania or White Flower Farm in Connecticut.

4. Have a guest speaker who produces perennials for resale speak to the class about the advantages and disadvantages of growing and selling perennials.
5. Visit a nursery/landscape trade show. Gather information and have conversations with as many agribusiness representatives as possible.
6. Visit flower shows held in the early spring. An example is cherry blossoms in Washington DC.
7. Search online to locate perennial flower distributors. How many are in your state?
8. Raise funds for your horticultural class and FFA chapter. Establish a contract with a seed company to collect milkweed seed pods. Harvest the seed for the seed company with the best contract.

Suggested Supervised Agricultural Experience (SAE) Activities

1. Raise several species of bamboo and market the bamboo to landscape contractors. Record all data to quality for an entrepreneurship SAE.
2. Design, plan, and install a prairie garden in your community. Record all hours and labor to conduct a community service SAE.

Additional Resources

www.prolandscape.com

www.prairienursery.com

www.bamboofarmingusa.com/

http://web.extension.illinois.edu/gardendesign/introduction.cfm

UNIT 27

Narrowleaf Evergreens

Objective

To identify and plant narrowleaf evergreens in the landscape.

Key Terms

balled and burlapped
bare root
berm
container stock
fascicles
heel-in
lath house
pot-bound

Competencies to Develop

After studying this unit, you should be able to:

- differentiate between the two types of narrowleaf evergreen leaves.
- list four reasons for using narrowleaf evergreens in the landscape.
- describe the primary purpose of a lath house.
- describe the proper fertilizer practices for narrowleaf evergreens.
- demonstrate the proper procedure for planting narrowleaf evergreens.
- calculate fertilizer's three active ingredients of nitrogen, phosphorus, and potassium.

Materials

- two different narrowleaf evergreen trees or tree branches
- several deciduous narrowleaf trees or tree branches
- 5-10-10 fertilizer
- round-point shovel
- nursery spade
- garden hose
- smartphone apps:
  - Leafsnap
  - GardenCompass
  - Spectracide
  - Pest Portal
  - AG PhD
  - Plantbid

Narrowleaf evergreens have long been popular landscape plants. This may be because they generally remain green year-round, thereby adding color to the landscape when other plants are dormant. Another advantage is that they are adaptable to various soil types and weather conditions.

Narrowleaf evergreens may have needlelike or scalelike leaves. Pines have needlelike leaves attached to the branch in bundles (known as sheaths or **fascicles**). For example, the white pine has five needles per bundle or sheath. These evergreens carry their needles throughout the year, with old needles dropped in the fall and new needles produced in the spring. Spruces have single needles attached to the branch, which are square, stiff, and sharp.

Narrowleaf evergreens with scalelike leaves have leaves arranged so that each scale overlaps another to form a flat spray. An example of a narrowleaf evergreen with scalelike leaves is the juniper.

TRY THIS Search for narrowleaf evergreens in your neighborhood. Take digital photos of various narrowleaf evergreens. Assemble the pictures into a digital collage or slideshow. Describe the purpose of each narrowleaf that you photograph.

Uses

The narrowleaf evergreen has long been a standard choice for foundation plantings because of its year-round green foliage. Figures 27–1 through 27–3 show several narrowleaf evergreens used as part of foundation plantings. Narrowleaf evergreens are also used extensively as boundary plants to accent borderlines of property (Figures 27–4A and B). The narrowleaf evergreen may also be used to screen a view of a less attractive area or to help control air currents. For example, in the winter months, a windbreak of narrowleaf

▲ **Figure 27–1** Mugho pine—*Pinus mugo mughus.*

▲ **Figure 27–2** Plume false cypress—*Chamaecyparis pisifera* 'Plumosa.'

▲ **Figure 27–3** Hemlock—*Tsuga canadensis.*

evergreens can cause the temperature on the protected side of the windbreak to be 5 to 10 degrees warmer than on the unprotected side (Figure 27–5).

GREEN TIP Narrowleaf evergreens are excellent resources to provide windbreaks in the microenvironment. For each foot of vertical height of the plant, it will give 7 feet of protection behind the narrowleaf evergreens.

GREEN TIP Go green by using trees to create a sound barrier. Many residential housing developments use trees as natural sound barriers. When located near a busy roadway, the plant foliage helps to absorb the sound of traffic.

▲ **Figure 27–4A** Blue Colorado spruce—*Picea pungens* 'Glacua' (winter).

▲ **Figure 27–4B** Blue Colorado spruce—*Picea pungens* 'Glacua' (spring).

▲ **Figure 27–5** Black Japanese pine—*Pinus thunbergii.*

Care

Narrowleaf evergreens are grown in nurseries. They are shipped three different ways: **bare root** (BR), **balled and burlapped** (B&B), or as **container stock** (C). Bare root narrowleaf evergreens are shipped while the plants are seedlings. When the seedlings arrive at their destination, the quality of the nursery stock must be maintained, and therefore the seedlings must be handled with care. Until the ground permits planting, the seedlings should be stored in a building that will maintain high humidity and a cool temperature range of between 33 and 40°F (0 and 50°C). The roots must be covered and kept damp with material such as straw, shingle tow, burlap, or sphagnum moss. Another good storage method is to **heel-in** the seedlings into well-rotted sawdust or soil outside. The roots must be prevented from drying out until they are planted in the ground.

Balled and burlapped narrowleaf evergreens may be moved at almost any time of the year except when the plant has shoots of new growth or when the ground is frozen. The new shoots must be allowed to harden before the plant is moved. If these plants are to be out of the ground for any length of time, it is best to use the heel-in process by mulching them heavily with well-rotted sawdust, pine bark, peat moss, straw, shingle tow, salt marsh hay, or hardwood bark to retain moisture around the roots.

To maintain healthy plants, narrowleaf evergreens should be watered regularly and stored in a lath house until they are planted on the site. A ***lath house*** gives protection from the sun by reducing the light available to the plant. This also reduces the transpiration rate (water loss) of the plant.

Planting Balled and Burlapped Plants

The prepared hole should be one and one-half to two times as large as the ball of soil around the tree being planted to allow room to set the tree in the hole and mix peat moss with the soil surrounding the hole. The ball of the tree is set 1 inch above the original soil line because the tree will settle somewhat in the hole. Remove the burlap, twine, and wires from the top third of the root ball. It is not necessary to pull the burlap out from under the ball, unless burlap made of plastic is used. Plastic will not rot and will restrict root development. Add prepared soil in and around the root ball until it is covered. When the hole is two-thirds filled with soil, fill it to the top with water and let it soak in. Then finish filling the hole with soil. With the remaining soil, form a saucer-like shape (known as a **berm**) around the tree to hold water that will be added later. A berm is used to conserve water and create a healthy, ecological system by managing runoff around the plant root system.

Planting Containerized Trees

Containerized trees are planted in the same way as balled and burlapped trees, except for the following differences.

1. The container is removed from the tree roots. Care must be taken not to break the root ball apart.
2. After the container is removed, check to see if the roots are **pot-bound** (the root ball will be excessively tight). If this has happened, then it is necessary to cut or loosen the roots to encourage growth from the root ball. This procedure stops the circular root growth

and allows the roots to make their growth into the soil. If the growth pattern remains circular, the plant growth will be restricted. In severe cases, trees can be killed if this growth pattern is allowed to continue.

Staking

Staking evergreens is important because it gives the tree needed support to allow the root system to become established and to prevent the tree from swaying in the wind. For trees 10 feet or less in height, the stakes should be 5 to 6 feet long and 1¾ inches square. Stakes should be made of wood that is capable of standing in the ground for at least 2 years.

To stake a tree, drive two stakes into the ground, with one of the stakes into the prevailing wind side of the tree. Drive the other stake directly opposite the first stake. Using a pliable 12- or 14-gauge soft steel wire and piece of garden hose or commercially prepared material (about 1 foot long), slip the wire through the hole in the hose or commercial sleeve and place it around the trunk of the tree about halfway up the tree. Pull the two ends of the wire around the first stake, making sure the section inside the rubber hose is around the trunk of the tree. Twist the wire around the stake. Repeat the process on the other stake. Using a pair of fencing pliers, twist the wire until it is tight enough to hold the tree in place. Repeat this procedure on the opposite stake, making sure the tree is secure from high wind and will grow straight.

Fertilizing

The general recommendation for fertilizing a narrowleaf evergreen is 3 to 6 pounds of 5-10-10 fertilizer per 100 square feet placed in a circle under the drip line at the outer end of the branches. Fertilize before new growth starts in the spring. Fertilizer with the analysis of 5-10-10 means 5 percent nitrogen, 10 percent phosphate, and 10 percent potash. If you apply 6 pounds per 100 square feet, you will apply 0.3 pounds of nitrogen (N), 0.6 pounds of phosphate (P_2O_5), and 0.6 pounds of potash (K_2O).

5 percent of 6 pounds = 0.3 pounds N

10 percent of 6 pounds = 0.6 pounds P_2O_5

10 percent of 6 pounds = 0.6 pounds K_2O

After applying the fertilizer, water it in with a garden hose sprinkler. Use evergreen fertilizer spikes according to the manufacturers recommendations. Organic fertilizers such as cottonseed meal or soybean oil meal may also be used. Although there is an advantage to the use of organic fertilizers in that there is no danger of burning the plant, they are usually more expensive.

Watering

Narrowleaf evergreens need to be watered every 10 to 14 days during the first year after planting. Thoroughly soak the soil to 6 inches or deeper when they are planted. Watering after planting will depend on the weather conditions of the geographic area.

Pruning

The two methods of pruning narrowleaf evergreens are the following:

- *Pruning* individual branches gives the plant a natural, informal appearance. (Large branches may be removed from specific locations on the plant.) Spreading junipers and *Taxus* should be pruned by removing the ends of long twigs. This gives an informal appearance to these plants.
- *Shearing* is similar to shaping a hedge. The ends of all small branches are clipped to shape the edges of the plant in straight lines, giving a more formal appearance. Hicks yew (Taxus 3 media 'Hicksii'), columnar juniper, and globe arborvitae may be sheared in this manner.

Characteristics of Specific Narrowleaf Evergreens

Figures 27–6, 27–7, and 27–8 list narrowleaf evergreens used in landscaping. The plants are listed in sequence according to their height. The general breakdown includes one group of those plants 3 feet or less in height; a second group of those plants ranging from 3 to 9 feet in height; and a third group of those plants greater than 10 feet in height.

The following information is given for each plant.

- The *hardiness* of each plant refers to one of the areas shown on the Hardiness Zone Map at the end of the text. It indicates where the plant can be grown in the United States.

NAME	HARDINESS	FOLIAGE COLOR	PERIOD OF INTEREST	LANDSCAPE USE	OTHER REMARKS
Erica carnea (Spring heath)	5	bright green	small, rosy red spikes in April	ground cover	Attractive spring colors.
Juniperus chinensis sargentii (Sargent juniper)	4	steel blue	blue berries in fall and winter	ground cover	Excellent for planting along seashore. Does well on steep banks to prevent erosion.
Juniperus horizontalis 'Wiltonii' (Blue rug juniper)	2	steel blue	blue berries in fall and winter	ground cover	Excellent rock garden.
Taxus baccata 'Repandens' (Spreading English yew)	4	dark green	fall and winter	foundation planting	Excellent plant; low maintenance; good color all year. Requires well-drained soil.
Taxus cuspidata aurencens (Dwarf Japanese yew)	4	light green	fall and winter	foundation planting	Compact plant that produces red berries. Requires well-drained soil.

Figure 27–6 Narrowleaf evergreens (3 feet or less in height).

NAME	HARDINESS	FOLIAGE COLOR	PERIOD OF INTEREST	LANDSCAPE USE	OTHER REMARKS
Chamaecyparis obtusa 'Nana'	3	deep dark green	all year	accent plant, specimen plant	Compact forrn; pyramidal shape.
Juniperus squamata 'Meyeri' (Meyer's juniper)	4	blue	all year	foundation planting	Needs good management and care in dry areas.
Juniperus virginiana 'Tripartita'	2	dark green	all year	foundation planting	Good in dry areas.
Pinus mugo mughus 'Compacta' (Mugho pine)	2	dark green	all year	foundation planting	Global shape; slow growing.
Picea abies 'Conica' (a) Dwarf Alberta spruce	2	light green	all year	foundation, planting, specimen plant, rock garden	Fine-textured plant.
(b) Bird's nest spruce	2	light green	all year		Requires well-drained soil.
Juniperus chinensis 'Pfitzerana' (Pfitzer juniper)	4	blue green	all year	foundation planting, screen plant	Plant in full sun. Requires well-drained soil. Control of bag worms necessary.
Juniperus chinensis 'Hetzii'	4	blue green	all year	screen plant, foundation planting	Plant in full sun; does well in dry areas.
Taxus × media 'Hicksii' (Hicks yew)	4	dark green	all year; fall fruit	foundation planting, screen plant	Plant in full sun. Excellent in formal gardens; columnar shape. Requires well-drained soil.

Figure 27–7 Narrowleaf evergreens (3 to 9 feet in height).

NAME	HARDINESS	FOLIAGE COLOR	PERIOD OF INTEREST	LANDSCAPE USE	OTHER REMARKS
Cypressocyparis leylandii (Leyland cypress)	4	light green	all year	screen or hedge plant	Fast grower, excellent windbreak; columnar shape.
Taxus baccata (English yew)	6	dark green	all year	screen or hedge plant, foundation planting on large buildings	Female plant has berries. Will stand shade. Good background shrub.
Taxus cuspidata (Japanese yew)	4	dark green	all year	screen or hedge plant, foundation planting on large buildings	Pyramidal form. Tolerates shade. Rapid grower. Produces red fruit on female.

Figure 27–8 Narrowleaf evergreens (over 10 feet in height).

- *Foliage color* is the color of the narrowleaf evergreen. Some types vary from a very light green to dark green.
- *Period of interest* is that time of year during which the plant is most attractive. This may be when it flowers, bears fruit, or changes foliage color. All of these create interest during a particular season of the year.
- *Landscape use* indicates how a particular plant is used in the landscape. The following are possible entries in this column:
  - *Ground cover* refers to plants, other than grass, that are used to cover the ground. Ground covers also have special uses, such as preventing erosion by holding steep banks in place (Figure 27–9).
  - *Foundation planting* refers to plants that are used around buildings to help accent and tie the building into the landscape (Figure 27–10).

▲ **Figure 27–9** Sargent Juniper—*Juniperus Chinensis Sargentii.*

▲ **Figure 27–10** Dwarf Hinoki False Cypress—*Chamaecyparis Obtusa* 'Nana.'

- *Rock garden* refers to plants used in gardens that are planted in a rocky environment, whether natural or artificially reproduced (Figure 27–11).
- *Specimen plants* refer to plants that are used alone for their own beauty or as an accent.
- *Screen plants* and *hedges* refer to those plants that confine certain areas, used to reduce noise or an unappealing sight.
- *Accent plants* are used to define borderlines.

Other remarks are given to help the individual make a more educated selection of plants.

Figures 27–12 through 27–18 identify narrowleaf evergreens by their leaf characteristics.

▲ **Figure 27–11** Blue rug juniper—*Juniperus horizontalis* 'Wiltonii.'

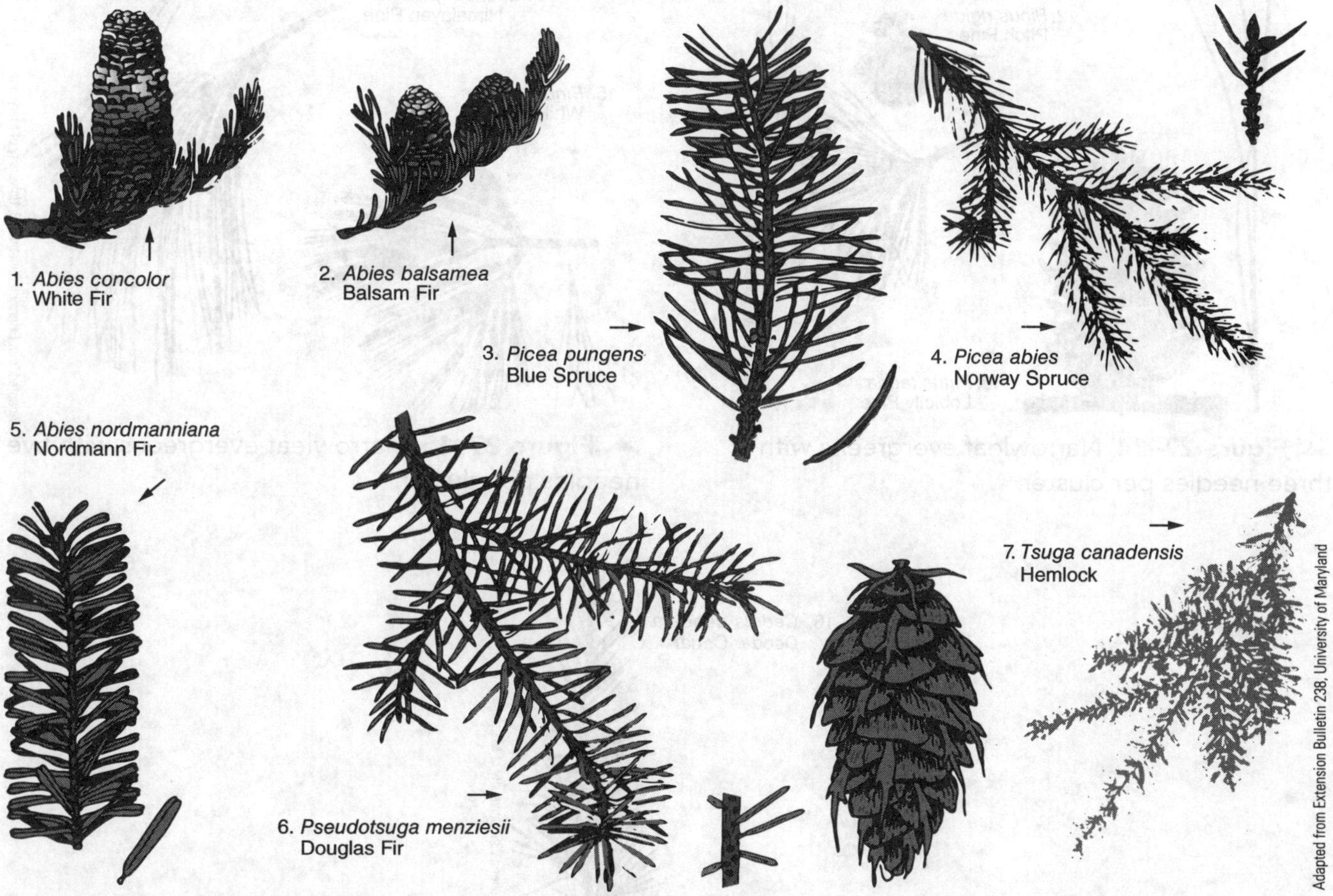

▲ **Figure 27–12** Narrowleaf evergreens with single needles.

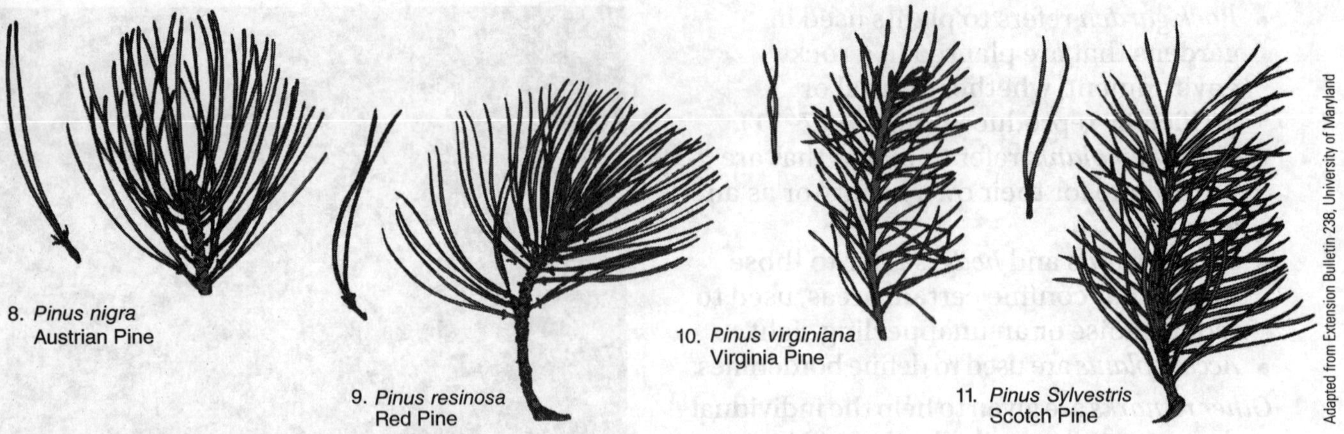

▲ **Figure 27–13** Narrowleaf evergreens with two needles per cluster.

▲ **Figure 27–14** Narrowleaf evergreens with three needles per cluster.

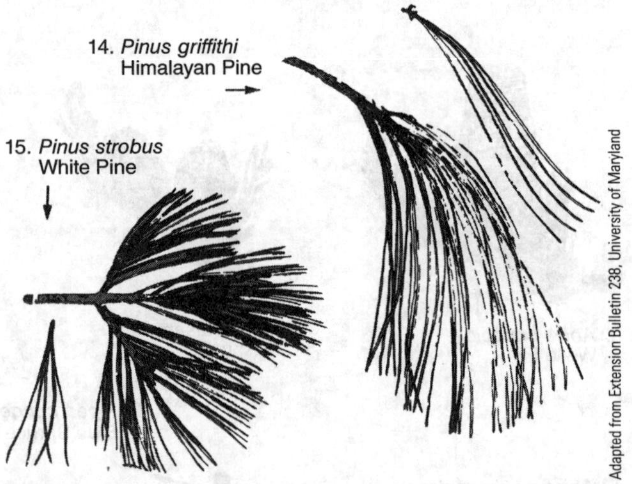

▲ **Figure 27–15** Narrowleaf evergreens with five needles per cluster.

▲ **Figure 27–16** Narrowleaf evergreen with more than five needles per cluster.

▲ **Figure 27–17** Narrowleaf evergreens that are deciduous. This group includes those evergreens that drop their needles in the winter.

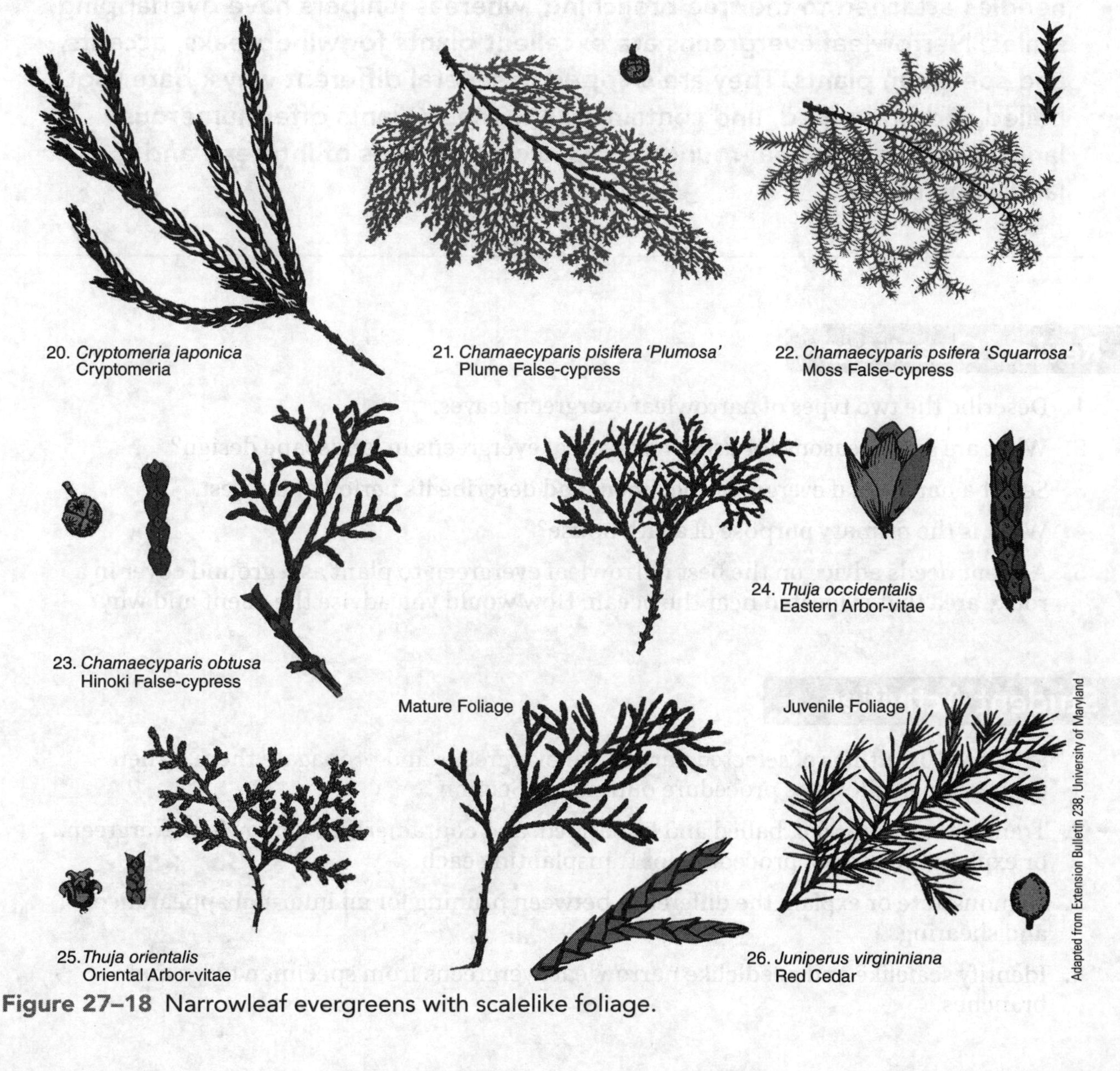

▲ **Figure 27–18** Narrowleaf evergreens with scalelike foliage.

SAFETY FIRST

OSHA's regulations ensure that employers and employees know about work hazards and how to protect themselves to prevent injuries in the workplace. For more information on occupational health and safety in the workplace, please visit the OSHA site at www.OSHA.gov.

Summary

Narrowleaf evergreens are very popular landscape plants. There are two types of narrowleaf evergreen leaves. Pines have two or more needles (leaves) in a fascicle. Spruces and firs have single needles attached to the tree branching, whereas junipers have overlapping scales. Narrowleaf evergreens are excellent plants for windbreaks, accents, and specimen plants. They are shipped in several different ways: bare root, balled and burlapped, and containerized. These plants offer numerous landscape qualities year-round: foliage color, periods of interest, and landscape use.

Review Questions

1. Describe the two types of narrowleaf evergreen leaves.
2. What are some reasons for using narrowleaf evergreens in landscape design?
3. Select a narrowleaf evergreen specimen, and describe its period of interest.
4. What is the primary purpose of a lath house?
5. A client needs advice on the best narrowleaf evergreen to plant as a ground cover in a rocky area that is located near the ocean. How would you advise the client and why?

Student Activities

1. Produce 10 cuttings of selected narrowleaf evergreens and propagate them. When propagating, follow the procedure outlined in Section 3.
2. Transplant a bare root, balled and burlapped, and containerized narrowleaf evergreen, or explain the correct procedure for transplanting each.
3. Demonstrate or explain the difference between pruning for an informal appearance and shearing.
4. Identify scalelike and needlelike narrowleaf evergreens from specimen twigs and branches.

5. Visit a local landscape nursery to learn the types of narrowleaf evergreens used in your geographic area.
6. Collect samples of narrowleaf evergreens. Preserve, label, and display the evergreen in the school for others to see.
7. Research evergreen production online. Write a two-paragraph summary on the status of the evergreen industry.

Suggested Supervised Agricultural Experience (SAE) Activities

1. Work the sales counter for a garden center that sells containerized trees and evergreens. Record your data to qualify for a placement SAE.
2. Work for a Christmas tree farm to qualify for an entrepreneurship SAE.

Additional Resources

http://ucanr.edu/sites/UrbanHort/files/80131.pdf

https://extension.unh.edu/resources/files/Resource000594_Rep616.pdf

http://pss.uvm.edu/ppp/articles/prunever.html

UNIT 28

Broadleaf Evergreens

Objective

To identify the cultural requirements, planting techniques, care, and uses of broadleaf evergreens in the landscape.

Key Terms

broadleaf
desiccation
drip line
raised beds
root ball
soilless
spreader sticker

Competencies to Develop

After studying this unit, you should be able to:

- list four ways in which broadleaf evergreens are used in the landscape.
- list four cultural requirements of broadleaf evergreens.
- describe the soil and fertilizer requirements of broadleaf evergreens.
- explain the procedure for transplanting broadleaf evergreens.
- list three pests that attack broadleaf evergreens and one control for each.

Materials

- branches and leaves of broadleaf evergreens grown in the local area
- one broadleaf evergreen properly balled and burlapped
- smartphone apps:
  - Leafsnap
  - GardenCompass
  - Spectracide
  - Pest Portal
  - AG PhD
  - Plantbid

The description of **broadleaf** evergreens is just as the name implies—*evergreen* plants (plants that hold their leaves all year) with broad leaves rather than the needlelike leaves of narrowleaf evergreens such as pines, yews, and junipers. Figure 28–1 illustrates some typical broadleaf evergreens.

Uses of Broadleaf Evergreens

Broadleaf evergreens are used in the landscape in several ways:

- Around foundations. The smaller-leaved broadleaf evergreens are commonly used as foundation plantings. Some of these are the Japanese holly, azalea, andromeda, dwarf English holly, and dwarf rhododendron.
- As specimen plants. A large rhododendron is beautiful as a singular specimen plant (Figure 28–2). American holly, some of the larger varieties of English holly, and magnolia also make beautiful specimen plants (Figure 28–3).
- As hedges. Privet, a broadleaf evergreen, is often used as a hedge. Japanese holly and boxwood are also used in hedges.
- Along sidewalks and entrances. Broadleaf evergreens are frequently used to lead the way to the home entrance. A privet hedge, if cut short, and azalea also work well as plantings along sidewalks.
- In woodland plantings. Azaleas and rhododendron make beautiful understory plants in open deciduous woods (Figure 28–4).

▲ **Figure 28–1** Leaves and twigs of three broadleaf evergreens: (left) American holly; (middle) English holly; (right) Japanese holly.

▲ **Figure 28–2** Rhododendron in full bloom, a beautiful specimen plant.

▲ **Figure 28–3** The American holly (*Ilex opaca*) is a beautiful landscape tree and also provides Christmas decorations.

▲ **Figure 28–4** Azaleas are used to add color to a spring landscape.

Figure 28–5 summarizes growing information and uses for some of the more common broadleaf evergreens.

NAME	SIZE AT MATURITY	SOIL AND SOIL pH PREFERENCE	PRUNING DATE	BLOOMING DATE	USES	OTHER REMARKS
Andromeda (*Pieris japonica*)	8'–10'	loam pH 5.0–6.5	after blooming in spring	early spring	foundation, specimen, center of group	Gives very early white blooms. Excellent for background. Hardy in Zones 6 to 9.
Azalea (*Rhododendron* varieties)	from creeping to 15' according to variety	loam pH 5.0–6.0	after blooming in spring	early spring through late spring depending on variety	foundation, beds, edging, woodland	Many variations in size, color, bloom date. Has a variety of uses. Should be sprayed for best results. Hardy in Zones 6 to 8.
Darwin Barberry (*Berberis darvinii*)	8'	broad range	anytime	spring flowering; berries in fall	hedge, foundation, specimen	Few thorns, bright orange flowers, purple berries. Needs full sun. Hardy in Zones 4 to 9.
Rosy Glow Barberry (*Berberis thumbergi* 'Rose Glow')	5'	broad range	anytime	spring flowering; berries in fall	hedge, foundation, specimen	Bright scarlet berries. Hardy in Zones 4 to 9.
Dwarf Crimson Pygmy Barberry (*Berberis thunbergii*)	4'	broad range	anytime	spring flowering; berries in fall	hedge, foundation, specimen	Rich purple foliage turns scarlet in autumn. Keep in full sun.
Warty Barberry (*Berberis verruculosa*)	4'	broad range	anytime	spring flowering; berries in fall	specimen, hedge	Yellow flowers in spring. Foliage white underneath. Hardy in Zones 4 to 9.
Boxwood, Common (*Buxus sempervirens*) (*B. sempervirens* 'Arborescens')	fast growing up to 15'	heavy loam pH 6.0–7.0	early spring	no flowers	hedge	Grows very fast. Full sun or shade.
Boxwood, English (*Buxus sempervirens*) (*B. sempervirens* 'Suffruiticosa')	slow growing up to 3'	heavy loam pH 5.0–7.0	early spring	inconspicuous	edging, walkways, topiary designs, inside corners at foundation	Suffers windburn in strong winter winds. Grows in full sun or shade. Grows slowly (1–2 inches each year). Hardy in Zones 6 to 8.
Camellia (*Camellia sasaqua*)	8'–15'	loam, high in organic matter pH 5.0–6.5	after flowering or early spring	autumn	specimen, border, hedge	White to pink single and double flowers. Will grow in full sun or partial shade. Hardy in Zones 7 to 9.

Figure 28–5 Common broadleaf evergreens. (*continued*)

NAME	SIZE AT MATURITY	SOIL AND SOIL pH PREFERENCE	PRUNING DATE	BLOOMING DATE	USES	OTHER REMARKS
(Camellia japonica)	15'–30'	loam, high in organic matter pH 5.0–6.5	after flowering	early spring	specimen	Dark green foliage; white to deep red single or double flowers. Needs sun. Hardy in Zones 7 to 9.
Firethorn (*Pyracantha coccinea* 'Lalandi')	20'	loam wide range of pH (6.0–8.0)	early spring	June	specimen	Many varieties. Difficult to transplant (use container plants). Orange fruit.
Holly, American *(Ilex opaca)*	20'–30'	sandy loam pH 6.0–7.0	winter holiday season	spring	specimen	Red berries in fall and winter; spiny leaves.
Holly, Chinese *(Ilex cornuta)*	from 4' to 15' depending on variety	sandy loam pH 6.0–7.0	throughout the year except late summer	spring	foundation, hedge, specimen	Many new excellent varieties. Red or yellow berries. Full sun or shade. Needs heavy fertilization.
Holly, English *(Ilex aquifolium)*	2'–10'	sandy loam pH 6.0–7.0	winter holiday season	spring	specimen	Glossy foliage. Yellow and red berries.
Holly, Japanese *(Ilex crenata)*	from 3' to 15' depending on variety	loam pH 5.0–6.0	throughout the year except late summer	early summer	hedge, topiary design, specimen, foundation	Small spineless leaves and black berries. Needs fertilizer each year.
Magnolia (*Magnolia grandiflora)* Southern magnolia	30'–50'	loam pH 6.0–6.5	after flowering	mid-summer	specimen	Native to the United States. 6" to 10" fragrant white flowers. Hardy in Zone 7 and in southern areas.
Sweet Bay Magnolia *(Magnolia virginiana)*	30'–50'	loam pH 5.5–6.5	after flowering	early summer	background, specimen	Evergreen in south, Zones 7 to 10. Deciduous in Zones 5 and 6. Hardy in Zones 5 to 10.
Nandina *(Nandina domestica)*	6'– 8'	loam pH 5.0–6.5	after flowering	mid-summer	shrub borders, front of taller evergreens	White flowers; red berries in autumn and winter. Hardy in Zones 7 and 8.
Privet *(Ligustrum japonicum)*	9'–18'	wide range	after flowering	June–July	hedge	Black berries. Hardy in Zones 6 to 7.

Figure 28–5 Common broadleaf evergreens. *(continued)*

NAME	SIZE AT MATURITY	SOIL AND SOIL pH PREFERENCE	PRUNING DATE	BLOOMING DATE	USES	OTHER REMARKS
Glossy Privet (*Ligustrum lucidum*)	8'–15'	wide range	after flowering	August–September	hedge, borders	Easy to grow. Hardy in Zones 7 to 10. Grows in shade.
Rhododendron varieties	3'–10' for most commercial varieties	varies with variety pH 5.0–6.0	early spring	spring	specimen, foundation	Requires mulch at all times and good drainage. High organic matter needed in soil. Hardy in Zones 4 to 8.
Viburnum (many varieties; select for your area)	5'–12' depending on variety	varies with variety	after flowering	spring	foundation, specimen	Fertilize each year. Attractive fruit, especially to birds.

Figure 28–5 Common broadleaf evergreens.

Cultural Requirements

Proper site selection and preparation of the soil is extremely important for creating favorable growing conditions for broadleaf evergreens. Cultural requirements, such as the amount of moisture, drainage, exposure, and sunlight, determine growth success.

- *Ample moisture* is required by most broadleaf evergreens for best growth. Organic matter holds moisture well. Sandy soils need more organic matter such as pine bark or sphagnum peat moss that could make up as much as 50 percent of the planting mix.

GREEN TIP Plant native species to ensure that appropriate fruits and berries are available for wildlife. Animals are an important aspect of our environment, and, in many cases, the health of our environment is closely associated with the health of wildlife.

- *Good soil drainage* is also necessary. If drainage is a problem, the planting may be raised above soil level. The addition of fine pine bark and sphagnum peat moss improves soil drainage by improving soil structure.
- *Mulching* is needed for the proper supply of moisture. Because these plants have green leaves year-round, a ready supply of moisture to the roots must be available at all times to replace moisture lost through the leaves due to transpiration. If the soil freezes in winter, the roots cannot pick up moisture, causing the plants to dry out. A mulch prevents the soil from freezing as deeply, and the roots are better able to supply lost moisture. In summer, a mulch keeps the soil cooler and saves moisture loss due to heat and drying winds. Mulching should *not* be done on boxwood and used at only 2 or 3 inches of coarse material on rhododendrons. The roots of these plants need a lot of oxygen.
- *A location that is sheltered from strong winds* is best for broadleaf evergreens. Cold winter winds and winter sun do the most damage by removing moisture from the leaves faster than the roots can replenish it in some cases. Drying out in winter is one of the most serious problems for broadleaf evergreens.

- *Enough sunlight* is needed to develop a full, compact plant. Plants tend to grow tall and leggy with too much shade. Full sunlight or partial shade is best for most of the broadleaf evergreens. For example, rhododendrons grow best in 40 percent shade. If plants do not flower well, they are probably not getting enough light.

Soil and Fertilizer

Any good loam soil is satisfactory for most broadleaf evergreens. An acidic soil is usually the ideal choice. The rhododendron, azalea, and Japanese andromeda require a pH of from about 5.0 to 6.0 for best growth. A soil pH higher than this results in an iron deficiency.

Fertilizer should be used in small amounts. Too much fertilizer, especially nitrogen, causes soft, fast growth and results in open, straggly plants that are unattractive. A slower-growing, compact plant makes a much better addition to the landscape picture. Also, if a fertilizer is applied in smaller amounts, the plant will not outgrow its space in the landscape as quickly. If all other cultural conditions are right, short twig growth and small, pale green leaves may indicate a need for fertilizer.

Fertilizer is added in the spring in a circular fashion around the plant at the **drip line**. (The drip line is an imaginary line directly below the outer edge of the branches.) A few handfuls carefully sprinkled around the plant are sufficient for small plants. For larger trees, use 1 to 1½ pounds per inch of tree trunk. Any type of fertilizer that is not alkaline may be used for "acid-loving" broadleaf evergreens; cottonseed meal is a safe fertilizer for these plants.

The varieties of Chinese and Japanese holly require more fertilizer than most broadleaf evergreens. For these evergreens, 2 to 3 pounds of 10-5-5 fertilizer per 100 square feet should be used. On individual plants, use 1½ pounds per 3 feet of height or spread of plant. Chinese holly grows best at a soil pH of 6.0 to 7.0.

Transplanting

Because these plants are evergreen, the demand for moisture is always high. The roots must always be in close contact with moist soil. For this reason, broadleaf evergreens are always moved with the **root ball**. This means that the soil is left intact around the roots and moved with the plant (Figure 28–6).

The hole for broadleaf evergreens should be wide enough so that 12 to 18 inches of space exists all the way around the root ball. The plants are always planted at the same depth or slightly higher than they were growing before; they are never planted deeper, as is done with many deciduous trees. No soil should be placed on top of the root ball (Figure 28–7). Be sure that the soil

▲ **Figure 28–6** This azalea was dug from the ground with the root ball intact in preparation for transplanting. Careful handling is important so that the soil is not broken loose from the roots.

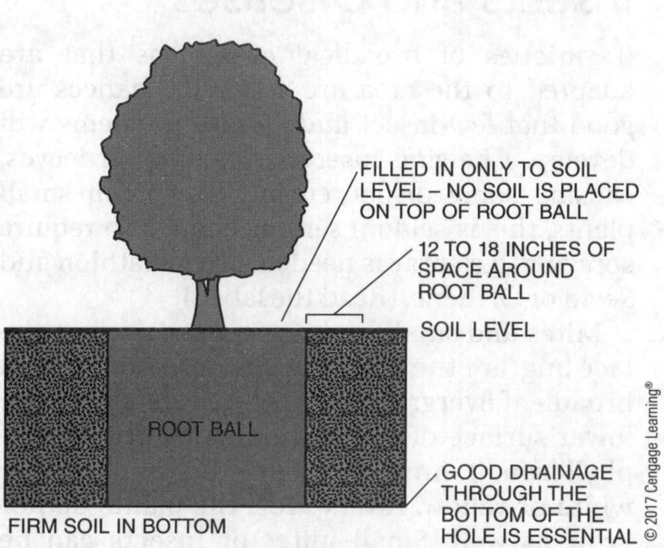

▲ **Figure 28–7** Transplanting a broadleaf evergreen.

in the bottom of the hole is firm so that the root ball does not sink after it is planted.

Plants must be planted in **raised beds** above the soil line in poorly drained or alkaline soils. Bring in enough good, well-drained topsoil of the proper pH or plant entirely in a **soilless** mix of two parts pine bark and one part peat moss. Raise the bed so the plant root ball sits on top of the existing, poorly drained soil.

Use your hands to firm the soil around and under the root ball, and water with a garden hose at low water pressure. Do not tramp with your feet or pack too tightly. Mulch around the plant with wood chips or pine bark to a depth of 2 to 3 inches and out beyond the branch ends.

Pruning

Broadleaf evergreens may be pruned for a special effect or cut back to thicken the plant. Pruning is most often done in early spring or immediately after blooming. Holly is often pruned during the winter holiday season, and the prunings are used for holiday decorations.

The rhododendron should be pruned at the point where a flush of growth stops and buds are more plentiful. This usually occurs at the point where branches extend from the plant.

Insects and Diseases

If varieties of broadleaf evergreens that are adapted to the area are selected, chances are good that few insect and disease problems will develop. Chewing insects may feed on leaves, causing some disfigurement. Except on small plants, this is seldom serious enough to require spraying. If a spray is needed, use malathion and Sevin or Orthene. (Read the label.)

Mites and small sucking insects, such as the lace bug, are the most serious pests that attack broadleaf evergreens. These insects attack the lower surface of the leaf and suck the chlorophyll from it. An attack causes leaves to appear white or yellow. Afterwards, the plants do not grow as well. Small mites or insects can be seen on the under surface of the plant's leaf with a 10-power microscope (one that magnifies the insects 10 times their normal size.) For control of these pests, spray with malathion if the pest can be hit with the spray. (Malathion will kill only if the insect comes in direct contact with the spray.) Orthene, a systemic, will kill without coming into contact with the pests and do a better job. However, systemics may be more toxic to humans and warm-blooded animals, so proper precautions must be taken. Less toxic pesticides are also available, such as Safer Insecticidal Soap and the new spray oils refined from cottonseed oil. For effective control, the insect's body must be covered with these materials.

Fungus diseases such as leaf spots, rusts, or dieback (caused by tips of branches being killed by a fungus) may be a problem. If these diseases are present, spray just after bloom and again in 10 days with Captan or Aliette. Some new systemic fungicides such as Subdue used as soil drenches give excellent plant protection. (Check the label for application information.)

Use a **spreader sticker** to apply all sprays to broadleaf evergreens. A spreader sticker is a chemical added to a pesticide that causes the spray to break into small droplets and spread evenly and thinly over the leaf surface. Shiny leaves are hard to get completely wet, and sprays tend to run off the plant without giving the proper protection. A spreader sticker helps to overcome this problem by causing the spray to stick to the plant more easily. Ordinary soap or detergent is acceptable as a spreader sticker if nothing else is available. Use only enough soap to produce light suds in the water. Many insect- and disease-resistant varieties require no spraying.

Most broadleaf evergreens are propagated by semihardwood cuttings. See Unit 7 for details.

TRY THIS Create an informational brochure describing the diseases and insects that affect broadleaf evergreens for homeowners caring for their landscape. Identify methods to control both diseases and insects.

Winterizing

Broadleaf plants usually suffer from winter burn (leaves turn brown) during the winter months in colder climates. Winter burn (**desiccation**), also called winter drying, results from the combination of cold drying winds, fluctuating temperatures from very cloudy to very sunny days, and frozen soil conditions. Broadleaf plants such as holly, rhododendron, azalea, camellia, boxwood, and nandina are susceptible to winter burn.

Spray materials are available to give broadleaf plants a protective coating to reduce the moisture from transpiring from the plants through the stomata (see Figure 28–8). This coating will stand on the plant for several months. The anti-desiccant should be applied when the temperature is above freezing for 24 hours. Properly watering the plants during dry periods in the late fall and placing protective barriers around the plants prior to winter are other ways to prevent winter burn.

Courtesy of Wilt-Pruf Products, Inc.

▲ **Figure 28–8** Wilt-Pruf guards against moisture loss year-round. It gives plant protection to evergreens, rhododendrons, boxwoods, azaleas, hollies, laurels, and other ornamentals.

SAFETY FIRST

OSHA's regulations ensure that employers and employees know about the work hazards and how to protect themselves to prevent injuries in the workplace. For more information on occupational health and safety in the workplace, please visit the OSHA site at www.OSHA.gov.

Summary

Broadleaf evergreens add a different texture to the landscape when used with narrowleaf evergreens, such as pines or spruces. They range in size from large trees, such as magnolias and American hollies, to dwarf small-leafed Japanese hollies. This range in size provides a selection for almost any landscape use. Azaleas and rhododendrons provide beautiful flowers in addition to attractive foliage, and everyone recognizes the red berries of the American holly in Christmas wreaths. Broadleaf evergreens provide beauty in northern climates with their dark green foliage.

Review Questions

1. Describe how you would plant a broadleaf evergreen in wet, poorly drained soil.
2. Which broadleaf evergreen should have no mulch over its roots? Why?
3. What are the main cultural requirements of broadleaf evergreens?
4. What are some reasons for choosing broadleaf evergreens over narrowleaf evergreens when designing a landscape?
5. How do you apply spray most effectively to broadleaf evergreens?

Student Activities

1. Draw up a chart such as the one shown in Figure 28–5. Insert local broadleaf evergreens that are different from the ones listed in the text. Supply the information in each column for the plants listed.
2. Tour the school grounds or part of the local community, identify various broadleaf evergreens, and note their use in the landscape. For each evergreen species found, use a smartphone tree identification app to confirm the species of evergreen.
3. Decide which broadleaf evergreens would be best in an area around your school. Draw a plan using these evergreens.
4. Research online for the current uses of the following plant species: azalea, rhododendron, and holly.
5. Explore the www.usda.gov website.

Suggested Supervised Agricultural Experience (SAE) Activities

1. Design and operate a landscaping business that specializes in created windbreaks using broadleaf evergreens to qualify for the entrepreneurship SAE category.
2. Conduct a research SAE that compares the function of different broadleaf evergreens in woodland planting.

Additional Resources

https://pubs.ext.vt.edu/426/426-607/426-607.html

http://web.extension.illinois.edu/cfiv/homeowners/021109.html

www.ext.colostate.edu/pubs/garden/07414.html

UNIT 29

Deciduous Trees

Objective

To properly position and plant deciduous trees in the landscape.

Key Terms

accent
containerized
contorted
deciduous
fastigiated
hardiness
monotony
privacy
tree wrap
weeping

Competencies to Develop

After studying this unit, you should be able to:

- list the six functions of trees in the landscape.
- select at least two specific trees to fulfill each of these six functions.
- demonstrate the proper planting technique for bare root, balled and burlapped, and containerized trees.

Materials

- deciduous or flowering trees, at least one of each in a container, bare root, and balled and burlapped
- round-point shovels for preparing the holes
- fertilizer
- peat moss
- mulching material (hardwood or pine bark)
- tree stakes, tree wrap, 12- to 14-gauge soft steel wire, small sections of garden hose
- a bar for punching holes to fertilize an established tree
- smartphone apps:
  - Leafsnap
  - LikeThat Garden
  - Spectracide
  - Plantbid

The Uses of Trees in the Landscape

Trees are often positioned in the landscape to serve a specific purpose. The following are some of the functions of trees in the landscape

- To provide shade. Shade trees keep temperatures inside houses cooler in summer as well as providing outdoor shade (Figure 29–1). A well-placed tree can reduce the summer room temperature of a frame house as much as 20 degrees in certain areas. To serve this purpose, trees may be planted as close as 15 feet to the house. Larger spreading trees, such as oaks, must be given more generous spacing.
- To frame the house. Trees are sometimes used to make houses the center of interest in the view from a street (Figure 29–2). Generally, the farther the house is set back from the street, the more effective the tree is as a framing device.
- To soften lines. A tree placed at the back of a home is effective in softening the lines of the house (Figure 29–3).
- To relieve bare spots. The end walls of many houses often have a bare look. This bareness can be relieved by planting a small or medium-sized tree near the side of the house (Figure 29–4). This creates a softening effect that a planting of shrubs alone may not accomplish.
- To screen an object. A single tree or a group of trees may be used to partially screen an object that would otherwise stand out too strongly in the landscape (Figure 29–5).
- To frame a view. Tall trees with an underplanting of shrubs or small trees may be used to frame a view from a house or terrace (Figure 29–6).
- To ensure **privacy**. A carefully positioned tree can create privacy by screening a porch or terrace (Figure 29–7).
- To **accent** the landscape. Flowering trees or trees with graceful and interesting branching habits (such as **weeping, fastigiated, contorted**) are often used as

Adapted from University of Maryland Extension Bulletin 183 by Robert L. Baker and Carl N. Johnson

▲ **Figure 29–1** To provide shade.

Adapted from University of Maryland Extension Bulletin 183 by Robert L. Baker and Carl N. Johnson

▲ **Figure 29–2** To frame the house.

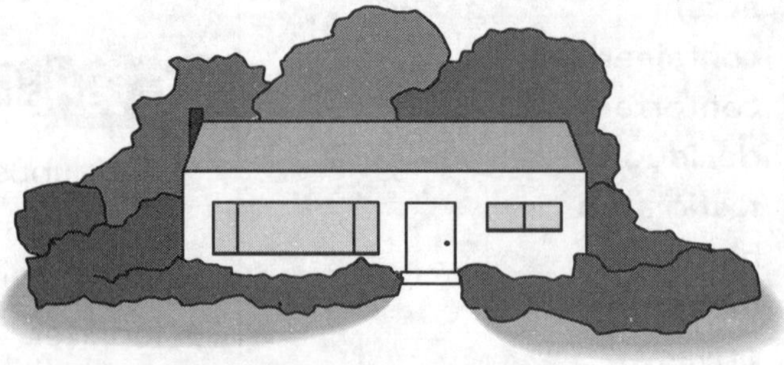

Adapted from University of Maryland Extension Bulletin 183 by Robert L. Baker and Carl N. Johnson

▲ **Figure 29–3** To soften lines.

Adapted from University of Maryland Extension Bulletin 183 by Robert L. Baker and Carl N. Johnson

▲ **Figure 29–4** To relieve bare spots.

Adapted from University of Maryland Extension Bulletin 183 by Robert L. Baker and Carl N. Johnson

▲ **Figure 29–5** To screen an object.

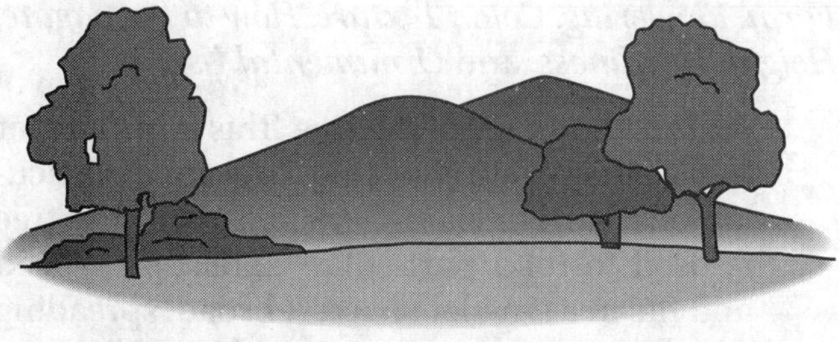

Adapted from University of Maryland Extension Bulletin 183 by Robert L. Baker and Carl N. Johnson

▲ **Figure 29–6** To frame a view.

Adapted from University of Maryland Extension Bulletin 183 by Robert L. Baker and Carl N. Johnson

▲ **Figure 29–7** To ensure privacy.

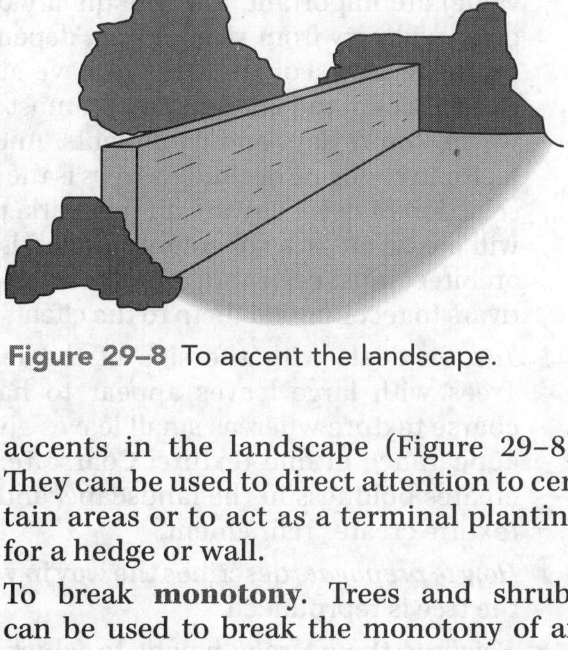

Adapted from University of Maryland Extension Bulletin 183 by Robert L. Baker and Carl N. Johnson

▲ **Figure 29–8** To accent the landscape.

accents in the landscape (Figure 29–8). They can be used to direct attention to certain areas or to act as a terminal planting for a hedge or wall.

- To break **monotony**. Trees and shrubs can be used to break the monotony of an obtrusive architectural feature or enclosure (Figure 29–9). The planting, however, should be placed off-center (use a zigzag pattern) so that the feature does not appear to have been cut in half. White brick walls, fencing, and other architectural means of enclosing property may be attractive in themselves, but they can become monotonous unless softened by planting materials.

Adapted from University of Maryland Extension Bulletin 183 by Robert L. Baker and Carl N. Johnson

▲ **Figure 29–9** To break monotony.

Adapted from University of Maryland Extension Bulletin 183 by Robert L. Baker and Carl N. Johnson

▲ **Figure 29–10** As windbreaks.

- As windbreaks. Tall evergreens are used as windbreaks, protecting the house from north and northwest winds in the winter (Figure 29–10). These should be placed about 50 feet from the house to be effective. (The rule of thumb for protection behind a windbreak is 7 to 10 feet of linear distance behind the windbreak for every 1 foot of windbreak vertical height. Therefore, a 10-foot *Pinus strobus* white pine would give 70 to 100 feet of wind protection behind the break.)

TRY THIS Create a technical landscape drawing that includes examples for each of the landscape uses listed in this section.

GREEN TIP Planting deciduous trees on the south, east, or west sides of your home lowers your heating and cooling expenses. In the summer, the trees shade your house from the hot sun. In the winter, the trees allow the warm sunlight into your home and raise your house temperature.

Multiple Use Trees

Some trees have multiple uses in the landscape. One of the best examples is the pecan tree *Carya illinoinensis*. Found throughout all the southern states, the pecan is very popular both as a nut tree and as a shade and large accent tree in the landscape. Its natural range is from Texas east to Alabama and from the Gulf Coast north to Iowa and southern Wisconsin. New cold-hardy selections have extended the planting area to include states as far north and east as New York.

The pecan is the largest member of the hickory family. Occasionally, trees are found that are 170 feet tall with trunk diameters of 6 feet, but a more general size is 90 to 100 feet tall with a diameter of 2 to 4 feet. The pecan tree grows best in the bottomlands with rich, moist soils where they are long-lived trees.

When used in the landscape, pecans need lots of space to spread out and up, and when given this space, they add a touch of symmetrical grace to the landscape. The fact that they bear nuts, which drop to the ground, restricts their use in lawn areas. They are best planted at some distance from the house where their stately form can be viewed.

> **GREEN TIP** Deciduous trees are important in a sustainable landscape environment. Deciduous trees are especially valuable for their many positive ecological aspects, such as their ability to cool the microelement and improve the soil conditions with a regenerative organic approach.

Tree Types

Figure 29–11 lists **deciduous** trees that may be used in landscaping for shade and ornamental flowering. The trees in this chart are separated according to their mature height, with small deciduous trees classified as those up to 35 feet in height; medium deciduous trees as those 36 to 75 feet; and large deciduous trees as those 75 feet and over in height.

The trees in the chart are classified according to various characteristics, under the headings *Form, Flowering, Color, Texture, How to Propagate, Height, Hardiness,* and *Ornamental Use.*

- *Form* is the shape of the tree. This is important to consider when choosing a tree for a particular location. For example, a tall, columnar tree is used to fill a particular landscaping need and fits in a smaller space. A broad, spreading tree requires more room and has a different appearance and use in the landscape.
- *Flowering and color.* If and when a tree flowers and the color of its blossoms are important factors in tree choice. Some trees have very large, attractive flowers whereas others have small, unnoticeable ones. Some trees vary in their leaf color from one season to the other, and others do not change at all. In some cases, a tree may be selected for its fall color.
- When landscaping with deciduous trees, it is important to consider foliage color display. Fall colors will change with the macroclimate of the area. Soil type, rainfall, and climate factors (e.g., temperature and sunlight value) are important, too. The quality of fall color will vary from year to year, depending on the season. For the trees to have attractive fall color, the autumn season must have warm, sunny days and cool nights. Another factor in choosing deciduous trees is the large selection of new cultivars on the market that will create an array of colors. The landscape architect must keep abreast of these new cultivars to recommend them to the client.
- *Texture* deals with the size of the leaves. Trees with large leaves appear to have a coarse texture whereas small leaves give an appearance of fine texture. Coarse texture creates boldness in the landscape, and fine texture creates refinement.
- *How to propagate* describes the way in which the tree is reproduced.
- *Height* is the average height to which that particular tree grows.
- ***Hardiness*** is the ability of the tree to live in a particular climate zone. This is influenced by the temperature, rainfall, and soil of that particular area. Some trees can tolerate lower temperatures than others. (In the chart, hardiness refers to US hardiness zones.) Hardiness zones give the average low temperature range for a particular area of the country.

SMALL DECIDUOUS TREES (UP TO 35 FEET IN HEIGHT)

NAME	FORM	FLOWERING	COLOR: SUMMER	COLOR: FALL	TEXTURE	HOW TO PROPAGATE	HEIGHT (CENTIMETERS)	HARDINESS	ORNAMENTAL USE	OTHER REMARKS
Acer palmatum (Japanese maple)	rounded moundlike	—	red to green	scarlet	fine	budding	209	5	specimen	Many varieties are available. Needs well-drained soil.
Acer ginnala 'Flame'(Amur maple)	rounded upright	purplish	green	scarlet	fine	seed	209	2	specimen, screening	Fragrant flowers.
Cercis canadensis (Redbud or Judas tree)	rounded	purplish pink mid-May	green	yellow	coarse	seed	359	4	border	Blooms same time as dogwood.
Chionanthus virginicus (Old man's beard)	rounded	white feathery late May	green	yellow	coarse	seed, softwood cuttings	309	4	specimen	Does well in moist, shady site.
Cornus florida (Flowering dogwood)	rounded	white mid-May	green	red	medium	seed	359	4	specimen	Does well in shade or full sun. Horizontal branching.
Cornus kousa (Chinese dogwood)	rounded	late May	green	red	medium	seed	309	5	accent, specimen	Bright red fruit. Excellent flower quality.
Crataegus crusgalli (Cockspur hawthorne)	rounded	small white	green	scarlet	medium	seed	359	4	border, specimen	Bright red fruit in fall. Needs sandy soil.
Crataegus oxycantha pauli (Paul's scarlet hawthorne)	dense rounded	double scarlet late May	green	—	medium	seed	359	4	border	Thorny.
Magnolia soulangeana (Saucer magnolia)	rounded open	white purple May	green	—	coarse open	seed, cuttings, grafting, layering	259	5	specimen, border	—

Figure 29–11 Common deciduous trees. *(continued)*

NAME	FORM	FLOWER-ING	COLOR SUMMER	COLOR FALL	TEXTURE	HOW TO PROPAGATE	HEIGHT (CENTIME-TERS)	HARDI-NESS	ORNAMEN-TAL USE	OTHER REMARKS
SMALL DECIDUOUS TREES (UP TO 35 FEET IN HEIGHT)										
Magnolia stellata (Star magnolia)	rounded open	double white to mid-April	green	yellow	coarse open	seeds, cuttings, grafting, layering	209	5	specimen	Best color appears in direct sun.
Malus floribunda (Japanese flowering crabapple)	dense	white early May	green	—	fine	T-budded, grafted	309	4	specimen	Flowers yearly.
Prunus cerasifera 'Thundercloud' (Thundercloud plum)	rounded	pink May	red	dark purple	medium	T-budded	209	3	specimen	Has dark purplish fruit.
Prunus serrulata (Kwanzan cherry)	vase	pink mid-May	green	—	coarse	T-budded	259	5	specimen	Many varieties are available. Showy display.
Prunus subhirtella 'Pendula' (Weeping cherry)	weeping	double pink early April	green	—	medium	grafted	259	5	specimen	Excellent as accent plant.
Syringa reticulata (Japanese tree lilac)	pyramidal	small creamy white	green	green/ yellow	medium	seed	309	4	specimen	Cherry-like bark.
MEDIUM DECIDUOUS TREES (36 TO 75 FEET IN HEIGHT)										
Aesculus carnea (Red horsechestnut)	pyramidal	pink to red spikes mid-May	green	brown	coarse	seed	759	3	specimen	—

Figure 29–11 Common deciduous trees. (*continued*)

MEDIUM DECIDUOUS TREES (36 TO 75 FEET IN HEIGHT)

NAME	FORM	FLOWERING	COLOR		TEXTURE	HOW TO PROPAGATE	HEIGHT (CENTIMETERS)	HARDINESS	ORNAMENTAL USE	OTHER REMARKS
			SUMMER	FALL						
Acer platanoides 'Crimson King' (Crimson King maple)	rounded dense	—	red-purple	red	medium	budding	509	4	accent	Brilliant foliage.
Acer platanoides 'Emerald Queen' (Emerald Queen maple)	upright oval	—	dark green	yellow	medium	budding	409	4	accent	Needs full sun.
Acer rubrum 'October Glory' (October Glory maple)	oval	—	dark green	red	medium	budding	509	3	accent	Excellent fall color.
Acer rubrum 'Red Sunset' (Red Sunset maple)	rounded	—	dark green	red	medium	budding	459	3	accent	Excellent fall color.
Betula pendula 'Laciniata' (Cutleaf weeping birch)	oval	—	bright green	yellow	medium fine	budding	509	2	accent	Weeping branches.
Betula pendula (Weeping European birch)	pyramidal	—	green	yellow	fine	seed	609	2	specimen	White bark in winter. Bronze birch borer can be a problem.
Carya illinoinensis (Pecan)	inverse pyramidal	—	green	yellow	medium	seed and grafting	1609	6	specimen	Excellent nut tree.
Cercidyphyllum japonicum (Katsura tree)	broad	—	green	yellow/ orange	medium/ fine	seed	609	4	accent	Excellent leaf and bark habit.

Figure 29–11 Common deciduous trees. (*continued*)

MEDIUM DECIDUOUS TREES (36 TO 75 FEET IN HEIGHT)

NAME	FORM	FLOWERING	COLOR: SUMMER	COLOR: FALL	TEXTURE	HOW TO PROPAGATE	HEIGHT (CENTIMETERS)	HARDINESS	ORNAMENTAL USE	OTHER REMARKS
Fagus sylvatica 'Atropurpurea' (Copper beech)	conical	—	dark purple	copper	medium/ fine	budding	709	5	accent	Excellent foliage.
Franklinia alatumaha (Franklinia tree)	rounded	white September	green	red	medium	—	369	6	border, specimen	Late fall flower.
Fraxinus pennsylvanica lanceolata 'Marshall' (Green ash)	rounded	—	green	yellow	coarse	seed	609	2	streets, border	Attractive form.
Gleditisa triacanthos "Shade-master" (Shademaster honeylocust)	rounded	—	light green	yellow	fine	budding	509	4	massing	Open branching.
Magnolia virginiana (Sweet bay)	rounded	white fragrant late May	green	green	coarse	seeds, cuttings, grafting, layering	609	5	specimen	Excellent flower quality.
Quercus virginiana (Live oak)	rounded wide-spreading	—	green	green	medium	seed	609	7	specimen, street	Evergreen in southern United States.
Oxydendrum arboreum (Sourwood or lily-of-the-valley tree)	rounded	white mid-July	green	red	medium	seed	759	4	specimen	Late summer flower; good fall color.
Sorbus aucuparia (European mountain ash)	pyramidal	clusters of white	green	reddish	medium	graft	609	5	specimen	Bright red berries appear in a cluster in fall.

Figure 29–11 Common deciduous trees. (*continued*)

NAME	FORM	FLOWERING	COLOR: SUMMER	COLOR: FALL	TEXTURE	HOW TO PROPAGATE	HEIGHT (CENTIMETERS)	HARDINESS	ORNAMENTAL USE	OTHER REMARKS
MEDIUM DECIDUOUS TREES (36 TO 75 FEET IN HEIGHT)										
Stewartia pseudo-camellia (Stewartia)	pyramidal	white camellia-like flower	green	purplish	medium	seed	609	5	specimen	Peeling bark in winter.
LARGE DECIDUOUS TREES (OVER 75 FEET IN HEIGHT)										
Acer platanoides (Norway maple)	rounded	small yellow before leaves	green	yellow	coarse	seed	1009	3	shade	Provides dense shade. Fast grower.
Acer rubrum (Red maple)	rounded	small red early April	green	brilliant red	medium	seed	1209	3	border, specimen	Several new varieties are available.
Acer saccharinum (Silver maple)	rounded	small green	green	green to yellow	medium	seed	1209	3	wildlife area	Fast growing; weak wooded.
Acer saccharum (Sugar maple)	rounded	green	green	yellow	coarse	seed	1259	3	specimen, shade	Used to produce maple syrup.
Fagus grandifolia (American beech)	pyramidal	—	green	bronze	medium	seed	909	3	specimen	Wildlife food.
Fraxinus americana (White ash)	rounded	—	green	yellow	coarse	seed	1209	3	border	Attractive form.
Ginkgo biloba (Ginkgo)	pyramidal	inconspicuous	green	yellow	coarse	grafted	1209	4	street, specimen, border	Female form has fragrant fruit. Fan-shaped leaf.

Figure 29–11 Common deciduous trees. (*continued*)

LARGE DECIDUOUS TREES (OVER 75 FEET IN HEIGHT)

NAME	FORM	FLOWERING	COLOR: SUMMER	COLOR: FALL	TEXTURE	HOW TO PROPAGATE	HEIGHT (CENTIMETERS)	HARDINESS	ORNAMENTAL USE	OTHER REMARKS
Gleditisia triacanthos 'Inermis' (Thornless honeylocust)	rounded open	greenish pealike June	green	yellow	fine	grafted	1209	4	lawn, specimen, shade	No thorns.
Juglans nigra (Black walnut)	rounded open	catkins	green	yellow	coarse	seed	1509	4	nut producing	Wood is used to make furniture.
Liquidambar styraciflua (Sweet gum)	pyramidal	greenish clusters	green	red to scarlet	medium	seed, leafy soft softwood cutting	1259	4	lawn, specimen, street, shade	Difficult to transplant. Good form. Corky ridges on stem.
Liriodendron tulipifera (Tulip tree)	rounded	greenish yellow tulip-shaped mid-June	green	yellow	coarse	seed	1809	4	specimen	Timber tree and fall color.
Magnolia acuminata (Cucumber tree)	pyramidal	greenish yellow early June	green	—	coarse	layering, seed, cuttings, grafting	909	4	specimen, border	Red to pink cucumber-shaped fruits.
Nyssa sylvatica (Black gum)	rounded	inconspicuous May-June	green	red	medium	seed	909	4	specimen	Small blue berries appear in midsummer.
Platanus occidentalis (Sycamore)	pyramidal	yellow-green	green	yellow-green	coarse	seed	1209	4	specimen	Natives tend to defoliate.
Populus nigra 'Italica' (Lombardy poplar)	narrow tall	—	green	—	medium	seed, hardwood cuttings	909	2	quick temporary screen	Short-lived tree.

Figure 29–11 Common deciduous trees. *(continued)*

LARGE DECIDUOUS TREES (OVER 75 FEET IN HEIGHT)

NAME	FORM	FLOWERING	COLOR		TEXTURE	HOW TO PROPAGATE	HEIGHT (CENTIMETERS)	HARDINESS	ORNAMENTAL USE	OTHER REMARKS
			SUMMER	FALL						
Quercus alba (White oak)	rounded	—	green	purple red	medium	seed	1509	4	specimen	Excellent lumber tree. Slow grower.
Quercus borealis (Red oak)	rounded	—	green	red	coarse	seed	769	4	specimen, street	Transplants easily.
Quercus coccinea (Scarlet oak)	rounded	—	green	bright scarlet	coarse	seed	1009	4	specimen, street	Difficult to transplant.
Quercus palustris (Pin oak)	pyramidal	—	green	scarlet	coarse	seed	1009	4	specimen	Very graceful. Excellent street tree.
Tilia americana (American linden)	rounded open	white June	green	—	medium	layering, seed, mound	1009	2	border	Wood is valuable for lumber.
Tilia cordata (Little leaf linden)	rounded open	white to yellow mid-July	green	yellow	medium	layering, seed, mound	909	3	shade	Excellent street tree.
Ulmus americana (American elm)	vase	yellow	green	yellow brown	medium	seed, softwood cuttings	1209	2	street	Possible problem with Dutch elm disease.

Figure 29–11 Common deciduous trees.

- *Ornamental use* describes common uses of the tree in the landscape. Specimen trees are grown solely for special characteristics they possess that create interest in a landscape. Street trees are those trees that are able to withstand the special conditions occurring along roadways and streets.

> **GREEN TIP** Tree leaves serve as environmental filters by trapping dust particles on the leaf surfaces of hairy or sticky leaves and absorbing some particles into the leaf pores. The dust particles on the leaves are flushed off by rainwater to the ground surface. In areas of heavy traffic, tree leaves absorb particles from the noxious gases found in exhaust fumes, such as ozone and nitrous oxide.

Figures 29–12 through 29–16 illustrate five deciduous trees commonly used in landscapes. Examine the trees and identify their physical characteristics, such as shape and density.

▲ **Figure 29–12** Kousa dogwood (*Cornus kousa*).

▲ **Figure 29–13** Maidenhair tree (*Ginkgo biloba*).

▲ **Figure 29–14** Weeping Japanese maple (*Acer palmatum*).

▲ **Figure 29–15** Stewartia (*Stewartia pseudocamellia*).

▲ **Figure 29–16** Japanese tree lilac (*Syringa reticulata*).

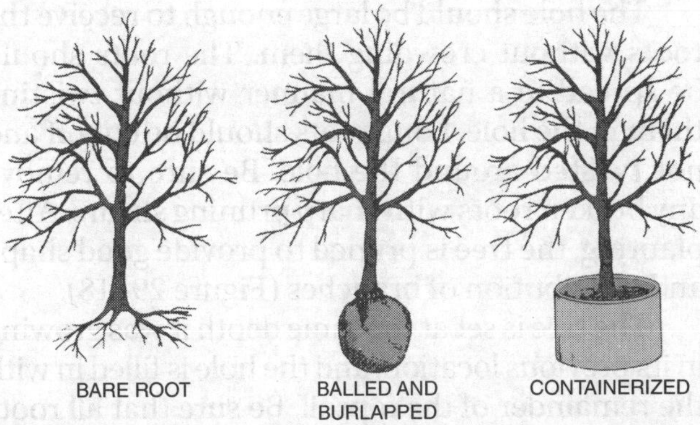

▲ **Figure 29–17** Ways in which deciduous trees are purchased.

Planting Deciduous Trees

Ways Trees Can Be Purchased

Deciduous trees may be purchased in three different forms: bare root (BR), balled and burlapped (B&B), or **containerized** (C) (Figure 29–17).

BARE ROOT Trees in this form should be purchased from a reliable nursery. The deciduous trees are dug in the nursery when the tree is dormant. No soil is left with the roots, hence the name bare root. Many nurseries purchase trees in this form for transplanting in the late fall or early spring (lining-out stock). It is best to work with only small bare root trees because they are easier to handle. Bare-root stock is usually less expensive. The cost of shipping large quantities of these trees is smaller as compared with balled and burlapped and container-grown stock.

BALLED AND BURLAPPED Trees in this form have a ball of soil around the roots. Almost all ornamentals can be transplanted in this form. The soil is held around the roots by burlap that is secured by twine and nails. Deciduous trees moved with a root ball larger than 20 inches should be drum laced to provide additional support for the root ball.

CONTAINER STOCK Trees in this form are planted in a container such as a basket or plastic or metal can. Many trees are grown today in this fashion because they can be transplanted at any time of the year without disturbing the root system and the grower has an increase in production per acre. Container stock becomes root-bound in the container if allowed to grow there too long.

Planting Bare Root Trees

Bare root trees are planted only when they are dormant. Because the soil is removed from the roots, the tree cannot replenish moisture rapidly enough to support leaves. Only after the roots have again made good soil contact can they supply the amount of water needed for a tree. The dormant tree requires adequate moisture to survive transplanting.

The bare roots should never be allowed to dry out before planting. Keep them covered with damp peat moss, sawdust, burlap, or some material so that the roots remain moist and away from the wind and sun. Some nurseries use a moisture-retention agent (e.g., Liqui-Gel) on the root system to prevent dry-out of the roots.

The hole should be large enough to receive the roots without crowding them. The roots should be spread in a natural manner without twisting them in the hole. Long roots should be cut off and not twisted around the hole. Be sure to remove any broken roots with sharp pruning shears. After planting, the tree is pruned to provide good shape and distribution of branches (Figure 29–18).

The tree is set at the same depth it was growing in its previous location, and the hole is filled in with the remainder of the topsoil. Be sure that all roots are covered, and work the soil around the roots as the hole is being filled. Firm the soil around the root system to hold the tree in correct growth position.

WATERING Use a garden hose (with the water running slowly) to settle the soil around the roots and to eliminate air pockets, which can cause the roots to dry unnecessarily. Water until the soil is saturated with water.

To ensure a good supply of water, form a ring (berm) around the base of the tree about 3 inches high. Fill the depression (saucer) with water (Figure 29–19). The water gradually soaks in and settles the soil around the roots. The berm also catches rainwater and keeps it from running off. Drive a stake on two sides of the tree and anchor the tree to the stake with a piece of rubber hose and wire only if necessary to hold the tree upright against the wind. Mulch with 2 to 3 inches of pine bark, hardwood bark, pine straw, or some other coarse material that will not blow away. Be sure to put the mulch about 1 inch away from the trunk of the tree. Never allow it to touch the tree; this discourages pests.

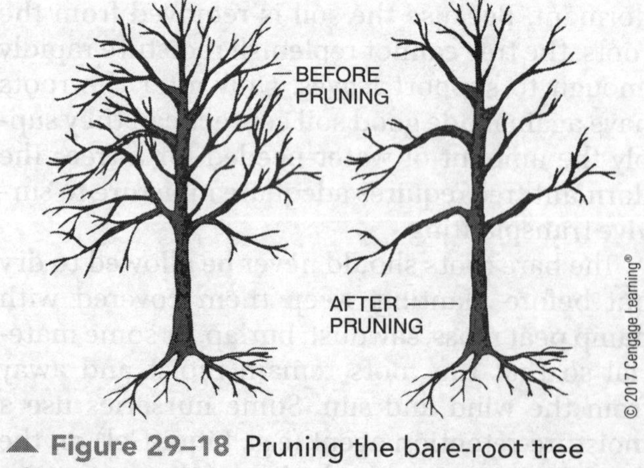

▲ **Figure 29–18** Pruning the bare-root tree after planting.

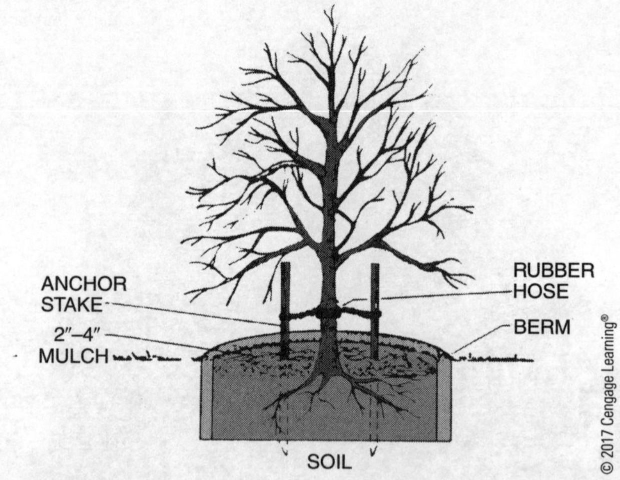

▲ **Figure 29–19** Formation of berm around base of tree.

Planting Balled and Burlapped Trees

Select the area in which the tree is to be planted. Dig a hole one and one-half to two times larger across than the ball of the tree to be planted. The hole should be dug so when the plant is set in the hole, the top of the ball sets slightly above the ground level. This allows for sinking of the tree root ball caused by settling. Set the tree in the hole, and loosen the burlap around the root ball (Figure 29–20). Fill the area around the root ball with soil. Follow the same procedure for watering as for bare-root stock.

Planting Containerized Trees

Containerized trees are planted in the same way as balled and burlapped trees except for the following differences.

1. The container is removed from the tree roots. It should be cut or otherwise removed carefully so that the soil is not broken from the tree.
2. After the container is removed, roots that may be growing around the outside of the root ball in a circular fashion are straightened out before planting. If the roots cannot be straightened, they are cut off at the edge of the root ball (called butterflying). This will stop the circular root growth, which tends to girdle root systems and restrict growth. In severe cases, it can kill trees.

▲ **Figure 29–20** Transplanting a balled and burlapped tree. After the tree has been placed in its new location, topsoil from the immediate area is distributed around the root ball.

Wrapping

It is necessary to wrap the trunk of the transplanted tree to prevent sunscald, bark damage, and water loss and to reduce chances of pest infestation on the trunk. The wrapping material should be 2- to 3-inch-wide burlap or commercial **tree wrap** paper. On large trees, it is best to wrap from the ground line up to and including several lower branches. This will allow excess water to escape from under the tree wrap material. Each turn of wrap spiral should overlap about one-half the width of the previous wrap. To hold the tree wrap in place use a light twine and lace up the trunk around the wrap. Note that tree wraps are used for temporary protection of trees with tender and thin bark during the winter months.

Staking/Guying

Staking or guying the tree prevents the wind from swaying it and loosening the roots. It also helps to keep the tree standing straight. Staking is not universally recommended as it was in the past. Research has shown that it causes weak trunks on young trees. For evergreens and trees planted in windy areas, however, it may still be necessary and is described here for use when needed. Some landscapers will still demand it, so you need to know how to do it properly.

PROCEDURE
Trees Braced by Staking

1. Staking must be completed within 48 hours of planting the tree. Place stakes parallel to walks and buildings.
2. Space stakes evenly and vertically on the outside of the tree ball, and drive firmly into the ground. Stakes must be driven at an angle and drawn to vertical. Note: Never drive a stake through the tree ball, as it will damage the tree's root system.
3. Cut pieces of hose long enough to loop around the trunk of the tree.
4. Place the hose around the trunk at the height required to provide optimum support. Then thread the wire through the hose, and pull both ends horizontally beyond the stake by about 2 feet.
5. Cut the ends of the wire, and then twist the wire at the rubber hose to keep it in place.
6. Wind both ends of the wire together around the stake twice, and then twist the wire back onto itself to secure. Cut off the excess wire. The wire should be within 2 to 6 inches of the top of the stake.
7. Follow steps 1–6 for each stake, keeping the tree straight at all times. There should be a 1- to 3-inch sway to the tree (the wires should not be pulled tight) for best establishment.

When staking, use a piece of rubber garden hose and wrap it around the tree trunk above the first set of branches. Then slip a piece of 12- to 14-gauge soft steel wire through the rubber garden hose. Twist the wire around and fasten it to each of the stakes (Figure 29–21). On large trees, the stakes should remain about 2 years. It is best

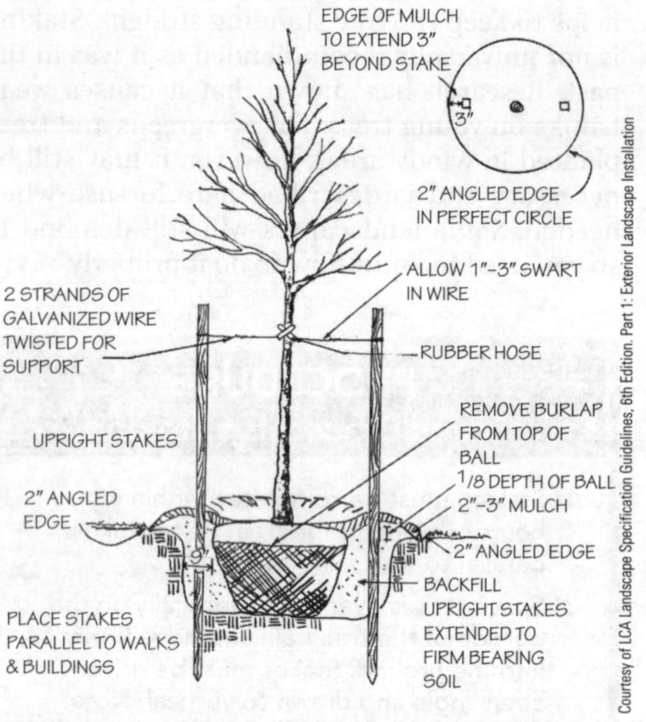

▲ **Figure 29–21** Bracing the tree by staking.

to remove the plant ties after this time because the hose may girdle the tree. As the tree grows older, it grows laterally, and if the ties are not removed, the tree will grow over the hose, thus weakening the tree. In a bad storm, the top of the tree could break completely off at the weakened point.

Fertilizing

When planting trees, it is recommended that a well-prepared soil be used for backfilling. A slow-release fertilizer such as Osmocote or MagAmp may be mixed in the soil at planting time according to manufacturer's recommendations. Use fertilizer tree stakes according to the manufacturers recommendations. Recent research shows that this may not be necessary on some tree species.

Another way to fertilize is to apply 2 pounds of a 5-10-5 fertilizer for each 1 inch of diameter of the tree trunk at a height 3 feet above ground level. Use a soil auger, wrenching bar, digging iron, or posthole digger to form holes that measure 15 to 24 inches deep and about 2 feet apart around the drip line of the tree (Figure 29–23). The fertilizer is distributed in equal amounts in the holes around the drip line of the tree. After the fertilizer has been applied, finish filling with soil the holes in which the fertilizer was placed.

No fertilizer is needed for the first growing season after planting. Use a slow-release fertilizer the second year after planting at the rate of 2 to 4 pounds of nitrogen per 1000 square feet per year. Use a commercial fertilizer with a 2:1:1 ratio with an analysis of 16-10-9 applied when the tree is dormant. The application of the fertilizer should come to the drip line of the tree. Determine the total square footage covered by the drip line of the tree. Use a soil auger, digging iron, or posthole digger to form holes 15 to 24 inches deep at 2 feet apart around the drip line of the tree.

Mulching

Mulch applied 2 to 3 inches deep is needed to retain moisture, control weeds, supply organic matter, and moderate the temperature for the roots. Mulching also improves the aesthetics of the area. Trees in the landscape may be mulched with very attractive materials such as hardwood bark, pine bark, coconut husks, and tanbark.

Mulch applied around deciduous trees should be spread evenly and neatly, 3 to 4 inches deep. When trees are mulched, it is important to keep the mulch away from the tree trunk, which can be done by running your hand down beside the tree trunk and pulling the mulch away from the trunk. This will help to prevent pests from entering the tree. The size of the mulched area under deciduous trees will vary according to personal preference. A rule of thumb is to apply mulch two-thirds of the distance from the tree trunk to the drip line of the tree.

Aftercare

The care after planting is important to ensure the proper growth of the newly planted tree. It is vital that the optimum quantity of moisture be in the soil at all times and critical to water the tree during the first year of growth (Figure 29–24). Never add less than 1 inch of water at any one time. Repeat every week from April through October if there is not sufficient natural rainfall to provide 1 inch of rain per week.

For larger trees that are staked or guyed, it is important to maintain support for the first year of planting. This will give the tree an opportunity to establish its root system for support. After the first year, the stakes or guying may be removed to prevent future girding of the tree trunk.

PROCEDURE
Trees Braced by Guying

1. Choose the correct size and number of stakes and size of hose and wire to guy the tree (Figure 29–22). The Landscape Contractors Association (LCA) provides Landscape Specification Guidelines for exterior landscaping that are helpful in determining the correct equipment (www.lcamddcva.org/). Guying must be completed within 48 hours of planting the tree.
2. Cut lengths of staking hose to extend 2 inches past the tree trunk when wrapped around.
3. Space stakes evenly on the outside of the tree saucer, and drive each firmly into the ground. Stakes must be driven at a 30-degree angle with the point of the stake toward the tree until 4 or 5 inches are left showing.
4. Place the hose around the trunk just above the lowest stout branch.
5. Thread the wire through the hose and past the stake, allowing approximately 2 feet of each of the two ends beyond the stake before cutting the wire.
6. Twist the wire at the rubber hose to keep it in place.
7. Pull the wire down and wind both ends around the stake twice. Twist the wire back onto itself to secure it before cutting off the excess.
8. Steps 1–7 are to be followed for each stake, keeping the tree straight at all times. There should be a 1- to 3-inch sway in the tree (the wires should not be pulled tight) for best establishment.
9. Flag the guy wires with surveyor's flagging tape or approved equal.

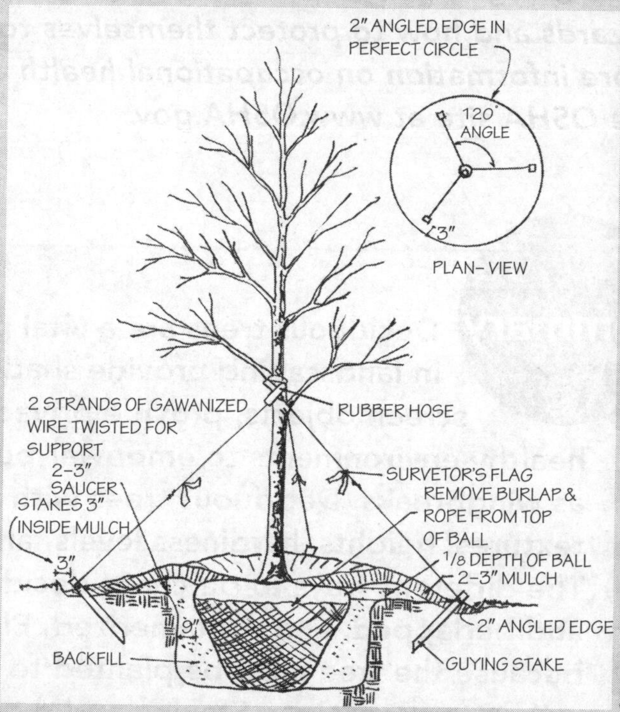

Courtesy of LCA Landscape Specification Guidelines, 6th Edition. Part 1: Exterior Landscape Installation

▲ **Figure 29–22** Bracing the tree by guying.

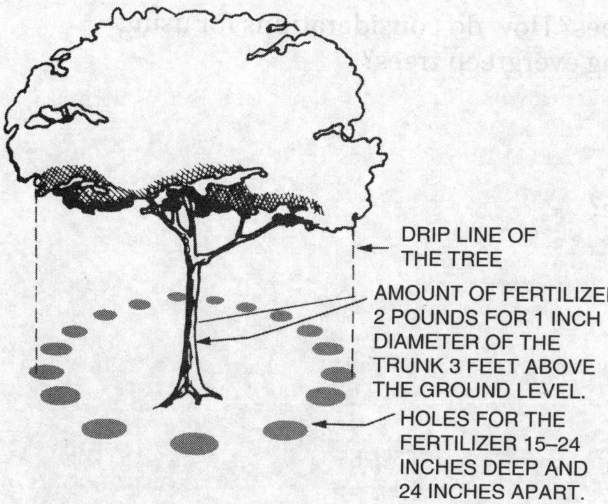

Source: US Department of Agriculture

▲ **Figure 29–23** Fertilizing the tree.

▲ **Figure 29–24** It is important to water after transplanting, especially during the first growing season.

SAFETY FIRST

OSHA's regulations ensure that employers and employees know about the work hazards and how to protect themselves to prevent injuries in the workplace. For more information on occupational health and safety in the workplace, please visit the OSHA site at www.OSHA.gov.

Summary

Deciduous trees are a vital part of landscapes today. Trees used in landscaping provide shade, frame the house, soften lines, screen objects, provide privacy, accent the landscape, provide healthy environments to empower public health for everyone, and serve as windbreaks. Deciduous trees with various flowering qualities, colors, textures, heights, hardiness levels, and ornamental uses are available. The different ways to purchase deciduous trees are bare root, balled and burlapped, and containerized. Planting for each of these is different because the tree must be planted to the original depth it was in the nursery or container. Staking and guying procedures are necessary for maintaining the stability of the tree. General maintenance (watering, fertilizing, and mulching) of the deciduous tree is important to ensure continued productivity of the plant.

Review Questions

1. Discuss at least six functions of trees in the landscape.
2. How do deciduous trees differ from evergreen trees? How do considerations for using deciduous trees in the landscape differ from using evergreen trees?
3. What is a multiple use tree?
4. Discuss the pros and cons of staking a tree.
5. What aftercare does a newly planted tree require?

Student Activities

1. Visit a local park or other landscaped area to practice tree identification. Use the Leafsnap app, or a nontechnical summer key, to identify deciduous trees. Identify and collect samples from at least 10 local trees. Preserve the leaves and organize them in a "leaf ID booklet," which includes an appropriate description of each tree.
2. Demonstrate the techniques used in the planting of a bare root, balled and burlapped, and container-grown tree.
3. Visit a local nursery in the spring to observe its bare-root tree planting operation.
4. Plant a landscape plan that uses balled and burlapped and container nursery stock. This work could be part of a local community service project.
5. Search online for the following terms: flowering trees; shade trees; deciduous trees.

Suggested Supervised Agricultural Experience (SAE) Activities

1. Design, plan, and install a landscape project for a local business, residence, or even your school. Record all hours, labor, and expenditures for an entrepreneurship SAE.
2. Participate in a placement SAE working for a forestry company that regularly works in the production of deciduous timber.

Additional Resources

http://bhort.bh.cornell.edu/tree/trees.htm

http://msucares.com/pubs/publications/p0146.pdf

www.ext.colostate.edu/pubs/garden/07418.html

http://ucanr.edu/sites/UC_Master_Gardeners/Local_Gardening_Articles_-_Info/Landscape_Trees/

UNIT 30

Deciduous Shrubs

Objective

To select, use, and care for deciduous shrubs.

Key Terms

adaptability
container-grown
deciduous
dormant
survival

Competencies to Develop

After studying this unit, you should be able to:

- identify at least five deciduous shrubs.
- list at least two uses of deciduous shrubs in the landscape.
- explain orally or in writing the three ways deciduous shrubs are purchased.
- outline the planting procedure for deciduous shrubs.

Materials

- one bare-root deciduous shrub
- one balled and burlapped deciduous shrub
- one container-grown deciduous shrub
- hardiness map
- round-point shovel
- nursery spade
- mulch (hardwood pine bark)
- 10-6-4 and 5-10-10 fertilizers
- hand pruners
- smartphone apps:
  - LikeThat Garden
  - Spectracide
  - Pest Portal
  - GardenCompass
  - Plantbid

Uses in the Landscape

Deciduous shrubs are used widely in the landscape to form borders, screens, background plantings, and foundation plantings. They offer an especially attractive feature in that they change in color as seasons change. Deciduous shrubs are selected for the various effects that are created by their flower color, their form, the fruit they bear, the color and shape of the stem, and the color and height of the foliage. Through careful selection, an interesting and attractive landscape design can be developed with deciduous shrubs, either alone or in combination with other plant materials. In today's landscape world, a myriad of new cultivars are available on the market.

The following are some important factors to consider when selecting deciduous shrubs for the landscape.

GREEN TIP Planting deciduous shrubs in the landscape helps reduce soil erosion, stabilizes soils, significantly reduces stormwater runoff, and shades landscapes and structures to help minimize water and energy use.

Period of Interest

A *period of interest* is that time of year in which plant materials create the most interest in the landscape. For example, flower color or fragrance of the flower might be a notable feature determining the period of interest. One example is the mock orange (*Philadelphus*). When the mock orange is in bloom, the fragrance is often apparent before the plant is seen. The period of interest for many shrubs is during the fall months, when they bear colorful fruit. An example is the cranberry bush (*Viburnum opulus*). It produces attractive red fruit in the fall that also provides food for wildlife in the winter.

GREEN TIP Deciduous shrubs provide oxygen, color, fragrance, and aesthetics to our environment in the forms of screening, borders, and special accents for background plantings.

▲ **Figure 30–1** *Spirea japonica* 'Lemon Princess.'

Foliage

The foliage of the plant has a definitive influence on its selection for a particular role in the landscape. For example, some shrubs, such as the oakleaf hydrangea (*Hydrangea quercifolia*), have large, coarse foliage. *Spirea japonica* 'Lemon Princess' has a bright chartreuse foliage color, which is very eye-catching (Figure 30–1). Others, such as white baby's breath spirea (*Spiraea thunbergili*), are very fine in texture. The two textures create a contrast in a landscape planting.

Height

When selecting deciduous shrubs, it is important to consider the height of the mature plant. For example, tall-growing shrubs should not be planted in foundation plantings where they may block windows.

Hardiness

The hardiness rating of a plant is based on the lowest minimum temperature that the plant can endure without noticeable damage to flower buds or wood. The **adaptability** of a plant to an area is further influenced by the amount of rainfall, wind velocity, exposure to sun, and soil type. Figure 30–2 gives essential information for choosing shrubs to fill a particular role in the landscape.

NAME	COLOR AND DATE OF FLOWERING	PERIOD OF INTEREST	HARDINESS	EXPOSURE	SOIL pH	PROPAGATION	REMARKS
DECIDUOUS SHRUBS LESS THAN 3 FEET TALL							
Chaenomeles japonica (Japanese quince)	orange to red	April	4	sun	6.0–7.0	cuttings	Border shrub. Grows in partial shade.
Cotoneaster horizontalis (Cotoneaster)	small, pink mid-June	fall	5	southern slopes	6.0–7.0	cuttings	Good for rock gardens. Low spreading branches.
Daphne genkwa (Daphne, blue daphne)	blue April	April	5	sun	6.5–7.5	cuttings	Difficult to transplant. Slow growing.
Deutzia gracilis (Slender deutzia)	white May	May	4	sun	6.0–7.0	cuttings	Showy, white.
Forsythia viridissima (Dwarf forsythia)	yellow April	April	5	sun	6.0–7.0	layering, cuttings	Rock garden plant.
Hypericum spp. (St. Johnswort)	yellow summer	summer	5	sun	5.5–6.5	cuttings	Many forms; yellow flowers throughout summer; dense.
Potentilla fruticosa (Cinquefoil)	yellow single rose mid-May	summer	2	sun	6.0–7.0	stem cuttings	Fine textured. Good border plant.
Spiraea bumalda 'Anthony Waterei' (Anthony Waterer spirea)	pink	June	6	sun	well drained 6.0 –7.0	cuttings	Blooms throughout summer. Compact and bushy
DECIDUOUS SHRUBS 3 TO 6 FEET TALL							
Abelia grandiflora (Abelia)	pink summer	summer	6	sun	6.0–8.0	cuttings	Flowers all summer. Good border or foundation plant.
Berberis thunbergi (Green leaf Japanese barberry)	yellow	fall, winter	6	dry sun	6.0–7.5	cuttings	Small hedge. Grows well under all conditions. Green foliage in summer; red foliage in fall.
Berberis thunbergi (Red leaf Japanese barberry) 'Atropurpurea'	yellow	fall, winter	6	dry sun	6.0–7.5	cuttings	Red all summer. Good specimen or hedge.
Buddleia davidii (Butterfly bush)	white, pink, and blue spikes	summer	6	sun	6.0–7.0	cuttings	Cut back to ground in spring. Fast-growing, arching plant.

Figure 30–2 Common deciduous shrubs. (*continued*)

NAME	COLOR AND DATE OF FLOWERING	PERIOD OF INTEREST	HARDINESS	EXPOSURE	SOIL pH	PROPAGATION	REMARKS
DECIDUOUS SHRUBS 3 TO 6 FEET TALL (*CONTINUED*)							
Euonymus alatus 'Compactus' (Dwarfwinged euonymus)	greenish May	fall	4	sun	5.5–7.0	cuttings	Brilliant red leaf in fall. Dense shrub; hedge plant.
Fothergilla monticola (Fothergilla)	white	fall, May	5	sun	5.0–7.0	cuttings	Fragrant white spring flowers. Quite stunning in Autumn. Nice in a shrub border.
Hydrangea quercifolia (Oakleaf hydrangea)	white mid-July	summer, fall	6	sun	moist soil 6.0–7.0	division, cuttings	Naturalistic effect when used to edge wooded lots.
Kerria japonica (Japonica)	yellow	winter, May		sun	6.0–7.0	division, cuttings	Green twigs in winter. Large perennial borders. Sometimes alled *Mt. Vernon shrub.*
Prunus glandulosa (Flowering almond)	double pink	early May	4	sun	6.0–7.0	cuttings	Double blossoms.
Rhododendron calendulaceum (Flame azalea)	yellow to red	late May	5	sun	4.5–5.5	layering, cuttings	Acid soil. Open growth habit.
Spiraea thunbergi (Baby's breath spirea)	white	April	6	sun	6.5–7.0	cuttings	Informal hedge. Fine textured.
Spiraea prunifolia (Bridalwreath spirea)	white	April	5	sun	6.0–7.0	cuttings	Arching branches; background plant
Spiraea vanhouttei (Vanhoutte spirea)	cascade of white, flat flowers	May	5	sun, shade	6.0–7.0	cuttings	Good hedge. Should not be sheared; best to prune to base.
Weigela spp. (Weigela)	pink to red	May	5	sun	6.0–7.0	cuttings	Abundance of flowers. Used in boundary plantings.
DECIDUOUS SHRUBS 6 TO 10 FEET TALL							
Aesculus parviflora (Bottlebrush buckeye)	white spikes	July	5	semishade	6.0–7.0	cuttings	Green mass of foliage in summer.
Elaeagnus umbellata (Autumn olive)	pale yellow	May	3	sun	5.0– 6.0	cuttings	Provides wildlife food. Pink-red fruit; silvery foliage. Used in border plantings. Very fragrant.

Figure 30–2 Common deciduous shrubs. (*continued*)

DECIDUOUS SHRUBS 6 TO 10 FEET TALL (*CONTINUED*)							
NAME	COLOR AND DATE OF FLOWERING	PERIOD OF INTEREST	HARDINESS	EXPOSURE	SOIL pH	PROPAGATION	REMARKS
Forsythia intermedia 'Spectabilis' (Border forsythia)	yellow	spring	5	sun	6.0–7.5	cuttings	Good in borders. Easy to transplant. Fast growing.
Ilex verticillata (Winterberry)	inconspicuous	winter	4	sun	5.5–6.5		Showy red fruit in winter. Grows well in swamps.
Kolkwitzia amabilis (Beauty bush)	pink	June	5	sun	6.0–7.5	cuttings	Boundary or specimen plant. Requires sun.
Ligustrum ovalifolium (California privet)	white	May	4	shade, sun	6.0–7.0	cuttings	Good hedge material.
Lonicera tatarica (Tatarian honeysuckle)	white to yellow	May	4	sun	6.5–8.0	cuttings	Red berries in June or July.
Philadelphus spp. (Mock orange)	white	June	6	sun	6.0–8.0 well drained	cuttings	Many varieties are fragrant.
Syringa chinensis (Chinese lilac)	lavender	April	5	sun	6.0–7.5 moderately drained	cuttings	Varieties available with white or red flowers.
Syringa vulgaris (Common lilac)	purple	April May	3	sun	6.0–7.5	cuttings	Powdery mildew is sometimes a problem. Suckers are pruned back to stimulate blooming.
Viburnum dentatum (Arrowwood)	flat white clusters	late May	3	wet to damp woods	6.0–7.5	cuttings	Excellent wildlife food source and border plant.
Viburnum macrocephalum (Chinese snowball)	large white flower heads	May	6	sun	6.0–7.5 rich soil	cuttings	Largest flower heads of all—snowballs.
Viburnum opulus (Cranberry bush)	white	May, fall	3	sun	6.0–7.5	cuttings	Park plantings. Red fruit. Purplish brown fall color.
Viburnum plicatum var. *tomentosum* (Double file viburnum)	white	May, fall	5	sun	6.5–7.5	cuttings	Flowers and fruit develop on upper side of branch.

Figure 30–2 Common deciduous shrubs.

Purchasing Plants

After particular shrubs are identified for use in the landscape, a source must be located. It is always best to purchase plants from a reliable nursery. Deciduous shrubs may be purchased as bare root (BR), balled and burlapped (B&B), or containerized (C).

Bare Root

A bare-root shrub is transported and transplanted with no soil around the roots. Only small shrubs should be purchased bare root. After plants reach 3 or 4 years of age or older, the **survival** rate is better if the plant is moved and planted with a root ball. Bare-root plants are moved only when they are **dormant**. This allows enough time for the roots to establish themselves prior to leafing out (followed by a high demand for moisture).

Balled and Burlapped

Balled and burlapped plants are purchased with a ball of soil around the roots. The root ball is wrapped tightly in burlap to hold the soil firmly around the roots and to prevent the soil from breaking apart and separating from the roots. The ball should be large enough to include most of the plant's root system or as large a ball as can be lifted and moved. Balled and burlapped plants can be transplanted any time during the year.

Container Grown

Container-grown shrubs are cultivated in various containers, such as pots, baskets, and tubs. The root system is damaged less during transplanting, so these plants can be transplanted at any time during the year. The roots of root-bound plants should be straightened after removal from the container. If twisted or curled roots cannot be straightened, they should be cut off at the edge of the ball to encourage the development of new roots as the plant continues to grow.

Planting Time

The best time to plant bare-root deciduous shrubs is when they are in their dormant stage. Balled and burlapped and containerized plants, on the other hand, may be transplanted any time during the growing season. This usually occurs in fall or early spring. In areas with mild winters, however, deciduous shrubs may be planted year-round. This includes areas south of Chesapeake Bay through the coastal plains of Texas and the coasts of Oregon and central California. Where winters are severe, spring is the best time to plant.

PROCEDURE
Planting

Note: Plant roots should never be allowed to dry prior to planting. Bare-root shrubs should be packed in moist peat moss or sawdust. Balled and burlapped and container-grown plants should be kept properly watered until planting time.

1. Dig a hole large enough to allow roots to spread in a normal manner without cramping or twisting. Dig deeply enough for the roots to sit on solid undug soil in the bottom and for the top roots to be at the same level covered in the nursery. The sides of the hole should be straight up and down and the bottom flat. This flat surface encourages straight growth of the root system.
2. Prune diseased and broken roots.
3. Plant the shrub at the same depth as that at which it grew before.
4. Work the soil in and around the roots, and apply 2 to 5 pounds per 100 square feet of 10-6-4 (foliage-type) or 5-10-10 (flowering-type) fertilizer on top of the soil.
5. Build up a slight berm around the outer edge of the hole to hold water.
6. Water well to settle the soil around the roots, eliminate any air pockets, and moisten the soil. Water regularly for the first year any time that the plant receives less than 1 inch of natural rainfall per week.
7. Mulch with chips or bark to a depth of 2 to 4 inches. Apply coarse material more heavily than fine material.
8. Prune the top to remove any broken branches and to help shape the plant.

Planting Site

Most shrubs prefer a well-drained soil. A good garden loam is the best soil type for most shrubs. Soil may be improved by the addition of organic matter to help hold moisture in dry soils and to improve drainage. Soil pH preference and the effects of exposure to sun and wind vary for each shrub type. The selection of the proper site is important because deciduous shrubs are long-lived plants. Deciduous shrubs should be fertilized in the spring with a commercial 10-6-4 fertilizer at 2 to 5 pounds per 100 square feet for foliage-type shrubs, and a commercial 5-10-10 fertilizer at 2 to 5 pounds per 100 square feet for flowering-type shrubs.

Holes for balled and burlapped and container-grown plants should be at least 3 times larger in diameter than the root balls to be planted. After the plant is placed in the hole, the container is cut away and removed or the burlap rolled down to expose the root ball as much as possible without loosening or breaking apart the ball. Soil is worked in around the root ball, and the plant is watered and mulched. Use tree/shrub fertilizer stakes according to the manufacturers recommendations.

Care of the Plant

TRY THIS After reading the following section, create a Standard Operating Procedure (SOP) to instruct other caretakers how to properly manage and care for deciduous shrubs. Display the SOP in a place where all people who care for the plants will notice it.

Disease and Insect Control

Some deciduous shrubs require spraying for control of insects or diseases. Consult your local agriculture experiment station or extension service for details. Most of the diseases that affect deciduous shrubs are caused by microscopic organisms such as bacteria and fungi. These small organisms infect parts of the plant structure such as roots, leaves, or stem.

Pruning

Pruning is used to thin out old or dead wood, to shape plants, and to control plant size.

(A) *Hydrangea quercifolia* 'Oakleaf hydrangea'

(B) *Exbury Azalea 'Gibraltar'* Azalea

▲ **Figure 30–3** Four deciduous shrubs used in landscapes. (*continued*)

(C) *Hibiscus rosa* 'Lord Baltimore.'

(D) *Viburnum carlesii* 'Korean Spice Viburnum.'

▲ **Figure 30–3** Four deciduous shrubs used in landscapes.

The following general rules apply to pruning shrubs:

1. Shrubs that bloom on wood grown the previous season should be pruned immediately after flowering. (Examples are forsythia and spirea.)
2. Shrubs that bloom on the current year's growth should be pruned in fall or early spring. (One example is the rose.)
3. Some shrubs require annual pruning to thin out old wood or wood that is shaded out and killed. (Examples are hydrangea, privet, and spirea.)
4. If shrubs send up shoots or suckers from the roots or at the base of the plant, at least some of them should be removed so that the growth on the plant does not become too thick.

Deciduous shrubs are a long-lived investment in the landscape. Their uses are as many as are the rewards in year-round beauty (Figure 30–3).

SAFETY FIRST

OSHA's regulations ensure that employers and employees know about work hazards and how to protect themselves to prevent injuries in the workplace. For more information on occupational health and safety in the workplace, please visit the OSHA site at www .OSHA.gov.

Summary

Deciduous shrubs are versatile plants that are widely used in landscapes to form borders and screens and for background plantings. Factors to consider in choosing deciduous shrubs include their period of interest with flower color, fragrance, foliage, height, and hardiness. Deciduous shrubs may be purchased in three conditions: bare root, balled and burlapped, and containerized. These plants are great to use in the landscape because they are long-lived and relatively easy to plant.

Review Questions

1. Discuss five deciduous shrubs that are commonly used in landscapes in your area.
2. How are deciduous shrubs used in landscape design?
3. Select one of the deciduous shrubs you identified in question 1, and describe its period of interest.
4. Discuss the different factors that contribute to a plant's adaptability.
5. Discuss the general rules that apply to shrub pruning.

Student Activities

1. Make 10 cuttings of selected deciduous shrubs in the area and propagate them. Observe the growth habit of the resulting plants. Practice pruning, fertilizing, and insect and disease control. If possible, transplant the shrubs to school grounds.
2. Design and develop a small rock garden using shrubs discussed in this unit that are common to the local area.
3. Ask a representative from a local nursery to speak to the class on the uses of deciduous shrubs in the landscape.
4. Search online to find current pricing for deciduous shrubs. Select five species of shrubs, and list the name and the price to purchase the plant from the nursery from the information you gather online.

Suggested Supervised Agricultural Experience (SAE) Activities

1. Participate in a placement SAE working for a local landscape contractor.
2. If land is available, establish an entrepreneurship SAE by growing, transplanting, and selling deciduous shrubs to the public.

Additional Resources

http://ucanr.edu/sites/UC_Master_Gardeners/Local_Gardening_Articles_-_Info/Landscape_Trees

http://extension.missouri.edu/p/G6830

www.ipm.iastate.edu/ipm/hortnews/2003/2-21-2003/shrubs.html

UNIT 31

Ground Covers

Objective

To select, establish, and maintain ground covers in the landscape.

Key Terms

broadleaf evergreen ground covers
deciduous ground covers
decomposed
erosion
functionally
ground covers
locality
narrowleaf evergreen ground covers
perimeter
prostrate

Competencies to Develop

After studying this unit, you should be able to:

- identify the three major types of ground covers.
- list three uses of ground covers.
- describe the cultural requirements of ground covers.
- identify five factors that must be considered when selecting ground covers.

Materials

- samples of different types of ground covers
- 5-10-5 synthetic fertilizer
- 6-2-2 cottonseed meal natural organic
- peat moss
- pine bark or hardwood mulch
- smartphone apps:
  - LikeThat Garden
  - Plantbid
  - Spectracide
  - Pest Portal
  - AG PhD
  - gro getter
  - Cailbrate My Sprayer

Ground covers are low-growing plants that cover the ground in place of turf. When used **functionally**, ground covers fill in bare spots in the landscape, help to prevent **erosion** of soil on steep banks, and fill in shady areas under trees where other plants have difficulty growing.

Types of Ground Covers

The three principal types of ground covers are broadleaf evergreens, deciduous plants, and narrowleaf evergreen plants. Most ground covers are perennial, but a few are annual plants. Figure 31–1 lists common ground covers and planting information for each. The three principal types are described here:

- **Broadleaf evergreen ground covers** retain their leaves year-round.
- **Deciduous ground covers** lose their leaves during the fall and winter season.
- **Narrowleaf evergreen ground covers** have needlelike or scalelike leaves. These plants retain their color throughout the year and thereby make excellent ground covers.

Uses

Ground covers play an important role in landscaping because they can be used in many areas where soil is not suitable for growing grass (Figure 31–2). These areas may be too steep, rocky, shaded, shallow soiled, or eroded for the proper growth of grass. However, ground covers are also used simply for the beauty they contribute to the landscape. Low-growing, dense ground covers that grow relatively slowly make attractive foreground plantings for shrubbery borders. They are also placed between plantings of broadleaf evergreens where they serve to keep the soil cool and shade the roots of certain plants, such as rhododendrons.

GREEN TIP Ground covers are low-growing plants that offer special stewardship in nature by conserving our soils on steep slopes, wet areas, and under the drip line of deciduous trees. They enhance the attractiveness of areas that are challenging in landscape design.

Selection

Before selecting ground covers for use in the landscape, the following questions must be answered:

- In what type of soil does the plant grow best?
- Is it suitable for the **locality**?
- How is it propagated?
- How long will it take for the plants to cover the area in which they are planted?
- What is the mature height of the plant?
- Are diseases and insects a problem?
- How expensive are the plants?

There may be other questions that require answering, depending on the particular landscape in question. Is the ground cover to serve a specific role in the landscape? If so, even more care must be taken when choosing the plant. If steep banks are to be covered for control of soil erosion, a plant that keeps its stem and plant top year-round should be selected. An evergreen would probably be best for this use.

Remember, also, that ground covers, like other plants, differ somewhat according to the soil type, moisture conditions, and amount of light they require. In a lightly shaded area, English ivy, *Pachysandra*, *Ajuga reptans* 'Chocolate Chip,' and *Vinca minor* are good choices. For a sunny bank, a **prostrate** juniper might be chosen. For planting in an area with continuous shade on the north side of the building, English ivy, *Ajuga reptans* 'Chocolate Chip,' or lily of the valley are wise choices. If the effect in the landscape is desired only for certain seasons of the year, select a plant that is most attractive during that season.

TRY THIS For each of the following environmental scenarios, identify the appropriate ground cover crop to use:

- a heavily shaded area
- a lightly shaded area
- a sunny bank or hill
- a steep bank to control erosion

NAME	COLOR & DATE OF FLOWERING	HEIGHT	LIGHT	FOLIAGE	PROPAGATION	HARDINESS	REMARKS
Ajuga reptans (Bugleweed)	blue to purple April or May	4″–8″	shade or sun	narrow leaves, 4″ long	seed and division	4	Gives dense coverage.
Arabis alpina (Rockcress)	white April to May	4″–10″	full sun	evergreen	seed and division	3	Spreads by creeping rootstock; useful in small areas.
Asarum caudatum (Wild ginger)	brownish purple June	6″–8″	shade	heart-shaped leaves with pungent taste	division	4	Grows best in moist soil that is high in organic matter.
Asperula odorata (Sweet woodruff)	white May to June	6″–8″	shade	fine	division	4	Scented leaves in whorls of six to eight.
Ceratostigma plumbaginoides (Plumbago)	blue summer	10″	shade or sun	green	division	5–9	Electric blue.
Chrysogonum virginianum (Golden star chrysogonum)	yellow May to June	4″–6″	sun	dense	division	4	Bright flowers.
Convallaria majalis (Lily-of-the-valley)	white spikes mid-May to mid-June	8″	shade	large oval leaves	division	2	Spreads by underground rootstock.
Cornus canadensis (Bunchberry)	small, yellow flowers May to June	9″	shade	evergreen	layering	2	Spreads by underground rootstock.
Coronilla varia (Crown vetch)	pinkish white June to September	1″–2″	full sun	compound leaves	division or seed	3	Excellent cover for steep slopes.
Dianthus deltoides (Maiden pink)	red to pink May to June	2″–6″	full sun	dense, glossy carpet	division or seed	2	Needs well-drained, acid soil
Epimedium grandiflorum (Barrenwort)	white, yellow, or lavender May to June	12″–15″	semi-shade	dense	division	3	Good in any soil.
Hosta undulata (Plantain lily)	lavender August	2″–3″	full sun to shade	dense, variegated	division	3	Dies to ground in fall.
Hypericum calycinum (Aaronsbeard, St. Johnswort)	bright yellow July to August	1″	semi-shade	dense after established	seed, division, or cuttings	6	Spreads freely by stolons.
Iberis sempervirens (Evergreen candytuft)	white late May	12″	full sun	narrowleaf evergreen	division, cuttings, or seed (roots easily)	4	Needs moist soil. Prune only after flowering in the spring.

Figure 31–1 Common ground covers. (*continued*)

NAME	COLOR & DATE OF FLOWERING	HEIGHT	LIGHT	FOLIAGE	PROPAGATION	HARDINESS	REMARKS
Lamiastrum (Yellow archangel)	yellow May	10″	sun, part shade	dense	division	4	Silver-edged foliage.
Liriope muscari (Lilyturf)	lavender to purple August	10″–15″	sun or shade	glossy	division	4	Spreads underground stem forming a mat or sod. Prune in the spring.
Lysimachia nummularia (Moneywort)	yellow summer	2″–3″	sun or shade	leafy, soft stem	division	3	Spreads quickly. May become pest in lawns if not controlled.
Ophiopogon japonicus (Dwarf lilyturf)	violet July to August	10″	sun or shade	glossy	divison	6	Forms a mat or sod.
Pachysandra terminalis (Japanese spurge)	small, white spikes early May	6″	shade	dark green, leafy	cuttings or division	4	Excellent with evergreens.
Sedum (Stonecrop)	yellow, white, or pink summer	2″–8″	full sun	evergreen	division or cuttings	3 or 4	Good for dry areas. Excellent new varieties.
Vinca minor (Periwinkle or myrtle)	blue violet or white late April	6″	partial to full shade	evergreen	division, root or stem cuttings	4	Needs fertile soil.
Arctostaphylos uva-ursi (Bearberry)	white to pink May	6″–12″	sun or partial shade	dark green	cuttings	2	Hard to transplant unless pot grown.
Calluna vulgaris (Heather)	white to red summer	4″–24″	sun	small needles	cuttings	4	Needs acid soil high in organic matter.
Cotoneaster adpressus (Cotoneaster)	white June	6″–12″	sun	small, dark green leaves	cuttings	5	Useful on banks.
Erica carnea (Heath)	rosy red early April	6″–12″	shade	small needles	cuttings	5	Needs loose, well-drained soil.
Euonymus fortunei (Wintercreeper)	none	4″	shade or sun	dark green	cuttings or division	5	Good on banks. Holds leaves most of winter.
Gaultheria procumbens (Wintergreen)	white mid-May	3″	shade	small, shiny green leaves	cuttings or division	3	Needs moist soil high in organic matter.
Hedera helix (English ivy)	small, greenish September	6″–8″	full or partial shade or full sun	dark green clinging vine	cuttings	5	Many cultivars available. Excellent plant.

Figure 31–1 Common ground covers. (*continued*)

NAME	COLOR & DATE OF FLOWERING	HEIGHT	LIGHT	FOLIAGE	PROPAGATION	HARDINESS	REMARKS
Helianthemum nummularium (Sunrose)	yellow, pink, or white June to July	6″–12″	full sun	evergreen	seed, cuttings, division	5	Does best in dry soil.
Lamium maculatum (Spotted dead nettle)	yellow summer	6″–8″	shade	green/white	division	3	Fast growing.
Mahonia repens (Creeping mahonia)	yellow early May	10″	sun	dull bluish green, leathery and spiny—like holly	root cuttings or division	5	Vigorous when established.
Mitchella repens (Partridgeberry)	white	4″–6″	shade	dark green	division or seed	5	Tolerant of dense shade and slow growing.
Paxistima canbyi (Paxistima)	small reddish early May	12″	sun or shade	small, dark green, and spreading	cuttings, layering, division	5	Needs fertile, acid soil.
Sarcococca hookeriana humilis (Sarcococca)	fragrant white mid-spring	1″–2″	light to heavy shade	glossy green	seed, division, cuttings	5	Prune in early spring to keep plant compact. Slow grower.
Teucrium chamaedrys (Germander)	white summer	12″	sun	green	cuttings	5–9	Knot garden.
Juniperus horizontalis (Creeping juniper)		12″–18″	full sun	summer—greenish blue winter—purple	cuttings	2	Excellent for banks and steep slopes.
Juniperus horizontalis 'Bar Harbor'		10″–12″	full sun	greenish blue	cuttings	2	Excellent for banks and steep slopes.
Juniperus horizontalis 'Douglasii' (Waukegan)		12″–18″	full sun	summer—steel blue winter—tinge of purple	cuttings	2	Excellent for banks and steep slopes.
Juniperus horizontalis 'Wiltoni' (Blue rug)		4″–6″	full sun	steel blue	cuttings	2	Excellent for banks and steep slopes.
Taxus baccata 'Repandens' (Spreading English yew)		36″	full sun to partial shade	deep dark green	cuttings	2	Excellent foliage and foundation plant. Expensive.

Figure 31–1 Common ground covers.

(A) *Vinca Minor* (Periwinkle)

(B) *Pachysandra terminalis* (Japanese spurge)

(C) *Galium odoratum* (Sweet woodruff)

(D) *Chrysogonum* (Goldenstar)

(E) *Lamiastrum* (Yellow Archangel)

▲ **Figure 31–2** Some commonly used ground covers. *(continued)*

Planting Ground Covers

When preparing the soil for ground covers, consider any special problems in the area. On a steep slope, very little can be done except to add topsoil if needed. No tilling or digging in of organic matter is possible on a very steep slope. If tilling can be done, add organic matter in the form of peat moss or rotted manure and 2 to 5 pounds of 5-10-5 fertilizer or 2 to 5 pounds of cottonseed meal spread evenly over each 100 square feet. Dig the materials into the top 6 inches of the soil before planting.

Ground cover plants are spaced according to their size, growth rate per year, and time

(F) *Lamium maculatum* (Spotted dead nettle)

(G) *Ceratostigma plumbaginoides* (Plumbago)

(H) *Hypericum calycinum* (St. Johnswort)

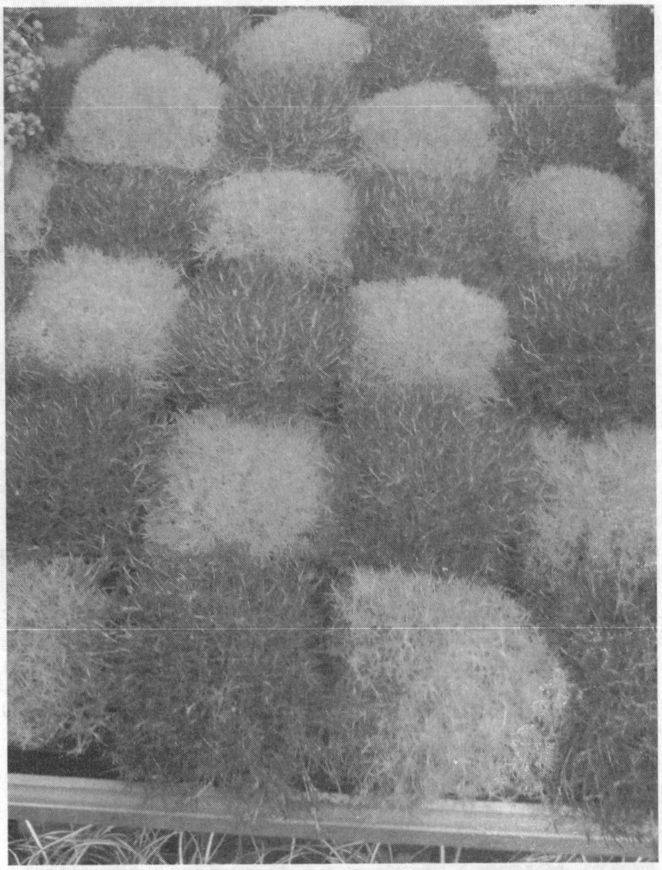

(I) *Santolina chamaecyparissus* (Lavender cotton)

(J) *Lamium macalatum* (Hermann's pride)

▲ **Figure 31–2** Some commonly used ground covers. (*continued*)

(K) *Liriope muscari* (Big blue)

(L) *Agave geminiflora* (Spaghetti strap agave)

▲ **Figure 31–2** Some commonly used ground covers.

allowed for plants to spread and cover the area. On steep slopes, the area may need to be covered more quickly. In this case, place the plants closer together. On more level areas, the plants may be spaced farther apart. This reduces original plant costs. It will, of course, take longer for the ground cover to grow together and completely cover the soil surface. Local nursery employees may be able to give spacing directions for the area to be planted. Water well after planting to settle the soil around the roots.

Caring for Ground Covers

Fertilizing

After the initial application at planting time, fertilizer should be applied only as needed to keep plants healthy. If fertilizing is necessary, it should be done in early spring in one of two ways: by means of a dry, granular fertilizer that is scattered over the area and watered in, or in the form of a soluble fertilizer that is mixed with water and siphoned through a garden hose and sprinkler system.

After ground covers are established, further soil cultivation is not necessary. However, it is good practice to mulch between newly positioned plants to aid in weed control and moisture retention until the ground cover has traveled over the entire soil surface. When mulching, cover the entire surface with the material before the plants are placed in the ground. This method is less time-consuming than if mulch were applied after the plants were installed. Push the mulch from the area where the hole is to be made, and pull it back around the plant as soon as the plant is set in the soil. Pine bark, hardwood bark, pine needles, well-rotted sawdust, peat moss, and **decomposed** leaves are suitable mulching materials for ground covers.

Pruning

Pruning is necessary only to confine the planting to the area in which it is desired. This is usually accomplished by mowing around the edges or otherwise cutting back the outer **perimeter** (edge).

Watering

Newly planted ground covers should be watered as needed during the first year. Any time the soil becomes dry or plants start to wilt, apply at least 1 inch of water to the entire planting area. If the proper plants are selected, watering should not be necessary after the first year.

Controlling Insects and Diseases

Insects and diseases are rarely a serious problem with ground covers. If control becomes necessary, identify the disease or insect, and spray with a recommended pesticide. (Be sure to use protective clothing as required by the pesticide container label.)

SAFETY FIRST

OSHA's regulations ensure that employers and employees know about work hazards and how to protect themselves to prevent injuries in the workplace. For more information on occupational health and safety in the workplace, please visit the OSHA site at www.OSHA.gov.

Summary

Ground covers are low-growing plants that cover the ground in place of turf. They have landscape value in special settings such as steep slopes, wet areas, and under the drip line area of trees. The different types of ground covers are broadleaf evergreens, deciduous, and narrowleaf evergreens. Ground covers add value to landscapes with additional flower color, foliage, and height. Planting ground covers and caring for them are important for maintaining an attractive landscape.

Review Questions

1. What are the three major types of ground covers used in landscape design?
2. What seven different factors should you consider when selecting a ground cover for a site?
3. What environmental benefits do ground covers provide?
4. When is it necessary to prune ground covers? How is pruning done?
5. A client would like to use Spreading English yew as a ground cover for a sunny area in front of her house but is concerned about the cost. What other plants might you recommend?

Student Activities

1. Draw a landscape plan depicting an area around the school or at home. Position ground covers where they could be used to best advantage. A class project might include actually planting and cultivating one type of ground cover.
2. Propagate ground cover cuttings in sand in 2 ¼-inch peat pots.
3. Collect samples of five different ground covers. For each one, identify the species and determine its purpose.
4. Calculate the amount of ground cover needed to cover 150 square feet.
5. Compete in the FFA Nursery/Landscape Career Development Event (CDE).

6. Google "Plant Ground Covers" (images), and select 10 that you like and would suggest to your client.
7. Search online for examples of landscape vines. Write a two-paragraph summary of the many types of landscape vines and their use in landscape design.

Suggested Supervised Agricultural Experience (SAE) Activity

1. Conduct a community-based SAE, in which you plant cover crops on a slope that is suffering from erosion. Record your erosion-control results as part of your SAE.

Additional Resources

www.hgtvgardens.com/ground covers/

http://urbanext.illinois.edu/ground covers/

https://classyground covers.com

UNIT 32

Bulbs

Objective

To force bulbs indoors and to use bulbs in the landscape.

Key Terms

acidic
alkaline
dormancy
embryonic
forcing bulbs
foundation
massing
natural
propagation
refrigerator

Competencies to Develop

After studying this unit, you should be able to:

- list four uses of bulbs in the landscape.
- describe the soil and fertilizer used in the flowering of bulbs.
- explain how planting depth and spacing of bulbs are determined.
- describe how to care for bulbs after they have flowered.
- list the steps in the forcing of bulbs.

Materials

- assorted bulbs and corms, including tulip, daffodil, crocus, hyacinth, amaryllis, begonia, calla, ismene, and gladiolus
- bulb planter and nursery spade
- bag of 5-10-5 fertilizer (organic or synthetic)
- sustainable bone meal
- cold frame or cold treatment storage
- bulb bed for demonstration
- greenhouse space, if available
- smartphone apps:
  - Weather Channel
  - Spectracide
  - Pest Portal
  - Calibrate My Sprayer
  - PGR Mix Master
  - Ortho Problem Solver
  - LikeThat Garden

Many people use the term *bulb* in a general way to refer to bulbs, corms, tubers, and rhizomes—all of which are structures containing an **embryonic** plant and the necessary stored food for plant growth. However, these structures are different in appearance and in their methods of **propagation**. Review Unit 9 for information on types of bulbs and descriptions of each.

▲ **Figure 32–1A** Flowering bulbs massed in the landscape add a burst of color early in the spring.

▲ **Figure 32–1B** Even a small bed of tulips accents a lawn area in the garden.

Uses of Bulbs in the Landscape

Many people purchase flowers that grow from bulbs, such as tulips, hyacinths, and daffodils, but are unsure about how to use them most effectively in the landscape. Bulbs are reliable, striking, and colorful, and they adorn the landscape with a beautiful image and an intriguing promise. Some landscape architects, such as the famous Wolfgang Oehme, believe that tulips look best when they are naturalized in the landscape. Just take a handful of bulbs and toss them in the area to be planted and plant them where they fall.

The benefits of bulbs are numerous:

- They create beauty in the garden early on in the growing season.
- They attract pollinators, such as bees, which provide a source of food in early spring.
- They provide an irresistible aromatic fragrance
- Many bulbs are off limits to pests. For example, deer will not graze on daffodils because they contain alkaloids that are toxic to them and therefore not palatable or attractive to them.

As a general rule, bulbs are most striking when they form a **massing** (grouping of color). This is usually more attractive than mixing various colors in the same bed. Bulbs are not attractive when planted in thin narrow rows. Circular groups or masses of a dozen or up to hundreds in one spot are better (Figures 32–1A and B).

> **GREEN TIP** Bulbs are used in sustainable landscapes to accent naturalistic settings such as rock gardens and wooded areas; they increase the area's attractiveness and give the environment "curb appeal."

In Wooded Areas

Bulbs grow well and produce beautiful color in **natural** wooded areas. Some bulbs grow well with evergreen ground covers in these areas. Growing taller than the ground covers, they give an excellent show of color. After they have flowered, the tops turn brown and die back. However, the ground cover remains to fill this area in the landscape until the next season's blooming of bulbs.

In Rock Gardens

Rock gardens are excellent places to use flowering bulbs. Bulbs that produce flowers that can grow low to the ground, such as crocus, dwarf daffodils, and species tulips, can be worked in very easily. To be most appealing, the flowering bulbs should be massed together. Using color masses throughout the garden accents the areas in which they are planted; this is an effective way to emphasize certain areas.

With Evergreen Shrubs

Many people use bulbs to add color around evergreen **foundation** shrubs. Bulbs planted in fall give early spring color to the landscape. Summer annuals can be planted to replace the color when the bulbs die back. When planting annuals, be careful not to damage the bulbs by digging too deeply.

As Cut Flowers

Tulips and narcissus are used extensively by florists in January, February, and March to create an early touch of spring in arrangements. When bulbs are to be used in these months, they must be forced to bloom out of season, a process which will be discussed later in this unit.

If bulb flowers are to be used as a cutting garden, the beds for cutting should be located in a special area so that when flowers are cut, the appearance of the landscaped area is not spoiled.

Selecting and Preparing Soils

Bulbs grow well in a good garden loam that is well drained. The organic matter content in the soil can be improved by adding sphagnum moss, leaf mold, garden compost, pine bark, or well-rotted manure. To apply these materials, spread them over the ground in a layer 3 to 4 inches thick and work in with a Rototiller the spring before planting in the fall.

Bulbs grow better if the soil pH is neutral, that is, neither **acidic** nor **alkaline**. If the pH value is below 6.0, ground limestone should be added at the rate of 10 pounds per 100 square feet to raise the pH to 7.0. A small amount of 5-10-5 fertilizer should be dug into the bottom of the bed.

Planting Bulbs

Bulbs such as the crocus, narcissus, hyacinth, and grape hyacinth are planted in the fall. Others, such as the dahlia, amaryllis, and gladiolus, are planted in the spring. It is important to consider the height of the various flowers because they are most attractive when they grow in sequence of height. When planting bulbs, use a bulb planter, nursery spade, or hand trowel.

Each type of bulb has a recommended planting depth and spacing (Figure 32–2). Spacing of the plants may be varied to achieve different effects. If a bed is to be showy, the bulbs are spaced closer together (Figure 32–3). As a general rule, bulbs should be spaced the same distance apart as their planting depth. Another approach, mentioned earlier and used by the renowned landscape architect and founder of the New American Garden movement Wolfgang Oehme of Washington DC, is to take a handful of bulbs and toss them into the garden area of desired choice. This gives a more natural appearance to the landscape. Water the bed thoroughly after planting.

GREEN TIP Bulbs are the easiest plants to use to enhance natural, dramatic landscapes. With extended flowering, a pleasing fragrance, a pallet of colors, bulbs form a natural pristine setting.

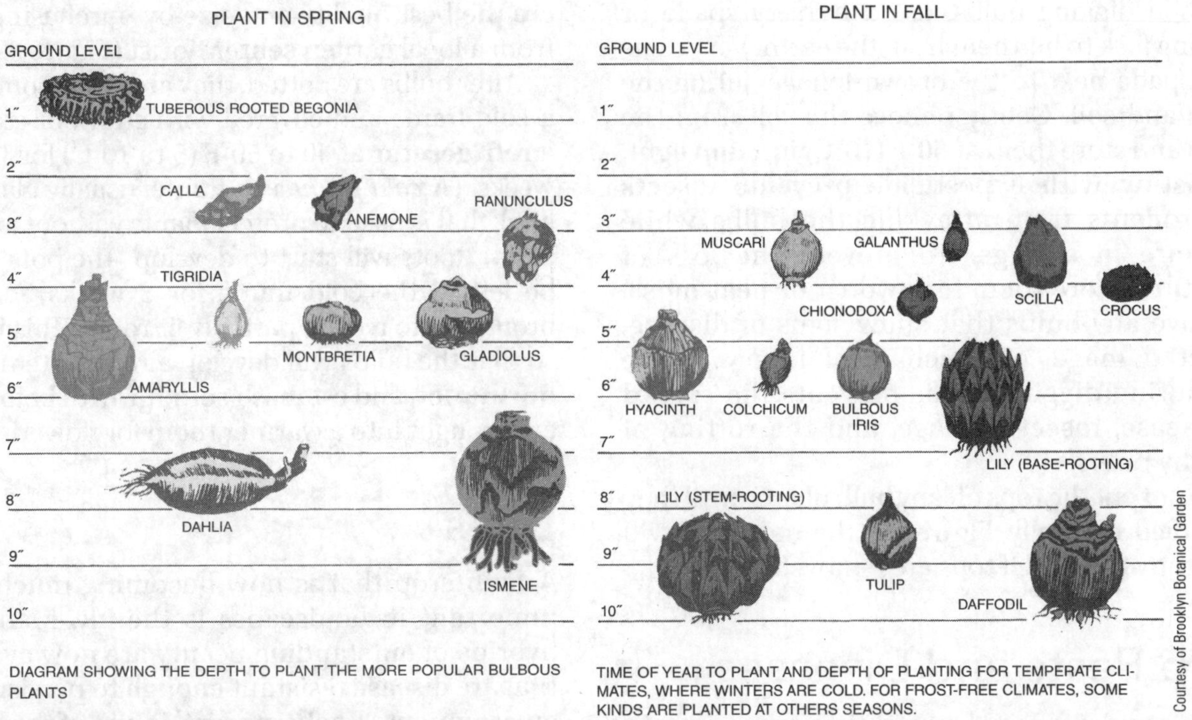

▲ **Figure 32–2** Planting depth of bulbs.

▲ **Figure 32–3** Tulip bulbs are planted 6 inches to center for maximum production of floral display.

Fertilizing Bulbs

Have the soil tested before adding any fertilizer to the garden being planted. Bulbs should be fertilized according to the recommendations of the soil test. This fertilizer should be covered with about 1 inch of soil before the bulbs are planted. Do not place the bulbs in direct contact with the fertilizer. The bulbs should then be completely covered with soil. Each spring after the first year, just before growth starts, provide a light application of a complete bulb food, such as bone meal 3-15-0 containing nitrogen and phosphorus, according to the soil analysis report.

Care after Flowering

Bulbs flowering in the spring should be dug *after the foliage turns yellow and dies back*. Tulip and hyacinth bulbs should be replanted every year to ensure high-quality blooming. There are now perennial tulips that do not require replanting. Daffodil, crocus, lily, and colchicum bulbs should be dug up every 3 to 5 years, thinned, and replanted.

Bulbs flowering during the summer should be dug after the first fall frost has killed the foliage. Some bulbs in this group are dahlia, tuberous-rooted begonias, calla, and gladiolus. It is important that the foliage of the bulbs has no green color in it. This condition indicates that the bulb has completed its growth cycle for the season, and all the food manufactured is stored in the bulb.

When digging bulbs, use a nursery spade or spading fork to lift them from the earth. Very carefully spade next to the brown foliage, lifting the bulbs and soil. Gently remove the soil from the bulbs and store them at 50°F (10°C) in a dim light.

Dusting with a pesticide prevents insects and rodents from attacking the bulbs while they are in storage. To prevent the loss of moisture, store them in sawdust or peat moss. Remove any bulbs that show signs of disease, insect damage, or mechanical injury (those cut during digging). This prevents the spread of disease, insect damage, and the rotting of healthy bulbs.

Do not cut the tops off any bulb plants until they have died naturally. Flowering the next year will be greatly reduced if tops are removed too soon.

Bulb Pests and Diseases

Bulbs require special protection from certain diseases and insects. Many pests can be controlled by maintenance of good cultural conditions. This is accomplished by removing weeds and other rubbish that provide a natural home for insects and disease organisms.

Chemical control is also effective and usually necessary. Applying chemicals before placing bulbs in storage is a good practice. Before applying any chemical, check the label to be sure it will not injure the bulbs. Consult local garden centers, nurseries, or hardware stores for information on pesticides available for use on bulbs.

Figure 32–4 lists the most common types of bulbs, pests and the damage they cause, and control methods that can be used.

Forcing Bulbs

Forcing bulbs, the procedure used to produce flowers, vegetables, and fruits out of season, is an interesting and challenging aspect of bulb production. This practice is used in the horticulture industry to make flowers available ahead of their typical blooming period. Individual gardeners also force bulbs to achieve the same affect in their homes. The fragrance and color of flowering bulbs such as tulips, daffodils, hyacinths, and crocus indoors during the late winter months can be very refreshing. When buying bulbs, be sure that they are the best quality and size by purchasing them from a local garden center florist (Figure 32–5).

After bulbs are potted, they are placed outside in a cold frame and covered with straw, or stored in a **refrigerator** at 40 to 50°F (5 to 10°C) for 10 to 12 weeks. (A *cold frame* is a frame, usually glass-covered, that is used to protect plants without artificial heat.) Roots will start to develop. The pots should be left in the cold frame for 8 weeks and then brought into a cool, partially lit room. This is done so that the bulbs will develop a root system, break **dormancy**, and establish some growth before they are brought into a warmer room for flowering.

Lilies

A bulb crop that is now becoming much more important in landscapes is the lily. Many new hybrids of outstanding beauty are now available that are disease resistant enough to persist in the environment. 'Enchantment' is one of the best to start with. Purchase others based on your own preferences, but look for varieties that are vigorous, disease resistant, and multiply from year to year. The lily bulb is soft and never dormant, so it must be handled carefully to avoid bruising; it also should never be allowed to dry out.

Notes about Bulbs

Consider the following important information regarding bulbs:

1. Always plant bulbs at the depth prescribed. Do not plant in shallow soil.
2. Always plant bulbs in well-drained soil.
3. Provide protection from mice where they are a problem. Mice love lilies and tulips especially. They will not eat daffodils or crown imperial (*Fritillaria imperialis*). Crown imperial may repel mice if interplanted with other bulbs. Cages made with 1/4-inch mesh wire and sunk into the bed with bulbs planted in them are also effective. The cage should be tight at corners and extend above the soil level for 1 or 2 inches.
4. All bulbs have planting instructions supplied with them when purchased. Be sure to follow these directions.

HOST, PEST, DISEASE	DAMAGE	CONTROL
Amaryllis		
Spotted cutworm Bulb mites Narcissus bulb fly Leaf scorch, red blotch	Feeds on flowers at night. Rotting bulbs. See Hyacinth. Decaying bulbs. See Narcissus. Reddish spots on flowers, leaves, bulb scales; stalks deformed.	Contact your local Extension Service for recommendations.
Gladiolus		
Thrips Botrytis and other flower blights Corm rots, scab Yellows (due to a soil fungus)	Leaves silvered, flowers streaked, deformed. Flowers, leaves, stalks spotted, then blighted. Lesions on corms, spots on leaves. Plants infected through roots, turn yellow and wilt.	Contact your local Extension Service for recommendations.
Hyacinth		
Bulb mites Aphids, several species Bulb nematode Soft rot	Minute; less than 1/25 inch, white mites in rotting bulbs. Leaves are curled; virus diseases may be transmitted. Dark rings in bulbs. Vile-smelling bacterial disease; often after mites.	Contact your local Extension Service for recommendations.
Iris (Bulbous)		
Tulip bulb aphid Gladiolus, iris thrips Leaf spot	See Tulip. Leaves russeted or flecked, flowers speckled or distorted. Light brown foliage spots with reddish borders.	Contact your local Extension Service for recommendations.
Lily		
Aphids (lily, bean, melon, peach, other species) Botrytis blight Mosaic and other virus diseases	Curl leaves, transmit mosaic and other virus diseases. Oval tan spots on leaves, which turn black, droop. Plants mottled, stunted.	Contact your local Extension Service for recommendations.
Narcissus		
Narcissus bulb fly Bulb nematode Basal rot Smoulder (botrytis rot) Scorch	Fly resembling bumblebee lays eggs on leaves near ground in early summer. Larva, fat, yellow maggot ½ to ¾ inch long, tunnels in rotting bulb. Dark rings in bulb. Chocolate-colored dry rot at base of bulbs. Plants stunted or missing; masses of black sclerotia on rotting leaves or bulbs. Yellow, red, or brown spots blight tips of leaves.	Contact your local Extension Service for recommendations.
Tulip		
Tulip bulb aphid Green peach, tulip leaf, and other aphids Botrytis blight, fire Cucumber mosaic Lily mottle viruses	Powdery white or grayish aphids common on stored bulbs. Transmit viruses to growing plants. Plants stunted, buds blasted, white patches on leaves, dark spots on white petals, white spots on colored petals, gray mold, general blighting. Small, shiny black sclerotia formed on petals; foliage rotting into soil and on bulbs. Yellow streaking or flecking of foliage. Cause broken flower colors, mottled foliage.	Contact your local Extension Service for recommendations.

Figure 32–4 Common bulb pests and diseases and their control.

▲ **Figure 32–5** Bulbs are graded according to size.

TRY THIS Create a display board showing different types of bulbs. Preserve the bulbs in wax paper or in small plastic bags and fasten them to the board. Clearly label each bulb and include a legible description of the bulb.

PROCEDURE
Forcing Bulbs

1. Identify bulbs to use for forcing. Select Number 1 (large) bulbs.
2. Select a potting medium that is well drained and high in organic matter.
3. Place drainage material (a stone or piece of clay pot) in the bottom of the pot.
4. Place a layer of the medium in the pot. Set the bulb in the pot, and fill in around the bulb with the growing medium so that the top of the bulb is exposed (Figure 32–6).
5. Water by setting the pot in a pan of water.
6. Be sure to label the pot correctly according to the variety of bulb.
7. Set the planted pot outside in a cold frame with a mulch of straw at a temperature of 50°F (10°C) from November 25 until January 15. Bulbs may be protected against disease by treatment with a fungicide. A refrigerator may be used for this cold treatment.
8. Remove the pot from the cold frame or refrigerator about January 15 (Figure 32–7), and place it in the greenhouse. Check for root development. Water well. Maintain a temperature of 60°F (16°C).
9. Tulip bulbs should bloom 5 weeks from the date of removal from the cold frame. Other plants may require more or less time to bloom.

▲ **Figure 32–6** The bulbs are set in a 6-inch azalea pot or bulb pan. Medium is added so that it almost covers the entire bulb.

▲ **Figure 32–7** After 8 weeks of cold treatment, the bulbs are brought inside to wait for blooming.

SAFETY FIRST

OSHA's regulations ensure that employers and employees know about work hazards and how to protect themselves to prevent injuries in the workplace. For more information on occupational health and safety in the workplace, please visit the OSHA site at www.OSHA.gov.

Summary

Bulbs are used in the landscape to enhance the garden with a massing of color. In the green industry, bulbs have a multitude of uses for gardens. In gardens with a naturalistic flair, bulbs can enhance a wooded area. Rock gardens are excellent locations for dwarf bulbs of crocus, daffodils, hyacinths, and grape hyacinths. Bulbs do well in soil with additional organic matter, and they need a complete fertilizer with a high content of phosphorus. When bulbs are planted, it is important to have correct spacing. It is best to set out all the bulbs before planting the entire bed. Bulbs have flat basal contractile roots. Bulbs can be forced for flower production. Precooled bulbs work best for forcing; cold storage for 15 to 16 weeks at 38°F is required. Monitoring root development of bulbs is important for the future production of flowers.

Review Questions

1. How are bulbs usually used in landscape design? What are some common bulbs used in your local area?
2. What type of soil works best for bulb planting?
3. Discuss the care of bulbs after the plants have flowered. Do some bulbs require different care than others?
4. Select a bulb that is commonly planted in your local area, and describe the pests and diseases to which it is prone.
5. Discuss some ways to prevent mice or other rodents from damaging bulbs.

Student Activities

1. Visit a local garden center, and note what types of bulbs are available. If possible, select and purchase bulbs for home or school use.
2. Force several types of bulbs, including tulips, daffodils, hyacinths, and crocuses. Develop a bulb-forcing schedule, and produce the flowers for sale on a certain date.

3. Under the supervision of the instructor, build a cold frame or use a refrigerator for use in producing and forcing bulbs.
4. Design a flower bed using bulbs for a home or school garden; include small plant markers that give brief descriptions of each bulb.
5. Attend a local garden show to observe the use of bulbs in the landscape. If possible, conduct an interview with the landscaper about the design of the plantings; share the interview answers with the class.
6. Cut and care for bulbs used in arrangements, and then create an arrangement that includes the bulbs.
7. Choose one type of bulb and research further information on its culture and uses.
8. Using the Internet, compile a list of different bulbs, and record their natural blooming dates. In this way, the sequence of their flowering can be determined.
9. Create a flower garden with bulbs, and use the Wolfgang Oehme method of naturalized planting.

Suggested Supervised Agricultural Experience (SAE) Activities

1. Prepare the soil properly, and conduct a home research SAE project planting flower bulbs such as tulips.
2. As a community-based SAE, adopt a section of roadway, and maintain the environment as well as plant and maintain bulbs to improve the aesthetic look of the roadside area.

Additional Resources

http://urbanext.illinois.edu/bulbs/planting.cfm

www.hort.purdue.edu/ext/ho-86.pdf

https://extension.umd.edu/hgic/flowers/bulbs

http://extension.missouri.edu/p/G6550

UNIT 33 Pruning Techniques

Objective

To select proper pruning techniques for specific plants and demonstrate their use in the landscape.

Key Terms

anvil pruner
espalier
heading back
hook and blade
horticulturist
lopper
thinning
topiary

Competencies to Develop

After studying this unit, you should be able to:

- list five reasons for pruning.
- describe four types of pruning.
- demonstrate pruning a stem at the proper angle.
- explain how the correct time to prune is determined.
- give three examples of each of the following types of plants, and describe the pruning technique used for each type.
  - deciduous spring-flowering shrubs
  - summer-flowering shrubs
  - broadleaf evergreens
  - conifers
- explain espalier pruning.
- describe topiary artistry.

Materials

- lopping shears
- lithium-powered chain saw
- anvil pruners
- hand pruners
- pruning saw
- stepladder
- protective equipment
- spring-flowering shrubs
- summer-flowering shrubs
- conifers
- broadleaf evergreens
- gas-powered hedge shears
- lithium-powered hedge shears
- lithium-powered pole saw
- pole pruner
- smartphone apps:
  - Leafsnap
  - LikeThat Garden
  - Spectracide
  - Ortho Problem Solver
  - Pest Portal
  - PGR Mix Master

Plants are pruned to accomplish the following:

- Remove dead, diseased, insect-infested, or broken branches. This keeps the plant healthy by removing any parts that might hinder further growth and stimulates new growth.
- Change the size or proportion of the plant. This is necessary when plants become overgrown in their landscape sites and sometimes occurs when the mature height and growth rate of the plant are not considered before it is placed in the landscape.
- Develop a special form or shape. Hedge pruning and topiary and espalier work are examples of this type of pruning. Topiary is a practice used in formal gardens in which plant material is trimmed into different forms and shapes. Espalier is a pruning method of training plant material to grow flat against a trellis or wall.
- Remove wilted or dead flowers and seed pods. Wilted or dead flowers give plants an unattractive appearance and should be removed. Seed pods use the plant's energy to develop seeds. Therefore, seed pods are removed to encourage further growth of the plant itself. For example, on rhododendrons and lilacs, seed pods develop rather than new shoots; the seed pods must be pruned to encourage new growth.
- Rejuvenate a declining plant. New growth is stimulated by pruning older wood. After pruning, the plant usually produces better-quality flowers and develops a better form (Figure 33–1).

(A) Shows an unpruned landscape rose. Note the length of the rose canes.

(B) Select the larger rose canes for removal to 10 or 12 inches above the ground level.

(C) Using lopper shears to remove the large rose canes.

▲ **Figure 33–1** Pruning a landscape rose (*continued*)

(D) Demonstrates the proper height to prune landscape roses.

▲ **Figure 33–1** Pruning a landscape rose

GREEN TIP Pruning regenerates the plant foliage, enhances the plant's ability to cool the microenvironment around the plant, and encourages new growth that gives cultural aesthetic appeal for the observer.

Equipment for Pruning

A few basic pruning tools are needed for all pruning (Figure 33–2). These tools must be kept clean and sharp so that when used, they will give a good, clean cut.

- Pruning saw. This folding pruning saw has a coarse teeth setting. The saw has a special setting to cut through the (live) green wood or dead wood without pinching the saw. The folding pruning saw blade closes into the handle for easier, safer, and better storage.
- Lopping shears. These shears have long handles to give more leverage for cutting larger branches. The type shown in Figure 33–2D is a scissor-action **lopper**.

(A) Pruning tools. From top to bottom: folding pruning saw, hook and blade pruners, anvil pruners.

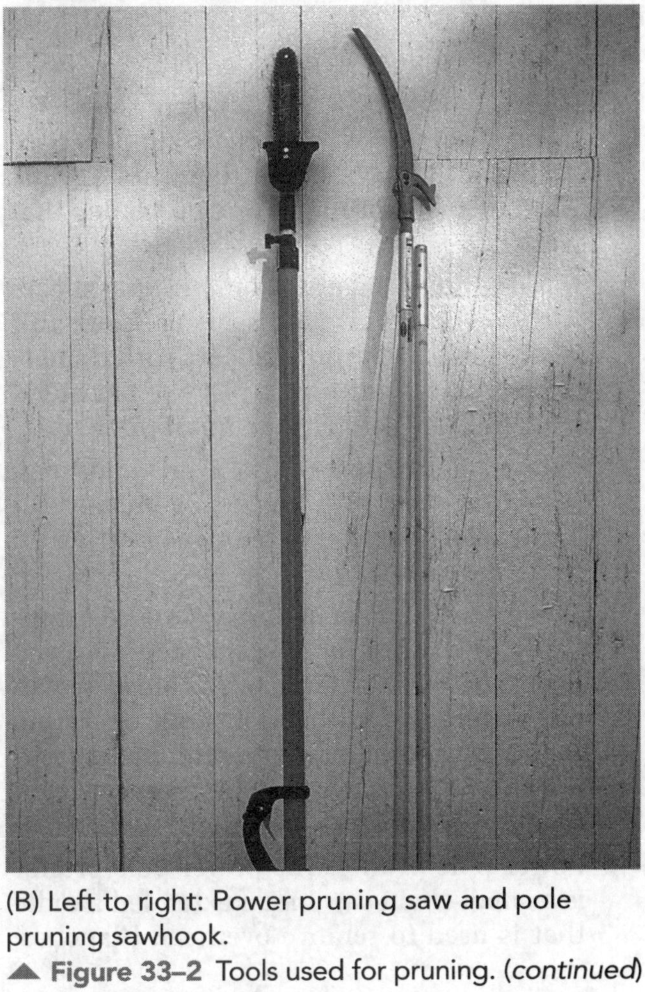

(B) Left to right: Power pruning saw and pole pruning sawhook.

▲ **Figure 33–2** Tools used for pruning. (*continued*)

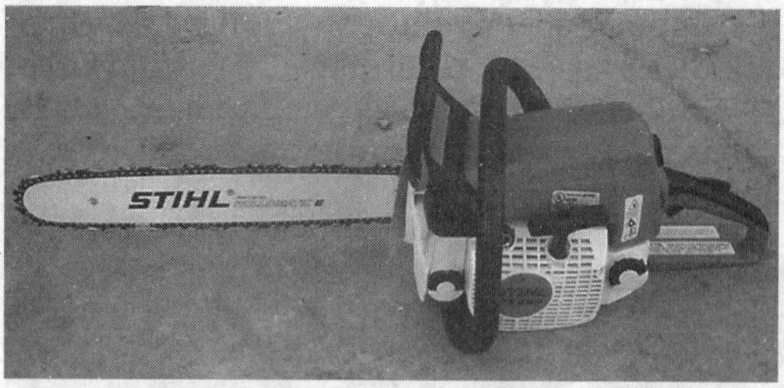

(C) Power chain saw.

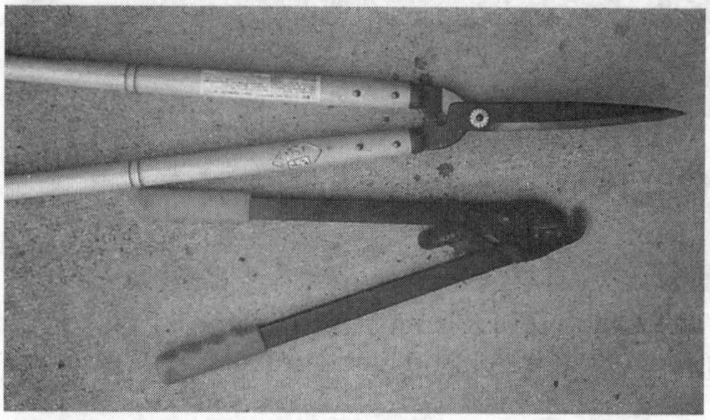

(D) Top: Anvil ratch loppers. Bottom: Japanese head shears.

▲ **Figure 33–2** Tools used for pruning.

- Hand shears. Hand shears are clippers consisting of a single blade that cuts against another piece of metal or two blades that work like scissors.
- Hand pruning shears. Hand pruning shears with two blades give a scissor cut (**hook and blade** pruner). Pruning shears with a single blade that cuts against a piece of flat metal (usually brass) is called an **anvil pruner**.
- Pole pruner. This tool has a saw and hook (which is used to pull, cut, or break dead branches). It is used to remove branches 10 to 12 feet overhead.
- Chain saw. A power chain saw is used extensively to remove large branches that are 3 inches or more in diameter. Chain saws can be dangerous—use caution while operating them. Also available are lithium battery-powered chain saws that are more environmentally friendly.
- Power pole pruner. The power pole pruner is a small chain saw on an extended handle that is used to remove overhead branches. It may be used to elevate lower branches to create improved vision. The new technology is the lithium battery-powered pole pruning saws that are more environmentally friendly.
- Power hedge trimmers. The power hedge trimmers are used to maintain the growth of leafy plants in the landscape. Power hedge trimmers are environmentally and mobile friendly, and are operational with battery, electric, or gasoline.

TRY THIS The instructor will display and demonstrate several pieces of pruning equipment. For each piece of equipment, identify the name of the tool and its purpose. After that is complete, demonstrate the safe use of each piece of equipment.

Types of Pruning

There are four basic methods of pruning. Before deciding which pruning method to use, consider the type of plant to be pruned and the desired finished effect.

Thinning

Thinning involves removing certain branches from a plant to open up the plant. This method is usually used when the **horticulturist** wants to keep the natural shape of the plant. Thinning allows more light to reach inner branches so that they can develop new growth more easily. The effect of thinning does not have to be noticeable (Figure 33–3).

Fruit trees are a good example of plants that are commonly thinned. These trees are pruned while the plants are dormant (in winter or dry season). All cuts are made to the collar of the limb so that the tree wound will heal properly (Figure 33–4). In pruning young fruit trees, it is important to thin limbs to leave about four to six lateral branches that are about 8 inches apart on the main trunk. The lowest branch should be about 2 feet from the ground. It is best to proceed in a spiral around the main trunk when pruning, keeping the lateral branches evenly spaced.

(A) An unpruned shade tree.

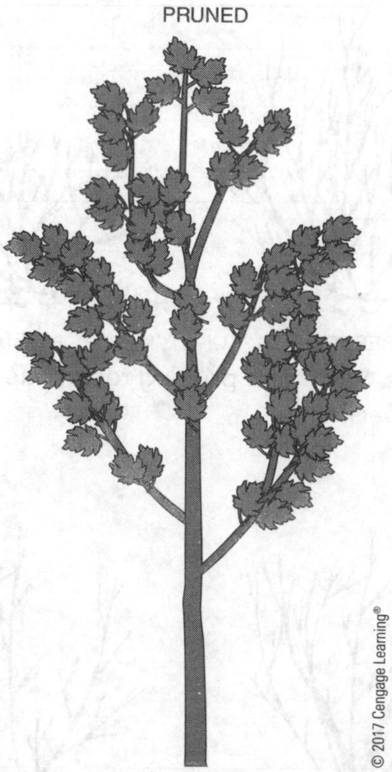

(B) The same shade tree properly pruned by thinning.

▲ **Figure 33–3** Thinning deciduous shade trees.

(A) Using a pole pruner to remove a lower branch.

(B) Pruning to the *collar* at the base of the branch.

▲ **Figure 33–4** Thinning involves removing certain branches from a plant to open up the plant.

Heading Back

Heading back is the removal of the end section of the plant branches at the same height (Figure 33–5). This causes new shoots to develop from dormant buds, therefore making the plant thicker. The plant will grow in the direction the bud is facing, so it is important to prune back to a live bud. The new growth will begin from that point. Heading back is usually considered the easiest of the four techniques discussed.

Renewal Pruning

In *renewal pruning*, old branches that are large and unproductive are removed. The oldest

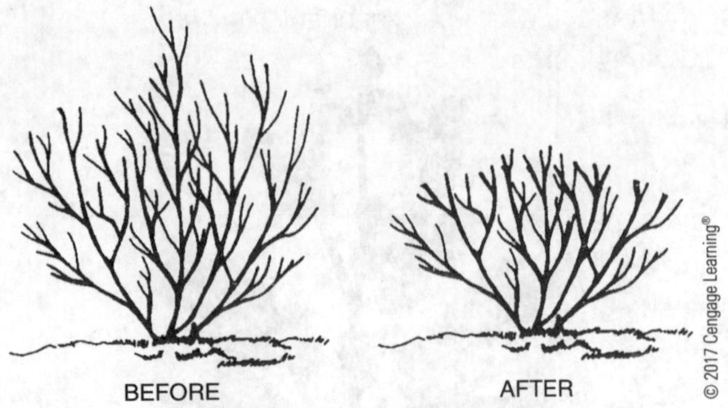

▲ **Figure 33–5** The pruning technique of heading back.

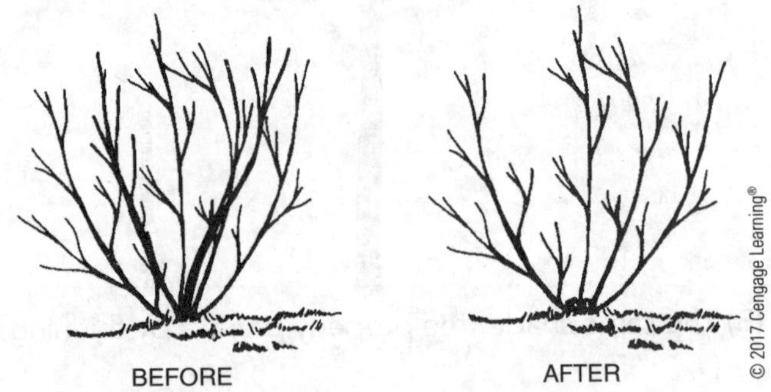

▲ **Figure 33–6** Renewal pruning before (left) and after (right).

branches are removed by cutting them back to ground level (Figure 33–6). This results in the development of new branches and better flowering on the new branches. Renewal pruning is usually effective on flowering shrubs.

Root Pruning

Root pruning, an important step in transplantation of plants, involves cutting off all the lateral roots of the plant with a sharp spade in a circle around the stem. In commercial nurseries, root pruners are mounted on tractors to allow for a more efficient operation. A general rule for root pruning is to cut in a circular motion around the plant so that 1 inch of stem diameter equals 10 inches of the circle diameter (Figure 33–7). It is best to root prune one growing season prior to transplanting the plant. This results in the formation of many small, fibrous roots. Early fall is the best time to perform root pruning.

▲ **Figure 33–7** The proper distance for root pruning deciduous trees.

How and When to Prune

Cutting at the Proper Angle

Cutting the plant stem at a 90-degree angle is an important aspect in the pruning process. Figure 33–8 shows how pruning cuts should be made. A cut that is made at too great an angle, as shown in Figure 33–8, leaves too great an exposed area for the plant to heal properly. It is also important that the cut be made the correct distance from the bud (Figure 33–9). Cutting too far from the bud leaves too much stub above the cut, causing the branch tip to dry out and die. It may also provide a home for insects, and results in an ugly appearance. Cutting too close to the bud may injure it.

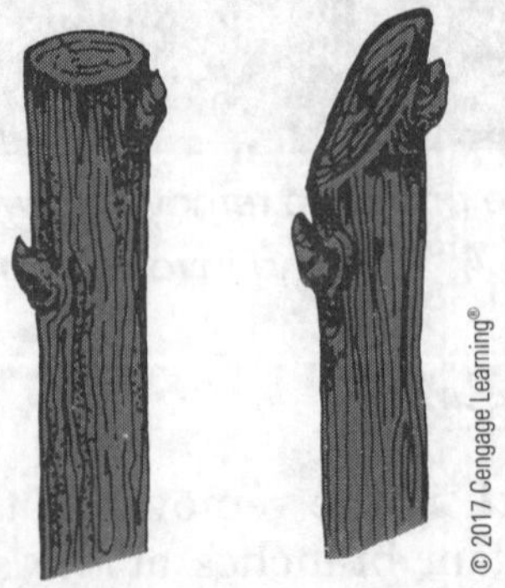

▲ **Figure 33–8** Pruning at the proper angle of the stem. The stem on the left has been cut at the correct angle; the stem on the right has been cut at too great a slant.

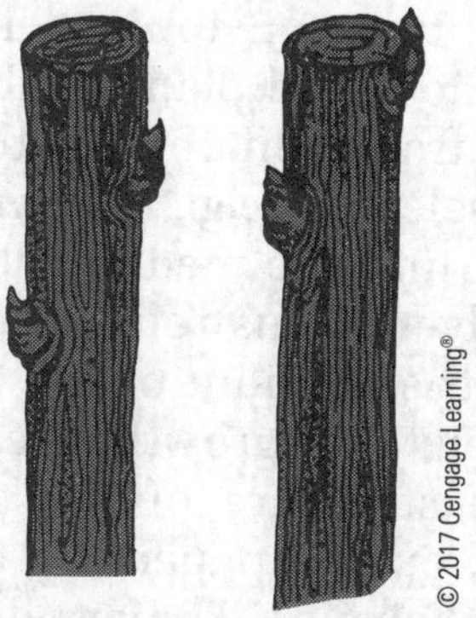

▲ **Figure 33–9** Pruning at the correct distance from the bud. On the left, the cut has been made too far from the bud; on the right, too close to the bud.

When to Prune

The time schedule for pruning usually depends on when the plant flowers or bears fruit. For example, if a plant flowers in the spring, it should be pruned after it flowers. This means that flower buds are developed during the previous season's growth. Remember that there must be flowering and fertilization of the flower buds before fruit will develop.

Shrubs that should be pruned after they flower are forsythia, deutzia, lilac, weigela, mock orange, bridal wreath, spirea, beauty bush, kerria, magnolia, sweet shrub, and Juneberry.

Plants that flower in the summer or fall are usually pruned just before growth begins in the spring. In this case, the flower buds develop on the new growth (current season's growth). Some examples of plants that may be pruned in summer or fall are abelia, crape myrtle, Russian olive, bush honeysuckle, roses, five-leaf aralia, Norfolk Island pine, bamboo, aucuba, heather, boxwood, bittersweet, croton, winter hazel, viburnum, blueberry, hemlock, and bald cypress.

Espalier Pruning

Espalier is a pruning process that trains the plant into a flat pattern onto a wall, fence, or trellis (Figure 33–10). The art of espalier was developed in medieval times. This technique will create artistry of design as well as only using a small area in the garden. Espalier has several other advantages in the garden: warmth from a building is an excellent source of winter protection, and the heat encourages flowering and improves fruit quality. Today, its use with dwarf fruit trees (i.e., Apple, *Malus* species and Pear, *Pyrus* species) gives needed support for the trees during the fruiting season. Other ornamental plants suitable for espalier include Blue atlas cedar (*Cedrus atlantica* 'Glauca'), Pyracantha (*Pyracantha* species), Japanese maple *(Acer palmatum)*, and vines such as English ivy *(Hedera helix)* and Boston ivy *(Parthenocissus tricuspidata)*. Different styles of designs are used for espaliering; for example, a more formal style would use the candelabra, circle, fan, or basket weave.

▲ **Figure 33–10** Espalier pruning of *Pyracantha coccinea* 'Laland.'

The process of training and pruning is ongoing. The time of pruning will vary with the type of plant used in the espalier. Fruit trees are pruned in late winter and early spring to encourage new growth. Pruning during June and July produces a dwarfing effect. From time to time, new growth shoots will occur and should be pruned to ensure the quality of the espalier.

Topiary Artistry

Topiary is the art of trimming and training plants into ornamental shapes. Topiary artistry has been practiced for centuries. Today, topiary art is used to enhance gardens with all types of shapes (Figure 33–11). An example of the scope of topiary is the display at the Disney parks, which ranges from numerous animated characters with phenomenal interest to a wreath pruned with miniature English ivy.

The technique of pruning topiary can be accomplished with hook and blade pruners or just a simple pinch between the thumb and index finger to remove tender shoots to shape the desired design for the topiary. In this type of topiary, live plants are planted into a container for active growth of the plant. The training starts with attaching the growing plant to a frame and maintaining the growth until it covers the wire frame.

▲ **Figure 33–11** A spiral topiary of a *Picea glauca* 'Cornica' (dwarf alberta spruce).

Another type of topiary involves using sphagnum moss and plant medium filled into the topiary frame. In this method, the plant material is selected (e.g., *Hedera helix* [English ivy]) to root into the medium. The English ivy will form adventitious roots at each node and attach into the medium of the topiary. As the plant develops new growth, it can be fastened to the base structure of the topiary with a hairpin (i.e., open a paper clip and cut into it to form the hairpin). Topiary displays may be used in gardens or interior settings to accent a focal point.

Plants with Special Pruning Needs

Some plants require special attention at pruning time. The following are some examples and their pruning requirements.

NANDINA Prune the oldest canes to the ground to start new growth. Prune to thin out the plant.

OLEANDER Prune the faded flower clusters. Remove the top branches to induce new growth and to control the size of the plant.

PRIVET Prune to control shape and size.

RHODODENDRON Pinch the flower truss after blooming to prevent the formation of seed pods. If plants are permitted to set seed, additional energy is required from the plant to produce new growth. After pinching, the plant will produce new growth faster and set more flower buds for blooming the following year.

HEMLOCK Prune in the early spring before new growth develops. This stimulates thicker growth and helps to control plant size and shape.

SPRUCE Prune to one-half of the new growth candles in August to result in a fuller tree. *Candles* are new growth on the ends of branches of pine species.

FIR Prune the new growth in August to one-half the length of the candle. Growth may be retarded by removing the tip bud in the spring before the candle develops. Usually, the only pruning done is to shape the tree. Firs have multiple buds on the stem for regrowth, whereas pine produces growth from the terminal end.

ANDROMEDA Prune to remove dead faded flower clusters in late spring. The new shoots should be pinched back to shape the plant and develop new growth.

AUCUBA Pinching back results in a thicker plant with more shape. If some of the leaves have turned black from winter burn, remove them.

ARBORVITAE Prune before new growth starts in the spring. Do not prune beyond the green leaves because this will result in a permanent brown area.

PINE Prune about one-half of each candle in June (Figure 33–12). This promotes thicker plant growth and the production of new side shoots.

AZALEA Prune after flowering to promote the production of new growth. Pinching back new top growth stimulates heavier blooming.

BOXWOOD Pruning boxwood is important to keep the plant compact and full. Prune in the fall/early winter by removing short branches with hand pruners. Remove diseased branches, but be sure to dip the hand pruners in a 70 percent alcohol solution between each cut to prevent the spread of the disease. Do not use hedge shears, because this causes the leaves to turn brown on the end where the leaf is cut.

HOLLY Prune various types of holly in December so that the branches can be used for holiday decorating. Select and cut the branches to develop the plant's shape and form. Holly may also be pruned in early spring before new growth begins.

MOUNTAIN LAUREL Prune to encourage growth of new shoots. Oddly shaped, tall, or leggy plants may be pruned to the ground to start new growth. When pruning, be sure that

(A) Pine tree with unpruned candles.

(B) Prune one-half of each candle.

(C) Pruning promotes thicker growth and production of new side shoots.

▲ **Figure 33–12** Pruning a pine tree.

▲ **Figure 33–13** Hand pruning a yew with scissor cut shears (Felco #2).

the natural shape and form of the plant are maintained.

YEWS Pruning yews is usually done in the late spring to remove any new growth in the process of heading back (Figure 33–13). Prune after the new growth has hardened. Yews are great plants to use for special design in topiary settings as well. Using the following pruning method allows one to maintain the natural shape of the yew cultivars; for example, Hicksii is upright in shape, while the English yew is more weeping and graceful. Yews may be pruned in the summer if necessary to control plant growth.

SAFETY FIRST

OSHA's regulations ensure that employers and employees know about work hazards and how to protect themselves to prevent injuries in the workplace. For more information on occupational health and safety in the workplace, please visit the OSHA site at www.OSHA.gov.

Summary

Pruning is an important aspect of the green industry. In this unit, five reasons for pruning plants and four types of pruning were discussed. Specialized pruning techniques, espalier and topiary, were also discussed. Timing is very important in the pruning process for survival of the plant as well as for controlling growth.

The proper use of hand and power tools in pruning is also important. You need to understand which tool is appropriate for the task at hand. Additionally, progress in the form of lithium battery-operated equipment has made tremendous impact on the green industry. These new tools have proven to be environmentally friendly and provide great service with aggressive management. Always remember to use caution when using or operating any tools or equipment to avoid injury.

Review Questions

1. Discuss five different reasons for pruning plants.
2. What are four different methods of pruning? Name a plant that is typically pruned using each of the four methods.
3. Explain why it is important to prune plants at the proper angle.
4. Select one plant that has special pruning needs, and explain what they are.
5. When is the best time to prune plants that flower in the summer or fall? Why?

Student Activities

1. Collect various branches of trees and shrubs. Take turns demonstrating the proper cutting technique to the rest of the class.
2. Under the supervision of your instructor, prune shrubs on the school grounds in accordance with the procedures specified in the text.
3. If possible, visit a local orchard to assist and learn during the pruning season.
4. Invite an arborist to speak to the class.
5. Have a demonstration on chain saw safety by a factory representative or certified instructor.
6. Develop a safety program (Standard Operating Procedures) for other students to follow while using pruning equipment.
7. Compete in the FFA Nursery/Landscape Career Development Event (CDE).
8. Research on the Internet the job availability of a career as an arborist or tree pruner.

Suggested Supervised Agricultural Experience (SAE) Activities

1. Work at a local orchard. Assist the manager in pruning all of the necessary trees to improve production for the next year to qualify as a placement SAE.
2. Research different types of landscape tools, and demonstrate their safe use to other students in your school and other schools to qualify as a community-based SAE.

Additional Resources

http://aggie-horticulture.tamu.edu/earthkind/landscape/proper-pruning-techniques/

http://extension.uga.edu/publications/detail.cfm?number=B949

http://umaine.edu/publications/2169e/

UNIT 34 Principles of Landscaping, Maintenance, Xeriscaping, Water Gardens, Irrigation, and Green Roofs

Objective

To apply the principles of landscaping, maintenance, and xeriscaping to an actual setting to understand the goals of the landscape profession.

Key Terms

architectural
canopy
cold hardiness
floral palette
hardscape paving
heat islands
heat tolerance
microclimate
pollutants
rain garden
residential
softscape paving
stormwater
subcanopy
subcontractor
transpiration
visualize
wind exposure
xeriscaping
xeric (waterwise)

Competencies to Develop

After studying this unit, you should be able to:

- describe the three major career fields within the landscape profession.
- list the main objectives of good residential landscaping.
- list the five principles of landscape design and examples of an application of each principle.
- list three hardscape paving and three softscape paving materials.
- list the technical procedures for landscape maintenance.
- calculate the volume of mulch needed to cover a landscape bed.
- list the basic concepts of xeriscaping.
- list ways the soil can be improved to conserve water.
- identify plants that can be used in a xeriscape setting.
- explain why mulches are important.
- explain what a water garden is.
- list the different types of irrigation systems used in the United States.
- list five green roof plants.
- list five reasons to use green roofs sustainable technology.
- use the CAD 3-D approach to landscape design.

Materials

- photographs or drawings of well-landscaped property
- basic landscape drawing tools
- landscape design hand tools
- A CAD landscape design system to demonstrate basic landscape design projects
- smartphone app: Plantbid

The Landscape Industry

The *landscape industry*, also known as the *green industry*, functions to improve our natural environment to meet the needs and desires of people. *Landscape architects* are professionals who integrate the disciplines of art and science and know how plants and landscape factors will react to the environment around them. They prepare designs that show what, where, and how objects are installed in the landscape and that provide for ease and comfort for use as well as appearance. In addition, today's landscape architects are using the new technologies of mechanized equipment and computers in conjunction with the design process, which is known as computer-aided design (CAD).

The three dimensional (3-D) approach to landscape design has opened another avenue to the presentation of landscape design. This procedure allows clients to view a photographic image of the landscape design, which improves the client's understanding of the landscape design and helps him or her visualize how the landscape will appear. This approach shows the mature landscape in scale with the area being landscaped and is a real enhancement tool for a landscape architect or designer presenting a landscape design to the client. The 3-D view affords the designer the opportunity to showcase background images of trees, shrubs, and other hardscapes in realistic color, texture, density, and form (see Figure 34–1).

Landscape features first take shape in the mind's eye of the landscape architect. The successful architect must be able to look at undeveloped, scarred, and ugly land and recognize its potential for becoming more attractive or functional for environmental and human uses.

Many states set high standards and legal requirements for those individuals calling themselves landscape designers or landscape architects. Often, 4 to 5 years of college training are required. State licensing may also be necessary. Only through training can the designer's full imagination be developed. Through on-the-job work experience, the designer can channel creativity and imagination in the development of practical design solutions.

The language of the landscape designer is *graphic art*. Through this art form (graphics), designers are able to reduce the actual

Courtesy of Drafix Software Inc.

▲ **Figure 34–1A** 3-D image before landscape installation.

Courtesy of Drafix Software Inc.

▲ **Figure 34–1B** 3-D image of how the complete landscape will appear.

Courtesy of Drafix Software Inc.

▲ **Figure 34–1C** 3-D image of special LED night lighting that conserves energy and enhances the environment of the completed landscape.

dimensions of the area to be landscaped to a size that can be illustrated to the client in the form of a reduced drawing. The landscape plan is a collection of graphic symbols that represents trees, shrubs, flowers, buildings, and other constructed materials necessary to the proposed landscape (Figure 34–2). When the land dimensions and the symbols are reduced in the same proportion, the drawing is said to be done to *scale*. Two examples of commonly used landscape scales are 1 inch on the drawing equal to 10 feet of real land (expressed as 1″ = 10′), and 1 inch on the drawing equal to 20 feet of real land (expressed as 1″ = 20′).

By means of landscape plans, designers are able to show their ideas to clients in a form that they can understand and discuss.

After the design has been approved by the client, it must be installed and brought into reality. *Landscape contracting* is the career field that deals primarily with the installation of landscapes. Landscape contractors are the main link between the design and implementation of a landscape, and, to be successful, they must have knowledge of plant materials, understand proper planting techniques, and have skills in engineering and building. Knowing how to lay a patio or build a brick wall is just as important as knowing how to plant a tree or shrub. The landscape contractor must be able to read and follow the scaled plans prepared by the landscape architect. If the landscape architect is responsible to the client for the actual development of the landscape, he or she may be the one who hires the landscape contractor to build the landscape. This practice is known as subcontracting. A sizeable landscape may involve numerous **subcontractors**, who are all responsible to the landscape architect. The landscape architect is, in turn, accountable to the client.

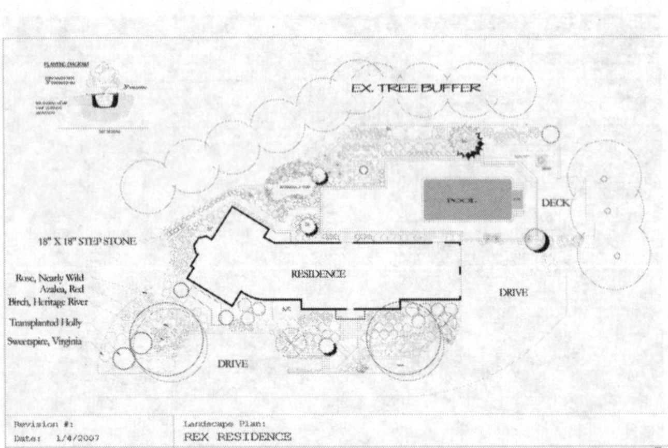

Courtesy of Drafix Software Inc.

▲ **Figure 34–2** A completed landscape design.

The care of the landscape after it has been installed is the job of the *landscape maintenance contractor*. Professionals in this career field are sometimes called gardeners or professional horticulturists. Some landscape maintenance contractors spend their time totally on a single landscape, such as a large estate, shopping center, cemetery, or park, whereas others serve numerous landscapes, accomplishing all or part of the maintenance tasks required by the landscape. Typical landscape maintenance tasks include lawn care (fertilizing, weeding, and mowing); tree and shrub care (fertilizing, pruning, and mulching); flower care; repair of walls, fences, walks, and drives; painting; and snow plowing. The fastest-growing part of the business is the management of landscapes after they have been installed.

The simplicity or complexity of the landscape design will make the maintenance of the landscape easy or difficult. It is vital that both the architect and the contractor keep maintenance requirements of the landscape in mind while developing the landscape. A design is weak if it contains so many high-maintenance features that the cost of upkeep is prohibitive.

Likewise, the landscape maintenance contractor must know how to read a landscape plan. As the landscape grows over a period of years, certain plants may need to be replaced. Others may require special pruning techniques. Flower plantings often require special care to obtain the desired color patterns year after year. A poorly trained or careless maintenance person can cause the best-designed landscapes to appear shabby.

Although there are different levels of involvement by various professionals in the landscaping business, all are interrelated. Neither the architect, the contractor, nor the landscape maintenance contractor can work without an appreciation for the role of others if landscapes that both serve and satisfy the client are to be created.

The Objectives of Residential Landscaping

The objectives of **residential** landscaping are evident in the definition of the term *landscaping*, which is to serve the needs and desires of people in development of the outdoor environment. Specifically, the goals of residential landscaping are to

- determine the exact landscape needs and desires of the homeowners.
- determine the capabilities of the land (site) to fulfill those needs and desires.
- develop the outdoor living areas of the landscape in a manner similar to the way indoor living areas are developed.
- design the landscape in such a way that maintenance practices do not exceed that which the homeowner is willing to do.
- keep costs within the budget of the homeowner.

The needs and desires of the homeowner are best determined by an interview. A direct conversation between the designer and the client promotes trust and confidence between the two parties. This helps the designer recognize the client's actual desires much more easily and allows for a personalized approach to landscaping. During the interview, questions such as the following should be answered:

1. How many family members are there? What are their ages?
2. How much does the family use the outdoor areas around the home?
3. Does the family entertain frequently? Large groups or small?
4. How much privacy from the neighbors and passing cars do they desire?
5. How much maintenance are they willing to do in the upkeep of the landscape?
6. Are there certain plants they are fond of, dislike, or are allergic to?
7. What service needs will the landscape be expected to accommodate (e.g., clothesline, trash cans, pets, vegetable garden, compost pile, or garden tool storage)?
8. Will the family be using the garden after dark?
9. How much does the family want to spend on the total development of the landscape? Do they want to spread the cost over several years?
10. Is the family willing to wait several years for the plants to reach maturity, or do they want large plants installed for an immediate effect?

Assessing the capabilities of the site requires a thorough *site analysis* by the landscape architect. Becoming familiar with the site helps the designer determine how easily and how many of the client's needs and desires can be met. The site analysis usually involves several visits to the property by the designer. A good site analysis often suggests additional possibilities for development that had not occurred to the homeowner but that might be of interest once identified.

The following are some of the things to look for and take note of during a site analysis:

- dimensions of the property
- topography of the site (how flat or rolling it is)
- quality of the topsoil and subsoil
- condition of the lawn areas
- types and condition of existing plants
- location of utility lines, meters, and utility easements
- good and bad views from the site
- locations of glass areas in the house and where they open onto in the landscape
- **architectural** style of the neighborhood
- environmental setting of the site and the neighborhood
- existing natural features such as streams, rock outcroppings, specimen plants, and wildlife habitat areas

The Concept of the Outdoor Room

When the design for a landscape is developed, it is helpful to **visualize** the outdoors in the same manner as the indoors. The home is a series of connecting living units called rooms. When the landscape designer sees the outdoor areas as rooms, the landscape becomes more familiar and easier to work with. Outdoor rooms have

▲ Figure 34–3 An outdoor room. Notice the materials that act as the walls, floor, and ceiling.

walls, ceilings, and floors, just as indoor rooms do (Figure 34–3). The main difference between indoor and outdoor rooms is in the materials used to construct them. Outdoor walls may be developed using shrubs, fences, brick or stone, exterior walls of buildings, or trellises. Outdoor floors may be the natural earth, sand, crushed stone, poured concrete, brick, decking, turfgrass, or many other similar surfacing materials. Outdoor ceilings can be developed with trees, awnings, canopies, or other overhead structures.

Trees are excellent materials for use in developing the outdoor ceiling (called a **canopy**). If an intimate patio setting is desired, a small tree branching overhead can create the low-ceiling effect (called a **subcanopy**). Also important to consider is the shade that the tree provides on a hot summer day. The southwest corner of a house will benefit greatly from the cooling effect of a medium to large tree planted nearby. Deciduous trees will allow the sun to pass through and warm the house with solar energy during the cold winter months.

The material selected for outdoor walls is usually determined by how much privacy and security is needed in the landscape. When total privacy or security is desired, the outdoor wall should be at least 6 feet tall and solid. When less privacy is needed, or security is not a factor, the outdoor wall can be lower and more open.

The main function of the outdoor wall is to define the shape and limits of the outdoor room. Plants or constructed materials used alone may accomplish this; however, the combination of natural and constructed materials often creates the most attractive effect of all (Figure 34–4).

There are several groups of materials from which to select for the outdoor floor. **Hardscape paving** includes paving brick, concrete, flagstone, tile, decking, and brick, among other materials. Pervious concrete, Figure 34–5, is another hardscape paving material that is used to create driveways, patios, and sidewalks. Because it is porous, pervious concrete allows the **stormwater** to drain off the hard surface and be contained in a collection reservoir. This stored stormwater may then be used for irrigating the landscape plants. This process conserves the stormwater runoff from polluting the groundwater. This sustainable concept

▲ **Figure 34–4** An outdoor wall constructed of natural stone materials.

▲ **Figure 34–5** Pervious paver hardscapes allows for environmental conservation of water by containing groundwater runoff.

is environmentally safe and valued to prevent contamination of the groundwater. Hardscape paving is expensive to install but inexpensive to maintain.

Softscape paving includes crushed stone, wood chips, marble chips, washed river gravel, pine needles, and other loose materials. Softscape paving has a moderate cost for both installation and maintenance. It requires periodic replacement of worn areas.

Turfgrass is the most popular of all surfacings. Turf has a low installation cost but a high maintenance cost because it must be cared for regularly. *Ground covers* are good for surfacing areas where no one will be walking, such as on slopes and directly under trees. Figure 34–6 describes the various outdoor room surfacings.

The type of material chosen for the outdoor floor is certainly affected by what the client can afford. However, there are other client needs to consider as well:

- Does the material absorb noise and/or dust?
- How well does it blend in with the rest of the landscape?
- Does it produce a glare from the sun?
- Does it track easily?
- Does it become slippery when wet?

TRY THIS Design, plan, and construct an outdoor classroom for your school. Acquire appropriate permission and a location from school administration, and then interview school staff and students to determine what needs the school has for inclusion in the outdoor classroom. Implement those needs in your design.

TYPE OF SURFACING	COST OF INSTALLATION	COST OF MAINTENANCE	DURABILITY
Hardscape paving	high	low	high
Softscape paving	moderate	moderate	moderate
Turfgrass	low	high	moderate
Ground covers	moderate	high at first	low

Figure 34–6 Comparison chart of various outdoor room surfacings.

The Principles of Landscape Design

Five basic principles guide the landscape architect or designer in planning the landscape's outdoor rooms:

- simplicity
- balance
- focalization of interest
- rhythm and line
- scale and proportion

The principle of simplicity is very important to the overall unity of the design. When this principle is correctly applied, the landscape is understood and appreciated by the viewer. The principle of simplicity is accomplished by repeating specific plants throughout the design and by massing plant types or colors into groups rather than spacing them so that each plant or color is seen separately. The fewer different objects there are for the eye to focus upon, the simpler the design will seem. Finally, straight-lined or gently curving bedlines around shrub plantings, rather than fussy, scalloped bedlines, add to the design's simplicity (Figure 34–7).

The principle of balance is applied by imagining the area of landscape placed on a seesaw. If properly balanced, the left side of the landscape should have no more visual weight than the right side. Balance may be either symmetrical or asymmetrical. *Symmetrical* balance is attained when one side of the landscape is an exact duplicate of the other side (Figure 34–8). This is a form of balance common in formal designs. It is sometimes applied to modern residential design, but more frequently asymmetrical balance is used. With *asymmetrical* balance, one side of the landscape has the same visual weight as the other side, but they are not duplicates (Figure 34–9).

The principle of focalization of interest recognizes that the viewer's eye wants to see only one feature as being most important within any given view. All other elements complement that important feature (the focal point) but do not compete with it for attention. When looking at a house from the street, the viewer's eye should go quickly to the front door. When sitting on the patio looking out across the backyard, the viewer may have no focal point at which to look unless one is created by the designer. Focal points may be created using especially attractive plants (*specimen* plants), statues, fountains, pools, and flower masses. Once created, all bedlines and plant arrangements should be designed to lead the eye of the viewer to the focal point (Figure 34–10).

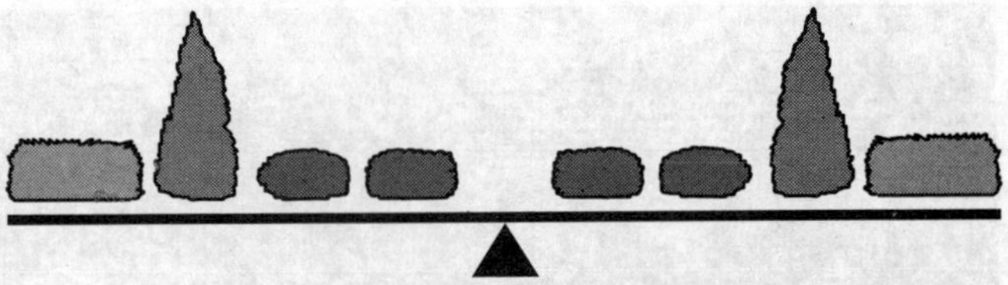

▲ **Figure 34–8** Symmetrical balance. Each side of the landscape duplicates the other.

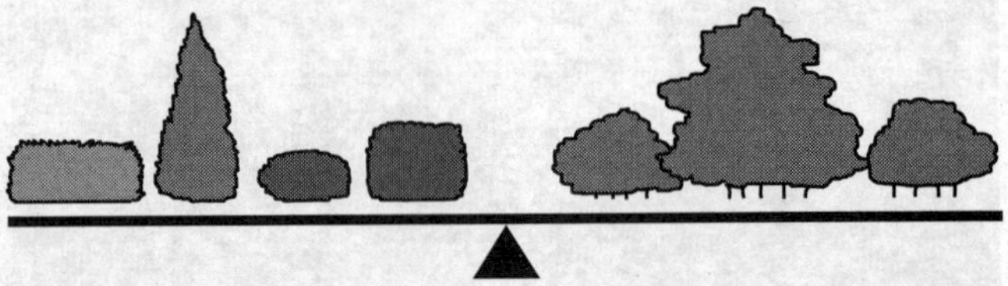

▲ **Figure 34–9** Asymmetrical balance. One side of the landscape provides as much interest as the other but does not duplicate it exactly.

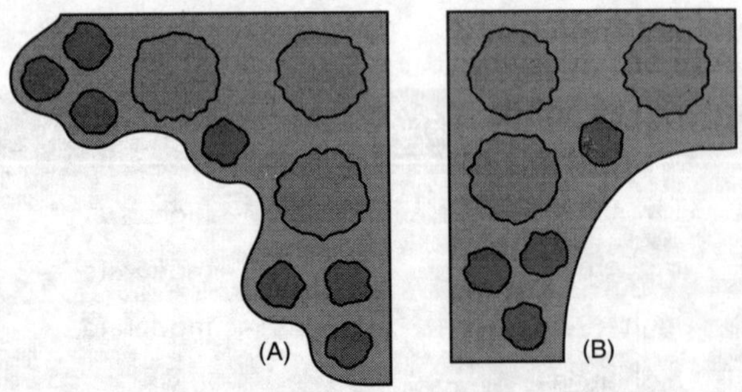

▲ **Figure 34–7** A fussy bedline (A) and a simple bedline (B). Bedline (B) was more thoughtfully planned and would be much easier to maintain.

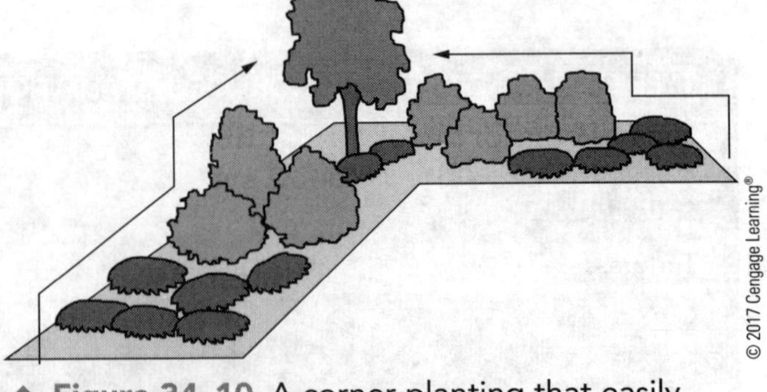

▲ **Figure 34–10** A corner planting that easily leads the eye to the focal point.

The principle of rhythm and line also contributes to the overall unity of the landscape design. This principle is responsible for the sense of continuity among different areas of the landscape. One way in which this continuity can be developed is by extending planting beds from one area to another. For example, shrub beds developed around the entrance to the house can be continued around the sides and into the backyard. Such an arrangement helps to tie the front and rear areas of the property together. Another means by which rhythm is given to a design is to repeat shapes, angles, or lines between various areas and elements of the design.

The principle of scale and proportion helps to keep all elements of the landscape in the correct size relationship without towering over the building when fully grown (Figure 34–11A and B). Plants selected for the landscape should add to human comfort in the setting. For example, plants and other materials used around a children's play area should be small so that the children can relate to them. In a world of giant adults, it is nice to feel as "tall as a tree," even when child and tree are only 3½ feet tall.

The principle of unity is the master principle of landscape design. Unity creates the flow among scale and proportion, balance, accent, rhythm, and simplicity (Figure 34–11C). The landscape design will complement the surroundings and create an aesthetic appeal that is pleasing and beautiful.

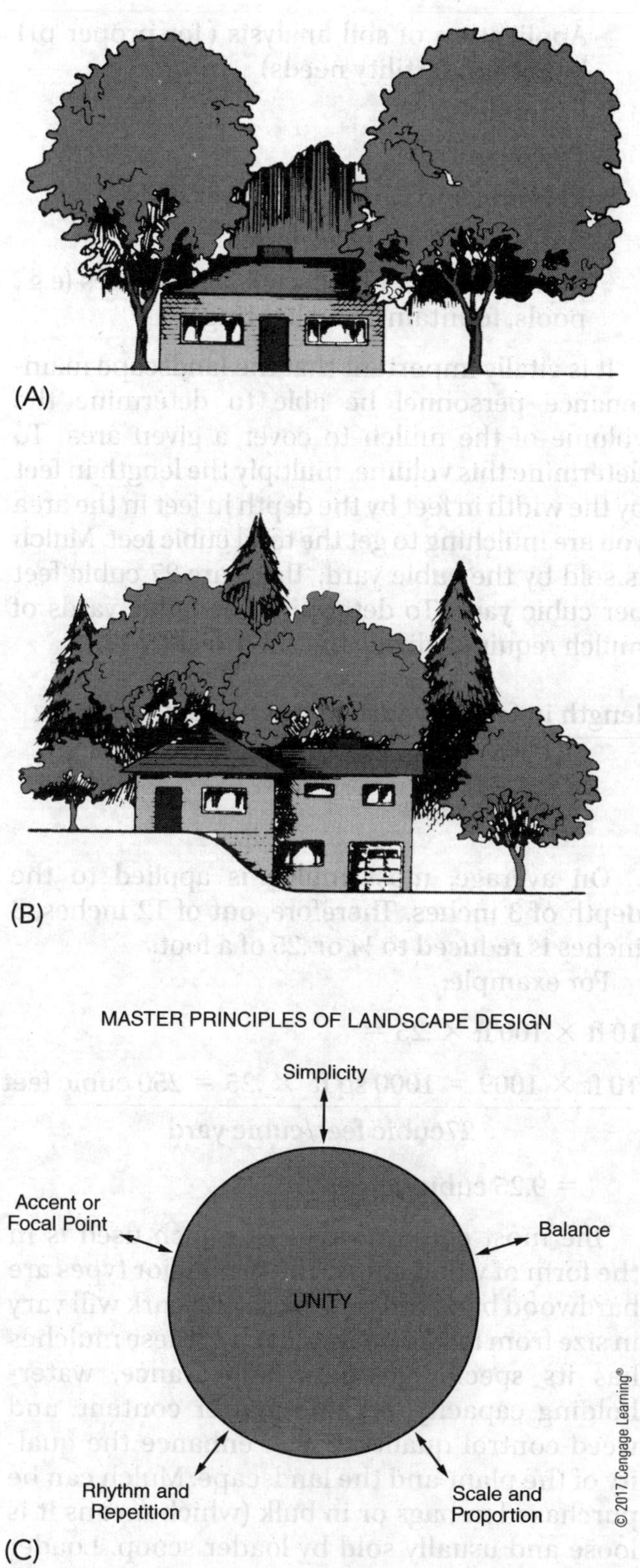

▲ **Figure 34–11** The trees in (A) overpower the structure to such an extent that it is dehumanizing and uncomfortable to view. The landscape in (B) is more carefully planned, with plantings in proportion to the structure. The master principles of landscape design are illustrated in (C).

Landscape Maintenance

Landscape maintenance is involved in caring for the grounds after the installation of the plant material, and it is one of the fastest-growing divisions of the green industry. Landscape maintenance is important because it maintains the aesthetics of the landscape area and the proper care of the plant materials and other landscape accessories (patios, pools, mulches, fertilizers, and organic/inorganic chemicals) throughout the year.

Landscape maintenance involves technical knowledge of the following skills:

- Replacement of plant material
- Application of mulch to proper depth around plants

- Application of soil analysis (for proper pH levels and fertility needs)
- Pruning
- Weed control
- Planting and caring for flower beds
- Proper mowing procedures
- Maintenance of landscape accessories (e.g., pools, fountains, and lighting)

It is vitally important that the landscape maintenance personnel be able to determine the volume of the mulch to cover a given area. To determine this volume, multiply the length in feet by the width in feet by the depth in feet in the area you are mulching to get the total cubic feet. Mulch is sold by the cubic yard. There are 27 cubic feet per cubic yard. To determine the cubic yards of mulch required, divide the cubic feet by 27.

$$\frac{\text{length in feet} \times \text{width in feet} \times \text{depth in feet}}{27 \text{ cubic feet/cubic yard}} = \text{cubic yards}$$

On average, most mulch is applied to the depth of 3 inches. Therefore, out of 12 inches, 3 inches is reduced to ¼ or .25 of a foot.

For example:

$$10 \text{ ft} \times 100 \text{ ft} \times .25 =$$

$$\frac{10 \text{ ft} \times 1009 = 1000 \text{ sq ft} \times .25 = 250 \text{ cubic feet}}{27\text{cubic feet/cubic yard}}$$

$$= 9.25 \text{ cubic yards}$$

The most common type of mulch used is in the form of wood chips. The two major types are hardwood bark and pine bark. The bark will vary in size from fine to coarse. Each of these mulches has its special aesthetic appearance, water-holding capacity, organic matter content, and weed-control quality, which enhance the quality of the plant and the landscape. Mulch can be purchased in bags or in bulk (which means it is loose and usually sold by loader scoop. Loader scoops vary in size, so it is important to check the volume of the bucket to ensure you are receiving the correct amount of mulch).

Landscape maintenance requires the use of hand tools and power tools Figures 34–12 and 34–13. Appendix C provides an illustrated list of common tools and their uses.

POWER TOOLS	USES
Zero-Turn Mower	Professional lawn-mowing equipment that increases mowing time efficiency.

POWER TOOLS	USES
Walk-Behind Mower	Professional lawn-mowing equipment.

POWER TOOLS	USES
Line and Blade Trimmer	To trim grass and plant material from areas unable to be mown with the walk-behind mower.

▲ **Figure 34–12** Power tools. *(continued)*

Lawn Edger	**To trim the grass along the edge of the sidewalk, driveways, and curbs.**	**Power Bed Edger**	**To accent a planting bed edge that separates the grassy area from the planting bed.**

▲ **Figure 34–12** Power tools.

HAND TOOLS		**USES**
Hand Trowel	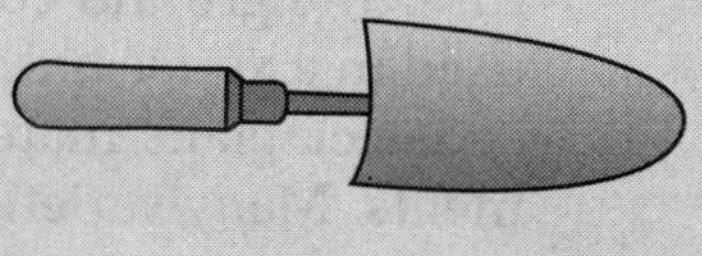	**Used to transplant seedling plants.**
Nursery Spade	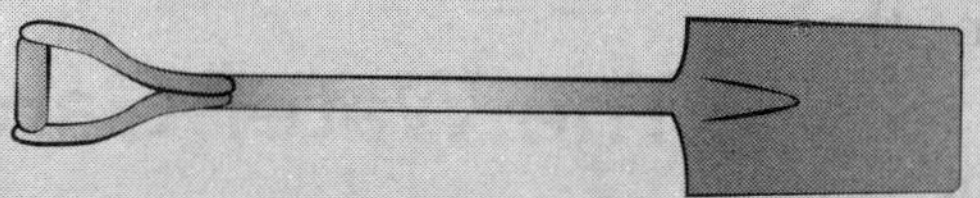	**Used for edging planting beds and digging nursery stock.**
Hand Pruners (Hook and Blade Pruners)	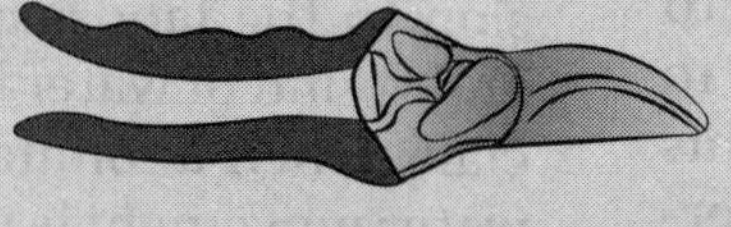	**Used to prune branches 3/4" to 1/2" in diameter.**
Pruning Saw	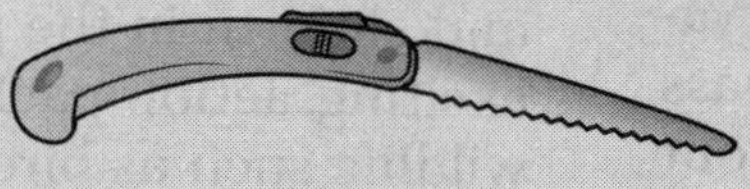	**Used to prune branches 1½' or greater in diameter.**
Pole Pruner	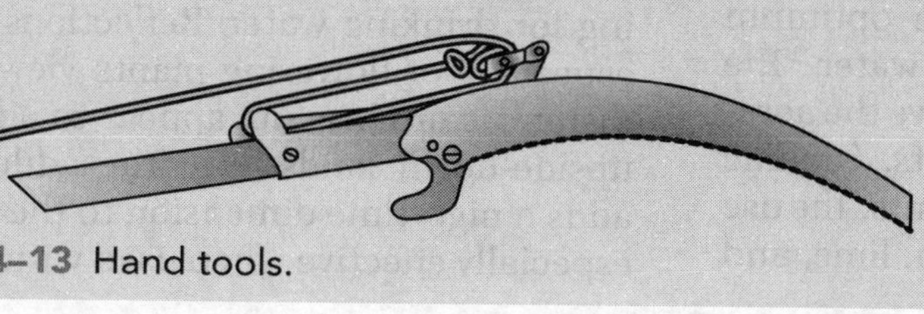	**Used to prune branches 1½" or greater in diameter 10' to 12' above the reach of the professional grounds maintenance worker.**

▲ **Figure 34–13** Hand tools.

Xeriscaping

Xeriscaping is used to practice water conservation in creative landscapes. The word *xeriscape* comes from the Greek word *xero* for "dry" and *scapes* for "dryscape." This application ensures water efficiency in all future landscape practices, especially for areas where water is scarce and expensive. The driest landscapes can come alive with texture and color by using waterwise (**xeric**) plants. This basic landscape practice has been used extensively in the southwestern United States. Over time, the demand for water has increased rapidly. Therefore, landscape practices must change to meet these new water requirements.

The basic concepts of xeriscaping involve the following applications of landscape principles:

- Good design and planning
- Improving the soil
- Turf areas
- Use of ground covers
- Low water use
- Plant selection
- Use of mulches
- Use of low-volume irrigation
- Proper landscape maintenance
- Reduced fertilizer use
- Reduced pest infestation
- Habitat and food sources for birds and butterflies
- Increased use of native plants in the landscape
- Conservation of a healthy environment

Good design and planning are important to proper xeriscaping because it is necessary to group plants in relation to their water needs. This grouping practice allows one to grow high-water-use plants and still maintain the water conservation needed. When designing a garden, use mass plantings with borders of lower plants in the front and taller plants in back (Figure 34–14).

Soil can be improved to provide optimum plant growth and conservation of water. The addition of organic matter will improve the aeration and water supply for the plants. Organic matter is an excellent source of nutrients. The use of soil amendments, such as gypsum, lime, and hydro-gels, aids in water retention and improved availability of water to the planted material.

Turf areas should be seeded in high-drought-tolerant grasses. The turf areas should be separate from other plantings and irrigated separately.

Mulches in the landscape are important to conserve moisture, reduce evaporation, keep the soil cool, and reduce weed growth. Mulches are organic and inorganic materials that provide aesthetic value to the landscape. A high portion of the water that falls is lost through evaporation or runoff. Mulching will reduce this water loss by 90 percent.

Low-volume irrigation is important to incorporate into the landscape. The use of water systems that deliver spray patterns for defined areas is necessary. The use of drip irrigation is a more direct way of pinpointing where the water is needed.

Proper landscape maintenance is needed to preserve the beauty of the landscape and conserve water. Weeding, pruning, pest control, and adjustments in the irrigation system help to conserve moisture in the landscape.

Ground covers give the landscape a wide variety of texture and color and retain soil moisture. Large turf areas transpire large quantities of moisture. Using ground covers in beds that are properly mulched will help to conserve moisture. The ground covers in Figure 34–15 may be used in a xeriscape setting.

Select plant material with low water requirements. Many varieties of plants survive well with minimal watering, including native and exotic plants, cultivated ground covers, trees, and shrubs.

The Water Garden

Water adds an entirely new and different dimension to the landscape (Figure 34–16). The sight and sound of water brings the garden alive, especially when a fountain, waterfall, or any form of water movement is used. Water plants, fish, frogs, and so forth in the pond also add a contrasting, exciting, action-packed dimension. Ponds attract wildlife such as birds, butterflies, and beneficial insects as well as other larger wildlife searching for drinking water. Reflections in the water, especially of flowering plants viewed from long distances or close up, appear to add a duplicate upside-down landscape. The addition of lights adds a nighttime dimension to the garden and is especially effective when used with waterfalls.

The area around a water garden can be made into an outdoor room with seats for resting and enjoying the sights and sounds. If placed near a patio or deck, the room is already formed and convenient for leisurely enjoyment of the garden. However, there is a caution with water gardens,

URBAN CONDITIONS	POOR/DRY SOILS	HEAVY CLAY SOILS
	Shrubs	
Acanthopanax sieboldianus	*Acanthopanax*	*Acanthopanax sieboldianus*
Aronia arbutifolia	*Acer ginnala*	*Aralia spinosa*
Berberis thunbergii and cvs.	*Berberis* × *mentorensis* and cvs.	*Aronia* spp.
Celastrus scandens	*Berberis thunbergii* and cvs.	*Berberis*, most
Chaenomeles spp. and cvs.	*Chaenomeles* spp. and cvs.	*Chaenomeles*, most
Chionanthus virginicus	*Cornus racemosa*	*Cornus*, shrub types
Cornus alba	*Cotinus coggygria* and cvs.	*Deutzia*, most
Cornus mas	*Cytisus* spp. and cvs.	*Elaeagnus*, most
Cornus stolonifera	*Elaeagnus angustifolia*	*Euonymus*, most
Elaeagnus umbellata	*Erica carnea*	*Forsythia*, most
Euonymus spp. and cvs.	*Genista lydia*	*Hamamelis* spp.
Forsythia spp. and cvs.	*Hamamelis* spp. and cvs.	*Juniperus*, most
Hamamelis spp. and cvs.	*Juniperus*, most	*Kolkwitzia amabilis*
Hibiscus syriacus and cvs.	*Kolkwitzia amabilis*	*Ligustrum*, most
Hydrangea spp.	*Ligustrum* spp. and cvs.	*Lonicera*, shrubby types
Ilex crenata and cvs.	*Myrica pennsylvanica*	*Myrica pennsylvanica*
Ilex glabra and cvs.	*Potentilla* spp. and cvs.	*Physocarpus opulifolius* and cvs.
Ilex × *meserveae* and cvs.	*Rhamnus* spp. and cvs.	*Pyracantha*, most
Leucothoe spp. and cvs.	*Rhus*, all	*Rosa rugosa*
Ligustrum spp. and cvs.	*Ribes alpinum* and cvs.	*Rhodotypos scandens*
Lindera benzoin	*Rosa rugosa* and cvs.	*Rhus*, most
Lonicera spp. and cvs.	*Viburnum lentago* and cvs.	*Ribes alpinus* and cvs.
Magnolia stellata and cvs.	*Yucca filamentosa* and cvs.	*Salix*, most
Mahonia aquifolium and cvs.		*Spiraea*, most
Myrica pennsylvanica		*Stephanandra incisa* 'Crispa'
Philadelphus virginalis		*Symphoricarpos* × *chenaultii* 'Hancock'
Physocarpus opulifolius		*Taxus*, most
Potentilla		*Thuja occidentalis* and cvs.
Pyracantha		*Viburnum dentatum*
Rhamnus spp. and cvs.		*Viburnum lentago*
Rhodotypos scandens		*Viburnum opulus* and cvs.
Rhus spp. and cvs.		*Viburnum prunifolium* and cvs.
Ribes alpinum and cvs.		*Viburnum sargentii* and cvs.
Rosa rugosa and cvs.		
Rosa species type		
Spiraea vanhouttei		
Syringa × *prestoniae*		
Syringa reticulata		
Syringa vulgaris		
Taxus spp. and cvs.		
Viburnum, most		
Wisteria sinensis and cvs.		
Yucca filamentosa and cvs.		
	Trees	
Acer × 'Celebration' *P.A.F.*	*Acer saccharinum*	*Acer campestre*
Acer ginnala	*Betula* spp. and cvs.	*Acer rubrum* and cvs.
Acer platanoides and cvs.	*Celtis occidentalis*	*Acer saccharinum*
Acer saccharinum	*Cercis canadensis*	*Betula nigra*
Chamaecyparis nootkatensis and cvs.	*Corylus corlurna*	*Crataegus* spp. and cvs.
Crataegus spp. and cvs.	*Elaeagnus* spp.	*Elaeagnus angustifolia*

Figure 34–14 Plant selection for landscaping. *(continued)*

URBAN CONDITIONS	POOR/DRY SOILS	HEAVY CLAY SOILS
Trees (*continued*)		
Elaeagnus angustifolia	*Eucommia ulmoides*	*Eucommia ulmoides*
Fagus sylvatica and cvs.	*Gleditsia triacanthos inermis* and cvs.	*Fraxinus pennsylvanica* and cvs.
Fraxinus pennsylvanica and cvs.	*Gymnocladus dioica*	*Gleditsia triacanthos inermis* and cvs.
Ginkgo biloba and cvs.	*Kerria japonica* and cvs.	*Koelreuteria paniculata*
Gleditsia triacanthos inermis and cvs.	*Koelreuteria paniculata*	*Liquidambar styraciflua* and cvs.
Koelreuteria paniculata	*Ostrya virginiana*	*Malus*, most
Liquidambar styraciflua and cvs.	*Picea* spp. and cvs.	*Ostrya virginiana*
Liriodendron tulipifera	*Platanus* × *acerifolia* 'Bloodgood'	
Malus spp. and cvs.	*Phellodendron amurense*	
Picea spp. and cvs.	*Pyrus*, most	
Pinus spp.	*Salix*, most	
Platanus spp. and cvs.	*Taxodium distichum*	
Populus spp., hybrids and cvs.	*Ulmus parvifolia*	
Pyrus calleryana and cvs.	*Ulmus* spp. and cvs.	
Taxodium distichum and cvs.		
Ulmus × *hollandica* and cvs.		
Ground Covers		
Ajuga reprans	*Aegopodium podagraria* 'Variegatum'	*Aegopodium podagraria* 'Variegatum'
Hedera helix	*Hedera helix*	*Ajuga* spp.
Juniperus spp.	*Liriope muscari*	*Euonymus fortunei* 'Colorata'
Liriope muscari	*Hosta* spp. and cvs.	*Hedera helix*
Ophiopogon japonicus	*Parthenocissus* spp.	*Hemerocallis*, all
Rosemarinus officinalis	*Rosemarinus officinalis*	*Houttunynia cordata* 'Chameleon'
Trachelospermum spp.	*Sedum*	*Juniperus*, low types
Vinca major		*Liriope muscari*
		Lonicera japonica 'Halliana'
		Pachysandra terminalis and cvs.
		Parthenocissus spp.
		Rhus aromatica and cvs.
Vines		
Bougainvillea spp.	*Bougainvillea* spp.	*Celastrus scandens*
Celastrus scandens	*Celastrus scandens*	*Wisteria* spp.
Clematis sp.	*Rosa banksiae*	
Ficus pumila 'Repans'	*Wisteria* sp.	
Gelsemium sempervirens		
Lonicera sempervirens		
Parthenocissus spp.		
Rosa banksiae		
Wisteria sp.		
Grasses		
Andropogon gerardi	*Andropogon gerardi*	*Andropogon gerardi*
Bouteloua curtipendula	*Bouteloua curtipendula*	*Bouteloua curtipendula*
Elymus virginicus	*Elymus virginicus*	*Elymus virginicus*
Koeleria macrantha	*Koeleria macrantha*	*Koeleria macrantha*
Panicum virgatum	*Panicum virgatum*	*Panicum virgatum*
Sporobolus heterolepis	*Sporobolus heterolepis*	*Sporobolus heterolepis*

Figure 34–14 Plant selection for landscaping.

GROUND COVERS	
Euonymus fortunei 'Colorata' Grasses, ornamental *Hemerocallis*, all *Juniperus*, low types *Lavandula* Lavender *Nepeta sp;cv* Catmint *Parthenocissus* spp. *Penstemon* Beardtongue *Rhus aromatica* and cvs. *Salvia superba cv* Sage	
SHRUBS	
Chamaebatiaria millefolium Fernbush	
GRASSES	
Bouteloua gracills Blonde Ambition Blue Grama Grass	
TREES	
Rhus aromatic Sumac	
APHANOSTEPHUS SKIRRHOBASIS	LAZY DAISY
Stachys byzantina	Lambs ear
Opuntia polacantha	Prickly pear cactus
Oenothera sp.	Evening primrose
Coreopsis sp.	Coreopsis
Dianthus plumarius	Dianthus
Centaurea Montana	Mt Bluet

Figure 34–15 Xeriscape plants.

▲ **Figure 34–16** Water gardens create the serenity of nature's beauty, sounds, and calmness.

especially large ponds: place these out of the path of small children, or add barriers to keep small children out of the water.

Water gardens with bridges or walkways through them allow the visitor to look down on top of the water surface. This view presents a clearer picture of water lilies and fish in the pond and creates in the visitor an intimate feeling of being part of the garden (Figure 34–17).

▲ **Figure 34–17** This water garden appears to be a small stream flowing under the bridge.

Water gardens can be sized and shaped to fit the landscape. They can vary from a small aboveground container on a patio or along a garden path to large ponds. Even the smallest container can have moving water in the form of fountains (Figure 34–18).

In formal settings, water gardens are usually shaped in rectangular, square, or circular forms and located in the center of the garden or in a straight line at the end of a path. They are edged in brick or flagstone. Informal water gardens tend to be designed in sweeping natural curves and generally do not have symmetrical balance. They are placed wherever they look good and are often edged with plants in an informal asymmetrical manner. Edges are treated so they blend into the surrounding garden and do not stand out as a dividing feature.

Locating the Water Garden

Most flowering water plants require at least 6 hours of full sun to flower well; full sun is better. In-ground ponds should be placed in full sun if possible. Aboveground containers, especially small ones, need some shade during the hottest part of the day to keep the water from overheating and damaging plants or fish.

The water garden should not be under trees that will shed leaves or needles into the pond. Tree roots can also damage in-ground ponds by tearing the liner or causing leaks.

▲ **Figure 34–18** This small water tub can be placed on a patio or tucked in between other plants in the landscape.

If waterfalls, fountains, or streams are to be created, the water garden must be near electricity to run the pumps needed to move the water. For safe operation use a ground fault interrupter (GFI) receptacle.

Wet areas with high water tables should be avoided in the placement of in-ground ponds. Groundwater can float ponds out of the ground.

Types of Ponds

In-ground ponds are lined with PVC or rubber liners. These liners make it easy to install a pond nearly anywhere and of any size or shape. Rubber liners last longer than PVC (at least 20 years compared with 10 years). All liners last longer if completely covered so sunlight does not hit them.

Preformed fiberglass or plastic rigid-sided liners come in a variety of sizes and shapes. They can be used both in-ground and aboveground and are easy to install. The pond, however, is limited in shape to those available for purchase. Flexible liners do allow for more flexibility in pond size and shape and are used to make small streams and waterfalls.

Aboveground containers come in many forms and types of material, from half barrels to concrete ponds, and they can be easily placed anywhere in the garden that provides some shade to moderate the water temperature in the container. Nearly any container that holds water can be used. Due to their small size, container gardens do not last that long and need to be emptied and cleaned out periodically. The smallest size should hold plants only. Fish require closer attention than plants, although goldfish, snails, and tadpoles may survive. These small containers should be attractive and placed so that they blend into the landscape. They may be hidden with plants or made of material, such as stone, that blends into the landscape.

Water Plants

It is obvious that the same landscaping principles used in the rest of the landscape also apply to the water garden. The major difference is working with different and often unfamiliar plants. Some plants for the water garden vary in size, shape, and texture from other landscape plants. The landscape effect is also variable because of these differences. Plants appear out of water and float on water as well as rise above it.

In the water garden, as in the land-based garden, it is important to select plants that grow to proper size for the size of the pond. For example, in a large pond, wide spread water lilies and lotus are a good choice. In a small tub container, garden dwarf water lilies are appropriate.

Plants must be adapted to your garden site. Tropical plants only flourish in warm climates. Many flowering plants need at least 6 to 8 hours of sunlight daily to flower well. Other plants require shade.

If the adjacent landscape is formal, the water garden and its plantings should be made in a balanced, symmetrical manner. If informal, an asymmetrical informal water garden is needed.

In sunny locations where color is needed, water lilies and lotuses are the most popular. There are hardy and tropical water lilies. Water lilies and lotuses require deep water (6 to 18 inches). Hardy water lilies come in colors of white, yellow, pink, and red. Tropical lilies also come in blue. Extending above the water level, water lilies make a spectacular show. Lotuses are not only spectacular with their white and pink flowers, but the white lotus is very fragrant. There are also water plants that grow well in shady areas, even ones that flower. Send for a water garden catalog to identify plants that grow well in your area. Also get advice from a water garden specialist before attempting to build a water garden. The details necessary to build a water garden are beyond the scope of this unit.

In planting the water garden, the gardener should use many of the same landscaping principles as used in the soil- or land-based garden. It is best to select plants for the water garden that blend into the surrounding landscape or contrast without clashing. For example, select flower colors that blend with or are the same as flowers growing outside the pond and that flower at the same time.

Foliage and growth habits should be similar or harmonious in their contrast. Low-growing water plants planted next to low growers on the pond bank blend in well. Tall water grasses, rushes, and cattails blend in well with ornamental grasses in the outside garden.

This type of planting ties the water garden into the rest of the garden. The edge of the pond must be hidden in the case of informal gardens. There should be a smooth transition so it is difficult to tell where the pond stops and the land garden begins. Formal gardens are tied into a walkway with stone or brick and may show a distinct edge.

Bog Plants

Many gardeners form bog areas around ponds to extend the water garden out into the landscape. Bogs are wet areas of soil or a sand and sphagnum peat mixture placed around the outside of the pond but not directly connected to the pond. If connected, silt and soil would wash into the pond from the bog area.

As mentioned earlier, moving water in the form of fountains, waterfalls, or small streams adds a feeling of tranquility or a sense of relaxation to the garden. The sound of water moving and the sight of it in motion bring life to what might be an otherwise uninteresting garden. Waterfalls are the most natural way of enjoying moving water in the landscape because they imitate nature.

Water plants, fish, frogs, and other water creatures make up an environment quite different from the ordinary flower bed and greatly enrich the landscape. The water garden definitely adds a new dimension in the landscape—one worth exploring and possibly adding to your garden.

Pondless Waterfalls

The pondless waterfall is a unique landscape feature to provide a wave of billowing sounds using nature's symphony in the landscape. The pondless waterfall allows water to be integrated into a garden tapestry by

- creating an accent and focal point in the garden setting.
- allowing a water garden while maintaining safety and security if small children are present.
- enabling the placement of a water garden in unique areas on the garden site.
- enhancing the garden with the natural sound of a babbling brook.
- providing an aesthetic value to the surrounding landscape.
- offering a good alternative for those who do not want a pond.

The site for the pondless waterfall should be in a setting with natural potential: the atrium of a commercial business, the interior garden in a hotel lobby, an area near a patio, or the front yard of a residence (Figure 34–19).

The pondless waterfall is designed to circulate the water from a basin that is 4 feet by 6 feet (Figure 34–20). This module unit contains about 150 gallons of water with a pump to circulate the water to a biofalls container at the head waters

▲ **Figure 34–19** Site selection for a pondless waterfall.

▲ **Figure 34–20** Installation of the underground module basin to pump the return water for the pondless waterfall.

of the waterfall, which is elevated at 3 to 4 feet above the basin area. The waterfall is usually designed to give a natural scenic exposure, imitating the soothing sound of a natural stream in the wilderness, including the presence of aquatic marginal plants (Figure 34–21).

Winterizing the Water Garden

The following instructions are useful for winterizing a water garden:

1. Stop fertilizing aquatic plants in September.
2. Protect the water garden from leaves falling into the water by covering it with netting.
3. Protect tropical lilies and other tender aquatic plants from frost damage.
4. In colder climates, use a floating heater to prevent the water garden from freezing.
5. Use a special circulating pump to provide additional oxygen for the fish.

▲ **Figure 34–21** Installed pondless waterfall.

The Irrigation Story

Water is the lifeline of plants and animals. With the demand for water resources increasing, consumers have developed greater respect for the conservation of quality water. Water is colorless, odorless, and tasteless. Water resources may come from a river, lake, spring, well, or a town, municipality, or city. The availability of these resources and their use depend on how far away they are for practical use and on the amount of seasonal rainfall for the geographic area.

Freshwater Use in the United States

According to the US Geological Survey (USGS) water use in the United States in 2010 was estimated to be about 355 billion gallons per day (Bgal/d), which was 13 percent less than in 2005. The breakdown by water use category is as follows:

- Irrigation: 38 percent
- Thermoelectric power: 45 percent
- Public supply: 14 percent
- Livestock, aquaculture: 1 percent
- Domestic (self-supplied): 1 percent
- Mining: 1 percent

Irrigation Systems

The USGS keeps records on the types of irrigation systems used in the United States (http://ga.water.usgs.gov/):

- Flood irrigation. Water poured onto the land flows through the fields.
- Spray irrigation. Water is sprayed from spray guns or sprinkler heads (Figure 34–22).
- Drip irrigation. Water is released slowly onto the crop through emitters made of small tubing (Figure 34–23).

Landscapers have adopted a variety of new irrigation practices and technologies for lawns in

© 2017 Cengage Learning®

▲ **Figure 34–22** Spray irrigation.

© 2017 Cengage Learning®

▲ **Figure 34–23** Drip irrigation.

recent years for the conservation of water and plant growth. An irrigation system as simple as a sprinkler connected to a garden hose or an ultramodern automatic irrigation system to enhance the landscape with greener lawns and beautiful gardens gives added value to a residence. Programmed irrigation improves application uniformity, reduces runoff, reduces the amount of water that will be leached below the root zone of the plants, reduces the groundwater pollution by pesticides, reduces the time needed to water, conserves water, and promotes plant growth. The automatic sprinkler system can increase the efficiency of using water in the landscape. The use of an irrigation system can deliver automatically the optimum volume of water used by the plants.

Basically, the automatic irrigation system has three major parts: sprinklers, valve controller, and computerized timer. The sprinklers distribute the precise water coverage of the area without gaps or runoff. Several basic engineered designs of lawn sprinklers are available for residential use. For smaller lawn areas, other ground covers, and shrubs, an affixed spray sprinkler or drip irrigation is used, whereas for larger areas, various rotary (pop-up spray heads) sprinklers are used.

Although several sprinklers are available, the rotary (pop-up) spray heads are more common in landscapes. They are installed underground and rise above the lawn surface to spray the water in a full circle pattern or a portion of a circular pattern with a 20- to 50-foot radius. The heads have an adjustment radius for various irregular areas. All sprinklers should be connected to the lateral lines with a Super Funny Pipe that allows you to position the sprinkler. In addition, this Funny Pipe acts as an extension cord between the sprinkler line and the sprinkler head and protects the water line from breaking if something hits the head.

The valve controller controls the water flow to the sprinkler. This controller directs the flow of water to each of the irrigation zones in the landscape.

Computerized timers are the brains of the irrigation system. The timer controls the actual time of day the sprinkler operates and the amount of water to apply.

Drip Irrigation (Subsurface Irrigation)

Drip irrigation has many uses in the horticultural field. It can be used to grow ornamental trees, shrubs, ground covers, flowering annuals, perennials, vines, fruit trees, vegetables, container gardens, and lawns.

In flower beds, water can be applied through drip irrigation, in which water is applied directly to the roots of the plants through small tubing with emitters in the irrigation tubing. These emitters are located at different intervals of the tubing, for example, at 18 or 24 inches apart. This tubing can be assembled quickly just by pushing together with no clamps.

The drip irrigation system supplies the water to the root zone of the plant slowly, which conserves water and allows more uniform application of the water. This system will enhance plant growth as well as the application of fertilizer. Water runoff is reduced to a minimum, which also reduces pollution of local water resources. From a management standpoint, each irrigated area must be controlled by paying close attention to the local soil survey maps of the type of soil in the area for optimum plant growth.

There are various factors one must consider when designing an irrigation system for a residential setting:

- Size and shape of the area.
- Soil type (e.g., clay soil holds more water than sandy soils due to its action exchange capacity). Refer to Unit 4 for review.
- Types of plants being grown.
- Available water resources.

SIZE AND SHAPE OF THE AREA When a system using pop-up sprinklers is designed, the spacing and rotation of the head must be considered to ensure even coverage of the ground area in a circular pattern. The area can be lined out to have the proper spacing of the tubing with the emitters for the correct coverage of the surface area.

SOIL TYPE The soil type determines the water-holding capacity of the soil. Soils with finer textures will have a higher water-holding capacity than soils that are more porous, such as sandy textured soils. Soil type plays a major role in the amount of water that is applied to the area and is reflected in the soil particles, which influence the growth of the plants.

TYPES OF PLANTS BEING GROWN Some plants are heavy users of water and therefore will withstand moist conditions for optimum growth in their life cycle. A great example of this type of plant is the holly (*Ilex*). At the other end of the spectrum are plants that use less water, the succulents. An example of this type of plant is sedum.

AVAILABLE WATER RESOURCES Water is a valued commodity, so it is important to use it efficiently. The resources for a residential home come from a city or town source, a well on the property, or a spring. The availability of the water will have a direct influence on the operation of an irrigation system.

To plan an irrigation system for a residential property, use a worksheet of graph paper and lay out the location of the house, sidewalks, garage, driveway, patios, and other structures. Identify the lawn, ground covers, and flower beds, and then calculate the square footage to be covered.

In selecting the sprinkler system, it is important to have sprinklers that will give the maximum spray radius for large areas so that fewer sprinklers and lower water flow values are needed. This will mean less trenching and fewer timer zones. Depending on the sprinkler selected, the radius of coverage will vary. To ensure the proper head-to-head location of the sprinklers, use a compass to draw the coverage to scale with circles, semicircles, and quarter-circles to represent the coverage (Figure 34–24).

The irrigation system should be winterized as follows:

1. Turn the water off at the main valve.
2. Drain the entire system.
3. Blow compressed air through the entire system to remove excess water. *Note*: Do not use over 50 psi on the system.

After the irrigation system is installed, it is important that water conservation be practiced:

1. As stated previously, use a rain-control sensor (turns the irrigation system off when it is raining).
2. Water early in the morning to reduce evaporation.

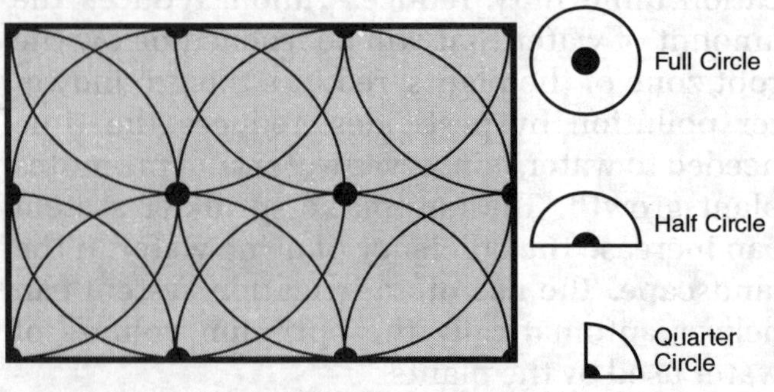

▲ **Figure 34–24** Demonstrates the proper head-to-head sprinkler coverage of irrigated area.

PROCEDURE

The basic steps of installations are as follows:

1. Use flags to indicate the location of the sprinklers in the design layout. Use marking spray paint to locate the irrigation water lines. The marking paint will make it easier to follow with a trencher to install the pipe.
2. Hook up to the water service from the house.
3. Trench in the main lines. (If trenching is done by hand, dig a trench 8 to 12 inches deep. In freezing climates, the line must be below the frost line. Cut the sod and lay it to one side, and dig out the soil and lay it to the other side; this will allow you to replace everything to its original condition. A power trencher may be used. It makes the job easier and saves time.)
4. Establish the zones to be irrigated, and connect the zone values to the main irrigation line.
5. Install the lateral irrigation lines of each zone, and connect the sprinkler heads at flags as the pipe is laid along the bottom of the trench and located according to the design.
6. Install the automatic timer in a convenient location out of the weather. Run each set of wiring harnesses to the proper zone values. Wire the zone values back to the automatic controller. This is a 12-volt system. Each zone will have a color code to follow.
7. Install the sprinklers zone by zone.
8. Check the system's operation zone by zone.
9. Install a rain-control sensor (turns the irrigation system off when it is raining).

3. Do not water at night because of disease problems.
4. Do not water in the heat of the day because the sun will evaporate the water before it can soak into the ground.
5. Water lawns more than planted beds.
6. Reduce the time of watering in shady areas.

GREEN TIP Green roofs create an environmental enhancement for inner-city and urban buildings to reduce the carbon footprint, conserve energy consumption, reduce storm runoff, and improve the aesthetic environment for our future generations.

Green Roofs

The green roof is another facet to enhance cities and urban environments for the future. Green roofs have been used for thousands of years, dating back to turf roofs and to the hanging gardens of ancient Babylon. In Europe, green roofs have been accepted, and in some cities, the government has even mandated their use. The rooftops of buildings in cities are large and expansive and in many cases are like deserts. Through implementation of green roofs, our cities and urban environments can rejuvenate and enrich the living environment.

Green roofs improve our environment by using plants that are drought tolerate to improve air quality where plants were not used previously (see Figure 34–25). By integrating living plants on the rooftop, additional oxygen is produced; water **transpiration** from the plants is released into the atmosphere; rainwater can be retained and used by the plants, reducing the volume of water that would end up as storm water; and reduced rooftop temperature creates more efficient use of the cooling and heating of the building. In addition, green roofs reduce the city and urban noise, absorb and filter airborne particles, reduce the city heat island effect, enhance the overall roof life longevity by two to three times, and create an aesthetic palette of nature for viewers.

In cities and urban areas, **heat islands** form as plants are reduced by construction of roads and buildings for the increased population. These structures absorb the sun's heat, causing surface temperatures and overall **microclimate** temperatures to rise.

Green roofs generate new methods of efficient energy use in buildings with the addition

BOTANICAL NAME	COMMON NAME	HEIGHT	BLOOM	FOLIAGE	BLOOMING TIME	PLANT TYPE	HARDINESS ZONE	EXPOSURE	REMARKS
Allium schoenoprasum 'Forescate'	'Forescate' chives	8–14″	Rose pink to bluish purple	Blue-green	May–June	Deciduous/Accent	3	Full sun to light shade	Low maintenance, Attracts butterflies
Allium cernuum	Nodding onion	12–16″	Lilac-pink flowers	Soft green	July–August	Deciduous/Accent	2	Full sun to moderate shade	Low maintenance, Attracts butterflies
Allium schoenoprasum 'Album'	White flowering ornamental chive	8–12″	White flowers	Blue-green	May–June	Deciduous/Accent	3	Full sun to light shade	Low maintenance, Attracts butterflies
Allium 'Millennium'	'Millennium' oranmental chive	12″	Iridescent pink-purple	Glossy green	July–August	Deciduous/Accent	5	Full sun to light shade	Low maintenance, Attracts butterflies
Allium senescens 'Glaucum'	Curly onion	4–12″	Pink	Twisted blue-gray	July–September	Deciduous/Accent	4	Full sun to light shade	Low maintenance, Attracts butterflies
Delosperma congestum 'Gold Nugget'	'Gold Nugget' ice plant	1″	Yellow flowers	Green	May	Evergreen/Accent	4	Full sun	Low maintenance, Attracts butterflies
Euphorbia myrsinites	Myrtle euphorbia	2–6″	Chartreuse-yellow	Blue-green	May–June	Deciduous/Accent	5	Full sun to light shade	Low maintenance, Attracts butterflies
Sedum acre 'Aureum'	Gold leaved goldmoss sedum	1″	Yellow flowers	Green	June–July	Evergreen	4	Full sun to light shade	Low maintenance, Attracts butterflies
Sedum album	White flowering sedum	1″	White flowers	Jelly-bean shaped	June–July	Evergreen	4	Full sun to light shade	Low maintenance, Attracts butterflies
Sedum album 'Coral Carpet'	'Coral Carpet' white flowered sedum	½″–1″	White flowers	Jelly-bean shaped	June–July	Evergreen	4	Full sun to light shade	Low maintenance, Attracts butterflies
Sedum album 'Green Ice'	'Green Ice' sedum	1″–2″	White flowers	Olive green	June–July	Evergreen	4	Full sun to light shade	Low maintenance, Attracts butterflies (introduced by Ed Snodgrass, Emory Farms)
Sedum album 'Red Ice'	'Red Ice' sedum	1″–2″	White flowers	Blue-green	June–July	Evergreen	4	Full sun to light shade	Low maintenance, Attracts butterflies (introduced by Ed Snodgrass, Emory Farms)
Sedum floriferum	Floific sedum	3″	Yellow flowers	Green to reddish-purple	June–July	Evergreen	3	Full sun to light shade	Low maintenance, Attracts butterflies
Sedum rupestre 'Angelina'	'Angelina' sedum	2–3″	Yellow flowers	Yellow	June–July	Evergreen	3	Full sun to light shade	Low maintenance, Attracts butterflies
Sempervivum 'Purple Beauty'	'Purple Beauty' hen and chicks	2–3″		Grayish-purple	July	Evergreen	5	Full sun to light shade	Low maintenance, Attracts butterflies
Sempervivum 'Ruby Heart'	'Ruby Heart' hen and chicks	2–3″		Silver-blue/rosy red centers	June	Evergreen	5	Full sun to light shade	Low maintenance, Attracts butterflies
Sempervivum 'Silverine'	'Silverine' hen and chicks	2–3″		Green/red tipped	June–July	Evergreen	5	Full sun to light shade	Low maintenance, Attracts butterflies

Figure 34–25 Plants for green roofs.

of the rooftop plantings during the growing season. Rooftop plants will increase the ability to cool down buildings in the city, urban settings, and industrial complexes with air circulation through the plants. In the winter season, rooftop plants will act as an additional insulation factor to conserve energy

Green roofs also make efficient use of stormwater runoff. Stormwater that is conserved with green roof gardens impacts our environment by reducing water runoff greatly. A green roof reduces the rainwater that runs off into the sewer systems that has **pollutants** from the environment (oil, salt, mercury), it reduces flooding of paved areas, it reduces the need for additional dry and wet drainage sediment ponds, and it reduces the amount of vehicular pollutants on paved areas that drain into the groundwater. Therefore, green roofs reduce the pollutants that will end up in the streams, rivers, and estuaries.

There are many factors to consider in the construction of green roofs, including an official certification of the architect or structural engineer (see Figure 34–26). Green roof plants have a shallow

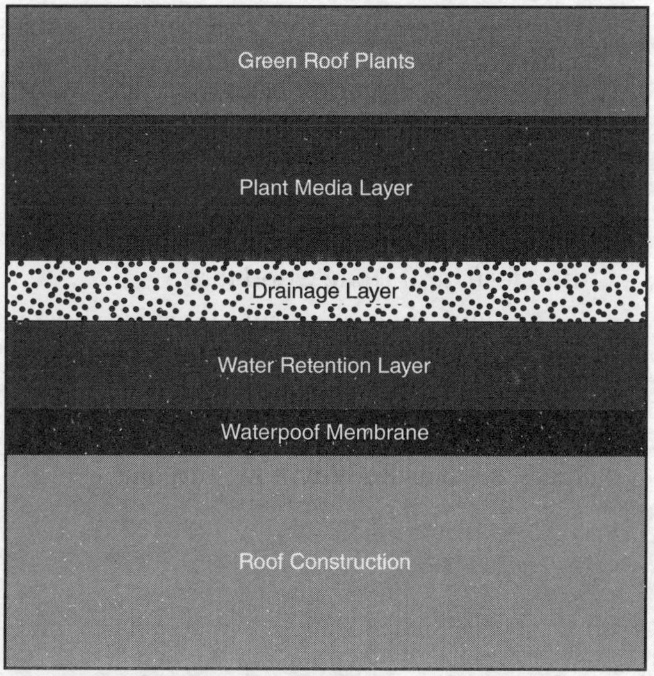

© 2017 Cengage Learning®

▲ **Figure 34–26** Cross-section diagram of green roof construction. Roof construction of the project requires the official certification of the architect or structural engineer for the correct load capacity to carry the weight of the entire green roof garden.

Courtesy of Live Roof, LLC, subsidiary of Hortech, Inc.

▲ **Figure 34–27** LiveRoof module ready to be installed.

root system, growing in media of 6 inches or less. They are low-growing plants with fine to medium texture; they are drought tolerant and can handle extensive harsh conditions. Green roof plants exhibit great **cold hardiness, heat tolerance**, and resistance to **wind exposure**. In the green industry, these plants are known as succulents, water-holding plants such as sedums, sempervivums, alliums, euphorbias, and delosperma that absorb the water resource for growth and release it into the atmosphere through their stomata that open during the night (a process known as CAM—crassulacean acid metabolism). Thus, green roof plants conserve water to enhance the environment, as well as create a natural habitat for birds to use plant seeds as a food source and a nesting ground for their young.

Green roofs help to absorb sound and reduce noise. The nature of the green roof media and vegetation can insulate the noise of trains, buses, trucks, and airplanes. The green roof plants may be deciduous or evergreen with a fine texture (small leaves) and a myriad of surface area to absorb, reflect, scatter, and diffuse sound (Figure 34–27).

Green roofs can be as beautiful as a well-manicured garden as viewed by looking down on to the garden. The extensive use of succulents with variegated leaves, showy leaf color, a vibrant **floral palette**, and foliage of deciduous, semi-evergreen, and evergreen plants, plus special accent plants that exhibit unique individual qualities catches the viewer's eye. Using the design principles of unity, scale, repetition, balance, and texture will create a magnificent natural enhancement of the environment (Figure 34–28).

▲ **Figure 34–28** LiveRoof at Haworth corporate headquarters, Holland, MI.

Green roofs' maintenance includes the following:

1. Perform an annual soil fertility test.
2. Check the soil pH, which should be 6.5 to 8.0.
3. Use slow-release sustainable organic fertilizers with an analysis of 14-14-14 at the rate of 20 pounds per 1,000 square feet.
4. Irrigate green roof plants in prolonged dry weather zones.
5. Perform weed control every 2 weeks to remove any weed seedlings and deadhead flower seedpods.
6. Replace any plants that have died to ensure continued coverage of the green roof area.
7. Check all roof drains to ensure they are clean and free flowing.
8. Remove any trash or debris in the area.
9. Keep an accurate record of the green roof maintenance.

Rain Gardens

The main goal of **rain gardens** is to help solve problems associated with the management of surface water drainage. For this reason, rain gardens have a very practical application in today's landscape world. They can range in size depending on the amount of surface water runoff. For example, rain gardens are typically created in natural depressions where rainwater runoff drains from the driveway, walkways, or roof onto graded areas. Rain garden design must consider the soil type that exists in area of the depression, and the plant selection should include native and wetland plants. These types of plants will filter the water runoff and create a beautiful garden in a section of the lawn that would normally be a wet, soggy area. The plants will also create a biofiltration area for the groundwater overflow on the property.

Selecting a diversity of native plants that will survive in very boggy, low, wet woodland or meadow-like conditions is important in rain garden design. Such plants can include the following:

- Native wildflowers
- Sedges (a grasslike plant)
- Most evergreen shade ornamental grass cultivars, including 'Blue Zinger,' 'Yushan,' 'Hobb,' 'Silk Tassel,' 'Oehme,' 'Evergold,' 'Everillio,' and 'Spark Plug,' as well as rushes from the *Juncus* species
- Ferns, such as Japanese painted ferns, autumn brilliance, Christmas fern, cinnamon fern, and Dixie wood fern
- Perennials, such as purple loostrife, hosta, bee balm, daylilies, Solomon's seal, bog rosemary, turtlehead, and blue iris
- Shrubs, such as elderberry and spicebush
- Trees, such as holly, willow, and oak.

SAFETY FIRST

OSHA's regulations ensure that employers and employees know about the work hazards and how to protect themselves to prevent injuries in the workplace. For more information on occupational health and safety in the workplace, please visit the OSHA site at www.OSHA.gov.

Summary

In this unit, the major career fields of the landscape profession were explored. An introduction to residential landscape design with applications of hardscapes, pondless waterfalls, water gardens, irrigation, and xeriscaping was provided. Landscapers have implemented water conservation practices for the irrigation of plants in today's world. The use of drip irrigation is increasing because it conserves water with optimum plant growth. In the selection of plants for each landscape display, water supply is an important consideration to ensure successful growth. And, bioengineering of green roofs is an important facet of the green industry to create sustainable biodiversity and enhance environmental improvements of the future.

Review Questions

1. What are the three major career fields in the landscape profession? What is the fastest-growing profession in the landscape industry today?
2. Discuss the objectives of residential landscaping.
3. Discuss five principles of landscape design and why they are important considerations when planning an outdoor room.
4. How does hardscape paving differ from softscape paving? In what situations would you choose one material over the other?
5. Discuss five reasons for landscaping an area with a rain garden.

Student Activities

1. Use cutouts of symbols on a felt board to practice arranging different types of plantings. Prepare the symbols (to represent various plantings and structures) beforehand. Try some arrangements that will (1) draw attention to a corner, (2) give partial screening, (3) give total screening, and (4) provide an interesting silhouette for the viewer's eye.
2. Invite someone who is in a landscaping profession (landscape architect, landscape contractor, or professional horticulturist) to the class to discuss his or her occupation. Ask the presenter to list all of the specific jobs that are employed during a large landscaping project.
3. As a class, install a planting around the school or in a nearby park. Discuss the function of the planting before the actual work is begun.
4. Search online for examples of outdoor rooms. Make a list of common characteristics to be included in an outdoor room.

Suggested Supervised Agricultural Experience (SAE) Activities

1. As a community-based SAE, design, plan, install, and maintain an outdoor room for the community.
2. Conduct an entrepreneurship SAE as a repair technician for landscaping equipment.
3. Participate in a placement SAE working for a landscaping company in which you maintain a landscape using all essential tools.

Additional Resources

www.LiveRoof.com

www.roofmeadow.com

www.lawnandlandscape.com

www.ext.colostate.edu/mg/gardennotes/413.html

www.the-landscape-design-site.com/principlesoflandscapedesign.html

SECTION 8

Sustainable Strategies for Lawn and Turfgrass Establishment and Maintenance

UNIT 35 Establishing the Lawn

Objective

To start a lawn by seeding, sodding, and sprig or strip planting.

Key Terms

calcium
carrier
damping-off
endophyte
erosion
germination
guttation
loam
nitrogenous
noxious weeds
pelletized
purity
recreational
stolonized
vegetative

Competencies to Develop

After studying this unit, you should be able to:

- list three reasons for establishing and maintaining a lawn.
- describe two methods of establishing proper drainage in a lawn.
- list three materials that are used to increase the organic matter in a new lawn.
- name eight items that must be included on a seed label.
- describe three ways turfgrass is started in the United States.
- demonstrate the five steps in seeding a lawn.

Materials

- Weed-free Composted Manure
- bag of lawn seed
- bag of lawn fertilizer, lime, and spreader
- sample of drainage tile
- samples of peat moss, well-rotted sawdust, and weed-free manure
- samples of commercial fertilizer, controlled-release fertilizers, and lime
- samples of cool-season and warm-season turfgrass
- several steel rakes
- rototiller
- straw
- lawn roller
- commercial mulching material
- smartphone apps:
  - AG PhD
  - Spectracide
  - Ortho Problem Solver
  - Pest Portal
  - Calibrate My Sprayer
  - PGR Mix Master

Sustainable lawns are a major part of most home landscapes. Basically, lawns are established for the following three reasons:

- They add beauty to the landscape. A well-kept lawn is very appealing and inviting.
- They are used as play areas for sports such as baseball, football, basketball, golf, or lacrosse, or for relaxation. Because sports are tough on lawns, it is important to select a lawn grass that can take wear.
- They provide an excellent cover to help control soil **erosion** while allowing the movement of air and water to roots of trees and shrubs in the soil below.

Turfgrass is a category of narrowleaf grass species that creates a uniform **vegetative** ground cover. A unique feature of turfgrass is that transpiration of water occurs on the tip of the grass blade, a phenomena called **guttation**. There are many turfgrass species used for lawns. Some provide an ideal surface for sports fields, parks, golf courses, and other **recreational** facilities. Each of these species has positive and negative aspects. To select the best grass species for your local area, it is best to follow the recommendations of the extension service of the land grant university or college in your state.

Turfgrass areas should be mowed regularly. Turfgrass serves a functional purpose by preventing soil erosion, but it has an aesthetic purpose as well. A properly managed turf area is very attractive. Turfgrass is able to withstand arduous use, and some species provide an ideal surface for sports fields and other recreational facilities. Turfgrass has two classifications: the cool-season turfgrass grown in the northern United States and the warm-season turfgrass grown in the southern United States.

> **GREEN TIP** Select an improved cultivar of the grass species to ensure high-quality turf performance in the future.

Soil and Grading

The first thing to consider in establishing a new lawn is the present condition of the soil. Is this an area in which the builder has graded off all the topsoil? Is the slope too steep to establish a lawn and mow it safely? Is the drainage adequate? These are all questions that must be answered before a lawn is established.

The builder usually establishes the rough grade. The lot is graded so that the land slopes away from the foundation of the house, to help prevent any water from entering the basement. After the rough grade is established, topsoil that the builder has set aside may be spread over the subsoil, which was kept in the rough grading. Six inches of topsoil should be spread evenly over the surface and tilled to loosen and break up clods (Figure 35–1).

The general slope for the lawn after the topsoil is spread should not exceed 15 percent, that is, no more than a 15-foot drop for every 100 feet of distance. Slopes greater than 15 percent are unsafe to mow. If a slope steeper than 15 percent cannot be avoided, it should be covered with plants that do not require mowing.

Drainage

Good drainage ensures a balance between the air and water in the soil. This, in turn, encourages proper root growth. Grasses can endure relatively wet soil but not bog or swamp conditions.

▲ **Figure 35–1** Topsoil being spread on an area that will be seeded to turfgrass.

There are two ways to establish proper drainage in lawns. One method is to install drainage tile about 3 feet below the surface of the soil to drain the subsoil. Another method is to make use of the slope of the land to drain surface water away.

Preparation of the Soil

As mentioned, 6 inches of topsoil should be spread over the rough graded subsoil. A good garden **loam** is the best medium for most grasses. If quality topsoil is not available, the organic matter content of the soil should be increased by adding well-rotted sawdust; weed-free, well-rotted manure; composted leaves and grass; or commercial peat moss at the rate of 6 cubic feet per 1,000 square feet of land. Work the material well into the soil with a rototiller. It may be necessary to remove stones or soil clods. Use a hand rake or a stone rake mounted on a tractor to remove stones. A good seedbed should have a firm and smooth, but not powder fine, surface texture.

GREEN TIP Productive soils will conserve water, encourage microorganisms to actively work to break down complex molecules for plant use, retain nutrients for use by lawn grass roots, and help control pollution of the environment.

Fertilizer

A complete starter fertilizer (18-24-6) with a high phosphorus content is recommended for use in establishing new lawns. Nitrogen promotes dark green growth in roots and blades, phosphorus develops a vigorous roots system, and potash improves hardiness and disease resistance. Some companies manufacture organic fertilizers (very insoluble matter that causes little or no water contamination in a blend of 100% organic materials mixed with minimal amounts of synthetic organics that offer economic savings) and inorganic fertilizers (synthetic).

Note that a number of states restrict, by law, the use of fertilizers containing phosphorus unless certain conditions are met. As a result, fertilizer companies are now offering "Phosphorus-Free" organic fertilizers as an alternative (Figure 35–2). Before establishing a lawn, it is important that you consult the laws governing your state regarding the use of fertilizer containing phosphorus content.

According to Phil Catron, president of NaturaLawn of America, there are various choices available when selecting a source for organic type fertilizers. These include, but are not limited to, the following:

- Class A biosolids (OCEANGRO, Milorganite, Hou-Actinite, etc.)
- Blood meals

Courtesy of Natural Alternative, Inc.

▲ **Figure 35–2** A total fertility program for establishing a lawn that is phosphorus-free.

- Bone meals
- Feather meals
- Soybean meals
- Corn gluten
- Activated composts (vegetable, leaf, etc.)
- Leather tankage
- Poultry litter

The benefits of using an organic fertilizer over an inorganic (synthetic) choice are many. Remember, the goal of lawn care is not so much to feed the turf plant directly but more to feed it indirectly by building up the soil's health, which in turn will feed the turfgrass.

Inorganic (synthetic) fertilizers share similar characteristics that include the following:

- High solubility makes them release very quickly.
- High salt indexes can burn (dehydrate) plants.
- Plant water needs and usage are increased with the synthetic forms of nitrogen.
- Excessive succulent growth can lower resistance to disease and adverse weather conditions such as drought.
- Loss of nitrogen is more readily apparent through leaching and/or vaporization.
- More frequent applications are required due to the "spike" release of the nutrients and the loss through leaching/vaporization.
- Many synthetic sources possess acidifying characteristics.

Organic fertilizer are slow-release forms of nutrients that help prevent a "flush" of growth, which means lawns do not need to be cut as frequently during high-growth months; they are also not soluble. Organic characteristics include the following:

- Less cell elongation in the plant from minimal flush growth helps keep disease incidence down.
- Negligible leaching and burning problems occur.
- Nutrients are provided to billions of soil organisms, which in turn help build the soil foundation. A strong soil foundation retains moisture better and assists the turf in not having to use up excessive carbohydrates.
- A stronger plant root system is developed that is better able to ward off pest and environmental problems.

Before applying fertilizer, the soil should be tested to determine the correct amount of fertilizer and lime to add. This can be done either with a special portable soil test kit or by sending a sample of the soil away to a soil-testing laboratory. When test results are returned from these laboratories, they should contain a recommendation regarding the amount and analysis of fertilizer to apply.

The results of soil tests sometimes indicate a need for the addition of lime to the soil. Lime reduces the acidity of soil and encourages root development. Different forms of lime are available for use on lawns, including ground limestone (calcium carbonate [$CaC0_3$]), burned lime (calcium oxide [CaO]), and hydrated lime (calcium hydroxide [$CaOH_2$]). When choosing a lime for use in a lawn, pick the one that gives the most **calcium** for the money. If the soil pH is too high or alkaline, sulfur or iron sulfate may be used to lower the pH.

Most lawn grasses grow best in a well-limed soil with a pH level from 6.0 to 6.5. If the soil test calls for lime, add the recommended amount evenly over the entire soil surface. Work it into the ground 4 to 6 inches before seeding.

Spreading the Fertilizer

The soil test will also identify any nutrient deficiencies and/or excesses in the soil. These may be addressed by the application of fertilizer. In 2013, 16 U.S. states had laws regulating the use and application of lawn fertilizer. It is therefore important to be aware of the laws governing fertilizer use in your state.

Fertilizer should be spread in two different directions on new lawns. Spread one-half of the recommended amount in a north-south direction and the rest in an east-west direction to better distribute the fertilizer.

Fertilizers are usually applied most efficiently by use of a spreader. It is important

▲ **Figure 35–3** Broadcast (rotary) spreader used for application of fertilizer or seeding grass.

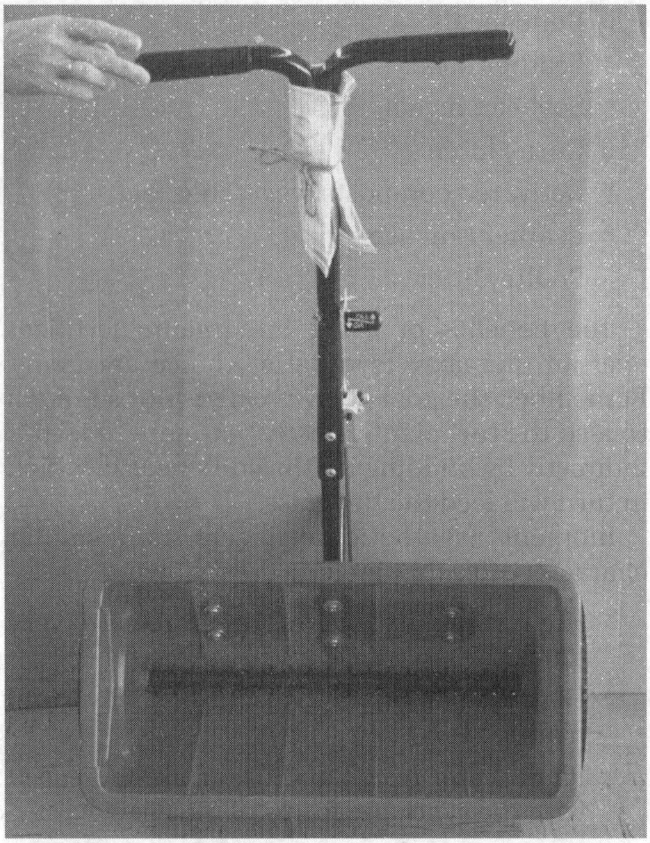

▲ **Figure 35–4** Conventional spreader (drop spreader).

that the fertilizer be applied uniformly. Two basic types of spreaders are available. Figure 35–3 shows a rotary spreader, which spins the fertilizer out over a relatively wide path. Figure 35–4 shows a conventional spreader (drop spreader), which spreads fertilizer in a strip as wide as the spreader. Both spreaders can be adjusted so that the proper amount of fertilizer is spread. The fertilizer is then worked into the soil surface with a hand rake (Figure 35–5).

Starting the Lawn

Lawns are started in one of two ways: (1) seeding or (2) vegetatively by sodding or plug, strip, or sprig planting.

Selecting the Seed

It is very important to purchase the best lawn seed. All lawn grass seed is required by law to have a label with the following information.

▲ **Figure 35–5** The fertilizer is worked into the topsoil with a hand rake.

NAME The seed package must give the correct name of all the seeds included in the package. Information on where this type of seed grows best—in a shaded area, partial shade, or full sun—is also provided.

PURITY **Purity** figures give the makeup of the seed by percentages of the different grasses included in the package.

PERCENT OF GERMINATION This figure indicates the percentage of seeds in the package that will sprout when planted. The higher this percentage, the more seed there is that will germinate. (A **germination** percentage of 85 means that 85 seeds out of 100 will germinate.)

OTHER CROP The percentage of other crop seed is important to consider. Crop seeds such as wheat, barley, orchard grass, and timothy are undesirable seeds for lawns.

INERT MATTER This is material that will not grow. It includes pieces of seed, sand particles, and chaff from the seed. All of these things add weight to the package, but they do not germinate. This type of material is undesirable.

WEED SEED A weed is any plant that is growing where it is not wanted. Weed seeds are not desirable in a lawn grass mixture. Some weed seeds are very small and difficult to remove from the desirable seed. Most packages of seed contain some weed seed, but the manufacturer is required by law to list the percentage present in each container.

NOXIOUS WEEDS Certain weeds have been declared by each state as **noxious weeds** because they are particularly difficult to control. In some cases, these weeds may be harmful. This material is listed on the package by a specific number, for example, one listing might read, "four garlic per pound."

YEAR TESTED Commercial seed must be tested every year for the correct germination percentage. The month and year that the seed was tested are sometimes given also.

COMPANY NAME The name of the company selling the seed and its correct address must be given on the label so that anyone having problems with the seed can contact the grower.

There are some much-improved varieties of cool-season grass seeds especially for use where heavy traffic problems exist. Some of the best are turf-type tall fescues that are drought and disease resistant, shade tolerant, and take a lot of wear. Check with your local lawn specialist for varieties best suited to your area.

Some grass varieties are resistant to insect attack. Recent research discovered that many of these grasses have a fungus growing in their leaves. This fungus, called an **endophyte**, does not harm the grass but protects it against insect attack. Insects will not eat an endophyte-infected plant. Endophytes are found in tall fescues and red fescue and are naturally transmitted through the seed. Researchers are attempting to place endophytes in other grass varieties to improve insect resistance. There is also genetic resistance to insects found in grasses. This is different from the resistance caused by endophytes.

Insect-resistant varieties should be used whenever they are available. The turf-type tall fescues Titan, Shenandoah, and Mesa are resistant because of endophyte infection.

The fine fescues that include creeping red fescue and chewing's fescue are environmentally friendly. They help to establish and maintain healthy attractive turf in shady areas. They are low-maintenance lawn grasses with minimal fertilizer and pesticide use needed; they tolerate dry, infertile acidic soils; and they require less water than other varieties of cool-season grasses. The unique characteristic of the fine fescues is that the grass blade folds into a tight "V" in bright sunlight to conserve moisture, provide a cooling effect, and reduce carbon dioxide.

Seeding

SOWING SEED Seed may be planted by hand or with a mechanical seeder. Read the label on the seed bag for the correct pounds of seed per 1,000 square feet, and set the spreader settings listed for the type of grass seed being sown. It is important to check the delivery rate before seeding the entire lawn. Check the seeder to be sure the open and closing control level is off. Fill the seeder hopper with the grass seed. During the seeding process, walk off with the seeder and turn it on to spread the seed. Sow 50 percent of the seed in a north-south direction, and 50 percent of the seed in an east-west direction. Sow one seeder width around the entire perimeter of the area being seeded. Then seed the inside area to have the correct coverage of the lawn area being sown. To obtain uniform distribution, the seed is mixed with small amounts of a **carrier** such as sand. The mixed material is divided into two equal parts; one part is sown in one direction, and the other part is sown crosswise to the first sowing. Figure 35–6 illustrates seeding with a mechanical seeder. Figure 35–7 lists seeding rates for cool-season grasses

▲ **Figure 35–6** Lawn seed is sown by hand or by means of a mechanical seeder. A rotary spreader is shown here.

GRASS VARIETY	POUNDS PER 1,000 SQUARE FEET
Tall fescues	6–8
Red or fine fescues	3–5
Bluegrass mixes	1½–2
Bluegrass fescue mix	2–3
Perennial ryegrass	5–6

Figure 35–7 Lawn seeding rates for cool-season grasses.

GREEN TIP The best time to seed a cool-season grass is in the late summer to early fall. This will ensure optimum conditions for soil moisture, favorable soil temperature, and less weed competition with the grass seed germination and seedling growth.

COVERING THE SEED The seed is lightly covered by hand raking. Large seeds are covered with ¼ to ⅜ inch of soil and small seeds with ⅛ to ¼ inch of soil. It is important that all seed is covered by and in close contact with the soil (Figure 35–8).

MULCHING Mulching with a light covering of weed-free straw, hay, or commercial **pelletized** fibers helps to hold moisture and prevent the seed from washing away during watering or rainfall. Straw also helps to hide the seed from birds. One (60- to 80-pound) bale to 1.5 bales of straw will cover 1,000 square feet of area. Mulches applied evenly and lightly may be left in place and the grass allowed to grow through. Peat moss or other fine material does not make satisfactory mulch. These materials become packed too tightly, resulting in the seed being planted too deep (Figure 35–9).

On terraced areas or sloping banks, cheesecloth, burlap, or a commercial mulching material (e.g., Curlex) helps to hold in moisture and

▲ **Figure 35–8** A garden rake is used to cover the seed.

▲ **Figure 35–9** Straw being spread over a newly seeded lawn.

▲ **Figure 35–10** The newly seeded lawn should be given at least 1 inch of water. Notice the newly seeded lawn is mulched with straw to shade and keep the soil moist. Seed comes up faster in moist soil.

keep the seeds in place. These products must be pinned with plastic staples to prevent it from sliding off the slope. The grass is able to grow through these mulching materials, which are biodegradable and provide additional sources of biomass to enhance seedling growth.

WATERING New seedlings should be kept moist until they are well established (Figure 35–10). After seeds have begun to germinate, they must not be allowed to dry out, or they will die. Avoid saturating the soil, however, as excessive moisture is favorable for the development of **damping-off**, a fungus disease.

Vegetative Planting

There are some grasses for which seed is not available, or the seed that is available does not produce plants that are true to type. These grasses must be planted by one of several vegetative methods, such as spot or plug sodding, strip sodding, sprigging, or stolonizing. Grasses planted by vegetative methods include zoysia, improved strains of Bermuda grass, St. Augustine grass, centipede grass, creeping bentgrass, and velvet bentgrass.

Whether plugged, strip sodded, sprigged, or **stolonized**, the planted material must be kept moist until well established. During the first year, light applications of a **nitrogenous** fertilizer every 2 to 4 weeks during the growing season helps to speed the spread of the grass.

Sodding

Sod consists of grass and grass roots in a thin layer of soil that is removed from the area in strips. It is rolled and transported to the area to be sodded (Figure 35–11). Unless good-quality sod is available and complete coverage is needed immediately, the expense of sodding is justified only on steep slopes or terraces where erosion may be a serious problem.

Sod should not be cut more than 1 inch thick. Sod that is cut ¾ inch thick will knit to the underlying soil faster than thicker sod. The sod pieces are laid in the same way as bricks are and fitted together as tightly as possible. After the first strip is laid, a broad board is placed on the sodded

▲ **Figure 35–11** Sod being installed. Notice rolls are not all started at the same place. This helps prevent slippage and separation of the sod surface.

strip. The board is knelt on and moved forward as the job progresses. This eliminates tramping on the prepared seedbed.

The planting area is prepared and fertilized in the same manner for sodding as for seeding. A roller is used to firm the seedbed after final hand raking and leveling. After the sod is laid, it is tamped or rolled lightly and topdressed with a small amount of topsoil to fill in any cracks between the pieces of sod. The back of a rake or a straw broom may be used for this job. It is important that the sod be kept moist until the roots have grown well into the soil.

SPOT SODDING OR PLUGGING *Spot sodding* or *plugging* is the planting of small plugs or blocks of sod at measured intervals. The plugs are spaced from 8 inches to 1 foot apart depending on how rapidly the area is to be completely covered. The closer the plugs are set, the faster the lawn will be completely covered with grass. The plugs are fit tightly into prepared holes (Figure 35–12). They are then tamped firmly into place. Zoysia grass is started in this way (Figure 35–13).

STRIP SODDING *Strip sodding* is the planting of strips of sod end to end in rows that are 1 foot apart. The sod strips should be 2 to 4 inches wide. Firm contact with surrounding soil is necessary.

SPRIGGING *Sprigging* is the planting of individual plants, runners, cuttings, or stolons at certain spaced intervals. The *runners* (sprigs of grass)

▲ **Figure 35–12** A turf plugger is sometimes used to make the holes into which plugs of grass are planted.

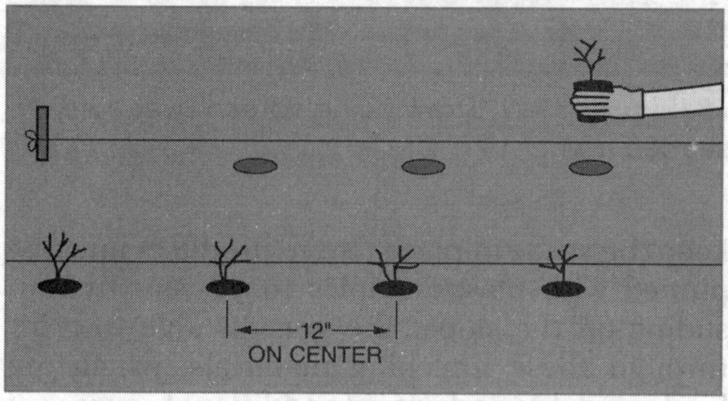

▲ **Figure 35–13** Zoysia grass plugs spaced 12 inches on center. A line is being used to keep the rows straight.

are obtained by shredding solid pieces of sod. The spacing of the runners is determined by how fast that particular grass grows, how fast the area is to be covered, and the amount of plant material available. Runners may also be planted end to end in rows (Figure 35–14).

STOLONIZING In stolonizing, shredded stolons are spread over the area with mechanized equipment. The spreading is followed by disking or rolling the planted area and topdressing with fertilizer. (*Topdressing* is the spreading of material on top of the soil.) Stolonizing is usually done only when large areas are to be planted or when the area to be planted has a highly specialized use, such as golf course putting greens.

▲ **Figure 35–14** Turf area planted with sprigs.

Large areas of Bermuda grass may be established by spreading shredded stolons with a manure spreader and disking lightly to firm into the soil. (*Disking* is the mixing of soil with a disc harrow.) This method requires 90 to 120 bushels of stolons per acre. Creeping bentgrass and velvet bentgrass are stolonized by spreading shredded stolons at the rate of 10 bushels per 1,000 square feet, topdressing with topsoil to a depth of ¼ inch, and rolling to firm the stolons into the topdressing.

TRY THIS Using the Internet, research the process of planting, harvesting, marketing, and installing sod. Use large distributors' websites and university websites for information. Include information about the modern equipment currently used in production (a field trip to a sod farm will provide even more information). Create a flowchart that identifies each step in the process of growing and marketing sod.

Choosing the Grass Type

When deciding which grass to plant in a certain area, the most important factors to consider are the climate conditions, such as the temperature of the area and the available moisture. Figure 35–15 shows a map of the continental United States, which is divided into areas suitable for the planting of certain grass types. Following is a list of the various grasses suitable for each of the climatic regions illustrated.

Region 1. Common Kentucky bluegrass, Merion Kentucky bluegrass, red fescue, and colonial bentgrass. Tall fescue, Bermuda grass, and zoysia grass in the southern portion of the region.

Region 2. Bermuda grass and zoysia grass. Centipede grass, carpetgrass, and St. Augustine grass in the southern portion of the region with tall fescue and Kentucky bluegrass in some northern areas.

Region 3. St. Augustine grass, Bermuda grass, zoysia grass, carpetgrass, and bahiagrass.

Region 4. Nonirrigated areas: Crested wheatgrass, buffalograss, and blue grama grass. Irrigated areas: Kentucky bluegrass and red fescue.

Region 5. Nonirrigated areas: Crested wheatgrass. Irrigated areas: Kentucky bluegrass and red fescue.

Region 6. Colonial bentgrass and Kentucky bluegrass.

Before planting, also check other sources, such as your instructor and local extension service, for varieties that are recommended for your locality.

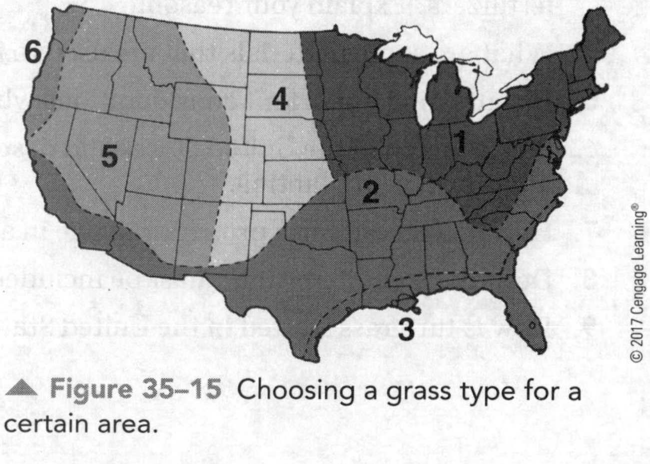

▲ **Figure 35–15** Choosing a grass type for a certain area.

SAFETY FIRST

OSHA's regulations ensure that employers and employees know about work hazards and how to protect themselves to prevent injuries in the workplace. For more information on occupational health and safety in the workplace, please visit the OSHA site at www.osha.gov.

Summary

Lawns are a major part of most residential home landscapes. An elegant lawn enhances the appeal and beauty of the landscape. The use of lawns for sports activities offers an opportunity for fun and relaxation. Lawns provide an excellent ground cover to help control erosion while allowing the movement of air and water to roots of trees and shrubs in the soil below. Successful establishment of a lawn involves several techniques: use of quality topsoil to ensure optimum growth of the turf; proper grading to provide correct drainage for movement of the water into the lawn; and seedbed preparation with addition of organic matter and fertilizer. Several ways to start a lawn are seeding, sodding, stolonizing, or sprigging. It is important to select the correct grass type for a particular area in the continental United States.

Review Questions

1. Discuss three reasons lawns are established.
2. Describe three unique characteristics of the fine fescue grasses.
3. If you were establishing a new lawn, would you use organic fertilizers or inorganic fertilizers? Explain your reason.
4. What are some materials that are used to increase the organic matter in a new lawn?
5. Discuss nine important considerations when starting a lawn by seeding.
6. Pick a region of the United States and describe the grass types that would be most appropriate for planting.
7. How do you establish proper drainage in a lawn?
8. Describe eight items that must be included on a seed label.
9. How is turfgrass started in the United States?

Student Activities

1. Create a display showing labeled samples of lawn seed.
2. Establish small yard-square plots of various lawn grasses. Practice leveling the area with a steel rake to establish a seedbed.
3. Practice using a lawn spreader to obtain uniform coverage of areas to be seeded. If the seed and the necessary ground are not available, sand can be used in place of seed and applied to marked areas of a parking lot.
4. Search online for information on sod production in your state. Identify the sod distributors and what regions they market their sod.
5. Visit a golf course and ask the turfgrass manager to describe where they purchase their sod and the process of installing and maintaining the sod on a golf course.

Suggested Supervised Agricultural Experience (SAE) Activities

1. Participate in a placement SAE by working for a landscape contractor installing sod on a client's lawn.
2. Work for a landscape company sowing seed for customer's lawns to qualify as a placement SAE.

Additional Resources

http://plantscience.psu.edu/research/centers/turf/extension/factsheets/lawn-establishment
https://store.extension.iastate.edu/Product/pm1072-pdf
www.ipm.ucdavis.edu/TOOLS/TURF/SITEPREP/seed.html
www.clemson.edu/extension/hgic/plants/landscape/lawns/hgic1203.html

UNIT 36 Maintaining the Lawn

Objective

To use proper lawn maintenance techniques.

Key Terms

- aeration
- formaldehyde
- gibberellic acid
- gluten
- granular
- parasites
- postemergence
- preemergence
- reel mower
- rotary mower
- selective
- sulfur coated
- thatch

Competencies to Develop

After studying this unit, you should be able to:

- list the seven factors of good lawn maintenance and explain each orally.
- describe the analysis of a good turf fertilizer.
- determine the best time to apply fertilizer to a lawn.
- demonstrate how to set the mowing height of a rotary mower.
- list the three causes of fungus disease in lawns.

Materials

- lime and fertilizer
- tape measure and enough string to lay out 1,000 square feet of lawn area
- one dull and one sharp mower blade
- knapsack sprayer for applying weed killer
- crabgrass weed killer (in combination with fertilizer)
- weed control
- rotary mower (lithium battery operated, electric, or gas powered)
- smartphone apps:
  - AG PhD
  - Spectracide
  - Ortho Problem Solver
  - Pest Portal
  - Calibrate My Sprayer
  - PGR Mix Master

After the lawn has been seeded or planted vegetatively, care must be taken to keep it healthy. Proper lawn maintenance is dependent upon seven factors:

- Planting the proper variety or species of grass
- Applying fertilizer and lime or sulfur at the proper time and in the proper form and amount
- Mowing to the proper height at the correct time
- Watering properly
- Using chemicals for weed, insect, and disease control if necessary
- Using the lawn in such a way that the traffic is not too heavy
- Providing the proper amount of light

Planting the Proper Variety or Species

Different varieties or species of grasses grow best under different temperature, light, and moisture conditions. If planting is to be done in a cool climate, for example, cool-season grasses should be selected. Some grasses, such as the creeping red fescues, grow better in shade than in full sun. They also grow better in dry areas. And some grasses are disease resistant and require less fertilizer. No grass does well in dense shade, however. Bluegrass grows well in cool climate areas receiving full sun and adequate moisture. Bermuda grass grows well in hot climates.

Another factor to consider when choosing a variety of grass is the way in which the lawn is to be used and the amount of traffic to which the lawn will be exposed. The more a lawn is used as a recreation or work area, the more stress there is on the grass. If heavy use is anticipated, select a hardy grass that can withstand more traffic. (The proper placement of walks and rest or play areas reduces traffic on grass plants.) Check with a local agricultural experiment station for the varieties of grass that grow best in your area.

Applying Lime, Sulfur, and Fertilizer pH Adjustment

Lime should be used whenever necessary to keep the soil pH in the proper range. Soil pH is the acidity or alkalinity of the soil and is measured on a scale from 0 to 14. Soils in the northeastern and northwestern United States are acidic, whereas soils in the southwestern United States are more alkaline. A pH of 6.0 to 6.5 is a good range for most grasses. Finely ground limestone is generally the best and least expensive form of lime to use. Late fall and winter are generally the best times to apply lime.

Lime moves down through the soil very slowly. Applications are spread on top of the soil as needed. Lime gradually works its way down into the soil at a rate of ½ to 1 inch a year.

Sulfur should be used on soils that are alkaline to adjust pH downward.

> ***TRY THIS*** Conduct a lime experiment. Designate three crop growing beds approximately 15 × 15 feet. In one grow bed, apply no lime; in the middle grow bed, apply ¾ of an inch of lime on the surface of the soil; and in the final grow bed, apply 1 ¾ inches of lime on the surface of the soil. Record the growth of the crops that are grown in each bed over the growing season. Compare the results, and prepare a report highlighting your findings.

Fertilizer

If a healthy lawn is to be maintained, annual applications of a turf fertilizer are needed (Figure 36–1). Nitrogen is leached from soil and must be replaced regularly.

Lawns require a fertilizer high in nitrogen (indicated by the first number on the fertilizer label). Of the three analysis numbers on fertilizer labels, the first number is always highest on high-quality maintenance turf fertilizers.

Many turf fertilizers supply nitrogen in an organic form. This form of nitrogen is released slowly and thus does not burn the grass. It also supplies nutrients over a longer period of time.

▲ **Figure 36–1** Power spreader and sprayer for lawn maintenance.

The urea **formaldehyde** form of nitrogen is often used as a slow-release fertilizer (**sulfur coated**) for turfgrasses. If an inorganic form of nitrogen that is released rapidly (such as nitrate of soda) is used, it must not be applied to wet, actively growing grass. Such an application would burn the grass badly.

Applying Fertilizer

Fertilizer may be applied by hand if uniform coverage can be obtained. However, it is better to use a spreader.

Fertilizer should be applied just before the active growing season. For cool-season grasses, this is in the very early spring or better in the fall. (Two fall applications, 6 weeks apart, are best.) These grasses grow in spring and fall and are either dormant or very slow growers in the hot, dry summer months. To fertilize them in summer would support weeds that thrive in hot weather and could use the fertilizer for growth. Where grass remains active all summer, fertilizer should be applied at least once during this time.

Care must be taken not to skip spaces or overlap areas with the fertilizer spreader. Difference in green shades and growth rate of the grass will result, ruining the appearance of the lawn. Fertilizer rates depend on things such as soil type and the amount of rainfall available in the area. Check local recommendations by contacting your local extension service for a lawn care calendar that gives the proper dates for fertilizing as well as spraying and seeding.

GREEN TIP Proper fertilizer use is one of the most important factors in maintaining a good lawn.

Mowing the Lawn

Two types of mowers are used for mowing lawns, the **reel mower** and the **rotary mower**. Rotary mowers may be powered by gas or by lithium battery, which is an environmentally friendly option. The type of mower used generally depends on individual preference; either one does a good job. Close mowing of ½ to 1 inch is done most efficiently with a reel mower.

Most lawns are cut too short because the homeowner believes that the grass looks better when short. A very short cutting reduces the leaf area of the grass to such an extent that it cannot manufacture enough food. Also, grasses that are cut too short encourage weed growth because the grass plants cannot kill weed seeds by "shading them out." Cool-season grasses should not be cut shorter than 2 to 3 inches. Warm-season grasses are cut shorter, from ½ to 1¼ inches, depending upon the grass variety. Because warm-season grasses grow rapidly during warm weather, they are better able to compete with weeds growing during the same season. Bermuda grass should be clipped to ½ to 1 inch and zoysia ¾ to 1¼ inches.

Lawns should be mowed often enough so that no more than one-third of the top is cut off in any one mowing (Figure 36–2). For example, if grass is kept at a height of 2 or 3 inches, only 1 inch should be cut off. This means that when the grass reaches 3 or 4 inches in height, it is time to mow again. When the grass is in periods of most active growth, mowing two or three times a week may

▲ **Figure 36–2** This lawn of creeping red fescue grass has been mowed to a height of about 2 ½ inches, ideal for this cool-season grass. When it reaches a height of 3 ½ inches, it will be ready for another mowing. Thus, no more than one-third of the leaf area is cut at any one mowing.

be required (Figure 36–3). This is a very important point to remember because if grass is allowed to grow to a height of 5 to 6 inches and is cut back to 3 inches in one mowing, most of the food-producing leaf blade is removed. The lawn may then appear yellow, an indication that it cannot manufacture enough food until more new leaves are formed. Constantly draining food reserves from the roots weakens the root system. Tall weeds also have a better chance to establish a root system if the lawn is not mowed often enough.

Cutting the lawn to the proper height requires adjustment of the mowing height of the mower blade (Figure 36–4). This is done by setting the mower on a flat, smooth surface and measuring from the surface up to the blade with a yardstick. Adjust the height of the mower blade according to the height of the grass being mowed.

CAUTION

Remove the spark plug wire before turning the mower over to measure the blade.

▲ **Figure 36–3** Current green industry mower used to manicure lawns.

▲ **Figure 36–4** Adjusting the mower height to the height required by the type of grass. Use a blade height indicator to measure the correct height of the mower blade. This blade will cut the grass to a height of 3 inches.

The mower blade should be kept sharp at all times so that grass blades are cut and not torn off. Jagged, split ends cause the lawn to appear uneven and provide open wounds for the attack of fungus diseases.

The new, improved mulching mowers cut grass blades into very fine pieces that do not have to be removed from the lawn. This saves mowing time and also reduces fertilizer needs by about 50 percent. As the grass blades rot, plant food elements are released into the soil.

Mowing Technique

The lawn should be mowed in a pattern such that it is cut in one direction at one mowing and at right angles to that direction in the next mowing. This helps to eliminate soil compaction and turf wear and gives the lawn a more even appearance. It may also help to reduce **thatch** buildup in the lawn. (*Thatch* is a layer of dead grass.) A slight overlap is necessary on each pass over the lawn. This overlap prevents missed areas and picks up any grass that the mower wheel may have only pushed down on the previous trip.

There is also the option of mowing to create aesthetically pleasing lawn striping or lawn patterns. This is a process in which grass blades are bent down with a roller or striping kits attached to the mower deck. The darker pattern results from the blades being bent toward you and a lighter pattern results when the blades are bent away from you.

If the lawn becomes too tall and it is necessary to cut off a large part of the grass blades, the grass should be collected in a catch bag attached to the mower or raked up and removed. Heavy accumulation of grass clippings shades out light and kills grass. It also mats down the grass and causes a buildup of thatch on the soil surface. Thatch reduces soil **aeration**, thus damaging roots and also providing a breeding place for insects and diseases. It is better to mow two or three times in close succession, taking off only 1 inch each time, until the desired height is reached.

New growth regulators sprayed onto the leaf blades of grass slow down the rate of growth of turfgrasses. This cuts down on the number of times the lawn must be mowed. One of these growth regulators, called PRIMO MAXX, works by temporarily inhibiting or stopping the use of **gibberellic acid** by the turfgrass. As explained in Unit 5, gibberellic acid stimulates cell division and lengthening, causing plants to grow taller. PRIMO MAXX reduces mowing frequency by half. Grass becomes greener in color and more compact. These plant growth regulators work best on lawns that have a single grass type. Mixed grasses tend to grow back unevenly; one grass type may regrow faster than others. PRIMO MAXX is relatively safe to use and breaks down in the soil into carbon dioxide and water within 3 months. Details on new growth regulators are beyond the scope of this text, but you may want to research them further.

Watering the Lawn

A great deal of water is required to give lawns the moisture they need. An inch of water equals about ½ gallon of water on each square foot of lawn. Unless an adequate supply is available, the lawn should not be watered at all. Shallow watering does more harm than good because it causes the grass roots to move to the surface to absorb water. Here, they are more easily dried out and require still more frequent watering. Surface roots are more easily torn loose from the soil during winter freezing and thawing. Frequent watering also encourages the growth of fungus disease on the grass blades.

At least 1 inch of water should be applied at each application. An inch of water penetrates a loam soil about 16 inches. To determine when enough has been applied by a sprinkler, set a rain gauge or any container with straight sides on the lawn on the area to be watered. When the container has 1 inch of water in it, 1 inch has been applied to the lawn.

Do not apply water faster than it can soak into the soil surface; runoff is wasted water. Do not water until the grass wilts and needs water, and then water only if you have enough water to do the job well.

A healthy lawn can become dormant and withstand a great deal of dry weather without being permanently damaged. The grass will become green and active again with the first good rain. The appearance may not be as pleasing when the grass turns yellow or brown, but little actual damage is being done.

Solving Problems in the Lawn

Controlling Weeds

If a lawn is heavily infested with weeds, chemicals should be used to eliminate the problem. However, development of a thick, healthy turf is the best way to guard against a serious weed problem. Weeds in a lawn are usually an indication of an unhealthy lawn caused by poor maintenance practices. When weeds are spotted, first check maintenance practices, and correct them if necessary. If weeds are still a problem, consider chemical control.

Two types of chemical weed killers are used on lawns. One is a **preemergence** weed killer, which kills germinating seeds and very tiny weed seedlings before the weeds become established. This is the type of weed killer that is applied in the spring to cool-season grasses for the control of crabgrass. It is often mixed with a fertilizer before application. Because crabgrass is an annual grass, the mature plant cannot be controlled without killing the lawn grass. However, the tender seedlings can be killed with small amounts of weed killer that are not strong enough to kill the lawn grass. Do not use weed killer on a newly seeded lawn. Corn **gluten** meal (organic herbicide), a totally safe material found at most feed stores, can be used to control weed seedlings. Applied in the spring and fall at 20 pounds per 1,000 square feet, it prevents weed seed from growing. It will also kill newly planted lawn grass seed. Corn gluten meal also acts as a slow-release nitrogen fertilizer.

Postemergence weed killers are applied after the weeds sprout and begin to grow. A sprayer is generally used to apply these types of weed killers (Figure 36–5). Notice that the sprayer is labeled to avoid using the container for any other purpose. Because many weed killers are **selective** and kill only certain types of plants, it is possible to select a chemical that does not kill grass but does kill broadleaf weeds. One chemical of this type is 2,4-D, which is used in lawns to kill many broadleaf weeds, such as dandelions and plantain. For specific weed identification and control, contact your local extension service to obtain bulletins written for the local area.

▲ **Figure 36–5** Preventive spraying to control weeds.

Lawn Diseases and Their Control

Most turf diseases are caused by parasitic plants called fungi. These **parasites** live in and on dead grass and in the soil where they attack the green grass and rob the soil of nutrients. This weakens and can kill the lawn grass plant. Fungus diseases are spread easily by mowing or simply walking through the lawn, especially if the grass is wet. The tiny seedlike spores spread rapidly.

For fungus diseases to cause a serious problem, the following must be in the area:

- grass plants on which the fungus can live
- fungus spores and a means of spreading them to the grass
- temperature and moisture conditions favorable for the growth of fungi

Disease Prevention

The best control measure for fungus is prevention. To reduce the chance of attack by fungus spores on lawns, follow these guidelines:

- Do not overuse nitrogen fertilizer.
- Add lime or sulfur as needed to maintain a pH of 6.0 to 6.5. (This helps to control thatch buildup.)
- Avoid thatch buildup by collecting clippings or by raking and removing thatch.
- Water only when necessary, and then water deeply. Do not repeat watering for 1 week. Water at night. Most fungus diseases require 14 to 16 hours of wet grass to become infectious or spread. The grass may be wet at night from dew, and if watering is done in the morning, the wet period is extended long enough for disease to spread.
- Mow frequently, and remove only one-third of the top growth at any one mowing. Cut grass to the proper height, and keep it within 1 inch of that height throughout the entire growing season.
- Keep trees pruned to allow sufficient light for good growth.

Insect Control

Insects can cause serious damage to lawns. Some insects attack the top of the plant, such as the chinch bug, sod webworm, and flea beetle. Serious damage is also done by insects that attack grasses below the soil level by eating roots and stems. Some below-the-surface feeders are the grubs (larvae) of the Japanese beetle and many other beetles and cutworms (Figure 36–6).

When insect attacks are serious, spraying or application of **granular** insecticides may be

▲ **Figure 36–6A** Checking the turf for diseases.

▲ **Figure 36–6B** Grub infestation in the soil in the grass root area.

necessary. Consult local extension service bulletins for pictures to assist in identification of the pests and control recommendations.

Thatch Control

Thatch is a layer of dead stems, leaves, and roots of grass that builds up on the soil surface and under the green leaf area of grass. A thick layer of thatch may prevent water from penetrating the soil, prevent proper soil aeration, and provide a breeding area for insects and disease. Heavy thatch buildup and reduced aeration of the soil result in shallow root systems.

The amount of thatch buildup varies somewhat according to the variety of grass that is planted. For example, Merion Kentucky bluegrass builds up thatch faster than common bluegrass. Bentgrass and some red fescues also tend to build up thatch.

Other factors that may cause thatch to build up include the following:

- not adjusting the mower blade properly, resulting in too much of the grass blades being cut, or not using one of the newer thatching mowers
- returning clippings to the lawn
- heavy fertilization
- heavy clay soil
- acidic soil (apply lime)

On small areas, thatch may be removed with a hand rake. A better method is to use a power-driven thatch machine; it is well worth the rental fee. Early spring is usually the best time to remove thatch from the lawn. Another solution to the problem of thatch is a light application of topsoil (⅛ inch). This buries the thatch, causing it to rot. However, the best thatch control measures are preventive. Proper mowing, fertilizing, liming, and removal of clippings if a mulching mower is not used usually provide adequate control.

Compacted Soil

One of the most needed lawn maintenance activities, especially for heavy soils, is aeration. As lawns get older or experience heavy pedestrian traffic, the soils get packed down. This closes soil pores and keeps out both air and water. The grass roots are robbed of two very necessary elements for good growth: oxygen and water.

The use of a machine known as a plugger/aerator is the best treatment for compacted soils (Figure 36–7). This roller-type machine pulls out plugs of soil about ½ to ¾ inch in diameter and up to 2 inches long (Figure 36–8). This leaves holes in the lawn soil for air and water to enter.

Pluggers/aerators come in small, walk-behind sizes and large, tractor-mounted sizes (Figure 36–9). In addition, compacted lawn soils may also be improved by the application of gypsum ($CaSO_4$ [calcium sulfate]). Gypsum will help to improve the soil aeration to allow for improved growth of the lawn grass, and it will take several years to improve the soil structure. Gypsum does not affect the pH of the soil, but is an excellent organic product to loosen compacted soils.

▲ **Figure 36–7** A plugger/aerator used for soil aeration. Notice the hollow tubes that pick up and push out plugs of soil to produce holes in the lawn.

▲ **Figure 36–8** A plug of soil removed from the lawn. It is about ¾ inch in diameter and about 1 ½ to 2 inches long.

▲ **Figure 36–9** A well-cared for lawn being aerated to stimulate deep root growth and moisture intake.

Loosened Grass

Rollers are used to press down grass that may have been loosened, pushed, or heaved out of the ground by freezing and thawing during the winter. Rolling should be done in early spring after the last freezing and thawing. The roller should be heavy enough to push the grass firmly against the soil. Too much weight may make the soil too compact and reduce aeration in the soil. This is especially true for heavy clay soils.

Moss

Moss sometimes grows in shady, wet areas of lawns. It is not a weed. Moss grows only where grass has died for some reason. There are four problem areas that can result in the growth of moss in the lawn as described in Figure 36–10.

PROBLEM	SOLUTION
too much shade	Reduce shade or plant a variety of grass that is shade tolerant.
poor growth of grass	Fertilize properly. (A thick turf prevents the growth of moss.)
acidic soil conditions	Test soil and add lime.
a wet area (moss will grow in soil too wet for grass)	Fill low, wet areas with topsoil or drain the area.

Figure 36–10 Conditions that may result in the growth of moss in the lawn.

SAFETY FIRST

OSHA's regulations ensure that employers and employees know about work hazards and how to protect themselves to prevent injuries in the workplace. For more information on occupational health and safety in the workplace, please visit the OSHA site at www.osha.gov.

Summary

Proper lawn maintenance is important to having a quality lawn. There are seven factors to maintaining a quality turf: select the proper variety or species of grass; apply fertilizer and lime or sulfur at the proper time and in the proper form and amount; select the proper mowing height at the correct time; use proper watering techniques; use chemical or organic substances for weed, insect, and disease control if necessary; use the lawn in such a way that the traffic is not too heavy; and provide the proper amount of light.

Review Questions

1. Describe six management practices that reduce or help to control turfgrass diseases.
2. What are the seven factors that are required for proper lawn maintenance? Are some of these factors more important than others?
3. Why would you apply gypsum to lawn soil?
4. What are the main causes of fungus disease in lawns? How do you prevent or control them?
5. Discuss mowing considerations for both cool-season and warm-season grasses. What mistakes do homeowners typically make when mowing the lawn?
6. How do you analyze a turf fertilizer?

Student Activities

The following activities are to be accomplished on an area of lawn assigned by the instructor.

1. Examine an area on school property, and make a list of problems found in the lawn. Using a smartphone or digital camera, take pictures of the identified problems. Write down possible reasons for the problem areas. and report to the class. Topics included in the report might be weed control, fertilizer needs, and insect and disease problems.
2. Develop a lawn maintenance and/or renovation schedule for the lawn area. Have the schedule approved by the horticulture teacher.
3. Locate a good source of lime or sulfur and turf fertilizer. Apply each to the lawn area at the proper time and rate using both a rotary and conventional fertilizer spreader. Calculate how much fertilizer to spread on 1,000 square feet, and set the spreader for that application. Check your accuracy by seeing how much fertilizer remains after the job or if you run out of fertilizer before the job is finished.
4. Adjust a rotary mower blade to the best mowing height for the grass to be mowed. Repeat with a reel mower.
5. Examine a sharp and a dull mower blade. Sharpen or observe the proper sharpening technique of the dull blade as demonstrated by the instructor. Balance the blade after sharpening.

> **CAUTION** Be careful of the sharp cutting edge.

6. Conduct a demonstration using PRIMO MAXX to illustrate the use of plant growth regulators on lawns to reduce the total number of mowings during the lush cool growing season in the spring.
7. Identify weeds in the lawn area. Determine whether a weed killer is needed for control and, if so, which type. Apply the chemical. Observe and record the results.

8. Visit a local golf course. Ask the greenskeeper to discuss the difference in equipment and technique when maintaining greens compared to fairways. Ask questions about the height of the grass and adjustment of the blade, as well as the maintenance schedule used to keep the greens in order.
9. Search the Internet for college programs that specialize in lawn and turfgrass management. Give a description of two available programs.

Suggested Supervised Agricultural Experience (SAE) Activities

1. Establish a lawn care business as an entrepreneurship SAE.
2. To qualify for a placement SAE, work for a professional sports complex as part of the turfgrass management team.

Additional Resources

http://publications.tamu.edu/TURF_LANDSCAPE/PUB_turf_Maintaining_Bermudagrass_Lawns.pdf

http://umaine.edu/publications/2243e/

http://ucanr.edu/sites/scmg/files/29791.pdf

www.turf.msu.edu/mowing-lawn-turf

UNIT 37

Renovating the Lawn

Objective

To select and use the proper renovation technique for a specific problem in the lawn.

Key Terms

aerator
herbicide
overseeder
renovation
Verticut

Competencies to Develop

After studying this unit, you should be able to:

- inspect a lawn area and determine whether it requires renovation.
- determine which of the four methods of renovation should be used on a lawn.
- use the step-by-step approach to renovate a lawn.

Materials

- grass seed
- organic/chemical fertilizer and lime, if needed
- weed killer and sprayer or other applicator
- garden rake
- thatch rake or Verticut machine
- lawn mower
- nursery spade or sod-lifting tool
- straw, burlap, or commercial mulching materials for mulching seeded areas
- overseeder (slicer)
- smartphone apps:
  - Weather Channel
  - Spectracide
  - Pest Portal
  - Calibrate My Sprayer
  - PGR Mix Master
  - Ortho Problem Solver
  - LikeThat Garden
  - AG PhD

When good maintenance practices are not carried out, or weather conditions have prevented the growth of healthy grass plants, it is time to consider lawn **renovation**.

Why Do Lawns Fail?

Altering Soils

Modifications to the landscape in urban and suburban areas can alter or negatively affect the soil contents. The process of excavating, grading, back-blading, and soil removal creates variability of soil materials. Materials in the soil, as well as soil compaction, can be different over a few feet because the building contractor may leave behind plastic sheeting, waste cement, nails, drywall, lumber pieces, and chemical pollutants. All of this matter results in mechanical changes to the soil texture, leaving little to virtually no organic matter, little or no topsoil, restricted drainage, and severe soil pH that is more extreme than most normal soils. Therefore, before renovating a lawn in this situation, it is necessary to perform a soil analysis at a soil-testing laboratory to determine the pH and major plant nutrients of nitrogen, phosphorous, and potassium. If hydrated lime is used to enhance the soil, it will serve as insect repellent without causing any harm to pollinators or plants. Therefore, in many situations, this scenario creates a modest challenge to change.

The Wrong Species or Variety of Grass

Be sure to check the type of grass best suited for the area. It is especially important to consider soil and light conditions and the amount of traffic on the lawn. Before purchasing seed, check the label for percentages of the different varieties of seed it contains. Good seed is expensive, but it saves money over a period of time because the lawn will require less reseeding. Seed cost is a small part of the total cost of establishing a lawn.

Improper Mowing

The lawn may have been cut too short. Some people prefer a lawn that is cut very short, but cool-season grasses cannot survive if mowed to heights of ½ to 1½ inches. Grass should not be allowed to grow to be 3 inches high and be cut back to 1 inch. Only one-third of the top should be removed at any one mowing. Remember that close mowing does not reduce the number of times the lawn needs to be mowed.

Improper Fertilizing

The most common mistakes in fertilizing are applying the fertilizer too late in spring for cool-season grasses or too late in fall for warm-season grasses. Fertilizer should be applied a few weeks before the grass begins its active growing stage both in the spring and fall and again in late fall for cool-season grasses. Use a slow-release fertilizer that is high in nitrogen. Apply fertilizer at the proper rate. It is important to follow the soil analysis recommendations to improve the efficiency of the macronutrients and micronutrients in the soil for lawns, other horticultural crops, and environmental health.

GREEN TIP Coffee grounds are a natural fertilizer that can be sprinkled directly on your lawn. Or you can sprinkle them near your plants and dig them in a little. Also, cold coffee is a ready-to-go natural liquid fertilizer.

Improper Watering

The most common mistake in watering is failure to apply enough water to soak into the soil to a depth of from 4 to 6 inches. This requires at least 1 inch of water at each watering. One inch of water penetrates 24 inches in sandy soil, 16 inches in loamy soil, and 11 inches in clayey soil. As a general rule, lawns should be watered no more than once a week and then only when the soil is dry.

Heavy Traffic

Too much traffic on lawns causes compaction of the soil, especially on those planted with a grass variety not known for its durability, and can result in bare spots. To solve this problem, reduce traffic or change the grass variety. The new, tall fescue grasses take heavy traffic. Lawn areas with heavy traffic can be corrected by using an **aerator** to remove cores of soil to open the surface area of the lawn area.

Excessive Shade

Dense shade and competing shallow roots of trees, such as maples, kill lawn grasses because the roots of maples are surface feeding and compete with the lawn grass root system. In these cases, the tree, and thus the source of the shade, must be removed, or a shade-tolerant grass planted instead. Other options are to mulch out to the drip line of the tree branch spread or landscape with other vegetative ground covers beneath the tree.

Soil Not Prepared Properly for Planting

A good topsoil, proper fertilizer, pH adjustment, and proper planting techniques are essential for a healthy lawn. Test the soil to determine fertilizer (macronutrients) and pH adjustment needs.

Infestation of Weeds, Diseases, and Insects

Weed, disease, and insect problems are usually the result of one or more of the preceding problems. A thick, healthy lawn turf is seldom overcome by insects, diseases, or weeds. (Refer to the list of smartphone apps in the Materials section for additional information.)

Thatch

When old grass leaves, roots, and stems accumulate on the soil surface faster than they rot, the buildup is called *thatch*. Thatch buildup on the soil surface prevents air from circulating in the soil—a very necessary condition for root growth. Thatch also causes water to run off rather than soak into the soil and harbors insects and disease organisms. To help speed up thatch rotting, maintain a soil pH of 6.2 to 6.5, and apply more fertilizer. Another solution is to remove the thatch with a thatch-removing machine or a special thatching rake.

GREEN TIP Use a commercial aerator. Aerate the lawn area to reduce the thatch in the lawn and improve air circulation and water and nutrient movement to the grass roots.

TRY THIS Collect several soil samples from a potential site for an urban garden. Bag and label the samples with appropriate information, and send them to your state land grant university for a detailed soil analysis. After the results are returned, determine whether the garden site will be productive using the existing soil structure.

Renovation Techniques

If the lawn still fails to prosper, renovation or rebuilding of the lawn must be considered. Lawns should be renovated at the time of year best suited for starting a new lawn, that is, just before the most active growing season for the type of grass being planted in the renovation. The entire lawn may be reseeded, resodded, or planted by other vegetative means such as strip planting or with plugs, or a technique known as spot patching may be used.

Weed-Infested Lawns

When more than one-third of the lawn grass plants are dead, and large bare spots exist, it may be better to kill the existing grass with a nonselective **herbicide** such as Roundup (glyphosate) and reseed the entire area (Figure 37–1). After the grass is killed, the entire lawn area may be overseeded. (Refer to Unit 35 for details.)

▲ **Figure 37–1** A section of lawn with less than 50 percent of the area producing healthy lawn grass. If the entire lawn suffers from such severe infestation, it is best to kill all vegetation and reseed.

PROCEDURE
Renovating a Thinly Covered Lawn with Less Than 50 Percent Turfgrass Stand

1. Perform a soil test of the lawn area to be renovated.
2. Weed control: Use a nonselective weed control herbicide. Contact the local extension service for the herbicide to use.
3. Lime the lawn area according to the soil test results.
4. Fertilize the lawn with a turf starter fertilizer (slow release, sulfur coated) according to soil test results.
5. Use an **overseeder** (slicer) (Figures 37–2A and 37–2B) to improve the present turfgrass situation. The overseeder (slicer) will slice the turf and soil vertically, which will remove thatch and loosen the soil. The overseeder (slicer) has a grass seed box attachment. Set the rate of flow according to the recommendation for the type of grass seed being used. The grass seed will make contact with soil to establish the future turf.
6. Lawn roller: Use a lawn roller to firm the seed to ensure good soil contact with the seed. The compression of the lawn roller will also create a capillary action of water movement in the soil. This attracts moisture to the seed for germination.
7. Use a commercial pelletized mulch to help hold the moisture close to the seed.
8. Irrigate the seeded area with at least 1 inch of water. Set up a rain gauge to determine the amount of water applied.
9. Follow all good maintenance practices.

▲ **Figure 37–2A** Overseeder (slicer).

▲ **Figure 37–2B** Overseeder vertical cutting blades.

Thinly Covered Lawns

If the lawn grasses cover the entire lawn area but do not produce a thick turf, the lawn may be salvaged as follows, providing good topsoil is present, and drainage is not a problem.

Spot Seeding

When there are isolated spots of dead or weak grass, the lawn may be renovated by a "patchup" of these spots. Any spot that measures 1 foot or larger should be treated. The same procedure is used to renovate these spots as when seeding a new lawn. The only difference is that the process is used on a much smaller area.

Spot Patching

When good sod can be taken from inconspicuous areas, small spots in the lawn may be patched up

▲ Figure 37–3 A sod repair to a small dead area in the lawn.

▲ Figure 37–4 Replacing sod at the correct level.

by digging out the weak area about 1 inch deep and replacing it with a piece of sod cut to fit the spot (Figure 37–3). For large spots, more than one piece of sod may be fitted into the hole. Tamp the sod so that it rests tightly against the loosened soil in the bottom of the hole. Fertilize and water well to stimulate new root growth. It is important that the sod knit well and quickly to the new soil area, and that it not be allowed to dry out until its roots have grown into the soil (Figure 37–4).

GREEN TIP Lawns create curb appeal in the sustainable landscapes that produce oxygen for our environment, reducing the carbon footprint. The ecological approach to lawn care is vitally important to help protect our environment. The organic substances are key products to help reduce the deterioration of our living space.

PROCEDURE
Renovating a Lawn with Greater Than 50 Percent Turfgrass Stand

1. Perform a soil test of the lawn area to be renovated.
2. Weed control: Use a broadleaf weed control herbicide. Contact the local extension service for the herbicide to use.
3. Mow the area to be renovated closely and remove the clippings by raking.
4. Dethatch: Remove thatch with a hand thatching rake or a mechanical dethatcher (Verticut), and remove the dead grass thatched out of the lawn area.
5. Aerate: Use a mechanical aerator to remove plugs of soil (refer to Figures 36–7 and 36–8 in Unit 36). Aerate the area in north-south and east-west directions to establish the seedbed.

6. Lime the lawn area according to the soil test results.
7. Fertilize the lawn with a turf starter fertilizer according to the soil test results.
8. Apply the turfgrass seed with the broadcast (rotary) seeder.
9. Use a drag mat to break up the aerator cores to ensure soil contact with the seed.
10. Lawn roller: Use a lawn roller to firm the seed to ensure good soil contact with the seed. The compression of the lawn roller will also create a capillary action of water movement in the soil, which attracts moisture to the seed for germination.
11. Use a commercial pelletized mulch to help hold the moisture close to the seed.
12. Irrigate the seeded area with at least 1 inch of water. Set up a rain gauge to determine the amount of water applied.
13. Practice proper maintenance techniques.

SAFETY FIRST

Lawn-care equipment must be operated in a productive and safe manner. The application of lawn organic/chemical fertilizers must be applied according to the soil test analysis recommendations. OSHA's regulations ensure that employers and employees know about the work hazards and how to protect themselves to prevent injuries in the workplace. For more information on occupational health and safety in the workplace, please visit the OSHA site at www.osha.gov.

Summary

Renovation of a lawn requires the use of the proper techniques to solve the problem. Why did the lawn fail? Was it due to use of the wrong grass seed, improper mowing, improper fertilizing, improper watering, heavy traffic, excessive shade, soil not prepared properly for planting, infestation of pests, or a thatch problem? Inspect the lawn area to determine the renovation procedure to use. If the lawn has less than 50 percent turfgrass stand, the most appropriate procedure is to overseed the lawn. If the lawn has greater that 50 percent of the turfgrass stand, then aerate the lawn and seed. Some small weedy areas of lawn may be replaced with sod to ensure a quality turf.

Review Questions

1. Discuss some of the main reasons why lawns fail to thrive.
2. Explain how you determine which renovation method to use on a lawn that has failed.
3. Why is aeration beneficial to lawn grass?
4. Why is it important to use an ecological approach to lawn care?
5. Describe the step-by-step approach to renovating a lawn.

Student Activities

1. Take a field trip to observe lawns in the local area. Note lawns that are in need of renovation and ones that appear to be healthy. Inquire about the present condition of the lawns and their maintenance schedules. Determine why the lawns are in varying conditions and report to the class.
2. Establish a 1,000-square-foot plot for use in practicing lawn renovation. Keep notes of the problems found, treatment, and results of treatment.
3. Invite a nurseryman, golf course superintendent, or other lawn specialist to the class to discuss the most common lawn problems in the local area and the methods used to correct them.
4. Note any soil drainage, shade, or nutrient deficiencies that exist in the local area. Create a lawn improvement plan to correct the problems that exist within the lawn.
5. Search the Internet using the following terms: lawn renovation; overseeding a lawn; fertilizing a lawn; aeration of lawns. Create a list of recommendations to renovate a lawn area.

Suggested Supervised Agricultural Experience (SAE) Activities

1. To qualify for a placement SAE, work for a lawn care and restoration business during the summer months.
2. Work for a landscape business that designs the drainage and foundation of a lawn area as a placement SAE.

Additional Resources

www.ext.colostate.edu/pubs/garden/07241.html

www.extension.umn.edu/garden/yard-garden/lawns/lawn-renovation/

http://plantscience.psu.edu/research/centers/turf/extension/factsheets/renovation

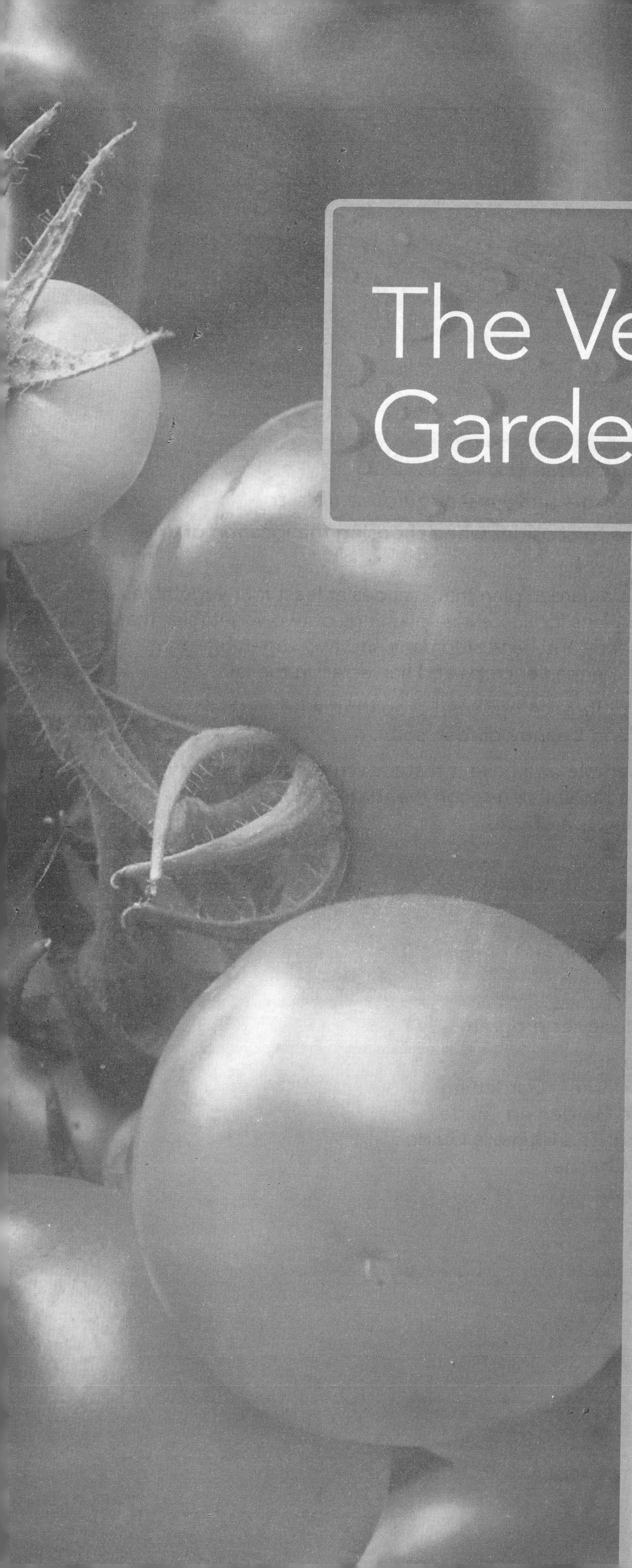

SECTION 9

The Vegetable Garden

UNIT 38 Planning and Preparing the Garden Site

Objective

To select and prepare a vegetable garden site.

Key Terms

annuals
banding
compost
crop rotation
germinate
hardiness
microorganisms
mulch
perennials
pH

Competencies to Develop

After studying this unit, you should be able to:

- list four items to consider when choosing the location of a vegetable garden.
- draw to scale a garden plan that includes at least four vegetables. Also include plans for successive plantings of two vegetables that are planted early and harvested early and two vegetables that are planted after earlier crops and harvested in the fall.
- explain the difference between preparing a heavy, clayey garden soil and a sandy garden soil.
- take a soil sample and have it tested. Write a fertilizer program according to the soil test recommendations for your garden plan with this type of soil.

Materials

- graph paper, ruler, and pencil
- seed catalogs
- computer designer program
- smartphone apps:
  - Vegetable Tree – Gardening
  - Vegetable Gardening Guide
  - Gardening: The Ultimate Guide
  - Weather Channel
  - Spectracide
  - Pest Portal
  - Calibrate My Sprayer
  - PGR Mix Master
  - Ortho Problem Solver

In accordance with the National Garden Association (NGA), of the 42 million households that grew some food products in 2013, 76 percent were vegetables, which was an increase of 19 percent over the past five years. No matter where you live, there is probably space for at least a small vegetable garden. Gardening is a good hobby that provides exercise, the satisfaction of growing some of your own food, and a savings on grocery bills. For the beginner, it is probably best to start with a small garden and work up in size as skill increases. The best techniques of planning and planting a garden are useless unless proper weed control, watering, and insect and disease control are practiced. Many of the skills developed in growing a small garden can be transferred to large-scale commercial vegetable production.

GREEN TIP Planning and preparing a garden site is an environmental dream come true. The use of the many natural products available gives gardeners the opportunity to help save our environment for future generations and to produce healthy soils, healthy food, healthy people, and a healthy planet.

Location

If irrigation of the garden is necessary, the vegetable garden must be located near a supply of water. More importantly, the location should have a healthy, loamy, well-drained soil and plenty of sunshine. Some vegetables will grow in partial shade, but most need full sun. A location in the shade of large trees is especially bad because the tree roots may extend into the garden and rob garden crops of needed plant food and moisture (Figure 38–1). If some of the garden plants must be grown in shade, leafy green vegetables and pumpkins suffer less from shade than other vegetables.

The garden should not be located on a steep hill where rainwater will run off rapidly because the runoff water is lost for use by the plants and also causes soil erosion. If the garden is located on a slope, form the rows along or around the hill instead of up and down the hill. This helps to prevent soil erosion and slows down water runoff, allowing it to soak into the soil.

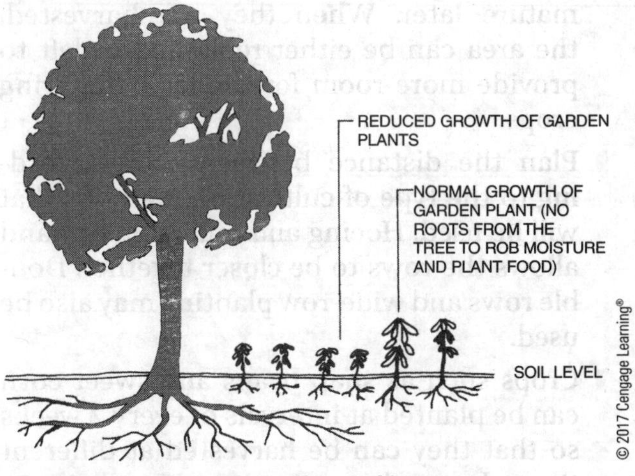

▲ **Figure 38–1** The effect of large trees on growth of garden crops. As shown in the illustration, tree roots usually extend farther than the limbs.

The Garden Plan

Deciding what to plant and how much to plant are important steps in vegetable gardening. First, list the vegetables you and your family like to eat. Decide if you want to plant enough to freeze or can for winter use, or only grow enough to eat during the summer. A garden plan saves time and work. Draw a sketch of the plot to scale, preferably on graph paper. Keep in mind that a small, well-kept garden gives better returns than a large, weedy one.

The following are some points to consider when planning the garden.

- Most garden vegetables are **annuals** that complete their life cycle in one year. A few are **perennials** that have roots living for many years, so they produce a crop each year.
- Plant perennials such as asparagus, rhubarb, horseradish, and sorrel together on one side of the garden where they will not interfere with the plowing and working of the rest of the garden.
- Group quickly maturing annuals together, or plant them between rows of crops that

mature later. When they are harvested, the area can be either replanted or left to provide more room for the later maturing crops.

- Plan the distance between rows according to the type of cultivation methods that will be used. Hoeing and mulching by hand allows the rows to be closer together. Double rows and wide-row planting may also be used.
- Crops such as snap beans and sweet corn can be planted at intervals of every 2 weeks so that they can be harvested at different times during the season.
- Replant areas where early crops, such as peas and lettuce, are harvested with fall crops, such as kale or turnips.

Beets, carrots, leaf lettuce, onions, early peas, radishes, and spinach can be planted in wide rows or in rows a foot apart (although 18 inches is preferable), but other vegetables must have more room. Plants set 18 inches apart with rows 2 feet apart is a good rule of thumb for most other vegetables except staked tomatoes, which must be planted about 3 feet apart in rows 3 feet apart. Sweet corn should be either in hills 3 feet apart each way or in rows 2 1/2 feet apart with plants spaced 1 foot apart. Vine crops (cucumber, melons, and squash) require 2 1/2 to 5 feet of space in all directions. For space savers, remember that cucumbers can be grown on a fence, and bush varieties of squash require less room than vine types. There are now new bush types of most melons, cucumbers, and squash.

Wide Row

A relatively new growing system for gardeners is the raised bed, wide-row system. Planting areas are raised above the existing soil level and the entire top of the wide row is planted in vegetable crops (Figure 38–2). The result is up to three or four times as much production per foot of row. Other advantages are that the raised beds dry out and warm up quicker in the spring, allowing earlier planting; there is deeper topsoil for plant roots, especially important for root crops such as carrots (Figure 38–3); and weeding is easier.

All of these spacing rules are for gardens that are hand cultivated. If a tractor or cultivator is

▲ **Figure 38–2** A garden laid out using beds for wide-row planting. The entire top of each of these beds will be planted with several rows of vegetable plants.

▲ **Figure 38–3** Carrots grown in a raised bed. Notice the carrots are much longer in the raised bed, thus producing more pounds of vegetables per foot of row.

to be used, more room must be allowed between rows. In a location with a considerable slope, make the rows across the grade to help prevent erosion during heavy rains. Use of mulch for weed control helps to control soil erosion on steep slopes, and no cultivation is necessary. Plant tall-growing vegetables on the north or west side of the plot to prevent shading of lower-growing varieties.

If your garden is tiny, a good garden plan is a half row each of radishes and leaf lettuce (to be followed by a succession planting of snap beans), a row of snap beans planted at the regular time, and a few tomato plants. With more space, plan to include snap beans; lima beans;

beets; broccoli; carrots; cabbage; a hill or two of cucumbers; greens such as spinach, Swiss chard, and leaf and head lettuce; a few early peas; and tomatoes. If there is plenty of room, include brussels sprouts, cauliflower, sweet corn, eggplant, melons, onions, peas, peppers, squash, and any other varieties of vegetables you especially like.

Crop Rotation

If possible, plan a **crop rotation** program for the garden. This will greatly reduce insect and disease problems. A 4-year rotation is best. The following areas are suggested: (1) a green manure crop, (2) high nitrogen feeders such as corn, (3) root crops, and (4) tomatoes, potatoes, eggplant, and other related plants. Plant corn or leafy vegetables such as lettuce after the green manure crop, and follow in sequence as numbered.

Mixed plantings of vegetables also help with insect and disease control. Bugs often zero in on plants by detecting odors associated with a specific plant. Mixed crops confuse bugs by the great number of odors or chemical stimuli.

This also works with underground pests. Researchers believe that the roots of pest-resistant varieties release compounds that repel or even kill insect larvae, protecting their roots as well as the roots of susceptible plants interplanted with them.

A garden guide from the W. Atlee Burpee Company is given in Figure 38–4 to help in the planning of a vegetable garden. The guide is designed to provide vegetables for four people.

Definition of Terms in Guide

- ***Hardiness*** is a measure of the crop's resistance to frost. See notes under the Growing Suggestions column.
- *Days to **germinate*** is the number of days from planting of the seed to the date on which the plant emerges.
- *Quantity to grow* is how much to plant for four. Increase or reduce to fit personal needs.
- *Days to harvest* refers to the number of days from planting of the seeds or plants in the garden to the date on which the crops are gathered.
- *Satisfactory pH range* is the soil acidity at which the plants grow best. Check soil with a soil test. Add lime or sulfur as recommended.
- *Uses* gives the best use for each vegetable.

The Garden Layout

Figure 38–5 shows a garden plan for spring and summer planting. Figure 38–6 shows a garden plan for the same garden later in the summer. Notice that the early maturing vegetables have been harvested and the same areas replanted with late summer and fall crops. Crops that were planted between tomatoes, such as broccoli, are harvested; however, no other crop is replanted in these places because the tomatoes need all the space.

> ***TRY THIS*** Create a vegetable garden design that would supply multiple families with produce. Draw and label each part of the design, including the spacing and varieties of vegetables that will be planted.

Garden Soil

The best garden soil is a healthy loamy one. It must have good drainage so that oxygen is available to roots, enabling the roots to penetrate the soil. A good supply of organic matter should be available to hold moisture and provide plant food. The soil should also have a good supply of plant food. The pH range (soil acidity) should be from 6.3 to 7.0 for most vegetables.

Recent research has definitely strengthened support for use of organic matter in the soil in the form of certified **compost**. Certified compost provides food for the many good microorganisms in the soil and ensures that the compost contains no detrimental contaminants, such as herbicides, or possible weed seeds. These beneficial **microorganisms** not only help control root diseases but also turn on the plant's defense system against leaf and stem diseases and insects. A healthy soil is alive with good microorganisms.

VEGETABLES AND TYPES	HARDINESS	DAYS TO GERMINATE	GROWING SUGGESTIONS	QUANTITY TO GROW	DAYS TO HARVEST	SATISFACTORY PH RANGE	USES
BEAN, snap bush and pole green, yellow	T	7–14	Sow bush types every 2 weeks until midsummer. Support pole types.	50 ft. bush; 8 hills pole	50–70	5.5–6.7	Fresh, frozen, canned. Vitamins A, B, C.
BEAN, bush shell red, white, green	T	7–14	Fava or English Broad Bean hardier than other types. Sow as early in spring as soil can be worked.	50 ft.	65–103	5.5–6.7	Fresh shell beans, or use dried for baking, soup, or Spanish or Mexican dishes. Vitamins A, B, C.
BEAN, lima bush and pole	T	7–14	Wait until ground is thoroughly warm before planting. Bush types mature earlier. Support pole varieties.	70 ft. bush; 8 hills pole	65–92	6.0–6.7	Fresh, frozen, canned, dried for baking. Vitamins A, B, C.
BEETS, red, golden, white	HH	10–21	For continuous harvest, make successive sowings until early summer. Do not transplant; this may cause forked or split roots.	25 ft.	55–80	6.0–7.5	Fresh, pickled, canned. Cook "thinnings" first, and tops later on for delicious greens. Vitamins A, B, C.
BROCCOLI	H	10–21	Plant again in midsummer for fall harvest. Grows best in cool weather.	25–40 plants	60–85*	5.5–6.7	Fresh, frozen. Vitamins A, B, C.
BRUSSELS SPROUTS	H	10–21	Pick lowest "sprouts" on stem each time; break off accompanying leaves but do not remove foliage.	25–40 plants	80–90*	5.5–6.7	Light frost improves flavor. Sprouts delicious fresh or frozen. Vitamins A, B, C.
CABBAGE, early, late, red, green	HH	10–21	Do not plant where any of the cabbage family grew the previous year.	25–40 plants	60–110*	5.5–6.7	Fresh, salads, coleslaw, sauerkraut. Winter storage. Vitamins A, B, C.
CARROT, long, short	HH	7–14	Short root types best for shallow or heavy soil. Plant again in midsummer for fall harvest.	25–30 ft.	65–75	5.2–6.7	Salads, relish, juice. Stews, soup. Vitamins A, B, C.

Note: See key on the last page of this figure.

Figure 38–4 Burpee vegetable garden guide. (*continued*)

VEGETABLES AND TYPES	HARDINESS	DAYS TO GERMINATE	GROWING SUGGESTIONS	QUANTITY TO GROW	DAYS TO HARVEST	SATISFACTORY PH RANGE	USES
CAULIFLOWER, white, purple	HH	10–21	Tie leaves over heads to whiten.	16–24	50–85*	6.0–6.7	Fresh, frozen, salad, relish. Vitamins A, B, C.
CELERY	HH	10–21	To whiten, mound soil up around mature stalks.	30–36	115–135*	5.5–6.7	Raw in salads and as relish. Cooked and creamed, soups. Vitamin A.
CHARD, red, white, stalked	HH	7–14	Pick frequently to encourage fresh leaves. Stands summer heat.	20–30 ft.	60	6.0–6.7	Cook leaves for greens; midribs and stalks like asparagus. Vitamins A, B, C.
COLLARDS	H	7–14	Easily grown, nonheading, cabbagelike leaves.	20–30 ft.	80	5.5–6.7	Cook leaves for greens. Popular in southern states. Vitamins A, B, C.
CRESS, garden and water	H	3–14	Sow in garden every 2 weeks for continuous supply. Also grows well on sunny windowsill. Grow watercress in moist, shady spots or along a shallow stream.	20–30 ft.	10–50	6.0–7.0	Salads, sandwiches, garnish, seasoning. Vitamins A, B, C.
CUCUMBERS, slicing, pickling	T	7–14	Grow on fence to save space. Keep picking to encourage new fruit.	8–12 hills	53–65	5.5–6.7	Salad, relish, pickles. Vitamin A.
EGGPLANT	T	10–21	Needs warm temperature—70 to 75°F for good germination. Pick fruits when skin has high gloss.	8–12 plants	62–75*	5.5–6.7	Delicious fried, sauteed, or in casseroles. Vitamin A.
ENDIVE	H	7–14	Grows best in cool weather.	20–30 ft.	90	6.0–7.0	Salad, greens. Hearts can be cooked and served with cream cheese or grated cheese. Vitamins A, B, C.

Figure 38–4 Burpee vegetable garden guide. *(continued)*

VEGETABLES AND TYPES	HARDINESS	DAYS TO GERMINATE	GROWING SUGGESTIONS	QUANTITY TO GROW	DAYS TO HARVEST	SATISFACTORY PH RANGE	USES
KALE	H	14–21	Mature plants take cold fall and winter weather. Frost improves flavor.	25–30 ft.	55–65	5.5–7.0	Chop young leaves for salads and sandwiches. Cook for greens. Vitamins A, B, C.
KOHLRABI	HH	14–21	Grow for spring or fall crop; thrives in cool weather.	16–20 ft.	55–60	5.5–6.7	Fresh, frozen; cooked like turnips. Vitamins A, B, C.
LEEK	H	14–21	Whiten and improve flavor by mounding soil around mature plants.	25–40 ft.	130	5.5–6.7	Fresh in salads. Cooked in soups, stews, or creamed.
LETTUCE, leaf	H	7–14	Make successive sowings in spring and another in late summer. Keep seedbed moist to get good germination for a fall crop.	25–40 ft.	40–47	6.0–7.0	Salad, sandwiches, garnish. Vitamins A, B, C.
LETTUCE, head	H	7–14	Needs cool weather in spring or fall to head well.	25–30 ft.	65–90	6.0–7.0	Salad, sandwiches, garnish. Vitamins A, B, C.
MUSTARD GREENS, fringed, smooth leaves	H	7–14	Grows as fall, winter, and spring crop in mild winter areas; spring and fall in north.	25–30 ft.	35–40	5.5–6.5	Greens. Vitamins A, B, C.
MELONS, cantaloupe, crenshaw, casaba, honeydew, watermelon	T	7–14	Very sensitive to frost. Black plastic mulch speeds maturity. Needs warm sunny weather when ripening for good flavor.	12–20 hills	75–120	6.0–6.7	Fresh, frozen. Ripe cantaloupes slip easily from stems. Ripe watermelons sound dull and hollow when tapped. Vitamins A, B, C.
OKRA	T	7–14	Needs hot weather to mature well. Pick pods young.	16–20 ft.	52–56	6.0–7.0	Soups, stews. Vitamins A, B, C.

Figure 38–4 Burpee vegetable garden guide. (*continued*)

VEGETABLES AND TYPES	HARDINESS	DAYS TO GERMINATE	GROWING SUGGESTIONS	QUANTITY TO GROW	DAYS TO HARVEST	SATISFACTORY PH RANGE	USES
ONIONS, yellow, white	H	10–21	Grows best in fine, well-drained sandy loam soil.	50–100 ft.	95–120	5.5–6.7	Fresh, salads, pickled. Vitamins B, C.
PARSLEY, curled or plain leaves	H	14–28	Attractive edging for flower garden; pot herb on sunny windowsill in winter.	10–20 ft.	72–90	6.0–7.5	Salad, garnish, seasoning. Dries or freezes well. Vitamins A, B, C.
PEA, dwarf, tall	H	7–14	Plant as early as ground can be worked.	40–100 ft.	55–79	5.5–6.7	Fresh, frozen, canned, dried. Vitamins A, B, C.
PEPPER, sweet, hot	T	10–21	Needs warm temperature—70 to 80°F for good germination.	8–10 plants	60–77*	5.5–6.5	Salad, stuffed, relish, seasoning. Vitamins A, B, C.
PUMPKIN, large, small bush, vine	T	7–14	For huge "contest" pumpkins, let only 1 or 2 grow per plant.	12–20 hills	95–120	5.5–6.5	Fresh, canned, frozen. Vitamins A, B, C.
RADISH, red, white, black	H	7–14	Make successive sowings until early summer; again a month before fall frost.	15–30 ft.	22–60	5.2–6.7	Relish, salad. Vitamins B, C.
RUTABAGA	H	14–21	Grows best in cool weather.	20–30 ft.	90	5.2–6.7	Fresh. Winter storage. Vitamins B, C.
SPINACH, crinkled, smooth	H	7–14	New Zealand and Malabar take hot weather; other varieties need cool weather.	20–40 ft.	42–70	6.0–6.7	Greens, frozen, canned. Vitamins A, B, C.
SQUASH, summer, bush, vine	T	7–14	Keep fruits picked so plants produce more.	8–12 hills	48–60	5.5–6.5	Fresh, frozen. Vitamin A.
SQUASH, winter, bush, vine	T	7–14	Black plastic mulch speeds maturity.	8–12 hills	80–120	5.5–6.5	Fresh, frozen, canned. Winter storage. Vitamin A.

Figure 38–4 Burpee vegetable garden guide. *(continued)*

VEGETABLES AND TYPES	HARDINESS	DAYS TO GERMINATE	GROWING SUGGESTIONS	QUANTITY TO GROW	DAYS TO HARVEST	SATISFACTORY PH RANGE	USES
SUNFLOWER	T	7–14	Use for screen plant. Protect maturing heads with bags to prevent bird damage.	25–50 ft.	80	6.0–7.5	Bird, poultry seed.
SWEET CORN, white, yellow	T	7–14	Plant in blocks of short rows for good pollination and well-filled ears.	50–100 ft.	63–90	5.2–6.7	Fresh, frozen, canned. Vitamins A, B, C.
TOMATO, red, pink, yellow	T	7–14	Hybrids especially need warm temperature—70 to 80°F for good germination.	16–20 plants	52–68*	5.2–6.7	Fresh, salad, canned, juice, pickled. Vitamins A, B, C.
TURNIPS, white, yellow	H	14–21	Grow best in cool weather.	20–30 ft.	35–60	5.2–6.7	Fresh, raw, or cooked. Leaves of some types for greens.

KEY
H — HARDY VARIETIES OF VEGETABLES AND FLOWERS—Tolerate cool weather and frost. Plant fall to early spring in zones G, H, I, J and 2–4 weeks before last killing spring frost in all other zones.
HH — HALF HARDY VARIETIES OF VEGETABLES AND FLOWERS—Tolerate cool weather and very light frost. For earlier maturity, varieties that transplant well can be started inside or in a cold frame 6–10 weeks before last expected light frost.
T — TENDER VARIETIES OF VEGETABLES AND FLOWERS—Cannot stand frost; plant in spring after last frost date.
* — Time from when plants are set into garden.

Figure 38–4 Burpee vegetable garden guide.

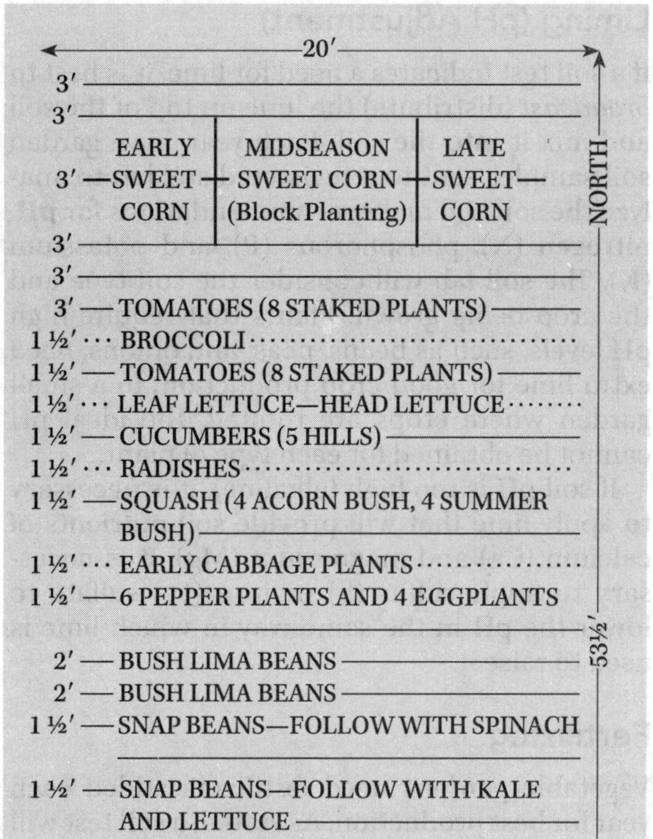

▲ **Figure 38–5** Suggested plan for a vegetable garden, 20 feet × 53½ feet, spring and summer. Corn is planted in blocks by variety for better pollination.

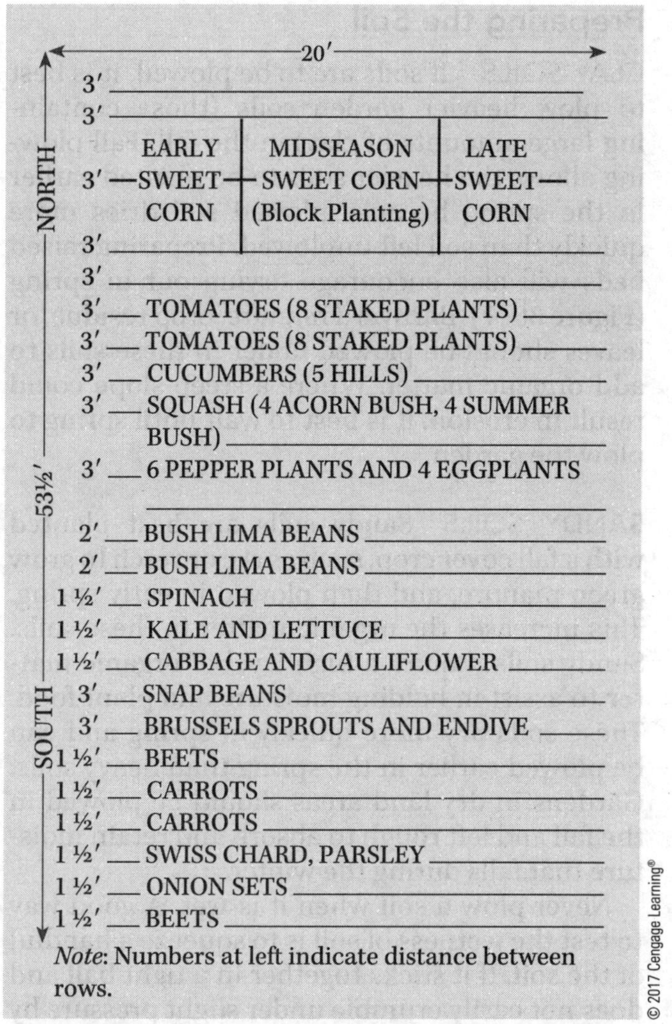

▲ **Figure 38–6** Suggested plan for a vegetable garden, 20 feet × 53½ feet, late summer and fall. This is how the garden in Figure 38–5 appears after quickly maturing varieties have been harvested and successive plantings have been made.

These good microorganisms kill disease-producing organisms or use up food and space in the soil so the disease-producing organisms have no space or food to live on.

A newer approach to gardening discourages tilling of the soil; rather the soil should always be covered with an organic **mulch**. As the mulch decays, it provides food for the good microorganism. As an alternative to yearly tilling, the garden should only be tilled when it is first started to get organic matter down into the soil if needed. After this first tilling, a mulch should be used to control weeds. The mulch can be pulled back to open rows for planting.

Preparing the Soil

CLAY SOILS If soils are to be plowed, it is best to plow heavier garden soils (those containing large amounts of clay) in the fall. Fall plowing allows the heavier soils to be planted earlier in the spring because plowed soil dries more quickly than soil left unplowed. Preparing raised beds will also encourage drying out in spring (Figure 38–7). Barnyard manure, crop residue, or leaves should be plowed under in these soils to add organic matter. Where a steep slope could result in erosion, it is best to wait until spring to plow the garden.

SANDY SOILS Sandy soils are best planted with a fall cover crop, such as rye or vetch to grow green manure, and then plowed in early spring. This increases the organic matter in these soils. Sandy soils require a high level of organic matter to assist in holding moisture and plant food. These soils dry more quickly in spring and can be plowed earlier in the spring than heavy soils. Gardens in dry land areas should be plowed in the fall and left rough to absorb and retain moisture that falls during the winter.

Never plow a soil when it is wet. A good way to test the wetness of soil is to squeeze a handful of the soil. If it sticks together in a tight ball and does not easily crumble under slight pressure by the thumb and finger, it is too wet for plowing or other working. When checking for wetness, check at the lower plow depth because soil may be dry on top and too wet on the bottom.

A soil test of the garden plot should be taken each year. Results of this test will show whether lime is needed to raise soil pH and how much of the major plant food elements are needed. Follow the recommendations given by the soil laboratory or the extension agent. Fall is a good time to take soil tests.

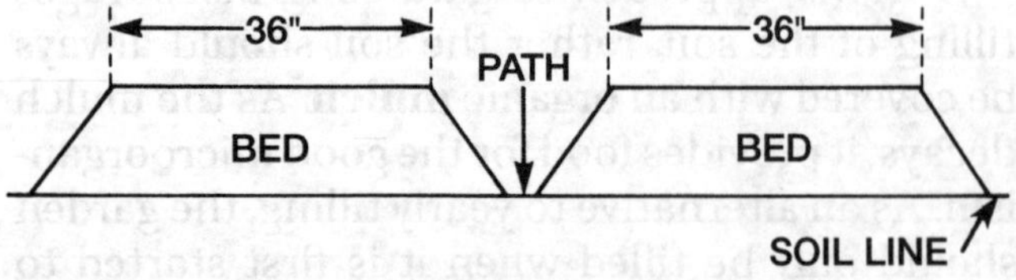

▲ **Figure 38–7** Raised beds allow earlier planting, especially in heavy clay soils. A 36-inch bed allows for wide planting or solid planting of small crops such as lettuce.

Liming (pH Adjustment)

If a soil test indicates a need for lime, it is best to *broadcast* (distribute) the lime on top of the soil and mix it into the soil. Each year, have garden soil samples sent to a registered soil lab to analyze the soil and make recommendations for **pH**, nitrogen (N), phosphorous (P), and potassium (K). The soil lab will consider the soil type and the crop being grown. Plants that require high pH levels, such as beans, peas, and onions, need extra lime for good crop production. In a small garden where crops are rotated, the ideal pH cannot be obtained for each type of plant.

If soil pH is too high (alkaline), it is necessary to apply lime that will provide soil nutrients of calcium (Ca) and magnesium (Mg). It is necessary to apply sulfur (S) or iron (Fe) sulfate to lower the pH in the same way in which lime is used to raise it.

Fertilizing

Vegetable gardens need fertilizers added each year for best production. As stated, a soil test will indicate the levels of nutrients available in the soil. From the soil test, the gardener can determine how much of each plant food element must be added to the garden soil.

To supply the major plant food elements, apply a good commercial fertilizer made for vegetable crops. The percentage of each of the three elements is given in the analysis on the bag. A 5-10-5 analysis indicates 5 percent nitrogen, 10 percent phosphate, and 5 percent potash. The plant food elements are always listed in this order. Good vegetable fertilizers are those with an analysis of 5-10-5, 5-10-10, 5-10-15, 10-10-10, and 10-6-4 Take a soil sample and have it analyzed by a certified soils laboratory to recommend the correct amount of N, P, K, and the pH level of the soil. Follow the recommendations of the soil test analysis.

In general, leafy vegetables and corn require larger amounts of nitrogen. The pod or fruit crops need more phosphorus fertilizer. Root crops need extra potash; therefore, it is important to have an analysis of the garden soil. Unless marked plant food deficiencies exist in the soil, it is not necessary to have special fertilizer for different crops in a small garden, which must be noted when filling out the soil analysis report sheet. In large commercial production, the analysis should match the crop needs as noted.

Applying Fertilizer

When compost and an organic mulch are used annually, no chemical fertilizer may be needed. Be sure to the follow the soil test recommendations from the soils lab.

BROADCASTING AND SIDE-DRESSING Vegetable crops may require fertilizer application as high as 5 pounds per 100 square feet. If this large an amount is added, it might be best to broadcast or spread evenly over the entire soil surface and mix half the amount into the soil before planting. Apply the rest later in the season as a *side-dress* on top of the soil on each side of the plants in the row, about 3 to 4 inches from the stem of the plant. (*Side-dressing* means placing fertilizer on the top of the soil around each plant.)

BANDING Another method for applying fertilizer, called **banding**, is to place the fertilizer in rows dug 3 inches from each side of the row of seeds or plants and slightly deeper than the depth at which the seeds are planted (Figure 38–8).

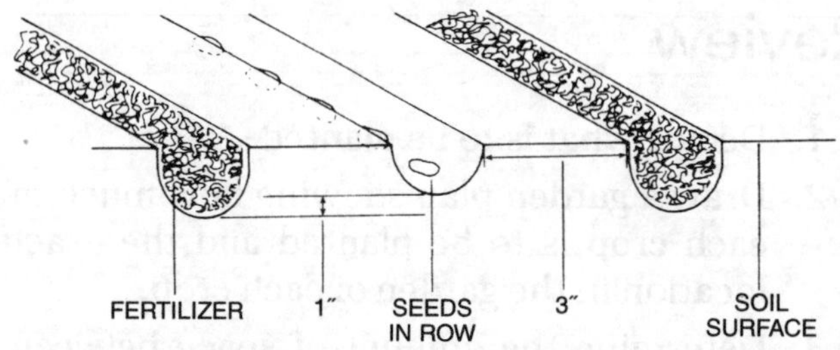

▲ **Figure 38–8** Banding fertilizer. The fertilizer is placed in bands 3 inches from each side of the seed row and 1 inch deeper than the depth of the seeds. Both the fertilizer and the seeds are covered with soil.

DIGGING IN If a soil tests very low in phosphorus and/or potash, it is best to dig in a fertilizer high in these elements. If applied on top of the soil, phosphorus and potash do not leach or wash down through the soil fast enough to be effective the same year. (If dug in one year, these elements remain in the soil for the following year.)

When fertilizer recommendations are given in pounds per acre, it is often difficult to translate the amounts into those for small rows. Figure 38–9 gives conversion figures for determining how much fertilizer to use on smaller plots.

WEIGHT OF FERTILIZER TO APPLY WHEN THE WEIGHT TO BE APPLIED PER ACRE IS:				
MEASUREMENT	100 POUNDS	400 POUNDS	800 POUNDS	1,200 POUNDS
Space between rows, and row length (in feet)	Pounds	Pounds	Pounds	Pounds
2 wide, 50 long	0.25	1.0	2.0	3.0
2 wide, 100 long	.50	2.0	4.0	6.0
2 ½ wide, 50 long	.30	1.2	2.4	3.6
2 ½ wide, 100 long	.60	2.4	4.8	7.2
3 wide, 50 long	.35	1.4	2.8	4.2
3 wide, 100 long	.70	2.8	5.6	8.4
Area (in square feet)				
100	.25	1.0	2.0	3.0
500	1.25	5.0	10.0	15.0
1,000	2.50	10.0	20.0	30.0
1,500	3.75	15.0	30.0	45.0
2,000	5.00	20.0	40.0	60.0

Figure 38–9 Approximate rates of fertilizer application per 50 or 100 feet of garden row and per 100 to 2,000 square feet of garden area, corresponding to given rates per acre.

Review

1. Decide what is to be planted.
2. Draw a garden plan showing how much of each crop is to be planted and the exact location in the garden of each crop.
3. Determine the amount of space between rows and whether or not interplanting between rows with early maturing crops is to be done.
4. Locate the garden close to water and in the sun on a loamy soil with good drainage.
5. Prepare the soil with organic matter by plowing under manure, compost, or cover crops.
6. Dig in fall or spring when the soil is dry enough to crumble or use the no-dig method.
7. Use lime or sulfur to correct pH. Dig into plowed or spaded soil.
8. Apply a good garden fertilizer or compost.

SAFETY FIRST

OSHA's regulations ensure that employers and employees know about the work hazards and how to protect themselves to prevent injuries in the workplace. For more information on occupational health and safety in the workplace, please visit the OSHA site at www.osha.gov.

Summary

Gardening is still one of the most popular and rewarding hobbies as well as a great source of nourishing food and exercise. It is important to plan a garden of proper size to furnish the volume of food needed and not too large to care for properly. Advances in organic soil management techniques now make it easier to produce healthy, pesticide-free produce. The organic gardeners were right—if you feed soil, the soil and its microorganisms feed and protect the plant.

Review Questions

1. Describe how to tell if a soil is too wet to plow.
2. What are the four most important items to consider when selecting a garden site?
3. Why is crop rotation important when growing vegetables?
4. Discuss how mulch is used in the vegetable garden.
5. What is the difference between broadcasting and side-dressing when applying fertilizer?
6. Discuss the difference in approaches you would take when preparing a heavy, clayey garden soil versus a sandy garden soil.

Student Activities

1. Draw a garden plan for a family garden using short-season crops.
2. Study a vegetable variety recommendation fact sheet from a local university or extension agent. Email an extension service in some other area of the country and request a fact sheet, or find one on the appropriate websites.
3. Study a garden seed catalog. Select and list the best vegetable varieties for your area. List the price of the seed to purchase.
4. Select a garden site to transform your plan into reality. Actually plant and care for the garden if possible. This may be a group or class project.
5. Research online to find examples of small-scale integrated pest management (IPM) plans for vegetable gardens. Select the one that will work for your designs.
6. For a great fund-raising activity, check out http://PowerfulPlants.net.

Suggested Supervised Agricultural Experience (SAE) Activities

1. Design, grow, and maintain a vegetable garden. Harvest and sell the vegetables at a farm stand to qualify for an entrepreneurship SAE.
2. Work for a vegetable farm selling produce at a farmer's market booth. Record your hours and income as a placement SAE.

Additional Resources

http://web.extension.illinois.edu/vegguide/

https://extension.umd.edu/learn/vegetable-garden-planning-2013

http://gardening.wsu.edu/vegetable-gardens/

www.gardening.cornell.edu/homegardening/scene0391.html

UNIT 39 Planting the Vegetable Garden

Objective

To demonstrate techniques for the planting of a small vegetable garden.

Key Terms

allelopathic
bag garden
bolting
cold frame
double cropping
floating row cover
high tunnels
phenology

Competencies to Develop

After studying this unit, you should be able to:

- list the best varieties of 10 vegetables for planting in the local area, including at least one disease-resistant variety for each.
- use the frost-free map and planting charts to determine planting dates for the 10 vegetables.
- list five vegetables that are direct seeded in the garden and five that must be seeded indoors for transplanting.
- describe the steps in seeding vegetables for transplanting outdoors.
- explain phenology in writing.

Materials

- listing of vegetable varieties recommended for the local area
- seed catalogs
- garden seeds
- containers for starting seeds to grow plants for transplanting (flats, Jiffy 7, etc.)
- garden rake, garden spade, garden hoe
- string for lining out rows
- garden hose
- aluminum foil or cardboard for cutworm collars
- 2.5 cu. ft. bag of topsoil
- smartphone apps:
  - Weather
  - Spectracide
  - Pest Portal
  - Calibrate My Sprayer
  - PGR Mix Master
  - Ortho Problem Solver
  - Vegetable Tree – Gardening Guide
  - Vegetable Gardening Guide
  - Gardening: The Ultimate Guide

The demand for fresh vegetables has grown at a phenomenal rate. According to the National Gardening Association (NGA), food gardening is at its highest level since more than a decade ago. In today's world, 50 percent of the people who garden grow their vegetables in containers. **Bag gardens**, or gardens grown in containers, offer gardeners the opportunity to grow produce on patios, rooftops, atriums, and foyers. A research study by Missouri Botanical Garden found that the top 10 vegetables grown in containers are beans, beets, chard, greens, eggplant, lettuce, peppers, radish, spinach, and tomatoes.

Selecting Varieties

One of the most important steps in the preparation of any garden is selecting the proper variety of vegetables to plant. For example, a smartphone app for vegetable garden seeds, online research on vegetable garden seeds, and seed catalogs may list dozens of varieties of sweet corn or tomatoes. Which is best for the area in which you live? An individual who sells seeds locally would probably be willing to offer advice concerning varieties. A local extension agent or horticulture teacher can also help to determine which varieties are recommended for the area. Consider the use of hybrid varieties for more vigorous plants and a higher yield.

It pays to purchase seed from a reputable professional and not to depend on home supplies. Homegrown seed may carry diseases, or cross-pollination may result in seeds that do not come true to variety. The seed from hybrid vegetables does not produce plants that are true to seed. The cost of seed is very low compared with the total cost of producing vegetables; it pays to start with the best.

Disease-resistant varieties should be used whenever a disease is known to be a problem in the area. The selection of disease-resistant varieties can minimize spraying and low yields at very little, if any, extra cost to the grower. Hybrid varieties are generally more vigorous and give higher yields for the same effort and cost. Hybrids should be used whenever possible.

Fresh seed should always be used. Read the date on the package to be sure that the seed was packed for the current year. Some seeds retain their vitality longer than others and may grow after being held over for 1 year. Percentage of germination will probably be lower than that listed on the package for first-year planting, however.

Vegetable seeds may be divided into three classes: (1) *short-lived* (do not germinate well after 1 or 2 years)—corn, leek, onion, parsley, rhubarb, and salsify; (2) *moderately long-lived* (often good for 3 to 5 years)—asparagus, beans, brussels sprouts, cabbage, carrot, cauliflower, celery, kale, lettuce, okra, peas, pepper, radish, spinach, turnip, and watermelon; and (3) *long-lived* (may be good for more than 5 years)—beet, cucumber, eggplant, muskmelon, and tomato. Any seed over 1 year old should be given a germination test to see how many seeds actually sprout vigorously. This should be done before any of the seeds are planted in the garden. *Always buy seeds that are dated for planting that year.*

Seed stored in a dry, cool, dark place remains good longer than seed that is stored in an open container in a warm, moist place. How seeds are stored helps to determine the length of active life and causes the figures given for the three classes of seed (short-lived, moderately long-lived, and long-lived) to vary considerably. These figures pertain to seed stored under good storage conditions.

Determining Planting Time

Planting seeds or transplanting plants at the proper time for the locality is very important. Temperatures can differ greatly between areas that are not many miles apart, causing the planting date for a particular vegetable to differ by as much as 1 or 2 weeks.

Vegetable crops may be grouped according to their hardiness (tolerance to cold). Figure 39–1 is a timetable listing the planting dates for some vegetable crops. The frost-free date in the spring of any locality is about the time that the oak trees in that area leaf out about 1 inch long or the American dogwood blooms.

Dr. R. J. Happ of the University of Vermont used the Persian Lilac as a planting guide. When the lilac was in its first leaf stages, cool-season crops such as lettuce, peas, and root crops could be safely planted; in its full bloom stage, warm-season plants such as beans and cucumbers also could be safely planted. By being observant, you can find plants in your area to use as frost-free guides.

COLD-HARDY PLANTS FOR EARLY SPRING PLANTINGS		COLD-TENDER OR HEAT-HARDY PLANTS FOR LATER SPRING OR EARLY SUMMER PLANTING			
Very hardy (Plant 4 to 6 weeks before frost-free date.)	**Hardy (Plant 2 to 4 weeks before frost-free date.)**	**Not cold-hardy (Plant on frost-free date.)**	**Requiring hot weather (Plant 1 week or more after frost-free date.)**	**Medium heat-tolerant (Good for summer planting.)**	**Hardy plants for late summer or fall planting except in the northern region (Plant 6 to 8 weeks before first fall freeze.)**
Broccoli	Beets	Beans, snap	Beans, lima	Beans, all	Beets
Cabbage	Broccoli	Okra	Eggplant	Chard	Broccoli
Collards	Cabbage	New Zealand spinach	Peppers	Soybeans	Brussels sprouts
Kale	Carrots	Soybeans	Sweet potatoes	New Zealand spinach	Cabbage
Leek	Cauliflower	Squash	Cucumbers	Squash	Collards
Lettuce	Celery	Sweet corn	Melons	Sweet corn	Kale
Onions	Chard	Tomatoes			Lettuce
Peas	Mustard				Mustard
Potatoes	Parsnip				Spinach
Radishes	Radishes				Turnips
Spinach					
Turnips					

Figure 39–1 Common vegetables grouped according to the approximate times they can be planted and their relative requirements for cool weather.

GREEN TIP Planting a vegetable garden provides the ultimate opportunity to learn and practice how to care for nature and harvest the reward of a productive resource to feed others.

Phenology

There are tables and charts with frost-free dates, but these dates vary from year to year. One very accurate method to determine planting dates is to watch outdoor plants and their response to climate.

All plants respond to degree days of temperature in starting growth, and most plants leaf out or flower when freezing weather is past. This relationship between climate and the life cycle of plants is known as **phenology**. Gardeners have been using these observations to guide them in time to plant for many years. Plants know best about what is going on around them. They integrate all of the environmental factors better than our computers do. For example, some seeds germinate only when the soil reaches a certain temperature.

A few examples of relationships among plants for determining planting time are as follows:

- Peas—when forsythia blooms
- Potatoes—when the first dandelion blooms
- Beets, carrots, broccoli, cabbage, lettuce, and spinach—when lilac is in first leaf
- Corn—when white oak leaves are the size of a squirrel's ear
- Beans, cucumbers, and squash—when lilac is in full bloom
- Tomatoes—when lily of the valley is in full bloom

It is important that the first planting of vegetables be as early in the spring as possible without danger of cold damage. Some vegetables can be planted before the frost-free date; others cannot. Those vegetables indicated for planting earlier than the frost-free date grow best in cool weather and cannot tolerate summer heat.

A gardener anywhere in the United States can determine the safe planting date for his or her particular area for different crops by using the information found in Figures 39–2, 39–3, 39–4, and 39–5. The maps show the average dates of the last killing frost in spring (Figure 39–2) and the first killing frost in fall (Figure 39–3). Specific planting dates are determined by plugging these dates into Figures 39–4 and 39–5.

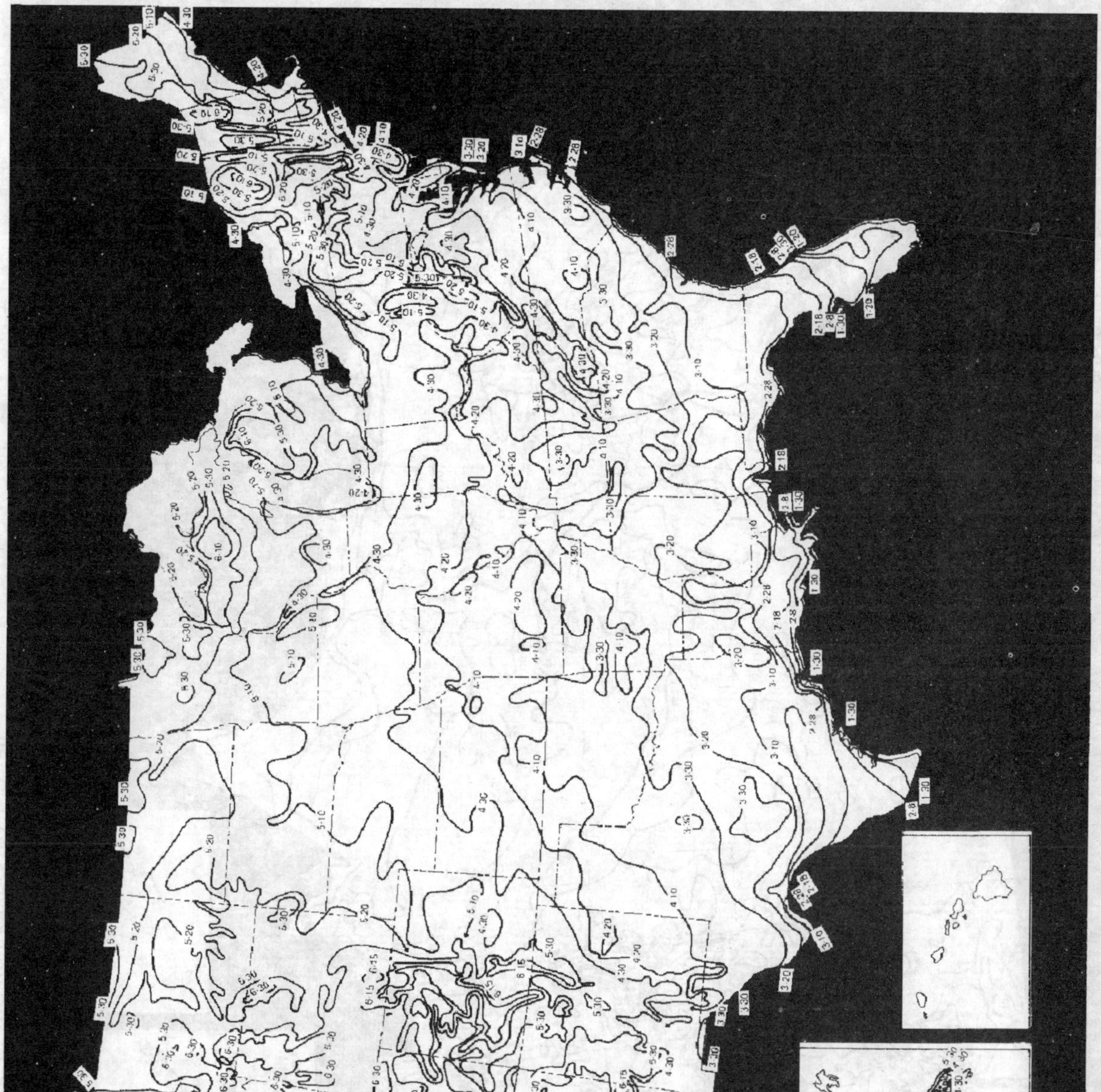

▲ **Figure 39–2** Average dates of the last killing frost in spring.
Note: Numbers such as 3-30 mean the third month (March), 30th day.

Source: US Department of Agriculture Home and Garden Bulletin 202.

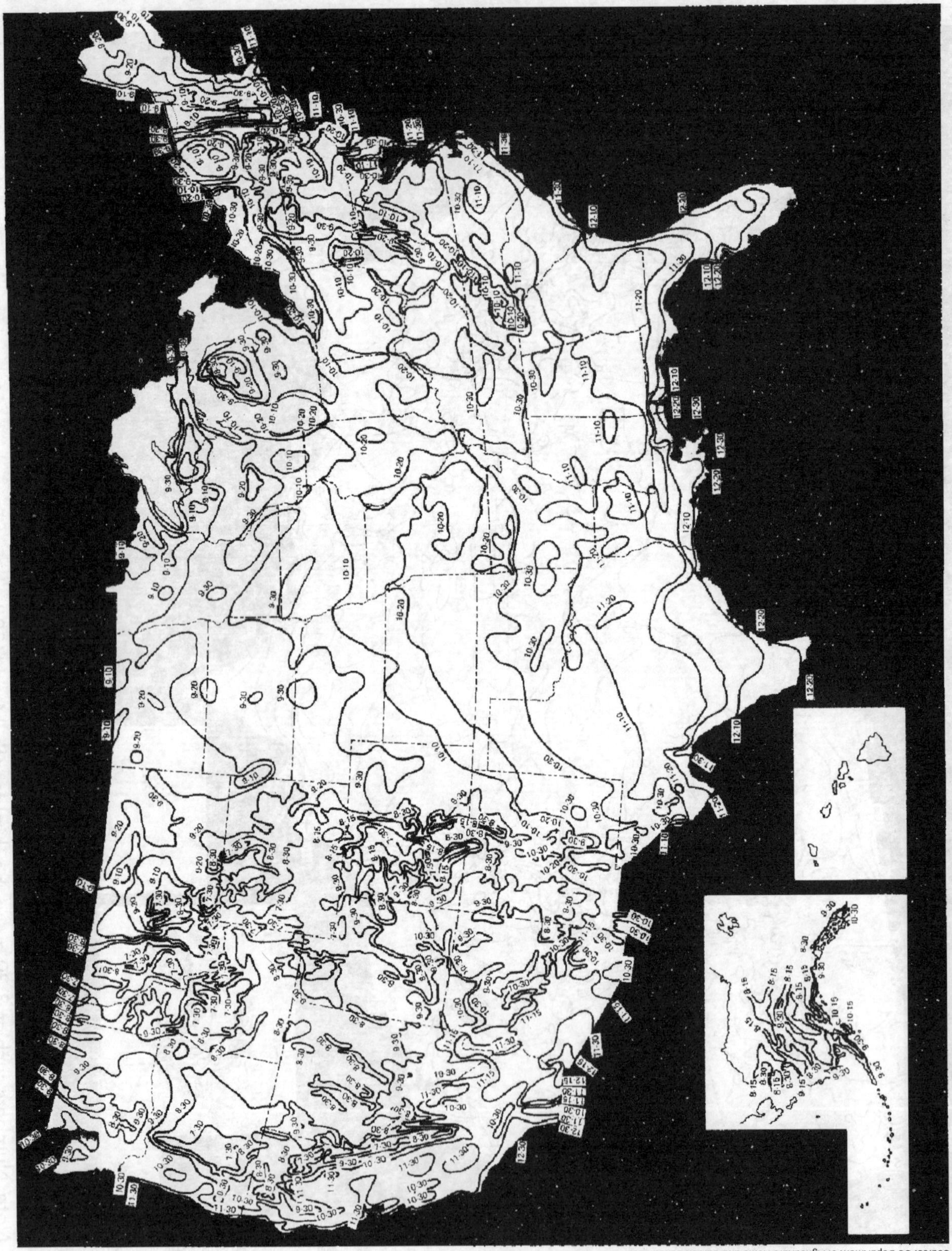

Source: US Department of Agriculture Home and Garden Bulletin 202.

Figure 39–3 Average dates of the first killing frost in fall.

Crop	Jan. 30	Feb. 8	Feb. 18	Feb. 28	Mar. 10	Mar. 20	Mar. 30
	SPRING PLANTING DATES FOR LOCALITIES IN WHICH AVERAGE DATE OF LAST FREEZE IS						
Asparagus[1]					Jan. 1–Mar. 1	Feb. 1–Mar. 10	Feb. 15–Mar. 20
Beans, lima	Feb. 1–Apr. 15	Feb. 10–May 1	Mar. 1–May 1	Mar. 15–June 1	Mar. 20–June 1	Apr. 1–June 15	Apr. 15–June 20
Beans, snap	Feb. 1–Apr. 1	Feb. 1–May 1	Mar. 1–May 1	Mar. 10–May 15	Mar. 15–May 15	Mar. 15–May 25	Apr. 1–June 1
Beet	Jan. 1–Mar. 15	Jan. 10–Mar. 15	Jan. 20–Apr. 1	Feb. 1–Apr. 15	Feb. 15–June 1	Feb. 15–May 15	Mar. 1–June
Broccoli, sprouting[1]	Jan. 1–30	Jan. 1–30	Jan. 15–Feb. 15	Feb. 1–Mar. 1	Feb. 15–Mar. 15	Feb. 15–Mar. 15	Mar. 1–20
Brussels sprouts[1]	Jan. 1–30	Jan. 1–30	Jan. 15–Feb. 15	Feb. 1–Mar. 1	Feb. 15–Mar. 15	Feb. 15–Mar. 15	Mar. 1–20
Cabbage[1]	Jan. 1–15	Jan. 1–Feb. 10	Jan. 1–Feb. 25	Jan. 15–Feb. 25	Jan. 25–Mar. 1	Feb. 1–Mar. 1	Feb. 15–Mar. 10
Cabbage, Chinese	(2)	(2)	(2)	(2)	(2)	(2)	(2)
Carrot	Jan. 1–Mar. 1	Jan. 1–Mar. 1	Jan. 15–Mar. 1	Feb. 1–Mar. 1	Feb. 10–Mar. 15	Feb. 15–Mar. 20	Mar. 1–Apr. 10
Cauliflower[1]	Jan. 1–Feb. 1	Jan. 1–Feb. 1	Jan. 10–Feb. 10	Jan. 20–Feb. 20	Feb. 1–Mar. 1	Feb. 10–Mar. 10	Feb. 10–Mar. 20
Celery and celeriac	Jan. 1–Feb. 1	Jan. 10–Feb. 10	Jan. 20–Feb. 20	Feb. 1–Mar. 1	Feb. 10–Mar. 20	Mar. 1–Apr. 1	Mar. 15–Apr. 15
Chard	Jan. 1–Apr. 1	Jan. 10–Apr. 1	Jan. 20–Apr. 15	Feb. 1–May 1	Feb. 15–May 15	Feb. 20–May 15	Mar. 1–May 25
Chervil and chives	Jan. 1–Feb. 1	Jan. 1–Feb. 1	Jan. 1–Feb. 1	Jan. 15–Feb. 15	Feb. 1–Mar. 1	Feb. 10–Mar. 10	Feb. 15–Mar. 15
Chicory, witloof					June 1–July 1	June 1–July 1	June 1–July 1
Collards[1]	Jan. 1–Feb. 15	Jan. 1–Feb. 15	Jan. 1–Mar. 15	Jan. 15–Mar. 15	Feb. 1–Apr. 1	Feb. 15–May 1	Mar. 1–June 1
Corn, sweet	Feb. 1–Mar. 15	Feb. 10–Apr. 1	Feb. 20–Apr. 15	Mar. 1–Apr. 15	Mar. 10–Apr. 15	Mar. 15–May 1	Mar. 25–May 15
Cornsalad	Jan. 1–Feb. 15	Jan. 1–Feb. 15	Jan. 1–Mar. 15	Jan. 1–Mar. 1	Jan. 1–Mar. 15	Jan. 1–Mar. 15	Jan. 15–Mar. 15
Cress, upland	Jan. 1–Feb. 1	Jan. 1–Feb. 15	Jan. 15–Feb. 15	Feb. 1–Mar. 1	Feb. 10–Mar. 15	Feb. 20–Mar. 15	Mar. 1–Apr. 1
Cucumber	Feb. 15–Mar. 15	Feb. 15–Apr. 1	Feb. 15–Apr. 15	Mar. 1–Apr. 15	Mar. 15–Apr. 15	Apr. 1–May 1	Apr. 10–May 15
Eggplant[1]	Feb. 1–Mar. 1	Feb. 10–Mar. 15	Feb. 20–Apr. 1	Mar. 10–Apr. 15	Mar. 15–Apr. 15	Apr. 1–May 1	Apr. 15–May 15
Endive	Jan. 1–Mar. 1	Jan. 1–Mar. 1	Jan. 15–Mar. 1	Feb. 1–Mar. 1	Feb. 15–Mar. 15	Mar. 1–Apr. 1	Mar. 10–Apr. 10
Fennel, Florence	Jan. 1–Mar. 1	Jan. 1–Mar. 1	Jan. 15–Mar. 1	Feb. 1–Mar. 1	Feb. 15–Mar. 15	Mar. 1–Apr. 1	Mar. 10–Apr. 10
Garlic	(2)	(2)	(2)	(2)	(2)	Feb. 1–Mar. 1	Feb. 10–Mar. 10
Horseradish[1]							Mar. 1–Apr. 1
Kale	Jan. 1–Feb. 1	Jan. 10–Feb. 1	Jan. 20–Feb. 10	Feb. 1–20	Feb. 10–Mar. 1	Feb. 20–Mar. 10	Mar. 1–20
Kohlrabi	Jan. 1–Feb. 1	Jan. 10–Feb. 1	Jan. 20–Feb. 10	Feb. 1–20	Feb. 10–Mar. 1	Feb. 20–Mar. 10	Mar. 1–Apr. 1
Leek	Jan. 1–Feb. 1	Jan. 1–Feb. 1	Jan. 1–Feb. 15	Jan. 15–Feb. 15	Jan. 25–Mar. 1	Feb. 1–Mar. 1	Feb. 15–Mar. 15
Lettuce, head[1]	Jan. 1–Feb. 1	Jan. 1–Feb. 1	Jan. 1–Feb. 1	Jan. 15–Feb. 15	Feb. 1–20	Feb. 15–Mar. 10	Mar. 1–20
Lettuce, leaf	Jan. 1–Feb. 1	Jan. 1–Feb. 1	Jan. 1–Mar. 15	Jan. 1–Mar. 15	Jan. 15–Apr. 1	Feb. 1–Apr. 1	Feb. 15–Apr. 15
Muskmelon[2]	Feb. 15–Mar. 15	Feb. 15–Apr. 1	Feb. 15–Apr. 15	Mar. 1–Apr. 15	Mar. 15–Apr. 15	Apr. 1–May 1	Apr. 10–May 15
Mustard	Jan. 1–Mar. 1	Jan. 1–Mar. 1	Feb. 15–Apr. 15	Feb. 1–Mar. 1	Feb. 10–Mar. 15	Feb. 20–Apr. 1	Mar. 1–Apr. 15
Okra	Feb. 15–Apr. 1	Feb. 15–Apr. 15	Mar. 1–June 1	Mar. 10–June 1	Mar. 20–June 1	Apr. 1–June 15	Apr. 10–June 15
Onion[1]	Jan. 1–15	Jan. 1–15	Jan. 1–15	Jan. 1–Feb. 1	Jan. 15–Feb. 15	Feb. 10–Mar. 10	Feb. 15–Mar. 15
Onion, seed	Jan. 1–15	Jan. 1–15	Jan. 1–15	Jan. 1–Feb. 15	Feb. 1–Mar. 1	Feb. 10–Mar. 10	Feb. 20–Mar. 15
Onion, sets	Jan. 1–15	Jan. 1–15	Jan. 1–15	Jan. 1–Mar. 1	Jan. 15–Mar. 10	Feb. 1–Mar. 20	Feb. 15–Mar. 20
Parsley	Jan. 1–30	Jan. 1–30	Jan. 1–30	Jan. 15–Mar. 1	Feb. 1–Mar. 10	Feb. 15–Mar. 15	Mar. 1–Apr. 1
Parsnip			Jan. 1–Feb. 1	Jan. 15–Feb. 15	Jan. 15–Mar. 1	Feb. 15–Mar. 15	Mar. 1–Apr. 1
Peas, black-eye	Feb. 15–May 1	Feb. 15–May 15	Mar. 1–June 15	Mar. 10–June 20	Mar. 15–July 1	Apr. 1–July 1	Apr. 15–July 1

[1]Plants.
[2]Included in this section because no appropriate coverage is available elsewhere.

Figure 39–4 Earliest dates and ranges of dates for safe spring planting of vegetables in the open. *(continued)*

Source: US Department of Agriculture Home and Garden Bulletin 202.

SPRING PLANTING DATES FOR LOCALITIES IN WHICH AVERAGE DATE OF LAST FREEZE IS

Crop	Jan. 30	Feb. 8	Feb. 18	Feb. 28	Mar. 10	Mar. 20	Mar. 30
Peas, garden	Jan. 1–Feb. 15	Jan. 1–Feb. 15	Jan. 1–Mar. 1	Jan. 15–Mar. 1	Jan. 15–Mar. 15	Feb. 1–Mar. 15	Feb. 10–Mar. 20
Pepper[1]	Feb. 1–Apr. 1	Feb. 15–Apr. 15	Mar. 1–May 1	Mar. 15–May 1	Apr. 1–June 1	Apr. 10–June 1	Apr. 15–June 1
Potato	Jan. 1–Feb. 15	Jan. 1–Feb. 15	Jan. 15–Mar. 1	Jan. 15–Mar. 1	Feb. 1–Mar. 1	Feb. 10–Mar. 15	Feb. 20–Mar. 20
Radish	Jan. 1–Apr. 1	Jan. 1–Apr. 1	Jan. 1–Apr. 1	Jan. 1–Apr. 1	Jan. 1–Apr. 15	Jan. 20–May 1	Feb. 15–May 1
Rhubarb[1]							
Rutabaga				Jan. 1–Feb. 1	Jan. 15–Feb. 15	Jan. 15–Mar. 1	Feb. 1–Mar. 1
Salsify	Jan. 1–Feb. 1	Jan. 10–Feb. 10	Jan. 15–Feb. 20	Jan. 15–Mar. 1	Feb. 1–Mar. 1	Feb. 15–Mar. 1	Mar. 1–15
Shallot	Jan. 1–Feb. 1	Jan. 1–Feb. 10	Jan. 1–Feb. 20	Jan. 1–Mar. 1	Jan. 15–Mar. 1	Feb. 1–Mar. 10	Feb. 15–Mar. 15
Sorrel	Jan. 1–Mar. 1	Jan. 1–Mar. 1	Jan. 15–Mar. 1	Feb. 1–Mar. 10	Feb. 10–Mar. 15	Feb. 10–Mar. 20	Feb. 20–Apr. 1
Soybean	Mar. 1–June 30	Mar. 1–June 30	Mar. 10–June 30	Mar. 20–June 30	Apr. 10–June 30	Apr. 10–June 30	Apr. 20–June 30
Spinach	Jan. 1–Feb. 15	Jan. 1–Feb. 15	Jan. 1–Mar. 1	Jan. 1–Mar. 1	Jan. 15–Mar. 10	Jan. 15–Mar. 15	Feb. 1–Mar. 20
Spinach, New Zealand	Feb. 1–Apr. 15	Feb. 15–Apr. 15	Mar. 1–Apr. 15	Mar. 15–May 15	Mar. 20–May 15	Apr. 1–May 15	Apr. 10–June 1
Squash, summer	Feb. 1–Apr. 15	Feb. 15–Apr. 15	Mar. 1–Apr. 15	Mar. 15–May 15	Mar. 15–May 1	Apr. 1–May 15	Apr. 10–June 1
Sweet potato	Feb. 15–May 15	Mar. 1–May 15	Mar. 20–June 1	Mar. 20–June 1	Apr. 1–June 1	Apr. 10–June 1	Apr. 20–June 1
Tomato	Feb. 1–Apr. 1	Feb. 20–Apr. 10	Mar. 1–Apr. 20	Mar. 10–May 1	Mar. 20–May 10	Apr. 1–May 20	Apr. 10–June 1
Turnip	Jan. 1–Mar. 1	Jan. 1–Mar. 1	Jan. 10–Mar. 1	Jan. 20–Mar. 1	Feb. 1–Mar. 1	Feb. 10–Mar. 10	Feb. 20–Mar. 20
Watermelon[2]	Feb. 15–Mar. 15	Feb. 15–Apr. 1	Feb. 15–Apr. 15	Mar. 1–Apr. 15	Mar. 15–Apr. 15	Apr. 1–May 1	Apr. 10–May 15

Crop	Apr. 10	Apr. 20	Apr. 30	May 10	May 20	May 30	June 10
Asparagus[1]	Mar. 10–Apr. 10	Mar. 15–Apr. 15	Mar. 20–Apr. 15	Mar. 10–Apr. 30	Apr. 20–May 15	May 1–June 1	May 15–June 1
Beans, lima	Apr. 1–June 30	May 1–June 20	May 15–June 15	May 25–June 15			
Beans, snap	Apr. 10–June 30	Apr. 25–June 30	May 10–June 30	May 10–June 30	May 15–June 30	May 25–June 15	
Beet	Mar. 10–June 1	Mar. 20–June 1	Apr. 1–June 15	Apr. 15–June 15	Apr. 25–June 15	May 1–June 15	May 15–June 15
Broccoli, sprouting[1]	Mar. 15–Apr. 15	Mar. 25–Apr. 20	Apr. 1–May 1	Apr. 15–June 1	May 1–June 15	May 10–June 10	May 20–June 10
Brussels sprouts[1]	Mar. 15–Apr. 15	Mar. 25–Apr. 20	Apr. 1–May 1	Apr. 15–June 1	May 1 –June 15	May 10–June 10	May 20–June 10
Cabbage[1]	Mar. 12–Apr. 1	Mar. 10–Apr. 1	Mar. 15–Apr. 10	Apr. 1–May 15	May 1–June 15	May 10–June 15	May 20–June 1
Cabbage, Chinese	(2)	(2)	(2)	Apr. 1–May 15	May 1–June 15	May 10–June 15	May 20–June 1
Carrot	Mar. 10–Apr. 20	Apr. 1–May 15	Apr. 10–June 1	Apr. 20–June 15	May 1–June 1	May 10–June 1	May 20–June 1
Cauliflower[1]	Mar. 1–Mar. 20	Mar. 15–Apr. 20	Apr. 10–May 10	Apr. 15–May 15	May 10–June 15	May 20–June 1	June 1–June 15
Celery and celeriac	Apr. 1–Apr. 20	Apr. 10–May 1	Apr. 15–May 1	Apr. 20–June 15	May 10–June 15	May 20–June 1	June 1–June 15
Chard	Mar. 15–June 1	Apr. 1–June 15	Apr. 15–June 15	Apr. 20–June 15	May 10–June 15	May 20–June 1	June 1–June 15
Chervil and chives	Mar. 1–Apr. 1	Mar. 10–Apr. 10	Mar. 20–Apr. 20	Apr. 1–May 1	Apr. 15–May 15	May 1–June 1	May 15–June 1
Chicory, witloof	June 10–July 1	June 15–July 1	June 15–July 1	June 1–20	June 1–15	June 1–15	June 1–15
Collards[1]	Mar. 1–June 1	Mar. 10–June 1	Apr. 1–June 1	Apr. 15–June 1	May 1–June 1	May 10–June 1	May 20–June 1
Corn, sweet	Apr. 10–June 1	Apr. 25–June 15	May 10–June 15	May 10–June 1	May 15–June 1	May 20–June 1	
Cornsalad	Feb. 1–Apr. 1	Feb. 15–Apr. 15	Mar. 1–May 1	Apr. 1–June 1	Apr. 15–June 1	May 1–June 15	May 15–June 15

[1]Plants.
[2]Included in this section because no appropriate coverage is available elsewhere.

Figure 39–4 Earliest dates and ranges of dates for safe spring planting of vegetables in the open. (*continued*)

SPRING PLANTING DATES FOR LOCALITIES IN WHICH AVERAGE DATE OF LAST FREEZE IS							
Crop	Apr. 10	Apr. 20	Apr. 30	May 10	May 20	May 30	June 10
Cress, upland	Mar. 10–Apr. 15	Mar. 20–May 1	Apr. 10–May 10	Apr. 20–May 20	May 1–June 1	May 15–June 1	May 15–June 15
Cucumber	Apr. 20–June 1	May 1–June 15	May 15–June 15	May 20–June 15	June 1–15		
Eggplant[1]	May 1–June 1	May 10–June 1	May 15–June 10	May 20–June 15	June 1–15		
Endive	Mar. 15–Apr. 15	Mar. 25–Apr. 15	Apr. 1–May 1	Apr. 15–May 15	May 1–30	May 1–30	May 15–June 1
Fennel, Florence	Mar. 15–Apr. 15	Mar. 25–Apr. 15	Apr. 1–May 1	Apr. 15–May 15	May 1–30	May 1–30	May 15–June 1
Garlic	Feb. 20–Mar. 20	Mar. 10–Apr. 1	Mar. 15–Apr. 15	Apr. 1–May 1	Apr. 15–May 15	May 1–30	May 15–June 1
Horseradish[1]	Mar. 10–Apr. 10	Mar. 20–Apr. 20	Apr. 1–30	Apr. 15–May 15	Apr. 20–May 20	May 1–30	May 15–June 1
Kale	Mar. 10–Apr. 1	Mar. 20–Apr. 10	Apr. 1–20	Apr. 10–May 1	Apr. 20–May 10	May 1–30	May 15–June
Kohlrabi	Mar. 10–Apr. 10	Mar. 20–May 1	Apr. 1–May 10	Apr. 10–May 15	Apr. 20–May 20	May 1–30	May 15–June 1
Leek	Mar. 1–Apr. 1	Mar. 15–Apr. 15	Apr. 1–May 1	Apr. 15–May 15	May 1–May 20	May 1–15	May 1–15
Lettuce, head[1]	Mar. 10–Apr. 1	Mar. 20–Apr. 15	Apr. 1–May 1	Apr. 15–May 15	May 1–June 30	May 10–June 30	May 20–June 30
Lettuce, leaf	Mar. 15–May 15	Mar. 20–May 15	Apr. 1–June 1	Apr. 15–June 15	May 1–June 30	May 10–June 30	May 20–June 30
Muskmelon[2]	Apr. 20–June 1	May 1–June 15	May 15–June 15	June 1–June 15			
Mustard	Mar. 10–Apr. 20	Mar. 20–May 1	Apr. 1–May 10	Apr. 15–June 1	May 1–June 30	May 10–June 30	May 20–June 30
Okra	Apr. 20–June 15	May 1–June 1	May 10–June 1	May 20–June 10	June 1–20		
Onion[1]	Mar. 1–Apr. 1	Mar. 15–Apr. 10	Apr. 1–May 1	Apr. 10–May 1	Apr. 20–May 15	May 1–30	May 10–June 10
Onion, seed	Mar. 1–Apr. 1	Mar. 15–Apr. 1	Mar. 15–Apr. 15	Apr. 1–May 1	Apr. 20–May 15	May 1–30	May 10–June 10
Onion, sets	Mar. 1–Apr. 1	Mar. 10–Apr. 1	Mar. 10–Apr. 10	Apr. 10–May 1	Apr. 20–May 15	May 1–30	May 10–June 10
Parsley	Mar. 10–Apr. 10	Mar. 20–Apr. 20	Apr. 1–May 1	Apr. 15–May 15	May 1–20	May 10–June 1	May 20–June 10
Parsnip	Mar. 10–Apr. 10	Mar. 20–Apr. 20	Apr. 1–May 1	Apr. 15–May 15	May 1–20	May 10–June 1	May 20–June 10
Peas, black-eye	May 1–July 1	May 10–June 15	May 15–June 1				
Peas, garden	Feb. 20–Mar. 20	Mar. 10–Apr. 10	Mar. 20–May 1	Apr. 1–May 15	Apr. 15–June 1	May 1–June 15	May 10–June 15
Pepper[1]	May 1–June 1	May 10–June 1	May 15–June 10	May 20–June 10	May 25–June 15	June 1–15	
Potato	Mar. 10–Apr. 1	Mar. 15–Apr. 10	Mar. 20–May 10	Apr. 1–June 1	Apr. 15–June 15	May 1–June 15	May 15–June 1
Radish	Mar. 1–May 1	Mar. 10–May 10	Mar. 20–May 10	Apr. 1–June 1	Apr. 15–June 15	May 1–June 15	May 15–June 1
Rhubarb[1]	Mar. 1–Apr. 1	Mar. 10–Apr. 10	Mar. 20–Apr. 15	Apr. 1–May 1	Apr. 15–May 10	May 1–20	May 15–June 1
Rutabaga	May 1–June 1	May 1–June 1	May 1–20	May 10–20	May 20–June 1		
Salsify	Mar. 10–Apr. 15	Mar. 20–May 1	Apr. 1–May 15	Apr. 15–June 1	May 1–June 1	May 10–June 1	May 20–June 1
Shallot	Mar. 1–Apr. 1	Mar. 15–Apr. 15	Apr. 1–May 1	Apr. 10–May 1	Apr. 20–May 10	May 1–June 1	May 10–June 1
Sorrel	Mar. 1–Apr. 15	Mar. 15–May 1	Apr. 1–May 15	Apr. 15–June 1	May 1–June 1	May 10–June 10	May 20–June 10
Soybean	May 1–June 30	May 10–June 20	May 15–June 15	May 25–June 10			
Spinach	Feb. 15–Apr. 1	Mar. 1–Apr. 15	Mar. 20–Apr. 20	Apr. 1–June 15	Apr. 10–June 15	Apr. 20–June 15	May 1–June 15
Spinach, New Zealand	Apr. 20–June 1	May 1–June 15	May 1–June 15	May 10–June 15	May 20–June 15	June 1–15	
Squash, summer	Apr. 20–June 1	May 1–June 15	May 1–30	May 10–June 10	May 20–June 15	June 1–20	June 10–20
Sweet potato	May 1–June 1	May 10–June 10	May 20–June 10				
Tomato	Apr. 20–June 1	May 5–June 10	May 10–June 15	May 15–June 10	May 25–June 15	June 5–20	June 15–30
Turnip	Mar. 1–Apr. 1	Mar. 10–Apr. 1	Mar. 20–May 1	Apr. 1–June 1	Apr. 15–June 1	May 1–June 15	May 15–June 15
Watermelon[2]	Apr. 20–June 1	May 1–June 15	May 15–June 15	June 1–15	June 15–July 1		

[1]Plants.
[2]Included in this section because no appropriate coverage is available elsewhere.

Figure 39–4 Earliest dates and ranges of dates for safe spring planting of vegetables in the open.

Crop	SUMMER OR FALL PLANTING DATES FOR LOCALITIES IN WHICH AVERAGE DATE OF LAST FREEZE IS					
	Aug. 30	Sept. 10	Sept. 20	Sept. 30	Oct. 10	Oct. 20
Asparagus[1]					Oct. 20–Nov. 15	Nov. 1–Dec. 15
Beans, lima				June 1–15	June 1–15	June 15–30
Beans, snap		May 15–June 15	June 1–July 1	June 1–July 10	June 15–July 20	July 1–Aug. 1
Beet	May 15–June 15	May 15–June 15	June 1–July 1	June 1–July 10	June 15–July 25	July 1–Aug. 5
Broccoli, sprouting	May 1–June 1	May 1–June 1	May 1–June 15	June 1–30	June 15–July 15	July 1–Aug. 1
Brussels sprouts	May 1–June 1	May 1–June 1	May 1–June 15	June 1–30	June 15–July 15	July 1–Aug. 1
Cabbage[1]	May 1–June 1	May 1–June 1	May 1–June 15	June 1–July 10	June 1–July 15	July 1–20
Cabbage, Chinese	May 15–June 15	May 15–June 15	June 1–July 1	June 1–July 15	June 15–Aug. 1	July 15–Aug. 15
Carrot	May 15–June 15	May 15–June 15	June 1–July 1	June 1–July 10	June 1–July 20	June 15–Aug. 1
Cauliflower[1]	May 1–June 1	May 1–July 1	May 1–July 1	May 10–July 15	June 1–July 25	July 1–Aug. 5
Celery[1] and celeriac	May 1–July 1	May 15–June 15	May 15–July 1	June 1–July 5	June 1–July 15	June 1–Aug. 1
Chard	May 15–June 15	May 15–July 1	June 1–July 1	June 1–July 5	June 1–July 20	June 1–Aug. 1
Chervil and chives	May 10–June 10	May 1–June 15	May 15–June 15	(2)	(2)	(2)
Chicory, witloof	May 15–June 15	May 15–June 15	May 15–June 15	June 1–July 1	June 1–July 1	June 15–July 15
Collards[1]	May 15–June 15	May 15–June 15	May 15–June 15	June 15–July 15	July 1–Aug. 1	July 15–Aug. 15
Corn, sweet			June 1–July 1	June 1–July 1	June 1–July 10	June 1–July 20
Cornsalad	May 15–June 15	May 15–July 1	June 15–Aug. 1	July 15–Sept. 1	Aug. 15–Sept. 15	Sept. 1–Oct. 15
Cress, upland	May 15–June 15	May 15–July 1	June 15–Aug. 1	July 15–Sept. 1	Aug. 15–Sept. 15	Sept. 1–Oct. 15
Cucumber			June 1–15	June 1–July 1	June 1–July 1	June 1–July 15
Eggplant[1]				May 20–June 10	May 15–June 15	June 1–July 1
Endive	June 1–July 1	June 1–July 1	June 15–July 15	June 15–Aug. 1	July 1–Aug. 15	July 15–Sept. 1
Fennel, Florence	May 15–June 15	May 15–July 15	June 1–July 1	June 1–July 1	June 15–July 15	June 15–Aug. 1
Garlic	(2)	(2)	(2)	(2)	(2)	(2)
Horseradish[1]	(2)	(2)	(2)	(2)	(2)	(2)
Kale	May 15–June 15	May 15–June 15	June 1–July 1	June 15–July 15	July 1–Aug. 1	July 15–Aug. 15
Kohlrabi	May 15–June 15	June 1–July 1	June 1–July 15	June 15–July 15	July 1–Aug. 1	July 15–Aug. 15
Leek	May 1–June 1	May 1–June 1	(2)	(2)	(2)	(2)
Lettuce, head[1]	May 15–July 1	May 15–July 1	June 1–July 15	June 15–Aug. 1	July 15–Aug. 15	Aug. 1–30
Lettuce, leaf	May 15–July 15	May 15–July 15	June 1–Aug. 1	June 1–Aug. 1	July 15–Sept. 1	July 15–Sept. 1
Muskmelon			May 1–June 15	May 15–June 1	June 1–June 15	June 15–July 20
Mustard	May 15–July 15	May 15–July 15	June 1–Aug. 1	June 15–Aug. 1	July 15–Aug. 15	Aug. 1–Sept. 1
Okra			June 1–20	June 1–July 1	June 1–July 15	June 1–Aug. 1
Onion[1]	May 1–June 10	May 1–June 10	(2)	(2)	(2)	(2)
Onion, seed	May 1–June 1	May 1–June 10	(2)	(2)	(2)	(2)
Onion, sets	May 1–June 1	May 1–June 10	(2)	(2)	(2)	(2)
Parsley	May 15–June 15	May 1–June 15	June 1–July 1	June 1–July 15	June 15–Aug. 1	July 15–Aug. 15
Parsnip	May 15–June 1	May 1–June 15	May 15–June 15	June 1–July 1	June 1–July 10	(2)
Peas, black-eye					June 1–July 1	June 1–July 1
Peas, garden	May 10–June 15	May 1–July 1	June 1–July 15	June 1–Aug. 1	(2)	(2)

[1]Plants.

Figure 39–5 Latest dates and ranges of dates for safe fall planting of vegetables in the open. (*continued*)

Source: US Department of Agriculture Home and Garden Bulletin 202.

SUMMER OR FALL PLANTING DATES FOR LOCALITIES IN WHICH AVERAGE DATE OF LAST FREEZE IS						
Crop	Aug. 30	Sept. 10	Sept. 20	Sept. 30	Oct. 10	Oct. 20
Pepper[1]			June 1–June 20	June 1–July 1	June 1–July 1	June 1–July 10
Potato	May 15–June 1	May 1–June 15	May 1–June 15	May 1–June 15	May 15–June 15	June 15–July 15
Radish	May 1–July 15	May 1–Aug. 1	June 1–Aug. 15	July 1–Sept. 1	July 15–Sept. 15	Aug. 1–Oct. 1
Rhubarb[1]	Sept. 1–Oct. 1	Sept. 15–Oct. 15	Sept. 15–Nov. 1	Oct. 1–Nov. 1	Oct. 15–Nov. 15	Oct. 15–Dec. 1
Rutabaga	May 15–June 15	May 1–June 15	June 1–July 1	June 1–July 1	June 15–July 15	July 10–20
Salsify	May 15–June 1	May 10–June 10	May 20–June 20	June 1–20	June 1–July 1	June 1–July 1
Shallot	(2)	(2)	(2)	(2)	(2)	(2)
Sorrel	May 15–June 15	May 1–June 15	June 1–July 1	June 1–July 15	July 1–Aug. 1	July 15–Aug. 15
Soybean				May 25–June 10	June 1–25	June 1–July 5
Spinach	May 15–July 1	June 1–July 15	June 1–Aug. 1	July 1–Aug. 15	Aug. 1–Sept. 1	Aug. 20–Sept. 10
Spinach, New Zealand				May 15–July 1	June 1–July 15	June 1–Aug. 1
Squash, summer	June 10–20	June 1–20	May 15–July 1	June 1–July 1	June 1–July 15	June 1–July 20
Squash, winter			May 20–June 10	June 1–15	June 1–July 1	June 1–July 1
Sweet potato					May 20–June 10	June 1–15
Tomato	June 20–30	June 10–20	June 1–20	June 1–20	June 1–20	June 1–July 1
Turnip	May 15–June 15	June 1–July 1	June 1–July 15	June 1–Aug. 1	July 1–Aug. 1	July 15–Aug. 15
Watermelon			May 1–June 15	May 15–June 1	June 1–15	June 15–July 20

FALL PLANTING DATES FOR LOCALITIES IN WHICH AVERAGE DATE OF FIRST FREEZE IS						
Crop	Oct. 30	Nov. 10	Nov. 20	Nov. 30	Dec. 10	Dec. 20
Asparagus[1]	Nov. 15–Jan. 1	Dec. 1–Jan. 1				
Beans, lima	July 1–Aug. 1	July 1–Aug. 15	July 15–Sept. 1	Aug. 1–Sept. 15	Sept. 1–30	Sept. 1–Oct. 1
Beans, snap	July 1–Aug. 15	July 1–Sept. 1	July 1–Sept. 10	Aug. 15–Sept. 20	Sept. 1–30	Sept. 1–Nov. 1
Beet	Aug. 1–Sept. 1	Aug. 1–Oct. 1	Sept. 1–Dec. 1	Sept. 1–Dec. 15	Sept 1–Dec. 31	Sept. 1–Dec. 31
Broccoli, sprouting	July 1–Aug. 15	Aug. 1–Sept. 1	Aug. 1–Sept. 15	Aug. 1–Oct. 1	Aug. 1–Nov. 1	Sept. 1–Dec. 31
Brussels sprouts	July 1–Aug. 15	Aug. 1–Sept. 1	Aug. 1–Sept. 15	Aug. 1–Oct. 1	Aug. 1–Nov. 1	Sept. 1–Dec. 31
Cabbage[1]	Aug. 1–Sept. 1	Sept. 1–15	Sept. 1–Dec. 1	Sept. 1–Dec. 31	Sept. 1–Dec. 31	Sept. 1–Dec. 31
Cabbage, Chinese	Aug 1–Sept. 15	Aug. 15–Oct. 1	Sept. 1–Oct. 15	Sept 1–Nov. 1	Sept. 1–Nov. 15	Sept. 1–Dec.1
Carrot	July 1–Aug. 15	Aug. 1–Sept. 1	Sept. 1–Nov. 1	Sept. 15–Dec. 1	Sept. 15–Dec. 1	Sept. 15–Dec.1
Cauliflower[1]	July 15–Aug. 15	Aug. 1–Sept. 1	Aug. 1–Sept. 15	Aug. 15–Oct. 10	Sept. 1–Oct. 20	Sept. 15–Nov. 1
Celery1and celeriac	June 15–Aug. 15	July 1–Aug. 15	July 15–Sept. 1	Aug. 1–Dec. 1	Sept. 1–Dec. 31	Oct. 1–Dec. 31
Chard	June 1–Sept. 10	June 1–Sept. 15	June 1–Oct. 1	June 1–Nov. 1	June 1–Dec. 1	June 1–Dec. 31
Chervil and chives	(2)	(2)	Nov. 1–Dec. 31	Nov. 1–Dec. 31	Nov. 1–Dec. 31	Nov. 1–Dec. 31
Chicory, witloof	July 1–Aug. 10	July 10–Aug. 20	July 20–Sept. 1	Aug. 15–Sept. 30	Aug. 15–Oct. 15	Aug. 15–Oct. 15
Collards[1]	Aug. 1–Sept. 15	Aug. 15–Oct. 1	Aug. 25–Nov. 1	Sept. 1–Dec. 1	Sept. 1–Dec. 31	Sept. 1–Dec. 31
Corn, sweet	June 1–Aug. 1	June 1–Aug. 15	June 1–Sept. 1			
Cornsalad	Sept. 15–Nov.1	Oct. 1–Dec. 1	Oct. 1–Dec. 1	Oct. 1–Dec. 31	Oct. 1–Dec. 31	Oct. 1–Dec. 31

[1]Plants.

Figure 39–5 Latest dates and ranges of dates for safe fall planting of vegetables in the open. (*continued*)

	FALL PLANTING DATES FOR LOCALITIES IN WHICH AVERAGE DATE OF FIRST FREEZE IS					
Crop	Oct. 30	Nov. 10	Nov. 20	Nov. 30	Dec. 10	Dec. 20
Cress, upland	Sept. 15–Nov. 1	Oct. 1–Dec. 1	Oct. 1–Dec. 1	Oct. 1–Dec. 31	Oct. 1–Dec. 31	Oct. 1–Dec. 31
Cucumber	June 1–Aug. 1	June 1–Aug. 15	June 1–Aug. 15	July 15–Sept. 15	Aug. 15–Oct. 1	Aug. 15–Oct. 1
Eggplant[1]	June 1–July 1	June 1–July 15	June 1–Aug. 1	July 1–Sept. 1	Aug. 1–Sept. 30	Aug. 1–Sept. 30
Endive	July 15–Aug. 15	Aug. 1–Sept. 1	Sept. 1–Oct. 1	Sept. 1–Nov. 15	Sept. 1–Dec. 31	Sept. 1–Dec. 31
Fennel, Florence	July 1–Aug. 1	July 15–Aug. 15	Aug. 15–Sept. 15	Sept. 1–Nov. 15	Sept. 1–Dec. 1	Sept. 1–Dec. 1
Garlic	(2)	Aug. 1–Oct. 1	Aug. 15–Oct. 1	Sept. 1–Nov. 15	Sept. 15–Nov. 15	Sept. 15–Nov. 15
Horseradish[1]	(2)	(2)	(2)	(2)	(2)	(2)
Kale	July 15–Sept. 1	Aug. 1–Sept. 15	Aug. 15–Oct. 15	Sept. 1–Dec. 1	Sept. 1–Dec. 31	Sept. 1–Dec. 31
Kohlrabi	Aug. 1–Sept. 1	Aug. 15–Sept. 15	Sept. 1–Oct. 15	Sept. 1–Dec. 1	Sept. 15–Dec. 31	Sept. 1–Dec. 31
Leek	(2)	(2)	Sept. 1–Nov. 1	Sept. 1–Nov. 1	Sept. 1–Nov. 1	Sept. 15–Nov. 1
Lettuce, head[1]	Aug. 1–Sept. 15	Aug. 15–Oct. 15	Sept. 1–Nov. 1	Sept. 1–Dec. 1	Sept. 15–Dec. 31	Sept. 15–Dec. 31
Lettuce, leaf	Aug. 15–Oct. 1	Aug. 25–Oct. 1	Sept. 1–Nov. 1	Sept. 1–Dec. 1	Sept. 15–Dec. 31	Sept. 15–Dec. 31
Muskmelon	July 1–July 15	July 15–July 30				
Mustard	Aug. 15–Oct. 15	Aug. 15–Nov. 1	Sept. 1–Dec. 1	Sept. 1–Dec. 1	Sept. 1–Dec. 1	Sept. 15–Dec. 1
Okra	June 1–Aug. 10	June 1–Aug. 20	June 1–Sept. 10	June 1–Sept. 20	Aug. 1–Oct. 1	Aug. 1–Oct. 1
Onion[1]		Sept. 1–Oct. 15	Oct. 1–Dec. 31	Oct. 1–Dec. 31	Oct. 1–Dec. 31	Oct. 1–Dec. 31
Onion, seed			Sept. 1–Nov. 1	Sept. 1–Nov. 1	Sept. 1–Nov. 1	Sept. 15–Nov. 1
Onion, sets		Oct. 1–Dec. 1	Nov. 1–Dec. 31	Nov. 1–Dec. 31	Nov. 1–Dec. 31	Nov. 1–Dec. 31
Parsley	Aug. 1–Sept. 15	Sept. 1–Nov. 15	Sept. 1–Dec. 31	Sept. 1–Dec. 31	Sept. 1–Dec. 31	Sept. 1–Dec. 31
Parsnip	(2)	(2)	Aug. 1–Sept. 1	Sept. 1–Nov. 15	Sept. 1–Dec. 1	Sept. 1–Dec. 1
Peas, black-eye	June 1–Aug. 1	June 15–Aug. 15	July 1–Sept. 1	July 1–Sept. 10	July 1–Sept. 20	July 1–Sept. 20
Peas, garden	Aug. 1–Sept. 15	Sept. 1–Nov. 1	Oct. 1–Dec. 1	Oct. 1–Dec. 31	Oct. 1–Dec. 31	Oct. 1–Dec. 31
Pepper[1]	June 1–July 20	June 1–Aug. 1	June 1–Aug. 15	June 15–Sept. 1	Aug. 15–Oct. 1	Aug. 15–Oct. 1
Potato	July 20–Aug. 10	July 25–Aug. 20	Aug. 10–Sept. 15	Aug. 1–Sept. 15	Aug. 1–Sept. 15	Aug. 1–Sept. 15
Radish	Aug. 15–Oct. 15	Sept. 1–Nov. 15	Sept. 1–Dec. 1	Sept. 1–Dec. 31	Aug. 1–Sept. 15	Oct. 1–Dec. 31
Rhubarb[1]	Nov. 1–Dec. 1					
Rutabaga	July 15–Aug. 1	July 15–Aug. 15	Aug. 1–Sept. 1	Sept. 1–Nov. 15	Oct. 1–Nov. 15	Oct. 15–Nov. 15
Salsify	June 1–July 10	June 15–July 20	July 15–Aug. 15	Aug. 15–Sept. 30	Aug. 15–Oct. 15	Sept. 1–Oct. 31
Shallot	(2)	Aug. 1–Oct. 1	Aug. 15–Oct. 1	Aug. 15–Oct. 15	Sept. 15–Nov. 1	Sept. 15–Nov. 1
Sorrel	Aug. 1–Sept. 15	Aug. 15–Oct. 1	Aug. 15–Oct. 15	Sept. 1–Nov. 15	Sept. 1–Dec. 15	Sept. 1–Dec. 31
Soybean	June 1–July 15	June 1–July 25	June 1–July 30	June 1–July 30	June 1–July 30	June 1–July 30
Spinach	Sept. 1–Oct. 1	Sept. 15–Nov. 1	Oct. 1–Dec. 1	Oct. 1–Dec. 31	Oct. 1–Dec. 31	Oct. 1–Dec. 31
Spinach, New Zealand	June 1–Aug. 1	June 1–Aug. 15	June 1–Aug. 15			
Squash, summer	June 1–Aug. 1	June 1–Aug. 10	June 1–Aug. 20	June 1–Sept. 1	June 1–Sept. 15	June 1–Oct. 1
Squash, winter	June 10–July 10	June 20–July 20	July 1–Aug. 1	July 15–Aug. 15	Aug. 1–Sept. 1	Aug. 1–Sept. 1
Sweet potato	June 1–15	June 1–July 1	June 1–July 1	June 1–July 1	June 1–July 1	June 1–July 1
Tomato	June 1–July 1	June 1–July 15	June 1–Aug. 1	Aug. 1–Sept. 1	Aug. 15–Oct. 1	Sept. 1–Nov. 1
Turnip	Aug. 1–Sept. 15	Sept. 1–Oct. 15	Sept. 1–Nov. 15	Sept. 1–Nov. 15	Oct. 1–Dec. 1	Oct. 1–Dec. 31
Watermelon	July 1–July 15	July 15–July 30				

[1]Plants.

Figure 39–5 Latest dates and ranges of dates for safe fall planting of vegetables in the open.

How to Use the Map and Tables

To determine the best time for spring planting of any vegetable in your locality, follow these steps:

1. Find your location on the map in Figure 39–2, and then find the solid line on the map that comes closest to that location.
2. Find the date shown on the solid line. This is the average date of the last killing frost. The first number represents the month; the second number represents the day. (Example: 5-10 represents May 10.) Note this date, and go to Figure 39–4.
3. Find the column with this date at the top. This is the only column you will use. It gives dates for all vegetable crops listed for your area.
4. In this column, locate the vegetable that you want to plant. The dates by the vegetable show the period of time during which the crop can be safely planted. The best time is on or soon after the first date. The second date is the latest date for planting. Planting time becomes less desirable as you move toward the second date.

Figure 39–5 is used with the map in Figure 39–3 in the same way to find planting dates for fall planting. A date about halfway between the two dates given in Figure 39–5 is usually best.

Along the northern half of the Pacific Coast, warm-weather crops should not be planted as late as the frost date and table indicate. Although frost comes late, very cool weather prevails for some time before frost, retarding the late growth of crops that prefer warm temperatures such as corn, beans, and tomatoes.

Planting Seeds

Assume that the garden area is prepared and soil pH and fertilizer requirements necessary prior to planting have been taken care of. The next step is to harrow or rake the garden in preparation for planting. The soil should be worked to break all large clods of soil and to form a firm seedbed. For a seed to germinate, there must be good contact with the soil so that moisture can be pulled from the soil into the seed. Large, open air spaces result in poor germination. *The smaller the seed, the more important it is to have a fine, firm seedbed.*

To be sure that the rows of seeds are planted straight, use a string stretched across the garden. To get the proper number of seed per foot or row, a wooden marker measured off in 1-foot lengths may be used (Figure 39–6). If there is no problem estimating distance, use of the marker pole is not necessary. To establish the proper distance between rows and the depth at which to plant seed, see Figure 39–7.

The planting row should be made at least deep enough to cover the seed at the proper depth. Small seeds, such as carrot and lettuce, should be covered with only ¼ to ½ inch of soil. The hoe handle can be used to form these shallow rows. For deeper planting, use the corner of the hoe blade or a wheel hoe. Read the directions on the seed package for the proper depth and distance apart to plant the seed, or use Figure 39–7 as a guide. Fertilizer may be banded on each side of the row.

A hoe or garden rake (Figure 39–8) is used to cover the seed. Try to cover the seed with fine soil that is free of lumps, stones, and clods. Cover all seeds in the row with the same depth of soil. Firm the soil around the seeds with the rake or hoe so that close contact is made between soil and seed. If the soil is a little wet when planting, do not pack or press soil down on seeds.

▲ **Figure 39–6** Using a line and marker pole, establish a straight row and the proper number of seeds per foot of row. Notice the even texture of the soil and the bean seeds.

	REQUIREMENT FOR 100 FEET OF ROW			DISTANCE APART		
				Rows		
Crop	Seed	Plants	Depth for Planting Seed (Inches)	Horse- or Tractor-Cultivated (Feet)	Hand-Cultivated	Plants in the Row
Asparagus	1 ounce	75	1–1½	4–5	1½–2 feet	18 inches
Beans:						
Lima, bush	½pound		1–1½	2½–3	2 feet	3–4 inches
Lima, pole	½pound		1–1½	3–4	3 feet	3–4 feet
Snap, bush	½pound		1–1½	2½–3	2 feet	3–4 inches
Snap, pole	4 ounces		1–1½	3–4	2 feet	3 feet
Beet	2 ounces		1	2–2½	14–16 inches	2–3 inches
Broccoli:						
Heading	1 packet	50–75	½	2½–3	2–2½ feet	14–24 inches
Sprouting	1 packet	50–75	½	2½–3	2–2½ feet	14–24 inches
Brussels sprouts	1 packet	50–75	½	2½–3	2–2½ feet	14–24 inches
Cabbage	1 packet	50–75	½	2½–3	2–2½ feet	14–24 inches
Cabbage, Chinese	1 packet		½	2–2½	18–24 inches	8–12 inches
Carrot	1 packet		½	2–2½	14–16 inches	2–3 inches
Cauliflower	1 packet	50–75	½	2½–3	1–2½ feet	14–24 inches
Celeriac	1 packet	200–250	⅛	2½–3	18–24 inches	4–6 inches
Celery	1 packet	200–250	⅛	2½–3	18–24 inches	4–6 inches
Chard	2 ounces		1	2–2½	18–24 inches	6 inches
Chervil	1 packet		½	2–2½	14–16 inches	2–3 inches
Chicory, witloof	1 packet		½	2–2½	18–24 inches	6–8 inches
Chives	1 packet		½	2½–3	14–16 inches	in clusters
Collards	1 packet		½	3–3½	18–24 inches	18–24 inches
Corn, sweet	2 ounces		2	3–3½	2–3 feet	drills, 14–16 inches; hills, 2½–3 feet
Cornsalad	1 packet		½	2½–3	14–16 inches	1 foot
Cress, upland	1 packet		⅛–¼	2–2½	14–16 inches	2–3 inches
Cucumber	1 packet		½	6–7	6–7 feet	drills, 3 feet; hills, 6 feet
Dasheen	5–6 pounds	50	2–3	3½–4	3½–4 feet	2 feet
Eggplant	1 packet	50	½	3	2–2½ feet	3 feet
Endive	1 packet		½	2½–3	18–24 inches	12 inches
Fennel, Florence	1 packet		½	2½–3	18–24 inches	4–6 inches
Garlic	1 pound		1–2	2½–3	14–16 inches	2–3 inches
Horseradish	Cuttings	50–75	2	3–4	2–2½ feet	18–24 inches
Kale	1 packet		½	2½–3	18–24 inches	12–15 inches
Kohlrabi	1 packet		½	2½–3	14–16 inches	5–6 inches
Leek	1 packet		½–1	2½–3	14–16 inches	2–3 inches
Lettuce, head	1 packet	100	½	2½–3	14–16 inches	12–15 inches

Source: US Department of Agriculture Home and Garden Bulletin 202

Figure 39–7 Quantity of seed and number of plants required for 100 feet of row, depths of planting, and distances apart for rows and plants. (*continued*)

	Requirement for 100 Feet of Row			Distance Apart		
				Rows		
Crop	Seed	Plants	Depth for Planting Seed (Inches)	Horse- or Tractor-Cultivated (Feet)	Hand-Cultivated	Plants in the Row
Lettuce, leaf	1 packet		½	2½–3	14–16 inches	6 inches
Muskmelon	1 packet		1	6–7	6–7 feet	hills, 6 feet
Mustard	1 packet		½	2½–3	14–16 inches	12 inches
Okra	2 ounces		1–1½	3–3½	3–3½ feet	2 feet
Onion:						
Plants		400	1–2	2–2½	14–16 inches	2–3 inches
Seed	1 packet		½–1	2–2½	14–16 inches	2–3 inches
Sets	1 pound		1–2	2–2½	14–16 inches	2–3 inches
Parsley	1 packet		½	2–2½	14–16 inches	4–6 inches
Parsley, turnip-rooted	1 packet		⅛–¼	2–2½	14–16 inches	2–3 inches
Parsnip	1 packet		½	2–2½	18–24 inches	2–3 inches
Peas	½ pound		2–3	2–4	1½–3 feet	1 inch
Pepper	1 packet	50–70	½	3–4	2–3 feet	18–24 inches
Physalis	1 packet		½	2–2½	1½–2 feet	12–18 inches
Potato	5–6 pounds, tubers		4	2½–3	1½–2 feet	10–18 inches
Pumpkin	1 ounce		1–2	5–8	5–8 feet	3–4 feet
Radish	1 ounce		½	2–2½	14–16 inches	1 inch
Rhubarb		25–35		3–4	3–4 feet	3–4 feet
Salsify	1 ounce		½	2–2½	18–26 inches	2–3 inches
Shallots	1 pound (cloves)		1–2	2–2½	12–18 inches	2–3 inches
Sorrel	1 packet		½	2–2½	18–24 inches	5–8 inches
Soybean	½–1 pound		1–1½	2½–3	24–30 inches	3 inches
Spinach	1 ounce		½	2–2½	14–16 inches	3–4 inches
Spinach, New Zealand	1 ounce		1–1½	3–3½	3 feet	18 inches
Squash:						
Bush	½ ounce		1–2	4–5	4–5 feet	drills, 15–18 inches; hills, 4 feet
Vine	1 ounce		1–2	8–12	8–12 feet	drills, 2–3 feet; hills, 4 feet
Sweet potato	5 pounds, bedroots	75	2–3	3–3½	3–3½ feet	12–14 inches
Tomato	1 packet	35–50	½	3–4	2–3 feet	1½–3 feet
Turnip greens	1 packet		¼–½	2–2½	14–16 inches	2–3 inches
Turnips and rutabagas	½ ounce		¼–½	2–2½	14–16 inches	2–3 inches
Watermelon	1 ounce		1–2	8–10	8–10 feet	drills, 2–3 feet; hills, 8 feet

Figure 39–7 Quantity of seed and number of plants required for 100 feet of row, depths of planting, and distances apart for rows and plants.

▲ **Figure 39–8** A garden rake is used to cover seeds evenly.

▲ **Figure 39–9** Seeds being planted in a flat to be grown indoors for later transplanting to the garden. Notice the labels placed for seed identification.

This causes a hard crust to form, and seeds will have difficulty breaking through the soil surface. Do not plant when the soil is wet. If a hard crust does develop, a gardening tool called the weasel can be used to break the soil's hard crust to enhance germination of the seed.

When using the wide-row system, small, easily covered seeds are planted by spreading them out over the top of the bed to the desired width and seed spacing and pulling dirt in from the edges to cover seeds. Several rows side by side can also be opened for larger seeds and covered the same as with the single-row system; the only difference is that the rows are closer together.

Planting Vegetable Plants

Mixing Crops for Insect Control

Mixed plantings reduce damage by insects. Mixing crops confuses insects because of the great number and variety of smells present. Insects that attack certain plants are attracted by the smell of that plant and show no interest in scents from other plants.

Mixing also works for underground pests. Roots of pest-resistant varieties may release compounds that inhibit or kill the insects' larvae. Sweet potato roots produce an **allelopathic** compound that reduces the growth of some weeds.

Seeding

Some vegetables are not directly seeded in the garden. These seeds are planted in flats or other containers before the frost-free date and later transplanted to the garden. Figure 39–9 shows a flat with rows marked off being seeded with vegetables for later transplanting. Soil in the flat should be a fine, carefully screened mixture of sand, soil, and peat moss or one of the soilless mixes. Seeds may also be planted in individual peat pots or Jiffy 7 peat pellets. Figure 39–10 shows a Jiffy 7 peat pellet planted with a single seed.

The length of time needed to reach transplanting size varies from vegetable to vegetable. A good general rule is to plant small seed in the flat or other container 6 to 8 weeks before plants are to be transplanted to the garden.

GREEN TIP Buy plants that are grown in peat pots, which are biodegradable once planted into the dirt. Plus, peat pots make planting easy and eliminate transplant shock.

After the seeds are planted, the flat or pot should be watered to settle the medium around the seeds. Close contact between the soil and seed is important for germination. Cover the flat or pot with newspaper, glass, or plastic to hold

▲ **Figure 39–10** Jiffy 7 pellet used to start individual plants from seed. (Left) Pellet before soaking in water; (Right) Plant ready for transplanting to the garden.

in moisture. The seedbed container should never be allowed to dry out because newly germinating seeds are easily killed by lack of moisture. Uncover the seedlings as soon as shoots appear through the soil. Water as often as necessary to keep the growing medium moist but not wet. Fertilize the seedlings as recommended on the fertilizer container label. Do not use too much fertilizer—new plants are easily burned. (Some soilless mixes have slow-release fertilizer in the mix, making additional fertilizer unnecessary.)

To determine which vegetable crops are usually planted in the garden as seeds and which ones may be transplanted to the garden as plants, see Figure 39–7. The column titled *Requirement for 100 Feet of Row* lists both the amount of seed and the number of plants for those crops that are transplanted to the garden as plants.

First Transplanting

If seeds are sown as directed on the package, two transplantings will be necessary. The first transplanting must be made soon after the seedlings develop the first pair of true leaves. At this first transplanting, each plant should be spaced about 2 inches apart in a flat or planted in individual peat pots or Market Packs. The soil mix used is the same mix the seeds were planted in—one-third soil, one-third sand or perlite, one-third peat moss, or use a commercial mix. A good soluble fertilizer should be used within a few days after transplanting and repeated according to directions. Some of the soilless planting mixes contain fertilizer, making the addition of fertilizer unnecessary.

PROCEDURE
Hardening Off in the Greenhouse

1. Place the plants where the temperature can gradually be lowered over a period of 10 days to 2 weeks. This slows the growth rate and toughens the plant.
2. Lower the humidity and allow the soil to become drier. This also slows growth and toughens the plant. Do not allow the plants to wilt, however.
3. If the plants are in flats, cut a square around the plants with a sharp knife about 10 days before transplanting. Use all the area of the flat, and make the blocks as big as possible. In the period of 10 days to 2 weeks, the plant will grow new roots in a tighter root ball and be able to withstand transplanting shock better.

Hardening Off

As the date nears for the plants to be transplanted to the garden, the conditions must be gradually changed to resemble as closely as possible the conditions of the outside garden. If plants are in a greenhouse in which temperature and humidity are high, steps must be taken to slow the growth rate to prepare the plant for chilling weather, drying winds, a shortage of water, or high temperatures. Tender, succulent growth must be hardened for the best growth of plants.

The hardening-off steps can be done in a greenhouse if all the plants it contains are being hardened off or in a special cold frame. **Cold frames** are usually used for hardening off. Permanent cold frames are usually constructed of concrete block or wood with a glass sash on top. A temporary cold frame is shown in Figure 39–11. Cold frames are never heated—the frame is opened enough during the day to keep the sun from raising the temperature too high and closed at night to hold heat in and prevent freezing of the plants. The cold frame is gradually opened more and more each day until the plants are hardened off and living in the outdoor environment. Most plants are then ready for transplanting. (Some plants, such as tomatoes, must not be transplanted to the garden until all danger of frost is past. On the other hand, other crops, such as cabbage, are more tolerant of cold weather and can be transplanted to the garden earlier.)

Figure 39–11 Cold frame to start and harden-off plants (to condition them for outdoors).

PROCEDURE
Transplanting Plants to the Garden

1. Water the plants to be transplanted the evening before or at least several hours before transplanting.
2. If plants are grown in peat pots, peel off the top half of the pot, and break off the bottom. This practice helps to speed the spread of roots into the garden soil, especially if dry weather follows transplanting. If roots are already growing through the pot, the bottom need not be broken.
3. If plants are grown in flats, cut the soil between the plants and lift them out with a putty knife or other small tool. Leave as much soil around the roots as possible. Handle carefully.
4. Check the distance apart at which the plants are to be placed, and set them immediately in the garden soil. Place the plants a little deeper in the garden soil than they grew in the original container. Some tall, spindly plants, especially tomatoes, may be planted deeper.
5. Pull soil toward the plant and firm slightly. Fill the hole only two-thirds full. Stop and water enough to settle the soil around the roots. Pull in enough dry soil to level the hole and cover the wet area. This prevents moisture loss and baking of the soil.

GREEN TIP Do not allow plant roots to be exposed to the damaging effects of the sun or drying wind. Such exposure will cause the plant's roots to dry out. The plant will wilt, and the roots will take longer to absorb water.

Second Transplanting

The vegetable plants are now properly hardened off and ready to transplant to the garden (Figure 39–12). If possible, transplant on a cloudy day or in late afternoon or evening. Plants will not wilt as much if they are not exposed to bright sunlight immediately after transplanting, resulting in a higher percentage of healthy, growing plants.

A strip of cardboard, aluminum foil, or newspaper wrapped around the stem and extending about 1 inch underground and about 2 inches aboveground will prevent cutworms from cutting off and killing the new plant (Figure 39–13).

Sustainable Farming

Some new products on the market now allow earlier outdoor planting of vegetables. These products provide both protection from the cold and protection against some insects, promoting earlier maturity and higher yields.

© 2017 Cengage Learning®

▲ **Figure 39–12** Transplanting a tomato plant to the garden. Notice the plant is in a peat pot so that the roots are not disturbed at planting time. Plant as deep as possible, placing the plant down in the hole with just the top leaves exposed. The peat pot should be completely covered to prevent drying out.

© 2017 Cengage Learning®

▲ **Figure 39–13** Placement of a cardboard collar around a tomato plant to prevent cutworms from cutting off the plant just above the soil level.

WATER-FILLED TEPEE The first of these products is a small item called the water-filled tepee (Figure 39–14). It is double-walled, made of plastic, measures 18 inches tall and 18 inches across, and holds 3/10 of a gallon (38 ounces) of water. It is open at the top. Placed around a newly planted plant, a tepee will control the temperature around that plant to a great degree because water is an excellent storage element for heat. In the daytime, the water warms up while keeping the plant inside from getting too hot. At night, the heat is released to keep the plant from freezing when temperatures are below freezing. Planting can be done a month earlier than normal, resulting in earlier maturity of crops and higher yields, especially for crops such as tomatoes.

Courtesy of Gardener's Supply Company

▲ **Figure 39–14** A water-filled tepee around a tomato plant. Tepees are very effective in protecting plants from both heat and cold.

Courtesy of Gardener's Supply Company

▲ **Figure 39–15** Lightweight floating row covers are very effective in protecting plants from heat, cold, insects, and rabbits.

FLOATING COVER Another new gardening product is the **floating row cover** (Figure 39–15). Made of very lightweight polyester material, the cover or blanket is placed over a row of plants and sealed with soil on both edges. As the plants grow, they lift up the lightweight blanket. The white material allows sunlight and rain to penetrate, holds in the heat at night, allows for only minor temperature changes, and may be left on until the plants get too large for the cover. Plants that need bees to pollinate flowers must be uncovered during flowering time.

Frost protection down to 27°F (23°C) is provided by the covers. Temperatures under these covers are from 1°F to 4°F warmer than outside temperatures, so planting can be done 1 to 2 weeks sooner. This is especially important if early market prices are higher, resulting in more profits.

Some plants remain covered until harvest if sprays for insects are not needed. Some insects are kept away from plants by these blankets, and sprays may not be needed. However, strong winds may be a problem, and a rye crop or other windbreak in narrow rows may be necessary.

The lightweight row covers have been tested at many state universities. Following are some of the results:

1. Increased yields on beans by as much as 40 percent
2. Increased yields and delayed **bolting** (see glossary) of lettuce, spinach, and Chinese cabbage
3. Protection against insects for cabbage, cucumbers, and melons
4. Protection from virus disease spread by aphids

Floating covers are being used both by home gardeners and large commercial growers.

Many gardeners are now using floating covers in wide-row planting. A row 1 to 3 feet wide is planted with closely spaced plants of the same variety. There is no wide spacing between rows; in fact, there are no rows in many cases. Seeds or plants are planted as close together as practical in a single wide row.

There are also other row covers that are placed over wire hoops stuck in the ground. They are applied in the same manner as the floating covers, but they also have a wire support to hold them up. For more information, contact your local agricultural experiment station.

High Tunnels

High tunnels are an effective method of protecting high-value crops from frost, pests, and environmental challenges. Considerable research at Pennsylvania State University, Department of Horticulture, has proven the economic value of planting both vegetables and small fruits, especially tomatoes and plug-grown strawberries, in high tunnels. High tunnels are greenhouse-like structures that lack a heating system built outdoors to grow vegetable plants and extend the growing season in both spring and fall. The high tunnel is a semi-circular or Quonset-shaped structure with a frame constructed of metal bows attached to metal posts that are driven into the ground. The structure is covered with polyethylene plastic, thereby using solar energy as a heat source. As the temperature rises, the plastic on the sides of the high tunnel can be rolled up for ventilation and to control the inside temperature for optimal vegetable production. Then, as the temperature cools off in the fall, the sides can be lowered to protect the tender vegetables from frost. The plants can be grown directly in the soil that is covered by the high tunnel. Other, more innovative approaches to high tunnel use include mounting the entire structure on a roller track so that it can be moved to new soil and used to grow another crop of vegetables. It is also possible to use a small tractor inside the high tunnel to cultivate the soil and prep it for a new vegetable crop. The economics of high tunnels offer another opportunity to improve the bottom line. Tall tunnels also reduce insect damage and lessen the need for pesticides.

TRY THIS Research the Internet to find information on high tunnel funding programs. Locate organizations that assist farmers to construct high tunnels and any funding or grant opportunities that could help a startup operation. Report to the class on how this funding can be used.

Growing Vegetables in a Bag

Bags or containers that grow vegetables offer more gardeners the opportunity to expand their resources and grow fresh vegetables and herbs in convenient coves, coveys, and walls. This method of gardening offers individuals ways to be creative, make something new and enduring, and create an environment that is soothing and harmonious.

PROCEDURE
Container Gardening

1. Purchase felt-like gardening bags and a 40 lb. bag of topsoil at the local garden center.
2. Fill the felt garden bag two-thirds full, level off, and plant the desired vegetables.
3. Place the bag in an area that receives at least 6 hours of sunlight per day, and water the plants as needed. The media should feel damp to the touch.
4. A general rule of thumb is to add 1 cup of 10-10-10 fertilizer and 1 cup of dolomitic lime ($MgCO_2$) per 2.5 cu. ft. of topsoil.

Double Cropping

Because some vegetables reach maturity earlier than others, consideration should be given to planting something in their place as they are harvested (called **double cropping**). Replanting the same crop and planting other crops are both possibilities. Vegetables may be grouped according to how long they grow in the garden and their use as replanted or successive crops (Figure 39–16).

CROPS OCCUPYING THE GROUND ALL SEASON	
asparagus	onions (from seeds)
beans, pole lima	parsnips
beans, pole snap	peppers
beets, late	pumpkins
carrots, late	rhubarb
corn, late	rutabagas
cucumbers	salsify
eggplant	squash
melons	tomatoes
okra	
SUCCESSIVE CROPS THAT MAY BE REPLANTED OR FOLLOWED BY OTHER CROPS	
beans, dwarf	peas
lettuce	radishes
mustard	spinach
parsley	turnips
EARLY CROPS THAT MAY BE FOLLOWED BY OTHERS	
beets, early	mustard
cabbage, early	onion sets
carrots, early	spinach
corn, early	turnips, early
LATE CROPS THAT MAY FOLLOW OTHERS	
beets, late	kale
Brussels sprouts	mustard
cabbage, late	peas, late
cauliflower	spinach
endive	turnips

Figure 39–16 By grouping vegetables in the garden according to their use as replanted or successive crops, harvesting is simplified.

Garden Tools

Very few tools are necessary for cultivation of a small garden (Figure 39–17). In most cases, these include the following:

- Spade or spading fork for digging up soil
- Steel bow garden rake for leveling and preparing the seedbed

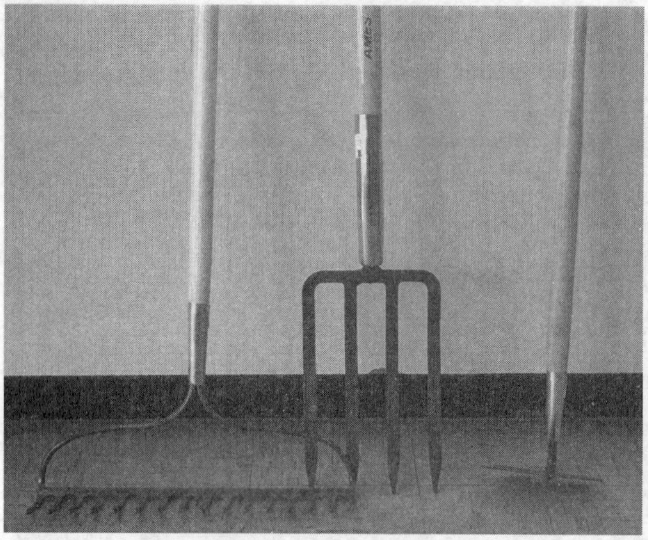

▲ **Figure 39–17** Essential tools for the home garden. From left to right: garden rake, fork spade, and garden hoe.

- Garden hoe for laying off rows and controlling weeds
- Strong cord for laying off rows
- Garden weasel for cultivation and aeration of the soil
- Wheelbarrow for hauling fertilizer and mulch
- Garden hose long enough to water the garden

For large gardens measuring about 2,000 to 4,000 sq. ft., a wheel hoe is helpful for saving labor. For gardens measuring 4,000 sq. ft., a rotary tiller rather than a rake is useful for preparing the soil for planting and cultivating to control weeds.

SAFETY FIRST

OSHA's regulations ensure that employers and employees know about the work hazards and how to protect themselves to prevent injuries in the workplace. For more information on occupational health and safety in the workplace, please visit the OSHA site at www.osha.gov.

Summary

Selecting the proper plant varieties and planting them in the garden at the proper date and in the proper way are most important factors for success. Plant too early, and the plants freeze. Plant too late, and crop yield is reduced. If plant varieties are not adapted to the local climate (hot, cold, and moisture), failure or lower production results. Proper choices and procedures greatly reduce the amount of work needed to produce a bountiful harvest.

Review Questions

1. List 10 common vegetables that are normally planted in a garden and then specify whether each should be direct seeded in the garden or transplanted to the garden as small plants.
2. What are the five steps in transplanting seedlings to the garden?

3. What is the biggest advantage of using row covers for vegetable plants?
4. Discuss two advantages of using high tunnels to grow vegetables.
5. What is the advantage of mixed plantings of vegetables, instead of solid beds of the same vegetable?
6. What is phenology?

Student Activities

1. Plant seeds of vegetable crops in flats for sale or for use in a home or school garden.
2. Plan and conduct an FFA-sponsored workshop to give a demonstration on the proper planting techniques for a home garden. Be sure to mention that the class has plants that are for sale.
3. Design and construct a temporary cold frame structure.
4. Search online to find information on the world seed bank. Describe how this program will ensure we will have viable vegetable seeds in the future.
5. As a fund-raising activity, raise vegetables in a school/community garden, and then use a certified kitchen to prepare premade and packaged salads and meals. Sell these meals to employees at large local businesses and offices.

Suggested Supervised Agricultural Experience (SAE) Activities

1. As an entrepreneurship SAE, raise vegetables in a private garden, then use a certified kitchen to prepare pre-made and packaged salads and meals. Sell these meals to the employees at large local businesses and offices.
2. As a community service SAE, raise vegetables in a local garden, can the vegetables to preserve them, and then donate the finished product to a local food pantry or those who may need assistance in the community.

Additional Resources

http://urbanext.illinois.edu/veggies/basics.cfm

http://content.ces.ncsu.edu/home-vegetable-gardening.pdf

http://cagardenweb.ucanr.edu/Vegetables/

UNIT 40

Caring for the Vegetable Garden

Objective

To care for a vegetable garden, producing a marketable-quality crop.

Key Terms

biopesticide
compost
crop rotation
cultivation
hairy vetch
microorganisms
mulch
organic
organic fertilizers

Competencies to Develop

After studying this unit, you should be able to:

- determine watering needs of vegetable plants and list three ways in which water is applied.
- determine the type and amount of fertilizer for a specific crop by using a soil test recommendation.
- establish pest control programs using mulches, cultivation, crop rotation, and beneficials.
- list five vegetables grown in the area, a pest that commonly attacks each vegetable, and one method of control for each.
- use organic material, such as compost and organic fertilizers, to reduce the need for synthetic fertilizers.

Materials

- rain gauge
- garden hose and sprinkler
- garden fertilizer
- garden tools (listed in Unit 39)
- pest control
- mulching material
- sprayer and/or duster
- PPE
- smartphone apps:
  - Vegetable Tree – Gardening Guide
  - Vegetable Gardening Guide
  - Gardening: The Ultimate Guide
  - Weather
  - Spectracide
  - Pest Portal
  - Calibrate My Sprayer
  - PGR Mix Master
  - Ortho Problem Solver

Planting the garden is only a small part of the work involved in growing vegetables. In planning the size of the garden, the time available for its care should be considered. A small, well-kept garden is much better than a large, weedy one. The best preparation, planning, and planting are of little value if the garden is not cared for properly.

GREEN TIP Caring for a sustainable vegetable garden is a value-added skill that integrates a multitude of organic facets to produce crops that are healthy for nature, people, and the planet.

Watering

All plants must be watered when they are transplanted to the garden. There may also be other times during the growing season when lack of rainfall makes watering desirable for best yield or even necessary for best plant growth.

When plants wilt for any length of time, production is reduced; watering helps to prevent serious damage to the plants at these times. If the plants do not recover from wilting soon after the sun is down, the need for water is even more serious.

Application of Water

If the garden is watered, it must be done properly—most plants are actually harmed by shallow watering. Soak the soil to a depth of at least 6 inches. This requires a great deal of water. To be sure the proper depth has been reached, dig down into the soil to see how far the water has penetrated. Another way to tell if enough water has been added is to place a can with straight sides or a rain gauge in the garden and measure how much water has been applied as a result of the sprinkling. The plants should receive no less than 1 inch of water per week. This amount includes any rainfall that may occur. Frequent, light waterings encourage shallow root systems and may actually reduce crop yield. Close row plantings of vegetables need watering less often because shading by plants keeps the soil cooler.

Water may be applied with sprinklers, with a hose, by trench, or by ditch irrigation. Ditch or trench irrigation can be used if the garden is fairly level.

The use of natural mulches 3 to 4 inches thick can reduce the need for water in the garden. A good **mulch** shades the soil and keeps it cool, reducing the rate of evaporation of water from the soil. Porous mulches also slow down runoff of rainwater. Plastic mulch does a good job in moisture retention and also warms the soil during the early season, encouraging growth. **Organic** mulches have an advantage by also adding plant nutrients to the soil.

Fertilizers (Additional)

By this time, some fertilizer should have been applied to the garden by broadcasting and plowing or digging in. Additional fertilizer application after the plants are growing is made by spreading the fertilizer around the plants (band application) or along the row (side-dress application) (Figure 40–1). If fertilizer was not applied to the garden earlier, the amount of fertilizer applied

▲ **Figure 40–1** Fertilizer being placed around a newly transplanted tomato plant.

after the plants have established growth will be increased slightly. In these cases, refer again to the soil test recommendations for the amount to apply. Never apply more than 3 pounds of fertilizer per 100 square feet in a single application.

Plants requiring high amounts of nitrogen, such as corn and leafy vegetables, respond best to applications of fertilizer to the soil surface, such as by side-dressing or band application. This is because nitrogen leaches down to the roots much faster than phosphorus or potash.

Use of **organic fertilizers** such as composted bovine manure can have a significant effect on insect control in the garden. Where composted bovine manure rather than chemical fertilizers is used, populations of flea beetles, aphids, caterpillars, and weed infestations are much lower. Better plant nutrition through minor elements or other materials added by the organic fertilizers apparently results in healthier plants. These results definitely lend support to the organic gardeners' claims that organic fertilizers can promote resistance to insect pests.

Organic Fertilizers

Generally, one feeding with organic fertilizers is adequate for the entire year. Some additional nitrogen may be needed. Organic fertilizers work with nature to increase the **microorganism** activity in the soil. Their nutrients become available to plants only when they are broken down by soil microorganisms. They are food for the microorganisms. Nutrients are released slowly, speeding up as the soil warms and plant growth increases. This results in more nutrients being available when plants need them most. Many chemical fertilizers tend to be available quickly after application, and plants cannot use all the nutrients before they are washed through the soil and run off into streams. Organic fertilizers also provide many, if not all, of the minor elements or micronutrients plants need for vigorous growth and good health.

Organic fertilizers are applied at a rate to add about 4 ounces of nitrogen per 100 square feet. Check the amount of nitrogen available in your fertilizer to determine the amount to apply. Organic fertilizers vary widely in nitrogen content. If plants are vigorous and healthy, an organic mulch or a 1- to 2-inch layer of **compost** may provide all the nutrients needed.

Seaweed or kelp contains many of the same benefits as an organic fertilizer and may be applied as a liquid to plant leaves on nutrient-deficient plants for a quick fix. Even healthy plants will generally increase yields with the use of seaweed.

Weed Control

Weed control is the single most important cultural practice in most gardens. All other effort is useless if weeds are allowed to overtake the garden. Weeds rob garden vegetables of water, plant food, and light. Some weeds help spread diseases, insects, and nematodes. Weeds are most easily controlled before they break through the ground or while they are quite small.

Cultivation

When weeds are small (less than 1 inch tall), shallow **cultivation** (turning the soil over 1 inch deep) will control the weeds and prevent damage to crop roots. Deep cultivation damages many crop roots and brings moist soil to the surface, which dries out and causes moisture loss from the soil. If weeds are allowed to grow taller than 1 inch, deep cultivation is needed to kill the weeds and, in the process, many vegetable plant roots are damaged.

The secret to good weed control is to kill the weeds when they are very small by using a hoe or other shallow cultivator. The oscillating hoe is an excellent tool for shallow cultivation. By sliding it back and forth just under the soil surface, small weeds are cut off and killed. This shallow cultivation is quicker, requires less muscle, and does not damage vegetable crop roots. As mentioned in Unit 39, a special type of hoe is used for wide-row planting.

As soon as the soil is dry enough to work after a rain, it should be hoed or cultivated to kill any weed seeds that have sprouted. This also leaves the soil surface loose to absorb later rainfall. Another advantage to the loose surface soil is that the weed seeds do not germinate and grow as quickly. Hoeing and cultivation should be done often enough to eliminate weeds while they are less than 1 inch tall. Hoeing or cultivation is unnecessary if weeds are not present, unless the soil has a hard crust on it that would prevent rainwater from soaking in.

Mulching

Mulching is another excellent method of weed control in the garden (Figure 40–2). Organic materials such as straw, leaves, sawdust, grass clippings, ground corncobs, and wood chips make excellent mulch. Apply several inches of mulch around plants to cover the entire soil surface. The application of materials that pack tightly, such as grass clippings, should be more shallow (2 inches deep) than coarse materials such as straw.

If weeds grow up through the mulch, the mulch is either not thick enough or not applied uniformly. It is important that the mulch keep all light from the soil surface to prevent weed seeds from germinating. For warm-season crops, such as tomatoes, melons, and beans, organic mulches are best applied only after the soil has warmed in late spring. Cool-season crops, such as cabbage, may be mulched earlier. Too early an application shades the soil, thereby preventing it from warming up in the spring and early summer.

GREEN TIP Be sure the material used for mulch does not contain weed seeds, otherwise your garden will be overrun with weeds once they germinate.

▲ **Figure 40–2** Straw mulch applied around a tomato plant. Notice that the mulch is thick enough to completely shade the soil.

When plant material is used for mulch, it is often necessary to apply extra nitrogen fertilizer. As the soil organisms rot the mulch, the soil is robbed of nitrogen, which is used as food by microorganisms. Sawdust uses a great deal of nitrogen in this rotting process, and fresh sawdust is best allowed to rot down for a year before use. To compensate, add extra nitrogen around the plants as a side-dress when organic mulches are used.

Organic mulches serve several purposes. They control weeds, conserve soil moisture, and add plant nutrients to the soil as they decompose.

Black polyethylene plastic makes an excellent mulch for vegetable gardens. The black color warms the soil and may be used early in the spring when plants are first planted in the garden. Polyethylene is available in rolls that may be rolled down the garden row on each side of the plants. Soil is placed over the edge of the plastic to hold it in place (Figure 40–3). Plants may be planted by punching a hole through the plastic and placing the plant in the soil below.

Mulches also conserve moisture in the soil, reducing the need for watering. If properly mulched, the garden need not be cultivated, saving a great deal of work in weed control.

GREEN TIP Because of their tunneling habits, mice and voles may become a problem under mulch that is applied too thickly. The mulch may also offer them an opportunity to chew seedlings or plants undetected. To prevent rodent damage, do not allow mulch to pile up and mow grassy areas near the garden regularly.

▲ **Figure 40–3** Black polyethylene plastic used as a mulch around tomato plants.

Recent research shows that the color of a mulch affects yield of certain crops. For example, US Department of Agriculture research has shown that the use of red mulch can increase tomato production from 32,921 pounds per acre to 37,057 pounds. Different colors of light, either reflected or directly applied, affect phytochrome, a light-sensitive pigment in plants. Red light increases yield in tomatoes, and white mulch increases potato yields as much as 25 percent. No difference in fertilizer or water use is required.

Straw mulch can help control Colorado potato beetle on either potato or tomato crops. For effective control, mulch before the plants come up in the case of potatoes or at transplanting time with tomatoes.

One of the most effective mulch materials for the garden is **hairy vetch**. This legume captures nitrogen from the air and stores it in the plant's roots. Therefore, it also serves as a source of fertilizer. Hairy vetch has many advantages over black plastic mulch when used with tomatoes, including higher yields and reduced fertilizer and pesticide use.

Dr. Arel Abdul-Baki, a plant physiologist at the Beltsville, Maryland, US Department of Agriculture Agricultural Research Center, found that tomatoes grown in vetch mulch yielded about 24 percent more fruit than those grown under black plastic mulch. Plant roots extended deeper into the soil under vetch and were able to survive dry soil conditions better. Under this system, the soil is not tilled at all.

Hairy vetch also helped control Colorado potato beetles on the tomatoes. Few beetles were found on plants mulched with vetch, whereas plants grown in bare soil and on black plastic were heavily infested.

Hairy vetch is a winter annual legume, planted in the fall. It is allowed to grow in the spring until the time to plant tomato seedlings. It is then cut down, and tomatoes are planted through it.

Chemical Weed Control

Modern herbicides provide dependable weed control where hand labor was once required. Herbicides require careful application because too much may injure the crop, and too little may not control the weeds.

Contact your local extension service or horticulture teacher for materials to use.

Application

A sprinkling can may be used to apply these herbicides. Determine the rate of application by applying only water first; then add the chemical and apply to the measured area. Generally, it is more desirable to make two trips over the area and walk in a different direction each time while applying one-half of the total amount of chemical. This technique provides a more uniform application. A similar application technique should be used when applying herbicides with hand sprayers. It is probably best *not* to use weed-killing chemicals in the small home garden.

CAUTION

Weed control chemicals are considered poisonous even though many of them are relatively nontoxic to humans. Read the label on the container, and follow all directions and safety precautions. Avoid contact with the skin and be sure to wash thoroughly after use. Store herbicides out of reach of children and pets and away from foodstuffs.

Thinning

Plants that are allowed to grow too close together compete with one another for things such as food and water. For this reason, it is sometimes necessary to remove a number of plants. Check directions on the seed package for thinning instructions. Never plant transplanted plants more closely together than recommended.

Insect and Disease Control

Insects and diseases can cause a serious decrease in production of the garden if not controlled.

Prevention

Some of the best control measures are preventive measures rather than cures.

- *Dispose of crop residue.* Dispose of all crop residue that may carry diseases or insects over to the next year's planting. Compost this material so that it rots before the next season.

- *Rotate crops.* **Crop rotation** is another good preventive measure. Avoid planting the same crop in the same area each year. A 3- or 4-year rotation is most effective.
- *Plant treated seeds.* Plant seeds that are treated to control disease as a good step in prevention of certain diseases.
- *Purchase resistant varieties.* Buy resistant varieties whenever possible. Many new varieties are disease resistant; some are insect or nematode resistant. A good example is the tomato Better Boy. It is labeled VFN, which means that it resists verticillium wilt, fusarium wilt, and nematodes.
- *Purchase healthy plants.* Plants that are vigorously growing because of good fertilizer and other cultural practices are more able to survive disease and insect attacks and produce good crops.
- *Practice mixed plantings.* See Unit 39.

If disease-resistant hybrids are planted and good cultural practices are followed, many garden vegetables will need no further insect or disease control. Corn, tomatoes, peas, beets, carrots, lettuce, onions, and turnips rarely need much additional help. Other vegetables, especially melons, cucumbers, beans, and cabbage, generally need some additional insect control. This varies with location and local problems. Use biological control agents (**biopesticides**) wherever possible (see Unit 17).

Chemical Control

Most garden insects can be effectively controlled with two relatively safe insecticides, malathion and Sevin. Most fungus diseases can be controlled with fungicides. These insecticides and fungicides can be mixed together and applied at the same time as a spray. Read the label for mixing, use, and safety precautions. Check to see how close to harvest date the sprays or dusts can be applied. The chemical must have time to break down and become harmless before the vegetables are eaten. Use of **compost** in and on the soil can reduce the need for chemicals.

CAUTION

Always apply pesticides only to those crops for which they are recommended. Apply only the amount that is specified on the label. Wash all treated vegetables before eating. Use *all* safety precautions.

Application

Many garden pesticides are applied as dusts using a hand duster. Dusts are available in diluted form and ready to apply. Sprays are often applied with a knapsack sprayer. Material that is sprayed is always diluted with water and should be applied on a quiet day when the wind is not blowing so that the chemical does not drift to other areas.

A spray schedule for the vegetable garden can be obtained at your local extension agent's office or from a state agriculture college extension service office.

Figure 40–4 lists vegetables and the insects that attack them.

GREEN TIP If a chemical must be used, read the recommendation on the pesticide label before purchase.

Harvesting

Harvest vegetable crops at the peak of quality (Figure 40–5). Some crops, such as sweet corn, peas, and asparagus, lose quality rapidly. With these vegetables, just a few days make a great difference in quality. Other crops, such as beets, cabbage, and pepper, retain their quality over a long period of time. See Unit 41 for more detail on harvesting.

GREEN TIP If you don't have your own garden, buy produce from your local farmers. Just look for locally grown foods at your grocery store or farmer's market. Many American meals travel about 1,500 miles to get to your plate by boat, truck, or plane. What does this mean? Wasted petroleum, and your fruit and vegetables are not their freshest and have lost many of their nutrients.

PLANT		PEST AND DESCRIPTION (Only the most damaging insects are listed for each crop in the chart.)	WHEN TO SPRAY
Asparagus	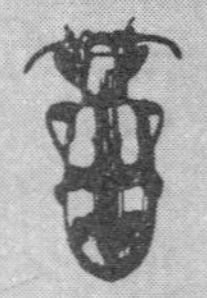 Asparagus Beetle	**Asparagus Beetle.** Metallic blue to black beetle with orange markings. Larvae olive green, ⅓ inch long. Feeds on leaves and shoots.	Contact your local extension agent's office.
Bean	 Mexican Bean Beetle	**Mexican Bean Beetle.** Feeds on underside of leaves. Adult is orange to yellow with 16 black spots on back. Larvae are fuzzy, ⅓ inch long.	Contact your local extension agent's office.
Beet		**Leaf Miner.** Adult is slender, gray, black-haired fly ¼ inch long. Larvae pale green or whitish. Leaves have blotches.	Contact your local extension agent's office.
Swiss Chard	 Beet Webworm	**Beet Webworm.** Larvae are yellow to green, with black stripe and numerous black spots on back. Up to 1¼ inches long.	Contact your local extension agent's office.
Broccoli		**Cabbage Looper.** Pale green worm, white stripes on back, that loops as it crawls. Up to 1½ inches long. Chews holes in leaf.	Contact your local extension agent's office.

Figure 40–4 Spray schedule for vegetables most often requiring spray. *(continued)*

PLANT		PEST AND DESCRIPTION (Only the most damaging insects are listed for each crop in the chart.)	WHEN TO SPRAY
Brussels sprouts Cabbage Cauliflower	 Cabbageworm	**Imported Cabbage Worm.** Velvet green, size up to 1½ inches long. Chews holes in leaves.	Contact your local extension agent's office.
Kale		**Aphid.** Small, green to powdery blue, soft. Causes curling of leaves.	Contact your local extension agent's office.
Kohlrabi		**Club Root Disease.** (swelling on roots causing wilting of plants)	Contact your local extension agent's office.
Celery	 Celery Leaf Tier	**Leaf Tier.** Larvae are greenish caterpillars ¾ inch long. Eats holes in leaves and stalks and ties leaves together with webs.	Contact your local extension agent's office.
Corn (Sweet)	 Corn Earworm	**Corn Earworm.** Larvae are green, brown, or pink worms; light stripes along sides and on back. Up to 1¾ inches long. Feeds on early corn shoot in season. Later feeds on kernels near top of ear.	Contact your local extension agent's office.

Figure 40–4 Spray schedule for vegetables most often requiring spray. *(continued)*

PLANT		PEST AND DESCRIPTION (Only the most damaging insects are listed for each crop in the chart.)	WHEN TO SPRAY
Cucumber Pumpkin Squash	 Cucumber Beetle	**Cucumber Beetle,** Flea Beetle, Leaf Hopper, Pickle Worm. Cucumber Beetle adult is yellow with black dots; Flea Beetle adult is black; Leaf Hopper adult is light green or brown; Pickle Worm adult is green. **Anthracnose.** Reddish-brown spots on leaves, tan cankers on stems, round sunken spots on fruit.	Contact your local extension agent's office.
Eggplant	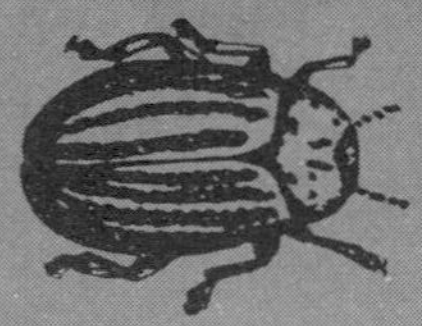 Colorado Potato Beetle	**Colorado Potato Beetle.** Adult is yellow, blackstriped, 3/8 inch long. Larvae are brick red, hunchbacked, 3/5 inch long. Eats leaves. Resistant to many chemicals.	Contact your local extension agent's office.
	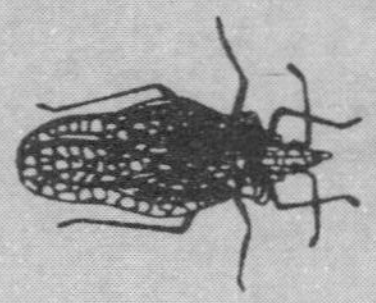 Eggplant Lacebug	**Eggplant Lacebug.** Adult is grayish to light brown; flat lacelike wings, 1/16 inch long. Nymph: yellowish, louselike; up to 1/10 inch long. Feeds on underside of leaves; leaves turn yellow and brown.	Contact your local extension agent's office.
Lettuce	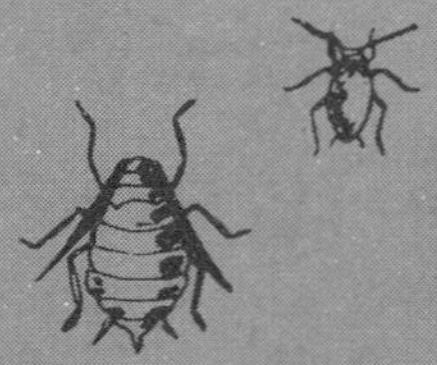 Aphid	**Aphid.** Same as broccoli. **Cabbage Looper.** Same as broccoli. Feeds on underside of leaves, leaving ragged holes.	Contact your local extension agent's office.

Figure 40–4 Spray schedule for vegetables most often requiring spray. (*continued*)

PLANT		PEST AND DESCRIPTION (Only the most damaging insects are listed for each crop in the chart.)	WHEN TO SPRAY
Muskmelon	 Aphid	**Aphid.** Same as broccoli. **Downy Mildew and Leaf Spot.** Brownish spots on leaves; leaves die. Fruits are not affected. **Mosaic (viruses).** Mottled green and yellow curled leaves.	Contact your local extension agent's office.
Okra	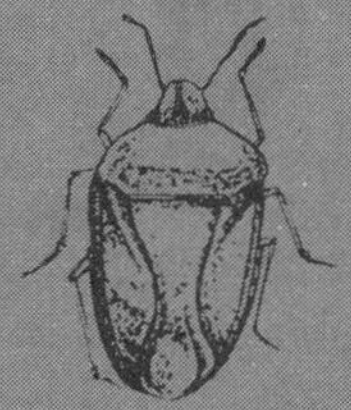 Southern Green Stink Bug	**Corn Earworm.** Same as corn. Eats holes in pods. **Stink Bug.** Black, green, or brown shield-shaped bug; $\frac{5}{8}$ inch long. **Wilt.** Causes yellow and wilted leaves.	Contact your local extension agent's office.
Onion	 Thrip	**Maggot.** White legless worm $\frac{3}{8}$ inch long. Burrows into young bulbs. **Thrip.** Adult is tiny, active yellow or brownish, winged insect. Larvae are white, same appearance as adult, only smaller. Sucks juices and causes blotches on leaves. Tips may turn brown and die.	Contact your local extension agent's office.
Pea	 Aphid	**Aphid.** Same as broccoli.	Contact your local extension agent's office.

Figure 40–4 Spray schedule for vegetables most often requiring spray. (*continued*)

PLANT	PEST AND DESCRIPTION (Only the most damaging insects are listed for each crop in the chart.)	WHEN TO SPRAY
Potato 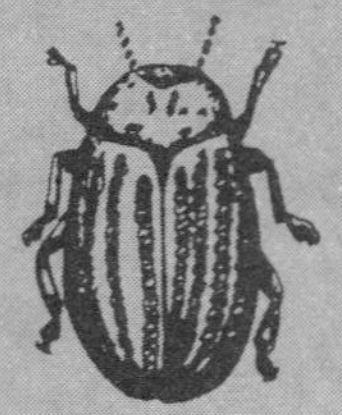 Colorado Potato Beetle 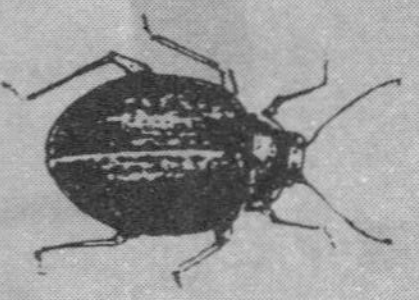 Potato Flea Beetle	**Colorado Potato Beetle.** Same as eggplant. **Flea Beetle.** Black, brown, or striped jumping beetles; about 1⁄16 inch long. Leaves appear to have been shot full of small holes. **Early Blight.** Leaves show small, irregular, dark brown spots. Fungus is carried in soil. **Late Blight.** Dark irregular dead areas on leaves and stems. Infected tubers rot in storage; plants may die in season.	Contact your local extension agent's office.
Tomato 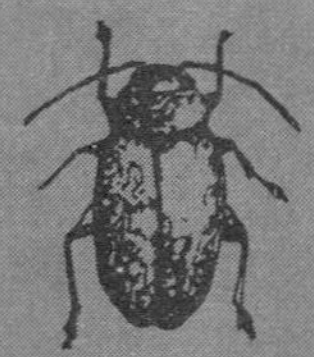 Flea Beetle	**Flea Beetle.** Same as potato. **Tomato Fruit Worm.** Larvae are green, brown, or pink; light green stripes along sides and on back; up to 1¾ inches long. Same as corn earworm. Eats holes in fruit and buds. **Fusarium and Verticillium Wilts.**	Contact your local extension agent's office.

Note: When organic practices of adding compost to the soil and mulching with a composted organic material are used, many plants will develop increased resistance to insects and diseases and not need chemical sprays. Use of natural controls such as beneficial bacteria in the soil and insect predators can also greatly reduce the need for chemical sprays. (See Unit 16.)

Figure 40–4 Spray schedule for vegetables most often requiring spray.

▲ **Figure 40–5** A well-cared-for garden planted using the wide-row system.

Gardening Notes—What's New

Consider the following gardening notes to stay up to date in your garden:

- *Tomatoes.* New varieties set fruit in hot or cold weather. 'Oregon Pride' and 'Oregon Star' are two examples for cool areas. 'Solar Set,' 'Sunfire,' 'Arkansas Traveler,' and 'Big Rainbow' are for hot areas.

  Tickle your tomato plants for 25 percent to 35 percent shorter, stockier plants. This means brushing your hand lightly over the tops of tomato seedlings for 1½ minutes twice a day while they are still indoors to emulate wind movement. This results in shorter plants, according to a study at the University of Georgia.

  Mulch with red plastic or hairy vetch.

TRY THIS Germinate and maintain tomato seeds in a school or community greenhouse. Maintain two groups of tomato plants. Conduct an experiment during the growing period to test the "tomato tickling" research from Georgia University. To test this theory, consistently brush or touch the top of tomato seedlings for 1½ minutes twice a day for one group of tomatoes, and leave the other group untouched. Record the growth of the two groups of tomato plants, and write a report detailing your findings.

- *Potatoes.* Produce 25 percent more with use of white mulch.

  Straw mulch can help to control the Colorado potato beetle if applied as soon as plants emerge.
- *Leeks.* Grow from bulblets faster than from seed.
- *Broccoli.* 'Emperor' variety is resistant to black rot.
- *Watermelon.* Seedless varieties are sweet and easier to eat (must also plant a pollinator).
- *Mixed plantings of vegetables.* This reduces damage by insects. Mixing confuses insects because of the mixed smell of chemicals given off by the different plants. They then have difficulty finding their favorite.

Some of the most destructive vegetable insects are pictured in Figures 40–6 through 40–9.

▲ **Figure 40–6A** The Colorado potato beetle larva.

▲ **Figure 40–6B** The Colorado potato beetle adult.

Source: US Department of Agriculture

▲ **Figure 40–7** The Mexican bean beetle, eggs, larva, pupa, and adult.

Source: US Department of Agriculture

▲ **Figure 40–9** Tomato hornworm. This one is parasitized by the branconid wasp. The white eggs attached to the worm will hatch, and the small wasp larvae will enter the worm and kill it.

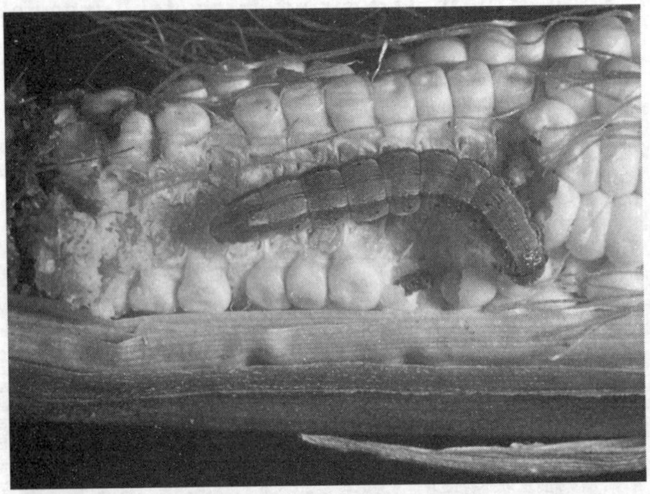

Source: US Department of Agriculture

▲ **Figure 40–8** Corn earworm.

▲ **Figure 40–10** A bag of the biological insecticide Dipel.

Review

In review, follow these steps when caring for a garden:

1. Choose fertile, well-drained soil and apply water and proper fertilizer to produce vigorous plants.
2. Plant crops adapted to the area.
3. Keep weeds under control.
4. Use only disease-free seed, certified if possible.
5. Treat seed with chemicals to protect against decay and damping-off.
6. Purchase disease-free plants from a reputable dealer.
7. Plant disease-resistant varieties whenever possible.
8. Destroy all plant refuse as soon as harvest is over to prevent carryover of diseases and insects. Compost if possible.
9. Use insecticides as a last resort and then only according to directions (Figure 40–10).
10. Select insecticides that do the least harm to helpful insects and the environment.

SAFETY FIRST

OSHA's regulations ensure that employers and employees know about the work hazards and how to protect themselves to prevent injuries in the workplace. For more information on occupational health and safety in the workplace, please visit the OSHA site at www.osha.gov.

Summary

Caring for the vegetable garden is a continuous task throughout the growing season. Control of insects, diseases, and weeds must be done on a timely and thorough basis. By using mulch, resistant varieties, proper planting techniques and soil preparation, and compost, time spent in the garden can be greatly reduced.

Review Questions

1. Discuss 10 key points for insect and disease control in the vegetable garden.
2. Select two vegetables, and name the pest that most commonly attacks them. Describe the pest, and discuss the chemicals used to control it. What other methods may be used to control it, beyond chemicals?
3. If fertilizer recommendations call for 50 pounds of actual nitrogen per acre, how many pounds of a 10-10-10 fertilizer would you need to apply to 1 acre?
4. Describe some methods of controlling weeds in the vegetable garden.
5. Discuss different methods for watering a vegetable garden. How do you determine if plants are receiving enough water?

Student Activities

1. Actually plant a garden. Demonstrate the techniques of planting and caring for the garden.
2. Work along with an experienced gardener.
3. Mulch tomatoes with a red mulch, and compare the yield against that with black plastic mulch.
4. Demonstrate how to compost garden refuse.
5. Search online for ideas on how to use compost in a vegetable garden bed. Choose one of the ideas you researched, and try it in your own garden experiment.
6. With the assistance and supervision of the instructor, use social media to advertise an FFA fund-raiser selling vegetables from a school or SAE garden.

Suggested Supervised Agricultural Experience (SAE) Activities

1. Work as an apprentice to a master gardener as part of a placement SAE.
2. Create a social media page to sell vegetables grown in an SAE garden to qualify as an entrepreneurship SAE.

Additional Resources

https://utextension.tennessee.edu/publications/documents/SP291-D.pdf

http://extension.arizona.edu/sites/extension.arizona.edu/files/pubs/az1435.pdf

http://web.extension.illinois.edu/vegguide/step05.cfm

UNIT 41

Favorite Garden Vegetables and Herbs

Objective

To cultivate and harvest five vegetables commonly grown in the local area.

Key Terms

determinate varieties
herbs
indeterminate varieties
site

Competencies to Develop

After studying this unit, you should be able to:

- list the requirements for planting, fertilizing, harvesting, and storing five garden vegetables that grow successfully in the local area.
- select five locally grown vegetables and plant and grow them to marketable quality.
- select and grow herbs to use for spicing up your favorite garden vegetables.

Materials

- seed catalogs
- seed
- planting site
- fertilizer
- minimal garden tools (spade, rake, garden weasel, hoe)
- duster or sprayer
- chemical sprays or dusts if necessary
- smartphone apps:
  - Vegetable Tree – Gardening Guide
  - Vegetable Gardening Guide
  - Gardening: The Ultimate Guide
  - Weather
  - Spectracide
  - Pest Portal
  - Calibrate My Sprayer
  - PGR Mix Master
  - Ortho Problem Solver

Notes on the Unit

Complete a soil test analysis for the lime and fertilizer recommendations. The two preplant methods of application mentioned for each vegetable are broadcasting and digging in and band application. Follow the soil test lab results recommendation for the correct fertilizer application.

Asparagus

Asparagus is a perennial and among the earliest of spring vegetables to appear. It grows best in areas where the soil freezes to a depth of at least a few inches in winter. Asparagus grows from a crown; the roots look like spiders with a long slender tapered root. The plant is dioecious, which means the male and female flowers are produced on separate plants. Male plants have several advantages over female plants: they appear earlier in the spring, they do not produce fruit and therefore do not produce unwanted volunteer seedlings, they have better disease resistance, and they are more productive.

The asparagus crop may be grown on almost any well-drained fertile soil. First, dig a trench 16 inches wide, laying aside the topsoil. Plow or spade lots of manure, leaf mold, rotted leaves, or peat into the subsoil to a depth of 14 to 16 inches (below the normal soil surface). Fill the trench with topsoil to within 6 inches of the natural soil level. Set the asparagus crowns and cover with 1 to 2 inches of soil. As the shoots grow during the first season, gradually replace the topsoil in the trench until it reaches the natural soil level. Set the plants at least 1½ feet apart in the row. Rows should be 4 to 5 feet apart.

Fertilizer

Follow the soil test analysis for the proper fertilization of the asparagus.

Harvest

Remove no shoots the first 2 years after planting. Each following year, remove all the shoots until July 1, when harvesting must be stopped to allow the plant to replenish root reserves of plant food. Harvest when the spears are 6 to 10 inches long. To harvest, break the spears off above soil level. Cutting below ground level reduces yield. Harvest frequently to prevent tough spears. In the fall, remove the dead plant tops and work them into the soil.

Use

Fresh, cooked fresh, frozen, canned.

Storage

Store by freezing or canning. Asparagus may be kept fresh for short periods of time at 32°F (0°C) and 85 to 90 percent relative humidity.

Lima Beans

Lima beans (Figure 41–1) are a warm-season crop and should not be planted until the soil and the air have warmed sufficiently. Limas are able to grow in most soils. In heavy soils that tend to turn crusty on top, the seed should be covered with sand rather than soil so that the seeds can push through the soil more easily. Another technique if the soil turns crusty on top is to use a garden weasel to break up the crusty soil so the seeds can emerge.

Lima beans require a growing season of about 4 months at a relatively high temperature. They need a richer soil than snap or green beans.

Limas grow both on poles and in bush form. The pole limas need some means of support such as poles or trellises. When poles are used, plant

Courtesy of W. Atlee Burpee & Company

▲ **Figure 41–1** Burpee's 'Fordhook' lima beans.

three or four plants per hill, and space the hills 2 to 3 feet apart. When planted to grow on trellises, space the plants 6 to 12 inches apart. Bush varieties are planted 3 to 4 inches apart in rows spaced 3 feet apart.

Fertilizer

Follow the soil test analysis for the proper fertilization of the lima beans.

Harvest

Harvest lima beans when the seeds are close to full size but before the color changes from green to white.

Use

Fresh cooked, frozen, canned, dried.

Storage

Limas remain fresh at 32°F (0°C) and 85 to 90 percent relative humidity. They can also be frozen, canned, or dried.

Snap Beans

Snap beans (Figure 41–2) should not be planted until the ground is warm. Successive plantings can be made every 2 to 3 weeks until 8 weeks before the fall frost date. Snap beans can be grown in a wide variety of soils, providing that the soil is well drained and not so heavy that it crusts, preventing seeds from emerging from the soil.

Fertilizer

Follow the soil test analysis for the proper fertilization of the green beans.

Harvest

Pick when seeds in the pods are one-third their mature size and when the pods are still young and tender.

Use

Fresh cooked, frozen, canned.

▲ **Figure 41–2** Snap beans.

Storage

Ideal fresh storage should be at 45 to 50°F (7 to 10°C) and 85 to 90 percent relative humidity. They can also be frozen or canned.

Beets

Beets (Figure 41–3) are very hardy vegetables and grow under a wide range of climatic conditions. They will not, however, tolerate severe freezing. Beets require a loose, loamy soil for best results.

> **GREEN TIP** Do not plant beets, cabbage, cauliflower, or brussels sprouts in the same area year after year because the same pests attack all these plants.

Beets are planted in rows as close as 16 inches apart for hand hoeing. Plants are spaced 3 inches apart in the row. Successive plantings may be made at 3-week intervals during the season. Seeds planted in hot, dry weather may not germinate well. Mulching or watering to keep the soil moist will help to resolve this problem. Beets do not grow well in acid soils. Check the soil pH, and add lime if needed to keep soil pH above 6.5.

© Brent Hofacker/Shutterstock.com

▲ **Figure 41–3** Red beet.

Fertilizer

Follow the soil test analysis for the proper fertilization of the beets.

Harvest

For tender beets, harvest when they are 1½ to 2 inches in diameter. A fall crop that is to be stored may be allowed to grow a little larger.

Use

Fresh cooked, canned, pickled.

Storage

Store fresh at 32°F (0°C) and 90 to 95 percent relative humidity for best results.

Broccoli

If a spring crop of broccoli is desired (Figure 41–4), plant seeds indoors and set plants in the garden in March or April. Set the plants 12 to 18 inches apart in rows 3 feet apart. Plants begin to yield sprouts about 10 weeks after planting. A fall crop may be direct seeded by sowing seed in June.

© al1962/Shutterstock.com

▲ **Figure 41–4** Broccoli.

Fertilizer

Follow the soil test analysis for the proper fertilization of the broccoli.

Harvest

The broccoli head, which is a cluster of flower buds, should be harvested before the buds begin to open. Small side clusters develop after the main head is cut and may be harvested over a long period of time.

Use

Fresh (as a greens crop), frozen.

Storage

For best results, store fresh broccoli at 32°F (0°C) and at 90 to 95 percent relative humidity.

Brussels Sprouts

Being hardier than cabbage, brussels sprouts can survive outdoors over winter in all the mild sections of the country. They may be grown as a winter crop in the southern regions of the United States, the same way as early and late cabbage is grown in southern sections. As the heads begin to form, break the lower leaves from the stem to give them more room. Do not remove top leaves.

Fertilizer

Follow the soil test analysis for the proper fertilization of the brussels sprouts.

Harvest

Remove the sprouts when they are 1 to 1¼ inches in diameter by cutting them close to the stem.

Use

Fresh, frozen.

Storage

See Broccoli.

Cabbage

Cabbage (Figure 41–5) is one of the most important of the home garden crops. It can be grown throughout almost the entire United States. For a spring crop, set the plants in the garden as early as the soil can be prepared. For fall crops, plant again in late summer.

Cabbage may be planted after early potatoes, peas, beets, spinach, or other early crops, or it may be set between rows of crops before these crops are harvested. Set plants about 1 foot apart in rows 3 feet apart. Plantings must be very shallow.

GREEN TIP Be sure to use methods for control of the cabbageworm to avoid losing your cabbage crop to these pests.

Fertilizer

Follow the soil test analysis for the proper fertilization of the cabbage.

Harvest

Harvest when the heads are firm, but before they split.

Use

Fresh (in salads and coleslaw), prepared as sauerkraut.

▲ Figure 41–5 Cabbage.

Storage

Fall cabbage may be stored in an outside pit or trench or in a cool vegetable cellar. Storage at 32°F (0°C) and at 90 to 95 percent relative humidity is best.

Carrot

Carrots (Figure 41–6) are cold-hardy. Seed may be planted as early in spring as the ground can be worked. Carrot seed does not germinate well in hot, dry weather. The soil requirements for carrots are very similar to those for beets. They prefer a deep, loose, sandy loam that is free of stones.

Space plants no closer than 1 inch apart in the row. Cultivate by pulling loose soil in toward the plant to keep the crown covered. This prevents the top of the carrot from turning green.

▲ Figure 41–6 Carrots, organic heirloom varieties.

Fertilizer

Follow the soil test analysis for the proper fertilization of carrots.

Harvest

Harvest carrots when the plants are 1 to 1½ inches in diameter, depending on the variety.

Use

Raw (alone or in salads), cooked fresh, canned.

Storage

Carrots must be dug and stored before frost kills the roots. Fresh carrots are stored best at 32°F (0°C) and at 90 to 95 percent relative humidity. They may also be stored outdoors covered with straw or other material to prevent freezing.

Cauliflower

Cauliflower (Figure 41–7) requires cultural conditions similar to those of cabbage. Cauliflower will not tolerate as much cold weather as cabbage, however. High temperatures cause poor-quality heads. Cauliflower grows in any type of well-drained, fertile soil. Set the plants 1½ to 2 feet apart in 3-foot rows. New cultivars of cauliflower have self-wrapping leaves to shield the snow-white, orange, green, or purple cauliflower heads. This is not the result of genetic engineering, but rather decades of selective breeding. It is interesting to note that the purple color in cauliflower is caused by the presence of the antioxidant group anthocyanins, which is found in red cabbage too.

▲ Figure 41–7 Cauliflower.

Fertilizer

Follow the soil test analysis for the proper fertilization of the cauliflower.

Harvest

Harvest when the heads are firm and compact. Do not allow the heads to become "ricey" appearing.

Use

Cooked, fresh.

Storage

Fresh storage at 32°F (0°C) and 85 to 90 percent relative humidity is best. May also be frozen and canned.

TRY THIS Using the Internet for research, determine the difference between genetic engineering and selective breeding of vegetables. Create a chart that describes the advantages and disadvantages of each method.

Celery

Celery is a cool-season vegetable grown as a spring or fall crop in the upper southern and in northern sections of the country. It is a good winter crop in lower southern regions.

Soil that is rich, moist, well drained, deeply prepared, and loose is a must for celery. Plenty of organic material is also necessary. About 10 weeks are required to grow healthy plants from seed to transplanting size. Set plants 18 to 24 inches apart for hand cultivation. For tractor cultivation, space plants 6 inches apart in rows 3 feet apart.

Fertilizer

Follow the soil test analysis for the proper fertilization of the celery.

Harvest

Harvest when large enough to eat.

Use

Fresh, alone, in salads and other dishes.

Storage

For early winter use, bank soil around the plants and cover the tips with straw or leaves to prevent freezing. May also be stored in a cold frame or cellar with high humidity. Storage at 34°F (2°C) with 90 to 95 percent relative humidity is best.

Celeriac

The cultivation of celeriac, known as turnip-rooted celery, is the same as that of celery. The root, rather than the top, is eaten.

Fertilizer

Follow the soil test analysis for the proper fertilization of the celeriac.

Harvest

Use the roots any time after they are big enough.

Storage

Store in the ground with straw or in cold storage.

Chard

See Beets. The only difference is that the plants grow larger and require spacing of at least 6 inches apart in the row. Chard needs a rich, loose soil. It is sensitive to soil acidity.

Chervil

Salad chervil is grown as parsley is. The seed must be bedded in damp sand for a few weeks before planting, or germination is very slow.

Turnip-rooted chervil is planted in the fall in southern regions and may not appear until spring. In northern regions, it is either planted in the fall or seeds are started in flats to grow transplants for spring planting. For cultivation, see Beets.

Chicory, Witloof

Witloof chicory is grown for both its roots and tops. It is a hardy plant that tolerates both hot and cold climates. A rich, loamy soil without much organic matter is best.

Sow seeds in rows in late spring. Thin so that they are kept 6 to 8 inches apart in rows. If sown too early, the plants go to seed, preventing later forcing.

Fertilizer

Follow the soil test analysis for the proper fertilization of the chicory.

Harvest

The tops are sometimes harvested when young.

Use

Top is used as cooking greens.

Storage

The roots are lifted in the fall and placed in moist soil in a warm cellar for forcing. Roots are covered with a few inches of sand. Under this covering, the leaves form a solid head known as a *witloof* (Dutch for "white leaf").

The Onion Group (Chives, Garlic, Leek, Shallot, Onion)

Onions thrive under a variety of climatic and soil conditions. A moist, warm climate is best.

The soil should be fertile, moist, and loose. It should also contain plenty of compost, have a fine texture, and be free of clods and foreign matter. Soil pH should be 7.0 for best results.

GREEN TIP Create your own compost by throwing in your vegetable waste instead of throwing it out; use it to replenish your garden. Compost enriches soil fertility by giving it vital nutrients, stimulating healthy root development, and improving your soil's texture, aeration, and water retention.

Onions are generally started as sets or as small dry onions grown the previous year from seed. Seedling onions have an advantage over sets in that they rarely go to seed and thus produce better bulbs.

Plant onion sets 3 inches apart in rows 1 foot apart. Plant 2 inches deep. Onion plants should have soil pulled away from them during cultivation to expose the onion; leeks should have soil pulled up around them.

Fertilizer

Follow the soil test analysis for the proper fertilization of the onions.

Harvest

Onion stalks may be harvested young for salad greens or left in the ground to mature. The mature onions are pulled for storage after the tops have died down.

Use

Fresh (in salads or alone), cooking, seasoning.

Storage

Store mature onions at 32°F (0°C) and in a relatively dry, well-aerated location at 70 percent relative humidity for best results.

Following is an explanation of members of the onion group only as they differ from onions.

Types of Onions

CHIVES Chives are perennials and should be planted where they can be left for more than 1 year. Divide and reset if plants become too thick. They are used primarily for seasoning.

▲ **Figure 41–8** Leeks harvested young for fresh consumption.

GARLIC Large bulbs must be separated before planting, as each contains about 10 small bulbs. Garlic is used primarily for seasoning.

LEEK Rather than forming a bulb, leeks produce a thick, fleshy, stemlike onion (Figure 41–8). Leeks are started from seed or small bulblets and harvested any time they are large enough to eat. Store like celery.

SHALLOT The shallot is a small onion. Its bulbs have a milder flavor than that of most onions. They seldom form seeds and are propagated by small divisions into which the plant splits during growth. The shallot should be lifted and separated, and the smaller ones should be replanted each year.

Collards

Collards are grown and used similarly to cabbage. They withstand heat better than other members of the cabbage group.

Collards do not form a true head, but rather a rosette of leaves that may be tied together for blanching.

The Sweet Corn Group (Sugar, Sugary Enhanced, Supersweet, Synergistic Hybrids)

Sweet corn (Figure 41–9) requires a great deal of garden space and is best used only in larger gardens. It is planted in 3-foot rows, 12 to 14 inches apart in the row, or in hills of three plants per hill spaced 3 feet apart. Also plant it at least 200 feet away from field corn to prevent cross-pollination.

Corn is a tender crop and should not be planted until the danger of frost has passed. The new supersweet varieties require warmer soil temperatures to germinate. Soaking seed in warm water increases the percentage of germination. It should not be planted in single rows. Plantings four rows wide are desirable to aid in pollination even if rows must be short. Sweet corn grows in almost any well-drained garden soil.

▲ **Figure 41–9** Sweet corn.

Fertilizer

Follow the soil test analysis for the proper fertilization of the sweet corn.

Harvest

Sweet corn is ready to eat when milk squirts from the kernel as the thumbnail is pushed into the kernel. The kernels should be full size. Pull and cook immediately. The sugar content of corn decreases rapidly after it is pulled from the plant. New, high-sugar varieties stay sweet much longer; the sugar in them is not converted to starch.

Use

Fresh as roasting ears, frozen, canned.

Storage

Store sweet corn by freezing and canning. Fresh ears are best held for a short time at 32°F (0°C) and at 85 to 90 percent relative humidity.

Types of Sweet Corn

SUGAR Labeled as SU, this is like the old-fashioned sweet corn that has a limited shelf life. The sweet corn needs to be processed right away to prevent the sugar content from converting to starch. The sooner it is processed or eaten, the higher the sugar content.

SUGARY ENHANCED Labeled as SE, this sweet corn has a gene that slows down the conversion of sugar to starch. This corn is sweeter and more tender than SU. The kernels are more tender and creamy than SH2 types.

SUPERSWEET Labeled as SH2, this sweet corn has a long shelf life with enhanced sweetness and kernels that are crisp and juicy.

SYNERGISTIC HYBRIDS These are a combination of 25 percent supersweet and 75 percent sugary enhanced. It has a longer shelf life than other varieties, plus the flavor and tenderness of the sugary enhanced.

Cucumber, Cucamelon, and Muskmelon (Cantaloupe)

Cucumbers, muskmelons, and cucamelons (Figures 41–10 and 41–11) are warm-season crops. Hot weather may be too severe and greatly lowers yields. In southern sections, they may be grown in the fall. A very fertile, loose soil high in rotted organic matter is necessary for best yields.

Seeds are planted in hills 6 feet apart, followed by thinning to four plants per hill, or in rows 6 feet apart with plants spaced 2 to 3 feet apart in the row. Cover seed with 1½ inches of soil. If soil is heavy and becomes crusty, cover seed with sand. Plants may be started indoors in peat pots and transplanted to the garden for earlier crops.

© MSPhotographic/Shutterstock.com

▲ **Figure 41–10** Cantaloupe (muskmelon) is very healthy as well as refreshingly tasty.

© julie deshaies/Shutterstock.com

▲ **Figure 41–11** Cucamelon.

GREEN TIP Control of the cucumber beetle is necessary for good crop yields. The beetle spreads bacterial wilt, which kills the plants.

Fertilizer

Follow the soil test analysis for the proper fertilization of the cucumbers or muskmelons.

Harvest

Cucumbers are harvested for fresh slicing while they are crisp and fresh and before any yellowing begins. Cucumbers used for pickling may be harvested at any size desired but before any yellowing begins. Pull all mature fruit off the plant, even if it is not wanted. Yield is reduced if fruit is allowed to remain on the vine.

Cantaloupe or muskmelon is harvested when the fruit separates from the stem with a slight tug or push with the thumb.

Use

Cucumbers are used in salads and pickling. Cantaloupe is eaten as fresh fruit.

Storage

Fresh storage should be at 45°F (7°C) and 85 to 90 percent relative humidity for best results.

Eggplant

The eggplant (Figure 41–12) demands warm weather and a warm soil. A growing season of from 100 to 140 days is needed for crop maturity.

Culture and fertilization are the same as those for tomatoes, except that eggplant may be planted closer together—18 inches apart in the row. Eggplant is a highly productive plant; six plants will furnish enough for the average family.

Fertilizer

Follow the soil test analysis for the proper fertilization of the eggplant.

Harvest

Harvest 42 days after the fruit begins to form and while the fruit is still glossy.

▲ **Figure 41–12** Eggplant.

Use

Cooked.

Storage

Eggplants should be stored like tomatoes.

Endive

For cultural requirements, see head lettuce under Lettuce. Endive is less sensitive to heat than head lettuce. In the south, it is a winter crop and in the north, a spring, summer, and fall crop. Tie leaves over the heart or shade and blanch the plant.

Fertilization

Follow the soil test analysis for the proper fertilization of the lettuce.

Harvest, and Use

See Lettuce.

Storage

Dig plants with a rootball attached, and place them in a cold frame or cellar where they will not freeze. Use as needed.

Kale

Kale is a member of the cabbage family. It is a hardy plant and lives over winter as far north as southern Pennsylvania. It is also resistant to heat and may be grown in summer. Kale is best grown as a cool-weather crop.

Plant in rows 12 inches apart, and keep plants thinned to 6 inches apart in the row. Sow seed in early spring or in early August for a fall crop. Seed may also be broadcast and raked in. Cover seed with only ½ inch of soil.

Fertilizer

Follow the soil test analysis for the proper fertilization of the kale.

Harvest

Kale may be harvested by pulling the entire plant from the soil or by cutting off the outside leaves as the proper size is reached. Old leaves are tough and stringy.

Use

Cooked as greens, fresh, frozen.

Storage

Keep cool (40°F; 4°C) and moist.

Kohlrabi

Kohlrabi is grown for its swollen stem. For cultural requirements and uses, see Cabbage.

Fertilizer

Follow the soil test analysis for the proper fertilization of kohlrabi.

Harvest

Harvest kohlrabi when it is young and tender.

Storage

Keep cool (40°F; 4°C) and moist.

Lettuce

Lettuce, a cool-season crop, is very sensitive to heat. In the southern United States, the growing of lettuce is continued into late fall, winter, and spring. In most other areas, it is a spring and fall crop.

Lettuce grows in any good garden soil. The soil pH should be between 6.5 and 7.0; check soil pH before planting. Head lettuce plants are grown indoors and transplanted to the garden. Leaf lettuce may be direct seeded. It may be planted very early because temperatures as low as 28°F (22°C) do not harm plants that have been properly hardened off. Set head lettuce plants 12 inches apart in rows 3 feet apart. Leaf lettuce plants may be planted much closer to each other.

Fertilizer

Follow the soil test analysis for the proper fertilization of lettuce.

Harvest

Harvest head lettuce when the heads are firm or slightly earlier. Leaf lettuce is picked before the leaves become tough. The older leaves should be used first.

Use

Fresh, in salads and other dishes.

Storage

Lettuce is stored at 32°F (0°C) and 90 to 95 percent relative humidity for best results.

Mustard

Mustard grows in almost any type of soil. It is a cool-season crop and matures rapidly. Prepare the same as a fresh-cooked greens dish, such as kale.

Fertilization and Storage

See Lettuce.

Okra

See Tomato.

Parsley

Parsley is hardy in cold weather but sensitive to heat. It thrives under the same conditions as kale and lettuce. It is important that parsley be planted early in the season. It grows in any good, loose garden soil. Because young parsley plants are slow to start, starting them indoors in flats and transplanting to the garden is best. Soaking seed overnight in water helps to speed up germination.

Plant parsley in rows 14 inches apart with plants 6 inches apart in the row.

Fertilizer

No special fertilizer is needed.

Harvest

Cut leaves as needed.

Storage

Store in the soil with mulch or in cool, moist conditions the same as other root crops such as carrots.

TURNIP-ROOTED PARSLEY This type of parsley is grown for the root and has much the same flavor as celeriac. The cultural conditions are the same as for parsley.

Peas

Peas (Figure 41–13) are a cool-season crop and can be seeded in the spring as soon as the ground can be prepared. A soil high in organic matter is best.

Plant in double rows 6 to 8 inches apart with just enough space between the rows to allow cultivation. Sow seeds 1 inch apart in the rows.

Fertilizer

Follow the soil test analysis for the proper fertilization of peas.

Harvest

Harvest when the pods are well filled but just before the seeds reach full size. Cook or process immediately after picking for best results.

Use

Fresh cooked.

▲ **Figure 41–13** Peas.

Storage

Peas are especially good fresh from the garden. If they must be stored, they remain freshest at 32°F (0°C) and 85 to 90 percent relative humidity. Freeze or can for permanent storage.

Peppers

The cultural practices for peppers (Figure 41–14) are the same as those for tomatoes, except that warmer soil is required before planting. Set plants 1½ to 2 feet apart in rows 3 feet apart.

Fertilizer

Follow the soil test analysis for the proper fertilization of peppers.

Harvest

Harvest peppers when they reach the desired size.

Use

Fresh (in salads and other dishes) and pickled.

▲ **Figure 41–14** Sweet pepper.

Storage

Store at 45 to 50°F (7 to 10°C) and at 85 to 90 percent relative humidity for best results.

Potato

The potato is one of the highest-producing vegetables in terms of food per area of land. A cool-season crop, potatoes are planted as early in spring as possible or in mid to late summer for a fall crop. Any loose, well-drained, fertile soil is suitable. The soil should be moderately acidic.

Plant small (1½ to 2 ounces), chunky rather than thin, cut pieces of potato with one or more eyes. Place the seed pieces 3 inches deep in the soil and space them 10 to 12 inches apart in rows spaced 3 feet apart.

Fertilizer

Follow the soil test analysis for the proper fertilization of the potatoes.

Harvest

Potatoes may be dug for immediate use as soon as they are large enough to eat. Do not dig potatoes that are to be stored until the tops are dead. Handle them carefully and avoid skinning. Do not leave potatoes exposed to light or they will turn green and be unfit for use.

Use

Cooked fresh.

Storage

Store potatoes at 45 to 50°F (7 to 10°C) in a well-ventilated area. Keep the storage area dark to prevent greening. If possible, curing at 50 to 60°F (10 to 16°C) in moist air before storage is an advantage.

Radish

Radishes (Figure 41–15) are hardy in cold weather but cannot tolerate heat. They may be grown in spring or fall. The crop matures from 3 to 6 weeks after planting, so a number of plantings can be made. A rich, loose, well-drained soil is best.

There are two types of radishes—the mild, small, quickly maturing type that reaches edible size in 20 to 40 days, and the large winter radishes that require 75 days for growth.

© pick/Shutterstock.com

▲ **Figure 41–15A** 'Cherry Belle' radish.

© pilipphoto/Shutterstock.com

▲ **Figure 41–15B** 'White Icicle' radish.

Fertilizer

Follow the soil test analysis for the proper fertilization of radishes.

Harvest

Both types of radishes are harvested as soon as they have reached a desirable size.

Use

Fresh (in salads).

Storage

Store as root crops are stored.

Rutabaga and Turnip

Rutabagas and turnips, cool-season crops, are similar in terms of soil and fertilizer requirements and ways in which they are used as food items. They are both grown in southern regions chiefly in the fall, winter, and spring; in northern parts of the country, they are grown mostly in the spring and autumn. Rutabagas grow best in northern regions; turnips are better for gardens south of Indianapolis, Indiana, or northern Virginia. A half ounce of seed of these vegetables allows for the broadcasting of 300 square feet.

Turnips require 60 to 80 days to mature; rutabagas require 90 to 110 days.

Fertilizer

If planted after early potatoes or other early crops that have been fertilized, no additional fertilizer is needed.

Harvest

Harvest as soon as the roots are large enough to eat.

Use

Fresh cooked.

Storage

Storage is the same as that for the other root crops—32°F (0°C) at 90 to 95 percent relative humidity.

Spinach

Spinach (Figure 41–16) is a hardy, cool-weather crop that can withstand winters in the southern regions and is an early spring and late fall crop in the northern regions. Seed should be planted as early in the spring as possible. Sow the seed in rows 12 to 30 inches apart (depending on the method of cultivation). Plants should be thinned so that they stand 2 to 4 inches apart in the row.

Spinach will grow in almost any well-drained soil. Check the requirement for spinach, and test the soil.

Fertilizer

Follow the soil test analysis for the proper fertilization of spinach.

Harvest

Harvest as leaves become mature enough to eat.

Use

Cook fresh (as greens) or eat fresh (in salads).

Storage

Spinach keeps freshest at 32°F (0°C) and 90 to 95 percent relative humidity.

New Zealand Spinach

New Zealand spinach is not related to common spinach. It is generally grown as a substitute for common spinach in hot weather. The plants are larger than common spinach and must be spaced 18 inches apart in rows that are 3 feet apart. Plant seed as soon as frost danger has passed.

Squash

Squash (Figure 41–17) is a hardier crop than melons and cucumbers. It is, however, a warm-season crop.

Summer squash, which includes all bush types, should be eaten while young and tender. This group consists of scallops, yellow straightneck, yellow crookneck, and the vegetable marrows such as cocozelle and zucchini. Summer squash plants are heavy producers.

Winter squash consists principally of vine types. They have hard rinds, making them easy to store. Winter squash includes varieties such as Boston marrow, table queen, Hubbard, delicious, and butternut.

Plant bush squash 2 to 3 feet apart in rows 3 to 4 feet apart. Plant vine squash 4 to 6 feet apart in rows 6 to 8 feet apart.

Fertilizer

Follow the soil test analysis for the proper fertilization of squash.

Harvest

Harvest summer squash while the skin is still tender. Winter squash must be allowed a longer time to mature and is harvested just before the first killing frost.

▲ **Figure 41–16** Spinach.

▲ **Figure 41–17** Summer squash.

Use

Fresh cooked and baked.

Storage

Store summer squash at 32 to 40°F (0 to 4°C) and at 85 to 90 percent relative humidity. Winter squash is stored at 50 to 55°F (10 to 13°C) and at 70 to 75 percent relative humidity.

Sweet Potato

Sweet potatoes (Figure 41–18) grow best in warm weather. They are grown in gardens as far north as southern New York and southern Michigan. A frost-free period of 150 days with relatively high temperatures is needed for their cultivation. Sprouts or slips should not be set out in the garden until all danger of freezing has passed. Home gardeners should purchase the sprouts for planting rather than attempting to grow them themselves.

Plant in rows 3 or 3½ feet apart, and space plants 12 to 15 inches apart in the row. Sweet potato plants are generally set on top of ridges. A well-drained, sandy loam soil that is moderately deep is best.

Fertilizer

Follow the soil test analysis for the proper fertilization of sweet potatoes.

Harvest

Harvest sweet potatoes just before the first frost. Allow the roots to dry before storage. Be careful not to bruise or injure the potatoes in any way.

▲ **Figure 41–18** Sweet potatoes.

Use

Fresh cooked.

Storage

Sweet potatoes should be allowed to cure for 1 to 2 weeks at a relatively high temperature (85°F). Storage is best at 50 to 55°F (10 to 13°C) and 85 to 90 percent relative humidity.

Tomato

The tomato is probably the most popular garden vegetable (Figure 41–19). It is a warm-weather crop and a highly productive plant. Tomatoes are planted in the garden only after the danger of frost is passed. Space the plants 3 feet apart in rows that are also 3 feet apart. Set the plants deeper in the garden than they were growing before transplanting. Plants are sometimes staked or caged for protection. A mulch is of great value in conserving moisture. Installation of a cutworm collar around each plant is necessary. Try the new wire caging method along with a heavy mulch for almost care-free tomato production. Tomatoes grow best in a well-drained loam soil.

Tomatoes are grouped into different production categories of "determinate" and "indeterminate."

Determinate varieties of tomatoes, also called mountain fresh tomatoes, are varieties that are bred to grow to a compact height of 3 to 4 feet. The tomato sets fruit on the terminal buds. The fruit set ripens in a 2-week period and dies. These

▲ **Figure 41–19** Tomatoes.

plants grow well in the garden setting and in a garden bag on a patio. Varieties included in this category are 'Roma,' 'Celebrity,' and 'Marglobe.'

Indeterminate varieties of tomatoes are also called vining tomatoes. They will grow and produce fruit the entire growing season, which is defined as the total number of frost-free days. The total frost-free days are determined by the total number of days from the last frost in the spring to the first frost in the fall. Vining tomatoes will flower, set new fruit, and ripen fruit throughout the growing season. They will grow to heights of 6 feet. These plants will need a sustainable management practice of staking and pruning throughout the growing season. Varieties such as 'Big Boy,' 'Better Boy,' 'Mountain Fresh,' 'Early Girl,' and selected heirloom species fall into this category.

Tomatoes today are being grafted to help enhance their resistance to soil-borne diseases and improve production of the heirloom plants.

Fertilizer

Follow the soil test analysis for the proper fertilization of tomatoes.

Harvest

Harvest tomatoes as they turn red but before the fruit becomes soft. For canning, the fruit should be fully red and fully ripe.

Use

Fresh (in salads), cooked fresh, canned, and for making catsup and tomato paste.

Storage

Ripe fruit is stored ideally at 50°F (10°C) and at 85 to 90 percent relative humidity. Green mature fruit stores best at 55 to 70°F (13 to 21°C).

Watermelon

Watermelons (Figure 41–20) require a great deal of garden space. They are more sensitive to cold than cucumbers or cantaloupes. Soil conditions are the same as those for cucumbers.

▲ **Figure 41–20** Watermelon.

Plant watermelons in hills 8 feet apart. Plant four to six seeds per hill and thin to two to three plants per hill. Mix about a bushel of organic material in the soil of each hill. (New bush types take much less garden space, and seedless varieties are easier to eat.)

Fertilizer

Follow the soil test analysis for the proper fertilization of watermelons.

Harvest

Watermelons are ready to harvest when the tendrils nearest the fruit die and the side of the fruit nearest the ground changes from white to creamy yellow.

Use

Fresh (as dessert or in salads); rinds may be pickled.

Herbs

Herbs are becoming an important part of the garden scene and fit in very nicely with our favorite garden vegetables. Not only do they have great garden appeal with their foliage and flowers, they are also practical plants with many uses (Figure 41–21). Most of our meals would be rather plain and be less interesting without herbs to spice things up. Herbs are also used for medicinal

▲ Figure 41–21A Curry.

▲ Figure 41–21B Oregano.

▲ Figure 41–21C Arugula.

purposes, but the emphasis here will be on their use to make our favorite foods taste better.

Planning the Herb Garden

As a decorative garden in the landscape, an herb garden may be designed like any flower garden. Formal or informal designs may be used. When its use is for food herbs, the herb garden may be planted as any other vegetable garden. The size of the garden depends on the need for herbs to spice food or the landscape effect desired. When planting for herbs used in cooking, a 4 × 6 foot plot will provide most of the fresh herbs needed during the summer months. If herbs are to be dried or frozen for winter use, double that size is required.

Placing plants with the same water requirements together is a good idea. Some perennial herbs with extensive root systems will require less water than most annual herbs.

Figure 41–22 lists the most widely used garden herbs and indicates whether they are annual, biennial, or perennial plants; how tall they grow; their light requirements; where to seed; and their use in cooking.

Note: Some of the herbs listed are not edible and are listed here to point out this fact and to present some other uses for herbs.

Site

Most herbs prefer a **site** where the sun shines at least 5 or 6 hours a day. In northern climates, more afternoon sun is tolerated, whereas in warmer climates, afternoon shade is an advantage. Fragrant herbs should be planted near paths or leisure areas where their fragrance can be enjoyed.

NAME (TYPE)	SUN	SOW	HT.	USE (COMPLETE INSTRUCTIONS ON EACH SEED PACKET)
Anise Hyssop (P) *Agastache foeniculum*	○	▭ ❋	2″–3″	Leaves used for tea. Prolific blue flower spikes. Zones 4–8.
Basil, Cinnamon (A) *Ocimum basilicum*	○	▭ ☼	18″	Cinnamon flavor and fragrance, used in sauces and salads.
Basil, Dwarf Bush Fineleaf (A) *O. basilicum*	○	▭ ☼	8″–10″	Very aromatic tiny leaves for salads and fresh tomatoes. Good for windowsill or container gardens.
Basil, Genovese (A) *O. basilicum*	○	▭ ☼	18″–24″	Sweet, extra-large leaves. Best for Italian pesto.
Basil, Green Bouquet (A) *O. basilicum*	○	▭ ☼	1′	Decorative edging plant with tiny, flavorful leaves. Use fresh or dried. BURPEE BRED
Basil, Green Ruffles (A) *O. basilicum*	○	▭ ☼	2′	Large, serrated, and quilted leaves. Enhances tomato dishes. BURPEE BRED
Basil, Lemon (A) *O. basilicum citriodorum*	○	▭ ☼	12″–18″	Silver green leaves. Lemony aroma and flavor for potpourris, teas, chicken, fish, vegetables, and herb vinegars.
Basil, Purple Ruffles (A) *O. basilicum*	○	▭ ☼	12″–18″	*All-America Winner.* Large and frilly purple leaves. Stronger flavor than Sweet Basil. BURPEE BRED
Basil, Siam Queen (A) *O. basilicum*	○	▭ ☼	20″	*All-America Winner.* More tender and more intensely flavored than Sweet Basil. Great in both Thai and Italian recipes.
Basil, Sweet (A) *O. basilicum*	○	▭ ☼	12″–18″	Classic herb, used in tomato sauce, pesto, and salads.
Basil, Sweet Dani (Lemon) (A) *O. basilicum*	○	▭ ☼	24″	*1998 All-America Selection.* Very strong lemon scent if touched. Excellent regrowth for multiple harvest. White flower spikes.
Borage (A) *Borago officinalis*	○ ◐	☼	18″	Young leaves used in salads and lemonade. Blue flowers can be candied.
Cat Grass (A) *Avena sativa*	○	☼	3″–4″	Fast-growing treat for cats.
Catnip (P) *Nepta cataria*	○ ◐	▭ ☼	18″	Favorite of cats. Green leaves are good for tea. Zones 3–9.
Celery, Cutting (B) *Apium graveolens*	○◐	▭ ❋	36″	Leaves used fresh in salads, soups, tomato juice, or garnish. Seeds used for seasoning.
Chamomile, German (A) *Matricaria recutita*	○	▭ ❋	2″–3″	Small white daisies are fragrant, used in brewing tea and making perfume or hair rinses.
Chamomile, Roman (P) *Chamaemelum nobilis*	○	▭ ❋	9″	Use small white daisies in potpourris, tea, and hair rinses. Zones 3–10.

Figure 41–22 Common garden herbs. (*continued*)

NAME (TYPE)	SUN	SOW	HT.	USE (COMPLETE INSTRUCTIONS ON EACH SEED PACKET)
Chervil, Curled (A) *Anthriscus cerefolium*	○	[tray] ❋	2′	Finely cut leaves for subtle seasoning in soups, fish, vegetables, cheese dishes, and herb butter.
Chives, Common (P) *Allium schoenoprasum*	○	[tray] ❋	12″–18″	Cut leaves for salads, potatoes, and soups. Good for growing indoors. Zones 3–10.
Chives, Garlic (P) *Allium tuberosum*	○	[tray] ❋	12″–18″	Flavor is midway between garlic and onion. Delicious in salads. spreads, and flavored vinegars. Zones 3–10.
Coriander or Cilantro (A) *Coriandrum sativum*	○	☼	18″–24″	Use pungent leaves (Cilantro) in Oriental and Mexican dishes. Seeds (Coriander) are used to make curry powder.
Cress Upland (A) *Barberea verna*	○ ◑	❋	3″–6″	Easy garden substitute for watercress. Harvest in 10–14 days.
Cumin or Comino (A) *Cuminum cyminum*	○	☼	6″	Seeds used to flavor curries and many Mexican dishes.
Dill (A) *Anethum graveolens*	○	☼	3′	Seeds flavor pickles; leaves enhance salads, soups, omelets, and vegetables.
Dill, Fernleaf (A) *A. graveolens*	○	[tray] ☼	18″	*All-America Winner.* Base-branching, dwarf plants ideal for containers.
Fennel, Florence (A) *Foeniculum vulgare azoricum*	○	❋	30″	Use fresh leaves in salads and with eggs, fish, and sauces. Bulbous stem can be steamed, grilled, or served raw.
Lavender, English (P) *Lavandula angustifolia*	○	[tray]	30″	Dried flowers and leaves repel moths and can be used in sachets. Fragrant flowers, aromatic leaves. Zones 6–9.
Lavender, Lady (P) *L. angustifolia*	○	[tray]	12″	*All-America Winner.* Smaller, hardier lavender for borders or pots. Blooms first year. Zones 5–9, north of Zone 5.
Lemon Balm (P) *Melissa officinalis*	○	[tray] ☼	2′	Use lemon-scented leaves in teas and potpourris. Zones 4–10.
Lovage (P) *Levisticum officinale*	○	[tray] ❋	3″–4″	Celery-flavored leaves for soups and sauces. Seeds are used in spreads, dressings, and pickling. Zones 3–9.
Majoram, Sweet (A) *Origanum majorana*	○	[tray] ☼	1′	Use fresh or dried leaves in salads, dressings, meat, sausage, lamb dishes, beans, and soups.
Mint, Curly (P) *Mentha aquatica var. crispa*	◑ ○	[tray]	2′	Small, curly leaves flavor cold drinks, teas, sauces, jelly, candy, fruit salads, and chocolate. Zones 3–10.
Mint, Spearmint (P) *Mantha spicata*	◑ ○	[tray]	18″–24″	Vigorous plants, leaves brighten iced drinks, sauces, vegetables, and lamb. Zones 3–10.

Figure 41–22 Common garden herbs. (*continued*)

NAME (TYPE)	SUN	SOW	HT.	USE (COMPLETE INSTRUCTIONS ON EACH SEED PACKET)
Oregano, Greek (P) *Origanum vulgare*	○	Start early indoors; Outdoors after all danger of frost	1′–2′	Widely used in Italian dishes, tomato sauce, pizza, fish, and salad dressing. Zones 5–10.
Parsley, Extra Curled Dwarf (B) *Petroselinum crispum*	○	Start early indoors; Outdoors after danger of heavy frost	10″–12″	Finely cut, curly leaves, compact plants.
Parsley, Plain or Single Italian (B) *P. crispum neapolitanum*	○	Start early indoors; Outdoors after danger of heavy frost	15″–18″	Plain, flat, deeply cut dark green leaves. Flavor more pronounced than Extra Curled Dwarf.
Parsley, Hamburg or Parsnip Rooted (B) *P. crispum tuberosum*	○	Outdoors after danger of heavy frost	15″–18″	Parsniplike root adds delicious flavor to soups and stews.
Pennyroyal (P) *Mentha pulegium*	○	Start early indoors; Outdoors after danger of heavy frost	6″	Leaves are used as an insect repellent. **Not safe for consumption**. Zones 6–10.
Rosemary (p) *Rosmarinus officinalis*	○	Start early indoors	2″–3″	Fragrant leaves flavor meats, poultry, and potatoes. Use in potpourris. Zones 7–10.
Sage, Common (P) *Salvia officinalis*	○	Start early indoors	1′–2′	Used in sausages, poultry, meat, bread, dressings, vegetables, omelets, and stuffing. Zones 4–8.
Savory, Summer (A) *Satureja hortensis*	○	Outdoors after all danger of frost	1′	The "bean" herb, used in dishes with beans, peas, and lentils. Excellent as a salt substitute.
Savory, Winter (P) *Satureja montana*	○	Start early indoors; Outdoors after all danger of frost	6″–12″	Leaves flavor beans, gravies, stews, dressings, soups, butter, and vinegars. Salt substitute. Zones 5–10.
Sorrel, French Soup (P) *Rumex scutatus*	○	Outdoors after all danger of frost	2′–3′	Small-leaved sorrel, prized for French soups and fish sauces. Zones 3–10.
Sorrel, Garden (P) *Rumex acetosa*	○ ◑	Outdoors after all danger of frost	2′–3′	Tangy foliage is tasty with salad greens. Zones 3–10.
St. Johnswort (P) *Hypericum perforatum*	○ ◑	Outdoors after all danger of frost	1′–3′	Lemon-scented yellow flowers used in dyes. Zones 3–9.
Tansy (P) *Tanacetum vulgare*	○ ◑	Start early indoors; Outdoors after danger of heavy frost	3′–4′	Fernlike leaves repel insects. Dried flower heads are ornamental. **Not safe for consumption**. Zones 4–10.
Thyme, Common (P) *Thymus vulgaris*	○	Start early indoors	6″–12″	Aromatic leaves season meats, poultry, stews, sauces, soups, and dressings. Zones 5–9.
Thyme, Creeping (P) *Thymus serphllum*	○	Start early indoors	2′	Creeping plants; same flavor as Common Thyme. Attractive small space ground cover. Zones 5–9.
Wormwood (P) *Artemisia absinthium*	○	Start early indoors	24″–30″	Ornamental silvery foliage. Leaves used to repel insects. **Not safe for consumption**. Zones 4–10.

KEY: Sun– ○ = Full sun; ◑ = Partial shade; ● = Full shade Type—A = Annual; B = Biennial; P = Perennial

KEY: Sowing— [indoor symbol] = Start early indoors; [frost symbol] = Outdoors after danger of heavy frost; [sun symbol] = Outdoors after all danger of frost

Figure 41–22 Common garden herbs.

GREEN TIP Using fresh green herbs will punch up the taste of your food.

Herbs are very similar to most other garden plants in their cultural requirements. They will grow in a wide range of soils, but as do most other plants, they grow better when their specific soil and cultural needs are met. Most seed packets list cultural information, which should be followed as closely as possible. A good, well-drained loamy soil high in organic matter with a pH of 6.5 to 7.5 is satisfactory for most herbs.

Soil

Where soils are wet, raised beds can be used to improve drainage. Adding organic matter to the soil also improves drainage.

Herbs require limited amounts of fertilizer. Too much nitrogen results in lush growth with fewer flowers and less aroma.

Plants and Planting

Purchase plants that you will use and enjoy from a reputable nursery. Remember, herbs are used to add spice and flavor to foods and for their beauty and fragrance in the landscape (Figure 41–23). Also remember there are annual, biennial, and perennial herbs. Annuals and biennials must be replanted or reseeded every year, whereas perennials return year after year. This information must be considered in planning where in the garden to plant herbs. Annuals should be planted where it is easy to rework the soil for seeding each year if necessary; some reseed themselves. Most biennials will reseed themselves, and, if planted two years in a row, they will set seed every year.

▲ **Figure 41–23** Herb garden.

Propagation

Herbs are propagated by seed, softwood cuttings, division, separation, or layering. Annuals and biennials are most often propagated by seed and by softwood cuttings taken during the growing season. Perennials may be propagated by seed, cuttings, separation, division, or layering. Herbs that should not be propagated from seed include selected cultivars of comfrey, garlic, most mints, and lemon thyme. See earlier units in this text for techniques of propagation.

Pruning

Pruning often takes the form of collecting herbs for use in cooking or as cut flowers. Harvesting for use in cooking is best done just before the plants bloom. Foliage usually has more flavor at that time.

Pruning is also done to remove dead foliage and to shape the plant.

Care and Maintenance

These maintenance steps are required for an herb garden:

1. Weed or cultivate as needed.
2. Fertilize with 5-10-5, 2 pounds per 100 square feet, when new growth shows in the spring and as needed to maintain active growth.
3. Water as needed, especially in the first year after planting.
4. In the fall, apply compost to beds (40 lbs. per 100 sq. ft., minimum).
5. Clean up dead foliage.
6. Apply a light, airy mulch in winter where freezing or thawing causes plants to heave up out of the soil.
7. After 3 or 4 years, if plants start to decline, replant the bed in early spring as soon as new growth is visible.
8. Because herbs are seldom bothered by insects or diseases, no treatment is required.

Fertilizer

Follow the soil test analysis for the proper fertilization of herbs.

Harvesting

Herbs can be used fresh as harvested or dried or frozen for future use. Harvesting should be done when foliage is at its peak for flavor or fragrance. Seeds should be mature.

Foliage is tied in small bunches and hung upside down to dry. Seed heads are spread out on an airy surface, and seed is shaken out when the seed heads are dry. Drying should be done in an airy, dark room with no sun.

Foliage will air dry in about 2 weeks under low-humidity conditions. A quicker method is to oven dry at a temperature between 77°F (25°C) and no hotter than 95°F (35°C). Artificial heat is definitely needed to dry roots.

For storage, leaves should be rubbed off the stem and either powdered in a blender or hand rubbed into small flakes. Store in an airtight glass container in the dark. Herbs should be used within a year of storage.

Some herbs are frozen for storage. Herbs that retain their flavor well when frozen are chervil, dill, lemon balm, large sweet marjoram, mints, parsley, rosemary, sage, thyme, and lemon verbena. Frozen herbs retain their flavor for 3 to 5 years, but they should be used sooner if possible.

SAFETY FIRST

OSHA's regulations ensure that employers and employees know about the work hazards and how to protect themselves to prevent injuries in the workplace. For more information on occupational health and safety in the workplace, please visit the OSHA site at www.osha.gov.

Summary We must select vegetables that grow best in the local environment and fertilize, harvest, and store them properly for marketable quality. We grow herbs for the garden appeal of their foliage and flowers and also for their use as spices.

Review Questions

1. Select five vegetables that are grown in the local area from the vegetables discussed in this unit. For each vegetable, list cultural requirements, including soil preparation and special needs of the plant, and how the vegetable is harvested.
2. If you were growing sweet corn in your garden, which cultivar would you select to grow and why?
3. Compare the plan for an herb garden with that of a vegetable garden. How are they similar and different?

Student Activities

1. Plant a vegetable garden as a class project. Examine a seed catalog, and select the varieties best suited for the area. (Local extension and experiment station bulletins give recommendations of varieties to plant.) Meet as a class the following fall to discuss reasons for the success or failure of the garden. Consider improvements for next year.
2. Visit a local farmers market. Make a list of the different varieties available. Include the size/quantity it is sold in, as well as the price.
3. Search online to determine which vegetables are most commonly cultivated in your area.
4. For a great fund-raising activity, find a reputable plant catalog company online, and conduct a plant sale using the catalogs provided by the fundraising company. Your FFA chapter will make a percentage of profit for every sale that is made using the catalog.

Suggested Supervised Agricultural Experience (SAE) Activities

1. Work at a local farm or country store that sells gardening and vegetable seeds and seedlings. Record your wages, hours, and skills gained for a placement SAE.
2. Raise and sell sweet corn at a roadside farm stand as an entrepreneurship SAE.

Additional Resources

http://ohioline.osu.edu/lines/vegie.html

http://urbanext.illinois.edu/containergardening/herbveggie.cfm

https://ag.arizona.edu/pubs/garden/mg/vegetable/herbs.html

SECTION 10

The Small Fruit Garden and Nut Crops

UNIT 42

Strawberries

Objective

To grow strawberries of marketable quality in the home garden and to become familiar with commercial strawberry-producing areas in the United States.

Key Terms

- crown
- day neutral
- green manure
- hill system
- matted-row system
- runners
- soil solarization

Competencies to Develop

After studying this unit, you should be able to:

- select a site for a strawberry planting and properly prepare the soil for planting.
- diagram four planting systems for strawberries.
- list the seven points to consider when selecting a strawberry variety to plant.
- list the seven steps in planting strawberries.
- explain the cultural practices of weed control, mulching, and fertilizing as they relate to the care of strawberry plantings.
- follow the 16 steps to better strawberry yields.

Materials

- extension bulletins on strawberry culture in the local area:
  - Northeast: Cornell University's Berry Resources and Pest Management Guidelines (http://gardening.cornell.edu)
  - Midwest: Ohio State University (http://ohioline.osu.edu/lines/fruit.html)
  - MidAtlantic: Pennsylvania State University Small Scale Fruit Production (http://extension.psu.edu/gardening/fphg)
  - South: Southern Region Small Fruit Consortium (www.smallfruits.org)
  - West Coast: University of California Agricultural Natural Resources (http://ucanr.edu)
  - Troubleshooting: Cornell Berry Diagnostic Tool (http://fruit.cornell.edu/berrytool/index.html)
- strawberry plants for identification of plant parts
- If a planting site is available:
  - plants
  - fertilizer
  - string
  - shovel
  - plant labels
  - mulcher
  - cultivator or hoe
  - tape measure
  - measuring sticks cut to proper plant spacing length
- smartphone apps:
  - Weather
  - Spectracide
  - Pest Portal
  - Calibrate My Sprayer
  - PGR Mix Master
  - Ortho Problem Solver
  - AG PhD

The Strawberry is a perennial plant (Figure 42–1). The leaves die back to the **crown** each fall in areas where winter temperatures drop below 23°F (25°C). The roots live from year to year, and new leaves and blossoms grow from the crown each spring. The new fruit is produced from these blossoms.

Runners are the means by which strawberries naturally propagate. Plants normally send out runners from early summer until fall. Each runner roots and forms new plants. Everbearing plants that produce berries all summer have fewer runners.

Strawberry Production

The strawberry is one of the first fresh fruits to be marketed in spring and is usually in high demand at this time. Strawberry plants are high producers on a per-acre basis and are considered to be a high-cash crop (Figure 42–2).

Strawberries are grown nationwide, but there are two large growing areas: (1) the Pacific Coast states, and (2) Florida and other plantings in the eastern United States. Approximately 65 percent of all strawberries grown in the United States are grown on the Pacific Coast. There are some major differences in cultural requirements between the two growing areas and within the Pacific Coast area.

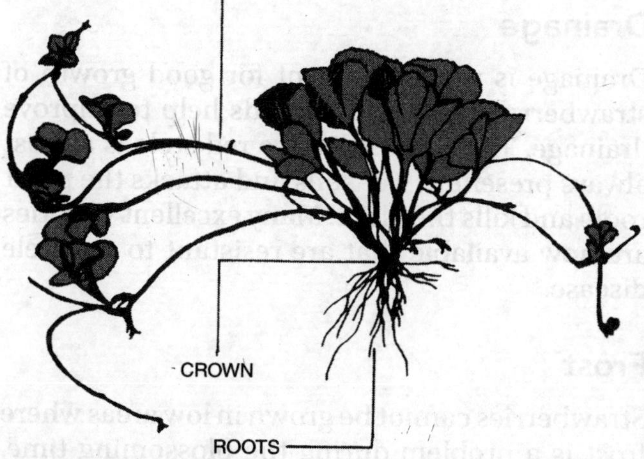

▲ **Figure 42–1** Parts of the strawberry.

▲ **Figure 42–2** Strawberries ready for market.

GREEN TIP Strawberries are loaded with vitamins, fiber and antioxidants to protect your heart, increase HDL (good) cholesterol, lower your blood pressure and guard against cancer.

Growing Areas on the Pacific Coast

California

Strawberries are grown in California by means of a **hill system**. The plants are kept in cold storage until summer, when they are set in the field. They are generally set in single or double rows. Production the first year is heavy.

Summer plantings are irrigated to keep salt accumulation in the soil minimal. Plantings are mulched with clear polyethylene.

Central Coast

This area includes Santa Cruz, Monterey, and Santa Clara counties. Plants are planted in early August and do not produce runners after the first year because of warm winters. Harvest season is from April until November.

Central Valleys (Near Fresno)

This area experiences much cold weather in winter and hotter-than-average weather in summer. The harvest season is April, May, and part of June. Plants are set in the fields in mid-July.

Santa Maria

This area includes Santa Barbara and San Luis Obispo counties. Harvest begins in April and continues through the summer and fall. Plants are set in the field in both winter and summer.

Southern California

This area includes the Oxnard coastal plain of Ventura County and the coastal regions of the counties of Los Angeles, Orange, and San Diego. Strawberries are harvested for only 1 year. Plants are set in the field both in summer and winter.

Oregon

The Northern Willamette Valley has the largest acreage of strawberries of any single area in the United States. Harvest is from May until early July. Many berries are also grown in the Hood River Valley.

Washington

The main strawberry-producing area in Washington is the Puget Sound region. Harvest is from early June until mid-July. Strawberries are also produced commercially in southwestern Washington.

Strawberry varieties grown in the Pacific Coast states are the spring- or June-bearing varieties. These varieties produce flower buds in late summer and bear flowers and fruit the following spring.

Because cultural requirements differ from area to area, specific recommendations should be obtained from the local extension service or a horticulturist. For a specific description of strawberry varieties, consult US Department of Agriculture Bulletin 1043, "Strawberry Varieties in the United States," available from local extension offices.

Growing Areas in the Eastern United States

This growing area includes Florida and all regions east of the Great Plains and north of the Southern Coastal Plains. Cultural practices are generally the same for this entire area.

In the eastern United States, dormant plants are usually set in early spring. Runners grow from the original plants starting in June, and a succession of new runners grow and take root around the original plant. Flowers are removed from the plants the first year, causing more vigorous growth. Fruit buds are formed in the fall and, by the end of October, can be seen in the crown. Fruit buds also develop in leaf axils of vigorous plants.

The number of leaves on a plant in the fall is a good indication of what the following year's production will be like. The more leaves there are, the more berries will result. The crop is generally harvested in late May and June. Most of these berries are sold fresh.

The following information applies to both the Pacific Coast areas and the eastern United States.

Selecting the Planting Site

Soil

Strawberries grow best in a loam or sandy loam soil. Clay loam or silty loam soils are very good to excellent for yields if well drained. If planted in a very sandy soil, plantings should be fruited (allowed to bear fruit) for only one season.

Supplying the soil with an abundance of organic matter is very important. This can be added by plowing under a crop such as rye or by applying compost. Care should be taken not to plow under weed seeds.

Soil solarization, which involves heating the soil by applying a clear plastic tarp for several weeks during a hot period of the year when the soil will receive the most direct sunlight, helps to kill weeds, diseases, nematodes, and insects while increasing essential nutrients for growing healthy plants.

Drainage

Drainage is very important for good growth of strawberry plants. Raised beds help to improve drainage. The fungus disease red stele is almost always present in wet soils and attacks the plant roots and kills the plant. Many excellent varieties are now available that are resistant to red stele disease.

Frost

Strawberries cannot be grown in low areas where frost is a problem during the blossoming time. Cold air collects in low areas in the same way that water does; frost damage in these pockets can be severe. Irrigation and row covers offer some protection.

Diseases

Avoid soils infested with red stele root rot or areas below infected fields where water may wash the disease into the planting site. 'Surecrop,' 'Earliglow,' and 'Allstar' are strawberry varieties that are resistant to the common types of red stele. These varieties may be grown where the disease is a problem. Check for other, more recently developed varieties for local areas.

If tomatoes, potatoes, or peppers have been grown in a soil recently, verticillium wilt is probably present. If this is the case, plant only a disease-resistant variety, such as the ones mentioned.

Soil Preparation

Soil preparation for planting strawberries should begin 1 or 2 years before the actual planting date. Soil tests should be made to determine soil pH or acidity. Strawberries grow best at a pH of 6.5 to 6.8. If necessary, lime or sulfur should be added to correct the soil pH at least 6 months or, better yet, 1 year before planting. This allows time for the lime or sulfur to dissolve and correct the soil pH before plants are set in the field. If magnesium is low, apply dolomitic limestone.

Fertile soils high in organic matter are necessary for top crop yield. Large quantities of composted animal manure, **green manure**, or cover crop should be plowed under to raise the organic matter level 1 to 2 years before planting. However, heavy green cover crops should not be plowed under just before planting strawberries—they should have time to partially decay so that large loose air pockets are not left in the soil.

In preparation for planting, the soil is plowed to a depth of 7 to 9 inches. If the soil test shows phosphorus and potassium to be low, both should be added to correct the deficiency before planting. Any phosphorus or potash should be either plowed under or mixed into the soil after plowing. The soil should be worked until it is firm and free of large clods and air pockets. The soil must pack firmly around the new plants.

Land that was in sod for years should not be directly planted with strawberries. Such soils may contain root-eating grubs that destroy new plants. It is best to plant a sod area with a cultivated crop, such as corn, for a year before planting strawberries.

Areas that contain nutgrass, quackgrass, Bermuda grass, Johnson grass, or other perennial weeds should be avoided. Contact the local extension agent for weed control issues.

Strawberries are being grown on raised beds more now than in the past. Because of this, the soil warms earlier in spring, drainage problems are lessened, the occurrence of root rot is reduced, and picking is easier. Use of the new floating row covers is also easier with raised beds.

Choosing Varieties to Plant

The selection of varieties to plant depends on climate, soil, and the purpose for which the crop is grown. The following are other factors to be considered.

- A variety that produces firm berries is needed for shipping to market.
- Large, softer-textured varieties for freezing have a bright red color, firm flesh, and a tart flavor.
- The fruit must ripen when the best market prices are available.
- If frost protection is not available, use plants that bloom late in the season to miss spring frosts that could destroy the crop.
- If planting for the wholesale market, varieties that are well known are planted, yielding enough volume to provide truckload lots. This ensures a better price for the fruit.
- Disease-resistant varieties should be used wherever possible.
- For the home garden and local roadside market, three or more varieties that ripen at different times are planted. This supplies fresh fruit over a longer picking season.

Day neutral everbearing strawberries are now a reality. These new varieties bear throughout the growing season, not just in spring and fall as many of the old everbearers did. They produce fewer runners and should be planted 6 inches apart each way, in double rows spaced 1 foot apart. For roadside markets, pick-your-own operations, and the home garden, this continuous fruiting is a big advantage.

Varieties now available are 'Tristar,' 'Tribute,' 'Tillikum,' 'Hecker,' 'Aptos,' and 'Brighton.' 'Tribute' and 'Tristar' are also disease resistant to red stele, verticillium wilt, and powdery mildew. 'Sweet Sunrise Strawberry' is a new high-yielding, June-bearing cultivar developed by USDA geneticist Chad Finn, Corvallis, Oregon.

Day neutral varieties require different management practices. They should be planted as early in spring as possible or in November or December in mild coastal and southwestern areas. Flowers should be pinched off for 6 weeks after planting, and all runners should be removed for best production. First fruit will be ripe about 30 days after blossom-pinching stops.

Day neutral varieties should be mulched to keep the soil cool and moist. This will maintain yields better in hot weather. The variety 'Tristar' is the best selection for hot weather.

Some growers treat day neutral varieties as annuals, harvesting for 1 year only. They can, however, be wintered over in many areas with adequate mulch.

Purchasing Plants

Purchase plants from a reputable nursery, and be certain they are disease and nematode free. Nursery plants should be certified true to variety and free of insects and signs of disease. Be sure that the plants are fresh upon arrival. The roots should appear bright and plump. If roots appear dry, soak the roots in water for 1 hour before planting or heeling in.

If plants cannot be planted immediately, either heel them in the soil outside or store them in cold storage with the roots packed in moist sphagnum moss or other packing.

Heeling in is done by making a V-shaped trench deep enough for the roots to spread out with the crowns at ground level. The plants are then placed along one edge of the trench so that the roots are separated, and the trench is filled in and packed around the roots. The plants can be kept in the trench until planted to the field (Figure 42–3).

If cold storage will be used, store plants that are not dormant at 40°F (4°C) and dormant plants at 32 to 36°F (0 to 2°C). When stored at the proper temperature, dormant plants can be kept for a longer period of time. When using cold

▲ **Figure 42–3** Heeling in. The plants are placed in the trench with the crown at soil level. The trench is then filled with soil and packed down.

storage, first wrap the plants in polyethylene plastic to keep them from drying out.

Planting Time

In the eastern United States, the best time to plant is early spring. Moisture is usually available from natural rainfall, and the temperatures are cool at this time. Pacific Coast regions are planted at differing times, as discussed earlier.

Planting Strawberries

Planting Systems

Five planting systems will be discussed in this section. The first three are generally used in the eastern United States. The other two systems are used by both eastern and western US growers.

HILL SYSTEM (No runners are allowed to grow.) In the hill system, plantings are made in double or four rows. Rows are spaced 12 inches apart, and plants in the row are 10 to 12 inches apart. A 24-inch space is left between the double rows (Figure 42–4). The double-row system allows for 29,000 plants per acre.

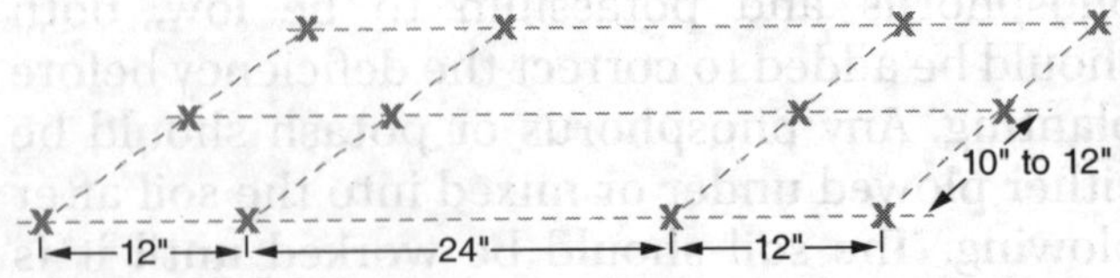

▲ **Figure 42–4** Double-row hill system. In this system, a 24-inch alley separates two rows that are 12 inches apart. Plants in the rows are 10 to 12 inches apart.

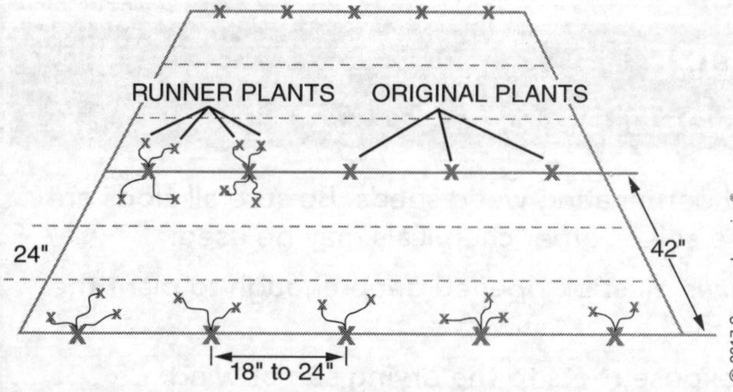

▲ **Figure 42–5** Spaced matted-row system. Plants are spaced 18 to 24 inches apart in rows 42 inches apart. Runners are allowed to grow so that plants are eventually spaced 6 inches apart. A 24-inch alley is left between runner plants.

The unwanted runners are generally cut off by machine and then thrown into the alley between rows where they are cut by sharp discs.

SPACED MATTED-ROW SYSTEM (Some runners are allowed to grow to fill in the rows.) In this system, plants are set 18 to 24 inches apart in the row, and a 42-inch space is left between rows. An acre allows for 6,225 plants if they are spaced 24 inches apart in the row and 8,300 plants if the 18-inch spacing is used (Figure 42–5).

MATTED-ROW SYSTEM (Most runners are allowed to grow.) Plant spacing for the **matted-row system** is the same as for the spaced matted-row system. The difference is that all runners are allowed to grow freely and fill in the rows. Plants become crowded in this system, and individual fruit size may be smaller than under the spaced matted-row system. This system is losing favor as narrow rows that can be mechanically tilled show advantages of lower cost and higher-quality fruit.

GREEN TIP Strawberries are often produced in a solid mat row with straw as an organic-matter mulch to control freezing and thawing of the soil that causes heaving of the roots. To control fruit rot, use the biocontrol method of using honeybees to spread a beneficial mold to the blossoms of the strawberries.

FOUR ROWS, DOUBLE ROWS, AND SINGLE ROW ON RAISED BEDS This is the most common method of planting in California. The system consists of raised beds spaced 2 feet apart. The two or four rows in each bed are spaced 8 to 12 inches apart, and plants are set in the rows 8 to 12 inches apart (Figure 42–6).

In the Pacific Northwest, plants are set in single rows on raised beds 39 to 42 inches apart with plants spaced 12 to 14 inches apart in the rows. Newer plantings are being made with 8-inch spacing in the row. Plants are often replanted each year with raised-bed plantings.

SINGLE-ROW PLANTING Recent research indicates that most strawberries will be grown on raised beds in single rows, with plants close together in the rows and rows 3 feet apart center to center. About 60,000 plants per acre will be planted.

In the eastern United States, a grower may plant in early spring and harvest a first crop in 60 days. Because runners are not needed with this type of close planting, plants will be allowed to flower the first year and bear fruit. The following spring will produce the largest crop, and several years of production may be expected from the planting. The introduction of new varieties, which produce few runners, will make this system more practical. These plantings often become narrow matted rows in time.

▲ **Figure 42–6** Strawberries planted in raised beds.

PROCEDURE
Planting Strawberries

1. Cultivate the soil immediately before planting to kill germinating weed seeds. Be sure all clods are broken up and that the soil is firm. Fumigation of the soil or other chemicals may be used.
2. Lay out rows according to spacing requirements. Rows must be opened deep enough to plant the roots straight down.
3. Keep the plant roots moist before planting. Never expose roots to the drying sun or wind.
4. Set the plant in the row at the proper depth with the crown just at ground level (Figure 42–7). This is very important. Plants set too deep smother and die; plants set too shallow dry out.
5. Fan the roots out and straighten them. If roots are too long, they may be trimmed to a length as short as 4 inches.
6. Pack the soil around the roots firmly. Do not cover the crown, and do not leave any roots exposed.
7. If possible, water immediately after planting.

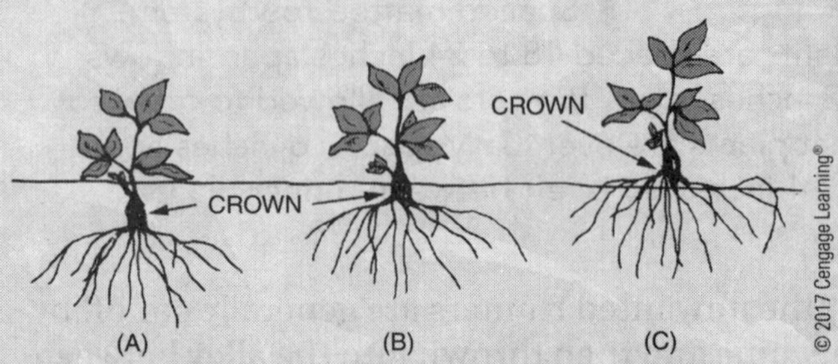

Figure 42–7 Plants set at different depths: (A) plant set too deep; (B) plant set at correct depth; (C) plant set too shallow.

Steps in Planting

Regardless of the system of planting, the general steps in the procedure are the same.

Many plants are set by machine in large operations. For small operations, a two-person team with a spade and buckets to carry plants is sufficient. One person inserts the spade in the ground and pushes it forward. The second person places a plant in the hole, after which the first person withdraws the spade and firms the soil around the plant with the foot.

Further watering will be needed in the eastern United States, and frequent irrigation during the summer is generally needed in all areas of California.

Care of the Planting

Floating Row Covers

A proven growing technique for strawberry production is the use of row covers. Made of lightweight materials (from 0.5 ounce to 2.0 ounces per square yard), row covers allow from 40 to 80 percent light transmission. Rainwater passes through, and temperature is modified under the covers. The covers are rolled out over rows of plants, and edges are covered with soil to hold them in place. No support is needed to hold them up off the plants.

These covers can help to get crops to market as much as 2 weeks earlier than normal in northern regions, when prices are higher. California growers also find that the covers provide some frost protection.

Dr. James Pollard, an associate professor of Plant Science at the University of New Hampshire, tested row covers on perennial strawberries. He found that floating row covers provide the following benefits:

- More fruit production because they induce plants to set more flower buds in the fall.
- Winter protection.
- Spring frost protection (Dr. Pollard recommends irrigation in addition because the degree of frost protection by row covers alone is not predictable [*uncover during flowering*].)
- Protection from virus diseases spread by aphids.

RESULTS OF TESTS BY DR. POLLARD The following are Dr. Pollard's test results:

- 1982–1983: Not much snow, low temperatures 215 to 220°F (223 to 229°C).

  Production of strawberries:

  With row covers—11,500 pounds per acre

  Mulched with hay—6,500 pounds per acre

  There were three severe frosts while the covers were still on. This yield with the covers is 75 percent greater than yield for the hay-mulched crop.

- 1983–1984: Mild winter, no spring frosts

  Production of strawberries:

  With row covers—23,000 pounds per acre.

  Mulched with hay—17,000 pounds per acre.

  Plants with the row covers produced 6,000 pounds or 35 percent more strawberries.

  Profit was calculated as follows:

Row cover cost:	$750.00 per acre
Cost to apply:	150.00 per acre
Total cost:	$900.00 per acre

At $.70 per pound for pick-your-own strawberries, even in the mild winter a profit was made.

6,000 pounds at $.70 per pound:	$4,200.00 per acre
Minus cost of row covers:	−900.00 per acre
Profit:	$3,300.00 per acre

Earliness to market with better prices per pound and reduced mulch costs from using hay or straw should result in even more profits.

Row covers must be removed at full bloom for fruit pollination, so frost protection is not a major benefit.

It appears that these new row covers show great promise. Contact your local agricultural college for more details.

Cultivation

Plantings should be cultivated within 6 weeks after they are placed in the soil. This loosens the soil and kills germinating weed seeds. Cultivation should be shallow (no more than 1 or 2 inches) and far enough away from the plant that the crowns are not exposed or covered too deeply by soil thrown by the cultivator. After this first cultivation, all commercial plantings should be treated with weed killers. Additional cultivation may be needed when the chemical weed control begins to fail, usually in about 2 months. Any plants that die may be replaced by placing runners in the location and allowing them to take root.

Weed Control

To obtain good yield of quality fruit, some method of weed control must usually be employed. Hand weeding is too expensive a method to use in strawberry production. The combination of mechanical cultivation and use of weed killers is the most practical method. Black plastic mulch is effective in controlling weeds, and it also conserves moisture and helps to control fruit rot.

In California, soil fumigation before planting is used to control weeds and nematodes. Because of the many different climatic and soil conditions that affect weed control with chemicals, no specific control measures are given here. Contact a local extension service or horticulturist for recommendations.

Mulching

Mulching is extensively used in eastern growing regions. Mulching helps to control freezing and thawing of the soil, which pushes plants out of the soil and exposes roots to drying. Mulching with natural material also helps in weed control if the mulch is applied so that it is thick enough (3 or 4 inches of straw) (Figure 42–8). It helps to conserve moisture and keeps the fruit clean. Small-grain straw makes the best mulch, but almost anything that does not contain weed seeds and stays in place is suitable. Black plastic also makes an excellent mulch except where excess heat builds up.

Clear polyethylene bed mulch is essential to successful strawberry growing in California. Plastic mulching takes much of the danger out of winter planting by increasing soil temperature as much as 10°F (−12°C). This results in earlier ripening of fruit, larger harvests, and longer harvest seasons. As mentioned earlier, new floating row covers are also being used.

▲ **Figure 42–8** A strawberry patch planted with solid mat rows of plants and mulched with straw for weed control.

Fertilizing after Planting

After planting, fertilizers should be used only in the late summer or early fall to increase the number of fruit buds in spring-bearing plants. Fertilizer containing nitrogen, applied in the spring, generally results in soft berries lacking in flavor and should not be used. Follow the soil test recommendations for optimum fruit production.

Frost Control

Planting on high ground prevents most frost damage to spring blossoms. If frost should occur during blooming time, an irrigation sprinkler system turned on during the time freezing temperatures exist will prevent the blossoms from freezing. Plants mulched with straw can be protected from frost by pulling the mulch back over the plants on nights when frost is expected. Row covers are also effective.

Winter Kill

In the eastern regions of the United States, freezing and thawing of the soil can heave newly transplanted plants out of the ground, resulting in drying of roots and death of the plants. A heavy mulch applied after the plants have been hardened off to winter conditions (usually after a temperature of 20°F [26°C] has been reached) will greatly reduce this loss. The plants should be completely covered to a depth of 3 or 4 inches. Floating row covers are also helpful.

Harvesting

All strawberries are picked by hand because no machine that can do the job properly is currently available. Poor harvesting techniques can ruin even the best berries. If handled roughly, berries are bruised. If stems are too short, they puncture other fruit in the container. In eastern growing areas, strawberries are almost always picked with a ½-inch stem attached. The berry should not be pulled from the vine; the stem should be pinched off instead, or the berry should be twisted between the thumb and index finger until the stem breaks. Pulling breaks the skin on the berry, and quality drops rapidly. For best flavor, allow fruit to ripen on the plant. In California, many berries are picked without stems for the freezing market. The fruit should be ripe for best flavor but not too soft. Pick-your-own operations, where the customer picks the fruit, have been the answer to many labor problems.

TRY THIS In small groups, create an instructional video that describes the best techniques to use when harvesting strawberries. Research the Internet for more information about harvesting. Share your completed video with the class.

Care of the Planting after Harvest

Weed Control

Weeds must be controlled after harvest either by mechanical cultivation and hand pulling, by mulching, or by chemical weed control. Failure to control weeds results in reduced crop production the following year.

Renewal (Revitalizing the Planting)

The root system of strawberries is often weakened while the fruit crop is maturing. To correct this situation, many growers mow the tops off the plants immediately after harvest. As soon as the tops dry, the area between the rows is dug up so that the row of plants is back to the narrow row width desired. Plants are thinned out within the rows with a hoe or spike-toothed harrow. Proper weed control must follow this operation.

Some strawberry plantings are allowed to produce fruit for only 1 year and are then plowed up and replanted.

Biocontrol of Fruit Rot

Gray mold is the most common cause of fruit rot of strawberries. A novel new control method is to use honeybees to spread another beneficial mold (*Gliocladium roseum*) to the blossoms of strawberries as the bees pollinate the flowers. A dispenser, fit in the beehive entrance, fixes fungus spores to the hair on the bee's body as she leaves the hive. Contact the local agricultural extension agent for proper control of gray mold.

Steps for Better Strawberry Yields

Follow these steps to improve strawberry yield:

1. Build up the level of organic matter in soils (a year ahead if possible).
2. Fumigate the soil for diseases, insects, and weeds.
3. Select locally adapted varieties. This is very important.
4. Plant as early as possible.
5. Plant on raised beds, especially in heavy soils.
6. Plant at the proper depth.
7. Space plants for maximum light (narrow rows).
8. Remove excess runners.
9. Fertilize in the early fall to encourage flower-bud formation.
10. Protect plants over the winter with mulch or row covers.
11. Protect plants from spring frost.
12. Provide bees for good pollination of blossoms.
13. Supply adequate moisture.
14. Spray for good disease and insect control.
15. Pick only ripe fruit and, in pick-your-own operations, insist that the customers pick only ripe fruit.
16. Renovate beds as soon as possible after the harvest.

SAFETY FIRST

OSHA's regulations ensure that employers and employees know about the work hazards and how to protect themselves to prevent injuries in the workplace. For more information on occupational health and safety in the workplace, please visit the OSHA site at www.osha.gov.

Summary

Strawberries are one of the first fresh fruits to be marketed in the spring. Commercially, strawberries are grown on the Pacific Coast and in the eastern United States and are considered to be a high-cash crop. With the new day-neutral varieties, such as Tristar, gardeners can have fresh berries for most of the growing season. Setting plants to the proper depth is very important to plant survival.

Review Questions

1. What type of use can best be made of the new day-neutral everbearing strawberries?
2. Discuss the four planting systems that are commonly used for strawberries, including the advantages and disadvantages of each.
3. What are seven points to consider when selecting strawberry plants? Why are these points important?
4. Discuss the advantages of mulching strawberry plants. Are there any disadvantages?
5. Discuss some steps to ensure better strawberry yields.
6. Discuss some methods of protecting strawberries from weeds and pests.
7. Explain weed control when growing strawberries.

Student Activities

1. Survey the internet to obtain creative ideals for a strawberry planter using the strawberry jar or vertical planters plant either one for a home project or school project.
2. Plan and care for a patch of strawberries at home or on the school grounds.
3. Visit a local strawberry grower.
4. Write in your notebook the 16 steps to better strawberry yields.
5. Search online for information about the irrigation of strawberries. Report on new technology being used to irrigate strawberries.
6. Using a smartphone app or Internet search, identify the major pests that destroy strawberries.

Suggested Supervised Agricultural Experience (SAE) Activities

1. Work for a local commercial strawberry producer, helping to harvest the crop for sale, to qualify as a placement SAE.
2. Grow your own patch of strawberries and sell in a roadside farm stand as an entrepreneurship SAE.

Additional Resources

www.hort.purdue.edu/ext/ho-46.pdf

http://pods.dasnr.okstate.edu/docushare/dsweb/Get/Document-1031/HLA-6214web.pdf

http://umaine.edu/publications/2067e/

UNIT 43
Blueberries

Objective

To plant and grow marketable blueberries.

Key Terms

chilling hours
green manure
highbush
lowbush
pollination
rabbiteye

Competencies to Develop

After studying this unit, you should be able to:

- list the six areas of the country in which blueberries grow naturally and the type of berry grown in each area.
- identify the two types of commercially grown blueberries and select the one best suited for growing in the local area.
- determine the soil conditions needed for the best highbush blueberry production by listing the three main soil requirements.
- draw to scale a diagram of a blueberry planting.
- plant, mulch, fertilize, spray, and prune one variety of blueberry.

Materials

- rooting frame to propagate blueberry plants
- blueberry cutting wood
- propagating knife, pruner, sprayer
- rooting hormone
- rooting medium (sphagnum moss and sand or perlite)
- soil test kit or a commercial soil testing laboratory
- graph paper
- computer with design capabilities
- smartphone apps:
    - Weather
    - Spectracide
    - Pest Portal
    - Leafsnap
    - Calibrate My Sprayer
    - PGR Mix Master
    - Ortho Problem Solver
    - AG PhD

Blueberries are now the preferred berry in the United States with an increase of 300 percent in consumption over the past decade, according to the US Highbush Blueberry Council. July is National Blueberry Month. Blueberries are one of the easiest and most rewarding small fruits to grow. If the soil pH can be adjusted so that it is low enough to support the acid soil requirement of blueberries, cultural practices should not be difficult. In addition to being a fruit plant, blueberries make an attractive landscape plant either as single plants or planted in a hedge. The spring blossoms are very fragrant, and the fall foliage is a beautiful red. The berries attract birds that come to feed on them.

It is much easier to cultivate blueberries in areas of the country where they grow naturally. Any commercial production is usually restricted to these areas. Figure 43–1 is a map showing the areas in the United States where blueberries grow in the wild. Many blueberries are harvested for sale from wild plants growing in the regions shown in the map. Blueberries can be divided into four categories:

- **Highbush** *blueberries (Vaccinium corymbosum)* are the most popular. Highbush varieties that will grow in US Department of Agriculture Plant Hardiness Zones 4 to 7 (northern highbush types) and Zones 7 to 10 (southern highbush types) are now available.
- **Lowbush** *blueberries (Vaccinium angustifolium)* grow in Zones 3 to 6.
- *Half-high blueberries (V. corymbosum × V. angustifolium)* grow in Zones 3 to 5. (These are hybrids between the highbush and lowbush types.) Hybridizing with the lowbush blueberry has produced some new commercially important blueberries.
- **Rabbiteye** *blueberries (Vaccinium ashei)* grow in Zones 7 to 9.

Only the highbush and rabbiteye blueberries are discussed separately in this unit.

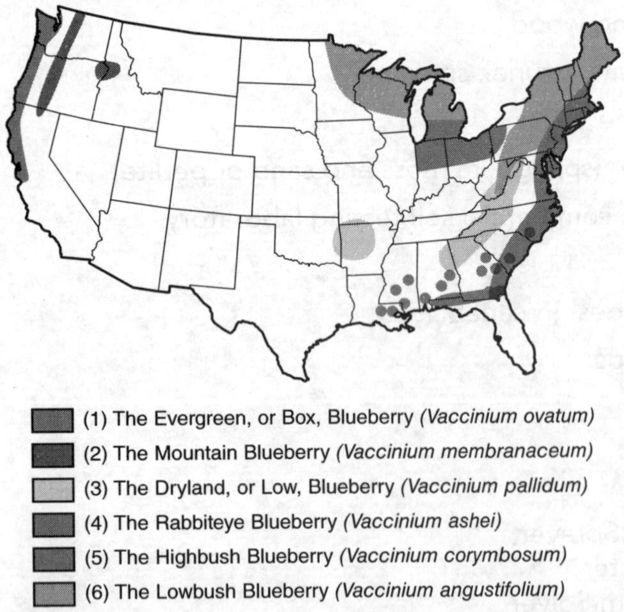

Source: US Department of Agriculture

▲ **Figure 43–1** Map of the United States, showing areas in which blueberries are extensively harvested.

GREEN TIP Blueberries are an excellent source of antioxidants, a natural health resource for humans. They are ranked as a superfood for health because in addition to antioxidants, they have high levels of vitamin C and potassium.

TRY THIS Research the Internet and study the health benefits of consuming blueberries in your diet. Create a bulletin board that highlights the health benefits of eating blueberries, and display it in your school.

Highbush Blueberries

Location

The highbush blueberry (Figure 43–2) grows naturally in an area from southeastern North Carolina to southern Maine and westward to southern Michigan. Highbush blueberries are planted where the soil is moist and acid. The berries have better flavor where the days are long and the nights are cool during the ripening season. Because blueberries bloom early, plantings should not be made in low areas where late spring frosts may kill the blossoms. Most plants

▲ **Figure 43–2** A fruit cluster of the highbush blueberry.

are not winter-hardy to temperatures below −20°F (−6°C). New hybrids, recently developed, are more cold-hardy and extend the growing area northward.

Highbush blueberries do require some winter cold to break the winter rest period and to blossom normally—about 1,200 hours of temperature below 40°F (5°C). They are not planted farther south than northern Georgia and northern Louisiana for this reason.

Soil

The highbush blueberry grows best in moist, acidic soil. A pH range of 4.6 to 5.0 is best. If soil pH is lower (more acidic) than this, add ground limestone to raise the pH. If the pH is higher than 5.0, add finely ground sulfur or fertilize with ammonium sulfate to lower the pH (make the soil more acid). Digging in sphagnum peat moss also lowers soil pH.

The soil must have good drainage. If water stands on the soil surface for several hours after a hard rain or for extended periods of time in winter or early spring, the area is not right for blueberries.

The best indication that blueberries may succeed in a soil is the presence of other native plants that grow best in acidic soil. If wild blueberries, huckleberries, azaleas, or laurel grow in an area, chances are good that cultivated blueberries will grow well there. A soil water table between 14 and 30 inches is ideal.

The soil should be high in organic matter. The plowing under of large amounts of compost or **green manure** crops 1 year before planting is very helpful. This material helps in moisture retention and in the development of good soil structure. Natural sand-peat mixtures make excellent blueberry soils. They drain well and the peat content holds moisture. The soil should be plowed and harrowed and rows established with a layoff plow for easier planting. (A layoff plow opens a trenchlike row in the soil.)

Planting

Plants are spaced 4 or 5 feet apart in rows that are spaced 9 or 10 feet apart. Spacing plants as such means that there are about 1,100 plants per acre. Two-year-old plants are generally used and are planted at the same depth at which they grew in the nursery row.

The planting is usually done as early in the spring as the soil can be worked. In New Jersey, some plants are set in the fall, and in eastern North Carolina, planting is done in late fall and winter.

Plant roots must not be exposed to drying wind or sun during the planting process. To prevent this, keep plants covered with moist burlap or some other covering. When planted, the roots of the plant are placed in holes big enough for them to fit in without bending or twisting around in the hole. Firm the soil around the plant roots and water if possible. Plants are cut back one-third to one-half at planting time by thinning out the weakest stems, not cutting off terminal ends.

Fertilizer

Soil pH or acidity should be adjusted, if necessary, 6 months before planting by mixing lime, sulfur, or iron sulfate into the soil, depending on which way the pH is to be adjusted. If the soil tests low in phosphorus or potash, these elements should also be plowed down or mixed into the soil before planting. Use a soil test to determine how much fertilizer is needed.

The fertilizer is broadcast around the plant to within 6 inches of the plant and out as far as the roots extend. This is usually 1 foot or more beyond the spread of the outer branches. Older plant roots extend farther out than younger plants. New shoot growth of 14 to 18 inches indicates proper fertilizer use.

Irrigation

The highbush blueberry is not drought resistant. During periods of dry weather, 1 to 2 inches of water per acre should be added every 10 days to 2 weeks during the picking season. Do not water any less than this amount at a single application. Shallow watering encourages shallow-rooted plants that fail to survive during periods of stress and winter months. Trickle irrigation under the mulch works well.

Mulches

Mulching with straw, sawdust, wood chips, or other material helps to conserve soil moisture and control weeds and erosion. These materials should be added deeply enough to shade the ground and to prevent the surfacing of weeds. A fine material such as sawdust should be applied only 3 or 4 inches thick, but coarse, open material such as straw should be up to 6 inches thick.

Mulches are the best weed control for blueberries because cultivation damages the shallow root system.

GREEN TIP Blueberries are productive when organic mulches are used in the planting bed. Researchers investigating the use of organic mulches in suppressing weed growth for blueberry plants have discovered that the most effective organic mulch is pine needles.

Pollination

Blueberries are partially self-fruitful or self-pollinating. But two or more varieties should be planted within two rows of each other to ensure good **pollination.** Bumblebees and other native pollinators do the pollinating. Wild insect pollinators are usually present in sufficient numbers to assure proper set of the fruit.

Pruning

Fruit is produced on wood of the previous season's growth. The largest fruit is produced on the most vigorous wood. This means that good vigorous growth (14 to 18 inches) must be maintained each year to produce a crop of large berries year after year.

Most varieties tend to bear too much fruit. Unless many of the fruit buds are removed, the berries are small, and there is not enough vigorous growth for the next year's crop. To prune varieties that are growing erect, thin out the plants in the center. To prune spreading varieties, remove the lower branches.

Heavy pruning should not be done in any 1 year. The heavier the pruning is, the smaller the crop will be. Too heavy pruning causes fruit to ripen earlier and over a shorter period of time. It is sometimes done to ripen berries earlier to obtain the highest price market.

Light pruning consists of thinning out the smaller, weaker branches and cutting some old stems to ground level each year. It is usually sufficient to ensure vigorous growth for the next year's crop and good-sized fruit during the current year (Figure 43–3). A vigorous plant requires less pruning than a weak, slow-growing plant. Vigorous plants tend to produce large berries without much pruning.

Very little pruning is needed until the third year after planting. After that, pruning should be done each year. Pruning done in the fall delays flowering in the spring by about 1 week. This provides protection against frost and does not change the harvesting date. Fall pruning should be done from Rhode Island southward.

Review

1. The center of the bush is kept open by removing old and weak wood.
2. Low-hanging branches are removed.
3. Most of the small slender branches are cut off at ground level, leaving only strong vigorous shoots and branches.
4. Some old branches are removed to ground level each year.

Propagation

Propagation of highbush blueberries is generally by hardwood cuttings (Figure 43–4). Cuttings 4 to 5 inches long are made from dormant shoots of the previous season's growth in winter. Only wood with leaf buds is used. If there are blossom buds on the tips of shoots, cut the ends off and discard them. Cuttings are usually

▲ **Figure 43–3** (A) A highbush blueberry bush before pruning. (B) The same blueberry bush after pruning. Notice that a number of large canes were removed to soil level. This stimulates new, vigorous shoots that are more productive.

grown in the nursery for 1 year after rooting and sold as 2-year-old plants for planting to the field.

Varieties

For use in western North Carolina, Maryland, and New Jersey:

'Atlantic'	'Bluecrop'	'Lateblue'	'Jersey'
'Herbert'	'Ivanhoe'	'Blueray'	'Elliott'

For use in eastern North Carolina:

'Berkeley'	'Croatan'	'Jersey'
'Morrow'	'Murphy'	'Wolcott'

For use in Michigan and New England:

'Bluecrop'	'Blueray'	'Collins'	'Coville'
'Earliblue'	'Jersey'	'Lateblue'	'Herbert'
'Elliott'	'Patriot'	'Early Bluejay'	'Northland'

For use in western Oregon and Washington:

'Berkeley'	'Bluecrop'	'Blueray'	'Collins'
'Coville'	'Darrow'	'Dixi'	'Herbert'

Most of these varieties are hybrids that tend to produce large fruit on vigorous plants.

Florida and Georgia are now getting into the early blueberry market with varieties such as 'Sharpblue' and 'Floridablue,' ripening about May 1. Fruit is being shipped to Michigan and other northern markets.

A fall-bearing variety is being developed in Minnesota. If the program produces a good commercial variety, a whole new blueberry market could be opened.

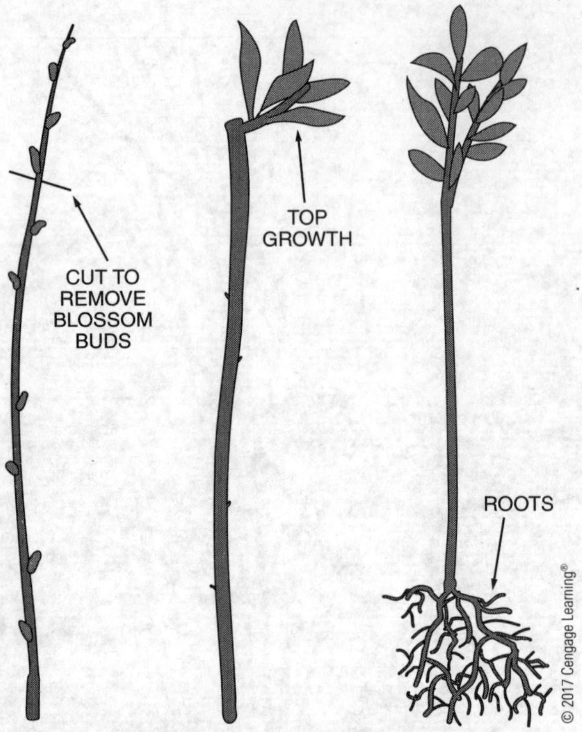

▲ **Figure 43–4** Propagation by hardwood cuttings. (A) shows the cutting wood and where it is cut to remove blossom buds. (B) shows the cutting as it begins to callus and establish top growth. (C) shows the well-rooted cutting.

Rabbiteye Blueberries

The rabbiteye blueberry has a natural range in southern Georgia, southern Alabama, and northern Florida. Rabbiteye blueberries are important mainly because they are able to grow in dry upland areas where the highbush will not, and they can be grown farther south because they require a very short rest period in winter. They are not as sensitive to soil conditions as the highbush and resist heat and drought better. The rabbiteye grows in areas from eastern North Carolina to central Florida and west to Arkansas and east Texas.

Culture

Rabbiteye varieties are planted in midwinter in rows 12 feet apart with plants spaced 6 to 12 feet apart in the row. Small plants are sensitive to fertilizer; none should be applied the first year.

Pruning is not generally done, but some pruning on older bushes is desirable. The older stems and smaller young shoots may require removal to prevent the bushes from becoming too thick. The rabbiteye blueberry is vigorous enough that pruning is not needed to produce strong new growth. The fruit ripens later and over a longer period of time than the highbush blueberry.

Plants are propagated by offshoots, or suckers, which grow up from the roots as far as 8 feet from the plant. They are dug up and cut free of the main root and grow in a nursery for 1 year before planting to the field. Propagation by softwood cuttings is also very successful.

Varieties

- 'Woodward' is the earliest and largest and has good flavor.
- 'Tifblue' has the lightest blue color and excellent flavor.
- 'Homebell' exhibits large fruit and very vigorous growth.
- Newer varieties such as 'Bonita' and 'Climax' have better flavor.

Many wild rabbiteye blueberries are harvested in the United States.

Newer Varieties

Southern blueberry growers can now grow southern highbush blueberries that are superior to the rabbiteye varieties. They are bred for heat tolerance, low winter **chilling hour** requirement, and superior fruit. All are self-pollinating, but two or more varieties planted together result in larger berries and crops.

The Florida Agricultural Experiment Station (FAES) has released a series of new plants. These are hybrids between the native lowbush Florida blueberry and the northern highbush blueberry. Starting in 1970 and continuing until 2003, 26 new blueberry varieties were released by the FAES. The most recent introductions are 'Sapphire,' 'Emerald,' 'Windsor,' 'Millennia,' 'Southern Belle,' 'Sebring,' 'Savory,' and 'Florida Rose.' These new plants have opened a large commercial market for early blueberries shipped north.

▲ **Figure 43–5** George Butler, owner of Butler's Orchards, and his daughter Susan harvest blueberries on their farm near Germantown, Maryland.

Harvesting (All Varieties)

For the fresh market, blueberries are generally picked by hand. In 8 hours, a picker can harvest 60 to 80 pints. Machine harvesting is used in some areas; the machines shake the berries into a net or hopper. Fruit harvested by machine must be sorted and cleaned. Most machine-harvested fruit is used for fruit that is processed before being sold.

Yields of blueberries range from 50 pints per acre on a 2-year-old planting to 6,000 pints on a 6-year-old planting. Blueberries are becoming a popular pick-your-own fruit (Figure 43–5).

Pest Control

Weeds

Weeds may be controlled by mulching, shallow cultivation, or use of weed killers. For weed killer recommendations for your area, see a local horticulturist or extension agent.

Diseases

North Carolina suffers the most severe losses of blueberries due to disease. Stem canker causes the most damage. Foliage diseases, stunt virus, stem blight, and root rot are also serious enough to consider a control program.

In New Jersey, the most serious diseases are stunt virus and mummy berry. In Michigan, *Fusicoccum* canker, *Phomopsis* canker, and mummy berry are serious problems. In the Pacific Northwest, mummy berry is difficult to control and is very serious.

GREEN TIP A well-drained soil and good sanitation help to control many diseases.

Insects (See Spray Schedule)

A spray schedule for the home blueberry grower is provided in Figure 43–6. More detailed information for commercial growers is available from the local extension service in US Department of Agriculture Bulletin 2254, "Commercial Blueberry Growing."

DISEASE OR INSECT AND WHEN TO CONTROL	DESCRIPTION	CONTROL
Scale insects (delayed dormant)	Small round spots on stem.	Follow the recommendations of the local agricultural extension agent for control of disease and insects.
Mummy berry (when buds first break)	Stems and blossoms blighted. As berries approach maturity, they become mummified and drop to the ground.	Follow the recommendations of the local agricultural extension agent for control of disease and insects.
Other fungus diseases (continue use of ferbam every 2 weeks)	Leaf, fruit, and stem fungus diseases.	Follow the recommendations of the local agricultural extension agent for control of disease and insects.
Plum curculio (2 weeks after bloom)	Dark brown beetles—¼ inch long, larva grayish white, legless, curved body, small brown head.	Follow the recommendations of the local agricultural extension agent for control of disease and insects.
Fruit worms (2 weeks after bloom)	Small gray moths lay eggs after bloom has fallen. Larvae are brownish red on top. Some are green on sides and underneath.	Follow the recommendations of the local agricultural extension agent for control of disease and insects.
Leaf hoppers (late May to June 5; repeat in 10 days)	Small, tent-shaped insects.	Follow the recommendations of the local agricultural extension agent for control of disease and insects.
Blueberry maggot (June 22; repeat every 10 days until harvest)	Adults look like small houseflies and are 3/16 inch long with a brown face and shiny black body, white on the side and rear of thorax. Maggots hollow berry by eating the flesh.	Follow the recommendations of the local agricultural extension agent for control of disease and insects.

Figure 43–6 Blueberry spray schedule.

SAFETY FIRST

OSHA's regulations ensure that employers and employees know about the work hazards and how to protect themselves to prevent injuries in the workplace. For more information on occupational health and safety in the workplace, please visit the OSHA site at www.osha.gov.

Summary

Blueberries are easy to grow if the soil is acidic (pH 4.6 to 5.0) and well drained. The plants are also useful as landscape plants. They make good hedge plants and provide good fall color. Blueberries also attract birds that enjoy the fruit. They are long-lived plants and require a minimum of care.

New varieties being developed will lengthen the blueberry season. This will result in fresh fruit availability extending into the fall.

Review Questions

1. Discuss the four categories of blueberries and which ones are used most frequently in commercial production.
2. Describe the soil requirements needed to grow blueberries.
3. Discuss the procedures for pruning both highbush and rabbiteye blueberries.
4. What are some of the insect pests that are harmful to blueberries?
5. Describe the procedures for fertilizing both highbush and rabbiteye blueberries.

Student Activities

1. Propagate blueberry plants from 10 hardwood cuttings. See Unit 8 for information on propagation of hardwood cuttings.
2. Take a soil test of a possible site for a blueberry planting at home or on the school grounds. Determine if the pH is suitable for growing blueberries. Report to the class on the findings. If suitable soil can be located, prepare the site for a blueberry planting using the plants propagated in the first activity.
3. If blueberries are cultivated locally, visit a local grower for tips on growing the crop.
4. Ask a local blueberry grower to demonstrate proper pruning techniques.
5. Search online for information on modern technology used in blueberry production.

Suggested Supervised Agricultural Experience (SAE) Activities

1. For summer employment and a placement SAE, work on the harvesting staff for a local blueberry producer.
2. Raise blueberries, harvest and market them at a farm stand as an entrepreneurship SAE.

Additional Resources

http://ohioline.osu.edu/hyg-fact/1000/pdf/1422.pdf

http://blueberries.msu.edu/growing_blueberries

http://content.ces.ncsu.edu/growing-blueberries-in-the-home-garden.pdf

UNIT 44 The Bramble Fruits

Objective

To outline a program for the propagation, cultivation, and maintenance of bramble fruit planting.

Key Terms

brambles
canes
erect
laterals
trailing
trellis
upright

Competencies to Develop

After studying this unit, you should be able to:

- list the five key points to consider when selecting a site for planting of bramble fruits.
- outline a soil-preparation plan for a bramble fruit planting.
- demonstrate the planting procedure for bramble fruits, including proper spacing of plants.
- outline a fertilizer and pruning program for bramble fruits.
- recognize insect and disease pests that infest brambles in the local area.
- list the seven measures for disease prevention in bramble fruit plantings.

Materials

- One of each of the following bramble fruits for identification purposes:
  - red raspberry
  - black raspberry
  - erect blackberry
  - trailing blackberry
- smartphone apps:
  - Weather Channel
  - Spectracide
  - Pest Portal
  - Leafsnap
  - Calibrate My Sprayer
  - PGR Mix Master
  - Ortho Problem Solver
  - Trees Pro
  - LikeThat Garden
  - Plantbid
  - AG PhD

The **bramble** fruits are native to North America and are some of the most delicious and useful of the small fruits (Figure 44–1). Although this unit addresses the blackberry, red raspberry, and black raspberry, most of the information also applies to the boysenberry, dewberry, loganberry, and tayberry, which are other **brambles**. Any difference in pruning or other practices concerning these brambles will be noted.

Location of the Planting

Good air circulation is important for bramble plantings. Because of this, a location that is higher than the surrounding area or with a slight slope should be chosen. A site that is above the level of the surrounding ground is not as subject to spring frosts that could kill the blossoms or cause winter injury of plants. However, brambles are not as subject to spring frost damage as strawberries and blueberries.

Do not plant in an area close to wild raspberry and blackberry plants. These wild plants may be diseased and will spread disease to the cultivated crop. Wild plants may be destroyed or removed from the area prior to planting.

Red raspberries should be planted at least 1,000 feet from black raspberries. Reds carry a virus disease that can be passed on to the black raspberries, causing severe damage.

Courtesy of Stark Brothers Nurseries and Orchards Co.

▲ **Figure 44–1** A blackberry branch with a typical fruit cluster ready for picking.

Brambles should not be planted in a soil that has been planted with potatoes, tomatoes, peppers, eggplant, or tobacco in the past 4 or 5 years. These plants may leave a fungus wilt disease in the soil that affects the planting of bramble fruits. The wilt organism can live only 4 or 5 years in the soil away from its host plant (a plant on or in which the disease lives).

Soil for Planting Site

The soil should have good drainage capacity and be high in organic matter. Subsoil drainage is more important than soil type. Organic matter increases the moisture-holding capacity of the soil, which promotes good growth but does not interfere with good drainage. The subsoil must be well drained with no hard layer that will stop roots from penetrating the soil or slow water movement through the soil. Plants that cannot establish a deep root system cannot produce a crop during dry weather. The brambles grow in a wide variety of soils, from sandy loams to clay loams. Red raspberry roots are finer and more sensitive to high soil temperatures.

Soil Preparation

If possible, begin preparing the soil for blackberries or raspberries 1 or 2 years before planting. Plow under green manure crops or animal manure to build up the organic matter level of the soil. The year the plants are to be set, plow the soil to a depth of 7 to 9 inches. Condition the soil by harrowing to break up all large clods and to level the soil surface.

A soil test should be taken before plowing. If lime or phosphorus and potash fertilizer are needed, they are applied and either plowed under or harrowed into the surface. Do not apply lime and fertilizer in the same application, however. It is probably best to plow under the fertilizer and harrow the lime into the surface after plowing. Lime may not be needed because bramble fruit grows well in a pH as low as 6.5 to 6.8.

GREEN TIP In preparation of the growing bed, it is necessary to plow down green manure crops and incorporate animal manures to raise the organic-matter content to enhance the soil.

The Planting Process

Always start with bare-root plants or tissue culture plugs that are certified to be disease free and true to variety. Caution should be taken to prevent the plant roots from drying out during the planting process. Start with wet roots, and keep the roots covered with wet burlap or plastic until planted. An additional precaution to help conserve moisture around the roots is to use a Viterra Agri-Gel and water slurry as a root dip, which will cling to the roots prior to planting. Do not place plants in the row so far ahead of the planters that they dry.

Establishing rows with a layoff plow at the proper distance apart makes the planting process easier. Place the plant in the row, and cover with soil, firming around the roots. The plants are set at the same depth in the ground at which they grew before. Red raspberries are planted 1 to 2 inches deeper than they grew in the nursery. Do not cover new sprouts that may have started to grow. Instead, pull soil up around them later, after the sprouts have had a chance to grow more. A spade may be used to plant the brambles (Figure 44–2). To plant, push the spade in the ground, pull it forward, place the plant in the opened slit spreading and keeping the roots straight, pull the spade out, step on the soil to force the slit closed, and firm the soil around the roots.

The 6-inch piece of cane on the new plant that is used as a handle is cut level with the ground and removed from the field. This is a disease-control measure. Tissue culture plants are disease-free.

▲ **Figure 44–2** Black raspberry tip plant being set in the ground. A spade is being used to make the hole in which the plant is placed. Note that the plant is held by the cane from the parent plant.

Planting Systems

Most of the bramble fruits are best grown in rows, either freestanding, on wire trellises, or staked to hold them up. Figure 44–3 gives recommended planting distances between rows of brambles.

The most effective **trellis** system for erect-growing brambles has two parallel wires, one on each side of the post. After harvesting fruit, the wires can be lowered to the ground until pruning of old **canes** is complete. The wires are then lifted to 24 to 30 inches high and fastened to each side of posts that are spaced 15 to 20 feet apart in the row. The new plant canes are placed inside the two wires. Wires may be wired together at intervals to prevent outward movement.

Trailing plants such as the boysenberry should have a two-wire trellis with the first wire 30 inches from the ground and the second wire

PLANT	DISTANCE BETWEEN ROWS*	DISTANCES BETWEEN PLANTS IN THE ROW
Black raspberry	7–10 feet	3–4 feet
Blackberry and Tayberry	7–10 feet	5 feet
Boysenberry (trailing types)	7–10 feet	8 feet
Purple raspberry	7–10 feet	3–4 feet
Red raspberry	7–10 feet	2–3 feet
Yellow raspberry	7–10 feet	2–3 feet

*Distance between rows varies with the method of cultivation that will be used. A small garden tractor works well at 7 feet apart; a large tractor requires 10 feet.

Figure 44–3 Planting distances for bramble fruits.

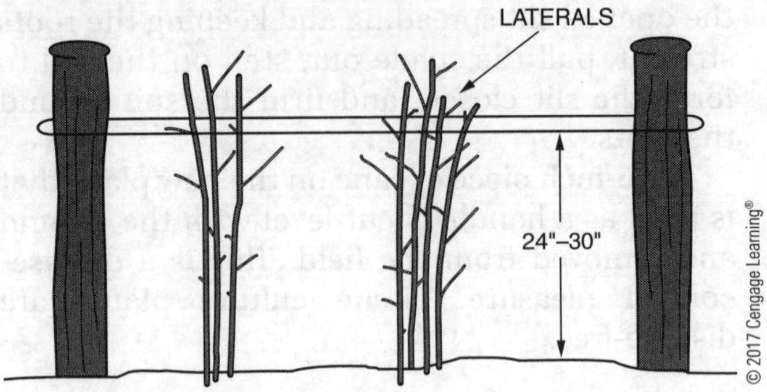

▲ **Figure 44–4** Two-wire trellis for black raspberry and erect-growing blackberry. The posts are set 15 to 20 feet apart.

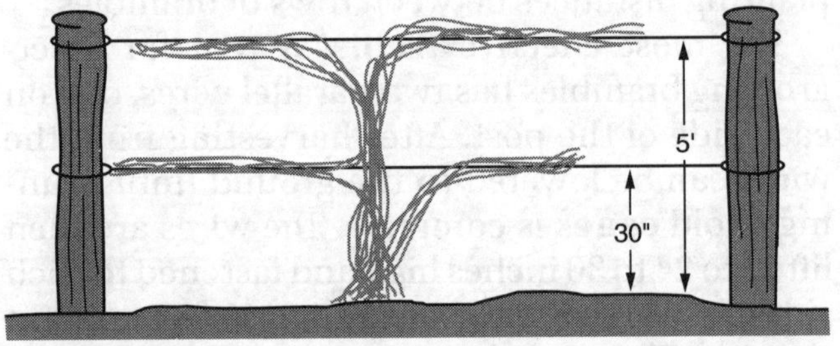

▲ **Figure 44–5** Two-wire trellis system used for trailing blackberries and boysenberries. Note new canes on the ground and fruiting canes tied to the two-wire trellis.

5 feet from the ground. The canes are tied to the wire with string. Some growers prune **erect** blackberries and raspberries back to an initial heading of 24 inches. Most canes of this height will stand without support. This practice is used commercially due to the cost of trellis construction. Some canes fall, and fruit is lost. If a trellis is not used, plants must be pruned shorter so that the canes are able to stand.

If trellis rows are 400 to 500 feet long, it is best to break the trellis system to provide a cross-lane through the center. Figure 44–4 illustrates a trellis system for erect blackberries or black raspberries. Figure 44–5 shows a two-wire trellis system for trailing blackberries or boysenberries.

Pruning and Training

The top growth of brambles lives for 2 years. New canes are produced each year from the crown or roots. Fruit is formed on wood of the previous season; these canes die soon after the fruit is harvested. The roots of plants are perennial and continue to live for many years.

Pruning and Training Freestanding Plants

PRUNING BLACKBERRIES DURING GROWING SEASON The first-season plants are pruned to a height of about 15 inches. After the first year, the height is increased to 24 inches because the plants are more vigorous and able to produce larger, stronger canes that stand better. Not all canes reach the correct size at the same time, so the planting should be checked once a week, and all canes 24 inches or longer should be headed back. Heading back causes the plants to form many side branches or **laterals**, which increases fruit production. Prune when no rain is expected for 48 hours as a disease-prevention practice.

PRUNING BLACKBERRIES IN EARLY SPRING Before buds swell in the spring, remove all canes that are ½ inch or less in diameter by cutting off at ground level. At least two canes must be left, but most plants should have four or five canes. Shorten all of the laterals to 18 to 24 inches in length. Fruit from pruned laterals is larger and of better quality.

PRUNING BLACKBERRIES AFTER HARVEST Immediately after the crop is harvested, prune all old canes that fruited by cutting off at ground level. This promotes new cane growth from buds at the base of the plant and helps to control diseases. At this time, continue the growing cycle by cutting back all new canes to the 24-inch height as soon as this height is reached. Purple raspberries are pruned the same way except that the new canes are headed back to 30 to 36 inches instead of 24 inches. As a laborsaving practice, commercial growers in Texas leave old canes for 3 or 4 years, and then mow off all tops and remove them from the field.

If plants are trained to grow on a trellis, the heading back of new canes should be kept about 12 inches higher than recommended for non-supported plants. Other pruning practices are the same.

New genetically thornless blackberries are now available. These varieties are easier to handle and pick fruit from and are nearly as productive as the thorny types; for example, 'Black Satin' and 'Smoothstem' are hardy to 5°F (−15°C), 'Thornfree' to 0°F (−18°C), 'Hull Thornless' to 10°F (−12°C), and 'Chester' to 12°F (−11°C). Check with your local agricultural research center for varieties best suited to your area.

PRUNING RED RASPBERRIES Red raspberries may be tied to a trellis or left freestanding. The canes are thinned to stand no closer than 4 to 6 inches apart, leaving about seven or eight strong canes per hill. Most red raspberries are grown in the Pacific Northwest.

PRUNING RED RASPBERRIES AFTER HARVEST All old canes are removed as with black raspberries. New canes are thinned, and all root suckers growing into the rows or between rows are removed.

The availability of the new 'Heritage' and 'Cherokee' fall-bearing red raspberries now makes it profitable to grow red raspberries as annuals to produce a fall crop, especially in states near 40 degrees latitude. Rows 20 to 24 inches wide are mowed to a height of 6 inches each spring. All canes are removed. New canes come up the same spring and produce a fall crop.

PRUNING UPRIGHT-GROWING BLACKBERRIES DURING GROWING SEASON All suckers that sprout between the rows should be pulled as soon as possible to prevent the planting from becoming too thick.

PRUNING UPRIGHT-GROWING BLACKBERRIES AFTER HARVEST All old canes that fruited are cut off at ground level and removed immediately after harvest. Commercial growers may want to leave old canes for 4 or 5 years and mow off everything for removal at that time to save labor costs. These canes should be burned or otherwise destroyed to help control diseases. Thin any new canes that surface, leaving four of the most vigorous canes per plant.

Blackberry canes with a great deal of fruit often fall due to lack of support. A wire trellis support may be needed. Tie canes to the wire with soft string, or place canes between wires in a two-wire trellis.

PRUNING SEMITRAILING BLACKBERRIES AND BOYSENBERRIES DURING GROWING SEASON Trailing blackberries are neither headed back nor summer pruned in any manner except to remove the old canes after fruiting. The new canes are allowed to grow along the ground and tied to a trellis in spring or fall after old canes are pruned out.

PRUNING SEMITRAILING BLACKBERRIES AND BOYSENBERRIES IN EARLY SPRING Laterals are shortened on all canes to 10 inches, and the long trailing vines are held together and tied to a two-wire trellis with soft string.

PRUNING SEMITRAILING BLACKBERRIES AND BOYSENBERRIES AFTER HARVEST Remove the old canes that have fruited immediately after harvest and destroy them.

Cultivation

Cultivate the bramble plants often enough to control weeds. Cultivation should be shallow—no more than 2 inches deep. Stop cultivation during harvest time to prevent fruit loss. Brambles should not be cultivated in the fall because this stimulates late growth, and plants do not harden off properly for winter.

Weeds may also be controlled with mulches and weed killers. Contact the local extension service for chemical weed control recommendations.

TRY THIS In a small group, research further planting and pruning techniques used to grow bramble fruits. Create a slideshow presentation that describes various planting systems and techniques, as well as pruning strategies. Include visual aids along with written information.

Fertilizing

As mentioned earlier, it is necessary to apply lime, phosphorus, and potash before planting. After planting, side-dress the bramble plants, but follow the recommendation of the local extension service office. Each year thereafter, apply fertilizer after following the soil test analysis results from the local extension service or a commercial soil-testing laboratory to ensure optimum berry production.

Harvesting

Berries are picked when they are fully ripe but still firm. To keep ahead of the fruit so that none gets too ripe, raspberries should be picked every 2 or 3 days. Hot weather ripens fruit more quickly and could cause picking dates to change. Blackberries should be picked every 3 or 4 days.

Only the firm berries are picked; all injured or soft berries are thrown away. Do not hold handfuls of berries during picking. This practice causes bruising of fruit that results in early spoilage and a product that cannot be marketed.

Place the fruit in the shade or some other cool spot as soon as it is picked. Heating in the container causes rapid spoilage. Berries picked in the morning when they are cool remain fresh longer than berries picked after the sun has heated the fruit.

Many growers are operating pick-your-own operations. This is one of the most economical systems for harvesting fruit if organized properly. Hand picking for the fresh market is very expensive, making up 50 percent of the total cost to produce the berries. Processed berries are often picked by machines. This process calls for special pruning and good disease control.

Average yields of fruit are about 6,000 pounds of blackberries per acre. Raspberries yield about 2,000 quarts per acre (Figure 44–6).

© David Kay/Shutterstock.com

▲ **Figure 44–6** A quart box of red raspberries.

Propagation

Black and purple raspberries are propagated by tip layering. Red raspberries are propagated from root suckers and root cuttings. The rooted suckers are dug up and separated from the parent plant and planted in early spring.

Root cuttings from red raspberries are taken in early spring. Pieces of root 2 or 3 inches in length are cut from the parent plant, covered with 2 inches of soil, and placed in the nursery. The roots send up shoots that are set in the field the following spring. When old plants are dug up, roots left in the ground send up many shoots that form new plants. This is an easy way to propagate red raspberries.

Upright blackberries are propagated from root suckers or root cuttings, as red raspberries are. Trailing blackberries are propagated by tip layering, just as black raspberries are. Tissue culture allows rapid propagation of virus-free plants of all the bramble fruits.

Varieties

There are many varieties of bramble fruits, and new varieties are constantly being developed. Some varieties grow best in colder climates; others survive better in southern regions of the United States. Recommended varieties for the local area may be obtained from the local agricultural extension service.

Insects and Diseases

Insects are not usually as destructive to the bramble fruits as diseases are. If control measures are used, attacks by disease organisms can be kept to a minimum without the use of an extensive spray schedule. Some spraying may be necessary, however, depending on local conditions and the needs of that particular year.

GREEN TIP New raspberry and blackberry cultivars show promise in the control of bramble viruses. These new cultivars are genetically resistant to some bramble viruses and aphids.

The following are recommended disease prevention measures.

1. Start with the proper site.
   - Do not plant black or purple raspberries in fields that were planted with tomatoes, potatoes, peppers, or eggplants.
   - Keep plantings of black and red raspberries separated by at least 1,000 feet.
   - Destroy all wild raspberry plants in or near the planting site 1 year before planting.
2. Choose disease-resistant varieties.
3. Buy and plant only disease-free plants. Tissue culture plants are safer for freedom from diseases.
4. Remove all diseased plants and burn them.
5. Remove old canes after harvest.
6. Keep the planting free of weeds and fallen leaves.
7. Use pesticides when necessary. Obtain a spray schedule from the local extension service or the US Department of Agriculture Farmers' Bulletin 2208, "Controlling Diseases of Raspberries and Blackberries."

SAFETY FIRST

OSHA's regulations ensure that employers and employees know about the work hazards and how to protect themselves to prevent injuries in the workplace. For more information on occupational health and safety in the workplace, please visit the OSHA site at www.osha.gov.

Summary

The bramble fruits—including raspberries, blackberries, and boysenberries—are native to North America and are some of the most useful of the small fruits. Their ease of culture makes them popular, especially the new thornless varieties.

Review Questions

1. What are five key points to consider when selecting a site for planting bramble fruits? Of these points, which ones are most important and why?
2. Discuss how to prepare the soil for a bramble fruit planting.
3. Describe disease-prevention measures for bramble fruits.
4. Describe how to fertilize, prune, and train one of the bramble fruits properly for optimal production.
5. How do bramble fruits propagate?
6. What insects or other pests typically harm bramble fruits in your local area? How are they prevented?

Student Activities

1. Visit a small fruit grower, and discuss all aspects of the operation with him or her.
2. Propagate and plant at least one bramble fruit. Practice proper pruning techniques in the program.
3. Search the Internet for information on blackberry production and raspberry production in your state. Report on the status of the bramble industry in your area.

Suggested Supervised Agricultural Experience (SAE) Activities

1. For summer employment and a placement SAE, work on the harvesting staff for a local bramble fruit producer.
2. Raise blackberries and raspberries, and then harvest and market them at a farm stand as an entrepreneurship SAE.

Additional Resources

http://extension.psu.edu/plants/gardening/fphg/brambles

www.gardening.cornell.edu/fruit/homefruit/6brambles.pdf

www.uky.edu/Ag/CCD/introsheets/hightunnelbrambles.pdf

UNIT 45 Grapes

Objective

To outline a program for the establishment and cultivation of a vineyard.

Key Terms

canes
renewal spurs

Competencies to Develop

After studying this unit, you should be able to:

- select a planting site for a vineyard and give reasons for its selection.
- prepare the soil and plant grapes in the selected planting site.
- describe a fertilizer program for a vineyard.
- list the characteristics of ripe grapes.
- list the procedures for weed control in vineyards.

Materials

- pruners
- canes from a grapevine for demonstration of proper pruning techniques
- all planting tools and materials for the construction of a trellis (if hands-on grape planting activity is possible)
- smartphone apps:
  - Weather Channel
  - Spectracide
  - Pest Portal
  - Calibrate My Sprayer
  - PGR Mix Master
  - Ortho Problem Solver

Grapes are rapidly becoming a popular homegrown fruit. Grapes have many uses and are very nutritious. They are consumed fresh, as juices and wines, as raisins, as jam and jelly, and as frozen products. Grapes are native to the United States. Many of the domesticated varieties have native wild grape ancestry in at least one parent. American fine wine grapes are native to central Asia. Grapes are grown commercially in most areas of the United States, except where the frost-free season is less than 150 days, in arid sections, and in areas having extremely high temperatures and very humid conditions. Some newer cultivars can be grown in central Florida and other warm areas.

> **GREEN TIP** Grapes are a valued health food. They are antioxidants and a source of manganese, vitamin B6, thiamin, potassium, and vitamin C.

Choosing a Planting Site

A level site or a steep slope that is higher than the surrounding area is important to reduce danger of frost damage, especially in the fall. The soil should have good drainage capacity and be high in organic matter.

A location to the south or east of a large body of water is favorable because water has a modifying effect on temperature. It protects against fall frost, delays early spring blooming and thus holds plants until the danger of spring frost is past, and keeps wood and buds from losing hardiness in spring.

Soil

Grapes grow well in most soil types, providing the subsoil allows deep rooting. *Hardpan* (compacted clayey soil), rock, and wet subsoil close to the surface are not suitable. A deep, fertile, well-drained loamy soil with a moderate amount of organic matter is best. A pH range from 5.5 to 7.0 is satisfactory.

Soils that are overfertilized, rich, or too high in organic matter produce a late-maturing crop with a low sugar content. Sandy or light soils tend to produce small crops of early-maturing fruit that is high in sugar content. A soil that falls somewhere between these two examples is best.

Soil Preparation

Grapes should not be planted directly into soil that has been in sod for years. Such land should be planted with a cultivated crop such as corn for at least 1 year before grapes are planted. The soil is plowed deeply and disc harrowed until it is in good planting condition.

After a soil test is done, any phosphorus or potash fertilizer needed should be plowed or disced into the soil before planting.

Planting

In southern regions of the United States, grapevines are planted in the fall as soon as they become dormant. North of Arkansas, Tennessee, and Virginia, vines are planted in early spring.

The ideal grape plant at planting time is 1 year old. The vines are planted at the same depth at which they grew in the nursery. Do not allow roots to dry out during the planting procedure. Immediately after planting, prune to a single stem with two or three buds remaining. Space plants 8 to 10 feet apart in the row, with the rows about 10 feet apart. Place plants directly under the trellis wire.

Grapes respond best to shallow cultivation (not over 2 inches deep). The vines should be cultivated soon after planting to kill germinating weeds. Cultivation is repeated as needed for weed control. The grape hoe, consisting of a manually operated blade mounted on a tractor, is an excellent tool for keeping the rows clear of weeds.

Trellis Construction

Grapes are trained to grow on a trellis so that the vines are given proper support. Generally, a two-wire trellis is used. Wooden, concrete, or steel posts are placed about 20 feet apart. The end posts must be securely planted at least 3 feet deep to remain firm. Line posts can be set 2 feet

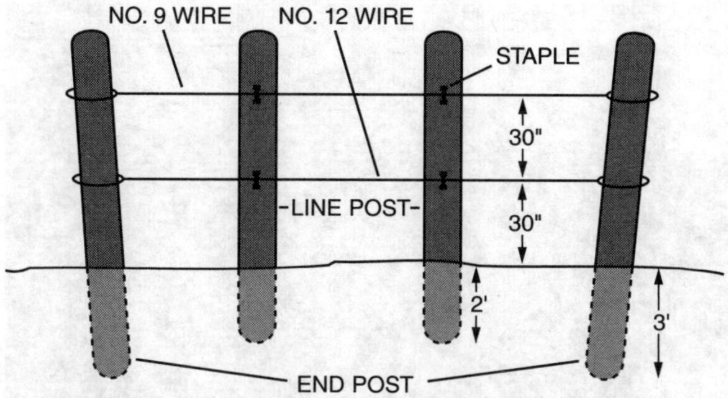

▲ **Figure 45–1** Typical two-wire grape trellis.

deep. Earth anchors set outside end posts are strongly recommended.

Number 9 steel wire is used for the top wire, and smaller number 12 steel wire is used for the bottom wire. The bottom wire is 30 inches from ground level, and the top wire is 30 inches from the bottom wire (Figure 45–1). The wires are fastened to the posts so that they slide through the staples or clips for easy tightening. The wires are tightened each spring before the vines are tied.

New planting and trellis systems are being evaluated as researchers look for higher production per acre and trellis systems that make it possible to prune mechanically.

> ***TRY THIS*** Research more trellis designs online, and then create your own blueprints for constructing a small trellis system at home or on school grounds. As a class, construct the trellis system according to the blueprint plan.

Training and Pruning

Grapevines are trained to fit the trellis construction previously described. Although there are several systems of training, the *four-arm Kniffin system* will be explained here. This system is used for training bunch grapes and works very well.

After the first growing season, the most vigorous cane is selected and tied to the top wire. It is then cut off above the top wire and all other **canes** are removed (Figure 45–2).

When pruning during the dormant season of the second or third year, select four vigorous canes

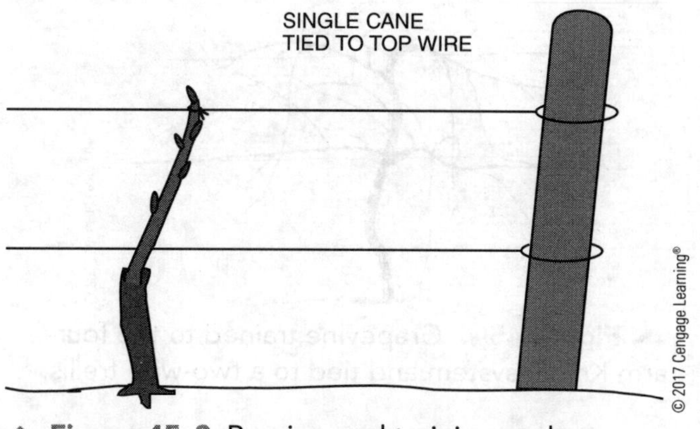

▲ **Figure 45–2** Pruning and training at the end of the first growing season.

for arms. The most fruitful canes are darkest in color, midsize, and on the outside where most light was received. Prune each of these canes to 8 or 10 buds in length. Select four other canes located as close to the arms as possible, and cut each back to two buds each. These short canes are called **renewal spurs**, and the buds on them will develop new canes to serve as the new arms for the vine the following year. All other canes are removed. Each year, this process of pruning is repeated. New arms are formed from spurs each year for the following year's crop. The old arms are removed during the pruning process. Fruit is always produced on new growth from the canes that are 1 year old.

The vines are pruned each year late in winter or early spring before growth begins. Most of the canes are removed each year, with the canes left as arms and renewal spurs being greatly shortened (Figure 45–3). This may seem to be very severe pruning, but it is essential for continued high yields and quality fruit.

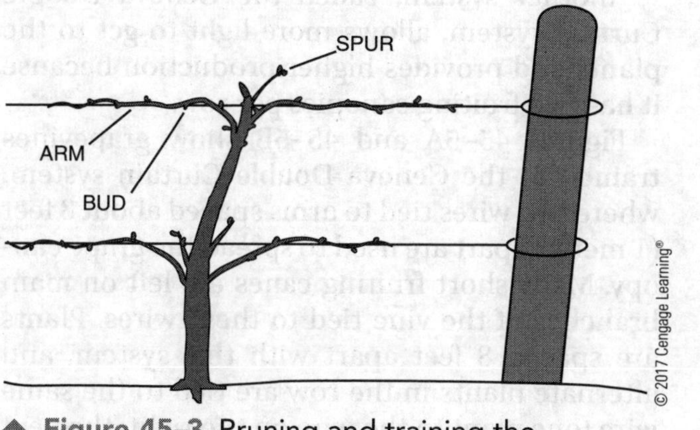

▲ **Figure 45–3** Pruning and training the second and third year.

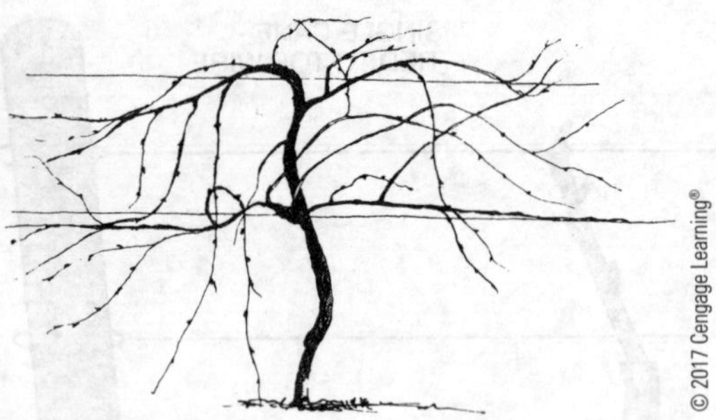

▲ **Figure 45–4** Grapevine trained to the four-arm Kniffin system and tied to a two-wire trellis.

The more vigorous a vine is, the more fruit it is able to produce. The number of buds left to produce fruit on the four arms as described earlier is from 32 to 40 buds total (8 to 10 on each arm), which is considered average. Weak vines have slightly fewer buds, and strong vines have more. A good rule of thumb is to leave 20 to 30 buds and then add 10 buds for each pound of 1-year-old wood pruned from the vine.

After pruning, the canes are tied to the trellis wire with string. The end of the arm is tied lightly just behind the last bud. That bud is rubbed off so that it does not grow and break the string as the cane grows. Other ties are made as needed to hold the arm to the wire. These are tied loosely so that the cane has room to grow and expand in diameter (Figure 45–4).

The four-arm Kniffin system is only one training system. Many others are also used. Mechanical harvesting requires different pruning and trellis construction. Plants must be cordon trained for mechanical harvesting. Canes are cut in horizontal and vertical planes, leaving many short canes and spurs instead of the long arms explained using the four-arm Kniffin system.

Another system, called the Geneva Double Curtain system, allows more light to get to the plants and provides higher production because it has two fruiting canopies per row.

Figures 45–5A and 45–5B show grapevines trained to the Geneva Double Curtain system, where two wires tied to arms spaced about 3 feet (1 meter) apart are used to spread the grape canopy. Many short fruiting canes are left on main branches of the vine tied to these wires. Plants are spaced 8 feet apart with this system, and alternate plants in the row are tied to the same wire (one plant to the wire on the right, the next plant to the wire on the left of the row, etc.).

▲ **Figure 45–5A** Vines trained to the Geneva Double Curtain system shown before pruning. Wooden spacers extend out on both sides of the posts to hold the two wires about 3 feet (1 meter) apart. Every other plant in the row is trained to the wire on the right side of the row and alternate plants to the wire on the left side. This gives a double curtain plant foliage.

▲ **Figure 45–5B** Vines pruned to shorten canes using the Geneva Double Curtain system.

In the near future, pruning may also be accomplished using new technology. Figure 45-6 shows an example of one of the robots being developed around the world to help mitigate the costs of seasonal labor in vineyards.

▲ **Figure 45–6** Robotic grapevine pruner.

Contact your local agricultural college for the latest information on grape pruning for your area.

Fertilizer

Fertilize grapes in early spring. Apply according to soil test recommendations. The fertilizer should be broadcast because grape roots tend to spread over a wide area.

Harvesting

Grapes are harvested when they are ripe. This stage is often difficult to detect because some grapes change color and appear to be ripe a month before they are actually ready to pick. The best test for table grapes is to taste them for sweetness or check the color of the seeds. Seed color changes from green to brown as the fruit ripens.

For grapes used in juices and wines, the percentage of soluble solids or sugar content is important. A refractometer or Balling hydrometer is used to determine ripeness. Proper acidity also ensures the best wine flavor and storability.

It is not always possible to harvest fruit when it is fully ripe. Some grapes crack or split open as ripening progresses and must be picked before they are fully ripe to prevent loss of fruit. Rotting in rainy weather or the danger of an early frost may also make early picking necessary.

Table grapes for fresh consumption are generally picked by hand, whereas juice and wine grapes are picked by machine. Figure 45–7 shows a grapevine with an abundant grape crop.

▲ **Figure 45–7** Grapevine with an abundant crop of fruit.

Weed Control

Weed control is generally accomplished through shallow cultivation and the use of the grape hoe. In one hoeing, the soil is pulled away from the wires toward the center of the rows. About 2 weeks later, the hoe is turned the other way, and the soil is pulled back toward the vines and under the trellis. The grape hoe removes most of the weeds and leaves only a few for hand hoeing. Chemical weed killers are also sometimes used. Check local recommendations from the state agricultural experiment station. Because the grape canopy is above the weed canopy, weed killers can be used more easily than with other small fruits.

Insects and Diseases

Grapes are sprayed for both insects and fungus diseases. At least three spray applications are needed for good control. Commercial growers spray up to eight times during the season. Consult your local county agricultural agent for the US Department of Agriculture bulletin on spraying grapes.

Major diseases that require control are black rot, powdery mildew, and downy mildew. Major insect pests are the flea beetle, leafhopper, berry moth, and Japanese beetle.

Propagation

Grapes are propagated commercially by hardwood cuttings. See the section of the text on propagation by hardwood cuttings for a description of this technique. If a disease-resistant rootstock is needed for certain varieties, plants are grafted or budded to a disease-resistant seedling rootstock.

SAFETY FIRST

OSHA's regulations ensure that employers and employees know about the work hazards and how to protect themselves to prevent injuries in the workplace. For more information on occupational health and safety in the workplace, please visit the OSHA site at www.osha.gov.

Summary

Grape growing is becoming popular in the United States. Grapes grow well in most types of soil and are trained to grow on a trellis to give the vines proper support. They are native to North America and grow very well there. Grape plants live for many years and represent a long-term commitment.

Review Questions

1. What are considerations related to selecting a planting site for a vineyard? What makes one site more appropriate than another site?
2. Describe how to prepare the soil before planting grapes.
3. Explain how to plant grapes after you have chosen a planting site.
4. Describe a fertilizer program for a vineyard.
5. Describe the characteristics of ripe grapes.
6. Discuss ways to control weeds in a vineyard.

Student Activities

1. Complete the following activity as a class project, using a section of school grounds or a local vineyard.
   - Plant and prune 1-year-old grapevines.
   - Fertilize a producing grapevine.
   - Prune the grapevine and tie it to the trellis.
   - Discuss the site selected.
   - Discuss weed control for the planting and carry it out, if necessary.
   - Observe the application of pesticides.
   - Determine when the fruit is ripe, and demonstrate proper picking techniques.
2. Ask a local grower to come and speak to the class about the culture of grapes as well as marketing grapes in the current food industry.
3. Search online for information on grape production in your state. Report on the variety of grapes that are predominantly grown in your state and why those grapes are grown more than others.

Suggested Supervised Agricultural Experience (SAE) Activities

1. As a placement SAE, work for a local viticulturist or grape production enterprise.
2. Grow a crop of grapes, and harvest and market them at a farm stand as an entrepreneurship SAE.

Additional Resources

www.extension.umn.edu/garden/yard-garden/fruit/growing-grapes-for-home-use/

http://extension.unh.edu/resources/files/resource000576_rep598.pdf

www.fruit.cornell.edu/grape/

UNIT 46

Nut Crops

Objective

To recognize four major nut crops in the United States and understand how they grow.

Key Terms

biannually
digging
harvesting
hulls
legumes
nuts
peg

Competencies to Develop

After studying this unit, you should be able to:

- list four important nut crops.
- understand where the major nut crops are grown.
- describe the plant type and size of each nut crop.

Materials

- professional grafting/budding tool
- budding knife
- budding rubber strips, poly budding tape, buddy tape
- nut tree understock material
- nut tree scions material
- laminar flow cabinet to perform tissue culture
- smartphone app:
    - Leafsnap

Nut Crops

A **nut** is generally defined as the seed and fruit of a plant that is encased in a hard shell or pod. Many, such as walnuts and pecans, grow on trees. However some foods that are referred to as nuts, such as peanuts, do not fit the botanical definition and are actually **legumes**, which are the fruit or seed of leguminous plants. This chapter reviews the most commonly produced nut crops, beginning with nuts that grow on trees and concluding with peanuts.

Tree nuts are an important part of our diet. In the United States, technological advancements have increased tree nut production. According to the 2012 USDA Census of Agriculture, tree nut total farms was 6971 with $17,940.202 value of nuts sold. While the total acreage of nut trees continues to decline according to the USDA, the market demand for nut crops accounts for a large segment of the edible horticultural crops. With new propagation methods, it now takes 2 to 3 years for nut trees to reach nut-bearing age, which is down from 5 to 6 years. Additionally, domestic and foreign demands for nuts have increased. In the United States, the per capita consumption of all tree nuts is more than 3.0 pounds (shelled basis). In 2012, US nut tree production generated more than $7 billion in value. The most profitable nut crops in the United States are almonds, English or Persian walnuts, pecans, peanuts, and pistachios.

California is the nation's number one producer of tree nuts. California orchards alone produce over 90 percent of US tree nuts. The United States also exports a large percentage of its nut crops annually, most notably to China. The US nut tree industry has been active in promoting the health benefits of nut trees as important sources of phytochemicals, which may help reduce a person's risk of developing some cancers and cardiovascular disease. The top tree nuts consumed in the United States are almonds, walnuts, and pecans.

In addition to the nuts listed previously, both hazelnuts and macadamia nuts are also grown commercially in the United States. The majority of hazelnuts are produced in the Willamette Valley of Oregon; worldwide, Turkey is the largest producer. Macadamia nuts are produced in Hawaii.

Pecans

Pecans are largely grown as a food crop. They are widely grown in the warm climates of Georgia and New Mexico. In 2012, the total US pecan crop was 302 million pounds. Georgia is the largest producer of pecans in the United States, producing 100 million pounds in 2012. Other top pecan producing states include New Mexico (65 million pounds), Texas (55 million pounds), and Oklahoma (25 million pounds). Pecans tend to bear **biannually**, meaning a large crop every other year.

Pecan trees are native to the United States and originated in central and eastern North America and the river valleys of Mexico. The nuts were an important source of food for Native Americans and early settlers. Today, pecan cultivars are a thriving nut crop grown largely in the southeastern, south central, and southwestern states. The pecan is the state tree of Texas, Mississippi, and Arkansas.

PLANTING Pecans have a chilling requirement of 150 to 250 hours below 45°F (7°C). Pecan trees are the largest of the hickories (Figure 46–1). They start bearing nuts when about 5 years old and reach full production in 15 to 20 years. Trees are wind pollinated, and it is best to plant two or more different cultivars close together.

In addition to the valuable nut crop, pecan shells are used to manufacture filler for plastic and veneer wood, and the trees produce lumber highly prized in furniture making.

▲ **Figure 46–1** Pecan tree.

Pecan trees require deep fertile soil with a pH of 6 to 6.5 and prefer some wind protection. They are relatively free of diseases and insect problems. Trees can reach heights of 100 feet, which is an important consideration when selecting a planting site.

Selected cultivars that have been grafted onto pecan seedlings will result in better nut crops. Small trees should be transplanted so the deep taproot is not damaged.

PRUNING Trees should be pruned to a central leader and only enough to remove dead and overlapping or too thick branches.

PROPAGATION Propagation is by whip and tongue grafting onto seedling pecan rootstock.

English or Persian Walnuts

English walnuts, also known as Persian walnuts, are large deciduous trees reaching a height and spread of 60 feet (18 m) (Figure 46–2). They are grown over a wide area of the United States but with a heavy concentration in California. Worldwide, China is the largest English walnut producer, followed by the United States. Walnut trees require a well-drained soil with a pH of 6.5 to 7.0. To flower and bear nuts, they annually require 500 to 1,000 hours of temperatures below 45°F (7°C).

PLANTING English walnuts are wind pollinated, and planting two or more varieties close together results in better nut production. Only named cultivars should be planted to ensure good quality nuts.

Walnuts are slow to become established. It is important to plant small trees so the long taproot is not damaged. Fall planting is best. Plant in a well-drained, moisture retentive soil.

Nuts are collected as soon as the outer **hulls**, or shells, open and should be stored in a cool, airy, dry area.

PRUNING Prune trees in winter to a central leader. Pruning should develop a well-balanced framework with wide forks or crotches between the tree trunk and branch. Little pruning is needed after this.

PROPAGATION Propagation is by whip and tongue grafting onto black walnut rootstock.

© schankz/Shutterstock.com

▲ Figure 46–2 English (Persian) walnut tree.

Almonds

The almond is a widely consumed nut. However, it is not really a nut, but rather the seed of a fruit belonging to the same family as peaches, plums, and so on. The nuts grow on a small tree, 15 to 20 feet (5 to 7 m) tall and 12 inches (30 cm) wide (Figure 46–3).

PLANTING Almonds grow only in areas with warm, dry summers and frost-free winters. Trees do have a chilling requirement of 300 to 500 hours of temperatures below 45°F (7°C). Trees should be planted in a sheltered, frost-free area with well-drained soil and a pH of 6.5. The small trees are spaced 20 feet apart.

© villorejo/Shutterstock.com

▲ Figure 46–3 Almond tree.

Almond flowers are insect pollinated, and trees are partially self-fertile, but cross-pollination using two or more different cultivars results in larger crops.

Trees should be mulched or cultivated to control weeds.

Almond trees produce nuts in 3 to 4 years. Harvest when the hulls start to crack. Clean and dry the nuts before storage.

PRUNING Nuts are born on 1-year-old wood. In summer each year, a number of old shoots that have cropped in previous seasons must be removed to encourage new growth for fruiting the following year.

PROPAGATION The propagation of nut crops is accomplished with asexual propagation by using the stem of the original plant, which is the exact genetic duplication of the original plant. For example, almonds are propagated by chip budding and tissue culture. Use almond seedlings for rootstock in dry areas, and use peach seedlings when grown on heavier, moist soils.

Peanuts

Peanuts are not actually nuts but the seeds of a legume. The USDA defines peanuts as both a field and grain crop and an oil seed crop. Peanuts grow on tender annual plants that originated in South America (see Figure 46–4). In the United States, Georgia produces 41 percent of peanut crops, followed by Texas, Alabama, North Carolina, Virginia, and Oklahoma. Different cultivars grow from 12 inches (30 cm) to 3 feet (1 m). The four cultivars commonly grown in the United States are 'Runner,' 'Virginia,' 'Spanish,' and 'Valencia.' They are tropical plants and need a frost-free site, growing best at temperatures between 68 and 80°F (20 and 30°C) and with a high 80 percent relative humidity.

PLANTING AND HARVESTING Seeds are sown in spring following the last frost when soil temperatures reach 61°F (16°C). Peanuts grow best in sandy, well-drained loam with a pH of 5.5 to 6.5. Fertilize with extra calcium, potassium, and phosphorous. Nitrogen needs are low because peanuts are legumes and can collect nitrogen from the air.

▲ **Figure 46–4** Peanut plant.

> ***TRY THIS*** Visit a peanut producer, or grow peanuts in your school program. Collect a sample of soil from the growing peanuts, and conduct a soil test. Determine whether the soil is in the right range for successful growth, and report on your findings.

Peanut flowers are self-pollinating. Flowers are fertilized above ground, and then the budding ovary (or **peg**) penetrates the soil where the peanuts develop. Pollination occurs at different times, causing peanuts to mature at different rates.

Peanuts are cultivated to hill up around roots when the plants are 6 inches tall. Soil should be kept loose, making it easier for the fertilized flowers to enter the soil. Irrigation is required during dry periods but not when the plants are in bloom.

Peanuts are ready for harvest in 17 to 20 weeks after sowing seed for upright types and 3 to 4 weeks later for prostrate types. 'Virginia' types may require more time, up to 25 weeks.

Peanuts are harvested with machinery in two stages. The first process, called **digging**, involves loosening the soil around the plants, cutting the taproots, and laying the plants on top of the soil where the peanuts are left to dry for 2 to 3 days. The second process, or **harvesting** (threshing, combining), separates the peanuts from the vines. Once harvested, the peanuts are placed in drying wagons to reduce moisture content by 8–10 percent before storage. After harvest, peanuts are placed in a cool, dry storage.

GREEN TIP

When peanuts are harvested, the peanuts are collected for drying, but the peanut vines can be left behind and mulched into the soil to restore nutrients.

SAFETY FIRST

OSHA's regulations ensure that employers and employees know about the work hazards and how to protect themselves to prevent injuries in the workplace. For more information on occupational health and safety in the workplace, please visit the OSHA site at www.osha.gov.

Summary

Nut crops produced in the United States include numerous varieties, but the four major crops are almonds, walnuts, pecans, and peanuts. Nut crops are mainly grown in the southern half of the country and the western states, specifically California. California is a major producer of almonds and walnuts, while Georgia produces the most pecans and peanuts. Nuts are popular for their health benefits.

Review Questions

1. Why are nuts a popular food in the United States?
2. What are the four most profitable nut crops in the United States?
3. Where do the majority of commercial nut crops come from in the United States?
4. In what ways do peanuts differ from other nut crops?

Student Activities

1. Have a local nut grower visit the class.
2. If interested in growing any of the nut crops, do some more research through your local extension services or at www.usda.gov and www.garden.org.
3. Visit a nut tree tissue culture lab. Report on the technology used in the process.
4. Have a tissue culture lab tech visit to speak to the class.
5. Research nut crop tissue cultures online, and write a report on the steps in the process of using tissue culture for reproduction.
6. Use the Leafsnap app to identify the genus of a nut tree.

Suggested Supervised Agricultural Experience (SAE) Activities

1. To qualify for a placement SAE, complete an internship at a tissue culture lab.
2. Work for a nut farm assisting in all aspects of production and harvesting as a placement SAE.

Additional Resources

http://smallfarms.ifas.ufl.edu/crops/fruits_and_nuts/nut_crops.html

www.gardening.cornell.edu/fruit/pdfs/nutpub.pdf

www.hort.purdue.edu/newcrop/NewCropsNews/94-4-1/nuts.html

SECTION 11

Holiday Crafts and Floral Designs

UNIT 47

Wreaths and Door Swags

Objective

To construct various seasonal decorative pieces, including the evergreen wreath, Della Robbia wreath, door swag, and kissing ball.

Key Terms

Della Robbia wreath
garland machine
greenery
Hillman wreath-making machine
lacquer
lotus pod
mistletoe
pinning machine
sphagnum moss
straw
swag
tuber

Competencies to Develop

After studying this unit, you should be able to:

- identify 10 different plant materials used in holiday decorations.
- construct an evergreen wreath, Della Robbia wreath, door swag, and kissing ball.
- list five different fruit and nut pods that can be used in the construction of a Della Robbia wreath.
- Use a smartphone to glean ideas on designing a holiday wealth or door swag.

Materials

- evergreen trees and shrubs used in wreath construction (white pine, Scotch pine, English boxwood, Douglas fir, noble fir, balsam fir, Norway spruce, Blue Colorado spruce, eastern red cedar, holly, rhododendron, and yew)
- seedpods of plants, pine cones, spruce cones, oak acorns, walnuts, shellbarks, pecans, filberts
- box wreath wire frame measuring 12-, 14-, 16-, or 18 inches
- Styrofoam ball, 4 inches in diameter
- 16-gauge and 22-gauge florist wire
- straw and sphagnum moss
- Hillman wreath frame, 10 to 36 inches in diameter
- smartphone apps:
  - Florista
  - My Flower Box Events

Holiday decorations, materials for their construction, and holiday plants play an important part in the wintertime business of nurseries, greenhouses, and garden centers. Some of the more popular holiday items are wreaths, door **swags**, and table centerpieces. These decorations usually consist of a base, some type of moisture-holding material, **greenery**, and decorative accessories that contrast with the plant material.

Evergreen Box Wreaths

To construct a foundation for an evergreen foliage wreath, a box wreath wire frame is used (Figure 47–1). After an appropriate frame has been chosen, it is filled with **straw** or **sphagnum moss**. This material is dampened to help hold it in place and keep the evergreen foliage fresh. After the straw or moss is packed in the wire frame, the frame is wrapped with a special plastic wreath wrap material (Figure 47–2). The wrap holds the straw or moss in place, keeps moisture in, and supports the greenery when it is punched into the frame.

The next step in the construction of a wreath is to select the foliage. The selection of foliage depends on personal preference, but the final design should be considered. Boxwood has been used in the construction of the wreath shown in this unit, but any type of greenery can be used in basically the same way.

Sprigs of the greenery in lengths of 4 to 6 inches are then punched individually into the wreath. This operation moves in one direction

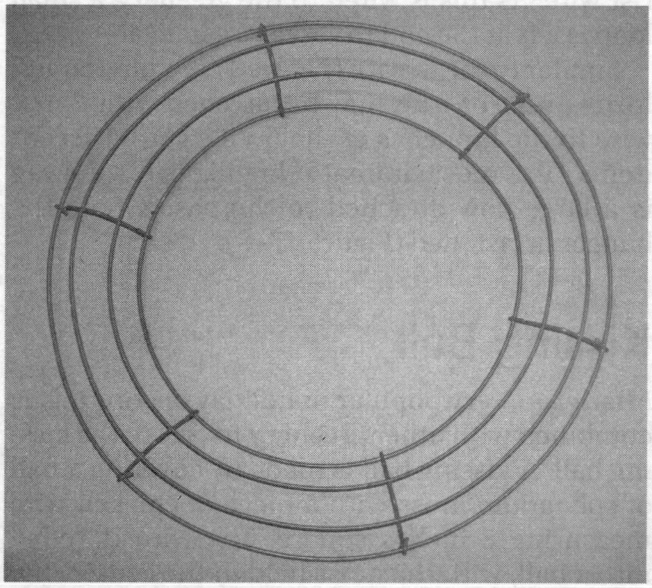

▲ **Figure 47–1** A box wreath frame. Frames are available from most florist supply companies in sizes of 10 to 36 inches in diameter.

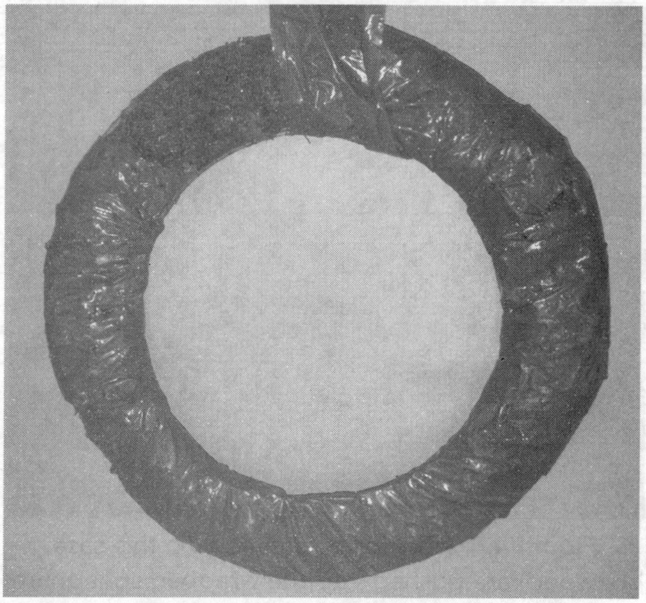

▲ **Figure 47–2A** The box wreath wire frame is filled with sphagnum moss and then wrapped tightly with plastic wreath wrap.

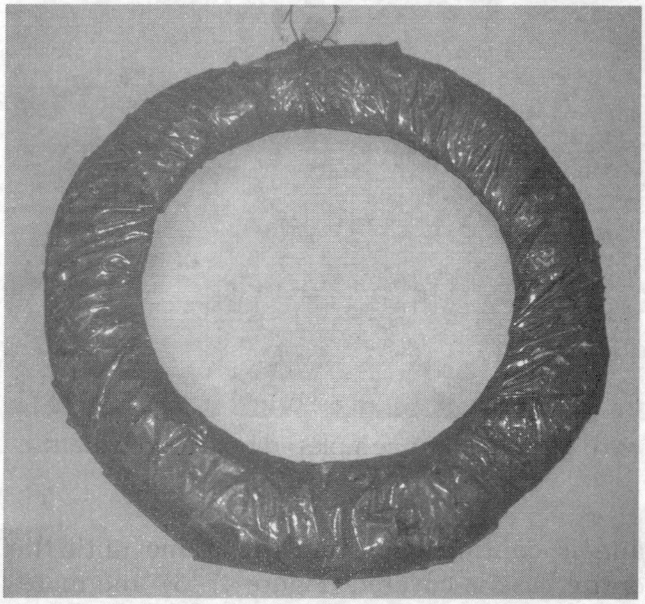

▲ **Figure 47–2B** A completed wreath form packed with sphagnum moss, with wire hanger.

▲ Figure 47–3 Sprigs of greenery (in this case, boxwood) are pushed into the wrapped sphagnum moss base until it is completely covered.

▲ Figure 47–4 A completed boxwood wreath.

and is continued around the frame until the entire base is covered (Figure 47–3). The material should flow smoothly and evenly for the best finished product. In other words, there should be no open spaces or sprigs that extend farther than the others. The completed wreath is shown in Figure 47–4.

Hillman Wreath Frame System

The Hillman wreath frame system consists of a 9-gauge wire wreath frame and a **Hillman wreath-making machine** (Figure 47–5). The wreath frame is available from most florist supply companies in sizes from 10 to 36 inches in diameter. Unlike the box wreath frame, the Hillman wreath frame requires no filler material (straw or sphagnum moss). The greenery is held in place by a system of crossbars that are clamped together using the Hillman wreath-making machine.

Evergreens such as white pine, blue Colorado spruce, Douglas fir, noble fir, balsam fir, or Scotch pine are generally chosen to construct the Hillman wreath. The greenery is pruned to 6- to 8-inch lengths and is gathered into bunches of four or five sprigs. For fullness, only those sprigs with terminal as well as lateral bud growth are used. Each bunch of sprigs is then clamped firmly into the wreath frame with the Hillman machine. This process is continued until the entire wreath is completed. After all of the greenery has been attached, accessories such as bows, cones, or ornaments may be added.

Door Swags

Door swags (Figure 47–6) are a popular decoration for entranceways to homes during the holiday season. Their construction is less complicated than that of the wreath. To make a door swag, several evergreen branches are simply grouped in an attractive arrangement that can be fastened to a door. The evergreen branches should be 24 to 36 inches long. If more than one branch is used, they may be tied together at the top with 16-gauge florist wire. As this is wired to the greenery, a small loop is left at the end to serve as a hanger.

Similar to the wreath (Figure 47–7), pine cones, **lotus pods**, or fruit may be fastened with florist wire to the branches of the door swag as accent items. The most dramatic highlight for a swag is a large bow attached to the base where the hanger is fastened (Figure 47–8).

Kissing Ball

Mistletoe, very popular in holiday decorating, is combined with other greenery to create the kissing ball. A kissing ball is made by covering a ball of sphagnum moss with a piece of chicken wire measuring 8 inches square. The wire is rolled into a ball, with the moss held in the center. The final size of the ball should be about the size of a softball. This serves as the foundation or base for the greenery. Other materials may be used, such as a Styrofoam ball or a large potato (**tuber**).

(A) The Hillman wreath frame system

(B) A student using a Hillman wreath-making machine to place the correct size bundles of greenery to give the wreath an even flow.

(C) Placing the finishing touches on the Hillman wreath.

(D) A completed wreath with a bow for a special accent.

(E) A completed wreath with a bow that has been woven and tucked to create a special holiday accent.

▲ **Figure 47–5** Hillman wreath making.

© 2017 Cengage Learning®

▲ **Figure 47–6** Door swags are constructed of several boughs of greenery.

© iStockphoto/stocknshares

▲ **Figure 47–7** Wreath design with fruit, seedpods, and leaves.

© 2017 Cengage Learning®

▲ **Figure 47–8** Bows, cones, or other ornaments are added for interest. The bow covers the cut ends of the boughs.

Della Robbia Wreath

A **Della Robbia wreath** is a wreath consisting of cones, pods, nuts, and fruits. It is constructed on a box wire wreath frame or a Styrofoam ring. A 10-inch ring makes an attractive candle-ring display, whereas the 16- to 18-inch rings form interesting door decorations.

The frame is placed face down on a table. The cones to be used for the wreath are then chosen and soaked in water for 30 minutes. This will cause the cones to close, making it easier to insert them into the frame. The cones will open again after they dry out. The cones that are used for the foundation of the wreath must be equal in size. The cones are inserted into the frame by pushing them in between the wires. Each cone is then wired in

PROCEDURE
Creating a Kissing Ball

1. Push a 16- to 20-gauge florist wire through a medium-sized potato to serve as the hanger for the kissing ball (Figure 47–9). Note that a medium-sized Styrofoam ball may also be used instead of the potato, but it will not provide moisture for the greenery.
2. Hang the potato or ball at about eye level or slightly lower. This makes insertion of the greenery much easier.
3. Cut the stems of greenery 4 to 5 inches long and at a 45-degree angle so the greenery will absorb moisture from the potato and be easier to stick into the potato (see Figure 47–10).
4. Insert greenery into the potato to cover the entire potato (see Figure 47–11). For best results, start at the side of the ball and work around the bottom in a circular pattern.
5. Continue to add greenery to cover the entire potato. The sprigs of greenery must be even in length for the best appearance.
6. Complete the kissing ball by adding decorations of ribbon loops and mistletoe (see Figure 47–12). Typically, a small piece of mistletoe is then fastened to the bottom of the ball with an attractive piece of ribbon or a bow.

▲ **Figure 47–9** A medium-sized potato can serve as the base for a kissing ball and provide moisture for the greenery.

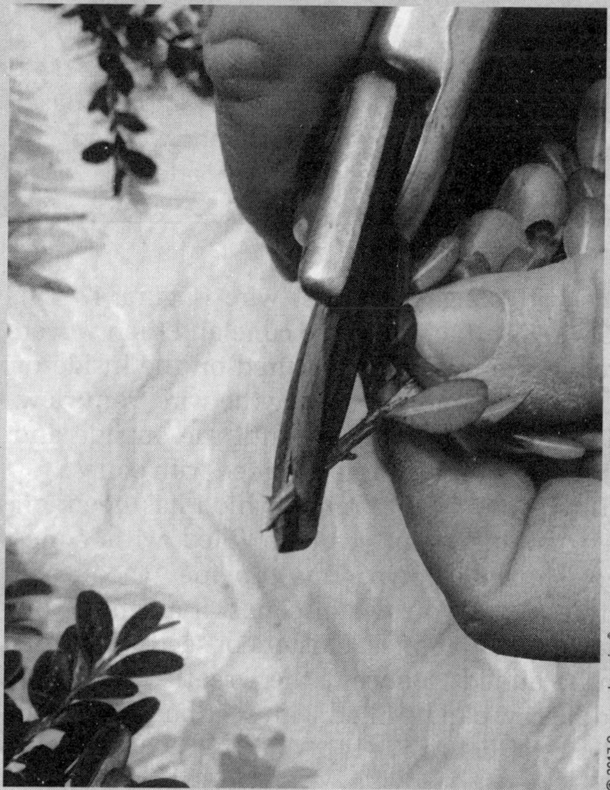

▲ **Figure 47–10** Cut the greenery at a 45-degree angle. This will make it easier to stick into the base and enable moisture to be drawn up through the greenery if using a potato for the base.

▲ **Figure 47–11** Insert greenery into the base, starting at the side of the base and working around the bottom in a circular pattern.

▲ **Figure 47–12** The completed kissing ball. A small piece of mistletoe may be fastened to the bottom of the ball with an attractive piece of ribbon or bow.

place with 22-gauge florist wire (Figure 47–13). The wire is attached to the cone and then woven through the frame and secured on the inside or back of the frame, concealing the wire from view. This process is continued until the outside and inside edges of the frame are lined with cones.

The center of the frame, which has been left open to this point, can be filled with cones placed upside down. This provides a contrast to the other two rows of cones. If a center row is formed with cones, clusters of seeds and nuts can be added to accent the wreath. The entire middle row can be filled with seedpods and nuts if desired (Figure 47–14).

TRY THIS Construct and sell evergreen and Della Robbia wreaths as a special fund-raising project for the class or school FFA chapter. Schedule the fund-raiser prior to and during the holiday season.

▲ **Figure 47–13** To line the inside and the outside of the wreath frame, the pine cones are pushed between the wires of the frame and secured with florist wire.

Garlands

Garlands are a tradition in America that was brought by the early settlers. This tradition

▲ **Figure 47–14** A completed pine cone wreath such as this one can be used for several years.

included garlands made of evergreen trees and shrubs (pine, holly, boxwood, fir, spruce, cedar, laurel, yew, and seedpods) adorning homes, churches, and businesses. Today, the use of artificial plastic and silk plant material can also be considered to extend the life of your holiday decor. This material is produced to appear as natural greenery.

The roping (garland) is a unique technique to create a graceful flow and give a festive touch. In the construction of the early roping, the base foundation was made with fox grapevine with the greenery fastened with twine. Today, the construction of the garland is made with a **garland machine** that uses fine-gauge florist wire to fasten the greenery to the twine.

In Figure 47–15, a garland machine is being used to design a garland of white pine. This machine uses black poly twine and feeds the green florist wire up through the half cup to twist the wire around the greenery and fasten it to the twine.

Also used in the design of wreaths, swags, garlands, and holiday centerpieces is the **pinning machine**. The pinning machine includes a metal pin; decorative accents are clamped on the opposite end of the pointed end of the pin, and then the pointed end pierces and fastens the adornments to the garland. The pinning machine shown in Figure 47–16 is hand-operated, but power-operated versions are also available.

▲ **Figure 47–15** The garland machine is used to attach decorative material such as evergreen clippings, seedpods, and fruit to a rope that will give flexibility of the arrangement to be draped or wrapped in a decorative fashion to adorn an entryway, mantel, table, or any creative focus the individual designs.

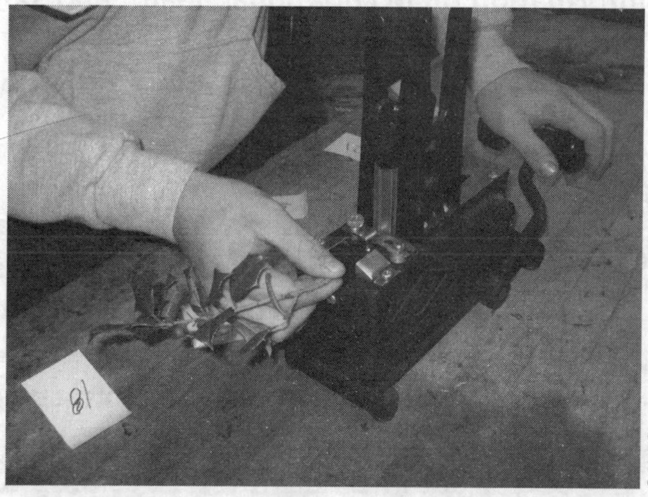

▲ **Figure 47–16** The pinning machine is used in the design of wreaths, swags, garlands, and holiday centerpieces to add special accents to the design for creative interest and special effects in the arrangement.

Locating Cones, Seedpods, Nuts, and Fruits

The best source of cones, seedpods, nuts, and fruit is a local florist supply house. However, field trips to the woods can often supply different types of seeds and nuts that are native to a particular area. For example, sweet gum balls, beechnuts, horse chestnuts, acorns, and shellbarks can be collected from trees in some areas. Rose hips, hollyhock seeds, and bamboo fruit might be found in your own backyard. A walk in the woods or a vacant lot might produce teasel, lotus, or milkweed pods.

GREEN TIP Wreath decorations were once homemade using naturally dried flowers, strings of popcorn or cranberries, apples, and sweets. Recreate those simpler traditions with unique, earth-friendly decorating tips. Reuse and recycle items in your home to add to your creation.

Fastening Seedpods and Nuts

To fasten seedpods or nuts to a wreath, they must first be drilled. The drilling must be done carefully because nuts often have a tendency to slip out of the driller's hand. Once the nuts or seeds have been drilled, they are fastened to the wreath with 22-gauge florist wire. Or, they may be glued into place with a hot glue gun. Observe caution when handling the glue gun.

The completed wreath can be sprayed with a clear **lacquer** for a shiny appearance. Clear lacquer is available in aerosol cans at most hardware stores. If a more natural or dull finish is desired, no lacquer is needed.

SAFETY FIRST

OSHA's regulations ensure that employers and employees know about the work hazards and how to protect themselves to prevent injuries in the workplace. For more information on occupational health and safety in the workplace, please visit the OSHA site at www.osha.gov.

Summary

In this unit, a variety of evergreen trees and shrubs of different aromatic senses, textures, and sizes have been used for the construction of a holiday wreath. In addition, various seedpods, cones, nuts, and fruits add special interest and beauty to a holiday wreath known as the Della Robbia wreath. Door swags and wreaths are popular holiday decorations. The basic mechanics of designing a wreath include the selection of a wreath frame with the desired base form and the choice of greenery and decorations. The use of a wreath-making machine will help to improve the efficiency of mass production of the wreaths for the holiday season.

Review Questions

1. Discuss 10 different plant materials that are commonly used in holiday decorations.
2. Describe the two main differences between constructing an evergreen wreath and constructing a door swag.
3. Describe five different cones, fruit pods, and nuts that can be used in Della Robbia wreaths.

4. Discuss some options for finding seedpods and other items to ornament a holiday decoration.
5. How do you keep wreaths and door swags looking fresh throughout the holidays?

Student Activities

1. Take a field trip to the local woods or fields to collect supplies for wreaths. Collect seedpods from trees and shrubs in the area. Cut branches of Scotch pine, white pine, and Douglas fir. (Note the distinct fragrance of white pine and Douglas fir.)
2. Visit a local florist supply house to examine materials available for holiday decorations.
3. Ask a local florist to speak with the class about special preparations for the holiday season and construction of holiday decorations.
4. Visit a library to examine magazines and pamphlets containing ideas for holiday decorations.
5. Attend a holiday craft fair to gain ideas to be used in the class.
6. Work with an elementary class to teach the younger students how to assemble holiday decorations.
7. Search online for instructional videos on garland and wreath-making.
8. Create your own instructional video on wreath-making. Share the video with the class.

Suggested Supervised Agricultural Experience (SAE) Activities

1. As a placement SAE project, work with a local florist to design and construct holiday decorations for sale. Record all earnings and time worked as documentation for your SAE project.
2. As an entrepreneurship SAE, design and construct holiday decorations and market those products for sale using social media for advertising.

Additional Resources

http://lancaster.unl.edu/hort/nebline/HolidayWreath.shtml

http://umaine.edu/publications/7012e/

www.hort.purdue.edu/ext/HO-22.pdf

UNIT 48

Creating Holiday Centerpieces

Objective

To construct and care for holiday centerpieces.

Key Terms

centerpieces
greenery
OASIS®
spongelike
variegated

Competencies to Develop

After studying this unit, you should be able to:

- construct a holiday centerpiece.
- list 10 plant materials that may be used for centerpieces.
- describe the procedures for keeping a centerpiece fresh.

Materials

- pieces of evergreen trees and shrubs
- floral foam (OASIS®)
- one candle
- 18-gauge florist wire
- florist bowl
- smartphone apps:
  - My Flower Box Events
  - Florista

Construction of Centerpieces

In many homes, **centerpieces** are an important part of the holiday season as they are used to decorate and create interest throughout the home. They are often found on dining room tables, dry sinks, and hutches. Candles, which have special meaning for many people, are usually part of the centerpiece design.

A variety of evergreens are suitable for use in centerpiece construction. Spruce, white pine, Scotch pine, Douglas fir, noble fir, balsam fir, juniper, yew, holly, and boxwood are some of the greens that may be chosen.

Keeping the Arrangement Fresh

To keep the centerpiece fresh during the entire holiday season, it must be watered. To water it, submerge floral foam in a tub of water deep enough to cover the entire arrangement. **OASIS®** floral foam is a good choice. Allow it to stand in the water until there are no longer any air bubbles coming out of foam. It is necessary to perform this procedure several times during the holidays to keep the **greenery** looking fresh.

PROCEDURE
Creating a Pine and Holly Centerpiece

1. *Preparing the base.* Select a half block of floral foam (OASIS®), a spongelike material, to be used as a base and to hold moisture for the greenery (Figure 48–1). To ensure that the foam is thoroughly saturated, soak it in a pan of water until there are no longer any air bubbles coming out of the floral foam.
2. *Inserting the pine.* Place the block on the design table. Take small clippings of white pine 4 to 5 inches long and push them in around the base of the floral foam. Be certain that a complete line of greenery is formed and that the pieces of pine are even in length.

CAUTION Never pull the greenery out of the OASIS® after it is in place; this causes air pockets that will prevent the greenery from absorbing the available moisture.

Start the next line of greenery about ¼ inch above the first line. This allows the design to flow smoothly. Continue until the line is completed.

Add additional rows until the greenery reaches the top of the floral foam block (Figure 48–2). Care must be taken not to insert the last row so that it is too close to the top of the block because floral foam crumbles and breaks apart easily.

3. *Positioning the candle and holly.* Place a candle in the center of the block. To be sure the candle stays in place, push a toothpick about ¼ inch into the center of its bottom. This acts as a balance and holds the candle firmly in the floral foam block. Insert sprigs of variegated English holly 4 to 5 inches long into the top portion of the floral foam. This particular type of holly has spines on the leaves and is very prickly, so this process must be done carefully. Add enough holly to cover the top of the block and make the centerpiece appear complete (Figure 48–3).

▲ **Figure 48–1** To begin construction of the centerpiece, select half a block of floral foam. Soak the floral foam in water until no air bubbles form. Set the soaked floral foam into a florist bowl container.

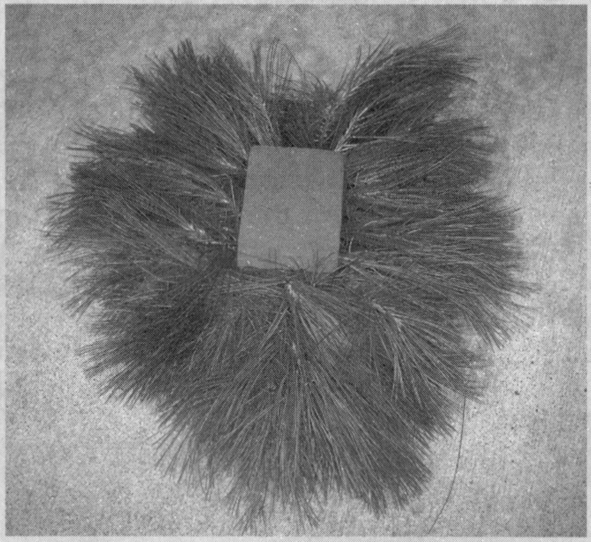

▲ **Figure 48–2** Pieces of pine 4 to 5 inches long are inserted into the sides of the floral foam block.

▲ **Figure 48–3** Placement of a candle and variegated foliage completes the centerpiece to create a special accent for the arrangement.

PROCEDURE
Creating a Miniature Holiday Tree Centerpiece

1. *Preparing the base.* Select a full block of floral foam (OASIS®) and soak it in water until there are no longer any air bubbles coming out of the OASIS®.
2. *Shaping the base.* Firmly set the full block of floral foam vertically into a standard florist bowl (Figure 48–4). Using a piece of 18-gauge florist wire, cut the floral foam to form a tapered cone by removing the corners as shown in Figure 48–5. Hold the florist wire firm while making these cuts. When finished, the block should appear as shown in Figure 48–6.

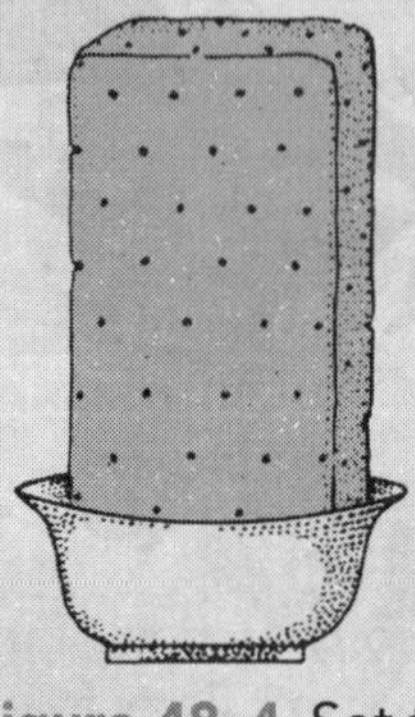

▲ **Figure 48–4** Set the floral foam firmly in the florist bowl.

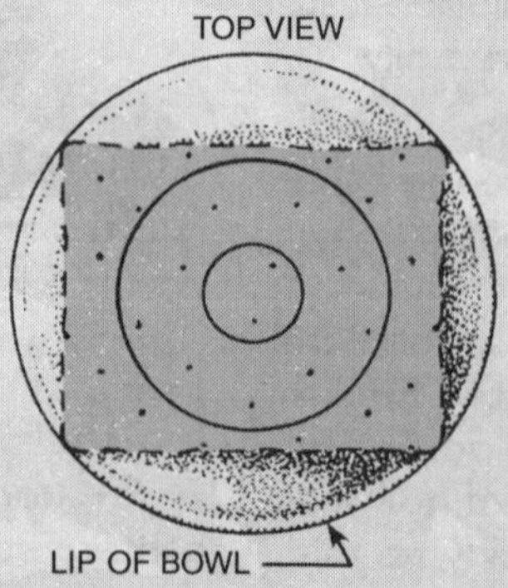

▲ **Figure 48–5** Top view of floral foam.

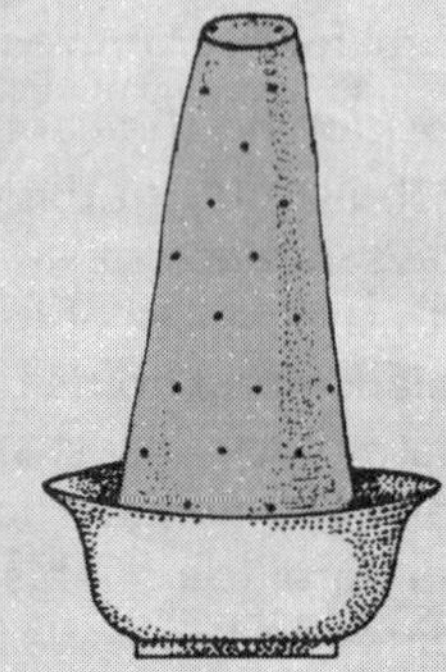

▲ **Figure 48–6** The cone-shaped floral foam.

▲ **Figure 48–7** The finished mini-tree.

3. *Preparing the greenery. Select your greenery from any of a number of narrowleaf or broadleaf evergreens. Prune sprigs to about 4 to 6 inches long.*
4. *Inserting the greenery. Beginning at the base of the floral foam cone, insert 4- to 6-inch sprigs evenly around the base, covering the container. As you move up the cone, you will have to keep pruning the sprigs of greenery so that the shape of your "mini-tree" tapers with its intended design.*
5. *Finishing the tree. Your finished tree should look like the one in Figure 48–7. Fill in with sprigs anywhere necessary and decorate as you desire. For an example of an undecorated and decorated mini-tree, see Figures 48–8 and 48–9.*

▲ **Figure 48–8** Undecorated holiday mini-tree.

▲ **Figure 48–9** Decorated holiday mini-tree with mini carnations, pom-pom chrysanthemums, and ribbon accents for a special effect.

TRY THIS As a class, use the Florista app through an iPhone or iPad to view videos on floral foam preparation. Use the techniques shared to prepare your own centerpiece base.

SAFETY FIRST

OSHA's regulations ensure that employers and employees know about work hazards and how to protect themselves to prevent injuries in the workplace. For more information on occupational health and safety in the workplace, please visit the OSHA site at www.osha.gov.

Summary

Designing a holiday centerpiece or miniature holiday tree is a unique experience. The basic materials for a centerpiece are the design bowl, floral foam, and plant materials or a candle of the designer's choice. A centerpiece must be watered to remain fresh during the entire holiday season.

Review Questions

1. Discuss 10 plant materials that can be used for centerpiece construction, including the advantages and disadvantages of each.
2. Describe techniques that are used to keep holiday centerpieces fresh.
3. How is floral foam (OASIS®) used in the construction of centerpieces, and how is it prepared before it is used?
4. What material is used to fill in the top of the centerpiece, and how long are the pieces that are used?

Student Activities

1. Take a walking field trip to a local area to collect greenery for the construction of centerpieces.
2. Design and make a centerpiece using the techniques in this textbook and from ideas found by searching online.
3. Set up a demonstration on how to water centerpieces after their construction. Present the demonstration at a local flower show or workshop.
4. After reviewing the suggested apps, select a design to make in class with the approval of your instructor.

Suggested Supervised Agricultural Experience (SAE) Activities

1. As a community service SAE, organize a group of volunteers to construct centerpiece decorations that will be sold as a fund-raiser for a local shelter. Track the amount of expenses, time spent, and donations raised to document your SAE project.
2. As a placement SAE project, work with a local florist to design and construct centerpiece decorations for sale. Record all earnings and time worked as documentation for your SAE project.

Additional Resources

www2.ca.uky.edu/hes/fcs/FACTSHTS/HF-LRA.093.PDF

www.howcast.com/videos/284-how-to-create-a-floral-centerpiece/

UNIT 49

Enhancements and Accessories for Floral Designs

Objective

To create attractive bows for decorative arrangements.

Key Terms

arrangement
bolt
cotton
decorator's bow
flocked
florist shears
plastic
satin
texture
velvet

Competencies to Develop

After studying this unit, you should be able to:

- select ribbon of appropriate size and color for a specific arrangement.
- construct a bow to be used on a wreath or door swag.

Materials

- 22-gauge florist wire
- wire cutters
- florist shears
- one bolt #40 red velvet ribbon (single or double sided)
- one bolt #9 red velvet ribbon (single or double sided)
- one bolt #5 red velvet ribbon (single or double sided)
- one bolt #3 red velvet ribbon (single or double sided)
- one bolt #1 red velvet ribbon (single or double sided)

The bow is an important accessory to many floral projects; it is used to lend a graceful touch to the design. Color, material, size, and quantity must be considered when selecting the ribbon used to make a bow.

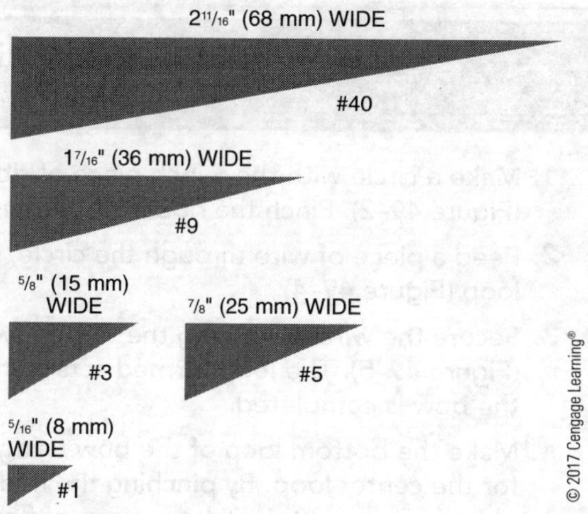

▲ **Figure 49–1** Ribbon sizes.

Bow color will depend on the **arrangement**, fashion trends, and individual tastes. Color should be a pleasing accent to the overall design. An almost infinite variety of ribbon colors are available. Patterns and styles, like color, will vary with fashion trends and individual tastes.

Many different materials are used to make ribbon today. Ribbon varies in **texture**, pattern, style, weight, and ease of workability. Because of these factors, certain designers have distinct preferences in the types of ribbons they use. Texture, for example, ranges from fine to coarse, such as a **satin** ribbon compared to a heavy burlap ribbon. Ease of workability varies with the texture and weight of the ribbon. Satin is regarded as the easiest ribbon with which to work. In addition, today's designer can choose from **velvet**, **plastic**, **cotton**, **flocked**, and burlap ribbon. Each has a different application in the designer's world.

Ribbon size follows an industry standard. The basic sizes used in this unit are #40 (2 11/16 inches wide) and #9 (1 7/16 inches wide). These sizes are shown in Figure 49–1. Ribbon is either single-faced or double-faced. The single-faced ribbon has a shiny side and a dull side. The double-faced ribbon is the same texture on both sides. Ribbon is available at local florist or craft shops where it is sold by the yard or by the roll, also called a **bolt**. To make the most economical purchase, it is important to check the total yardage on the bolt of ribbon.

Many arrangements, in addition to those used during holiday celebrations, require decorative bows. There are two basic types of decorative bows to use for holiday arrangements. The first type is a pinch bow that is made by loops of ribbon in a circular fashion. The other, a decorator's bow, is in the shape of a bow tie. A well-made bow adds an attractive finishing touch and can be made easily with practice.

Decorator's Bow

The bow demonstrated here is known as a **decorator's bow**. It is made of velvet ribbon that is 2 11/16 inches wide (#40 ribbon).

The materials needed for the sample bow are #40 velvet ribbon and 22-gauge florist wire. From the velvet ribbon, cut one piece 6 inches long (for the center of the bow), one piece 18 inches, one piece 20 inches, and one piece 30 inches long (for loops). Use three pieces of florist wire, each 18 inches long.

TRY THIS Create a display board showing the various sizes, styles, materials, and colors of available ribbons. For each ribbon on the display board, describe the appropriate use of that style of ribbon.

PROCEDURE
Making a Decorator's Bow

1. Make a circle with the 6-inch piece of ribbon, allowing the ends to overlap about an inch (Figure 49–2). Pinch the ribbon together at the center where the two ends overlap (Figure 49–3).
2. Feed a piece of wire through the circle, pulling it tight around the crushed, overlapped side of the loop (Figure 49–4).
3. Secure the wire by turning the ribbon two or three times while holding the ends of the wire (Figure 49–5). The loop formed acts as the center of the bow and covers all exposed wires when the bow is completed.
4. Make the bottom loop of the bow using the 20-inch piece of velvet. Overlap the ends as shown for the center loop. By pinching the middle of the loop together, form a figure eight with the ribbon (Figure 49–6). Take a piece of florist wire and wire the loop in the center, holding the bow tight. Twist the wire by turning the ribbon two or three times (Figure 49–7).
5. Form the top loop of the bow with the 18-inch piece of ribbon. Overlap the ends as before, and then form a figure eight as in Step 4. Crush the loop in the center, and wire it with the last piece of wire. Twist the ribbon two or three times to secure the wire.
6. Take the two larger loops and place the smaller of the two on top of the larger one (Figure 49–8).
7. Use the 30-inch piece of ribbon to form the tails of the bow. Fold the ribbon in half to determine the middle. Crush the ribbon together at this point so that it can be easily wired (Figure 49–9). Place the tails on top of the two loops from Step 6.

▲ **Figure 49–2** Make a circle with the 6-inch piece of bow, allowing ends to overlap an inch.

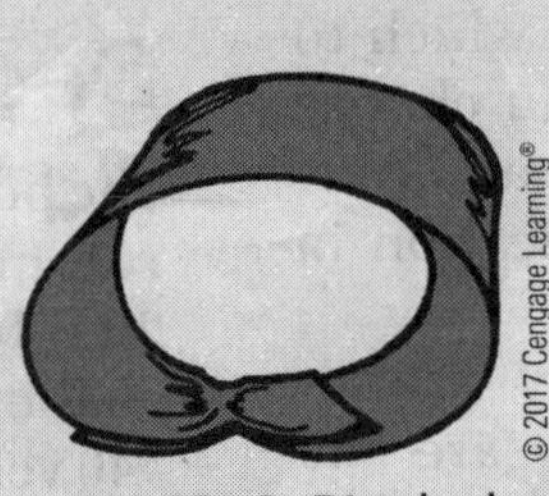

▲ **Figure 49–3** Pinch the center of the ribbon where the two ends overlap.

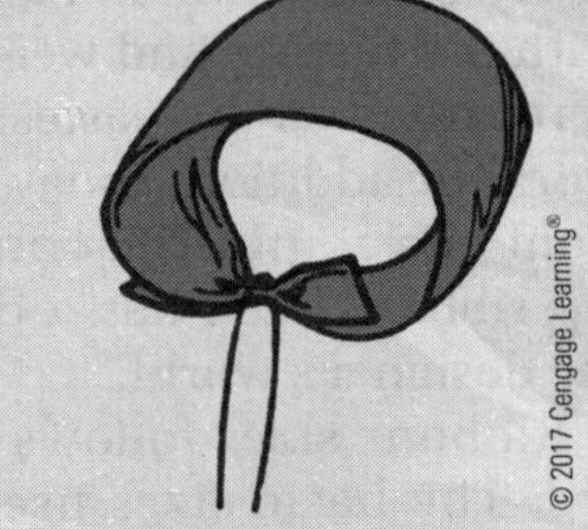

▲ **Figure 49–4** Feed a piece of wire through the circle, pulling it tight.

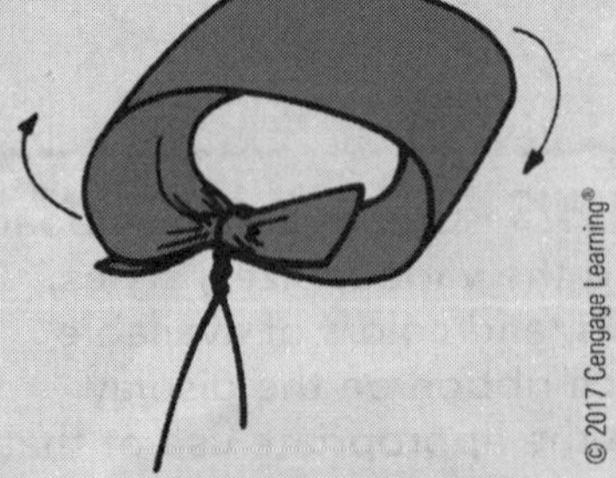

▲ **Figure 49–5** Secure the wire by turning the ribbon two or three times while holding ends of wire.

▲ **Figure 49–6** Using the 20-inch piece of velvet, make the bottom and center of the loop. Pinch the middle of the loop together to form a figure eight.

▲ **Figure 49–7** Wire the loop in the center, and twist the wire by turning the ribbon two or three times.

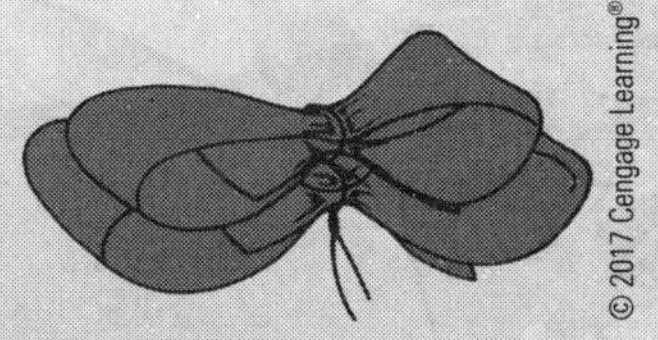

▲ **Figure 49–8** Place the smaller loop on top of the larger one.

8. With the wire attached to the center loop formed in Steps 2 and 3, wire the entire bow together. Pull the center tight to form the completed bow (Figure 49–10).
9. The tails of the bow should be two different lengths to create interest. The ends of the ribbon may be cut with **florist shears** on a diagonal or in a fishtail design for an interesting variation (Figure 49–11).

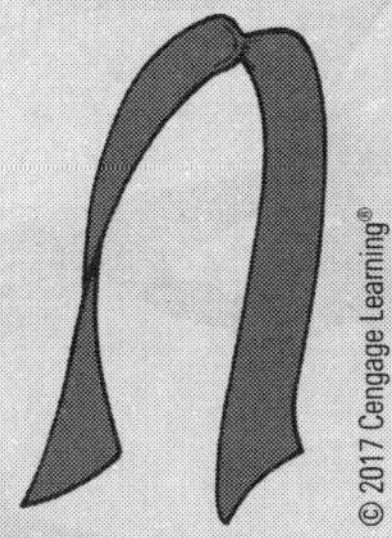

▲ **Figure 49–9** Use the 30-inch piece of ribbon to form the tails of the bow. Fold in half, and crush in the center to easily wire.

▲ **Figure 49–10** Wire the entire bow together, and pull center tight.

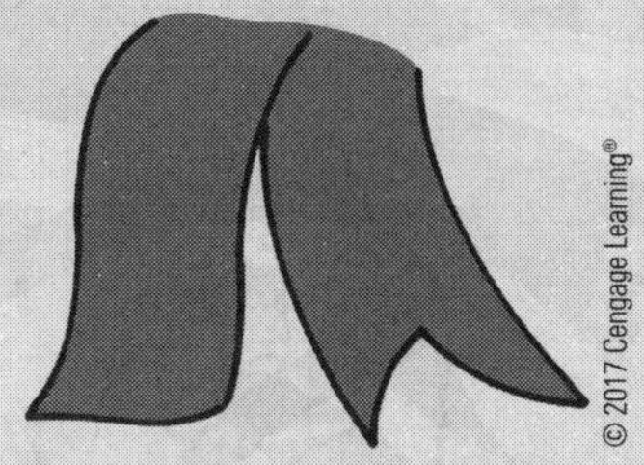

▲ **Figure 49–11** Cut the ends as desired.

PROCEDURE
Making a Pinch Bow

1. Take a piece of 22-gauge florist wire, and tape the center 3 inches of the wire with florist tape. Folding the wire in half, work the folded wire between the thumb, index finger, and third finger, making an inverted "U" shape. Put the wire aside.
2. Hold the ribbon between the thumb and index finger with the shiny side of the ribbon facing you.
3. Take one end of the ribbon, and make a loop about 3 inches long, shiny side up.
4. Pull ribbon through and make a loop on the opposite side the same size as the first loop.
5. As the second loop is made, pinch the bow in the center. Continue the same procedure until there are three loops on one side and two loops on the opposite side. As the ribbon is turned to make the third loop, pull the ribbon up into the center of the bow to make the bow appear fuller. Repeat previous steps until the entire amount of the ribbon is used.
6. Now, hold the bow firmly between the index finger and the thumb. The florist wire that you made ready earlier is used to fasten the bow together. This wire is 18 inches long and coated green to blend better with arrangements.
7. With your free hand, pick up the piece of previously folded wire. Pull the wire up around the pinched center of the bow. Take the opposite end of the florist wire and run it through the loop in the wire, pulling it tightly to form a completed bow.

Pinch Bow

To make the pinch bow, use about 3 yards of #9 florist ribbon. Refer to Figure 49–12 while following the procedure.

Ribbon Accents

Ribbon loops with tails are used to accent designs with color and texture throughout the design to create a sense of unity. Ribbon loops with tails enhance and add pizzazz by using spray glitter of the designer's desired color.

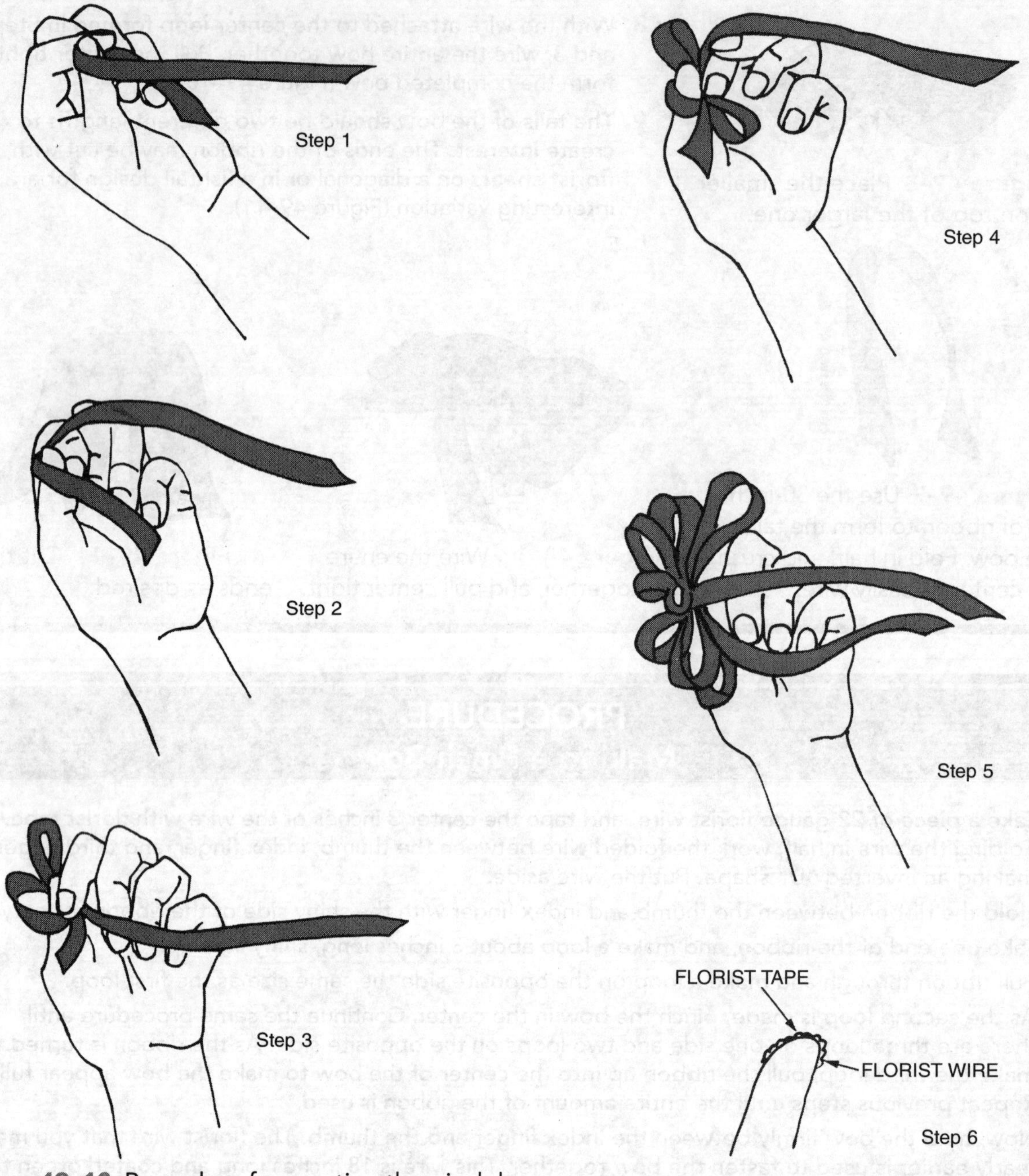

▲ **Figure 49–12** Procedure for creating the pinch bow.

PROCEDURE
Designing Ribbon Loop with a Tail

1. Select the desired size of ribbon. In most situations, a #3, 5, or 9 ribbon works well.
2. Cut the desired size of ribbon to 7 inches long.
3. Holding the end of ribbon in the right hand, fold the loop, and pull the tail up behind the loop.
4. Using 20-gauge florist wire, cover the base of the loop and tail with florist tape. (Florist tape is sticky on both sides and will stick to itself as the florist wire is covered.)
5. Insert the loop and tail design into the arrangement for a special holiday accent. (See Figure 49–13.)

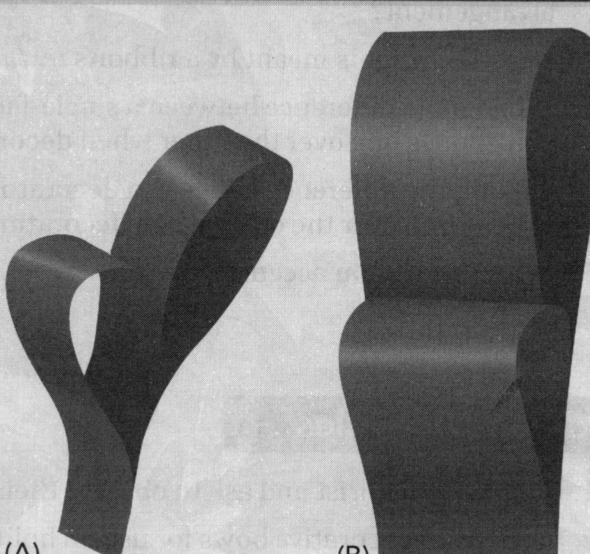

(A) (B)

▲ **Figure 49–13 A and B** The side and front view of a completed ribbon loop with tail.

SAFETY FIRST

OSHA's regulations ensure that employers and employees know about work hazards and how to protect themselves to prevent injuries in the workplace. For more information on occupational health and safety in the workplace, please visit the OSHA site at www.osha.gov.

Summary

Bows are important accessories for floral designs. When adding a bow, the designer needs to understand different sizes of ribbon, select a color, select single-faced or double-faced ribbon to accent the design, and know the mechanics of assembling the bow. The process of construction will depend on the type of bow being designed. Wider ribbon is used to make a decorator's bow than for a pinch bow.

Review Questions

1. How do you select the appropriate ribbon size and color when decorating an arrangement?
2. Describe what is meant by a ribbon's *texture*, and give some examples.
3. What is the difference between a single-faced and a double-faced ribbon? When would you choose one over the other when decorating an arrangement?
4. Explain the difference between a decorator's bow and a pinch bow. When would you choose one over the other when decorating an arrangement?
5. What is a ribbon accent?

Student Activities

1. Visit a local florist and ask to observe their procedure for making bows.
2. Construct decorative bows for use on holiday decorations. Consider selling the bows as an FFA fund-raising project.
3. Prepare and participate in the FFA Floriculture Career Development Event (CDE). This is usually offered at the state and national levels.
4. Create two bows of different styles based on the "Procedures" presented in the unit.
5. Use the smartphone apps My Flower Box Events and Florista to view examples and videos of floral designs.

Suggested Supervised Agricultural Experience (SAE) Activities

1. As an entrepreneurship SAE, prepare premade floral bows and then retail them for sale using Internet advertising.
2. To qualify for a placement SAE project, work for a local florist preparing arrangements and floral accessories. Record all earnings and time worked as documentation for your SAE project.

Additional Resources

www.baxtercountyfair.org/pdf/Principles_of_Floral_Arrangement.pdf

www.dillonfloral.com/

www.joann.com/home-and-floral-decor/floral-and-supplies/containers-and-fillers/

UNIT 50

Floral Designs

Objective

To create a floral design.

Key Terms

accent
axis
balance
color wheel
complement
focal point
harmony
line flower
proportion
repetition
rhythm
scale
symmetrical
unity

Competencies to Develop

After studying this unit, you should be able to:

- identify the types of flowers used in floral designing.
- list and describe the nine basic principles of floral design.
- identify the six basic floral design shapes.
- design a circular floral arrangement.

Materials

- fresh floral material (carnations, baby's breath, baker's fern)
- floral container
- floral foam (OASIS®)
- florist wire and wire cutters
- sharp florist knife
- florist shears
- smartphone apps:
  - My Flower Box Events
  - Florista

Basic floral design is an art that has played an important role throughout history and is a part of many functions in our lives. Because it is a specialty, floral design requires both specific skills and tools (Figure 50–1). It is also very functional. Floral design is used at weddings, funerals, and other formal and informal functions (Figure 50–2). Flowers help to adorn any event at which they are used.

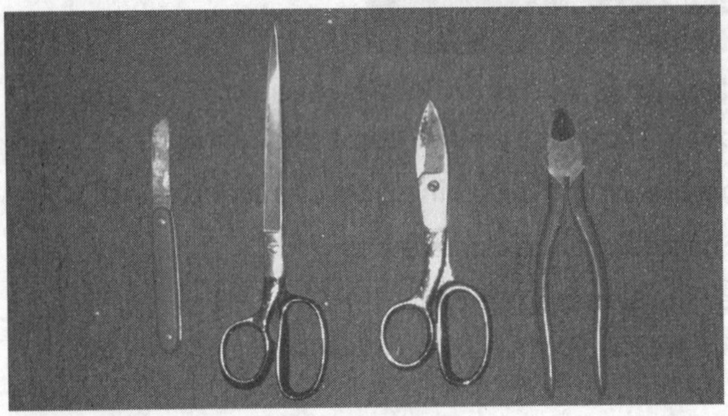

▲ **Figure 50–1** Florist tools. (Left to right) florist knife, ribbon shears, florist shears, diagonal cutters.

Today, the florist industry has broadened opportunities for those people who are interested in a career in the florist business. The industry has expanded to malls, supermarkets, airports, and hotels to meet the ever-growing needs of individuals who recognize the value of flowers in their everyday lives. The florist trade association and wholesalers of floral products offer employment opportunities as well. It is important that people seeking employment in these fields have a basic skill and knowledge of the florist industry.

Basic Floral Design

In basic floral design, one must be familiar with the factors that affect the design. When designing an arrangement, it is necessary to use the following basic principles of design that serve as the guidelines for floral design: balance, focal point, proportion, scale, accent, repetition, rhythm, harmony, and unity.

Design Principles

Balance is achieved in a floral design when flower size and container flow together and **complement** each other. The arrangement has visual weight on each side of the **axis** to make it

(A)

(B)

(C)

▲ **Figure 50–2** Some floral designs. (A) Cascading wedding bouquet. (B) Asymmetrical right angle informal design. (C) Symmetrical design.

stand alone. If the arrangement is **symmetrical**, the floral design could be divided in half and have equal visual weight on each side of the arrangement. This type of design is used to create a more formal arrangement. In asymmetrical design, the arrangement is offset, creating a visual weight of being heavier to one side of the axis. This design allows the individual to use a more natural effect or informal approach to designing.

The **focal point** is important in some designs because it creates the accent and interest. This will catch the eye of the viewer. Focal points are created by the use of unusual flowers, by grouping flowers of one color, or by using a larger flower. The focal point must complement the entire arrangement.

Proportion in an arrangement is important to keep relative size, color, and texture of the flowers to create a pleasing arrangement. The general rule of thumb for floral designs is that they be at least one-and-one-half to two times the height or width of the container (Figure 50–3). To create the proper proportion in the arrangement, the points farthest from the focal point should have the flower buds. In designing toward the focal point, flowers in full bloom are used.

Scale is a principle that must be taken into consideration. The size of the flowers, where the arrangement is to be used, placement of the arrangement, and overall size of the completed design are all factors of scale. The design must be in proportion to the surrounding area.

Accent is used in a design to draw attention to the design. Many designers use the focal point to create this emphasis. Accent is that point of interest that catches one's attention and enhances the design.

Repetition is used to accentuate the flowers and colors used in a design. Repeating the use of material or color throughout the design can be very effective.

Rhythm creates a sense of continuity among the shapes, colors, and textures. It creates a feeling in the viewer that everything flows together in the arrangement, thus giving a feeling of motion.

Harmony is created in a floral design when all the parts flow together to give a total, completed look to the arrangement.

Unity is created when all of the arrangement flows together and is not segmented. For a look of unity, the flowers in an arrangement must complement each other through color, size, and shape.

As noted, color plays an intricate role in design and can factor into other design principles such as unity, rhythm, and repetition. The **color wheel** (Figure 50–4) is a helpful tool for designers and can help learners understand the theory behind applying color to floral arrangements.

Color in arrangements is enhanced by the relationships of the primary, secondary, and tertiary colors.

- P—Primary colors (red, blue, and yellow are sets of colors that are combined to create a range of colors)
- S—Secondary colors (colors made by mixing two or more primary colors, such as green or orange)

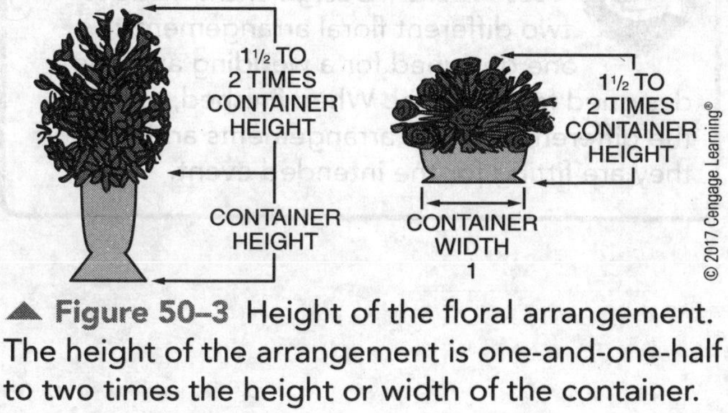

▲ **Figure 50–3** Height of the floral arrangement. The height of the arrangement is one-and-one-half to two times the height or width of the container.

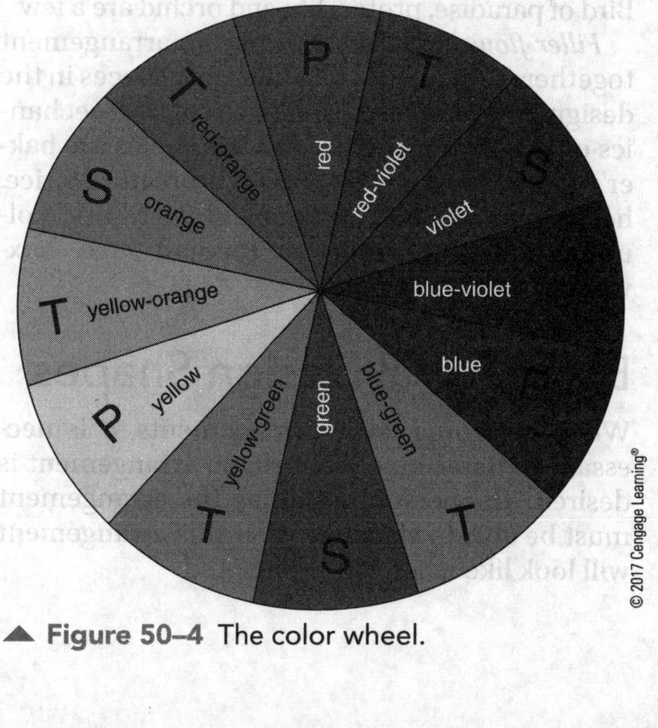

▲ **Figure 50–4** The color wheel.

- T—Tertiary colors (colors made by mixing either one primary color and one secondary color or two secondary colors, such as yellow-orange or blue-violet)

When selecting the flowers and foliage for the desired design, one must consider the desired reaction of the viewer. The warm reds, oranges, and yellows create a feeling of excitement, while the cooler blues, greens, and purples create a sense of calmness. If constructing boldness is desired in the design, use opposite colors on the color wheel (i.e., purples and yellows). If one desires a more subtle arrangement, use colors next to one another on the color wheel (i.e., blues and purples).

Design Materials

In floral design, there are four basic types of design materials: line, mass, form, and filler flowers.

Line flowers include snapdragons, liatris, delphinium, foxglove, ligularia, loosestrife, Bells of Ireland, and gladiolus. In most cases, these flowers are used to establish the outer framework of an arrangement.

Massing flowers have the flower head on the terminal end of the stem. These flowers are used in the arrangement to create all of the design principles and to accent the focal point. Some examples are roses, carnations, chrysanthemums, zinnias, coneflowers, tulip, shasta daisy, Michelmas daisy, and asters.

Form flowers have a unique shape or form. They may be used to create the accent or the focal point. Bird of paradise, protea, lily, and orchid are a few.

Filler flowers are used to tie the arrangement together. They help to fill any empty spaces in the design. They are also used to cover the mechanics of the arrangement. Some examples are baker's fern, asparagus fern, baby's breath, statice, heather, huckleberry, ornamental grasses, columbine, astilbe, lavender, Russian sage, boxwood, croton, and palm.

Basic Floral Design Shapes

When designing floral arrangements, it is necessary to determine what shape arrangement is desired. The person designing the arrangement must be able to visualize what this arrangement will look like when it is completed.

Basic floral design shapes include round (circular), triangle, horizontal, crescent, Hogarth's (S) curve, and right angle (Figure 50–5).

Designing a Floral Arrangement

A good, basic design to start with is the circular design that has a rounded form. It has many applications and can be used in various settings such as a centerpiece or coffee-table arrangement or in a foyer in front of a mirror where it is viewed from all sides.

In the construction of the circular design, select massing flowers to establish the outline and form the circular shape of the design. A circular design does not have a focal point because it is viewed from all sides. To have the correct proportion of the design, use the general rule of thumb of one-and-one-half to two times the height or width of the container. The stem length of the massing flowers will be determined by this rule. Arrange the massing flowers to establish the basic outline of the design. With the basic outline established, arrange the other massing flowers so they are within the outline to give a rounded appearance.

Filler flowers have several purposes: to tie the arrangement together, to fill in any voids in the arrangement, and to provide the greenery to cover the floral foam and container. Filler flowers and greenery provide more fullness and give the arrangement a completed appearance.

> ***TRY THIS*** Design and construct two different floral arrangements: one designed for a wedding and one designed for a funeral. When finished, explain the differences in the arrangements and why they are fitting for the intended event.

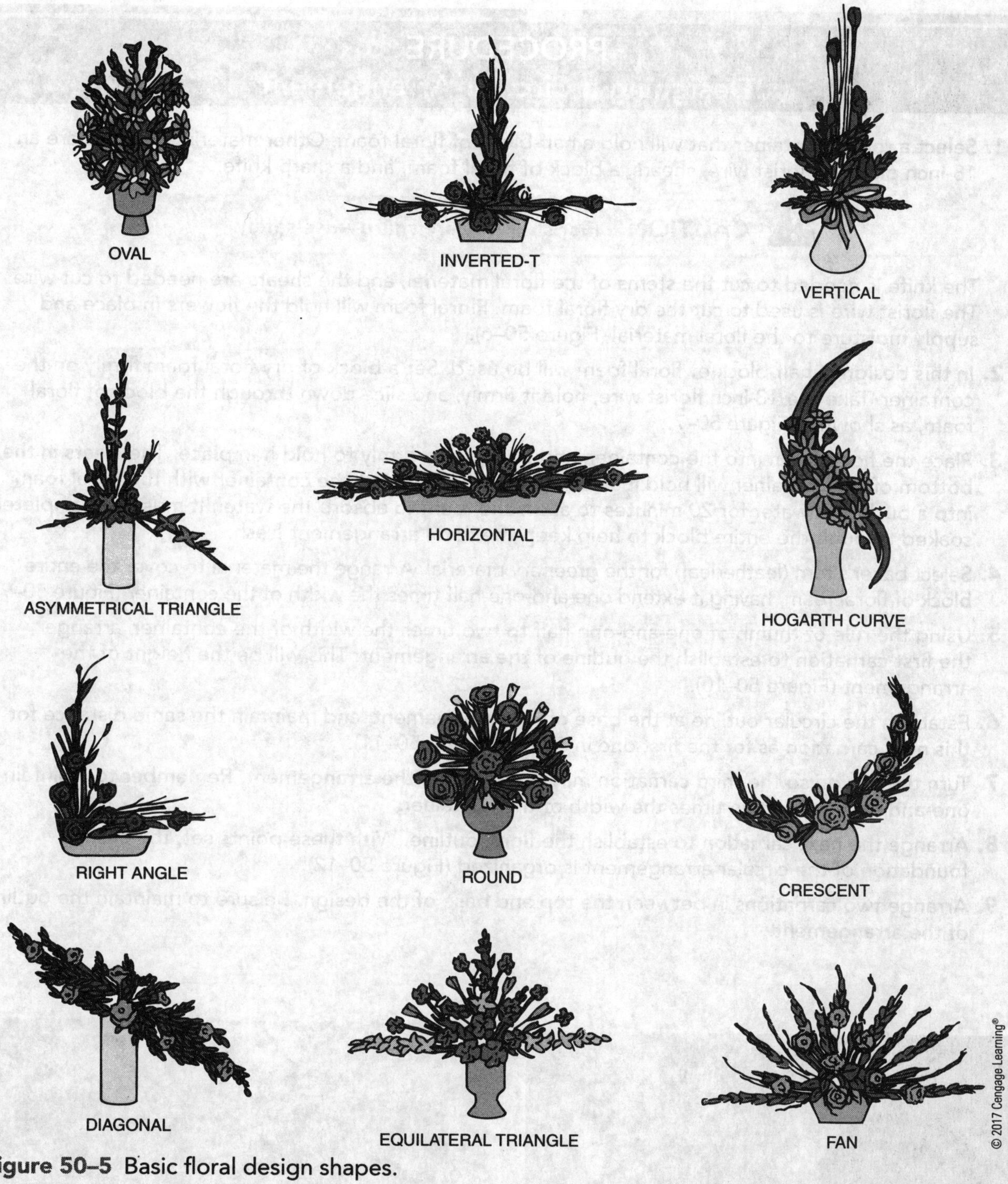

▲ **Figure 50–5** Basic floral design shapes.

PROCEDURE
Designing a Circular Arrangement

1. Select a round container that will hold a half-block of floral foam. Other materials needed are an 18-inch piece of florist wire, shears, a block of floral foam, and a sharp knife.

CAUTION Florist knives are sharp—use safely.

The knife is needed to cut the stems of the floral material, and the shears are needed to cut wire. The florist wire is used to cut the dry floral foam. Floral foam will hold the flowers in place and supply moisture to the floral material (Figure 50–6).

2. In this design, a half-block of floral foam will be used. Set a block of dry floral foam firmly on the container. Take the 18-inch florist wire, hold it firmly, and slice down through the block of floral foam, as shown in Figure 50–7.
3. Place the floral foam into the container, and push down firmly to hold it in place. The cleats in the bottom of the container will hold it securely (Figure 50–8). Set the container with the floral foam into a bucket of water for 20 minutes to allow the foam to absorb the water. It must be completely soaked through the entire block to help keep the floral arrangement fresh.
4. Select baker's fern (leatherleaf) for the greenery material. Arrange the material to cover the entire block of floral foam, having it extend one-and-one-half times the width of the container (Figure 50–9).
5. Using the rule of thumb of one-and-one-half to two times the width of the container, arrange the first carnation to establish the outline of the arrangement. This will be the height of the arrangement (Figure 50–10).
6. Establish the circular outline at the base of the arrangement, and maintain the same distance for this next carnation as for the first one in Step 5 (Figure 50–11).
7. Turn the design so the third carnation may be placed in the arrangement. Remember to maintain one-and-one-half to two times the width of the container.
8. Arrange the next carnation to establish the floral outline. With these points set, the basic foundation of the circular arrangement is organized (Figure 50–12).
9. Arrange two carnations in between the top and base of the design. Be sure to maintain the outline of the arrangement.

▲ **Figure 50–6** The mechanics of a floral arrangement include the following materials: container, floral foam, shears.

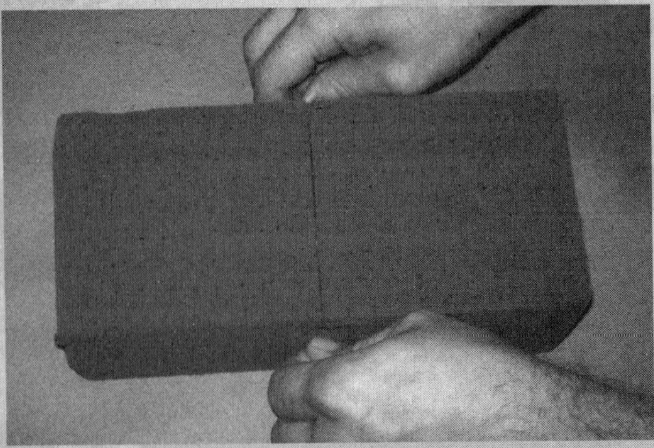

▲ **Figure 50–7** Cutting dry floral foam with florist wire to fit the container.

▲ **Figure 50–8** This container has cleats in the bottom to hold the floral foam firmly in place. The floral foam is placed in the container dry and then soaked with water.

▲ **Figure 50–9** Baker's fern used as a greenery material extends one-and-one-half times the width of the container to form a foundation.

▲ **Figure 50–10** Establishing the height with the first carnation added to the arrangement.

▲ **Figure 50–11** Establishing the width of the floral design at the base of the arrangement.

10. Add the remaining carnations to the arrangement until the circular shape of the arrangement is filled. It is important to keep space in between each carnation to allow their natural beauty to radiate. The greenery should extend out around each carnation.
11. To complete the design, filler flowers (such as baby's breath) are arranged in between the carnations. This will create a completed arrangement (Figure 50–13).

▲ **Figure 50–12** Placing the carnations at the base establishes the circular design. Continue the circular pattern, maintaining space between each flower, to finish the arrangement.

▲ **Figure 50–13** Baby's breath is added to tie the arrangement together.

SAFETY FIRST

OSHA's regulations ensure that employers and employees know about work hazards and how to protect themselves to prevent injuries in the workplace. For more information on occupational health and safety in the workplace, please visit the OSHA site at www.osha.gov.

Summary

Basic floral design is an art that has played an important role throughout history. Floral arrangements are a part of many functions today. The florist industry plays a major role in the horticultural field for employment and gross sales in the total economy. In basic floral design, it is important to understand and employ the basic design principles of balance, focal point, scale, proportion, accent, repetition, rhythm, harmony, and unity. A designer must understand the basic types of design materials, line flowers, massing flowers, form flowers, and filler flowers, as well as the basic floral design shapes, round (circular), triangle, horizontal, crescent, Hogarth's (S) curve, and right angle.

Review Questions

1. Discuss the different types of flowers that are typically used in floral designing.
2. List and give examples of the nine basic principles of floral design.
3. Describe the six different shapes that are used in floral design.
4. How would you use a color wheel to create a design that conveys excitement?
5. Describe the four basic types of design materials that are used in floral design.

Student Activities

1. Visit a local florist and observe the designing of various arrangements.
2. Attend a flower and garden trade show to observe competitive events of floral designers.
3. Have a local florist come to class as a guest speaker.
4. Prepare and participate in the FFA Floriculture Career Development Event (CDE). This is usually offered at the state and national levels.
5. Put on floral demonstrations or workshops at local fairs, malls, shopping centers, and garden club meetings.
6. Make a display board showing the basic floral design shapes commonly used by florists.
7. Read florist trade association magazines to keep up to date on current trends in the florist industry.
8. Design floral arrangements for fund-raising projects.
9. Visit floral design schools at the college level, and sit in on a class to determine if this is a career interest.
10. Use the smartphone apps My Flower Box Events and Florista to view examples and videos of floral designs.

Suggested Supervised Agricultural Experience (SAE) Activities

1. As an entrepreneurship SAE, prepare floral arrangements and then retail them for sale using Internet advertising.
2. To qualify for a placement SAE project, work for a local florist preparing arrangements for customers. Record all earnings and time worked as documentation for your SAE project.

Additional Resources

http://ati.osu.edu/floral-design-marketing

www.morrisville.edu/programsofstudy/schoolofagandnr/horticulturefloraldesign/

www.wku.edu/floralshop/

UNIT 51

Corsages and Boutonnieres

Objective

To create a corsage and a boutonniere.

Key Terms

boutonniere
calyx
clutchwire
consultation
corsage
corsage pin
cotton wad
dexterity
Floraltape
florist tape
hairpin
hook
Parafilm
piercing
podocarpus
ribbon
salal

Competencies to Develop

After studying this unit, you should be able to:

- list the various floral materials used to make corsages and boutonnieres.
- select the correct wiring procedure for the different flower types.
- tape the flowers correctly.
- demonstrate the four wiring procedures used with the appropriate flowers.
- select the correct size ribbon for the construction of a corsage.
- feather a carnation flower.

Materials

- floral shears
- floral tape
- floral wire
- floral knife
- ribbon
- corsage pins
- boutonniere pins
- cut flowers
- filler flowers/greenery
- smartphone apps:
  - LikeThat Garden
  - Florista
  - Flower Slideshow
  - Season Flowers

Corsages and boutonnieres have been a special part of the florist business for years. They are popular for many occasions, such as school proms, dances, holiday parties, and weddings.

Many special techniques that require more **dexterity** than other forms of floral design are used in the construction of corsages and boutonnieres. With practice, a designer can create unique arrangements that are charming and beautiful.

Corsages

Corsages should be created with the wearer in mind; they must be complementary to the individual. The size of a corsage is a major consideration. For example, a bird of paradise worn by a flower girl in a wedding is totally out of scale and proportion to her size. Sweetheart roses or miniature carnations are better choices. Therefore, the designer must use good judgment when selecting the flowers for a corsage.

The corsage should be coordinated with the individual's apparel. In many situations, the wearer will have a **consultation** with a designer regarding the type and color of flowers desired, the **ribbon** to use, and the position at which the corsage will be worn. For example, a wristlet corsage could be used when a pin-on shoulder corsage would not be appropriate because the wearer's dress is strapless. When constructing a corsage, the designer should keep in mind that standard practice calls for flowers to be displayed in their natural growing position.

Construction of a Corsage

To make an attractive corsage, the designer must first select a high-quality flower. Next, the designer must wire the flowers. Wiring is necessary because the wire allows the designer to place the florets in the correct position in the design, whereas natural stems make the corsage too bulky.

When wiring corsages, always use florist wire. This wire is available in a standard length of 18 inches and a wide range of gauges (thicknesses). The lower the gauge number, the thicker the wire will be. A range of 22- to 32-gauge wire is good for corsage construction.

The florist wire is used for several reasons:

- To hold the flowers in the correct position in the design
- To make arranging the flowers in the corsage easier
- To prevent the flower heads from breaking off the natural stems
- To reduce the bulkiness of using natural stems
- To make adding accessories such as a bow, puff, or flocked leaves easier

Methods of Wiring Flowers

There are four methods of wiring flowers to make a corsage or boutonniere: the **hook**, **piercing**, **clutchwire**, and **hairpin** methods. Each is described in the Procedures that follow.

Taping the Flower

Florist tape is also used in the construction of corsages and boutonnieres to cover all exposed florist wire, stems, and calyxes. Florist tape is available in ¼-inch, ½-inch, and 1-inch widths. The most commonly used width is ½ inch. Florist tape also comes in a variety of colors. Although green is the most common color, tape also comes in orchid, brown, black, blue, white, red, yellow, purple, and orange.

The tape is applied to the stem by holding the wired flower between the thumb and forefinger of one hand and applying the tape to the stem by twisting the wired flower with the other hand, pulling the tape tightly around the stem. This will create a stickiness, causing the tape to adhere to the stem as the wired floret is wrapped. The tape should be applied evenly, covering all wire and exposed stem. The excess wire on the stem should be left for the time being, as it will be helpful when the construction process begins.

Florist tape is sold under brand names such as **Floraltape** and **Parafilm**.

PROCEDURE
Hook Method

This method is used for composite (daisy-like) florets. Typical flowers wired this way are asters, black-eyed Susans, daisy chrysanthemums, and any flower with a short tough calyx (Figure 51–1).

1. Cut the stem to within ½ inch of the calyx. Strip any leaves off the stem.
2. Insert the end of a 6-inch piece of 22-gauge wire in the stem and up through the flower.
3. Bend the end of the wire above the flower into a hood ¼ inch long.
4. Slowly and gently pull the wire back through the flower until it catches in the top of the calyx and is not visible in the flower center.
5. Tape the flower, starting high on the calyx and continuing down the stem and wire, using a smooth, twisting motion. (The procedure for taping flowers is given in detail later in this unit.)

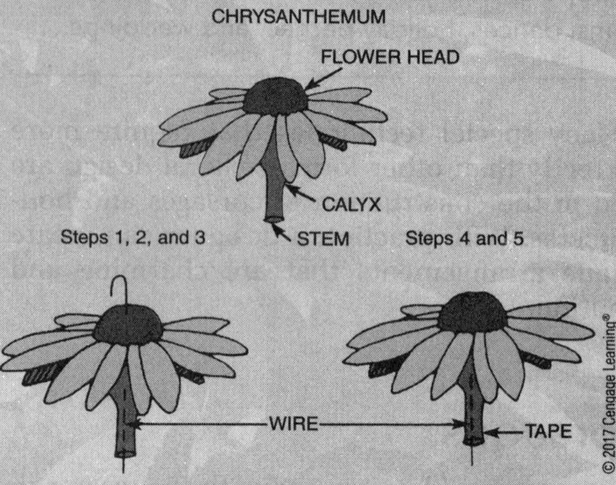

▲ **Figure 51–1** The hook method of wiring flowers.

PROCEDURE
Piercing Method

This method is used for flowers with a large calyx, such as roses and carnations (Figure 51–2).

1. Remove the stem ½ inch below the calyx, and pierce the calyx halfway down, pulling the wire through to one-half the length of the wire.
2. Bend the wire down on both sides. Do not wrap the wire around the stem. Wrapping will make the stem too bulky.
3. Tape the flower, starting high on the calyx, pulling and taping just under the petals. Continue taping both the flower and the wire until all of the wire is covered. If the flower is not evenly taped, retape to get a smooth, tight wrap around the flower and the wire.

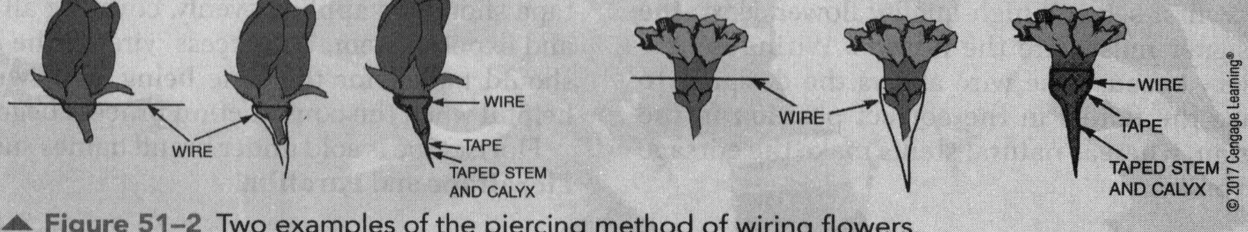

▲ **Figure 51–2** Two examples of the piercing method of wiring flowers.

PROCEDURE
Clutchwire Method

This method is used on gladiolus florets, camellias, and garden, football (standard), and spider mums (Figure 51–3).

1. Take 8 inches of wire and fold it in half. Place the fold around the calyx and/or stem of the flower.
2. Wrap one-half of the wire around the flower's stem or calyx and around the other half of the wire several times, until all the wire is used.
3. Tape the flower, starting at the calyx and working down until all the wire is covered.

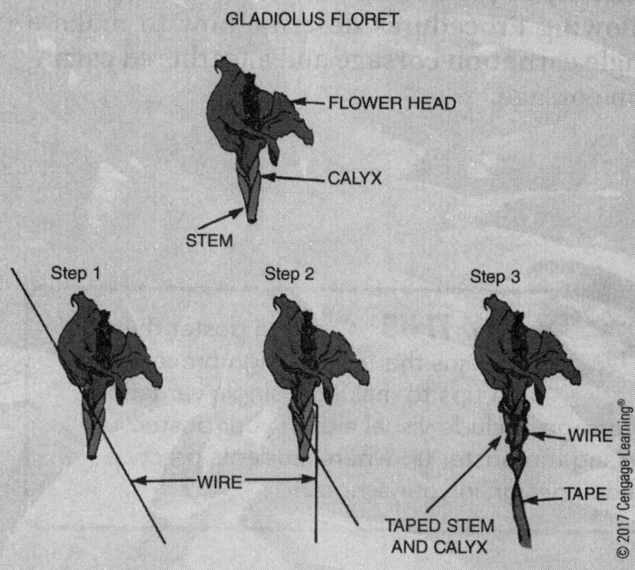

▲ **Figure 51–3** The clutchwire method of wiring flowers.

PROCEDURE
Hairpin Method

This method is used with flowers that have a tubular floret, such as the *Stephanotis* (Figure 51–4).

1. Take 8 inches of florist wire and bend it in half. Place a small wad of moist cotton in the bend to support the flower.
2. Insert the wire down through the floret, keeping the cotton wad in place.
3. Gently pull the florist wire down until the cotton wad is firmly fixed in the flower trumpet. Twist the florist wire around the calyx. Tape the floret into place.

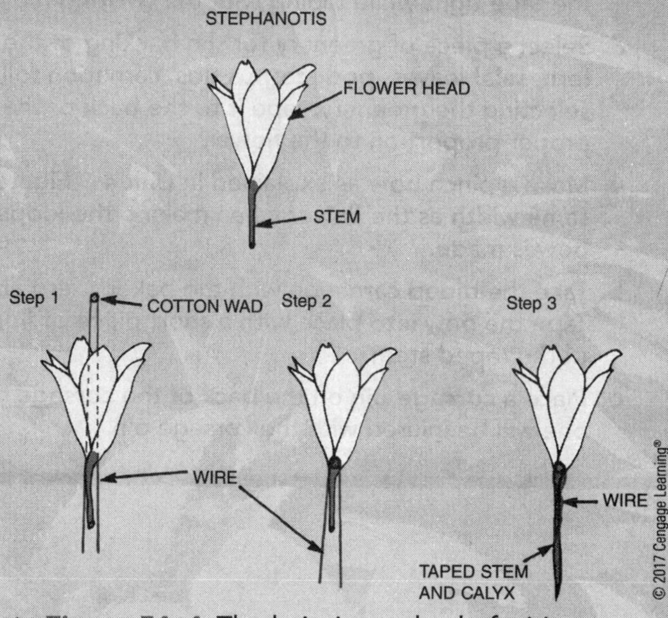

▲ **Figure 51–4** The hairpin method of wiring flowers.

Techniques of Corsage Making

Corsage construction is an area of the florist business that tests one's creative ability and dexterity to produce an attractive corsage. The following Procedures describe how to make a single carnation corsage and a feathered carnation corsage.

TRY THIS Create a poster that explains the basic design procedure and tips to making a single carnation corsage. Include visual aids in your poster. Hang the poster up where students practice floral design in your school.

Boutonnieres

The **boutonniere** is a single flower or small bouquet pinned to a man's lapel or in his lapel buttonhole to add a touch of color to the man's clothing for numerous special occasions.

Flowers for boutonnieres are selected based on the individual's preference. Flowers most commonly used for boutonnieres are the carnation, rose, chrysanthemum, and *Stephanotis*. Other flowers available on a seasonal basis are the bachelor's button and lily of the valley.

The technique of making a boutonniere is simple. The first step is to select a flower of the highest quality and freshness. Wire the flower with the proper florist wire, and tape it smoothly and evenly. Use a 20- to 22-gauge wire for roses, carnations, and chrysanthemums, and a 24-gauge wire for *Stephanotis*. Following are the steps to use for making a rose boutonniere.

PROCEDURE
Making a Single Carnation Corsage

1. Select a top-quality standard carnation. The color choice is based on the individual's preference. Remove the stem ½ inch below the calyx.
2. Using a piercing method, wire the flower with an 8-inch piece of 20- to 22-gauge florist wire. Tape the flower by holding the flower in one hand and twisting it as the tape is applied to the calyx stem and the wire. Using ½-inch florist tape, cover all exposed stem and wire. Remember to keep the tape tight while taping with the twisting motion. Leave the stem about 4 inches long.
3. Select a piece of greenery for the backing of the corsage. Baker's fern, asparagus fern, sprengeri fern, salal leaves, podocarpus tips, carnation foliage tips, or ribbon leaves may be used. After selecting the greenery, tape it to the back of the carnation with the ½-inch tape. Be sure it is in the proper proportion to the flower.
4. Make a pinch bow as explained in Unit 49 (Figure 49–12). In the bow, be sure the ribbon is the same width as the flower when making the loops. Personal taste will also dictate how large the bow is made.
5. Take the taped carnation with the baker's fern and add the bow to the underside of the carnation. Tape the bow into place with a short piece of florist tape. Using florist shears, remove about 1 inch of the taped stem.
6. Place a corsage pin on the back of the corsage by inserting it vertically into the stem. This way no one will be injured with the corsage pin.

PROCEDURE
Making a Feathered Carnation Corsage

1. To create this corsage, select two large carnations. To feather a carnation bloom, use a sharp florist knife to cut *one* of the blooms in half. Using 9-inch pieces of 24-gauge florist wire, wrap the remaining calyx of one of the flower halves. (*Note:* hold the flower firmly as the wire is twisted around the calyx.) Tape the carnation calyx and wire. Repeat the same process for the other half of the flower.
2. Take the other standard carnation flower and split the calyx about ½ inch on each side by using a sharp florist knife to allow the flower to open more widely. This flower will be used as the focal point in the corsage. Using the piercing method at right angles to the cuts in the calyx, wire and tape this flower.
3. For greenery in this arrangement, choose three short tips of carnation foliage. Using three pieces of 4-inch, 28-gauge florist wire, tape a foliage tip to each of the individual wires.
4. Design the corsage by laying out the three flowers as taped along with the three individual carnation foliage tips (see Figure 51–5).
5. Take one of the feathered carnations and one of the carnation foliage tips and press them together. The stickiness of the tape will hold them sufficiently until you can put them together with a short piece of florist tape.
6. Add the next feathered carnation, moving down the main stem of the corsage. Press this flower into place and tape it in the main stem.
7. Add the two remaining foliage tips below the last feathered carnation and on each side of the stem. Again, press the taped foliage tips to the stem and tape them into place with a short piece of florist tape.
8. Now add the focal point, the standard carnation, taping it at the base of the two feathered carnations.
9. Select a bow that will harmonize with the corsage, and tape it below the last flower. Trim the long stems, but be sure to leave them straight to balance the corsage (Figure 51–6).

▲ **Figure 51–5** Design for a feathered carnation corsage.

▲ **Figure 51–6** Completed feathered carnation corsage.

PROCEDURE
Making a Rose Boutonniere

1. Select a high-quality flower for making the rose boutonniere (Figure 51–7).
2. Using the piercing method, attach a 6-inch piece of 20- to 22-gauge wire to the stem.
3. Bend the florist wire down next to the stem.
4. Select a piece of greenery to use on the back of the boutonniere. In this case, use baker's fern. Be sure to keep the greenery in proportion to the size of the rose, as it is used to complement the rose.
5. Tape the baker's fern, rose stem, and wire all in one step. Be sure the tape is applied tightly and evenly down the stem. After taping, cut off the excess stem. The stem must be kept in proportion to the flower. Leave the stem straight because it gives a more natural appearance on the lapel.
6. Mist the boutonniere, and attach a boutonniere pin with the sharp point parallel to the stem. Be sure it will not injure the wearer. Place the boutonniere in a small bag or box.

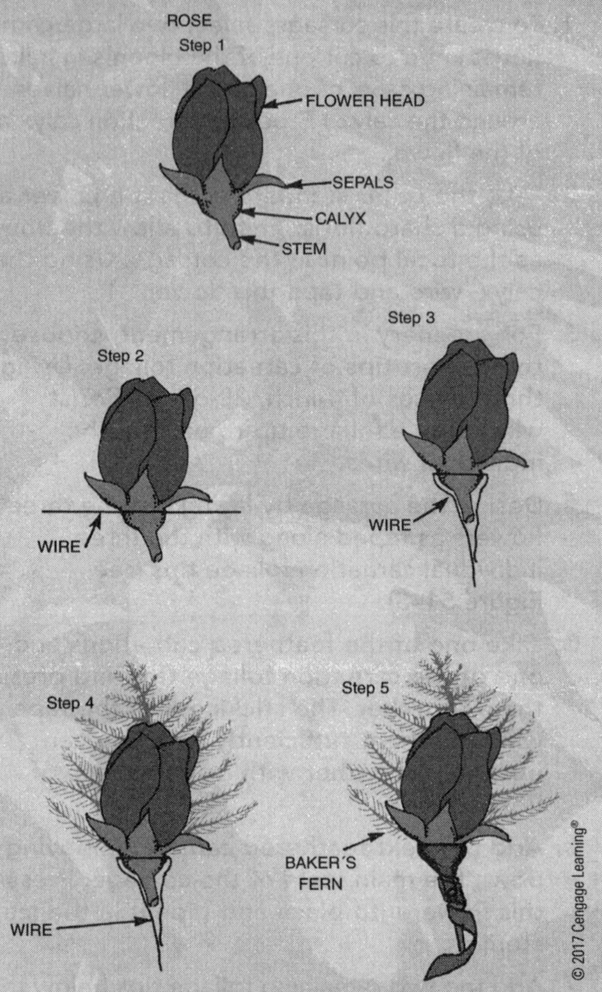

▲ **Figure 51–7** Steps in making a rose boutonniere.

SAFETY FIRST

OSHA's regulations ensure that employers and employees know about work hazards and how to protect themselves to prevent injuries in the workplace. For more information on occupational health and safety in the workplace, please visit the OSHA site at www.osha.gov.

Summary

Corsages and boutonnieres are an important part of the florist business. They are popular for school proms, dances, holiday parties, and weddings. Corsages and boutonnieres must be created with the individual in mind to coordinate with the individual's apparel. In the construction of corsages and boutonnieres, the designer must understand that each flower shape requires a different wiring process.

Review Questions

1. What are the five floral materials used to make corsages and/or boutonnieres?
2. There are four methods of wiring flowers for boutonnieres or corsages. Give an example of flowers for each method.
3. What does it mean to feather a carnation? Describe how to do this, and explain when it is done.
4. How do you select the appropriate ribbon size for a corsage?
5. Describe the procedure for taping flowers. Why is this done?

Student Activities

1. Visit a local florist to observe the making of corsages and boutonnieres.
2. Consider as a class fund-raising project the making and selling of corsages and boutonnieres, perhaps around a holiday or for a formal school dance (e.g., the prom).
3. Prepare and participate in the FFA Floriculture Career Development Event (CDE). This is usually offered at the state and national levels.

Suggested Supervised Agricultural Experience (SAE) Activities

1. As an entrepreneurship SAE, prepare corsages and boutonnieres, and then retail them for sale using Internet advertising as well as local publications.
2. To qualify for a placement SAE project, work for a local florist preparing corsages and boutonnieres for customers for events such as weddings and proms. Record all earnings and time worked as documentation for your SAE project.

Additional Resources

www.wku.edu/floralshop/boutsandcorsages.php

http://extension.ucsd.edu/studyarea/index.cfm?vAction=singleCourse&vCourse=ART-80002

http://learn.org/articles/Florist_Training_Programs_and_Degrees_Online_Program_FAQs.html

SECTION 12

Business Operations

UNIT 52 Beginning a Horticultural Business

Objective

To demonstrate the process of how to create a horticultural business.

Key Terms

accounts receivable
attitude advertising
classic partnership
closed corporation
collateral
corporation
dividends
fixed assets
immediate response advertising
limited partnership
partnership
private corporation
Safety Data Sheet (SDS)
sole proprietorship
vendor
working capital

Competencies to Develop

After studying this unit, you should be able to:

- describe and compare three legal forms of business operation.
- explain the value of a market survey.
- list sources of capital for new and established businesses.
- list factors to be considered in choosing a business site.
- draw up a physical plant layout plan and a staff organization chart.
- outline the major laws and regulations that affect horticulturists.
- describe the methods, values, and limitations of advertising.
- list characteristics of effective advertising.
- list characteristics of effective displays.

Materials

- advertisement samples
- photos of displays
- Web site samples
- market survey
- sample business plan

The dream of owning your own horticultural business is a challenging and rewarding path, and having an interest in growing a significant business is important. Horticultural business owners today must have desire, drive, dedication, commitment, and skills in the ownership of their business. When starting a business, it is desirable to have experience in sales and sales management, and entrepreneurship, as well as significant net worth.

Why Have Your Own Business?

First of all, note that most people do not own their own business; they never will, nor do they want to. Still, they have interesting and rewarding careers.

For those who do operate or seek to operate their own business, a desire for one or more of the following may be the motive:

- The opportunity to exercise leadership and make all major decisions that impact the operation of the business (i.e., the chance to be your own boss)
- The opportunity to build a business around your own interests and personal strengths
- Job security, insofar as this ever exists
- Certain tax advantages
- Money (however, the perception that the owner of a business makes the most money is not always correct)

Conversely, the advantages of nonownership include the following:

- Lack of major responsibility for decisions that affect the future of the business
- The opportunity to pursue a career interest without the problems of ownership
- Freedom to change jobs and relocate as the desire or opportunities arise

There are reasons why self-employment appeals to certain people, and there are also personal qualities that make business ownership satisfying. Leadership skills and a fondness for both the industry and people are vital to successful operation. Liking plants is not enough. Not everyone is equally comfortable with workers and consumers or tolerant of their behavior under myriad conditions. The owner sets the tone for the way customers are treated and the spirit that exists among the staff. Someone who prefers working alone or is unusually shy might be better advised not to seek a career as an owner.

GREEN TIP In your new business, be sure to create a position in the green market. Advertise, produce, or provide products and services that are environmentally friendly.

Forms of Business Organization

After the decision is made to begin a business or to reorganize an existing firm, another decision must be made: how the business will be legally organized. Three types of organization are common: the sole proprietorship, the partnership, and the corporation. Each type has certain advantages.

Sole Proprietorship

The **sole proprietorship** is the simplest form of doing business. It is the easiest to begin and the most private. After assuring that no zoning laws or local ordinances are being violated, the horticulturist can simply obtain the necessary licenses and insurance, set up shop, hang out a sign, plant a field, load a lawnmower on a truck, and be in business. The sole proprietor puts up the capital, hires and directs all employees, reaps all profits, absorbs all losses, and is personally and totally responsible for all debts. There is no separation between the sole proprietor's personal finances and those of the business. A sole proprietorship may be advantageous if you have substantial personal assets. However, should the new venture fail, your personal assets may be lost.

Sole proprietorships have other disadvantages. One is that the business can be limited by the weaknesses of the owner. The business usually grows at the outset and then reaches a plateau due to the proprietor's inability to direct its growth further. Another disadvantage is that the business ends with the death of the owner. If the death is premature and unexpected, and no provision has been made for the sale or continuation of the business, liquidation may be the only recourse.

In the ornamental horticulture professions, sole proprietorships are common. Retail flower shops, garden centers, and small landscape maintenance firms are frequently organized in this form, at least at the beginning.

Partnership

A **partnership** is a business engaged in by two or more persons. It is also a common form of business in ornamental horticulture that works best when the partners have interests and abilities that complement rather than duplicate each other. For example, a flower shop partnership might involve one partner with an interest in design and another who prefers sales and customer contact. A nursery could be owned by one partner who looks after promotion, sales, and administration, and another who oversees outdoor field production. Should the partners decide to add a greenhouse range for propagation, they might take in another partner to provide capital and/or take charge of the greenhouse operation.

In modern partnerships, all agreements between partners are in writing before the business begins. The terms of agreement should include the percentage of original capital to be provided by each partner, the responsibilities assigned to each partner, the business objectives of the partnership, and provision for the dissolution of the partnership in the event a partner dies or wants to leave the business. While such a contract, prepared in advance, will not ensure that partners work together compatibly, it will help to expose potential areas of misunderstanding or disagreement before the pressures of the work arena complicate them.

In the **classic partnership**, the two partners contribute the same amounts of capital and divide the work equally. They each draw the same weekly salary, and at the end of the year, they divide the profits equally. In such a situation, each partner is an agent of the business, empowered in the eyes of the law to act on behalf of the business, to contract with other firms, receive credit, and obligate the company. Both partners become liable for the actions of either partner.

There are variations on the classic partnership. In one, a single partner provides most or all of the capital, and the other(s) provide the technical and administrative skills to operate the business. Another variation is the **limited partnership**, where one or more partners perform some function, such as partial capitalization, but limit their claims on the profits or liabilities to a share agreed to in advance. Where liability is limited for some of the partners, at least one partner must agree to be legally responsible for all the liabilities of the company. Limited partnerships usually restrict the right of the limited partner(s) to obligate the firm.

Corporation

Corporations are the third form of business operation. A corporation is the most public method of doing business and more governed by state and federal laws than are sole proprietorships or partnerships. More expense is involved in establishing a corporation, but it is not excessive. The corporation is given its legal life by the state, similarly as cities and villages. Therefore, to the state, the corporation becomes a legal entity, separate from its owner and managers. The profits of the corporation are taxed independently of the incomes of the company's owners. The liabilities of the corporation are also legally separate from the personal debts of its owners. A corporation can file or be the defendant in a lawsuit without affecting the personal finances of the owners.

Because the corporation is a separate entity, it has greater flexibility than noncorporate businesses. Management decisions may be easier to make if the business is not tied to the personalities of one or two owners. Some corporations are completely managed by nonowners. Therefore, the ownership of a corporation can change while the business continues without interruption.

The ownership of a corporation is obtained through the purchase of shares, each share having a particular monetary value. When major decisions must be made, requiring a vote of the owners, each shareholder is entitled to a vote proportionate to the number of shares owned. At the end of the fiscal year, the profits of the corporation are paid to the shareholders on the basis of the number of shares owned. The profits are termed **dividends**. If the corporation does not make a profit, no dividends are paid, and the value of the stock may be reduced.

There are several types of corporations, two of which are common in the ornamental horticulture industry. A **private corporation** sells its stock publicly and can be owned by many people, often as an investment. A **closed corporation** does not sell its stock publicly. It is often owned by the members of a single family. If additional capital is needed to finance expansion of the business, either corporation may sell more stock.

Operating decisions are made by a board of directors, elected by the stockholders. Stockholders cannot obligate the company. Because most ornamental horticulture corporations are closed corporations with few stockholders, the election of directors, their titles, and the conduct of meetings are usually informal. The incentive to incorporate is usually the desire to limit the liability of the owners.

Market Surveys

The desire to start a business is not enough to ensure its success, even when the owner has enough capital to get the business started properly. Hard work, great talent, and the support of family and friends will not guarantee the survival of a business. Simply stated, the success of a business depends on enough people wanting the products or services of the business and having the money to pay for them.

The geographic area from which a business attracts most of its customers is termed its market. For garden centers, flower shops, retail nurseries, lawn maintenance firms, and landscape contractors, the market area may be limited to a city or county. For wholesale nurseries, mail-order nurseries, landscape architects, and propagators, the market area can often extend beyond state lines. The safest way to begin a business is to study the market beforehand. An analysis known as a market survey can answer the following questions:

- What is the extent of the market area? If deliveries are to be made, how far can the truck travel before the cost of delivery becomes prohibitive? How far must customers travel to reach the business? Are they more likely to use the telephone to call in orders or to come in personally?
- Is there an established desire for the products or services of the business, or must one be created?
- How strong is the economy of the market area? In the opinion of some people, ornamental horticulture products and services are not essentials. Do the local people have sufficient funds to support their desire for what the business is selling?
- How many similar businesses are already established in the market area? A pie can only be cut into so many pieces before they become too small to satisfy. A market area will only support a certain number of similar firms profitably. A saturated market area is not the place to begin a new operation.
- How good is the existing competition? It is difficult to gain an edge against a competitor that enjoys a good reputation in the community and backs it with competent performance and fair prices.
- Can qualified employees be attracted and retained? In upper-income market areas, there may be a scarcity of workers, and workers may be unable to move into the area at the wages a new business can pay.

If possible, the would-be owner should let someone else do the market study. Wishful thinking and ego can get in the way of an objective analysis. A hopeful newcomer may be sure that customers will beat a path to the door of a new business when someone finally does it right. The facts may indicate that the business would be better located in a different area. In a saturated market area, someone must eventually close each time a new business opens. The chances are good that it will be the new business.

> ***TRY THIS*** Conduct a market survey in your area for a lawn maintenance firm. Use sources such as the census, state and county websites, USDA data, and personal experience to identify the type of customers in the area and their needs. Write a report that determines if the lawn maintenance firm would have an appropriate customer base in your area, and explain your findings.

Business Plan

Before sources of capital can be identified for a potential start-up business, generally some form of a business plan will be required. The following are some general components to include in a business plan:

- mission statement
- description of the company and product
- market analysis
- description of the management team
- marketing plan
- strengths and weaknesses analysis
- cash flow statement
- revenue projections
- conclusion

A business plan is essential for organized management of the company and also serves as a document to prove to a lender that you have a well-developed plan for success.

Sources of Capital

For someone considering a new business, the initial excitement of plans and dreams can be quickly replaced by the realities of financing. Where does the money come from to start a business? How much is necessary to start and support the business until it begins to show a profit?

Because there is no predetermined period of time after which a business is guaranteed to survive, capital must be discussed in terms of initial funding and subsequent funding necessary to sustain the business. Capital is needed for two reasons. One is to purchase **fixed assets** such as real estate, coolers, furniture, equipment, and similar items. The other is to pay wages, purchase supplies, and pay utility bills. Such funds are termed **working capital.**

There are three ways to obtain capital for a business:

1. Use your own money or that of other people willing to invest their money in your business.
2. Borrow from lending institutions such as banks, insurance companies, and loan companies; from persons seeking to buy into the business as limited partners; or from other companies that will market your future crop and are willing to pay you in advance (termed contract growing).
3. Return the profits of the firm to the business for use as capital.

Each method of raising capital has features that may make it the appropriate or necessary choice.

Using Your Own Money

You should use your personal savings as much as possible in starting a new business for two main reasons. First, there is no interest charge when you use your own money. If money is also invested by family or friends, it can be with the written understanding that no interest will be paid on the initial investment but that they will share in the profits for a predetermined period of time. If interest is to be paid on the money of friends or family, it may be less than that charged by a bank. The second reason for using personal savings is that it is evidence of your commitment to the business. Such evidence is important when credit is sought later from other sources. Certainly, the healthiest financial situation is to have enough capital available from personal sources that the borrowing of working capital is only necessary during peak seasons, and the loans are of short duration.

Borrowing Money

For the novice as well as the established horticulturist, borrowing may be the only way to obtain capital, but it is also the most expensive way. In addition, it requires you to reveal a good deal of information about your background and personal finances, precise details on how the money is to be used, and in-depth records of the financial history of the business. Lending institutions tend to be conservative and cautious when making loans, especially to new businesses with no financial history, hence their inquiring attitude. Regardless of the institution, it will require **collateral**, or something of value to be guaranteed as security for the loan. Big ideas, enthusiasm, and a college degree are not collateral, although the personal credentials of the borrower can lend credibility and promote confidence. Usually accepted as collateral are such things as land, buildings, vehicles, equipment, crops, or **accounts receivable**.

Several kinds of loans are available. Many financial institutions can be used to acquire a business loan. Several institutions are also dedicated to agricultural financing. Farm credit can be acquired through the USDA-FSA, the Farm Credit System (FCS), the Farmers Market Promotion Program (FMPP), commercial banks, or private investors usually referred to as "Angel" investors.

CHARACTER OR SIGNATURE LOANS These are short-term loans granted to businesspeople with good credit ratings. Obtaining working capital for 30 to 90 days requires only the borrower's signature if the bank approves a line of credit in advance. This kind of loan is reserved for those who are well established and successful. A young person starting out will seldom qualify for a character loan, but it is the kind of loan all should aspire to attain.

TERM LOANS These long-term loans are repaid over a period of years, usually by the month. This is an expensive loan due to the interest rate charged. Term loans may be made entirely by a private lending institution or by the federal Small Business Administration (SBA) in association with a bank. The SBA supports small businesses by granting a percentage of loans to borrowers that in its opinion deserve the chance for success. Through the agency's support, the borrower obtains capital at a rate of interest lower than what the bank alone would charge.

ACCOUNTS RECEIVABLE LOANS These loans are obtained by established businesspersons by pledging money owed to the business from customers. This is a short-term loan that is repaid as the customers pay their bills.

LOANS ON LIFE INSURANCE These loans can be obtained using the cash value of the borrower's life insurance as collateral.

LIMITED PARTNERSHIP LOANS These loans are made by individuals who provide capital in return for a share of the profit; this type of capital is termed equity funds. Most owners do not want strings attached to the loan, particularly attempts by the lender to influence management decisions, so the terms of the loan need to be spelled out in writing.

CONTRACT GROWING LOANS Made to growers by firms that will eventually market their crops, these loans provide the grower with the money necessary to produce a particular crop while ensuring the lender of a guaranteed price. Recently, contract growing has been used by growers to ensure that they will have a market for their crops and thus avoid expensive overproduction.

AVAILABLE GRANTS Several grants are available for agribusinesses at all stages of business development. The most common grants target start-up businesses and young business owners. Several grants and low-interest loans are offered through the USDA and the Farm Service Agency (FSA). Examples of some of the programs available are the Farm Loan Program, Rural Development Loan Program, the Beginning Farmers and Ranchers Program, Specialty Crop Block Grant Program, the Farmers Market Promotion Program, and the Organic Certification Cost Share Program. Each type of grant has very specific qualifications for people applying for grant funding. Research into the details of each grant is important to determine if the grant would apply to your business venture.

Reinvesting Profits into the Business

Retaining the profit that a business makes rather than paying it out to the owners is an important method of raising interest-free capital. Although it is not enough for a developing and expanding business, it can be of major significance for a successful, established firm that needs capital to replace inventories and depreciated equipment, or to pay taxes and other operational costs. To gain the greatest benefit from this source of capital, the administrator of a business must be attentive to money owed the company. The longer customers withhold payment, the more it costs the business because an unpaid account represents money tied up in expended materials and services that is unable to be used.

Selecting the Business Site

One of the initial uses of capital is to obtain a site for the new business. The opportunity to select from among several sites is not always available. When it is, however, numerous factors must be considered.

SIZE The size of the site must be evaluated in terms of how much area the business needs at the beginning and how much expansion is planned if the business is successful. A field nursery requires more area than a garden center, and both need more space than a retail flower shop. A landscape designer does not require much space to operate but would need a lot more to expand into a contracting business. Small sites that offer no room for expansion should be avoided if the horticulturist foresees the business eventually growing and needing larger production or storage facilities. Conversely, large sites should be avoided by small retail businesses. Taxes and other costs of maintaining a large site can be good reasons for not purchasing a piece of property that is larger than will be needed now or in the future.

NATURAL FEATURES Natural features of the site and adjacent areas are of special concern to production firms. Fertile, rock-free soil of good texture and drainage is necessary for a field nursery. Good quality air, available water, and ample sunshine benefit all production nurseries as well as greenhouses. Level terrain that is easily accessible and areas that are protected from harsh extremes of weather are also of importance for any production firm.

Garden centers, landscape nurseries, flower shops, and maintenance firms may benefit from the presence of shade trees, bodies of water, rock outcroppings, and other natural features that contribute to the aesthetics of a site.

ZONING REGULATIONS Zoning regulations determine what use can be made of a site. Some areas of a community may be zoned only for a residential development and perhaps for churches and schools. Other areas may be zoned for heavy industry, and still others for light industry or commercial purposes. Zoning ordinances may restrict expansion, regulate architectural style, or even specify landscape features. The zoning of an area can also affect the rate of taxation on the land and buildings. Careful investigation of the zoning patterns of a community and the frequency of zoning changes can prevent a site that seems otherwise perfect from becoming a liability.

UTILITIES Utilities must be on the site or easily installed to avoid high costs of hookups. Nurseries and greenhouses need adequate sources of fresh water in large quantities. Everyone requires electricity, and special wiring or special forms of electricity may sometimes be needed. Oil, gas, or other fuel suppliers must be nearby and reliable.

ACCESS Access to the site is of prime concern. The type of access depends on the type of business. A retail operation needs to be close to main streets and highways so that suppliers can make deliveries. Growers who wholesale their plants must be near highways and railways if the plants are to reach their markets. Equally important is access to the business for customers. A landscape nursery needs to be on a major road near the community of clients. A flower shop or garden center has the best chance of attracting impulse buyers if it is located along a major road where traffic is heavy and access is convenient. However, if traffic is too hectic, as at certain intersections, motorists may avoid stopping. One-way streets can also be bad sites if homeward-bound traffic is passing in the opposite direction on another street. Where walk-in traffic is an important source of customers, the store could benefit from a location along a pedestrian mall, in a shopping center, or even in a major hotel lobby.

COMPATIBILITY Compatibility of the new business with other businesses around it will frequently determine whether customers are attracted to the operation. Other businesses generate customer traffic. Because most retail flower shops, garden centers, and nurseries benefit from people willing to linger and shop, it is advantageous to be located among other businesses that cater to shoppers. Convenience stores that cater to a dash-in and dash-out trade do not make compatible business neighbors. Stores that serve customers who remain in their cars, such as banks and fast-food restaurants, make especially poor nearby neighbors.

MERCHANTS' ASSOCIATIONS Merchants' associations are common in retail areas and are often very helpful to new and established businesses alike. The members of active associations often share the expenses of group advertising, holiday decorations, security systems, remodeling, group insurance plans, and other items of collective interest. At times, as necessary, they can even be effective for political lobbying. In evaluating a potential retail site, it is worthwhile to ask nearby merchants about the quality and effectiveness of the local merchants' association. Lack of such an association may indicate an area on the decline.

LEASE AGREEMENTS Lease agreements are usually required when land or buildings are rented rather than purchased. A lease should not be signed until the business operator has thought carefully about the future. If a long-term lease is sought, the agreement may need to permit expansion when needed. If another site is likely to be preferable when capital availability permits it, a short-term lease may be preferable. Any responsibilities of the landlord should be written into the lease agreement. Maintenance responsibilities are especially important, and the period of time for their accomplishment should be specified. If possible, an agreement that vacant properties nearby managed by the same property owner will not be leased to competing businesses should also be written into the lease.

COST OF ACQUISITION Cost of acquisition is the final factor in the evaluation of a site. For some would-be business owners, it may be the most important factor of all. The cost of the site should be weighted carefully against its advantages and disadvantages to determine if the price is fair or if another site is a better value.

Buying an Established Business

An alternative to beginning a business from the ground up is to purchase an established operation. In many cases, the cost of entry into the profession can be reduced if an established business can be purchased at a sacrificed price.

A horticulture operation will usually be put up for sale for one of three reasons:

1. The owner has reached the age of retirement and has no one to carry on the business.
2. The owner has become ill and is unable to carry on the business.
3. The business has been unsuccessful.

The first two situations can provide excellent opportunities to acquire a successful business at a reasonable price. The third situation does not provide such an opportunity. An unsuccessful operation may have a bad location, obsolete or improper equipment, or a history of poor products or service that leaves it with a bad reputation. Such obstacles are difficult to overcome.

When a successful business is purchased, a certain following of established customers (termed goodwill) comes with the purchase. Having established customers is an important asset, but not one that can have a value (and price) assigned to it easily. Sellers usually try to inflate its value; buyers should resist paying much for something that is so difficult to appraise.

When considering the purchase of an established business, the future owner should look at its past performance as an indication of how it will do in the future and also assess the potential of the business under new ownership. The seller should be asked to supply profit-and-loss statements for the preceding 5 years. These should be taken to an accountant and studied carefully. The earnings, the increase or decline of sales, the salaries of employees, and the budget for advertising during the 5 years can all provide a picture of the past performance of the company. In attempting to determine how the business will do under new ownership, there must be a strong indication of steady growth over the next 5- to 10-year period. The business must promise not only a good salary for the owners but a return on the invested capital. If the projection by the accountant does not favor both salaries and profits, it would be best to keep looking for a more suitable business.

If and when a good business is found for purchase, the selling price must be evaluated in terms of the factors that it represents:

- Equipment will seldom be brand new. Its value at the time of purchase will be greater if it is newer. If the equipment is old and has been fully depreciated, it may be a liability.
- Existing inventory is not always usable. It may not be what the buyer would have stocked and may contain outdated products that have been rejected repeatedly by customers. The inventory values on the books of the seller should not be accepted without investigation. Items may be carried at full value when in reality they are unsalable and nearly worthless.

- Accounts receivable are monies owed to the seller that will be owed to the buyer after the sale is finalized. If a large percentage of those receivables is long-standing bad debts, they should be discounted in all or in part.
- Goodwill, as already mentioned, has a vague value and should not represent a large percentage of the sale price.
- Use of the existing name may be part of the sale price or may have no value assigned to it. It may be advantageous to use the established name for a while, especially if the firm has a good reputation, but the advantage should be carefully assessed. It is difficult to assign a value to reputation.

Using Space Efficiently

Once acquired, the site must be put to use in a manner that will maximize the return on capital. Space must be assigned to each of the major areas of the business. For example, a retail business such as a garden center usually requires areas for the following:

- customer sales and product display (indoors and outdoors)
- inventory storage
- plant holding
- parking
- employee use
- climate control

In addition, production operations such as greenhouses or nurseries require areas for the following:

- plant growth
- equipment and supply storage
- plant grading and packaging
- shipping and receiving

Each area can be further divided into subareas. Examples of these subareas include the following:

- Customer sales and product display:
  - space for shelves of products
  - display cooler for cut flowers
  - display greenhouse for foliage plants and potted flowers
  - outside sales yard for woody plants, bedding plants, lawn ornaments, and bulk materials
  - restrooms
  - check-out area
  - landscaped setting for the business
- Powerful plant tags with QR Inventory:
  - storage
  - warehouse for bulk materials and equipment
  - coolers
  - storeroom inside the retail centers
- Plant holding:
  - heeling-in/overwintering
  - acclimatization
  - overwintering, bare-root storage
- Parking:
  - customer parking
  - employee parking
  - access roads
  - space for snow to be piled
- Employee use:
  - offices
  - workrooms
  - lockers and rest rooms
- Climate control:
  - central heating
  - air conditioning
  - windbreaks
- Plant growth:
  - greenhouses
  - nursery fields
  - lath houses
  - propagation beds
- Equipment and supply storage:
  - garage
  - fuel tanks
  - small engine repair shop
  - tool shed
- Plant grading and packaging:
  - grading and packaging room
  - overwintering, bare-root storage
- Shipping and receiving:
  - maneuvering space for trucks
  - loading dock
  - access roads
  - space for snow to be piled

The use of space in a business must be analyzed at several levels. First, there is the allotment of space to selected subareas. For example, parking lots must be large enough to hold the number of vehicles expected to be there at one time on a busy day at peak season. Assigning too little space to the lot will result in lost sales and annoyed customers. However, assigning too much space to the lot may deprive other areas of the space they need. Similarly, customer sales areas must allow enough room for comfortable browsing through a diversified array of products. Allowing too much space between products will limit the number of items that can be displayed, and crowding too much in will make the shopping area look like a bargain basement. Some florists and landscape design firms assign space to customer consultation areas. Although such areas are nice to have, the space might be used more profitably if the retail area is small.

Use of space must also be analyzed in terms of inventory control. Quantity buying is generally regarded as wise because it results in lower unit costs. However, it also requires more space to store the inventory and ties up capital until the inventory is sold. In small shops where the storage area is small, quantity buying can result in limited product offerings.

Space efficiency is also important in the production of plants. Production schedules must be planned so that bench space is available when it is time to expand the spacing of a crop. Planning must also foresee when a bench will be empty so that another crop can be started as soon as possible. Heating a partially empty greenhouse drives up the grower's cost of operation. Equally costly and inefficient is a partially harvested nursery field if no new crop is planted. With land as expensive as it is, few growers can afford inefficient use.

Organization Plans

It is a good idea to plan the layout of the business on paper to determine how big each area and subarea will be as well as how each is to relate to the other. This physical plant layout plan should be made while the business is still in the developmental stage. It allows the owner to think about the business in an organized way and also provides a beginning point for rethinking the layout if expansion is needed later.

Proper sizing of each area and ease of circulation are the keys to a successful layout (Figures 52–1 and 52–2). Examples of sizing have already been given. Circulation patterns are equally important. Both the customers and the employees must be considered. Customers will seek the most direct route from their cars to the products they came to buy. They do not want to pass through work areas or go past the loading dock in the process. They are generally receptive to passing other products on the way to the product they want, however. Meanwhile, employees do not want to be routed through the customer sales area to get to the restroom. Neither do employees want to take their coffee break in the sales yard.

A staff organization plan can also be useful when a business is starting out and each time that expansion or reduction in the staff causes duties to be reassigned (Figures 52–3 and 52–4). A staff organization plan helps employees understand their roles within the company. This reduces conflict over who has responsibilities in a particular area. No business is too small to benefit from an organization plan.

Building the Inventory

After a business is established and has earned a good credit rating with suppliers, it will have little difficulty obtaining materials and supplies as needed. A new business may encounter some problems at the outset, however. First, suppliers are wary of extending credit to someone who has no credit history. Second, if shortages develop, suppliers are likely to service their established customers first, leaving newcomers with unfilled orders.

The best way to begin the purchase of materials and supplies is to order small quantities that can be turned over quickly and paid for promptly. Large purchases tie up capital too long and can create credit problems. Also, many items are seasonal (such as holiday items) or highly perishable (such as flowers and bedding plants), and these should only be purchased in quantities that can be turned over quickly.

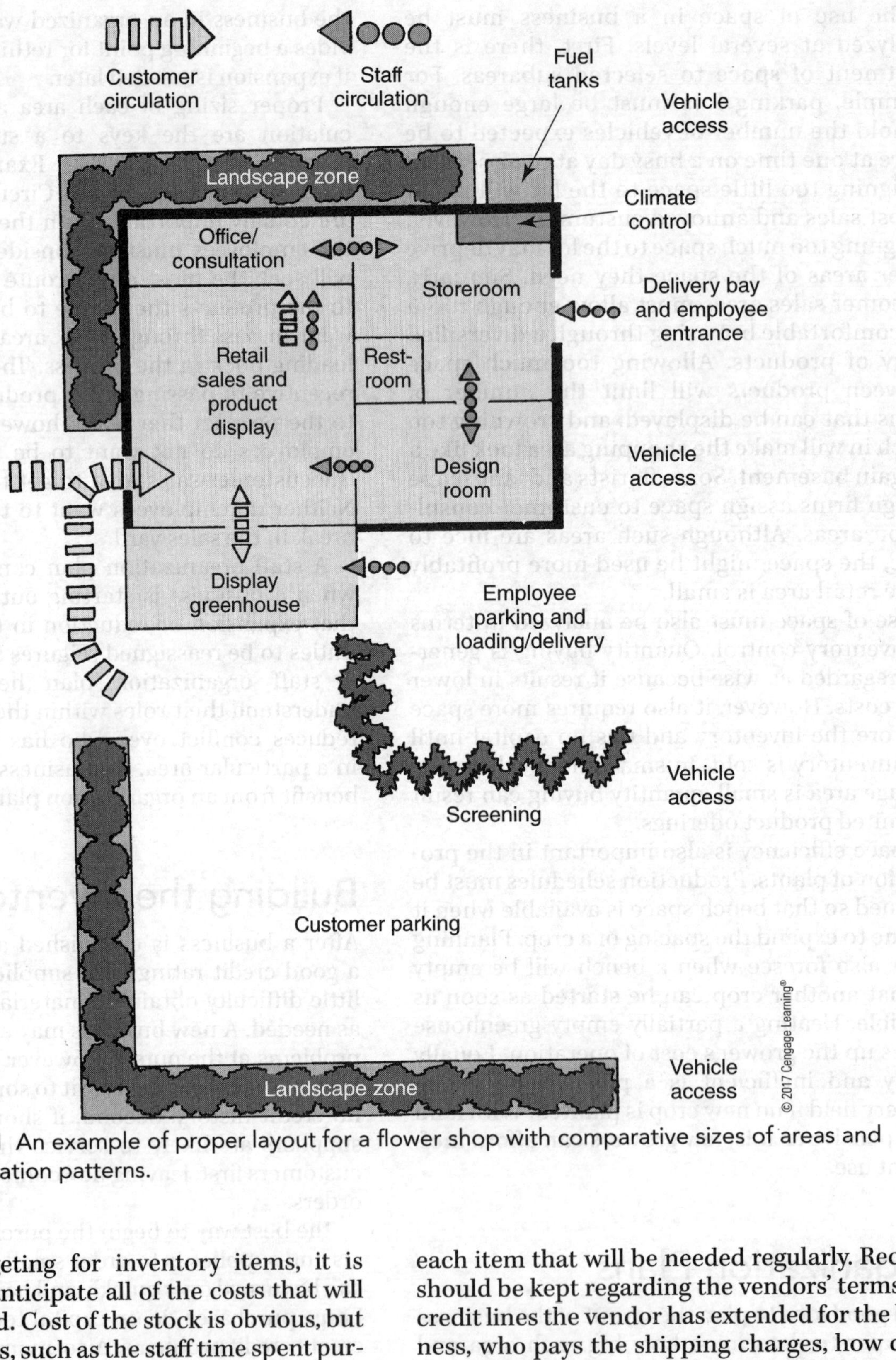

▲ **Figure 52–1** An example of proper layout for a flower shop with comparative sizes of areas and predicted circulation patterns.

When budgeting for inventory items, it is important to anticipate all of the costs that will be encountered. Cost of the stock is obvious, but there are others, such as the staff time spent purchasing, receiving, and storing the stock, and the cost of keeping stock control records.

To ensure an uninterrupted supply of materials, a reliable source (**vendor**) should be found for each item that will be needed regularly. Records should be kept regarding the vendors' terms, the credit lines the vendor has extended for the business, who pays the shipping charges, how delivery is made, and the time usually required to fill orders. It is also desirable to establish cordial relationships with the salespeople who represent the vendors in your area.

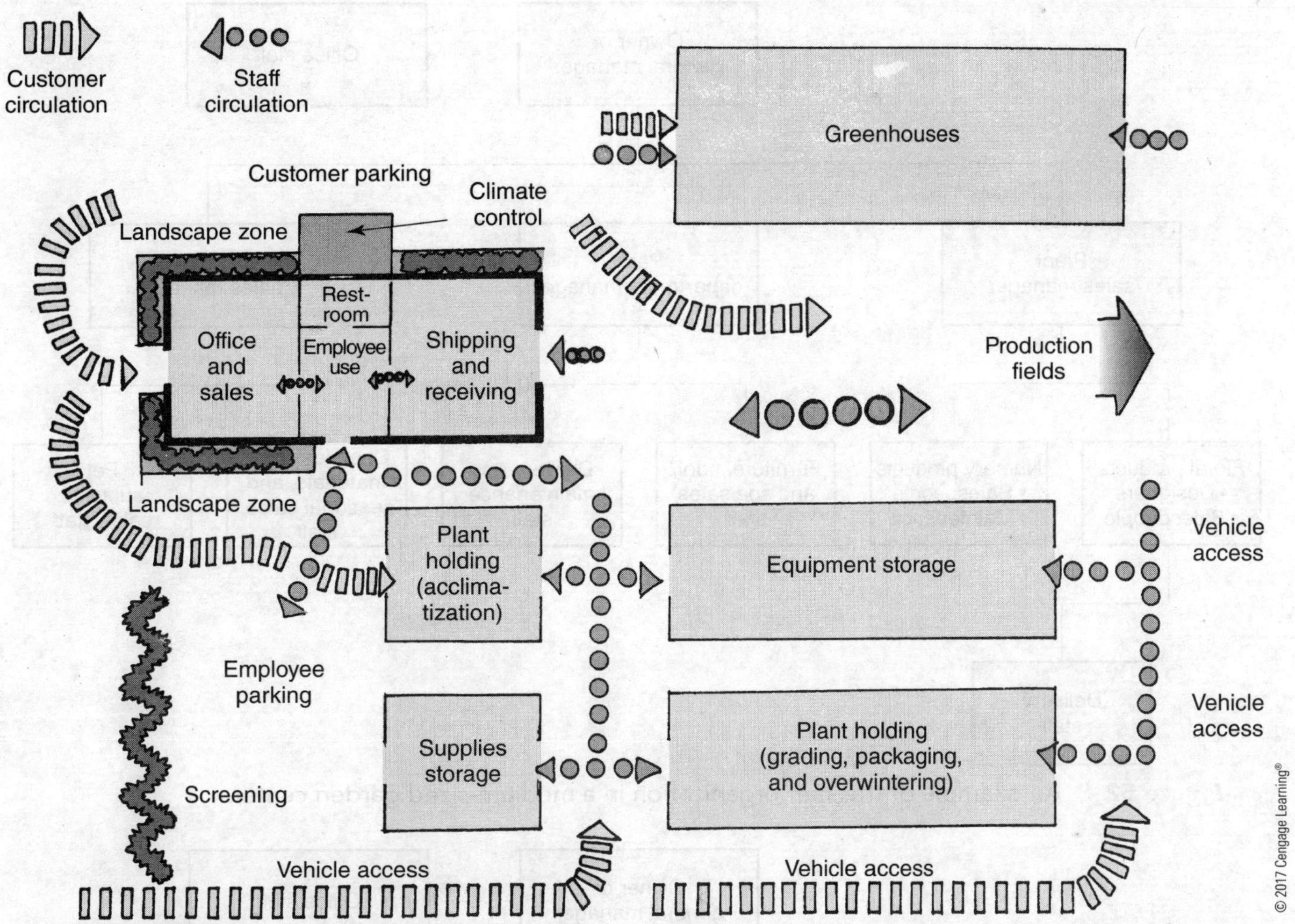

▲ **Figure 52–2** An example of proper layout for a production nursery showing predictable circulation patterns. Conflicting intersections are minimal.

Laws and Regulations

Like all businesses, horticultural businesses are subject to regulation by all levels of government. Horticulturists may have an above-average number of laws and regulations to comply with, especially if they are involved in plant production. Some of the laws and regulations are described in Figure 52–5. An agricultural business will usually have an attorney on retainer or an attorney who is familiar with the business and easily available for consultation.

Several laws govern the treatment and safety of employees of a business. The owner and manager should be knowledgeable about OSHA regulations and how to keep the working environment safe. Clear labels and instructions help to make sure employees have proper information about the materials and equipment being used. Using up-to-date equipment is also important to maintain safety. When any type of chemical or hazardous substance is used at a business, a **Safety Data Sheet (SDS)** (formerly known as Material Safety Data Sheet [MSDS]) should be available on-site to keep records and information about the handling of the substance. An SDS is a document that contains information on a chemical product's potential hazards (health, fire, reactivity, and environmental) and how to work safely with it. Maintaining a binder of SDSs is an essential starting point for the development of a complete health and safety program. Proper protocol for handling and storing hazardous substances should always be followed by the management and staff. A good safety record makes all aspects of business management easier and more efficient.

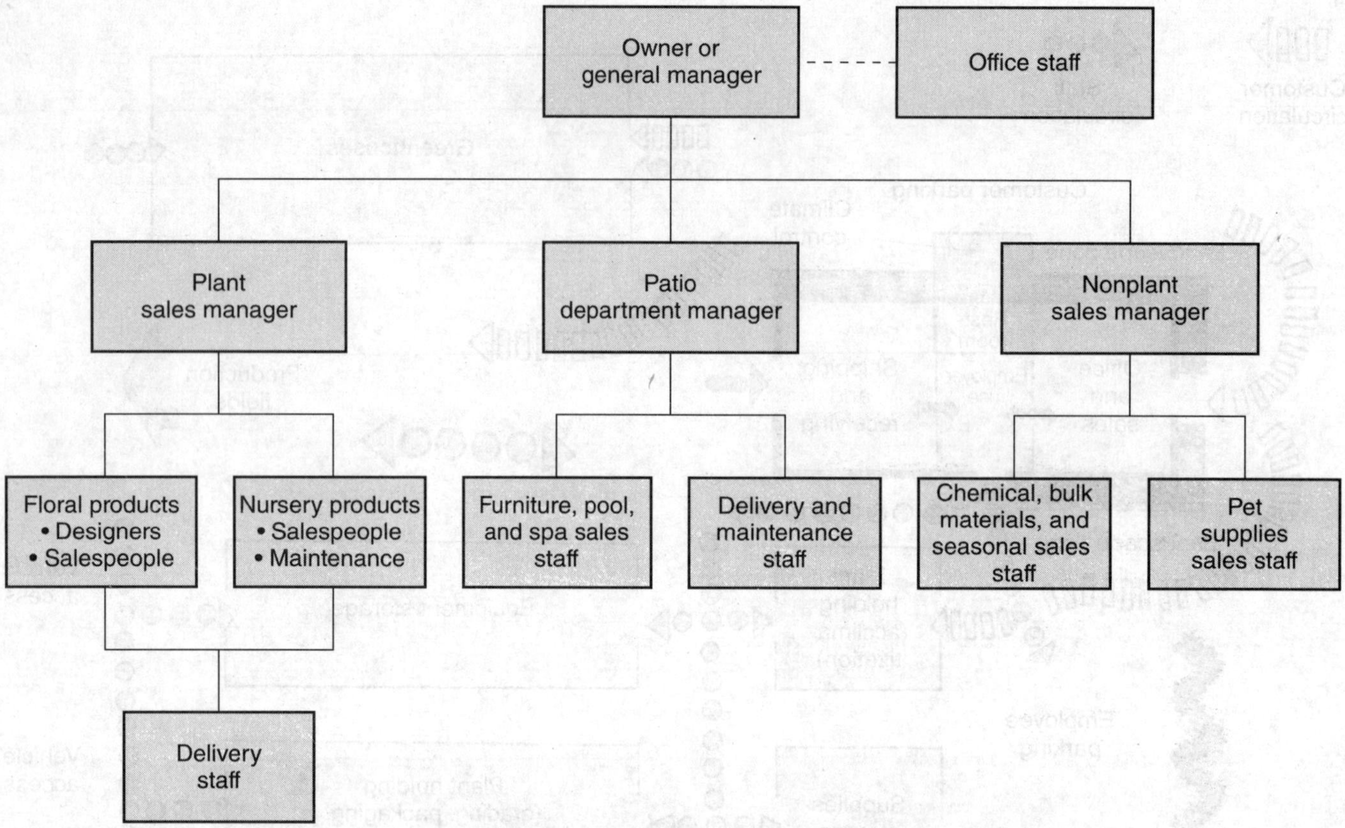

▲ **Figure 52–3** An example of the staff organization in a medium-sized garden center.

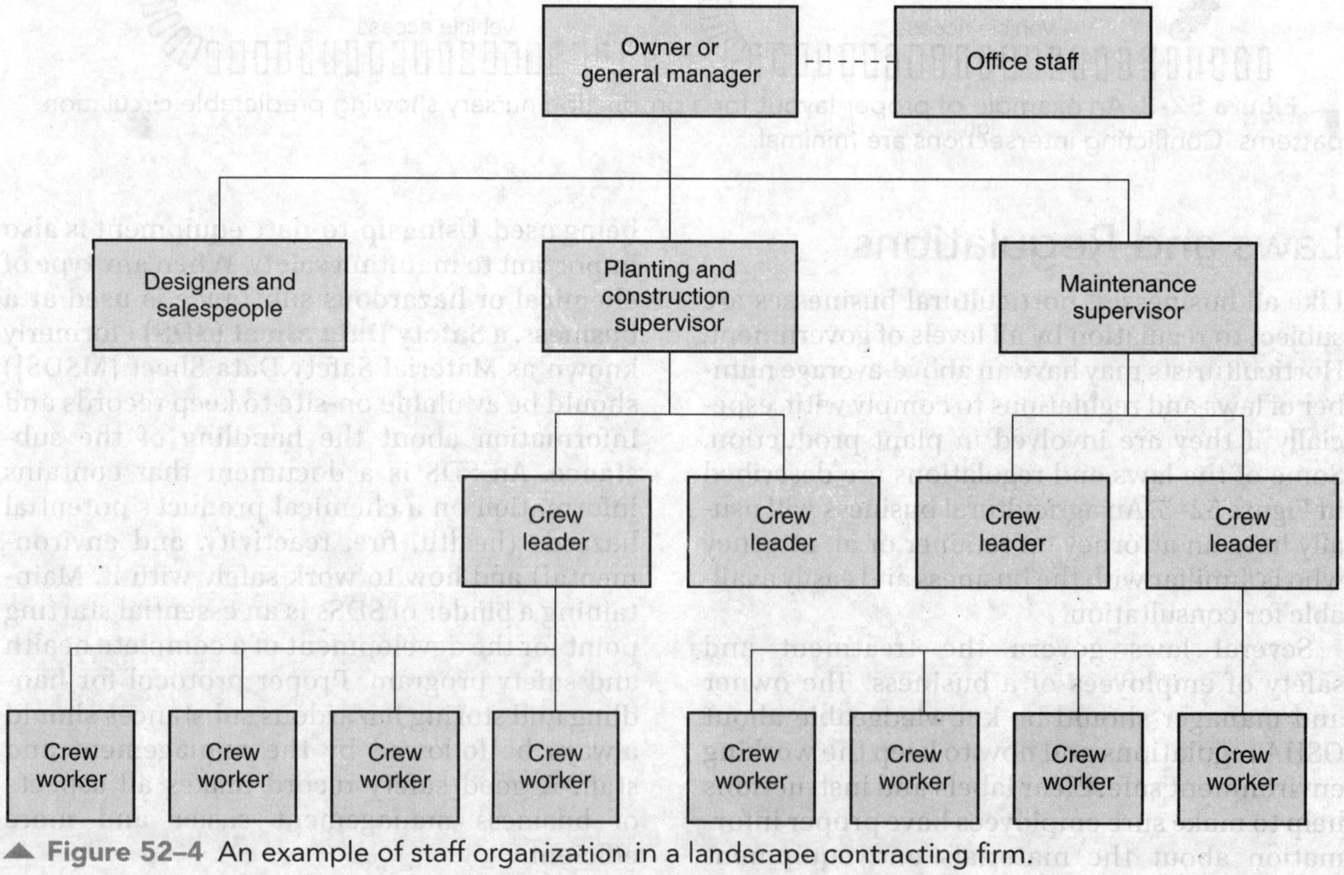

▲ **Figure 52–4** An example of staff organization in a landscape contracting firm.

LAW OR REGULATION	PURPOSE	EMPLOYER COST	EMPLOYEE COST	COMMENT
Deceptive practice or product regulations	To protect the consumer from unrealistic descriptions and claims for products and services.			Regulations such as these may be consumer protection laws or industry-initiated trade practices.
Fair Labor Standards Act	To establish a minimum wage for all workers; to establish standards for payment of overtime work; to prevent abuse of child labor.	Cost borne by employer		Equal pay must be given for equal work on jobs that require equal skills and responsibilities.
Family and Medical Leave Act	Permits employees to take time off, unpaid, for up to 12 weeks for maternity, child adoption, or serious illness.			The leave is unpaid, but all benefits must be fully restored when employees return.
Farm Labor Contractors Registration Act	To ensure fair treatment of migrant workers. They must be accounted for through regular payroll accounts and must be provided by each employer with information about their work area, housing, insurance, transportation, wage rates, and charges.	Cost borne by employer and paid to a farm labor contractor		Nurseries are most likely to be affected.
H2-B program	Ten-month visas are issued to legal aliens who want to work only temporarily in the United States. It permits a company to employ needed foreign laborers during a peak season without the need to retain them during the off-season.	Must pay the prevailing labor wages	Visas cost $750 each	Alien employees must have their visas recertified each year by the federal INS.
Immigration Reform and Control Act	All employers must verify that every new hire is either a US citizen or authorized to work in the country.			Civil and criminal penalties are applied to employers who knowingly recruit, hire, or retain unauthorized aliens.
Occupational Safety and Health Act	To ensure safe and healthful working conditions for employees.	Cost borne by employer	No cost, but required to comply with the law by using all safety devices properly	All employers with at least 11 employees are covered by the law. The law is enforced by the states.
Pesticide application certification (Environmental Pesticide Control Act)	To limit the use of toxic pesticides and to ensure that they are used only by persons knowledgeable about their safe and proper use.	Cost (minimal) borne by applicator		Once certified, the applicator must stay up to date by acquiring a number of recertification points by a certain date. Points are awarded for participation in various professional meetings.

Figure 52–5 Laws and regulations affecting horticulturists. *(continued)*

LAW OR REGULATION	PURPOSE	EMPLOYER COST	EMPLOYEE COST	COMMENT
Plant Patent Act	To protect the right of the inventor or discoverer of a plant to be the sole asexual propagator of the plant.	May increase cost of production for a grower		Plant patents are valid for 17 years.
Plant Pest Act	To regulate by quarantine the entry and distribution of plants and plant products into and across the United States to protect against foreign plant pests.	Fees charged for import/export permits and nursery inspections		Quarantines may control the movement of plants between countries, between states, or within individual states.
Price and Services Act	To ensure that all commercial customers are treated equally by a seller by forbidding selective prices or services.	No cost, but employer must ensure compliance		The use of printed wholesale price lists is a common means of proving compliance with the law.
Social Security	To provide financial assistance to persons after their retirement; to provide income for surviving dependents after a worker's death.	A percentage of a base amount of employee's wage (subject to change over time)	An equal amount is withheld from paycheck	Current trends are for both the percentage and the base amount to increase regularly to meet the expenses of the system.
Unemployment compensation	To provide financial protection for workers who lose their jobs for reasons other than firing.	3.4 percent of the first $6,000 of wages (subject to change over time)		
Warranties Act	To define the terms "full warranty" and "limited warranty" as used by tradespeople.	Replacement cost borne by employer		Horticulturists are not required to offer warranties, but if they do, they must either replace a product completely or in part in compliance with the terms of their stated warranties.
Workmen's compensation insurance	To provide compensation to employees injured on the job.	Cost borne by employer		Coverage is required for all businesses having at least one employee in a hazardous occupation or at least three employees.

Figure 52–5 Laws and regulations affecting horticulturists.

Advertising the Business

The purpose of advertising is to focus public attention on items or services that are for sale. Advertising is related to, but different from, promotion, publicity, and public relations. All are needed for the successful growth of any business.

A modern horticultural business must advertise to retain old customers and add new ones. To be productive, advertising must generate both sales and profits and generate them in quantities greater than could be expected without advertising. Customer traffic and sales volume are not enough to make advertising successful. It must make profits for the business.

There are two types of advertising, each with a different objective. The purpose of **immediate response advertising** is to make customers reach for the telephone or come into the store soon after reading or hearing the advertisement. Such advertising must offer merchandise or services at a time when customers are receptive to them. For example, flowers and special arrangements are of predictable interest to consumers during holiday times and should be advertised by florists just prior to the holidays (Figure 52–6). Home landscaping is on the minds of property owners in the spring more than in the winter. The timing of the advertising is important for maximum return on the advertising dollar.

The purpose of **attitude advertising** is to build the reputation of the business. Done on a regular basis, attitude advertising seeks to implant in customers' minds an understanding of the services or products available from the business along with a sense of its status or reliability (Figure 52–7). Successful attitude advertising will motivate customers to seek out the business in the future.

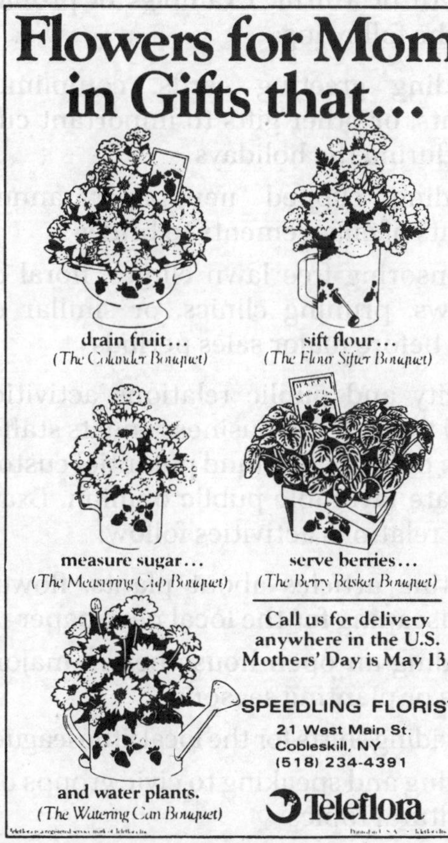

▲ **Figure 52–6** This example of immediate response advertising was mailed to all regular customers of a flower shop prior to Mother's Day to promote the shop and feature the holiday specials of its wire service affiliation.

▲ **Figure 52–7** This example of attitude advertising was run in the newspaper in early spring. It states clearly what the firm does and suggests quality and reliability.

The Advertising Media

The media through which a business can advertise are known to nearly everyone: newspapers, magazines, billboards, telephone directories, radio, television, Internet websites, and social media outlets (e.g., Google Forms). In addition, small businesses can use direct mail, e-mail advertisements, handbills, display windows, county fairs, flower and garden shows, and a variety of other means.

Advertising is a cost of doing business and must be budgeted in advance. The business owner must know which media will result in the greatest return on the money spent. The point to remember is that advertising must result in measurable profits. Therefore, establishing appropriate advertising media outlets is important for horticultural businesses. A retail florist may find that a weekly advertisement in the local newspaper builds public awareness of the shop (attitude advertising) and also increases the sale of weekly specials (immediate response advertising). Garden centers, nurseries, and landscapers may use the local newspaper, television and radio stations, billboards, and even e-mail advertising during the spring and fall when customers are thinking about their home landscapes. Such advertising is usually attitudinal, yet it can evoke immediate response if it features specific products or services at promotional prices.

Wholesale nurseries, large landscape contractors, and landscape architects gain little benefit from newspaper or broadcast advertising. They must reach a customer population that often extends across state lines. Magazines, particularly trade journals, are one means of reaching their clientele. Exhibition at industry conventions and trade shows is another means of putting the business before the proper audience. Direct mail and e-mail advertising is also used. As much as possible, journal advertisements, exhibitions, direct mail, and e-mail should induce an immediate response, although attitudinal advertising is valuable and commonplace in this market also.

Telephone directory advertising can vary in its value to the business. In a small community, most citizens are familiar with the few horticultural operations in the area and will just require the number to contact them. Their choice of one over the other is not likely to be influenced by the presence or size of the advertisement in the yellow pages. In a larger population, customers may use the directory to select a flower shop, garden center, retail nursery, landscaper, or lawn care company. Even so, with an increasing number of customers in various demographic populations using the web as a search tool, advertising dollars may reap a better return on investment if spent on building and designing a website for the business than on printing advertisements in paper directories.

Immediate response advertising has a short-term value. It must motivate the customer to seek out the advertised product or service the same day, the same week, or the next at the latest. The value of such advertising is both direct and indirect. Directly, the items advertised must sell in excess of what would have been sold without advertising and generate enough revenue to pay for the advertising while leaving a profit on the advertised items. Indirectly, the customers that advertising attracts to the business may make additional impromptu purchases. A new business operator should study the response of consumers to determine which medium or combination of media is most effective for immediate response advertising.

The value of a particular medium may depend on the items being advertised. For example, a flower shop might invest in a series of 30-second radio commercials. If prom corsages are advertised, a classical AM station will not reach the teenage market as effectively as an FM rock or pop station might. Advertising prom corsages could generate a profit if done at the right time and through the right medium. The single greatest value of any advertising is to let customers know that you have a good supply of what they want at the time they want it. Advertising to create a demand for a new product is often beyond the capabilities of a small business and must be left to manufacturers and trade organizations.

Advertising has definite limitations. It can never fully compensate for a poor business location or inadequate parking facilities. The quality of the merchandise or service must be high, even when the price is reduced.

Promotion, Publicity, and Public Relations

A complete advertising program must include promotion, publicity, and public relations. Often, no clear distinctions can be drawn among them. In general, promotion is the range of activities whose purpose is to establish goodwill or further the growth of a firm. Examples of promotions include the following:

- Sending greeting cards, complimentary plants, or other gifts to important customers during the holidays
- Sending clipped newspaper announcements of engagements to brides
- Sponsoring free lawn clinics, floral design shows, pruning clinics, or similar events just before major sales periods

Publicity and public relations activities are intended to bring the business and its staff to the attention of the public (and potential customers) to generate favorable public opinion. Examples of public relations activities follow:

- Writing articles about plants, flowers, or landscaping for the local newspaper
- Holding an open house before major holidays or planting seasons
- Providing shirts for the local little league team
- Joining and speaking to civic groups on horticultural topics
- Sponsoring a local FFA chapter event
- Using social media such as Facebook for promotions
- Sending regular e-mail newsletters to customers

The return on the dollars spent in these areas is not easily measurable, and so the budget should not lump them together with advertising. To do so would complicate and distort the interpretation of the effectiveness of the advertising program.

Budgeting for Advertising

At the outset, the business owner must decide what percentage of the anticipated sales revenue will be spent on advertising. The percentage will vary with the financial condition of the business, the amount of competition in the market area, and the nature of the products or services sold. By multiplying the total anticipated sales for a year of operation by the percentage allotted for advertising, a dollar amount can be assigned to the advertising program for the year.

Some business owners determine how much to spend each month by correlating the percentage of advertising dollars with the percentage of total sales that each month contributes. For example, if June contributes 12 percent of the annual sales, then 12 percent of the advertising budget is spent for June sales. Some months are naturally high-volume sales months, however. Christmas sales in a flower shop are an example. The sales for the month of December may be 35 percent of the annual sales, yet only 20 percent of the advertising budget may be needed to generate the sales.

An advertising budget should also allocate funds to special, nonrecurring public relations or promotional events such as the store's grand opening, historical commemorations, or the sale of an entire truckload of plants obtained at a special price. Special-event advertising should be added to the regular advertising budget to arrive at the total allocation.

The US Small Business Administration (SBA), in its "Advertising Guidelines for Small Retail Firms," recommends that a monthly record of advertising expenditures be kept to determine if the advertising budget is being spent as planned. The chart in Figure 52–8 can be used to keep such a record.

	MONTH		YEAR TO DATE	
Account	Budget	Actual	Budget	Actual
Media				
Newspapers				
Radio				
Television				
Literature				
Internet/Social Media				
Other				
Promotions				
Exhibits				
Displays				
Contests				
Advertising expenses				
Salaries				
Supplies				
Travel				
Stationery				
Postage				
Subscriptions				
Entertainment Dues				
Total				

Figure 52–8 Advertising expense and budget record.

Writing Descriptive Copy

The best way to obtain advertising copy that will make customers call or stop in is to hire a professional advertising agency. For smaller businesses that may not be able to afford a large professional advertising agency, there are many online freelance sites where an owner or manager can contract the services of an advertising professional on an hourly basis or per item fee. Of course, the owner or manager may also choose to prepare the copy himself or herself or provide the necessary information to the advertising staff of the newspaper, website, or radio or television station. Therefore, a new business owner should have a basic understanding of how to construct a good advertisement.

A lot of money is spent on nonproductive, even counterproductive, advertising. Humor that falls flat or wears thin after repeated exposure is not good advertising. Advertisements that do not contain information about specific products or services are unlikely to result in increased sales. Omitting prices or allowing customers to suppose them higher than they really are is not productive. Advertisements that criticize a competitor or make unbelievable claims for a product or service may harm the reputation of the business.

To prepare honest and descriptive advertising copy that sells, remember the following points:

1. Study the buying patterns of the customers, and advertise to appeal to them. For example, bronze mums are properly advertised in the fall but are not what customers are seeking during the Christmas season. Hardy bulbs for fall planting should be advertised then, not during the winter, when they are obvious leftovers.
2. Advertise representative products and services. For example, avoid featuring low-cost items if the business normally carries a better line of merchandise.
3. Give customers good value for their dollars. Promoting broadleaved evergreens in an area that is too cold to ensure their survival will not create goodwill. It is better to feature plants of predictable hardiness.
4. Be definite about the price of what is advertised. Avoid giving price ranges that make it hard to determine what an item will cost. Featuring flowering shrubs for $5 to $30 is too vague. Pricing them at $5 and up is even worse. Tie the price to a size or quantity so that customers know what they will get for the money.
5. Describe the benefits to the customers of purchasing the product or service. The size or length of the advertisement will determine how many benefits can be described, but the most obvious benefit, as well as the one most likely to motivate the customer, should be mentioned. For example, a lawn service company can advertise benefits such as "rich, green, weed-free turf" (an obvious benefit) and "more free time to spend enjoying the summer" (a strong motivator).
6. Suggest a need for the item. This can often be incorporated into the description of benefits.
7. Encourage the customer to act immediately. Close the advertisement with a statement that motivates customers to pick up the telephone or stop by the store while it is still fresh in their minds.
8. Keep the layout of printed advertisements simple. Use a heading in large type and an illustration to catch the reader's eye, follow with the copy, and close with the price and store name, address, and hours.
9. Develop a recognizable format for a series of advertisements. Although specific items may change, the look or sound of the advertisement should strike a familiar note. The use of standardized graphics can help to create the desired identification.
10. Keep broadcast advertisements conversational. Avoid an impression of aloofness.
11. Devise a way to measure public response to the advertisement. Use of a code word in a printed advertisement allows a count of the number of persons responding. Radio and television advertisements can offer some special benefit to customers who use a key word when placing their orders. Figure 52–9 is an example of a good advertisement for ornamental horticulture through the use of a QR code.

Courtesy of MasterTag

▲ **Figure 52–9** A MasterTag is a great example of how to advertise in horticulture through the use of a QR code.

© iStockphoto/RapidEye

▲ **Figure 52–11** A prominent display of fertilizers is placed where customers will pass by as they enter and leave this garden center.

Using Displays to Advertise

The business itself may be its own best advertisement. A customer who comes to buy one item may be encouraged to buy others if they are displayed freshly and attractively (Figure 52–10). Such displays are known as point-of-purchase advertising. Florists, garden centers, and retail nurseries may use or feature live materials in their displays. Landscapers and lawn care services may use color photographs of completed properties. Other displays may feature bottles, boxes, or bulky bags of chemicals, seeds, and fertilizers (Figure 52–11). The following

© Lynn Watson/Shutterstock.com

© Elnur/Shutterstock.com

▲ **Figure 52–10** Attractive displays can spark customer interest in other items. Shown here are two examples of fresh and attractive displays in retail flower shops.

techniques will help make in-store displays more successful.

1. Keep the display simple. Feature either a single product or service or a limited choice of related products or services. Complexity in the display can create indecision in the customer's mind.
2. Suggest a use for the product. This may establish a need for the item in the customer's mind.
3. Keep the display small. Display a few of the items, but avoid a mass display. Create a qualitative image rather than a quantitative one.
4. Keep the display complete. If customers remove items from the display to purchase, replace the items quickly.
5. Coordinate the display with other advertising. The display may feature merchandise advertised in the newspaper or on the radio or television.
6. Keep all plant material in peak condition. Do not let wilted blossoms, yellow leaves, wilted foliage, or broken branches suggest merchandise of inferior quality.
7. Neatness counts. All signs should be neatly lettered, nonfaded, and free of spattered water, soil stains, and dust.
8. Light the displays. Show windows offer opportunities for effectively lit displays even after the store closes for the day.
9. Change the displays frequently. Regular customers will cease to notice them if they remain unaltered for too long.

SAFETY FIRST

OSHA's regulations ensure that employers and employees know about work hazards and how to protect themselves to prevent injuries in the workplace. As mentioned earlier in the unit, the owner and manager of a business should be knowledgeable about OSHA regulations and how to keep the working environment safe, including labeling and instructions to make sure employees have proper information about the materials and equipment being used as well as providing up-to-date equipment to maintain safety. Additionally, when any type of chemical or hazardous substance is used at a business, a Safety Data Sheet should be available on-site to keep records and information about the handling of the substance. For more information on occupational health and safety in the workplace, please visit the OSHA site at www.osha.gov.

Summary

Ornamental horticulture businesses may be structured as sole proprietorships, partnerships, or corporations. Each has features that make it appealing, but the corporation offers the greatest legal protection to the owners. Also, a corporation may offer the greatest opportunities for diversification and growth of the business.

No one should begin in business without a careful market survey to determine the probability of success in the location under consideration. The study should analyze the range of the market area, the need for the business, the economy, the competition, and employee availability.

A business requires capital to purchase fixed assets and to keep the business operating (working capital). Capital can be obtained by using personal savings, borrowing from lending institutions, or reinvesting profits into the business.

In selecting a site for the business, the buyer should consider factors such as size, natural features, zoning regulations, utilities, access, compatibility of the new business with other businesses nearby, strength of merchants' associations, lease agreements, and the cost of acquisition. If purchasing an established business, the buyer should consider factors such as the equipment included and its condition, the existing inventory and its market value, the accounts receivable, goodwill, and use of the existing name.

Once acquired, the business site must be allocated to different uses. Proper sizing and ease of circulation are the keys to a successful layout of the physical plant. A staff organization chart can help employees understand their roles within the company.

When starting out, materials and supplies should be ordered in small quantities that can be turned over quickly and paid for promptly. The less capital tied up in inventory, the better.

Advertising seeks to focus public attention on items or services that are for sale. To be productive, advertising must generate sales and profits in quantities greater than would be expected without advertising. There are two types of advertising: immediate response advertising and attitude advertising. A complete advertising program must also include promotion, publicity, and public relations.

This chapter also includes a summary of the laws and regulations that affect horticulturists.

Review Questions

1. Discuss the similarities and differences among a sole proprietorship, partnership, and corporation.
2. Why is it important to perform a market survey prior to investing money and time in a new business?
3. Discuss three ways to obtain capital for a business.
4. Discuss important factors to consider before choosing a location for a new business.
5. Describe some of the major laws and regulations that affect horticulturists. Why is it important to be aware of these laws before starting a new business?
6. Discuss some methods of advertising. What makes some methods more effective than others?
7. Describe an effective display. What features make it effective?

Student Activities

1. Select an item or service that would be sold by the type of ornamental horticulture business that interests you. Prepare a digital radio advertisement for it. Specify the time of year the advertisement should be featured. Play your audio recording for the class.
2. Prepare a brief essay on why you would or would not like to own your own business. Discuss the factors that you think will contribute most to your job satisfaction and how ownership or nonownership would affect them.
3. Create a rough draft of a business plan for a flower shop in an urban city. Include the components mentioned in the text.
4. Search online for business management courses that are offered for entrepreneurs.
5. Use Google Forms and develop a survey to determine interest in the purchase of horticulture products from the school horticultural department.

Suggested Supervised Agricultural Experience (SAE) Activities

1. To qualify for a placement SAE, work in the business office of a local garden center.
2. As an entrepreneurship SAE, manage and operate your own lawn maintenance business operation.

Additional Resources

www.morrisville.edu/programsofstudy/schoolofagandnr/horticulturebusinessmanagement/

www.clemson.edu/majors/horticulture

www.beginningfarmers.org/farm-business-planning/

www.fsa.usda.gov/Internet/FSA_File/fsa_br_01_web_booklet.pdf

UNIT 53 Financial Operations of a Horticultural Business

Objective

To determine considerations for the financial operation of a horticultural business.

Key Terms

actuating
asset
balance sheet
buying function
cash flow statement
distribution function
financing function
inventory report
long-term plan
net worth
profit and loss statement
promoting function
selling function
short-term plan

Competencies to Develop

After studying this unit, you should be able to:

- describe steps in planning a business venture.
- state five basic functions performed in the operation of a small business.
- select a product or service for a personal or group enterprise.
- determine the basic functions performed by small-business managers.
- analyze the outcome of a business venture.
- use small-business financial records.
- analyze the benefits of self-employment versus other types of employment.

Materials List

- variety of publications with pictures
- materials to make a collage or bulletin board
- Internet access

Operating Businesses

Five basic functions are performed in the operation of both a small business and a sales project. These functions represent the basic steps in moving a product or service from the supplier to the consumer. They are the buying function, selling function, promoting function, distribution function, and financing function.

Buying Function

The **buying function** involves selecting a product or service to be marketed or sold for profit. The selection of a product or service is based on thorough marketing research, which determines consumer (customer) needs and wants. It also determines who, if anyone, is already promoting the product or service.

Selling Function

The **selling function** includes studying the product or service to determine the reasons that customers will want and need the product or service. This function also involves developing suitable customer approaches. Planning sales presentations, determining methods of overcoming objections, and planning for the close of a sale are all part of this function (Figure 53–1).

Courtesy of the National FFA Organization

▲ Figure 53–1 Effective selling and promotion is absolutely essential to the success of most businesses. How a product is displayed is important. The flowers at this greenhouse are eye-catching and visually appealing, which is sure to attract customers.

Promoting Function

The **promoting function** involves developing a plan to identify ways to make potential customers aware of the product or service to be offered. Examples of promotional activities include newspaper, news websites, blogs, social media, TV, radio, and outdoor advertising.

Distribution Function

The **distribution function** involves physically organizing and delivering the selected product or service (Figure 53–2). For products, this includes receiving, storing, and distributing the merchandise. Many of the same activities are associated with services, particularly distributing to the customer.

Financing Function

The **financing function** includes obtaining capital for the initial inventory, recording sales,

© Pi-Lens/Shutterstock.com

▲ Figure 53–2 Product distribution may be made to retail outlets or to other businesses, homes, farms, and many other destinations.

maintaining inventory, computing profit or loss, and reporting the results of the venture. A venture is an activity with some risk involved.

Selecting a Product or Service

The first function in the process of establishing a new business or developing a fund-raising project is the selection of a product or service (buying function). This involves analyzing potential products/services to determine the potential demand (number of customers). The analysis of potential products or services should answer the following questions:

- Who are the potential purchasers of the proposed product or service? Customers can often be viewed in groups who have similar interests.
- On what basis does each group make decisions on the product or services? Typical responses to this question include price, quality, and continuing service after the sale.
- What are the competing products or services? How do they compare with your product or service relative to price, quality, and service?
- What is the total estimated demand for your product or service?
- How much profit would you make on the sale of this product or service? (Multiply the number of units sold by your markup per unit, and subtract any expenses you incur in the purchase, sale, and delivery of the product or service.)

Organization and Management

The role of a venture's organization and management is to make things happen so the venture can achieve its goals. To accomplish this, managers work mainly with data and people. Managing involves getting all the parts of the business—including personnel, marketing strategies, finances, and records—to function together to achieve the venture's goals.

No two managers have jobs that are the same. The jobs are shaped by the type of venture and the personality of the individual manager. However, managers perform the same functions:

- planning work
- organizing people and resources for work
- actuating work
- controlling and evaluating work

Planning

When managers make plans, they set objectives or goals. They establish policy for the business, and they recommend strategies to achieve the goals. They develop procedures, methods, and/or programs to support policies and implement strategies leading to achievement of goals. Plans must be constantly reviewed and updated. No matter how thorough the plans, they do not guarantee success. In planning, managers must make short-term and long-term plans. **Short-term plans** are accomplished in several days or weeks. **Long-term plans** are accomplished over several months or years.

Organizing

After a plan is developed, the work must be organized. The manager must identify the people needed to carry out the plan and arrange for the necessary equipment and supplies. Staff development may also be required.

Actuating

Actuating means simply putting the plan into action. The manager needs to inform employees of the plan. All persons involved must understand their roles. Actuation also includes motivating people to work efficiently and effectively and to want to get the job done.

Controlling and Evaluating

Managers must carefully control implementation of plans after the work begins. The quality and quantity of all results must be evaluated. If the results are satisfactory, work can continue. If problems arise, changes must be made, and alternate plans may need to be developed. Managers must make adjustments in personnel,

© Hurst Photo/Shutterstock.com

▲ **Figure 53–3** The manager must be familiar with all of the functions and able to work with all of the employees of the business.

equipment, policies, or procedures whenever necessary (Figure 53–3).

In a very small venture, the same person may carry out all management functions. As a venture grows and its goals and objectives become more involved, additional managers may be required. When they are hired, organizational lines of responsibility must be developed and job descriptions written to identify the functions that are performed by each manager.

> **GREEN TIP** Use recycled equipment or buy only energy-efficient office products and equipment such as energy-saving lightbulbs. Many companies manufacture green-friendly products, and the cost is reasonable. Your use of electricity decreases, and therefore the business expenses decrease.

Small-Business Financial Records

Financial records are invaluable tools to the entrepreneur and manager. Good financial management allows the manager to maintain control of the business venture and to increase profits or reduce losses.

Financial records can reveal which items are selling and which are not. These records can also show the extent of success of each person on the sales force. Financial records should indicate the amount of inventory on hand and how much has been sold. How much profit is made and the venture's total value at any time can also be computed using appropriate financial records.

© RAGMA IMAGES/Shutterstock.com

▲ **Figure 53–4** Financial records are fundamental management tools.

The small-business owner or entrepreneur needs and uses detailed financial records. Financial records for items such as buildings, fixtures and equipment, credit, debt, and return on investments are examples of records maintained by small businesses. However, balance sheets, profit and loss statements, and inventory and sales reports may be useful for a group project, as well as for a small business. Some typical financial records are discussed in the following sections (Figure 53–4).

Balance Sheet

A **balance sheet** is like a photograph of the business at a given time. It shows the assets, liabilities, and owner's investment on a particular date. The equation for the balance sheet is the following: assets = liabilities + net worth. **Assets** are anything the venture owns, including cash on hand, equipment, and inventory. Liabilities are both current and long-term debts. **Net worth** is the owner's investment in the business, including profits as they occur. A sample balance sheet is shown in Figure 53–5.

> ***TRY THIS*** Obtain records from a sample or real business, and create a balance sheet using the appropriate data. Examples of financial information can be found on the Internet or from a real business that is willing to share its financial records.

BALANCE SHEET

	Date:	________
ASSETS		
Current:	Cash	________
	Merchandise	________
	Accounts Receivable	________
Fixed:	Land	________
	Building	________
	Machinery	________
	Equipment	________
Other:		
	1.	________
	2.	________
	3.	________
TOTAL ASSETS:		________
LIABILITIES		
Current:	Notes Payable	________
	Accounts Payable	
Noncurrent:		
	Debts more than one-year maturity	________
TOTAL:		
NET WORTH:		________
TOTAL LIABILITIES AND NET WORTH:		________

Figure 53–5 A balance sheet is a statement of a business owner's assets, liabilities, and net worth.

PROFIT AND LOSS STATEMENT

For period ending ________

TOTAL SALES	$75,000
COST OF GOODS SOLD	
Beginning Inventory	55,000
(Plus) Purchases	10,000
(Less) Ending Inventory	15,000
TOTAL COST OF GOODS SOLD	50,000
(Beginning Inventory plus Purchases minus Ending Inventory)	25,000
GROSS PROFIT	25,000
(Total Sales minus Total Cost of Goods Sold)	
EXPENSES	
Salaries	18,000
Payroll Taxes	2,500
Rent	4,500
Advertising	1,000
TOTAL EXPENSES	26,000
NET PROFIT (LOSS) (before taxes)	(1,000)
(Gross Profit minus Total Expenses)	

Figure 53–6 Sample profit and loss statement.

Profit and Loss Statement

The **profit and loss statement** projects costs and other expenses against sales and revenue over time. The five basic sections in a profit and loss statement are (Figure 53–6):

- Total Sales
- Cost of Goods Sold
- Gross Profit
- Expenses
- Net Profit

Cash Flow Statement

A **cash flow statement** describes the availability of funds for the purposes of running a business (Figure 53–7). When a positive cash flow exists, the income that is received by the business exceeds the expenses, and there is cash on hand. A negative cash flow indicates that the expenses that are generated by the business exceed the income at a given time. Nearly every business deals with periods of negative cash flow, and it is at these times that an operating loan may be needed to tide the business over until income is received. The cash flow statement allows a business owner to estimate the need for credit and to establish when the credit will be needed. The cash flow statement is intended to help the business manager and the loan officer determine how much operating cash is needed, when it is needed, and when it will be repaid.

Inventory Report

The **inventory report** includes how many units of each product are on hand, how many were

INVENTORY REPORT

Item No.	Description	Beginning Inventory	Items Sold	Ending Inventory	Total Cost Beg./Inv.	Total Cost End/Inv.	Total Sales	Total Profit
000	Tick Tack Tote	45	45	0	33.75	.00	45.00	11.25
001	Brown Tray	32	32	0	25.60	.00	48.00	22.40
002	Fish Tray	20	20	0	15.00	.00	30.00	15.00
003	Lion Tray	6	6	0	4.80	.00	9.00	4.20
004	Guitar Tray	8	8	0	6.40	.00	12.00	5.60
005	Asst. Animal Tray	34	34	0	25.50	.00	51.00	25.50
006	Elephant Tray	10	10	0	5.00	.00	15.00	10.00
007	Screwdriver Set	22	22	0	11.00	.00	33.00	22.00
008	Brush & Shoehorn	36	36	0	32.40	.00	81.00	48.60
009	Fully Auto. Umbrella	9	9	0	20.70	.00	31.50	10.80
010	Auto. Umbrella	8	8	0	16.00	.00	24.00	8.00
011	Dad's No. 1 Keychain	200	190	10	90.00	4.50	190.00	104.50
012	Grandpa Keychain	100	79	21	45.00	13.95	79.00	47.95
013	Dad's Pad	48	30	18	28.80	10.80	45.00	27.00
014	Mini Screwdriver Set	75	75	0	60.00	.00	150.00	90.00
015	Dad's Plaque	72	72	0	28.80	.00	108.00	79.20
016	Tic Tac Toe	36	36	0	21.60	.00	36.00	14.40
017	Dad's Pen	1	1	0	.80	.00	1.50	.70

Figure 53–7 Sample cash flow statement.

sold, each item's cost, total sales, and profit. The inventory report can be used to make decisions such as reducing the price of slow-selling items, reordering other items, and learning which items are yielding the best profits. The use of computers at checkout stands allows this report to be updated each time a sale is made. Some software is also capable of generating new orders for products as the inventory for a particular item is reduced to a preset level.

SAFETY FIRST

OSHA's regulations ensure that employers and employees know about work hazards and how to protect themselves to prevent injuries in the workplace. The owner and manager of a business should be knowledgeable about OSHA regulations and how to keep the working environment safe. For more information on occupational health and safety in the workplace, please visit the OSHA site at www.osha.gov.

Summary

The horticulture business is a dynamic, fast-growing business that offers entrepreneurs myriad avenues to explore success in a highly specialized form of business. Operating a business involves many categories of management skills such as planning, organizing, actuating, controlling, and evaluating. Accurate financial records help business owners evaluate and make decisions for the future of the business.

Review Questions

1. Discuss the different steps that are required when planning a business venture.
2. Describe the five basic functions that are performed in the operation of a small business.
3. What basic functions do small-business managers perform?
4. Discuss the advantages and disadvantages of self-employment versus other types of employment. If you had to select one, which would you choose and why?
5. Why is an inventory report important?

Student Activities

1. Write the Terms to Know and their definitions in your notebook.
2. Develop a digital collage illustrating business opportunities in horticulture.
3. Seek election as your local FFA chapter treasurer, and maintain the business records of the local FFA organization.
4. Create an individual business plan using the information you have learned from this unit. Discuss with your parents and trusted advisors what steps you would need to take to make your business plan a reality.
5. Search the Internet for business plan templates. Choose one that is appropriate for your business idea.

Suggested Supervised Agricultural Experience (SAE) Activities

1. Conduct an entrepreneurship SAE as a repair technician for landscaping equipment. Prepare a business plan, income statements, balance sheets, and profit and loss statements for your SAE records.
2. Participate in a placement SAE working for a horticultural company in which you help to maintain the financial records.

Additional Resources

http://smallbusiness.chron.com/develop-strategic-marketing-plan-horticultural-firm-44874.html

http://horticulture.oregonstate.edu/content/business-planning

http://ucanr.edu/sites/placernevadasmallfarms/files/143084.pdf

UNIT 54 Human Relations: Personnel Management and Customer Sales

Objective

To outline a program to create effective personnel managers and salespersons.

Key Terms

customer service
good salesperson
part-time
personnel management
seasonal

Competencies to Develop

After studying this unit, you should be able to:

- define and state the objectives of personnel management.
- list the characteristics of a good personnel manager.
- list the characteristics of a good salesperson.
- describe an effective sales procedure.

The Importance of Human Relations

The owner or manager of a horticulture business is at the top of a human relations triangle that includes the employees of the business and the customers of the business (Figure 54–1). No business can operate for long with a serious problem between any members of the triangle. The manager is responsible for ensuring that such problems do not arise and, if they do, resolving them. Employees must be kept happy with their work and productive on behalf of the company. Customers must be pleased with the products and services they obtain from the company and equally pleased with the people who wait on them, answer the telephone, deliver to their homes, install or maintain their landscapes, or respond to their complaints. The two groups will be discussed separately, but the dependency of the business on a compatible relationship between employees and customers is obvious.

Personnel Management

The direction of workers in a manner that brings out their best efforts and attitudes on behalf of the business is termed **personnel management.** Although the owner of the business is the foremost personnel manager, other persons in

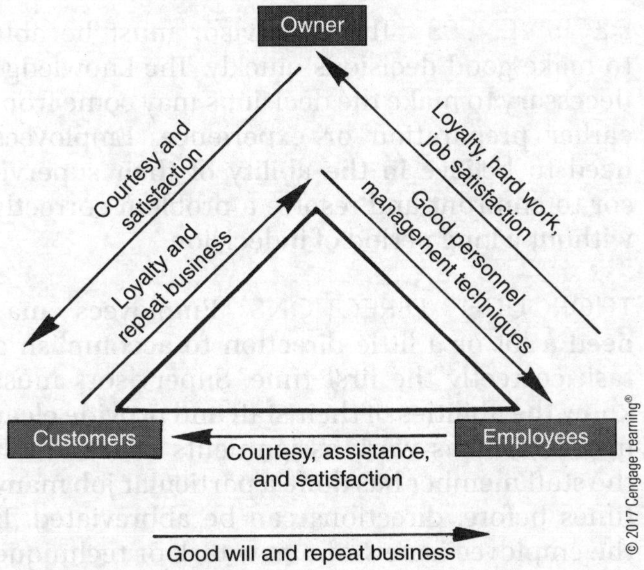

▲ **Figure 54–1** The triangular relationships within a business.

supervisory or leadership positions must also develop good personnel management skills.

Much of what constitutes good personnel management seems to be somewhere between common sense and the Golden Rule. In addition to being an application of courtesy and friendliness in the workplace, personnel management requires good leadership skills.

Setting Realistic Expectations

The supervisor of any group of employees needs to understand what those employees can be expected to accomplish. Expectations may change as employees gain experience on the job.

Certain job skills are more physical; others are more intellectual. Some require charm and personality; others require the ability to be decisive and inspire customer confidence. A supervisor usually has different expectations of a college-trained permanent employee than of a part-time high school student. In any case, a supervisor's ability to get along with all employees begins with the ability to set realistic expectations.

Depending on its type and size, a horticulture operation may have up to three basic categories of employees. A supervisor will probably have different expectations of each.

CAREER-DIRECTED These employees can be expected to have personal goals for advancement. They may want to stay with the company for many years or they may intend to start their own business after a few years of on-the-job training. They may have technical training or other education that is useful to the business. They can be taught new skills or be given increasing responsibilities and be expected to learn quickly. Career-directed employees need frequent and varied rewards, including periodic salary or wage increases. They usually have opinions and suggestions for improvements that should be listened to and respected.

PART-TIME These employees' career goals usually do not involve advancement in the business. The job may be a way to supplement family income, a way to fill empty time, or a way to earn money for a short-term objective such as schooling, a new car, or a vacation. **Part-time** employees may have education or practical experience to bring to the business. They may possess the maturity and skills needed to staff the business over weekends or during vacation periods. They often feel as much loyalty to the company as career employees but do not seek or expect major leadership responsibilities.

TEMPORARY SEASONAL **Seasonal** employees are hired full-time when there is more business than the permanent staff can handle. Examples are temporary sales help hired in a flower shop at Christmas time, field laborers hired at harvest time in a nursery, and planting crewmembers hired in the spring by a landscaper. These employees are hired for their ability to perform a few skills competently and quickly, such as balling-and-burlapping, operating heavy equipment, or duplicating wire-service arrangements. The employees may be students on vacation, migrant field workers, or others who have no long-term employment goals with the company. Money and work experience are their major motivations. Loyalty to the company and an interest in its growth and success are not common in seasonal employees. They may require closer supervision than other employees.

GREEN TIP Train your employees on how to go green. Start a green team to research more ways to make the horticultural business more eco-friendly.

A Diverse Workplace

The faces of America's horticultural workforce are increasingly diverse as more and more companies are meeting their employment needs with nontraditional workers. Workers from Mexico and Central America have filled the majority of field positions in American nursery, arborist, and landscape firms since the mid-1980s. They are an increasingly important human resource in the turf industries as well. For many Hispanic workers, their seasonal employment in America is valued for the higher wages they can earn compared to in their home country. They also value the opportunity it gives them to return home during the off-season. As skilled migrant workers learn English, they gain the opportunity to advance within their companies to fill positions of greater responsibility and leadership. Asians and Asian American workers are also increasingly present in the green industry workforce. By the mid-point of the twenty-first century, citizens of Asian heritage are predicted to represent 25 percent or more of the population, so their presence in the shops, sales yards, and offices of horticultural firms can be expected.

The incorporation of immigrant workers into a traditional American workforce presents a challenge to those who lead and supervise. Cultural differences, complicated by language difficulties, can build barriers between employees and reduce productivity. An effective leader must develop methods of addressing, correcting, and surmounting these differences and difficulties. Bilingual individuals have an advantage because they are able to manage employees who speak languages other than English.

Other Qualities of Good Leadership

In addition to realistic expectations for each employee, effective supervision requires an understanding of what motivates workers to respond positively to their work and their supervisors. Most people respond in a positive manner to a supervisor whose competence they respect, whose expectations are fair and impartial, and whose attitude is upbeat. Consider the following qualities of good leadership.

DECISIVENESS The supervisor must be able to make good decisions quickly. The knowledge necessary to make the decisions may come from earlier preparation or experience. Employees need to believe in the ability of their supervisor to confront and resolve a problem correctly without a long period of indecision.

THOROUGH DIRECTIONS Employees may need a lot or a little direction to accomplish a task correctly the first time. Supervisors must know the abilities of their staff and provide clear instructions as work assignments are made. If the staff member has done a particular job many times before, directions can be abbreviated. If the employee is new or a new task or technique is being introduced, the supervisor must take longer to explain and demonstrate what is to be accomplished and the quality that is expected. Posting directions or Standard Operating Procedures (SOPs) in the workplace can help remind employees of workplace procedures.

COMPETENCE Employees seldom perform at a level they believe is beyond the capability of their supervisor to recognize and appreciate. Therefore, supervisors must perform the tasks that they assign frequently enough to retain and refresh their own skills. The supervisor may need to take over a task in the event that an employee is suddenly absent; therefore, they need to be able to perform those tasks as well.

FAIRNESS Supervisors must allocate work among their staff equitability. Two workers with the same competencies and receiving the same wage or salary should be expected to perform equally. Otherwise, feelings of resentment develop. Fairness implies equal treatment for all, whether in the choice of vacation weeks, the assignment of new tools, or the requirement to work over holidays. If benefits accrue with seniority, this should be understood by all employees at the time of their hiring. Partiality or bias, real or perceived, can be demoralizing to a staff or crew.

UNDERSTANDING A good supervisor is predictable. Employees must be able to anticipate the consequences of performing in a way contrary to established company policies. Certain inflexibility is necessary, or all discipline will quickly disappear. However, the enforcement of company policies must be tempered by understanding when circumstances warrant it. If a trusted and proven employee suddenly begins performing carelessly or erratically on the job, something is probably happening outside of work to explain it. A supervisor who points out the problem to the employee privately and without a tone of reprimand is likely to be effective and appreciated. It is seldom necessary to become involved with the employees' personal problems. Pointing out how the symptoms of the problem are being manifested on the job is often enough.

RESPECT Supervisors who expect respect from their staff must demonstrate their own respect. Many employees have insights and abilities developed through years of experience on the job. Others have knowledge attained from recent classroom experience. The employee's point of view should be sought by the supervisor whenever possible. At times, it may be offered without solicitation. If the viewpoint has merit, it should be used. If inappropriate, the employee's viewpoint should at least be listened to courteously and an explanation given as to why it is inappropriate. Using an employee's suggestion or asking an employee for an opinion is not a sign of weakness or incompetence in a supervisor—it is merely good sense.

SINCERITY All supervisor-employee relationships benefit if the staff perceives that the supervisor is sincerely interested in their well-being. Concern for the employee as well as for the company can be exhibited in numerous ways, such as asking about employees' children or vacations, or whether they are satisfied with their work. Equally important is that the employee understands the supervisor cares about the business, the satisfaction of the customers, and the quality of work.

PRAISE Everyone appreciates recognition. When an employee or group of employees does a job well, the supervisor should recognize the accomplishment. It may be as simple as telling the employee privately that the work was well done, for example: "Nice job. You got that done in record time, and it looks great." It may be more public, such as complimenting a crew or crewmember in the presence of peers. Then it is best to praise the action or the results to avoid embarrassing individuals or implying criticism of other staff members. For example, a supervisor might say, "We've had a productive week. Everything went well and that job at the Johnson home really made some money for us because it went faster than we expected. They've already called to say how pleased they were." The crew responsible has been praised, and the other workers have not been compared or criticized.

REWARD In return for their continuing good performance, employees expect their status within the company to improve. The most expected reward is an increase in salary or wage, or a bonus. Surprisingly, monetary reward is usually short-lived in terms of employee satisfaction. We all tend rapidly to adjust our way of living to the new income level and look for our next reward. When a business's financial circumstances do not permit monetary reward of deserving employees, or when the supervisor has no authority to reward an employee with money, other forms of reward can be offered. These may even be more long-lasting sources of worker satisfaction than financial rewards. Deserving employees may be the subject of public relations news articles. Others may be presented with service awards or prizes. Some companies recognize the "employee of the month" by posting the employee's photo in a prominent location. Other companies recognize employee excellence through titles, such as assistant grower or crew leader. Something as simple as a handwritten note from the supervisor to the employee acknowledging the latter's performance can be an effective reward.

Qualities to Avoid

Certain qualities in a supervisor may cause negative feelings to develop in an employee or group of employees. The probable result of negative feelings between workers and their supervisor is low worker morale, reduced productivity, and even theft or vandalism. Although no supervisor can expect to perform flawlessly, an effort should be made to avoid the following characteristics.

INFLEXIBILITY Company rules must be applied fairly to all employees, but little is gained by a policy of "no exceptions for anyone at any time." If the rules penalize a worker for tardiness, most employees will see the need for it and the logic behind it. Employees who oversleep may expect to have their paycheck adjusted to reflect the lost time. However, if an employee arrives late once because of a flat tire, enforcement of the rule will probably be seen as needless rigidity. "They don't care about me" is a difficult attitude to reverse once implanted in an employee's mind.

PARTIALITY In the close working environment of many horticultural businesses, friendships form easily. If personal associations between supervisors and employees outside of work cause partiality at work, however, resentment is bound to result. Joking with some employees or spending time outside of work with some, but not all, members can be interpreted as partiality. Such an interpretation can be as harmful to worker morale as if favoritism actually exists. It must be guarded against.

CONDESCENSION Training a new employee or assigning a new task to an old employee usually requires the supervisor to give instructions. They must be delivered in a manner that respects the employee's intelligence, skills, and insights. Talking down to employees by oversimplifying the directions or giving shallow answers to serious questions insults their intelligence.

SUBJECTIVE CRITICISM No one likes being criticized or reprimanded, yet every supervisor must do so on occasion. The objective of properly applied criticism must be to prevent the employee from feeling personally offended or embarrassed while assuring that the mistake in behavior or performance is corrected. If the criticism is directed to an individual, the individual should be spoken to privately so that no other workers overhear. Whether the criticism is directed toward an individual or a group, the supervisor should remain objective. Criticizing the actions or the outcome rather than specific persons allows employees to retain their own sense of worth and avoids a long-term negative reaction to the supervisor. All criticism should be delivered in a positive manner without anger or profanity. Emotional reprimands are much more likely to be taken personally.

DOUBLE STANDARDS Employees have the right to expect that policies set for them will be followed by those in supervisory positions as well. Supervisors are responsible for exemplifying the positive qualities and attitudes that they desire or require of their workers.

INDIFFERENCE The small-business atmosphere of most horticultural enterprises requires that supervisors give frequent positive reinforcement to their workers. When a supervisor inspects the work done by an employee or crew, the supervisor is expected to react, not merely walk on. Equally damaging to working relationships can be a supervisor who displays no concern when births, deaths, illnesses, or other personal events occur in the lives of employees. "They don't care about me" is an attitude that can be avoided by a few outward expressions of concern from a supervisor.

Complicating Factors

Even in businesses in which the owner and supervisors are sincerely interested in their workers and make a concerted effort to be good managers, problems still occur. Assuming that an employee does not dislike the supervisor, but that they are still not communicating effectively, it could be for one of the following reasons:

- Age differences. When the employee and supervisor are of widely different ages, either may be guilty of believing that they have nothing in common. Young people may tend to believe that the growing pains of life are known only to them. Older employees can be intimidated by the strength or boldness of younger ones or may have no patience with the inexperience of young workers, forgetting that they were once that young and inexperienced themselves.
- Educational differences. Less educated employees may feel intimidated by a better educated, articulate supervisor. Young supervisors recently graduated from college can expect to be regarded suspiciously by the employees they direct. If the workers are appreciably older, more experienced, or have seniority with the firm, the young supervisor may be on the receiving end of reverse snobbery.
- Sex differences. Equality and respect in the workplace are essential for all employees. Verbal communication and behavior between male and female employees must remain professional and appropriate. A manager can destroy workplace respect by making comments or gestures that make a member of the opposite sex feel uncomfortable. Any such complaints should be taken seriously and dealt with according to company policy.
- Ethnic and background differences. The background and ethnic heritage of all employees should be respected and accepted. Disrespecting one's heritage will make for a very negative workplace environment. Any complaints of racial or ethnic harassment need to be appropriately resolved according to company policy.
- Position or status differences. Some employees are not comfortable around authority figures, no matter how hard the supervisor tries to develop a good relationship. Supervisors who are promoted from the ranks may find that their former friends on the staff suddenly act differently toward them for this reason alone.

If one or more of these factors leads to complications of the human relations within a firm, it may or may not affect the workers' productivity, but it will affect their job satisfaction. As much as possible, a good supervisor or manager will seek to break down barriers that stand in the way of communication. Workers seldom initiate the action.

Reconciliation of differences requires that supervisors seek out and emphasize the many areas that they and the employees have in common. Focusing on areas of common interest while tolerating differences that do not affect the business can eliminate, or at least minimize, the number of problems that arise due to common differences.

TRY THIS Create and write down a workplace scenario that could cause relational problems at a business. Describe actions or qualities of a manager that could help solve or avoid the situation.

Matching the Employee with the Job

If the company has an up-to-date organizational chart, then each position in the company should have a job description developed specifically for it. Before placing a new or current employee into that job, management personnel should determine if the person has the right combination of personal and professional skills to be successful. Both the employee and the company suffer when there is not a proper match between the person and the needed job performance.

PERSONS WITH A ...	MIGHT PERFORM SUCCESSFULLY AS ...
Positive, aggressive personality	Sales representatives, recruiters, personnel trainers
Positive, reserved personality	Team players, crew workers, department managers
Negative, aggressive personality	Groundbreakers to open new territories or cut through red tape
Negative, reserved personality	Employees whose work is not directly tied to a team, such as shop mechanics or accountants

Figure 54–2 A match between personality types and job functions can contribute to a highly functional business.

Whether measured during the hiring process or observed on the job, the personality of an employee can be categorized. At the risk of oversimplification, but to illustrate the point, we all may know people who are characteristically positive and upbeat, just as we may know others who are usually reserved and often negative about life or situations. We can all think of people we have known who are aggressive, take-charge types, and others who are more comfortable as followers or working alone. To exemplify the match between personality types and job performance, consider the following possibilities described in Figure 54–2.

Although these examples oversimplify the complexities of personality, they serve to illustrate the concept of fitting employees to the job that offers them the best opportunity to enjoy their work and perform it in a way that benefits all members of the customer-company-employee triangle.

Customer Relations

The other major group of people in the human relations triangle is the customers. They are the people who keep the business in existence. Their goodwill must be cultivated as carefully as the horticulture crops they purchase. Goodwill and sales are intimately related business objectives, and the sales staff must be trained to promote them simultaneously. Personnel on the sales staff must have the right personal relations skills to be successful in the job.

Characteristics of Customers

The type of customer to be served will depend first on whether the business is retail or wholesale. Retail sales are likely to be directed to homeowners and other nonhorticulturists who will purchase items in comparatively small quantities. Wholesaling usually means that the customers will be other industry personnel.

Although there are differences in the ways products and services are marketed to the retail and wholesale markets, wholesale and retail buyers are not as different as might be expected. Some know exactly what they want and only need assistance obtaining it. Others browse, surveying the entire inventory to generate an idea of what, if anything, to purchase. In between is the customer who definitely plans to purchase an item or two but is willing to purchase more if something looks interesting.

The second factor determining the type of customer to be served is the economic class toward which the business is directed. If the business is known for costly products and services, certain types of customers will shy away, and others will seek it out. If low-priced merchandise is promoted, the same is true. A business located in a middle-income neighborhood can expect to serve homeowners and amateur gardeners who seek good-quality and familiar merchandise at affordable prices. A more affluent clientele will expect top-quality and often unusual merchandise with attentive service, and they will expect to pay a premium price in return.

In any case, two motivations bring a customer to a horticulture business: need and desire. Need is the most direct motivation; it has an urgency about it. As soon as the customers believe they can afford the item or service, they move to satisfy the need by going to the shop or calling on the telephone. New homeowners might regard the establishment of a new lawn and a foundation planting as a need for their property. They may have the desire for a maintenance firm to care for the landscape, but they may not be sure they can afford it. Still they may call for an estimate.

Characteristics of Good Salespeople

Customers deserve assistance from a **good salesperson** who possesses the characteristics described in the following sections.

FRIENDLINESS Customers should be greeted soon after walking through the door or into the sales yard. If the customer's name is known, it should be used. Not every customer warrants a handshake, but it can be appropriate with a regular customer. Salespeople need to smile easily and frequently and present themselves as approachable to the customer. Personal problems, boredom, or business problems must not be transferred to the customer in the form of a negative personality, a short temper, or a short attention span.

HELPFULNESS Customers should be given a short period of time to get their bearings in the store or sales area and to look around. Browsers are easily spotted, usually by their slow pace and meandering ways. Customers who know what they want usually seek out a salesperson soon after entering.

When approaching a customer, too many salespeople ask, "Can I help you?" Such a question is unnecessary and has also become a cliché. A better approach to a customer is "How may I help you?" The customer's response can then determine further action or inaction by the salesperson.

KNOWLEDGE Salespeople should be familiar with the plants and products that are being offered for sale. They should know the answers to questions that customers are likely to ask; for example, "How big does this plant get?" "What are the light requirements?" "Is it safe to spray this around children and pets?" In addition to responding to customer questions, the sales staff should be able to suggest appropriate products and services to supplement those initially sought by the customer. This should increase sales volume along with customer satisfaction.

HONESTY The salesperson must be honest in describing how a product or service will fulfill the client's need or desire. Delivery of the product when promised and starting the service on time are also reflections of the salesperson's honesty in the mind of the customer.

PHYSICAL PRESENTATION AND ARTICULATE SPEECH Salespeople must be dressed to meet customers and create a good impression on behalf of the firm. Overdressing can make the customer uncomfortable; a sloppy or soiled appearance can discourage customers. Conservative dress is usually the wisest approach in business. Equally important is the salesperson's ability to speak clearly, accurately, with good grammar, and without profanity, slang, or verbal clichés. Even conversations meant to be private between employees can be heard by customers and should maintain appropriate language.

COURTESY Salespeople must be polite and positive with customers, even if they do not return the courtesy. It takes little effort to say please and thank you but it leaves an impression of caring and civility with the customer. Addressing customers by name or as sir or ma'am tells them that their importance to the success of the business is recognized.

Making the Sale

A greeting and offer of assistance should be made to customers soon after they enter the store, greenhouse, or sales yard. When the first contact is by telephone, the offer of assistance is immediate. The salesperson must then determine the needs of the customer and set about to satisfy those needs. The approach must be positive so that the customer feels good about making the purchase. The features of the product or service must be described at whatever level the customer requires for complete understanding. Obviously, the description used for a retail customer is different from that used for a wholesale horticultural buyer. The salesperson would also attempt to interest the customer in products or services that support or are related to the initially needed item. The attempt should not be high pressured or appear mercenary. Emphasis should be placed on how the customer will benefit from the purchase, not on how the company or salesperson will profit. The customer should be encouraged to ask questions, and the salesperson should answer them directly. Where the sale is potentially sizeable, the salesperson should be prepared with information regarding financing possibilities. If a service is being sold rather than a product, the salesperson must be skilled at helping the customer visualize the service. For example, photographs of typical lawns or landscapes or illustrations of plants or flowers can be shown, or quick sketches can be done as the client watches.

Successful sales result in satisfied customers who (1) get what they came for and perhaps even more, but (2) do not spend more than they should have spent, and (3) will return again for future purchases.

Every Employee Is in Customer Service

One of the most critical concepts to convey to the employees of a horticultural enterprise is that everyone is involved in **customer service** (Figure 54–3). While salespeople and on-site foremen readily grasp the concept, others who do not have direct contact with customers may not understand. The leaders and managers of the business must convey to all employees that whether answering the telephone, digging a plant in the field, repairing a chainsaw in the shop, moving the cup on a golf green, or unpacking a case of cut flowers, each is contributing to the ultimate satisfaction of the person who will receive the product or service in its final form.

Career Development

Two FFA Career Development Events (CDEs) that build the skills mentioned in this unit are the Job Interview CDE and the Agricultural Sales CDE. The National FFA Job Interview CDE is designed for FFA members to develop, practice, and demonstrate skills needed for seeking employment in the agriculture industry. Each part of the event simulates, as closely as possible, real-world activities that will be used by real-world employers. The Job Interview CDE is developed to help participants in their current job search (for SAE projects, internships, and part-time and full-time employment). Therefore, the participants need to create a cover letter, résumé, and references that reflect their current skills and abilities and must be targeted to a job for which they would like to apply. They will then perform a live interview with a judge to determine the quality of their interview preparation,

▲ **Figure 54–3** Successful customer service is only achieved when the employees care about the customer's needs.

The purpose of the National FFA Agricultural Sales CDE is to evaluate skills that are essential for an individual to be successful in the agricultural sales field. The process of selling agricultural products is essential for production and marketing of agricultural products. The following are skills that will be developed through this CDE:

- Develop verbal, written, and interactive communication skills.
- Discuss features and benefits of a product.
- Identify potential customer objections.
- Demonstrate knowledge of proper product use.
- Identify prospective customers through marketing data.
- Introduce the product to prospective customers.
- Develop a sales call that determines and addresses customers' needs and objections.
- Understand the basic business structure necessary to sell and deliver a product.
- Attempt to close the sale by asking for customer's buying decision.
- Establish and build customer confidence in the product.

Preparing and competing in either of these CDEs will prepare students to find a career and be successful in a horticultural business. These are lifelong skills that FFA members use throughout their entire careers.

SAFETY FIRST

OSHA's regulations ensure that employers and employees know about work hazards and how to protect themselves to prevent injuries in the workplace. The owner and manager of a business should be knowledgeable about OSHA regulations and how to keep the working environment safe. For more information on occupational health and safety in the workplace, please visit the OSHA site at www.osha.gov.

Summary

Human relations in a horticultural business involve the management, the employees, and the customers in an interdependent association.

Personnel management is the direction of workers in a manner that brings out their best efforts and attitudes on behalf of the business. Good supervisors are able to set realistic expectations of their employees, taking into account each worker's individual abilities and limitations, educational and work experience, and potential. Employees may be career-directed, part-time, or temporary seasonal.

Most people respond in a positive manner to a supervisor who exemplifies decisiveness, competency, fairness, sincerity, understanding, and respect for the employees, offering praise and reward when warranted and delivering directions thoroughly. Employees can also be expected to react negatively to a supervisor who frequently displays inflexibility, partiality, a condescending attitude, indifference, or sarcasm, or one who gives subjective criticism or applies double standards in adherence to company policies.

Communication can be complicated by differences of age, education, sex, ethnic background, or status within the company. The supervisor must usually seek to open the channels of communication when problems arise, rather than waiting for the problems to resolve themselves.

Because sales are intimately related to customer goodwill, the sales staff must be trained to promote both simultaneously. Good salespeople offer a customer friendliness, helpful assistance, knowledge of the materials and services sold, honesty, and courtesy, and they exhibit good grooming and proper grammar when speaking. Successful sales are those that result in satisfied customers who got what they came for and perhaps even more, did not spend more than they could afford, and will return again for future purchases. The FFA provides several Career Development Events that prepare students for careers in the horticultural business sector.

Review Questions

1. Define personnel management.
2. Discuss the characteristics of career-directed employees, part-time employees, and temporary seasonal employees.
3. Discuss the ways in which customers may differ depending on the type of horticulture business.
4. Discuss the different qualities that make a good and poor supervisor.
5. Describe the procedures for effective selling. How do the qualities of the salesperson help determine a positive outcome?

Student Activities

1. Assume that you are preparing to conduct a training session on customer service for selected employees of the company. Attending the training session will be the receptionist, accountant, delivery person, sales representative, designer, and shop mechanic. Prepare a brief essay explaining how you would convince each of them of the important role he or she plays in customer service.
2. Prepare and participate in the Agricultural Business Management CDE through the FFA organization.
3. Prepare and participate in the Agricultural Sales CDE through the FFA organization.
4. Conduct a mock sales situation between staff and a customer to demonstrate the proper treatment of a customer. Prepare and present with another student in the class.

Suggested Supervised Agricultural Experience (SAE) Activities

1. As a placement SAE, work as an intern in the human relations department of a large horticultural business.
2. Work with a retail manager to learn how to become a retail manager at a garden center.

Additional Resources

www.growcom.com.au/_uploads/15640Human_resource_and_indstrial_relations.pdf

www.lakeland.cc.il.us/as/as/academicprograms/model_pdfs/02%20TechPrgrms_022814WEB%2060.pdf

UNIT

55 Business Communications

Objective

To demonstrate appropriate forms of business communication.

Key Terms

letterhead
salutations
website

Competencies to Develop

After studying this unit, you should be able to:

- select and explain the reason for the most appropriate form of communication for specific business needs.
- compose a correct and effective business letter.
- place a business telephone call correctly.
- answer a business call correctly.

The Purpose of Business Communications

Although there are various types of horticultural businesses, most are very customer oriented. Unlike many other businesses in which the final recipients of the company's products are seldom seen by the persons who produced them, the customers of florists, nurseries, greenhouses, and landscape firms are frequently seen and many times known by employees of the company. Even the suppliers and competitors of horticulture companies are often familiar names and faces to members of the staff. That familiarity is an asset because it minimizes misunderstanding. With a smile, a handshake, a specific inflection of the voice, or the reply to a question, a person is able to convey or clarify a message. Take away one or more of those attributes of the one-to-one contact, and problems may arise. Without an accompanying smile or the touch of a handshake, a written agreement can sound suspicious. Without the opportunity to ask a question, a mistake can be made in filling an order. Without understanding the sensitivity of a person to the propriety of how a message is sent, an unintentional affront can be felt. Therefore, business communications become extensions of the people who represent the company. They must be done with clarity and professionalism if they are to come close to simulating the personal contact that they replace.

The Technology of Business Communication

Today's horticulture business makes full use of the technology that has swept across the country in recent decades. Computer systems, high-speed networks, online management software, digital billing and payment programs, smartphone apps, social media with virtual communities and networks, and many other advancements have changed the way businesses operate. Telephones answer themselves and both send and receive messages automatically. They identify callers before being answered. They have escaped the constraints of wires and can now be found in cars and trucks, back pockets, and in the center of conference tables. Their fiber optic lines can be used to send and receive faxes of printed documents. E-mail enables businesses to contact suppliers and customers as well as to respond to their queries. The Internet gives every company the opportunity to showcase its products and services to a worldwide audience as well as to receive orders and payment electronically. Still, it must be remembered that a voice mail system cannot answer a question, a cell phone cannot differentiate between a friendly call and a nuisance call, and an e-mail cannot smile or shake a hand. With the many advancements in communication technology, there comes the danger of overlooking the personal connection; however, maintaining that personal connection is vitally important, even with the convenience of technology.

Written Communications

Business correspondence is done both on paper and electronically. Because both forms can be saved, copied, and forwarded, they have certain features in common. They also have differences in the appropriateness of their use.

Using E-Mail

Few innovations compare to e-mail in both its popularity as a form of written communication and its convenience. Within seconds of sending a message, it appears in the e-mail inbox of its recipient. It enables the sender to avoid the often annoying experiences that accompany voice mail systems and/or protective receptionists who run interference for their employers. However, there are several precautions to remember when using e-mail in business. First, e-mail is seldom a suitable substitute for a written letter if and when that letter is contracting or obligating the business. Second, the immediacy of its impact depends on how often the recipient checks for new messages. Third, because it is electronic, the message can be forwarded almost anywhere in the world without the knowledge of the original writer. What is written in an e-mail is documented and can be used in a positive or negative way. Wording of e-mails needs to be carefully considered before sending the message. Always re-read an e-mail before sending it. Finally, the impersonal nature of electronics devalues the sincerity of certain types of messages in the minds of some people. For example, if the recipient of the e-mail message receives an overload of junk mail messages, he or she may not recognize the importance of one sent by a company in need of a timely reply and does not open it to read. Also, e-mail should never be used for thank-you notes or messages that have an emotional tone, such as congratulatory or sympathy notes. Those necessitate a handwritten note sent through the USPS. Some of the personal aspects of life and business must retain the human touch.

TRY THIS With the permission of the instructor, create a formal business e-mail that is intended to attract a new customer to the business. Send the e-mail to your instructor for review.

Writing Business Letters

Unlike conversations, business letters are permanent records. When a request, an offer, or an order is made in writing, it becomes a form of contract. Prepared with care, a letter expresses clearly and correctly the message of the writer. Prepared carelessly, the letter can cause error, allow for omission or misinterpretation, or unintentionally offend.

As a tool of business, a letter must possess the following characteristics:

- correct mechanics
- conventional format
- polite and friendly tone
- thorough coverage of its subject(s)
- accuracy
- brevity

Correct Mechanics

All business correspondence should be typed on good-quality white- or cream-colored paper with a printed **letterhead**. The letterhead should not occupy more than one-fifth of the page and should contain the name, address, and telephone number of the company. It may also contain a logo, piece of line art, fax number, and e-mail and/or website address. Care should be taken to ensure that the letterhead is not overcrowded or gaudy. The purpose of stationery is communication, not advertising.

The majority of business letters in America today are prepared on a computer using a word processing program. All modern word processing programs have many different formatting options and even contain embedded templates for documents such as business letters. An understanding of appropriate formatting is necessary to prepare a business letter. The most common basic style used is full block style, which aligns the components of the letter along the left margin.

Proper form requires that a business letter have the following components, placed and spaced as noted (Figure 55–1).

DATE The date line varies in its location depending on the format and length of the letter. It is typed beneath the letterhead leaving enough space to balance the letter without crowding at the top or bottom.

INSIDE ADDRESS The full name and address of the person(s) and firm you are writing to should be entered against the left margin at least four lines under the date line. The inside address should not extend beyond the center of the page.

The name and title or position of the person you are writing to should be included if you know it. Customers are usually addressed as the following:

Mr.	Miss
Mrs.	Ms.
Mr. and Mrs.	Messrs.

A title follows the name in the inside address. For example:

Mrs. Irene Clark, Branch Manager
Green-Gro Supplies
213 Carriage Road
Chicago, IL 12345

SALUTATION If a business letter had a smile and a handshake, it would be the salutation, which is spaced two lines below the inside address and punctuated with a colon. It should

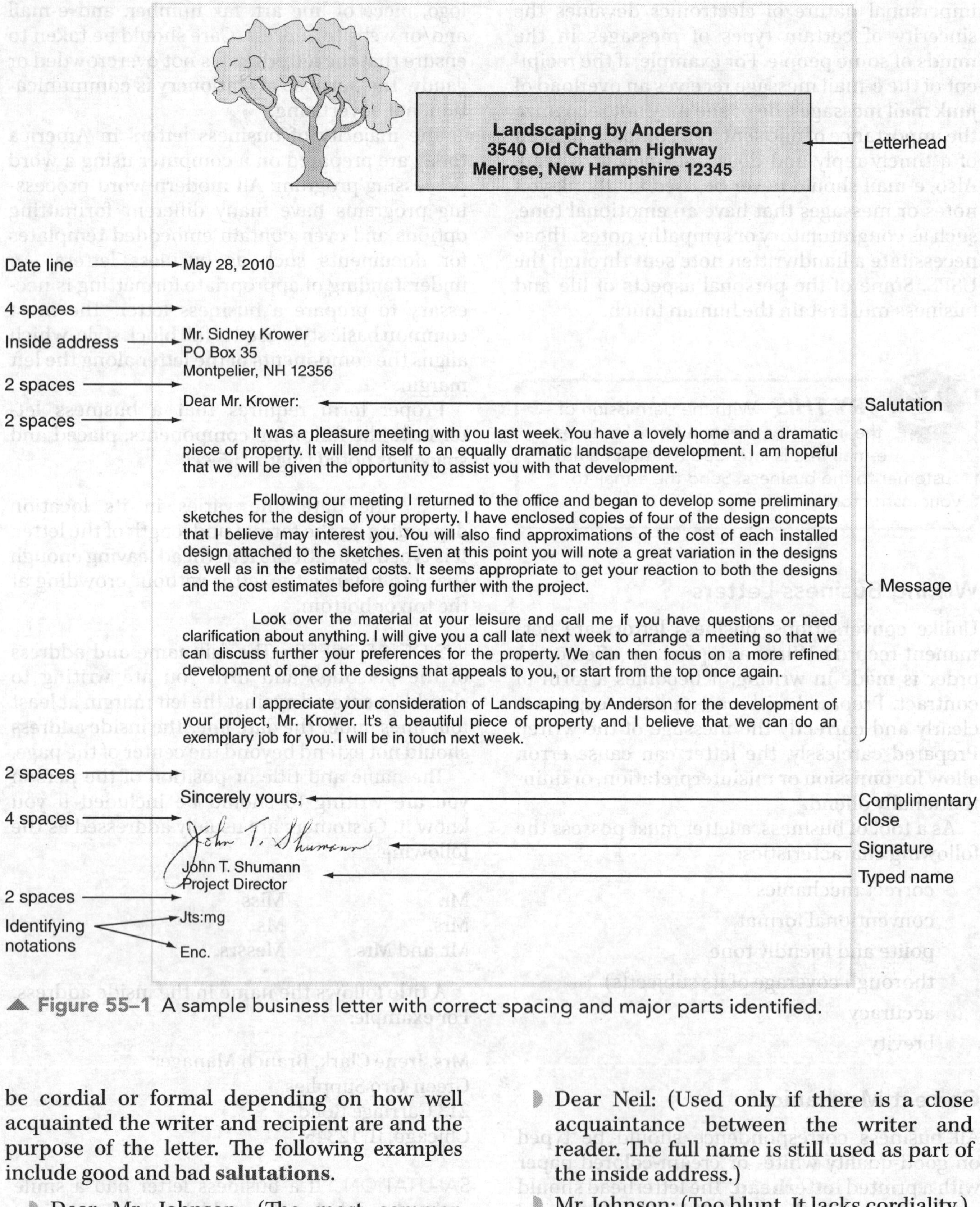

Landscaping by Anderson
3540 Old Chatham Highway
Melrose, New Hampshire 12345

May 28, 2010

Mr. Sidney Krower
PO Box 35
Montpelier, NH 12356

Dear Mr. Krower:

It was a pleasure meeting with you last week. You have a lovely home and a dramatic piece of property. It will lend itself to an equally dramatic landscape development. I am hopeful that we will be given the opportunity to assist you with that development.

Following our meeting I returned to the office and began to develop some preliminary sketches for the design of your property. I have enclosed copies of four of the design concepts that I believe may interest you. You will also find approximations of the cost of each installed design attached to the sketches. Even at this point you will note a great variation in the designs as well as in the estimated costs. It seems appropriate to get your reaction to both the designs and the cost estimates before going further with the project.

Look over the material at your leisure and call me if you have questions or need clarification about anything. I will give you a call late next week to arrange a meeting so that we can discuss further your preferences for the property. We can then focus on a more refined development of one of the designs that appeals to you, or start from the top once again.

appreciate your consideration of Landscaping by Anderson for the development of your project, Mr. Krower. It's a beautiful piece of property and I believe that we can do an exemplary job for you. I will be in touch next week.

Sincerely yours,

John T. Shumann

John T. Shumann
Project Director

Jts:mg

Enc.

▲ **Figure 55–1** A sample business letter with correct spacing and major parts identified.

be cordial or formal depending on how well acquainted the writer and recipient are and the purpose of the letter. The following examples include good and bad **salutations**.

- Dear Mr. Johnson: (The most common business salutation. It is both cordial and formal.)
- Dear Neil: (Used only if there is a close acquaintance between the writer and reader. The full name is still used as part of the inside address.)
- Mr. Johnson: (Too blunt. It lacks cordiality.)
- Dear Mr. Neil Johnson: (Stiff and unnatural. When the customer is addressed with a title

such as Mr., the first name is not included in the salutation.)

- Neil Johnson: (Blunt and stiff. It lacks both courtesy and cordiality.)
- Dear Neil Johnson: (This is sometimes a good solution when people's names give you no clue to their sex, and you do not know them personally.)

BODY The message to be conveyed is in the body of the letter, which begins two spaces beneath the salutation. To fill the page most attractively, the top and bottom margins should be the same, as should the side margins. Brief letters of only two or three sentences usually look best if lines are double spaced, but most letters are single spaced. A double space may be left between each paragraph.

COMPLIMENTARY CLOSE Two spaces beneath the body of the letter is the courteous ending of the communication. The first word is capitalized, and a comma is used at the end. Its degree of formality should correspond with the rest of the letter. If the salutation has been formal, such as "Dear Mr. and Mrs. Andrews," then a similar closing is appropriate. If the letter concerns an overdue account, the closing should not be chummy. Consider the following examples:

- Sincerely,
  Sincerely yours,
  Yours sincerely,
  (used in most regular business correspondence)
- With best regards,
  Cordially,
  (used when a tone of informal friendliness is desired)

One correspondence gimmick seen too frequently is running the last sentence of the letter into the complimentary close. For example:

In the hope that we can be of service to you in the future, I remain

Sincerely yours,

The final sentence of the body should be independent of the complimentary close.

SIGNATURE The signature is handwritten. It should be legible, not a scrawl, despite what you may have seen. Four spaces are allowed for the signature. If the name of the firm is included, it should be typed in capitals two spaces beneath the complimentary close and above the signature. The name of the writer is typed below the signature. If the writer has a title and chooses to use it, it may be added beneath the typed name. If the name of the firm precedes the signature, then the firm assumes responsibility for agreements or statements made in the letter. Obviously, not everyone corresponding on business stationery is empowered to speak for the company, so the form of the signature is important. For example:

- Sincerely,
  CHEMGREEN LAWN SERVICE
  Todd R. Martinson
- Sincerely yours,
  STACEY LANDSCAPE CONTRACTING
  Linda K. Stacey
  President
- With best regards,
  Terry A. Forsyth
  General Manager

NOTATIONS Business letters usually carry one or more abbreviations to identify the person who composed the letter, the secretary who typed it, and other persons who will receive copies of it and as a reminder that materials were enclosed with the original letter. For example:

- JEI: Identifying initials placed in the lower-left corner of the last page of the letter. They are either aligned opposite the typed signature or title or placed two spaces lower. The first set of initials belongs to the person who composed the letter. The second set belongs to the secretary or typist.
- Enc. or Enc. (3): An abbreviation of the word enclosure is used to indicate that something has been included with the letter. If more than one item is enclosed, the number of items may follow in parentheses. This notation is made two lines beneath the identifying initials and against the left margin.
- cc or xc: The initials cc mean carbon copy, and the initials xc mean exact copy. They are followed by the name(s) of the person(s) receiving copies of the original letter.

The notation is spaced two lines beneath the enclosure notation or, if there are no enclosures, two lines beneath the identifying initials.

- P.S.: Postscripts are sometimes used in sales letters to focus additional attention on a point made in the body of the letter. As such, they are a gimmick. They should not be used in formal correspondence to add a point forgotten in the original letter. In such a case, the letter should be rewritten and typed again. Postscripts follow two lines beneath any of the above notations.

ENVELOPE Two envelope sizes are commonly used in business correspondence: 3⅝ × 6½ inches and 4⅛ × 9½ inches. The smaller size is frequently used for billing, and the larger is used for letters. Other sizes exist for other purposes such as when folding is undesirable or when oversized enclosures are included. The usual business envelopes should be of the same color and quality of paper as the stationery they contain. Some have a cellophane window to allow the inside address on the letter to show through, thereby eliminating the need for typing on the envelope. If a windowed envelope is not used, the full name and address of the individual or firm you are writing to must be typed, single spaced, on the envelope (Figure 55–2). Business envelopes should also have your firm's return address printed in the upper left corner.

The smaller envelopes require that 8½ × 11-inch stationery be folded three times before insertion. Larger envelopes usually require only two folds in the letter. In the two-fold technique, the paper is folded in thirds lengthwise (Figure 55–3). In the three-fold technique, the paper is first folded in half lengthwise, then in thirds perpendicular to the fold.

Setting the Tone of the Letter

Nearly all correspondence sent by horticulturists goes to suppliers, customers, potential customers, or members of the profession, and requires the same courtesy and concern that would be extended in a face-to-face meeting. Because there is no opportunity for facial expression or voice inflexion to support or explain the words, the letter must be phrased especially carefully.

Courtesy is a quality of every good letter and can be incorporated even when firmness is required, as with a collection letter for an overdue account.

Conversational phrasing will allow the letter to be read easily and with understanding. Business letters need not sound like insurance policies or tax forms, even when legal matters are their subject.

▲ Figure 55–2 A correctly addressed envelope. The typing is centered between top and bottom and on the right side of center.

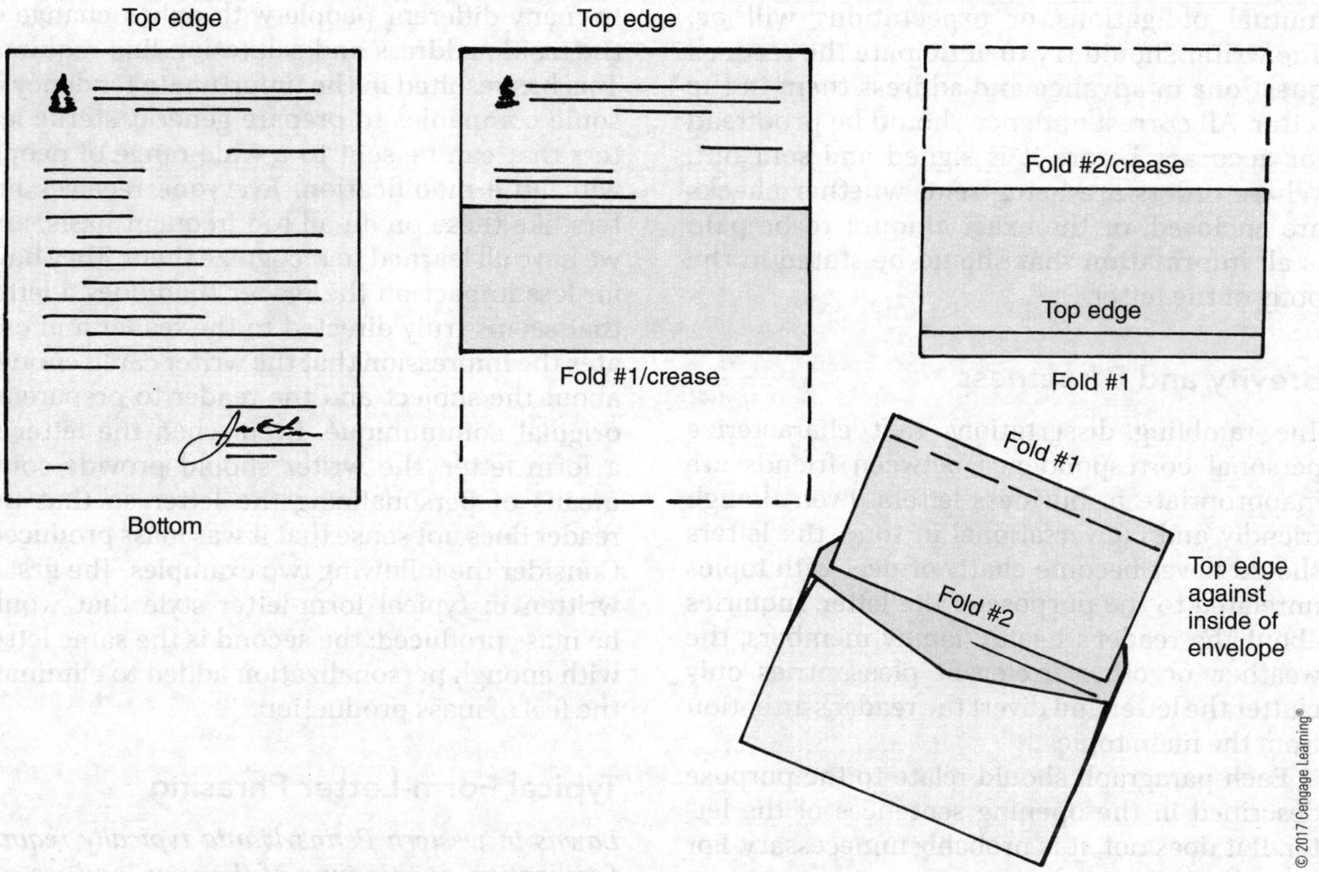

▲ **Figure 55–3** Correct preparation of a two-fold letter and placement into an envelope.

Concern for the reader rather than the writer will encourage a better reception. For example, the words "Due to a backlog of work, we will be unable to install your new lawn until sometime in May" create certain disappointment in the customer, who is not concerned about the backlog of work but is concerned about the lawn. This customer may seek another firm to provide the satisfaction desired. A better way to say the same thing is, "We appreciate your contacting us to install your new lawn and will be there during the first week of May to do the job." Disappointment is not built into the phrasing, and the importance of the customer is stressed. Similarly, the words "Please drop in any time from 9 A.M. to 5 P.M. Tuesday through Saturday so that we can be of service" are more inviting than "Closed Sunday and Monday. Open Tuesday through Saturday."

Thoroughness and Accuracy

In a conversation, the listener can ask the speaker to clarify or complete statements that are not fully explanatory. In a business letter, no such opportunity exists. Therefore, the writer must be certain that the topic of the letter is covered thoroughly and accurately. Failing to state a price or stating an incorrect price may lose a customer or obligate the firm to a loss.

The opening paragraph of the letter should state the purpose of the letter and define the subject(s) to be covered. If a review of the background is needed, it should be included in the first paragraph and kept as brief as possible. For example:

- "I am writing because we have had no response to our earlier requests for payment of your overdue account."
- "Last fall, you requested an estimate of the cost for design and installation of your patio. If you are ready to proceed with the project this spring, we are eager to be of service."

Each paragraph in the letter should aid the reader's understanding of what the writer is requesting or proposing, and what their mutual obligations or expectations will be. The writer should try to anticipate the reader's questions in advance and address them in the letter. All correspondence should be proofread for accuracy before it is signed and sent out. Where orders are being sent, whether checks are enclosed, or the exact amount to be paid is all information that should be stated in the body of the letter.

Brevity and Directness

The rambling dissertations that characterize personal correspondence between friends are inappropriate in business letters. Even though friendly and conversational in tone, the letters should never become chatty or deal with topics unrelated to the purpose of the letter. Inquiries about the reader's health, family members, the weather, or other irrelevant pleasantries only clutter the letter and divert the reader's attention from the main topic.

Each paragraph should relate to the purpose described in the opening sentences of the letter. If it does not, it is probably unnecessary. For example:

- "I am interested in purchasing a new Ajax Tree Digger, Model R17. I saw it demonstrated recently at a trade show in Wichita and need additional information about its capabilities.
- The Tree Digger should enable our firm to double our digging operation without any increase in personnel. As you know, every dollar saved these days is important.
- Please have a sales representative call on me here at the nursery. Because I need the machine in time for spring digging, I shall appreciate your prompt attention to my request."

The second paragraph is of no importance to the purpose of the letter, which is to obtain more information about the tree digger. It should be eliminated.

Personalizing the Letter

Word processing technology permits a quick and easy correction of mistakes and omissions. It also makes it possible to send identical letters to many different people with only a change of the inside address and salutation line required. This has resulted in the unfortunate tendency of some companies to prepare generic, sterile letters that can be sent to a wide range of people with little modification. Everyone receives letters like these on an all too frequent basis, and we have all learned to recognize them. They have far less impact on the reader than does a letter that seems truly directed to the reader and creates the impression that the writer cared enough about the subject and the reader to prepare an original communiqué. Even when the letter is a form letter, the writer should provide some means of personalizing the letter so that the reader does not sense that it was mass produced. Consider the following two examples. The first is written in typical form-letter style that would be mass produced; the second is the same letter with enough personalization added to eliminate the feel of mass production.

Typical Form-Letter Phrasing

Lawns in western Pennsylvania typically require fertilization at this time of the year, and we are prepared to do the job. If you believe we can be of service to you, please give us a call at (827) 555-1234 for a free estimate.

Personalized Phrasing

It is time to fertilize the lawns of Pittsburgh, and we want to help make your Grandview Avenue residence a showplace of your neighborhood. Mr. Jones, please give us a call at 555-1234 so that we can provide you with a free estimate.

While the differences are slight, the second example inserts the reader's name, city, and street, easily lifted from the mailing address. It also provides a local telephone number without an area code. The point to note is that the writer needs to be careful that business letters don't lose their personal touch and become stiff and formal in tone.

Telephone Communication

Use of the telephone for business transactions has advantages and disadvantages. Like business letters, the telephone does not permit eye-to-eye contact or a smile and handshake between the horticulturist and the customer or business associate. The telephone allows questions to be asked when clarification is needed that the business letter does not. However, a telephone call cannot be proofread as a letter can. What is said, in haste or error, is heard immediately. Also, a letter does not have a tone or irritating voice pitch that can create a negative impression in the listener's ear. Neither does it proceed too rapidly or slowly, requiring time-consuming repetition for total understanding.

Certain types of horticultural operations depend on the telephone as a sales tool more than other operations do. For example, retail florists do a large percentage of their order taking by telephone. A garden center or lawn care firm is likely to use the telephone more often for answering customer questions and receiving requests for estimates. In certain situations, horticulturists and their employees are more likely to initiate the phone calls. In other situations, they tend to receive them. In either case, the voice on the telephone becomes the personality of the firm. It must be pleasant and interesting to hear, and the person behind the voice must be knowledgeable, interested in the person at the other end of the line, professional in manner, friendly, attentive, and courteous. The voice should be directed into the mouthpiece of the telephone, and the receiver should not drop beneath the chin. Persons who are naturally monotonic should practice developing a more expressive voice, possibly through the use of a tape recorder, if they want to make effective use of the business telephone.

Placing a Business Call Correctly

To place a business call correctly requires preparation. The caller must have a clear understanding of what the call seeks to accomplish, what points are to be covered, the sequence in which they will be covered, what information is to be obtained or released, and what commitment is to be sought or offered on behalf of the firm. Irrelevant chatter should be avoided on the assumption that the receiver of the call is busy and will not want to waste time.

The following is a checklist for making a business telephone call:

- Be certain of the number (including area code) before dialing.
- Have the name of the person and firm you are calling in mind.
- Allow ample time for answering. Ten rings are usually recommended by most telephone companies.
- Identify yourself and your firm immediately. For example: "Good afternoon. This is Brian Miller of Miller's Greenhouses."
- Ask to speak to a specific person if possible. Otherwise, briefly identify the reason for the call, and ask to speak to someone who can assist you. For example: "I'm calling about the possibility of carrying the Ajax line of giftware in our shop. May I speak with someone on your sales staff, please?"
- Provide information and ask questions in an organized, sequential manner, and at a rate of speed that permits the listener to comprehend and record as necessary.
- In closing, summarize the major items discussed during the conversation to ensure mutual agreement and understanding of what the call accomplished.
- Close the conversation in a polite and friendly manner as briefly as possible after the purpose of the call has been accomplished.

A business call should not be made when other things require the caller's attention. The call should wait until you can proceed without distraction or interruption. If some unexpected event makes it necessary for you to leave the line for a moment, arrange to call back at another time because no one enjoys being put on "hold." Hold lines are too often overused and may alienate more business than they serve.

Answering a Business Call

An incoming call is much less predictable than an outgoing call. It may be a customer placing an order, making an inquiry, or registering a complaint; a supplier offering information about an order; or a garden club seeking a guest speaker. Equally unpredictable are the telephone skills of the callers. They may offer their questions or information in a disorganized or emotional manner. Dealing professionally with all callers, including rude or irate callers, requires the ability to listen, comprehend, and respond with a voice that can convey friendliness and concern.

The ability to listen, hear, and understand is not as common as you might think. Employees assigned to answer business calls must often learn to listen more attentively than they usually do. Otherwise, the telephone conversation can become a series of needless requests for repetition.

In preparation for receiving business calls and making them as productive as possible, the following points are worth noting:

- Be knowledgeable about the business, its inventories and services, its pricing structure, the amount of time required to accomplish customer requests, and the dates by which they can be accomplished.
- Have a pad of order forms, a blank pad of paper, a price list, and sharpened pencils by the phone.
- Answer the phone as quickly as possible and by no later than the third ring.
- Identify the business first, and then give your own name. If not too cumbersome, add an appropriate salutation. For example: "Good afternoon. Larry's Lawn and Garden. This is Janice." "Fountain View Florist. Rod Johnson speaking." "Good morning and thanks for calling Gateway Nursery. This is Barry Byrne."
- Determine the name of the caller at the outset of the conversation. If callers do not identify themselves, ask. If a name or firm is not given distinctly, ask for a repetition.
- Listen intently, shutting out other distractions, to determine the purpose of the call, the needs or concerns of the caller, and the disposition of the caller (e.g., friendly, upset, angry, or confused).
- Avoid interrupting the caller. Resist the urge to complete sentences for a slow talker or direct the thinking of an uncertain customer before you fully understand the situation.
- Ask questions to ensure that you understand the caller's message clearly. Do not make assumptions.
- Take notes to ensure that you remember important points and that commitments made are honored.
- Be courteous and positive, even if the caller is rude, profane, or irritating. Use the caller's name occasionally to personalize the conversation. Even if the conversation does not lead to an agreement, it can be conducted courteously.
- Price must be discussed if a request for merchandise or services is the purpose of the call. Never assume that a customer will pay a price unless it has been discussed and agreed upon.
- Review the major points of the conversation (e.g., the item[s] ordered, specific preferences for colors or style, date needed, when workers or sales representatives will call, and the price agreed on).
- Close the conversation courteously and allow the customer to hang up first. Nothing is gained from an extended conversation. It merely ties up the phone lines and delays the next business call.

Using Telephone Technology Properly

Alexander Graham Bell would be amazed at how his invention has evolved. Telephone services are a vital part of every horticulture business today. They link employees to customers, employees to other employees, and businesses to suppliers. Services are offered by a diversity of providers that compete against one another with a wide range of enticements that promise to make a company faster, more efficient, more productive, more cost effective, and so forth. Blind acceptance of these claims by a business can lead to the belief that customers will be better served after the latest techno-wonder system is in place. Often that is not the case. The reality is overdependence on automated phone systems can actually turn some customers away. Automated phone systems can be very helpful for a business that fields many calls at the same time. It identifies the nature of the call and directs the caller to the correct personnel. Any automated answering system should link the customer to a person as quickly as possible. Using a cell phone should be avoided for important business calls. When receiving a call that is made incomplete by loud crackles and missed words due to poor signal, customers have a right to be annoyed. Here are some tips on using telephone technology in a business setting:

- Avoid answering machines during business hours whenever possible. Provide sufficient and trained staff who can answer and address customer concerns. If a hold line is being used frequently, a larger staff is needed to do the job properly. Also, recognize that there is no value in having the telephone answered by a person who is unprepared to offer any assistance to the caller. "I'm sorry, Ms. Andrews is out. No, I don't know when she will be returning. I'm sorry; I don't know anything about that. Would you like to be connected to her voice mail?"
- If a caller must be placed on hold while the receiver searches for something, make the wait brief, and do not play music in the listener's ear.
- Avoid overdependence on answering services. The most they can offer is message-taking, and they have no knowledge of the business.
- Calls to customers made from cell phones are subject to breakups and should be conducted on a clear line at a time when the caller can give full attention to the customer and the purpose of the call.
- When making a call, be prepared for the possibility of having to leave a message on a machine. Too many callers stutter and stammer instead of leaving a meaningful message. The two most important parts of the message are often the parts delivered most hastily: the caller's name and return phone number. The name should be delivered slowly and spelled if necessary. The telephone number should be delivered the same way. Most likely the listener is trying to transcribe the name and number. If delivered too quickly or slurred, the call may never get returned.
- Fax messages use the telephone lines as well. Many businesses have preprinted forms that contain information similar to the letterhead of stationery. Before sending a fax, the sender should be certain that the message identifies the sender of the fax as well as the intended recipient. The return fax number of the sender should also be included if not part of the preprinted form.

Websites

If a company does not have a web page, it is both unusual and a sign to many people that it is behind the times. In an attempt to convince itself and its customers that it is modern and relevant, many green industry businesses develop web pages. If done well, the **website** can provide the viewer with a positive first impression of the business, explain its products and services, and offer examples of past accomplishments, with accompanying endorsements from satisfied former customers. Some websites will also provide a way for viewers to contact the company as well as place orders and make payments (Figure 55–4).

If constructed properly, the website can create a positive impression of the company in a way that no yellow page directory can match. However, many websites are of little use to the viewers, either because they are so abbreviated in their content that they offer little more than some colorful photos or because they are cluttered with insignificant material such as photos from the last employee picnic or lengthy listings of the credentials of employees whom

▲ **Figure 55–4** This florist is taking a computerized floral order. The system is linked to a remote station that allows the designer to complete the order.

the viewer neither knows nor cares about. Most Internet visitors to a company website have arrived there with an interest in learning quickly and concisely what the company does and how to access those products or services, and they may have questions about cost, scheduling, or other related concerns. The website needs to be easy to navigate and offer the information being sought, without undue and irrelevant clutter. It also needs to be kept current. Opening a web page in October and finding a greeting like "Spring is almost here and we can't wait to work with you!" is certain to convey a negative impression. Posting time schedules or price lists can also be dangerous if they are not kept current. Also having an e-mail "Contact Us" connection that encourages the viewer to send a message to the company is fine, unless no one at the company monitors the e-mail, and thus the potential customer is not responded to in a timely manner.

Websites should be developed by people who are trained and experienced. Most green industry companies would be wise to outsource the project to a professional website designer. Once developed, the site should be monitored frequently for accuracy and relevancy. Its organization should be logical for use, perhaps with multiple directories to help users find what they need and avoid what they are not interested in. Websites should also download quickly, which may necessitate simplifying some of the graphics.

If the website is to be truly communicative to present and potential customers, then designing it correctly is only the beginning. Once developed, it must then be managed continuously to attract new customers, communicate with current customers, and convey a positive impression of the company; otherwise, it is a pointless and unnecessary expense for the company.

The Language of Business Communications

As stated earlier, the horticulturist who communicates most effectively incorporates a friendly and conversational tone into letters and calls, and emphasizes the positive rather than the negative. In addition, horticulturists should learn to describe their products and services in ways that evoke enthusiastic and receptive responses. For example, a customer may call a flower shop before Christmas to order "something for the front door." A salesperson can suggest "an evergreen door swag" or "a cluster of fresh pine boughs with accents of red holly berries and pine cones, tied with a red velvet bow." Both descriptions are truthful and refer to the same product, but the second enables the customer to visualize the product better. In a similar fashion, a landscaper can describe a proposed patio as "800 square feet in area, made of brown brick, with a built-in grill and surrounding plantings" or as "800 square feet of patio living area done in brick of a warm earth tone and incorporating cooking facilities and cooling shade." The second description has the important quality of evoking the anticipation of use.

SAFETY FIRST

OSHA's regulations ensure that employers and employees know about work hazards and how to protect themselves to prevent injuries in the workplace. The owner and manager of a business should be knowledgeable about OSHA regulations and how to keep the working environment safe. For more information on occupational health and safety in the workplace, please visit the OSHA site at www.osha.gov.

Summary

Letters and telephone conversations are extensions of the ornamental horticulture business and must be handled professionally.

Business letters are the most permanent form of communication. Their permanency is their greatest advantage. A good business letter is written in correct form and conventional format, in a polite and friendly tone, with thorough and accurate coverage, and as briefly as possible.

Proper form requires a letterhead that is restrained and that contains the firm's name, address, and telephone number; a date line; inside address; salutation; body (message); complimentary close; signature; and notations.

Use of the telephone for business communication permits questions to be asked when clarification is needed. The development of an expressive voice is important to good telephone communication. The voice must be pleasant and interesting to hear, and the person behind the voice must be knowledgeable, interested in the caller, and professional in manner.

To place a business call correctly, the caller should know in advance what the call seeks to accomplish, what points are to be covered, the sequence of coverage, the information sought or to be released, and the commitment to be sought or offered on behalf of the firm.

Answering a business call requires good listening skills coupled with the ability to comprehend and organize information or questions as offered by a caller.

The Internet offers additional means of communicating with present and potential customers. The use of e-mail is fast and appropriate for certain types of communications, but it may not be appropriate for business communication that is emotional. Websites offer the opportunity for companies to display their products and services and communicate with customers in a way that suggests they are up to date and skilled. For that reason, websites should be carefully designed to be helpful to users and not waste their time or convey a negative image of the company.

The advances in communication technology offer many new variations on the traditional ways of communicating with customers and suppliers. Businesspeople need to select the most appropriate techniques and use them in a way that enhances their services to their customers.

Review Questions

1. Discuss situations when it is more appropriate to send a business letter instead of an e-mail and vice versa.
2. Describe the components of a business letter.
3. Why is it important to use correct mechanics and proper tone when composing a business letter?
4. What are some ways to personalize a business letter?
5. What are some important considerations regarding using an answering machine or an answering service?
6. Describe an effective website.

Student Activities

1. Assume that you are the general manager of Dakota Gardens, a retail nursery at 1504 W. Purdue Road in Bismarck, North Dakota. Write a letter to a customer, Mr. Tracy Kostecki, at 2600 Ball Avenue in Bismarck. The letter concerns his overdue account of $187.56, which has gone unpaid for 90 days. You have written to him twice already and tried to reach him by telephone, but he has not been available and has not returned your calls. Prior to this indebtedness, he has been a good and valued customer who has purchased and paid for much larger amounts of merchandise, always promptly. Your normal procedures are to turn bad accounts over to a collection agency after 90 days.
2. Practice your telephone skills by working with another person in the simulation of a business call. Before beginning, make up a realistic premise. Select a firm name and type of business. Select a customer or vendor name and problem. Make several calls, varying the problem (e.g., an order, a complaint, or a request for a bid). Sit facing in the opposite directions or on opposite sides of a screen that prevents any contact except by voice. The local telephone company may be able to provide a telephone simulator training kit for greater realism. School telephone systems also provide a way to conduct practice calls. If possible, record the conversations, and play them back for critiques.
3. Prepare for, and participate in, the FFA Agricultural Sales Career Development Event.
4. Prepare for, and participate in, the FFA Agricultural Communications Career Development Event.
5. Use the Internet to research examples of business letters and templates that could be used to construct an appropriate business letter.

Suggested Supervised Agricultural Experience (SAE) Activities

1. To qualify for a placement SAE, work in the business office of a local garden center. Responsibilities should include phone and e-mail communications.
2. As an entrepreneurship SAE, manage and operate your own lawn maintenance business operation. Record the communication strategies used to interact with customers.

Additional Resources

http://hbin.tamu.edu/greenhouse-firms/

www.safnow.org/

http://catalog.k-state.edu/preview_program.php?catoid=13&poid=3341

UNIT 56

Horticulture in the Age of Technology

Objective

To introduce the various types of technologies and marketing used in the horticulture industry.

Key Terms

CAD (computer-aided design)
digitizer
drones/unmanned aerial vehicles (UAVs)
sensory input devices
social media
video imaging

Competencies to Develop

After studying this unit, you should be able to:

- discuss the benefits and limitations of modern technology in the green industry.
- explain the major types of computer programs currently popular in the industry.
- explain other technological advances that now serve the industry.
- project future benefits of technology to the industry.
- describe three key social media outlets used in horticultural marketing in the millennial age.
- list the benefits of horticultural marketing with Quick Response codes (QR codes).

Technology is not new. It is defined by time. Throughout history, each generation has witnessed distinctive and unprecedented technological breakthroughs thought impossible by earlier generations. The horse and buggy was retired by the horseless carriage. The earthbound passenger train was swept away by the airplane, which has since been surpassed by the gravity-defying rocket. Communications have progressed from jungle drums to wireless cell phones. Entertainment has evolved from parlor games to surround-sound home theater systems. Only slightly less dramatic has been the transformation of American business. Where there were once labor-intensive offices filled with secretaries, file clerks, accountants, keypunch operators, typewriters, filing cabinets, mimeograph machines, and a computer the size of a refrigerator that hardly anyone knew how to use, there is now something strikingly different. Fewer people, operating fewer machines with expanded capabilities, have brought a new look to the office workplace. The computer is now the central feature of every office and most desktops.

The Benefits and Limitations

Like so many advances that preceded it, the computer age has offered many benefits, with a few accompanying drawbacks. Modern computers are much smaller than their first-generation ancestors and have many more capabilities. They can process and store vast amounts of data in a variety of forms. They can accelerate calculations, project the effects of changes, and even predict the changes. They can allow instant recall of historical data, react to sensors, correct spelling and grammar in multiple languages, create graphic illustrations from numerical data, activate environmental systems, turn drawings into electronic impulses, and restructure them as drawings in far distant places. They even provide a source of constant communication with access to an Internet connection.

The Most Common Uses of the Computer

Most businesses in America and the developed nations have embraced computer technology and are quick to accept and incorporate each new advancement. The companies of the green industry are no different. The explosive growth of the Internet gives companies in the most remote locations access to the nation and the world. Current uses of the computer in the offices of American horticulture firms include the following:

- correspondence
- customer accounts
- accounts payable
- payroll
- taxes
- mailing lists and labels
- inventory records
- accounts receivable
- billing
- sales analysis
- e-mail
- office newsletters and promotional mailings to customers

Beyond the office, industry-specific operations include the following:

- greenhouse environmental control
- soil and media mixing
- labeling of plants
- labor saving equipment
- landscape specification preparation
- quantification of materials for construction
- plant selection
- irrigation system design and graphics
- landscape design and graphics
- cost estimating
- technical calculations such as cut and fill
- geographic information systems
- weather data
- Internet access to catalogues and current prices of suppliers
- web page promotion of the company, its products, and services
- social media communications

Word Processing

Word processing programs give office staff members the opportunity to prepare correspondence and reports that have a highly professional appearance featuring varied font styles, sizes, and even colors. Form letters can be personalized with the recipient's name and other specifics that increase the probability of it being read rather than discarded. Spelling errors and incorrect grammar are minimized as software becomes increasingly user-friendly and responsive to human frailties. Computers have also replaced the filing cabinet, with large capacities to store documents on the internal hard drive as well as on external storage devices such as flash drives or on cloud-based services. The use of e-mail can send simultaneous and paperless memos or copies to a number of people in different offices, buildings, and cities. Chapter 55 dealt with the use of technology in business communications.

GREEN TIP Go paperless. Use social media as your primary means of corresponding; advertise on the Internet; purchase paperless faxing software; and keep records on USB drives or other data devices. Save our forests.

Numerical Calculations

With the capability of software programs to accomplish mathematical functions, prepare spreadsheets, and convert numerical data to bar graphs, pie charts, and other graphic interpretations, the office computer can fill many additional roles. Computers are commonly used to track inventory, accounts receivable, and accounts payable; prepare estimates, billings, and paychecks; maintain time records; and prepare tax reports and other financial statements.

Computer technology allows garden centers and nurseries to update their inventories when plants and other materials are purchased. One scan by a computerized register generates a printed receipt for the customer and simultaneously deducts the item from the inventory.

Landscapers are able to quantify materials from designs and convert the data into precise bids in a fraction of the time required by conventional measuring and counting methods. Last-minute changes to design specifications can be quickly input and the bid updated. Less time is required for calculations that in the past were time-consuming, tedious, and subject to human error. At the same time, the accuracy of the calculations has probably improved because, with the use of computers, a company can keep close track of labor hours, material costs, overhead costs, and equipment usage. Early software estimating and inventory programs were directed at a generic business market. More recently, a number of specialized programs have reached the market specifically targeting the nursery, landscape, lawn care, irrigation, or florist companies. They range in sophistication and in their usefulness to the user. It seems to be generally agreed by industry practitioners that the best software is that which is sufficiently complex to meet their needs and that can be customized to fit their individual business. They do not want to modify the business to conform to inflexible software. In the short span of time that the green industry has been using commercially developed estimating, inventory, and other office practice software, numerous products have fallen in and out of favor. There are two reasons that most companies change from one program to another. Either they find a new program that is a better fit with the way they do business, or they outgrow the capabilities of their current software to track the financial activity of their expanding business. One of the most successful niches developed by enterprising software companies has been the creation of specific software for specific companies, rather than attempting to customize an existing program for a buyer.

Input devices for numerical calculation programs include a keyboard, a scanner, and a **digitizer** (Figure 56–1). The digitizer permits direct measurement of lengths, areas, volumes, and perimeters from drawings, so it is essential for the conversion of drawn data to numerical equivalents. The output device for numerical calculations is usually a printer.

Graphic Visualization

Several types of computer graphic visualization systems enjoy popularity among horticulturists, most of which are adaptations of programs developed earlier for other industries.

▲ **Figure 56–1** A handheld stylus is used to select commands and symbols on a digitizer or menu tablet.

Computer-aided design (CAD) systems are accelerated drafting systems, designed to do what centuries of draftsmen have done with their T-square and pencils. Computer drafting has increased mainly because of a PC-based CAD software program named AutoCAD (a registered trademark of AutoDesk, Inc.). Due to the necessity for hardware, software, and user training, AutoCAD is an expensive investment for a company. Moreover, because it is a general system, marketed to a variety of users, it requires a third-party, discipline-specific software interface to customize it for a specific type of design and drafting. A growing number of landscape and irrigation design-specific software programs are available, and more are to come. However, the one most commonly used by landscape architects and landscape designers is LANDCADD.

Developed by a group of practicing professionals and released commercially in 1984, LANDCADD is marketed by LANDCADD Inc. of Franktown, Colorado. Since its inception, the program has advanced through many generations of revisions and currently offers modular programs that permit a company to purchase only those programs that it needs, avoiding the requirement of paying for ones that it will not use. At the heart of the software system is the Site Planning and Landscape Design Module. It permits creation of two- and three-dimensional plan views, elevations, and perspectives of landscape features and entire sites. Textures, colors, enlargements of special areas, lettering, and a variety of other graphic features are possible. When used by a skilled operator, the drawings can rival hand-drawn plans. As the design develops, the plant list is automatically created and tabulated. Designers can select from the symbols that come with the program or develop and add their own to further customize the system. Figure 56–2 illustrates a landscape plan developed using LANDCADD. Three-dimensional drawings of buildings, terrain, and grading plans are also possible using the system's Planning and Design Module. With the rapid growth of irrigation services and the increased number of companies offering them, the Irrigation Design Module is widely used (Figure 56–3). Other modules permit material measurement, cost estimating, and plant selection.

It warrants repeating that LANDCADD is but one company now offering these discipline-specific software programs. Figure 56–4 illustrates a drawing done using different software. Future advances by competing companies will ensure a free market, innovative research and development, responsiveness to the needs of the green industry, and fair prices to the industry consumers.

The input device for computerized drawing is usually a mouse (Figure 56–5), or a digitizing tablet and stylus. The tablet and stylus permit the designer to place drawings on the monitor's screen in a manner remotely similar to that of the traditional draftsman.

The output device for CAD system drawings is usually a plotter (Figure 56–6). It provides a print of quality, including colored lines, while also permitting larger sizes than possible on the printers of conventional word processors.

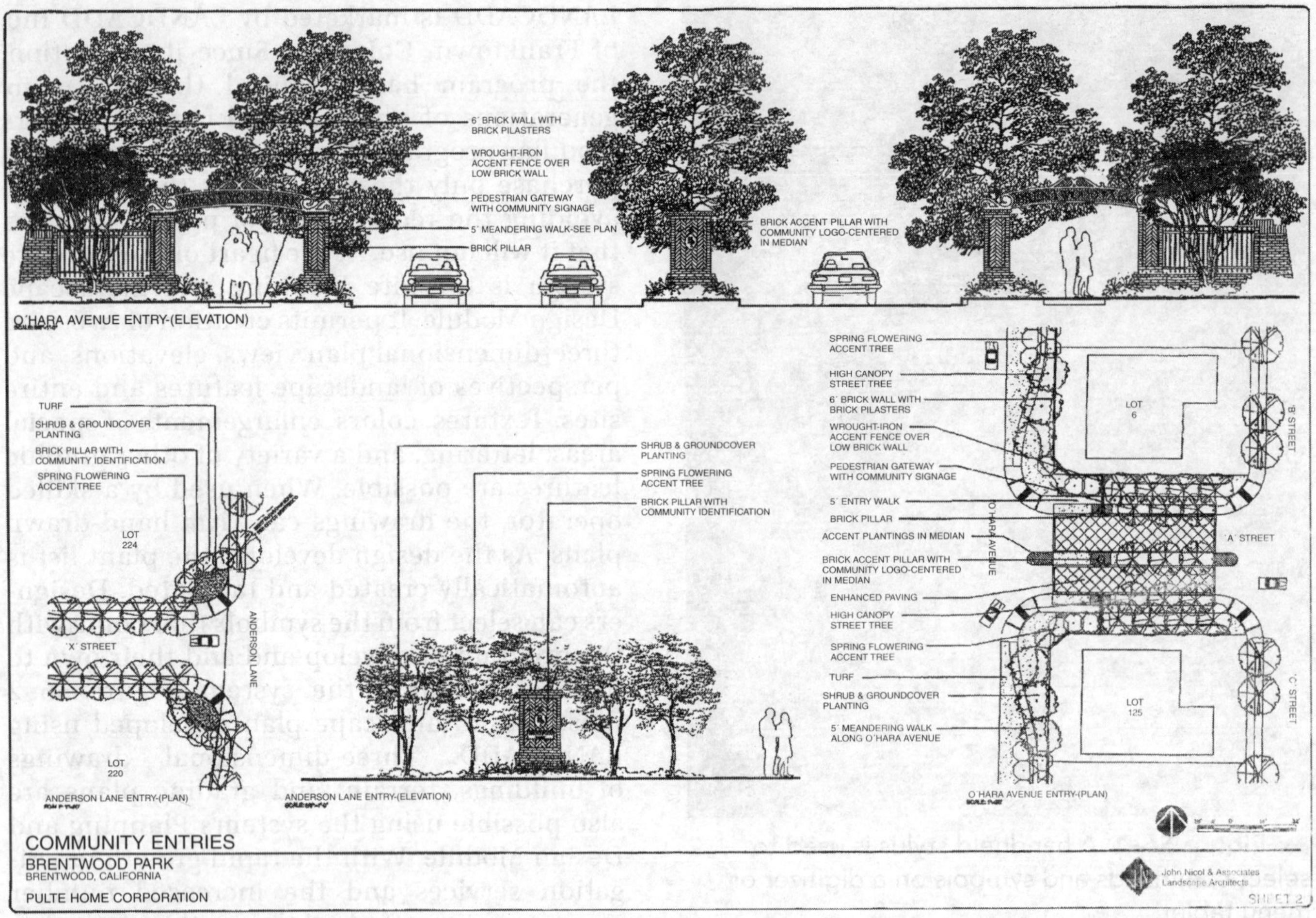

Courtesy of Eagle Point Software Corporation

▲ **Figure 56–2** An example of a plan using LANDCADD to produce the design and lettering. Trees were added using a paste-up technique.

> ***TRY THIS*** Use a CAD system to make a landscape design drawing that includes common landscape components. Print the drawing and present to the instructor for review.

Video imaging is a computerized technique that superimposes a landscape concept over a photographic image of the undeveloped landscape. Most commonly used by residential landscape designers to illustrate their proposals for areas around a client's home, the designer begins with one or more photographs of the building captured with a conventional camera, a digital camera, or a video camera. One or more selected images of the building are then entered into the system's memory. Thereafter, the image can be brought up on a computer monitor used to superimpose color photo images of plants, pavings, pools, lawn, walls, fences, and other landscape elements over the building. The system also permits the designer to modify the photo to illustrate the impact of removing existing features, changing colors, or even remodeling the architecture. The resulting image gives customers a full-color suggestion of how their landscape will look before it is actually installed, which is an excellent sales tool (Figure 56–7).

Several good video imaging systems are available on the market. PRO Landscape Design Software, by Drafix Software, Inc., will create a "life-like" before and after image of your proposed landscape using a digital photograph of a house or building. Video imaging is less a design system and more a graphic technique because it does not create scaled drawings. It relies on the eye of the user to create the proper perspective (Figure 56–8). After the graphic image is composed on the monitor's

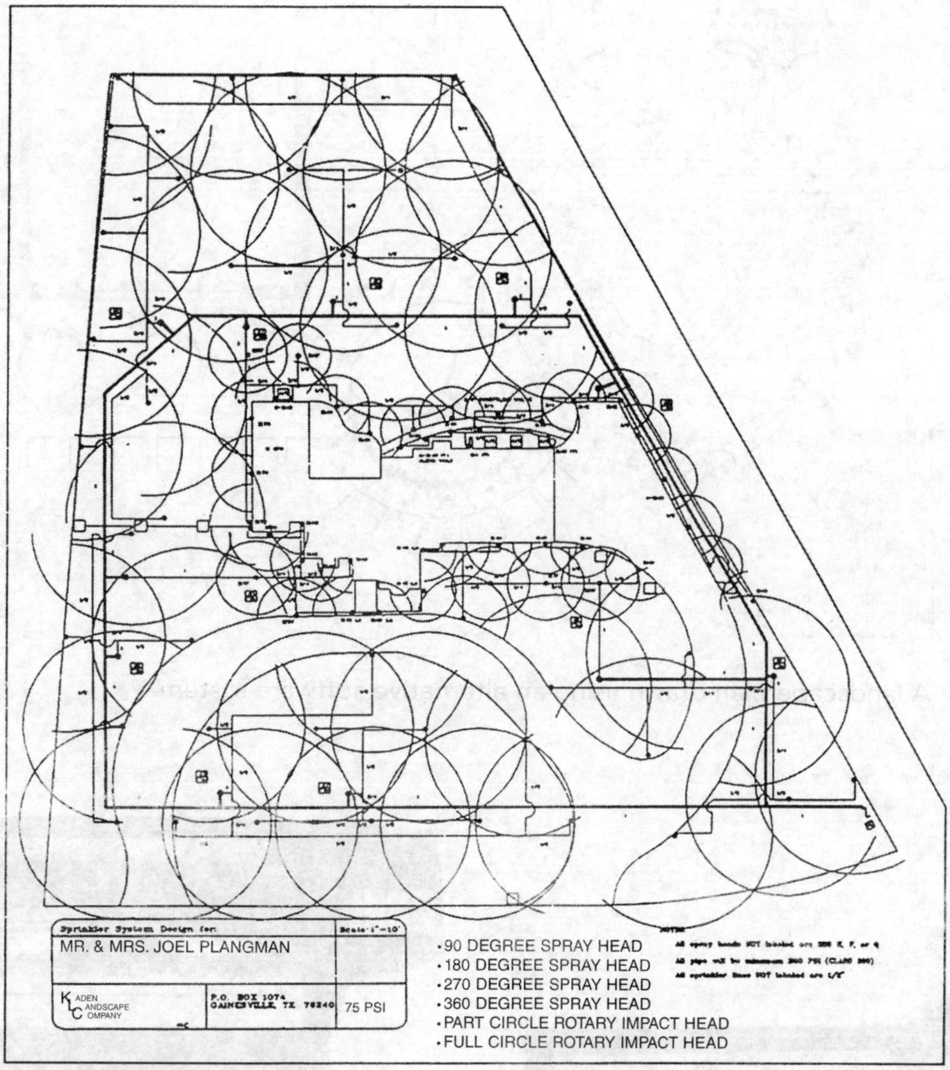

Courtesy of Eagle Point Software Corporation

▲ **Figure 56–3** A LANDCADD irrigation design.

screen, it can then be reproduced as a color print or as videotape for presentation to a client audience of just 1 person or of 101 people. The PRO Landscape Design Software and others are able to interface with CAD systems such as LANDCADD to apply color and realistic surface textures to the line drawings produced by those systems.

Both CAD drawings and video images can be saved, reproduced, transmitted electronically, and made available to others with the ease and immediacy of e-mail.

Graphic design using the computer is now available to any person or company willing to invest in the software. Products range from expensive systems used by commercial artists to simple, inexpensive clip art programs. Coupled with the photo scanning capabilities of the computer, the graphic design programs permit companies to prepare their own newsletters, advertising layouts, and promotional mailings. Customers and employees can be remembered on their birthdays or other occasions with personalized greetings. Clients can be shown before-and-after photos of their landscapes. Garden centers and flower shops can replace the amateurish felt marker signs with computer-prepared signing that looks as if it came from a custom print shop.

Other Uses of Modern Technology

The development of the technology of today and tomorrow has been either made possible or accelerated by the ongoing evolution of computers. The scientific breakthroughs of the space

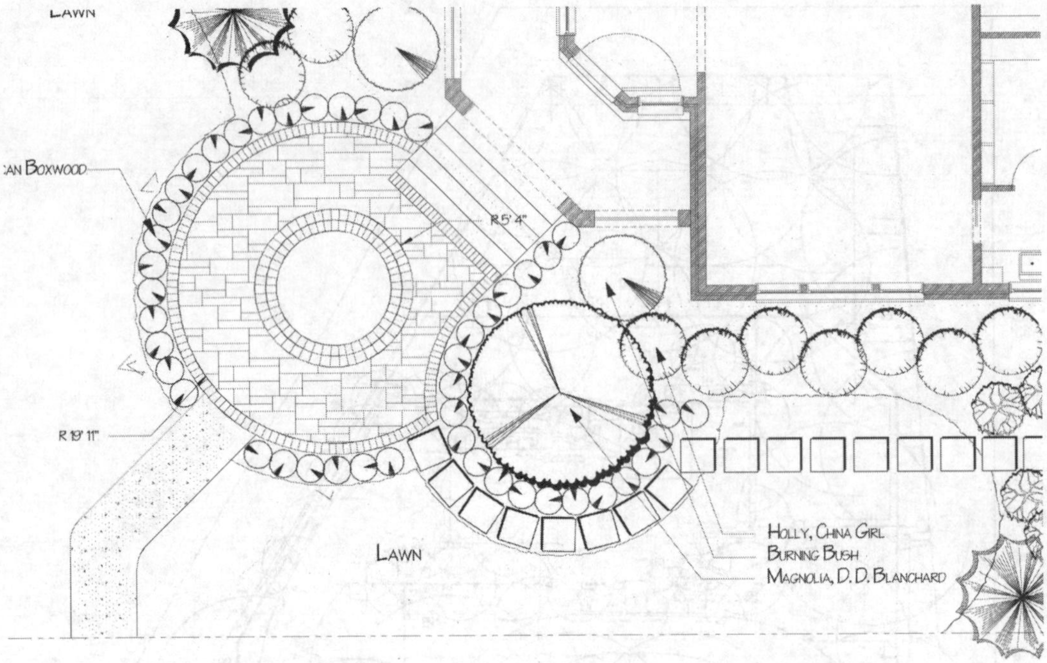

Courtesy of Drafix Software Inc.

▲ **Figure 56–4** A landscape plan drawn using an alternative software system.

© Chuck Rausin/Shutterstock.com

▲ **Figure 56–5** Computers have replaced traditional drafting tools in many offices.

© Michal Vitek/Shutterstock.com

▲ **Figure 56–6** An A- through E-size dual-mode plotter.

program have ushered in an era of nearly instant communications that link people and companies worldwide. Researchers have machines, apparatuses, and techniques available for their use that early pioneers in science could have never envisioned. Growers now press buttons to mix soil rather than lift shovels. They let sensors determine when to water rather than sticking

Courtesy of Drafix Software Inc.

▲ **Figure 56–7A and B** Before-and-after examples of computer imaging.

Courtesy of Drafix Software Inc.

▲ **Figure 56–8A** Computer-savvy skills using 3D landscape design are the future technology in horticultural sales. This shares with the client how the landscape will appear upon completion.

their fingers into the soil. Golf course superintendents consult their on-site weather station to determine the feasibility of irrigating portions of the golf course with the controls on the smartphone and the advisability of applying chemicals or preparing work schedules. No listing of the ways that computer-based technology is used throughout the green industry could ever be complete or current, but consider the following examples as representative.

The Greenhouse Industry

Enlisting the speed and alertness of computers and smartphones to help grow plants in greenhouses has been made possible by the development of **sensory input devices**. Independent of human

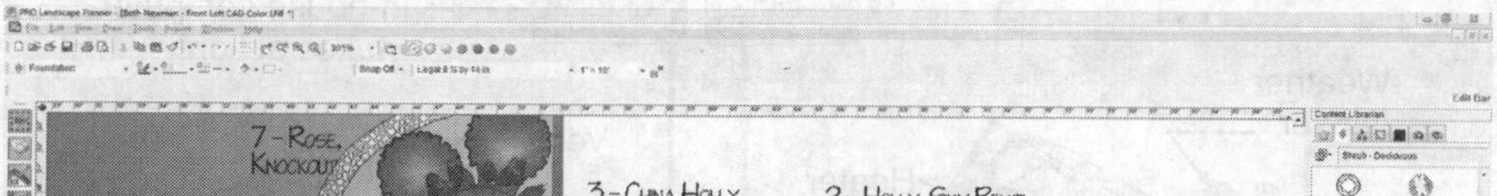

Courtesy of Drafix Software Inc.

▲ **Figure 56–8B** Computerized 2D-landscape design color rendering.

commands, the sensors can detect and monitor temperatures, light, humidity, carbon dioxide, wind velocity and direction, soluble salt levels, pH, and many other growth-related factors. The computerized control system has three main functions:

- Sensing. The system objectively measures environmental factors using a series of independent sensors. The sensors are located both inside and outside the greenhouse range. They are not restricted to single locations within a house, which makes them a significant improvement over older, analog control systems.
- Decision making. An independent computer monitors the sensors, analyzes the data they provide, and compares the existing environmental description against the grower's programmed instructions.
- Taking action. The system activates equipment within the greenhouse to implement the decisions made.

Computerized environmental control systems are component systems whose parts are linked together by coaxial cables and microprocessors, and whose data can be made available to the grower at any time, either in graphic displays or in digital or printed form by means of the computer monitor. Figure 56–9 illustrates a typical computerized environmental control system in a greenhouse range.

Unlike older greenhouse control systems, which are basically reactive systems (e.g., the temperature drops below a minimum, and the thermostat turns on the heat full force to warm the house), computerized systems can be programmed to respond almost continuously to the integrated sensory information. The result is a greenhouse environment that remains nearly constant, changing more slowly than the reactive systems. Temperature extremes, crop drying and watering, humidity levels, light conditions, and other growth factors adjust gradually. The assumption is that this stability results in less traumatic growing conditions and thereby improves plant growth. Logical though the assumption is, researchers are just beginning to evaluate the new technology. As computerized production technology continues to develop, it can be anticipated that such lags between manufacturer claims and independent confirmation will be common.

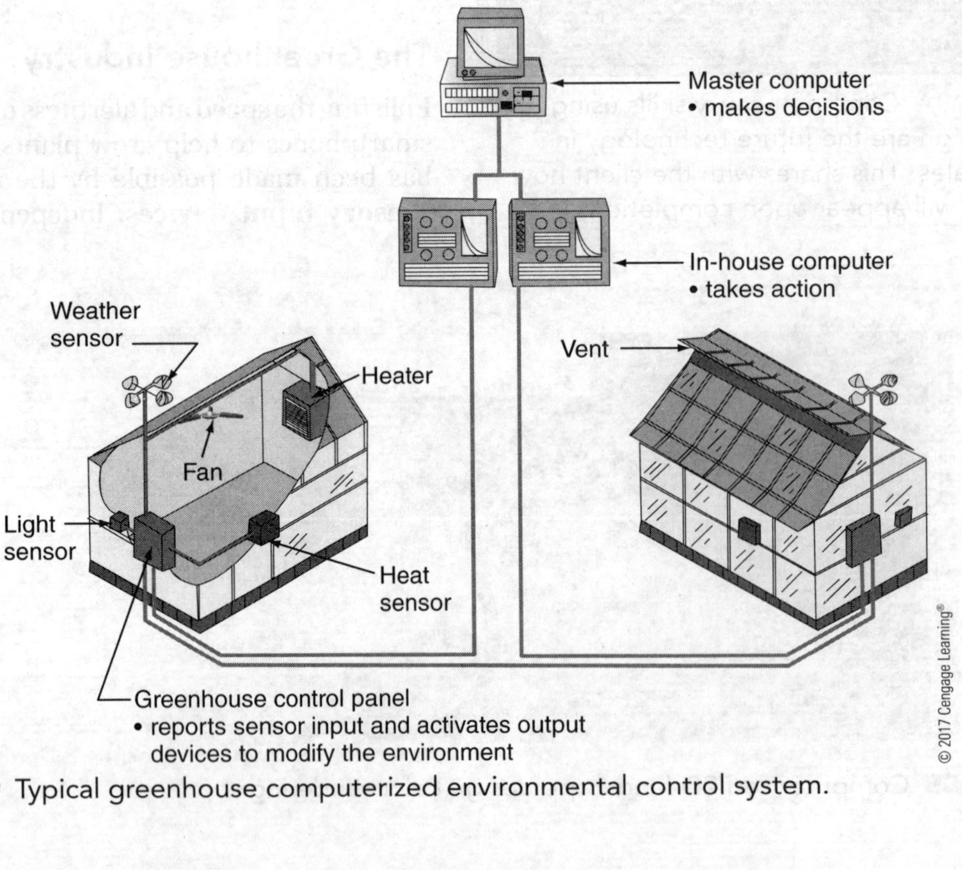

▲ **Figure 56–9** Typical greenhouse computerized environmental control system.

Representative greenhouse environmental control system manufacturers are Argus Controls in Canada, Priva Computers Inc. in the Netherlands, and Wadsworth Control Systems in Arvada, Colorado. It is interesting to note that none of the companies incorporate the word "greenhouse" in their name, probably recognizing (and hoping) that their computer-operated systems have a more comprehensive adaptability to other and future green industry operations.

The Retail Florist Industry

The green industry business that remains most tied to its handcrafting roots is retail floristry. However, even it has become high-tech in comparison to its recent past. While worldwide wire service is not new, the method and speed of transmission are as current as the latest telecommunications technology can make them. Florists across the country and around the world are linked by means of numerous cooperative agreements that enable customers to make a call to their local flower shop and send flowers thousands of miles away for arrival that same day. While FTD, Teleflora, and other established wire services still dominate, other companies such as Sears Roebuck, 1-800 FLOWERS, and Hallmark stores are now successfully attracting wire orders and transmitting them through their own network of affiliated shops. A customer can now order products electronically using a smartphone and never have verbal communication with the seller.

Even local business is often given a technical look as the florist, while designing an arrangement in the traditional way, answers the telephone and speaks with a customer using a headset rather than a handheld receiver, so that work is not interrupted. The same flower shop may also elect to advertise its services on the Internet or link itself with other products or companies that will bring its name before a much greater audience than ever before possible. They may also receive and process orders electronically.

The Landscape Industry

In addition to the uses of the computer as an aid to drawing, calculating, and preparing graphic presentations, the landscape industry has adapted other forms and types of technology for its benefit. Multiple-use machinery is gradually replacing equipment that could only dig a hole, cut a trench, or perform other single functions. Wireless communication systems permit faster, easier, and more frequent communication between people in the field, in the office, and on the road somewhere in between. Customer concerns can receive quicker attention. Problems can be resolved sooner. Confusion and errors are reduced. Industry trade associations are offering expanded services to their members using the interactive capabilities of the Internet and e-mail. Some companies are even able to let customers calculate their own maintenance estimates using interactive websites whereby the customer enters specific measurements and other data about property from home that is paired with prices preset by the company. The result is a do-it-yourself estimate generated by customers that saves the company time, particularly the time lost doing estimates that are not accepted. While this is not yet widely practiced, the potential is there for other types of customer-company interaction. More common, and gaining in usage, are computer-generated estimates that can be done on-site by project directors and others, working with handheld systems that can calculate and print out as the customer waits. By reducing the turnaround time between a customer's query and the company's reply, additional business is generated.

Landscape architects are using the data provided by geographic information systems (GIS) to ascertain the suitability of near and distant sites for varying types of development. They are also able to receive and transmit data electronically between their offices and those of associated professionals who are also involved in their projects. Computer systems can now convert flat plans on a monitor into video walk-throughs of the proposed design

The Turf Industries

Today, the development of a programmable unmanned mower has the potential to provide horticulturists with the ability to manicure lawns without manpower, thus increasing productivity (Figure 56–10). The application of these robotic mowers will vary. Some are targeted at residential homeowners who may not have the time to

▲ Figure 56–10 Husqvarna automatic mower.

maintain their lawns. However, turf companies have access to the usefulness of this technology in their businesses.

Advances in maintenance equipment efficiency and irrigation technology have made operations more cost-effective for lawn care services, golf course operators, and landscape contractors.

The application of current and near-future research has resulted in bioengineered grasses that are more resistant to environmental stress and/or require fewer mowings to maintain their desired appearance.

Probabilities and Possibilities for the Future

The businesses, scientists, educators, practitioners, and customers of the green industry can anticipate further changes in how ornamental horticulture is practiced and presented in the near and distant future. The most high-profile changes will probably result from further advances in computer and telecommunication technology. Faster and easier access to an ever-expanding marketplace will give consumers greater choices. A customer in Boston may choose to order orchids directly from Hawaii rather than the local flower shop, and get the same quality, price, and prompt delivery. A nursery grower may bypass the local wholesale supplier and order needed supplies directly from their manufacturers. Salespeople may spend more time in face-to-face conversations with customers on television monitors than in their offices or shops. Voice commands may soon replace, or at least supplement, drawn plans in the offices of landscape designers and architects. Plant breeders can be expected to respond to increasing environmental concerns with accelerated efforts to produce plants that require less maintenance, less fertilization, less water, and are more resistant to pests and other stress factors. The greater movement of people, plants, and plant products around the world can be expected to bring new pests into regions of the planet that are vulnerable to rapid, invasive damage. That will introduce new challenges for arborists, landscapers, and others.

Consolidations and mergers are already occurring throughout the landscape, lawn care, and arborist industries. At least in the short term, those mergers can be expected to continue, and the practice may widen to include the florist and nursery industries as well. As large regional or national corporate identities are established within the green industry, they mirror what is happening nationally in other industries such as property management. Gaining the advantages of improved purchasing power, higher-profile advertising, state-of-the-art technology, and easier access to the centers of power where decisions are made, such companies offer formidable competition to local independent companies that may offer a comparable or even better product or service but cannot get the attention of the decision makers who may not be local.

Horticultural Marketing in the Millennial Age

Marketing strategies have changed to meet the needs of the millennial generation (ages 18–34), which is the fastest growing segment of garden clients. With the integration of smartphone apps, Internet websites, social media, and QR codes, the use of technology in the horticultural world has become a plus as it can make a horticultural operation more efficient and can speak directly to this young, growing segment of customers.

Online resources are widely accepted as a source of information for the purchase of horticultural products. Many people use **social media** to increase their knowledge about products and services and also use it to their advantage to select desired products that best meet their wants and needs. Horticultural marketing specialists have to be influential in communicating the value of garden plants, tools, and an outdoor living lifestyle to these savvy consumers. These specialists handle this monumental task by sharing the economic and environmental benefits of plants and outdoor living.

In a recent presentation, Sid Raisch, Horticultural Advantage consultant, said, "69% of the customers research the product online and then buy in the store." According to Annie Kemp, author and horticulturist, the key social media resources of the horticultural industry are Facebook, YouTube, LinkedIn, and Pinterest:

- Facebook is one of the most popular social networks that connects people virtually. Customers who use Facebook can get daily updates of promotions or products, make comments, ask questions, and interact with company staff.
- YouTube allows anyone to upload videos that are informational and can instruct viewers on how to complete a specific type of landscape project.
- LinkedIn is a professional networking site that engages professionals and enables them to access knowledge and opportunities. Horticultural professionals can use this resource to make contacts with others in the business and share ideas and strategies.
- Pinterest is an online wish list and scrapbook of ideas. Many horticultural ideas and inspirations can be identified through this website.

These media resources also offer business owners the opportunity to showcase their products as well as share information. For example, a nursery may inform customers about its selection of available plants or feature unique color combinations. A grower might post details about sustainability related to shade versus sun or discuss plant life cycles, growing media, and general care for optimal esthetics.

GREEN TIP The use of ornamental edibles, or plants that both look attractive and taste good, is all the rage among Millennials. Some examples of popular edible plants used in landscaping include varieties of kale, tomatoes, and blueberries.

QR CODES QR codes are being used in the horticultural world to market products. Through the technology of a smartphone, horticultural customers are able to retrieve pertinent information related to specific plants by using a QR code. Each individual QR code includes a picture of the plant and text information such as sunlight exposure, moisture requirements, life cycle classification, planting instructions, growth habit, height and spacing, hardiness zones, and heat zones. Additionally, it will include general tips and possibly video.

Gerry Giorgio, a manager at MasterTag, a company that provides horticultural care tags to growers, provides some tips on adding information to your QR codes that is relevant and useful:

- Offer new information that can't be found on the physical plant tag.
- Make sure landing pages are mobile-ready for easy viewing on a smartphone.
- Keep your message short. People are moving and don't have time.
- Consider a more visual message. Images are good; video is better.
- Be sure you have sufficient signage directing people to scan the codes, as they may not be obvious at first.

Drones

Drones, or **unmanned aerial vehicles (UAVS)**, are becoming frequently used in agriculture. Because drones can take aerial photos, orchards and vineyards can use them to identify problem areas throughout growing areas and focus precise effort to correct problems or deficiencies in the crops. Drones are also becoming more affordable as time goes on, so small producers may be able to use this technology as well.

As one example, in recent research, Dr. John Nowatziki of North Dakota State University presented data from the first unmanned drone system test in agriculture in the spring of 2014. His research is to use the drone system to monitor plant deficiencies in nitrogen, iron, and other plant health, as well as to help identify weeds.

The Flip Side of Technology

Just as the development of automobiles negated the need for buggy makers, the use of computers has changed the staffing needs of businesses. The incorporation of new technology into horticulture companies has created some problems. The one most common to all types of businesses is the impact on the workforce. Most upgrades in technology necessitate the retraining of some employees, which may or may not involve an expense. Some technical advances also displace workers or require their replacement with better-educated workers who expect higher salaries and who may not be easily recruited. Older workers, long in their service and loyalty to a company and knowledgeable of existing technology, may resist accepting the new knowledge and techniques. New technology excites some people and is perceived as a threat by others.

As the current and future generations of American horticulturists train to work with the new technology, fewer people will remain to perform the labor-intensive entry-level and field tasks that will still constitute a large portion of the production services performed by green industry companies. The employee shortage is already a major concern in all sectors of horticulture. The new technology may herald an era of increased sophistication in the ways horticulturists present their products and services, while simultaneously provoking a shortage of personnel to do the work.

SAFETY FIRST

OSHA's regulations ensure that employers and employees know about work hazards and how to protect themselves to prevent injuries in the workplace. The owner and manager of a business should be knowledgeable about OSHA regulations and how to keep the working environment safe. For more information on occupational health and safety in the workplace, please visit the OSHA site at www.osha.gov.

Summary

Technology is rapidly expanding throughout the industries of ornamental horticulture. Operations common to all businesses, as well as those distinctive to the nation's flower shops, greenhouses, nurseries, arborists, and landscape and lawn care firms, are now routinely aided by computers. Computer-centered machines, equipment, and software programs are changing how plants are grown, inventories are monitored, plans are drawn, estimates are prepared, communications are transmitted, advertising is prepared and delivered, and customers are serviced. The advances in technology and their applications to ornamental horticulture seem certain to continue into the future. Those advances will bring opportunities and challenges to industry professionals as they strive to balance the benefits of improved quality, craftsmanship, speed, client satisfaction, or other measures against the costs of new equipment, technically trained employees, and the shortage of entry-level laborers that may result.

Review Questions

1. What are three characteristics of each of the social media used in the horticultural industry today?
2. How do you perceive social media impacting the horticultural industry of tomorrow?
3. What are the benefits and limitations of modern technology in the green industry?
4. Do you think the use of drones will increase in the horticultural industry? How do you see them being used?

Student Activities

1. Conduct research on the Internet, and then create a report on the use of modern technology in the horticultural industry.
2. Find examples of horticultural businesses that use social media to promote or educate people about their business.
3. Ask a local florist how technology impacts their business.
4. Use a smartphone app to scan a QR code on a horticultural product. Record the information that is displayed.

Suggested Supervised Agricultural Experience (SAE) Activities

1. To qualify for a placement SAE, work in the communications department of a large garden center.
2. As an entrepreneurship SAE, manage and operate your own lawn maintenance business operation. Manage a website and social media outlet to advertise your business.

Additional Resources

www.bidwell-training.org/majors/horticulture-technology/

www.hort.purdue.edu/newcrop/originhorttech.pdf

http://thompsonschool.unh.edu/hort/greenhouse-sales

UNIT 57 Leadership Development in Horticulture

Objective

To develop basic leadership skills.

Key Terms

- 4-H
- adjourn
- amend
- Boy Scout
- business meeting
- career success
- citizenship
- Cooperative Extension System
- courage
- enthusiasm
- executive meeting
- extemporaneous speaking
- FFA (the National FFA Organization)
- gavel
- Girl Scout
- integrity
- knowledge
- lay on the table
- leadership
- lead
- loyalty
- main motion
- manage
- minutes
- motion
- order of business
- parliamentary procedure
- personal growth
- plan
- point of order
- premier leadership
- presiding officer
- refer
- secretary
- selflessness
- stewardship
- tact

Competencies to Develop

After studying this unit, you should be able to:

- define leader and leadership.
- explain why effective leadership is needed in agriscience.
- list some characteristics of good leaders.
- describe the opportunities for leadership development in FFA.
- demonstrate positive leadership skills.

Materials

- paper, and pencil or pen
- iPad, laptop, smartphone
- bulletin board materials
- Internet access

What is **leadership** in horticulture? Horticulture has been described as a broad and diverse field that is not just plants or supplies and services; not just professions or products, processing, and distribution; not just mechanics or forestry; and not just renewable natural resources or production. It is the impact that one has on others through their profession.

Leadership Defined

Leadership may be defined as the capacity or ability to solve problems and to set a direction. To **lead** is to show the way by going in advance or guiding the actions or opinions of others. To do this in horticulture, you must have **knowledge** of technical information and people. You must know how to organize and **manage** activities. Most jobs are too big for one person. We can do only part of what needs to be done. Therefore, we need to use the help of others.

A leader uses the knowledge and skills of others to achieve a common goal. For instance, a quarterback on a football team will use leadership skills to coordinate the team players and communicate a united plan to achieve a touchdown. Similarly, the wise batter in baseball will hit the ball in a place that not only gets the batter on first base, but permits other runners to advance around the bases. A properly placed hit supports the common goal of achieving runs. A single base hit, where everyone advances and no one gets out, may be the best for the team. Conversely, a line drive that doesn't quite make the fence may result in a third out, with no chance for other players to score.

Why Leadership in Horticulture?

Agriculture is a highly organized industry that involves people and complex processes. Leadership skills are necessary whenever people are assembled (Figure 57–1). Those who teach in horticulture are part of a team of agricultural education teachers, principals, supervisors, community advisory groups, and others (Figure 57–2). Those in horticulture are typically part of teams consisting of the manager, office staff, sales representatives, field personnel, and board of directors. Those in horticulture may be the owner, manager, spouse, children, hired help, or neighbors who assist at times. The manager of a horticulture business must **plan** (think through, determine procedure, assemble materials, and train staff to do a job). After a job is planned, it may be accomplished through management. To

Courtesy of the National FFA Organization

▲ **Figure 57–1** A leader must learn to coordinate the work of all members of a team to avoid having a team member working on something that has already been completed.

Courtesy of the NYS FFA

▲ **Figure 57–2** Agribusinesses rely on team efforts to achieve goals. Here, FFA members work together to explore new agribusiness technologies for future demands in the horticulture industry.

Courtesy of the National FFA Organization

▲ **Figure 57–3** Managers are leaders working together with others on a day-to-day basis.

manage is to direct people, resources, and processes to reach a goal. A manager uses leadership skills continuously in working with others on a day-to-day basis (Figure 57–3).

GREEN TIP Teach and lead your employees about sustainable plants to increase the sustainable use and conservation of native wildflowers, plants, and landscapes to stay green! Volunteer in your community by cultivating urban vegetable gardens for food, planting shrubs and flowers, or enhancing the grounds of shelters or other nonprofit locations. Donate new seeds, plants, or fruits and vegetables for a local food drive or community garden. Continue to focus on sustainable horticulture to create a cleaner world.

A landscaper is someone who plans, plants, builds, or maintains outdoor ornamental plants and landscape structures. Sometimes working alone and sometimes working with others, the landscaper is a leader in many respects. When developing a plan for a customer, the landscaper exerts leadership. The landscaper knows the name, function, and performance of each plant. This information is used to develop an acceptable plan. Because the customer has personal ideas about landscaping, a professional must consider these ideas in the plan. It may test the landscaper's leadership skills to lead the customer to an acceptable plan with plants that survive the climate and conform to acceptable landscape practices.

The landscaper and other horticulture personnel may be called on as officers or members of professional organizations, such as AmericanHort, American Horticultural Society, or the Farm Bureau, to give testimony before legislators or other public officials on the need for laws, regulations, or other actions that affect their work. It may be quite possible to end the day presiding over a meeting or taking **minutes**, the instant written record of the meeting or hearing (notes), at a professional or civic meeting.

Horticulturalists, as well as all employees in the agricultural industry, have a responsibility to be good stewards of our natural resources. **Stewardship** is the careful and responsible management of something entrusted to one's care. Members of the horticultural industry have a sense of responsibility to make decisions that are sustainable and positive for the future of our natural resources.

For the horticulture student, the need for leadership skills is also apparent. Having confidence to participate fully in class is important (Figure 57–4). To meet prospective employers and conduct a supervised agricultural experience program requires leadership skills. To participate as a member or serve as an **FFA (the National FFA Organization)** chapter officer directing the functions of the student organization develops many leadership skills (Figure 57–5). Functioning in the community, participating in group meetings, and making friends readily all require leadership skills. Good leadership skills greatly improve individual marketability in the working

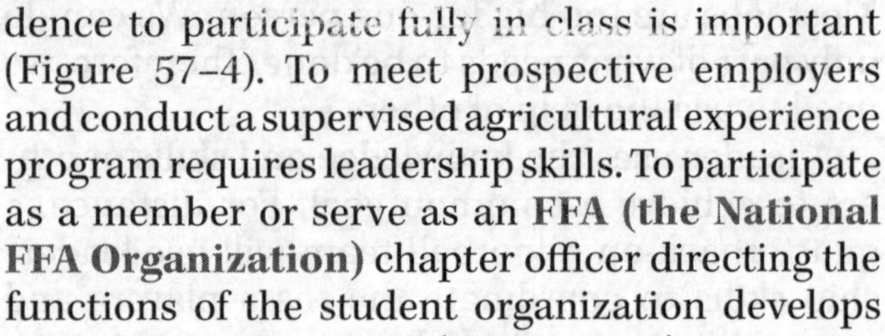

Courtesy of the National FFA Organization

▲ **Figure 57–4** Participation in class discussions is the first step toward developing qualities of leadership.

▲ **Figure 57–5** Many horticultural youth experience leadership training as members of the FFA and are recognized for their horticultural skills and competence through which they earn scholarships for their college education.

world. Leadership also improves the performance of employees and encourages them to stay with the company for a long period of time.

To be a good citizen, you must earn your way in life without infringing on the rights of others. Useful **citizenship** uses leadership to promote the common good in society. Citizenship means functioning in society in a positive way.

Traits of Good Leaders

Good leaders must have **integrity** (honesty). Without it, others cannot trust an individual with the power to manage or control, even in minor things. A leader must have knowledge, which means familiarity, awareness, and understanding. Good leaders are dependable and have the **courage** (willingness to proceed under difficult conditions) and initiative to carry out personal and group decisions. To lead, one must demonstrate the initiative to carry out personal and group decisions. To lead, one must also communicate, which requires good speaking and listening skills. A good leader is also able to delegate responsibilities to others to demonstrate confidence in those with whom they work.

In working with others, **tact**, or the skill of encouraging others in positive ways, is useful. Similarly, a sense of justice to ensure the rights of others is important. **Enthusiasm**, or energy to do a job and inspiration to encourage others, is useful. **Selflessness** means placing the desires and welfare of others above yourself. It, too, is an important quality for good leadership. These traits encourage **loyalty**, which results in reliable support for an individual, group, or cause. These and other traits are achieved through effective leadership development.

Leadership Development Opportunities

Modern schools provide extensive opportunities for students to develop leadership skills. Students develop leadership in school organizations, athletics, and in classroom and laboratory situations. Some become leaders at home, on the job, or in community organizations.

4-H Clubs

The **Cooperative Extension System** is an educational agency of the U.S. Department of Agriculture (USDA) and an arm of your state university. It provides educational programs for both youth and adults that include personal, home and family, community, and agriscience resources development. The Cooperative Extension System also sponsors **4-H** clubs. The 4-H network of clubs is directed by Cooperative Extension System personnel to enhance personal development for youth and provide skill development in many areas, including agriscience. The four Hs in 4-H stand for head, heart, hands, and health. These provide the basis for the 4-H pledge: "I pledge my head to clearer thinking, my heart to greater loyalty, my hands to larger service, and my health to better living for my Club, my community, my country, and my world."

Scout Organizations

Girl Scout and **Boy Scout** organizations provide opportunities for leadership development and skill development in agriscience and other areas. Scouts focus heavily on outdoor activities and provide excellent leadership development and natural resources skills. They provide recognition through a system of merit badges, which are earned by learning skills and obtaining experiences in many areas, including agriscience.

FFA

The National FFA Organization is dedicated to making a positive difference in the lives of students by developing their potential for **premier leadership, personal growth**, and **career success** through agricultural education. Students under the age of 21 who are enrolled in a systematic program of agricultural education are eligible for membership in the FFA.

AIM AND PURPOSES The FFA is part of the horticulture curriculum in most schools where agriculture and horticulture programs are offered. As an important teaching tool, the FFA serves as a laboratory for developing leadership and citizenship skills. These, in turn, are helpful in learning horticulture skills. The primary aim of the FFA is the development of agricultural leadership, cooperation, and citizenship. The specific purposes of the FFA may be paraphrased as follows:

- To develop competent and positive agricultural knowledge and leadership
- To develop awareness of the global importance of horticulture and its contribution to our well-being
- To strengthen the confidence of FFA members in themselves and their work
- To promote the intelligent choice and establishment of a successful career
- To stimulate development and to encourage achievement in individual horticulture experience programs
- To improve the economic, environmental, recreational, and human resources of the community
- To develop competencies in communications, human relations, and social abilities
- To develop character, train for useful citizenship, and foster patriotism
- To build cooperative attitudes among horticulture students
- To encourage wise use and management of resources

Courtesy of the National FFA Organization

▲ Figure 57–6 The FFA emblem contains five important symbols.

- To encourage improvement in scholarship
- To provide organized recreational activities for horticulture students

THE EMBLEM The FFA emblem contains five major symbols that help demonstrate the structure of the organization (Figure 57–6):

Eagle—The emblem is topped by the eagle and other items of our national seal. The eagle was placed in the emblem to represent the national scope of the organization. It could also represent the natural resources in horticulture.

Corn—Corn is grown in every state in the United States. The cross section of corn in the emblem represents the connection that agriculture and the FFA create between people, the same way that every kernel of corn is touching another kernel. It reminds us of our common interest in horticulture, regardless of where we live.

Owl—The owl represents knowledge and wisdom. Use of this symbol in the emblem recognizes the fact that people in horticulture need a good education and that education must be tempered with experience to be of greatest usefulness.

Plow—The plow has been used to represent work–labor–effort. These qualities are needed to cause things to happen and to get results in horticulture.

Rising Sun—The rising sun is a symbol of a new era in agriculture and the progressive nature of horticulture. It is symbolic of the need for workers in horticulture to cooperate and work toward common goals.

The FFA emblem may be constructed one symbol at a time. When assembled and dissembled in this manner, it is a good device to help others understand the FFA.

THE COLORS The official FFA colors are blue and gold. The shade of blue is national blue. The shade of gold is the yellow color of corn. Therefore, the colors are called national blue and corn gold.

MOTTO The FFA motto contains phrases that describe the philosophy of learning and development in horticulture:

- Learning to Do
- Doing to Learn
- Earning to Live
- Living to Serve

"Learning to Do" emphasizes the practical reasons for study and experience in horticulture. It also suggests ambition and willingness to productively use the hands, as well as the mind. "Doing to Learn" describes procedures used in horticulture instruction at the doing level. Experiencing results from doing is the most permanent result of learning. "Earning to Live" suggests that FFA members intend to develop their skills and support themselves in life. And "Living to Serve" indicates an intention to help others through personal and community service.

SALUTE The Pledge of Allegiance to the American flag is the official FFA salute. The words of the pledge are the following: "I pledge allegiance to the flag of the United States of America and to the Republic for which it stands, one nation under God, indivisible, with liberty and justice for all."

DEGREE REQUIREMENTS The FFA has four degrees, which indicate the progress a member is making: Greenhand, Chapter, State, and American degrees. The Greenhand and Chapter FFA degrees are awarded by the FFA chapter in the agriculture or horticulture department of the school. The State FFA degree is awarded by the State Association. The American FFA degree is awarded by the National FFA.

The Greenhand degree is so named to indicate that the member is in a learning modality. He or she is developing basic skills through FFA participation and by studying the principles of horticulture. To receive the Greenhand degree, the member must meet the requirements spelled out in the current FFA *Official Manual*. In general, the requirements for the Greenhand degree are the following:

- Be enrolled in an agricultural education course
- Have satisfactory plans for a Supervised Agricultural Experience (SAE) program
- Recite the FFA creed, motto, and salute
- Describe the FFA emblem, colors, and symbols
- Explain the FFA Code of Ethics and proper use of the FFA jacket
- Have satisfactory knowledge of the history of the organization and of the Chapter Constitution and Program of Activities
- Know the duties and responsibilities of members
- Own or have access to a copy of the *Official Manual* and *FFA Student Handbook*
- Submit a written application for the Greenhand degree

The requirements for the other three degrees help the member learn and grow professionally from one level to the next in the organization. The *Official Manual* contains the exact requirements for all degrees, details of membership, and chapter operation for the FFA. This information can also be found at the national FFA website (www.ffa.org).

CAREER DEVELOPMENT EVENTS The FFA sponsors competitive Career Development Events (CDEs) for a wide range of career interests. The first level is in the local chapter at the school. The second level is the district and regional level within the state. The third level is the state association level, and the fourth is the national level. Local FFA advisors determine which events are appropriate for students in their programs. The competitions that are conducted at each level should reflect the content of the instructional programs.

PHILOSOPHY FOR NATIONAL FFA CAREER DEVELOPMENT EVENTS Students are important customers of agricultural education and the FFA who recognize quality and value in products and activities. When provided an opportunity to fashion their educational experiences, they generally make wise decisions based on needs. Perceptions, images, and opinions of others influence students. They value change based on their perceived personal needs as well as the needs of others. They sometimes value change for the sake of variety. Adults are concerned about the experiences of students and want to help organize experiences that will meet the future needs of students while accomplishing the purposes of agricultural education and the National FFA Organization. The National FFA Organization assumes the leadership role in developing and continuously improving relevant FFA CDEs. Although the National FFA Organization should be aware of the needs of state associations and should react to those needs, it should help initiate opportunities that reflect relevant and emerging technology. National FFA CDEs should be developed with significant input from FFA members, teachers, partners, respective industry sponsors, and others involved in agricultural education. National CDEs reflect instruction that currently takes place in the entire agricultural education program, including classroom instruction, laboratory instruction, individualized instruction, and/or supervised agricultural experience. However, it is appropriate for the national organization to develop CDEs and awards that stimulate instruction in emerging areas that reflect both current and future community, national, and global workforce needs. The authority for ensuring the relevance of an FFA activity is ultimately vested in the National FFA Board of Directors. The national organization promotes and facilitates CDEs. CDEs that include team activities are based on cooperation and teamwork while recognizing the value of competition and individual achievement. The role of CDEs is to motivate students and encourage leadership, personal growth, citizenship, and career development. Students are recognized for achievement in CDEs with awards, public recognition, and scholarships. Quality standards are used as a basis for achievement. Recognition for achievement is reflective of the total effort required by the chapter/team/individual and should take place at all levels of participation. The National FFA Organization also encourages accessibility and provides opportunities for achievement and recognition for students with diverse backgrounds. High expectations should be consistently communicated to those who are involved in CDEs and awards. The following are the CDE categories:

- Agricultural Communications
- Agricultural Issues Forum
- Agricultural Technology and Mechanical Systems
- Agricultural Sales
- Agronomy
- Creed Speaking
- Dairy Cattle Evaluation and Management
- Dairy Cattle Handlers' Activity
- Environmental and Natural Resources
- Extemporaneous Public Speaking
- Farm Business Management
- Floriculture
- Food Science and Technology
- Forestry
- Horse Evaluation
- Job Interview
- Livestock Evaluation
- Marketing Plan
- Meats Evaluation and Technology
- Milk Quality and Products
- Nursery/Landscape
- Parliamentary Procedure
- Poultry Evaluation
- Prepared Public Speaking
- Veterinary Science

Some FFA CDEs require contestants to know how to grade agricultural products, such as eggs, meats, poultry, fruits, and vegetables. Other events require students to evaluate live animals, such as beef and dairy cattle, horses, poultry, sheep, and swine. Some require mechanical abilities, such as welding, plumbing, electronics, irrigation, surveying, engine troubleshooting and repair, painting, woodworking, and general tool use.

Some events, such as Floriculture and Horticulture, require knowledge of the art and science of floral arrangement. Some CDEs include Forestry, Vegetables, and Nursery/Landscape competitions, which require students to identify plants, plant materials, landscape design, equipment identification, operation and safety, plant cloning, nurserymen skills, landscape design presentation

Courtesy of the National FFA Organization

▲ **Figure 57–7** FFA Career Development Events provide motivation to students to develop leadership skills through public speaking and by using parliamentary procedure to conduct the business of the organization.

orally and visually, problem solving with pests and client intervention, insects, and diseases. Land Judging (conducted by soil and water districts) involves evaluating the soil and land and recommending appropriate management practices. **Parliamentary procedure** teams demonstrate their ability to conduct meetings using their knowledge of correct parliamentary practices. These procedures are used to open meetings, conduct business, close meetings, and write minutes according to acceptable practice in the real world. Various types of speaking contests help individuals to sharpen their speaking skills (Figure 57–7).

Each of these skill events is organized to ensure that the participants are competing as individuals and as teams of three or four participants. An example of a chapter achievement is the National Chapter Award. This competition involves most or all of the students in the horticulture program, and encourages them to make valuable improvements within the program, school, and community.

Public Speaking

Oral communication skills are important for good leadership. Effective leaders must speak with individuals, committees, small groups, and in large forums. The ability to relax, speak clearly, and state what is pertinent to the subject at hand is useful. These skills are developed by applying basic principles of speech preparation and organization.

Speeches may be prepared or extemporaneous. **Extemporaneous speaking** is delivering a speech with little time for preparation. Extemporaneous speaking is a real-life skill that is used daily in horticulture careers. The ability to speak extemporaneously is enhanced by learning to deliver prepared speeches. Both prepared and extemporaneous speaking are used to teach public speaking skills in the FFA.

The National FFA Organization sponsors three different types of speech competitions: Creed Speaking, Prepared Public Speaking, and Extemporaneous Public Speaking. Each event provides opportunities for students to stand before an audience or judges and deliver a speech in a public setting.

Creed Speaking

The Creed Speaking event is for students who are first-year members of the FFA. It requires the speaker to repeat the FFA Creed from memory. The emphasis is on accuracy and delivery. When the speaker is finished, the judges will ask him or her questions. The speaker explains what the creed means to him or her and answers all questions.

The Creed Speaking event gets students started in speaking without undue concern over the content of the speech. It is a good way to help students succeed before they are expected to prepare the written content of a speech. After students gain confidence, they often become motivated to participate in other speech contests.

Prepared Speaking

The Prepared Speaking contest provides an opportunity for a student to research an agricultural topic and develop his or her own ideas. The content of the speech must be original, not copied from someone else. The length of the speech should be 6–8 minutes, and points are deducted if the speech is too long or too short.

Part of this speech competition is based on the quality of the written manuscript. The neatly typed manuscript (double spaced) is given to the judges ahead of time. Questions are developed from the manuscript. Each contestant responds to the questions for 5 minutes following the delivery of the speech. Contestants are judged on quality, effectiveness, accuracy of the written manuscript, speech delivery, and responses to the questions.

Extemporaneous Speaking

Extemporaneous speaking is a valuable skill for life. This skill is used every day by most horticulture professionals. Salespeople use extemporaneous speaking skills to negotiate sales. Agricultural educators use these skills to teach classes or to teach individuals how to deal with problems and issues. Agricultural executives and administrators use extemporaneous speaking skills to convince their employees and stockholders to support their leadership and business plans. Nearly everyone can benefit from learning extemporaneous speaking skills.

Extemporaneous Speaking competitions require students to gather original documents and materials in a notebook or file, but no written preparation of a manuscript may be done before the competition. Each competitor draws for a speech topic, and he or she is allowed to prepare for 30 minutes using only the materials that were assembled earlier. The judges are allowed 5 minutes to ask questions after the speech has been delivered.

Planning the Speech

A speech should have at least three sections: the introduction, body, and conclusion. The plan should clearly identify the sections.

INTRODUCTION The introduction indicates the need for and importance of the speech. It should be carefully planned and spoken with confidence. The introduction may be in the form of statements or questions. If the introduction is not to the point, does not fit the occasion, or is not delivered in a spirited manner, the audience may not listen to the rest of the speech. The introduction might be only a few lines long, but it must capture the attention of the audience.

BODY The body of the speech contains the majority of the information. It should consist of several major points that support one central theme or objective. Each major point is supported by additional information to explain, illustrate, or clarify the point. It is best to write the body of the speech in outline form (Figure 57–8).

SPEECH OUTLINE

INTRODUCTION

Honorable judges, instructors, and fellow students
1. Rabbits, cows, plants, and plows—WE STILL NEED THEM!
2. When the family farm goes under, a piece of America goes under with it

BODY

The 2010 census revealed...
1. The midsize farm is likely to be the true family farm
   - Owned and operated by the family
   - Family receives benefit from their work
2. Some believe the family farm is a relic of the past!
   a. Press fascination with bankruptcy sales
   b. Farms not in view from interstate highways
   c. Small population on farms
   d. Animal rights groups and unionizing efforts
3. The family farm endures in the United States
   a. Better managed farm businesses buy the weaker farms
   b. Disease epidemics threaten specialized operations
   c. Farm retailing is on the rise
   d. Small farms are becoming legitimate
4. Farm bankruptcy must be minimized
   a. Farm failure affects general businesses
   b. We will miss fresh farm produce
   c. We will see more pollution
   d. We will depend more on imports
5. Are there remedies? Yes!
   a. Remove politics from marketing
   b. Recognize farmers as astute business people
   c. See the total industry of agriculture
   d. Tax farm land for farm use
   e. Increase agriculture land preservation programs

CONCLUSION

Indeed...
1. General George Washington nearly lost the continental army at Valley Forge from lack of food, shelter, and clothing.
2. It can happen to us if we don't maintain a healthy farm situation in the United States today
3. Don't give away our most valuable resource—the ability to feed, clothe, and shelter ourselves!

Figure 57–8 An outline of a winning speech on the importance of family farms.

An outline will help direct the thought and delivery of the speech. After outlining the speech, a carefully worded narrative may be written to help the speaker fully develop the content of the speech. However, it must be emphasized that a speech should be given from an outline to avoid the temptation to memorize the speech. Memorization of a speech has two serious pitfalls. First, there is the danger that it will sound like someone else's words and lack authenticity. Second, if the line of thought is lost during delivery, the speaker may not be able to find the location in the narrative. This can greatly damage the quality of the speech.

CONCLUSION The conclusion should remind the audience of the major theme or central points of the speech and briefly restate them. The conclusion should leave the audience feeling like they want to take action to implement or adopt what you have said. Some speeches call for action, whereas others call for changes in attitude or perception. The more powerful speeches move people to action. The words needed to do such a big job must be carefully planned.

Giving the Speech

Giving the speech can be fun and provide much satisfaction (Figure 57–9). However, this fun and satisfaction does not come easily. The speaker must prepare ahead of time and practice the speech extensively. Practicing the speech until the content becomes familiar helps speaking become nearly automatic. Recite your speech out loud to yourself several times. Then practice in front of a mirror to observe facial expressions, posture, and gestures. Finally, give the speech in front of others, and invite them to make suggestions to improve the delivery (Figure 57–10).

Courtesy of the National FFA Organization

▲ **Figure 57–9** Speaking in public can be a pleasant and exciting experience after you have learned the proper way to organize and present your thoughts. Speaking is a skill that is learned only through practice as you stand and speak in front of an audience.

TRY THIS Complete the following tasks to develop your speaking skills:

- Choose a topic for a speech, conduct research on the topic, and write an organized speech.
- Ask your teacher to read and make suggestions on the content of your written speech.
- Learn the content thoroughly through repeated thought and practice.
- Record the speech and observe the sound, speed, power, and effectiveness of your voice. Make corrections to improve the delivery.
- Practice the speech in front of a mirror to observe posture, hand gestures, and facial expressions. Your posture should be erect and natural, with hands at your sides or resting lightly on the edges of the podium. Your hands should be used occasionally for gestures that emphasize a point, show direction, or indicate count.
- Ask your teacher for a score sheet for judging speeches. Deliver the speech in front of a trusted person who can check your delivery against the score sheet and provide suggestions for improvements. This may be a friend, relative, or teacher.
- Deliver your speech in front of your class for experience and suggestions.
- Anticipate possible questions the judges may ask and prepare the answers.
- Make some statements in your speech that you are well prepared to defend. They may lead to questions from the judges for which you are well prepared.
- Ask your teacher to critique your speech for final approval.
- Deliver your speech in front of civic groups and/or in FFA public speaking competitions.
- Record your speech on video for later critique and review.

National Public Speaking Contest — ***Judge's Score Sheet***

PART I. FOR SCORING CONTENT AND COMPOSITION

Items to Be Scored	Points Allowed	Points Awarded Contestant 1	2	3	4	5	6	7	8	9	10	11	12	13
Content of Manuscript	200													
Composition of Manuscript	100													
Score on Written Production	300													

PART II. FOR SCORING DELIVERY OF THE PRODUCTION

Items to Be Scored	Points Allowed	Points Awarded Contestant 1	2	3	4	5	6	7	8	9	10	11	12	13
Voice	100													
Stage Presence	100													
Power of Expression	200													
Response to Questions	200													
General Effect	100													
Score on Delivery	700													

PART III. FOR COMPUTING THE RESULTS OF THE CONTEST

Items to Be Scored	Points Allowed	Points Awarded Contestant 1	2	3	4	5	6	7	8	9	10	11	12	13
Score on Written Production	300													
Score on Delivery	700													
TOTALS	1000													
* Less Overtime Deductions, for each minute or major fraction thereof	20													
* Less Undertime Deductions, for each minute or major fraction thereof	20													
GRAND TOTALS														
Numerical or Final Placing														

* From the Timekeeper's record.

Explanation of Score Sheet Points

Part I: For Scoring Content and Composition

1. *Content of the manuscript* includes:
   Importance and appropriateness of the subject
   Suitability of the material used
   Accuracy of statements included
   Evidence of purpose
   Completeness and accuracy of bibliography
2. *Composition of the manuscript* includes:
   Organization of the content
   Unity of thought
   Logical development
   Language used
   Sentence structure
   Accomplishment of purpose-conclusions

Part II: For Scoring Delivery of Production

1. *Voice* includes:
   Quality
   Pitch
   Articulation
   Pronunciation
   Force
2. *Stage presence* includes:
   Personal appearance
   Poise and body posture
   Attitude
   Confidence
   Personality
   Ease before audience
3. *Power of expression* includes:
   Fluency
   Emphasis
   Directness
   Sincerity
   Communicative ability
   Conveyance of thought and meaning
4. *Response to questions* includes:
   *Ability to answer satisfactorily the questions on the speech that are asked by the judges indicating originality, familiarity with subject, and ability to think quickly.
5. *General effect* includes:
   Extent to which the speech was interesting, understandable, convincing, pleasing, and held attention.

*NOTE: Judges should meet prior to the contest to prepare and clarify the questions asked.

Figure 57–10 A score sheet for evaluating speeches.

Parliamentary Procedure

What is parliamentary procedure, and why are so many people familiar with it? Why is it important? Why should horticulture students be interested in learning parliamentary procedure? Parliamentary procedure is a system of guidelines or rules for conducting meetings. Most Americans who are influential in their communities are familiar with parliamentary procedure.

Parliamentary procedure is used to guide the meetings conducted by city councils, school boards, church groups, commissions, professional organizations, and civic organizations, such as garden clubs, Lions, Rotary, and Ruritan clubs. Horticulture students should be interested in learning this procedure so they can have their opinions heard and influence decisions that affect their lives. Parliamentary procedure is important because it permits a group to do the following:

- Discuss one thing at a time.
- Hear everyone's opinion in a courteous atmosphere.
- Protect the rights of minorities.
- Make decisions according to the wishes of the majority of the group.

Requirements for a Good Business Meeting

A good **business meeting** is a gathering of people working together to make wise decisions. Wrong decisions cause unhappiness, loss of income, inefficiency in business and social activities, injury, and other problems. The most common outcomes of poorly run business meetings are waste of time and lack of results. Meetings run by groups of individuals who know and use parliamentary procedure are smooth, efficient, orderly, and focused, and such meetings accomplish much more than poorly organized meetings (Figure 57–11). Some requirements for a good business meeting follow.

Effective Presiding Officer

A **presiding officer** is a president, vice president, or chairperson who is designated to lead a business meeting. He or she should be committed to the goals of the organization and should want to lead the group in making good group decisions. A good presiding officer must know and use proper parliamentary procedure.

Courtesy of the National FFA Organization

▲ **Figure 57–11** Membership in the FFA provides opportunities for students to lead class discussions and to learn to conduct the business of an organization by taking turns as the presiding officer.

Competent Secretary

A **secretary** is a person elected or appointed to take notes and prepare minutes of the meeting, which refers to the official written record of a business meeting. Minutes should include the date, time, place, presiding officer, attendance, and motions discussed at the meeting. They should be written clearly, include all actions taken by the group, and be kept in a permanent secretary's book.

Informed Members

Informed members are members who are active in the organization and want to be part of the group. They give previous thought to issues to be discussed and gather useful information about the issues. They share these thoughts with others in the meeting. This permits everyone to have the benefit of the best thinking in the group and permits the best decisions to be made. Effective members know and use parliamentary procedure to bring out important points of discussion and to advance the agenda of the meeting in an orderly manner.

A Comfortable Meeting Room

The meeting place must be comfortable and free from distractions. A moderate temperature and good lighting are essential. Members should be seated so they can hear and see each other. Seating people at a table or in a circle works well for small groups. Large groups must rely on a good sound system for the presiding officer to be heard, and members must speak clearly with good volume. Good public speaking skills and thorough knowledge of parliamentary procedure help the members conduct effective meetings.

Courtesy of the National FFA Organization

▲ **Figure 57–12** The presiding officer and each member should take time to learn how to use the gavel as a tool to conduct the business of an organization.

Conducting Meetings

THE ORDER OF BUSINESS The **order of business** refers to the items and sequence of activities conducted at a meeting. The order of business is usually made up by the secretary. This generally grows out of an **executive meeting**, which is a meeting of the officers to conduct the business of the organization between regular meetings. They may also consider what needs to be discussed by the total membership at the regular meeting. The essential items in an order of business include the following:

- Call to order
- Reading and approval of minutes of the previous meeting
- Treasurer's report
- Reports of other officers and committees
- Old business
- New business
- Adjournment

Other items or activities that are frequently included in orders of business are programs, speakers, or entertainment.

Parliamentary Practices

USE OF THE GAVEL The **gavel** is a wooden mallet used by the presiding officer to direct a meeting (Figure 57–12). It is used to call the meeting to order, announce the result of votes, and adjourn the meeting. It is also used to signal the members to stand, sit down, or reduce the noise level of the group. The gavel is a symbol of the authority of the office of president or chairperson, and it should be respected by all attending the meeting.

In some organizations, such as the FFA, a system of taps is used to signal the audience to do certain things. In FFA meetings, the gavel is used as follows:

- One tap means the outcome of or decision about the item under consideration has been announced by the presiding officer.
- Two taps means the meeting will come to order, members should sit down if standing, or members should be quiet except when recognized.
- Three taps means members should stand up.

Obtaining Recognition and Permission to Speak

For a meeting to be orderly, members must speak one at a time and in some logical and fair sequence. The presiding officer is regarded as the "traffic controller," and calls on members as they request to be recognized according to certain rules. To be recognized, the member should raise a hand to get the presiding officer's attention. The presiding officer should call the member by name; then the person should stand and address the presiding officer as Madam or Mr. Chairperson, or Madam or President. The individual should then proceed to speak. Both

SUMMARY OF MOTIONS					
Motion	**Debatable**	**Amendable**	**Vote Required**	**Second**	**Reconsider**
PRIVILEGED					
Fix time which to adjourn	No	Yes	Majority	Yes	Yes
Adjourn	No	No	Majority	Yes	No
Recess	No	Yes	Majority	Yes	No
Question of privilege	No	No	None	None	Yes
Call for orders of the day	No	No	2/3	None	No
INCIDENTAL					
Appeal	Yes	No	Majority	Yes	Yes
Point of order	No	No	None	No	No
Parliamentary inquiry	No	No	None	No	No
Suspend the rules	No	No	2/3	Yes	No
Withdraw a motion	No	No	Usually none	No	No
Object consideration of question Negative vote only	No	No	2/3	No	Yes
Division of the question	No	Yes	Majority	Yes	No
Division of the assembly	No	No	No	No	No
SUBSIDIARY					
Lay on table	No	No	Majority	Yes	No
Previous question before vote	No	No	2/3	Yes	Yes
Limit debate	No	Yes	2/3	Yes	Yes
Postpone definitely	Yes	Yes	Majority	Yes	Yes
Refer to committee	Yes	Yes	Majority	Yes	Yes
Amend	Yes	Yes	Majority	Yes	Yes
Postpone indefinitely	Yes	No	Majority	Yes	Yes vote only
Main motion	Yes	Yes	Majority	Yes	Yes
UNCLASSIFIED					
Take from table	No	No	Majority	Yes	No
Reconsider	No	No	Majority	Yes	Negative vote only

For more details on parliamentary procedure, see a parliamentary procedure book such as *Robert's Rules of Order.*

Courtesy of the National FFA Organization

▲ **Figure 57–13** Parliamentary skills are useful in FFA and other organizations.

the presiding officer and the members should understand the correct classification of motions (Figures 57–13 and 57–14).

Presenting a Motion

A **motion** is a proposal, presented in a meeting, to be acted upon by the group. To present a motion, the member raises his or her hand and is recognized by the presiding officer. Then the member states, "Madam/Mr. President, I move that . . ." (and continues with the rest of the motion). The words "I move" are important to say when beginning the motion. Otherwise, you will be regarded as incorrect in your usage of good parliamentary procedure. For a motion to be discussed by the group, at least one other member must be willing to have the motion discussed. That second individual expresses this willingness by saying, "Madam/Mr. President, I second the motion."

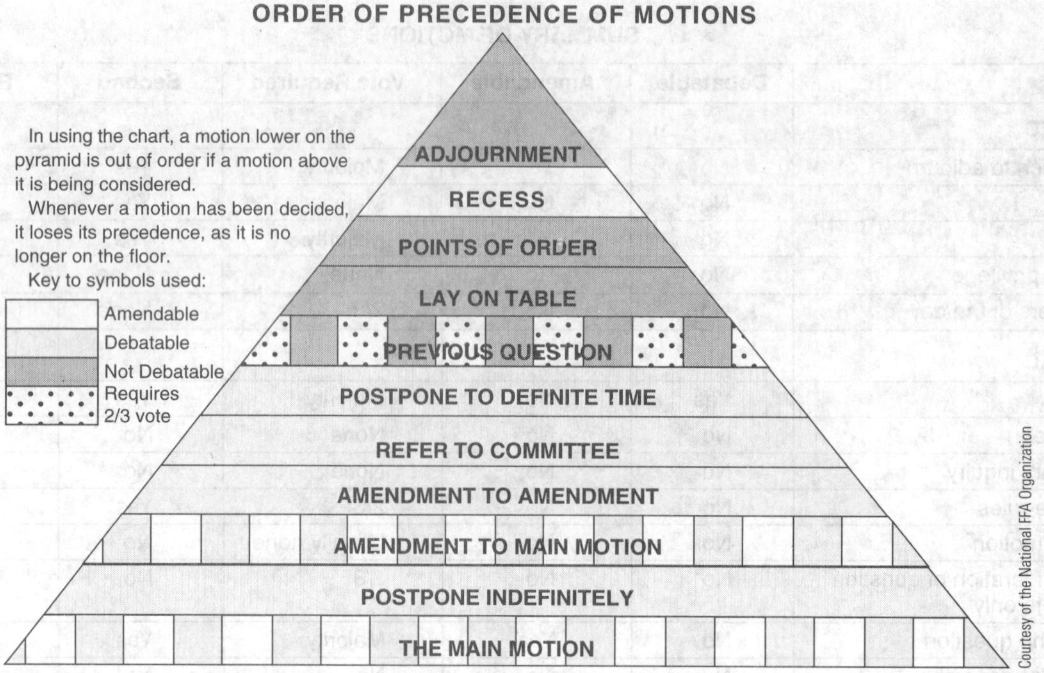

▲ **Figure 57–14** Correct order of precedence of motions.

Some Useful Motions

There are dozens of motions, but a few basic motions are generally known and widely used:

- **Main motion**. A basic motion used to present a proposal for the first time. The way to state it is to obtain recognition from the chairman and then say, "I move . . . ."
- **Amend**. A type of motion used to add to, subtract from, or strike out words in a main motion. The way to present an amendment is to say, "I move to amend the motion by . . . ."
- **Refer**. A motion used to refer to a committee or person for finding more information and/or taking action on the motion. The way to state a referral is to say, "I move to refer this motion to . . . ."
- **Lay on the table**. A motion used to stop discussion on a motion until the next meeting. The way to table a motion is to say, "I move to table the motion."
- **Point of order**. A procedure used to object to some item in or about the meeting that is not being presented properly. The procedure to use is to stand up and say, "Madam/Mr. President, I rise to a point of order!" The presiding officer should then recognize the member by saying, "State your point." The member then explains what has been done incorrectly.
- **Adjourn**. A motion used to close a meeting. The procedure is to say, "I move to adjourn."

The FFA *Student Handbook* provides a listing of additional motions and explains how to use parliamentary procedure for more effective meetings. Horticulture students are encouraged to develop effective leadership skills. The ability to work effectively as a member of a group is essential for all. The ability to function as a chairperson or officer creates more opportunities to serve and influence the communities in which we live. The development of self-confidence is essential. Self-confidence is a product of knowledge and skill. Therefore, each student should strive to learn to speak well, to function in groups through parliamentary procedure, and to use the opportunities in FFA for personal growth and development.

SAFETY FIRST

Being the owner and manager of a business means also being a leader. Being a good leader requires exemplifying safe practices and ensuring the safety of all employees. OSHA's regulations ensure that employers and employees know about work hazards and how to protect themselves to prevent injuries in the workplace. Therefore, the owner or manager of a business should be knowledgeable about OSHA regulations and how to keep the working environment safe. For more information on occupational health and safety in the workplace, please visit the OSHA site at www.osha.gov.

Summary

Leadership is paramount to any organization's success. The skills developed are a vital asset to each individual who is involved with an organization that has a supreme focus. The National FFA Organization is dedicated to making a positive difference in the lives of students by developing their potential for premier leadership, personal growth, and career success through agricultural education. The success of the 4-H program is based on the 4-H family, their involvement, commitment, and the many volunteer hours given by adult leaders who contribute in any way they can by sharing their time and talents. A well-rounded student will be involved in as many youth leadership organizations as possible.

Review Questions

1. How would you define leadership?
2. Why is effective leadership needed in agriscience?
3. What are some characteristics of good leaders?
4. Discuss some of the opportunities FFA offers for leadership development.
5. How would you demonstrate positive leadership skills?

Student Activities

1. Write the Terms to Know and their definitions in your notebook.
2. Make a list of the many ways that you exercise leadership in your family, school, and community.
3. Develop a poster board showing the symbols of the FFA emblem.

4. Develop a handout or flyer illustrating the purposes of the FFA. Include the FFA colors.
5. Write down five CDEs and five proficiency awards for which you would like to try out. Discuss these with your classmates and teacher. Determine what tasks are required for each event.
6. Participate in the Extemporaneous Speaking CDE at the local level.
7. Ask your teacher to let you use a gavel and direct a mock class or FFA meeting to learn parliamentary skills.
8. Form a parliamentary procedure team consisting of you and your classmates, and then demonstrate various parliamentary skills to the class.
9. Under the supervision of your teacher, develop an appropriate solution for a school or district issue using parliamentary procedure. Nominate and appoint a small committee to present your idea, in the form of a main motion, to the principal or school board.
10. Consider running for a leadership position in your FFA chapter. Find which offices are available, and make a list of the requirements.
11. Attend an FFA activity you have never participated in. This could be a local chapter, district, state, regional, or national activity or convention.
12. Search online using the following terms: leadership defined; leadership traits; leadership skills; National 4-H Clubs; National FFA Organization; Boy Scouts of America; Girl Scouts of the United States of America; FFA CDEs; how to give a speech; how to use parliamentary procedure.

Suggested Supervised Agricultural Experience (SAE) Activities

1. To qualify for a placement SAE, intern with the staff of 4-H or another youth leadership organization.
2. As a community service SAE, organize and lead the community to restore the landscape of a local public area.

Additional Resources

www.ffa.org

www.4-h.org

www.scouting.org

Appendix A Scientific and Common Names

Abelia grandiflora—Abelia
Abies balsamea—Balsam Fir
Abies concolor—White Fir
Abies nordmanniana—Nordman Fir
Acer ginnala—Amur Maple
Acer palmatum—Japanese Maple
Acer palmatum 'Atropurpureum'—Japanese Maple
Acer platanoides—Norway Maple
Acer platanoides 'Crimson King'—Crimson King Maple
Acer platanoides 'Emerald Queen'—Emerald Queen Maple
Acer rubrum—Red Maple
Acer rubrum 'October Glory'—October Glory Maple
Acer rubrum 'Red Sunset'—Red Sunset Maple
Acer saccharinum—Silver Maple
Acer saccharum—Sugar Maple
Achillea millefolium—Yarrow
Adiantum cuneatum 'Croweanum'—Maidenhair Fern
Aechmea fasciata—Bromeliads
Aeschynanthus lobbianus—Lipstick Vine
Aesculus carnea—Red Horsechestnut
Aesculus parviflora—Bottlebrush Buckeye
Ageratum houstonianum—Ageratum
Aglaonema 'Emerald Bay'—Emerald Bay Aglaonema
Aglaonema simplex—Aglaonema
Agropyron cristatum—Crested Wheatgrass
Agrostis palustris—Creeping Bentgrass
Agrostis tenuis—Colonial Bentgrass
Ajuga reptans—Bugleweed
Alliums schoenoprasum—'Forescate' Chive
Aloe vera—Medicine Plant
Alyssum saxatile—Golddust
Ananas comosus—Pineapple
Anchusa italica—Alkanet
Andropogon gerardi—Big Bluestem
Anemone pulsatilla—Windflower
Anthemis tinctoria—Golden Daisy
Antirrhinum sp.—Snapdragon
Aphelandra squarrosa—Zebra Plant
Aquilegia alpina—Columbine
Aquilegia canadensis—Columbine
Aquilegia sp.—Columbine
Arabis alpina—Rockcress
Araucaria excelsa—Norfolk Island Pine
Araucaria heterophylla—Norfolk Island Pine
Arctostaphylos uva-ursi—Bearberry
Armeria maritima—Sea Pink
Asarum caudatum—Wild Ginger
Asparagus sprengeri—Asparagus Fern
Aspidistra elatior—Aspidistra
Asplenium nidus—Birds Nest Fern
Aster alpinus—Hardy Aster
Astilbe arendsii—False Spirea
Aucuba japonica—Japanese Aucuba
Axonopus affinis—Carpet Grass
Beaucarnea recurvata—Ponytail Palm
Begonia evansiana—Hardy Begonia
Begonia socotrana—Begonia
Berberis thunbergil—Japanese Barberry
Betula pendula—Weeping Birch
Bougainvillea spectabilis—Bougainvillea
Bouteloua curtipendula—Sideoats Grama
Bouteloua gracilis—Blue Grama
Browallia sp.—Browallia
Buchloe dactyloides—Buffalo Grass

Buddleia davidii—Butterfly Bush
Buxus sempervirens—Common Boxwood
Calendula sp.—Pot Marigold
Callistephus chinensis—Aster
Calluna vulgaris—Heather
Camellia japonica—Camellia
Campanula calycanthema—Bellflower
Campanula medium—Canterbury Bells
Campsis radicans—Trumpet Vine
Canna indica—Canna Lily
Carex morrowii variegata—Sedge
Carya illinoinensis—Pecan
Cedrus atlantica 'Glauca'—Blue Atlas Cedar
Cedrus deodara—Deodar Cedar
Celastrus scandens—American Bittersweet
Celosia sp.—Cockscomb
Cerastium tomentosum—Snow-in-Summer
Ceratostigma plubaginoides—Plumbago
Cercidyphyllum japonicum—Katsura Tree
Cercis canadensis—Redbud
Chaenomeles japonica—Flowering Quince
Chamaecyparis obtusa 'Nana'—Dwarf Hinoki Cypress
Chamaecyparis pisifera 'Plumosa'—Plume False-Cypress
Chamaecyparis pisifera 'Squarrosa'—Moss False-Cypress
Chasmanthium latifolium—Wild Oats
Chionanthus virginicus—Old Man's Beard
Chlorophytum comosum—Spider Plant
Chrysanthemum maximum—Shasta Daisy
Chrysanthemum morifolium—Chrysanthemum
Cissus rhombifolia—Grape Ivy
Clematis hybrida—Clematis
Clematis 'Jackmani'—Clematis
Clematis montana 'Nellie Moser'—Nellie Moser Clematis
Cleome speciosa—Spiderflower
Codiaeum variegatum—Croton
Coleus blumei—Coleus
Convallaria majalis—Lily of the Valley
Coreopsis lanceolata—Tickseed
Coreopsis sp.—Coreopsis
Cornus canadensis—Bunchberry
Cornus florida—Flowering Dogwood
Cornus kousa—Chinese Dogwood
Coronilla varia—Crown Vetch
Cortaderia—Pampas Grass
Corylus avellana—Filbert
Cosmos sp.—Cosmos
Cotoneaster adpressus 'Praecox'—Cotoneaster
Cotoneaster congesta—Cotoneaster
Cotoneaster dammeri—Lowfast Cotoneaster
Cotoneaster dammeri 'Skogholmen'—Skogholmen Cotoneaster
Cotoneaster horizontalis—Rock Spray Cotoneaster
Crassula argentea—Jade Plant
Crataegus crus-galli—Cockspur Hawthorne
Crataegus oxyacantha 'Pauli'—Paul's Scarlet Hawthorne
Crataegus phaenopyrum—Washington Hawthorne
Cryptomeria japonica—Japanese Cryptomeria
Cupressocyparis leylandil—Leyland Cypress
Cyclamen persicum—Cyclamen
Cynodon dactylon—Bermudagrass
Cytisus 3 'Praecox'—Warminster Broom
Dahlia variabilis—Dahlia
Daphne odora—Daphne
Delphinium belladonna—Delphinium
Delphinium elatum—Delphinium
Delosperma congestum—'Gold Nugget' Hardy Iceplant
Delphinium sp.—Larkspur
Deutzia gracilis—Slender Deutzia
Dianthus sp.—Pinks
Dianthus barbatus—Sweet William
Dianthus caryophyllus—Carnation
Dieffenbachia 'Exotica'—Dieffenbachia
Digitalis purpurea—Foxglove
Dimorphotheca sp.—Cape Marigold
Dizygotheca elegantissima—Aralia
Dolichos lablab—Hyacinth Bean
Dracaena fragrans 'Massangeana'—Corn Plant Dracaena
Echeveria secunda—Echeveria (Cat and Kittens)
Echinacea purpurea—Cone Flower
Elaeagnus umbellata—Autumn Olive
Elymus virginicus—Wild Ryegrass
Epimedium grandiflorum—Barrenwort
Eremochloa ophiuroides—Centipede Grass
Erica carnea—Heath
Euonymus alatus—Winged Euonymus

Euonymus fortunei 'Argenteo marginatus'—Wintercreeper Euonymus
Euonymus fortunei 'Carrieri'—Wintercreeper Euonymus
Euonymus fortunei 'Colorata'—Wintercreeper Euonymus
Euonymus fortunei 'Vegetus'—Wintercreeper Euonymus
Euonymus japonicus—Euonymus
Euonymus kiautschovicus—Spreading Euonymus
Euphorbia myrsinites spurge—Myrtle Euphorbia
Euphorbia pulcherrima—Poinsettia
Fagus grandifolia—American Beech
Fagus sylvatica—Beech
Fatshedera lizei—Fatshedera
Festuca arundinacea—Tall Fescue
Festuca glauca—Blue Fescue
Festuca rubra—Creeping Red Fescue
Ficus benjamina 'Midnight Braid'—Ficus Midnight Braid
Ficus benjamina 'Monique Corkscrew'— Ficus Monique Corkscrew Ficus
Ficus elastica 'Decora'—Rubber Plant
Ficus lyrata 'Little Fiddle'—Little Fiddle Ficus
Ficus pumila 'Repans'—Creeping Ficus
Fittonia sp.—Fittonia
Fothergilla monticola—Fothergilla
Franklinia alatamaha—Franklinia
Fraxinus americana—White Ash
Fraxinus pennsylvanica lanceolata 'Marshall'—Green Ash
Fritillaria imperialis—Fritillaria
Fuchsia excorticata—Fuchsia
Gaillardia grandiflora—Gaillardia
Gaillardia sp.—Gaillardia
Gardenia jasminoides—Gardenia
Gaultheria procumbens—Wintergreen
Gelsemium sempervirens—Carolina Jasmine
Ginkgo biloba—Ginkgo
Gladiolus grandiflorus—Gladiolus
Gleditsia triacanthos 'Shademaster'—Shademaster Honeylocust
Gomphrena globosa—Globe Amaranth
Gynura sarmentosa—Velvet Plant
Gypsophila paniculata—Baby's Breath
Hedera helix—English Ivy
Hedera helix 'Baltic'—Baltic English Ivy
Helianthemum nummularium—Sunrose
Helianthus sp.—Sunflower
Helichrysum sp.—Strawflower
Hermerocallis sp.—Daylily
Hibiscus moscheutos—Mallow Rose
Hibiscus rosea 'Lord Baltimore'—Lord Baltimore Hibiscus
Hibiscus syriacus—Rose of Sharon
Hippeastrum sp.—Amaryllis
Hosta undulata—Hosta
Hoya carnosa—Hoya
Hyacinthus orientalis—Hyacinth
Hydrangea quercifolia—Oakleaf Hydrangea
Hypericum sp.—St. John's Wort
Iberis sp.—Candytuft
Ilex aquifolium—English Holly
Ilex crenata—Japanese Holly
Ilex cornuta 'Nellie Stevens'—Nellie Stevens Holly
Ilex crenata 'Convexa'—Convexleaf Japanese Holly
Ilex crenata 'Helleri'—Helleri Holly
Ilex crenata 'Hetzi'—Hetz Holly
Ilex crenata 'San Jose'—San Jose Japanese Holly
Ilex crenata 'Stokes'—Stokes Holly
Ilex glabra—Inkberry Holly
Ilex × *meserveae*—Meservea Holly
Ilex opaca—American Holly
Ilex verticillata—Winterberry
Impatiens wallerana—Impatiens
Imperata—Japanese Blood Grass
Ipomoea sp.—Morning Glory
Iris sp.—Iris (Flag)
Juglans nigra—Black Walnut
Juglans regia—English Walnut (Persian Walnut)
Juniperus chinensis 'Hetzii'—Hetz Juniper
Juniperus chinensis 'Pfitzeriana'—Pfitzer Juniper
Juniperus chinensis 'Sargentii'—Sargent Juniper
Juniperus conferta—Blue Pacific Juniper
Juniperus horizontalis 'Douglasii'—Waukegan Juniper
Juniperus horizontalis 'Plumosa'—Andorra Juniper
Juniperus horizontalis 'Wiltoni'—Blue Rug Juniper
Juniperus squamata—Meyeri Juniper
Juniperus virginiana—Eastern Redcedar
Kalanchoe blossfeldiana—Kalanchoe
Kalanchoe pinnata—Kalanchoe

Kalmia latifolia—Mountain Laurel
Kerria japonica—Mt. Vernon Shrub
Koeleria macrantha—Prairie Junigrass
Koelreuteria paniculata—Golden Rain Tree
Kolkwitzia amabilis—Beauty Bush
Lagerstroemia indica—Dwarf Crape Myrtle
Lamium maculatum—Spotted Dead Nettle
Lantana camara—Lantana
Larix decidua—European Larch
Lathyrus sp.—Sweet Pea
Leucothoe catesbaei—Leucothoe
Leucothoe fontanesiana—Leucothoe
Leymus—Blue Wild Rye
Liatris pycnostachya—Gayfeather
Liatris spicata—Purple Gay Feather
Ligustrum japonicum—Privet
Ligustrum lucidum—Glossy Privet
Ligustrum ovalifolium—California Privet
Lilium longiflorum—Lily
Limonium sp.—Statice
Limonium sinuatum—Statice
Lippia citriodora—Lemon Verbena
Liquidambar styraciflua—Sweet Gum
Liriodendron tulipifera—Tulip Tree
Liriope muscari—Liriope
Lobelia sp.—Lobelia
Lobularia sp.—Sweet Alyssum
Lolium multiflorum—Annual Rye Grass
Lolium perenne—Perennial Rye Grass
Lonicera japonica—Japanese Honeysuckle
Lonicera sempervirens—Honeysuckle
Lonicera tatarica—Tatarian Honeysuckle
Lupinus polyphyllus—Lupine
Lysimachia nummularia—Moneywort
Macadamia integrifolia—Macadamia
Magnolia acuminata—Cucumber Magnolia
Magnolia grandiflora—Southern Magnolia
Magnolia stellata—Star Magnolia
Magnolia virginiana—Sweet Bay Magnolia
Magnolia × 'Soulangeana'—Saucer Magnolia
Mahonia aquifolium—Oregon Holly-Grape
Mahonia bealei—Leatherleaf Mahonia
Mahonia repens—Creeping Mahonia
Malus floribunda—Japanese Flowering Crabapple
Maranta leuconeura—Prayer Plant
Matteuccia struthiopteris—Ostrich Fern
Matthiola sp.—Stock
Metasequoia glyptostroboides—Dawn Redwood
Mirabilis sp.—Four-o-Clock
Miscanthus floridulus—Giant Miscanthus
Mitchella repens—Partridgeberry
Musa maurelii—Banana
Myosotis sp.—Forget-Me-Not
Nandina domestica—Heavenly Bamboo
Narcissus jonquilla—Narcissus (Daffodils)
Nephrolepis exaltata—Boston Fern
Nicotiana sp.—Flowering Tobacco
Nyssa sylvatica—Black Gum
Ophiopogon japonicus—Dwarf Lilyturf
Osteospermum ecklonis—African Daisy
Oxydendrum arboreum—Sourwood
Pachysandra terminalis—Japanese Pachysandra
Paeonia sp.—Peony
Panicum virgatum—Switch Grass
Papaver nudicaule—Iceland Poppy
Papaver orientalis—Oriental Poppy
Papaver sp.—Poppy
Parthenocissus sp.—Boston Ivy
Parthenocissus tricuspidata—Boston Ivy
Paspalum notatum—Bahia Grass
Paxistima canbyi—Paxistima
Pelargonium hortorum—Geranium
Pennisetum sp.—Dwarf Fountain Grass
Pennisetum glaucum sp.—Pearl Millet
Pentas lanceolata sp.—Pentas
Peperomia obtusifolia—Peperomia
Peris japonica—Japanese Andromeda
Petunia hybrida—Petunia
Phalaenopsis amabilis sp.—Phalaenopsis Orchid
Phalaris arundinacea 'Picta'—Ribbon Grass
Philadelphus tomentosus—Mock Orange
Philodendron cordatum—Heartleaf Philodendron
Phlox paniculata—Summer Phlox
Phlox sp.—Phlox
Phlox subulata—Moss Phlox
Physalis alkekengi—Chinese Lantern
Picea abies—Norway Spruce
Picea abies 'Nudiformis'—Bird's Nest Spruce
Picea omorika—Serbian Spruce
Picea pungens—Colorado Spruce
Picea pungens 'Glauca'—Blue Colorado Spruce
Pieris floribunda—Mountain Andromeda

Pieris japonica—Andromeda
Pilea cadierei—Pilea
Pinus griffithi—Himalayan Pine
Pinus mugo 'Mugo'—Mugho Pine
Pinus nigra—Austrian Pine
Pinus resinosa—Red Pine
Pinus rigida—Pitch Pine
Pinus strobus—White Pine
Pinus sylvestris—Scotch Pine
Pinus taeda—Loblolly Pine
Pinus virginiana—Virginia Pine
Pittosporum tobira—Pittosporum
Platanus occidentalis—Sycamore
Platycerium bifurcatum—Staghorn Fern
Plectranthus australis—Swedish Ivy
Poa annua—Annual Blue Grass
Poa pratensis—Kentucky Blue Grass
Poa trivialis—Rough Blue Grass
Podocarpus macrophyllus—Yew Podocarpus
Polygonum aubertii—Silverlace Vine
Populus nigra—Lombardy Poplar
Portulaca sp.—Rose Moss
Potentilla fruticosa—Cinquefoil
Primula polyantha—Primrose
Prunus cerasifera 'Thundercloud'—Thundercloud Flowering Plum
Prunus dulcis—Almond
Prunus grandulosa—Flowering Almond
Prunus serrulata—Oriental Cherry
Prunus subhirtella 'Pendula'—Weeping Cherry
Pseudotsuga menziesii—Douglas Fir
Pyracantha coccinea 'Lalandi'—Firethorn
Pyracantha sp.—Pyracantha
Pyrethrum roseum—Painted Daisy
Quercus alba—White Oak
Quercus borealis—Red Oak
Quercus coccinea—Scarlet Oak
Quercus palustris—Pin Oak
Quercus virginiana—Live Oak
Rhododendron minus 'Carolinianum'—Rhododendron Carolinianum
Rhododendron pericly-menoides—Azalea Nudiflora
Rhododendron 'Exbury Hybrid'—Deciduous Azalea
Rhododendron 'Gibraltar'—Deciduous Azalea Gibraltar
Rhododendron 'Hino Crimson'—Hino Crimson Azalea
Rhododendron 'Hinodegiri'—Hinodegiri Azalea
Rhododendron 'P.J.M.'—P.J.M. Rhododendron
Rhododendron 'Purple Gem'—Purple Gem Rhododendron
Rhododendron prunifolium—Deciduous Azalea
Rhododendron schlippenbachi—Schlippenbachi Rhododendron
Rhododendron sp.—Azalea
Rhododendron × 'Catawbiense'—Catawba Rhododendron
Rosa banksias—Lady Bank Rose
Rosa hybrida—Hybrid Tea Rose
Rosmarinus officinalis—Rosemary
Rudbeckia fulgida 'Goldstrum'—Black-Eyed Susan
Saintpaulia ionantha—African Violet
Salpiglossis sp.—Giant Velvet Flower
Salvia sp.—Scarlet Sage
Sansevieria hahni—Hahn Snake Plant
Sansevieria trifasciata—Sansevieria (Snake Plant)
Sarcococca hookeriana 'Humilis'—Sarcococca
Scabiosa sp.—Pincushion Flower
Schefflera actinophylla—Schefflera
Schefflera digitata—Umbrella Tree
Scindapsus aureus—Pothos
Sedum sp.—Stonecrop
Sempervivums sp.—Hens and Chicks
Senecio cruentus—Cineraria
Sinningia speciosus—Gloxinia
Sorbus aucuparia—European Mountain Ash
Spiraea bumalda 'Anthony Watereri'—Anthony Waterer Spiraea
Spiraea japonica 'Lemon Princess'—Lemon Princess Spirea
Spiraea prunifolia—Bridalwreath Spirea
Spiraea vanhouttei—Vanhoutte Spirea
Sporobolus heterolepis—Prairie Dropseed
Stenotaphrum secundatum—St. Augustine Grass
Stephanandra incisa—Cutleaf Stephanandra
Strelitzia reginae—Bird of Paradise
Syngonium sp.—Nephthytis
Syringa chinensis—Chinese Lilac
Syringa vulgaris—Common Lilac

Tagetes sp.—Marigold
Taxodium distichum—Bald Cypress
Taxus baccata 'Repandens'—Spreading English Yew
Taxus cuspidata aurescens—Dwarf Japanese Yew
Taxus cuspidata 'Capitata'—Japanese pyramidal yew
Taxus × media 'Hicksi'—Hicks Yew
Taxus × media 'Nigra'—Anglojap Yew
Teucrium chamaedrys—Germander
Thuja occidentalis—American Arborvitae
Thuja orientalis—Oriental Arborvitae
Thunbergia sp.—Black-Eyed Susan Vine
Tilia americana—American Linden
Tilia cordata—Little Leaf Linden
Tillandsias sp.—Air Plant
Tithonia roundifolia—Mexican Sunflower
Trachelospermum jasminodes sp.—Star Jasmine
Tropaeolum sp.—Nasturtium
Tsuga canadensis—Canada Hemlock
Tulipa hybrida—Tulip
Ulmus americana—American Elm
Verbena × *hortensis*—Verbena
Veronica spicata—Speedwell
Viburnum carlesii—Korean Spice Viburnum
Viburnum dentatum—Arrowwood
Viburnum macrocephalum—Chinese Snowball
Viburnum opulus—Cranberry Bush
Viburnum plicatum 'Tomentosum'—Doublefile Viburnum
Vinca major—Large Leaf Vinca
Vinca minor—Periwinkle
Viola tricolor—Pansy
Weigela sp.—Weigela
Wisteria floribunda—Wisteria
Wisteria sinensis—Wisteria
Wisteria sp.—Wisteria
Yucca filamentosa—Desert Candle (Yucca)
Zebrina pendula—Wandering Jew
Zinnia sp.—Zinnia
Zoysia japonica—Zoysia Grass
Zygocactus truncatus—Christmas Cactus

Appendix B

Professional, Educational, and Trade Organizations for Horticulture and Related Areas

Agronomy, Soil

AMERICAN SOCIETY OF AGRONOMY
5585 Guilford Road
Madison, WI 53711-5801
(608) 273-8080
http://www.agronomy.org

CROP SCIENCE SOCIETY OF AMERICA
5855 Guilford Road
Madison, WI 53711-5801
(608) 273-8080
http://www.crops.org

SOIL AND WATER CONSERVATION SOCIETY OF AMERICA
945 SW Ankeny Road
Ankeny, IA 50023-9723
(515) 289-2331
http://www.swcs.org

SOIL SCIENCE SOCIETY OF AMERICA
5585 Guilford Road
Madison, WI 53711-5801
(608) 273-8080
http://www.soils.org

WEED SCIENCE SOCIETY OF AMERICA
810 East 10th Street
Lawrence, KS 66044
(800) 627-0326
http://www.wssa.net

Botany

AMERICAN BOTANICAL COUNCIL
6200 Manor Road
Austin, TX 78723
(512) 926-4900
http://www.herbalgram.org

AMERICAN PUBLIC GARDENS ASSOCIATION
351 Longwood Road
Kennett Square, PA 19348
(610) 708-3010
http://www.publicgardens.org

BOTANICAL SOCIETY OF AMERICA
PO Box 299
St. Louis, MO 63166
(314) 577-9566
http://www.botany.org

Business Management

UNITED STATES SMALL BUSINESS ASSOCIATION
409 3rd Street, SW
Washington, DC 20416
(800) 827-5722
http://www.sba.gov

Educational Organizations

ASSOCIATION OF CAREER AND TECHNICAL EDUCATION
1410 King Street
Alexandria, VA 22314
(800) 826-9972
http://www.acteonline.org

NATIONAL ASSOCIATION OF AGRICULTURAL EDUCATION
300 Garrigus Building
University of Kentucky
Lexington, KY 40546-0215
(800) 509-0204
http://www.naae.org

NATIONAL FFA ORGANIZATION
PO Box 68960
6060 FFA Drive
Indianapolis, IN 46268-0960
http://www.FFA.org

Fertilizers, Chemicals

FERTILIZER INSTITUTE
425 Third Street, SW
Suite 950
Washington, DC 20024
(202) 962-0490
http://www.tfi.org

Flowers, Plants

AMERICAN CAMELLIA SOCIETY
100 Massee Lane
Fort Valley, GA 31030
(478) 967-2083
http://www.americancamellias.com

AMERICAN ORCHID SOCIETY
10901 Old Cutler Road
Coral Gables, FL 33156
(305) 740-2010
http://www.aos.org

AMERICAN RHODODENDRON SOCIETY
PO Box 525
Niagara Falls, NY 14304
(416) 424-1042
http://www.rhododendron.org

AMERICAN ROSE SOCIETY
8877 Jefferson Paige Road
Shreveport, LA 71130
http://www.rose.org

HOLLY SOCIETY OF AMERICA
PO Box 803
Millville, NY 08332-0803
(856) 825-4300
http://www.hollysocam.org

Government Agencies

USDA ANIMAL AND PLANT HEALTH INSPECTION SERVICE
1400 Independence Avenue SW
Washington D.C. 20250
(844) 820-2234
http://www.aphis.usda.gov

USDA NATURAL RESOURCES CONSERVATION SERVICE
1185 New Litchfield Street
Torrington, CT 06790
(860) 626-8258
http://www.nrcs.usda.gov/wps/portal/nrcs/site/national/home/

Horticulture

AMERICAN HORTICULTURAL SOCIETY
7931 E. Boulevard Drive
Alexandria, VA 22308
(703) 768-5700
http://www.ahs.org

AMERICAN SOCIETY FOR HORTICULTURE SCIENCE
1018 Duke Street
Alexandria, VA 22314
(703) 836-4606
http://www.ashs.org

HORTICULTURE RESEARCH INSTITUTE, INC.
525 9th Street NW, Suite 800
Washington D.C. 20004
(202) 789-2900
http://www.hriresearch.org

Landscaping

AMERICANHORT COLUMBUS, OHIO
2130 Stella Court
Columbus, OH 43215 USA
(614) 487-1117

AMERICAN SOCIETY OF LANDSCAPE ARCHITECTS
636 Eye Street, NW
Washington, DC 20001-3736
(202) 898-2444
http://www.asla.org

ASSOCIATION OF PROFESSIONAL LANDSCAPE DESIGNERS
2207 Forest Hills Drive
Harrisburg, PA 17112
(717) 238-9780
http://www.apld.com

IRRIGATION ASSOCIATION
8280 Willow Oaks Corporate Drive, Suite #400
Fairfax, VA 22031
(703) 536-7080
http://www.irrigation.org

NATIONAL ASSOCIATION OF LANDSCAPE PROFESSIONALS
950 Herndon Parkway
Suite 450
Herndon, VA 20170
(703) 736-9666
http://www.landscapeprofessionals.org

PROFESSIONAL GROUNDS MANAGEMENT SOCIETY
720 Light Street
Baltimore, MD 21230
(410) 223-2861
http://www.pgms.org

WASHINGTON, DC
525 9th St NW, Suite 800
Washington, DC 20004 USA
(202) 789-2900
http://www.americanhort.org

Nurseries

AMERICAN WHOLESALE NURSERIES, INC.
565 N. Service Road
Dix Hills, NY 11746
(631) 673 2800
http://www.americanwholesalenurseries.com

ORNAMENTAL GROWERS ASSOCIATION
4 E Wilson St # A
Batavia, IL 60510
(630) 879-0520
http://www.ogaoni.com

PERENNIAL PLANT ASSOCIATION
3383 Schirtzinger Road
Hilliard, OH 43026
(614) 771-8431
http://www.perennialplant.org

Outdoor Living

CALIFORNIA REDWOOD ASSOCIATION
818 Grayson Road, Suite 201
Pleasant Hill, CA 94523
(925) 935-1499
http://www.calredwood.org

GOLF COURSE SUPERINTENDENTS ASSOCIATION OF AMERICA
1421 Research Park Drive
Lawrence, KS 66049
(800) 472-7878
http://www.gcsaa.org

KEEP AMERICA BEAUTIFUL, INC.
1010 Washington Boulevard
Stamford, CT 06901
(203) 659-3000
http://www.kab.org

NATIONAL SPA AND POOL INSTITUTE
2111 Eisenhower Avenue
Alexandria, VA 22314
(703) 838-0083
http://www.nspi.org

Pest Control

ASSOCIATION OF NATURAL BIOCONTROL PRODUCERS
PO Box 1609
Clovis, CA 93613-1609
(559) 360-7111
http://www.anbp.org

ENTOMOLOGICAL SOCIETY OF AMERICA
3 Park Place, Suite #307
Annapolis, MD 21401-3722
(301) 731-4535
http://www.entsoc.org

NATIONAL PEST MANAGEMENT ASSOCIATION
10460 North Street
Fairfax, VA 22030
(800) 678-6722
http://www.npmapestworld.org

WEED SCIENCE SOCIETY OF AMERICA
810 East 10th Street
Lawrence, KS 66044
(800) 627-0326
http://www.wssa.net

Power Equipment, Parts

OUTDOOR POWER EQUIPMENT INSTITUTE, INC.
341 S. Patrick Street
Old Town Alexandria, VA 22314
(703) 549-7600
http://www.opei.org

POWER TOOL INSTITUTE
1300 Sumner Avenue
Cleveland, OH 44115-2851
(216) 241-0105
http://www.powertoolinstitute.com

Seeds

AMERICAN SEED TRADE ASSOCIATION
1701 Duke Street
Alexandria, VA 22314
(703) 837-8140
http://www.amseed.org

Soil Conditioners

CANADIAN SPHAGNUM PEAT MOSS ASSOCIATION
2208-13 Mission Avenue, St. Albert
Alberta, Canada T8N 1H6
(780) 460-8280
http://www.peatmoss.com

MULCH AND SOIL COUNCIL
7809 N FM 179
Shallowater, TX 79363
http://www.mulchandsoilcouncil.org

PERLITE INSTITUTE, INC.
4305 N. Sixth Street
Suite A
Harrisburg, PA 17110
(717) 238-9723
http://www.perlite.org

Trees

AMERICAN FORESTS
1220 L Street NW, Suite # 750
Washington, DC 20005
(202) 737-1944
http://www.americanforests.org

AMERICAN SOCIETY OF CONSULTING ARBORISTS
9707 Key W Ave #100
Rockville, MD 20850
(301) 947-0483
http://www.asca-consultants.org

INTERNATIONAL SOCIETY OF ARBORICULTURE
PO Box 3129
Champaign, IL 61825-3129
(888) 472-8733
http://www.isa-arbor.com

NATIONAL CHRISTMAS TREE ASSOCIATION
16020 Swingley Ridge Road, Suite #300

Chesterfield, MO 63017
(636) 449-5070
http://www.realchristmastree.org

SOCIETY OF AMERICAN FORESTERS
5400 Grosvenor Lane
Bethesda, MD 20814-2198
(866) 897-8720
http://www.safnet.org

TREE CARE INDUSTRY ASSOCIATION
136 Harvey Road, Suite #101
Londonderry, NH 03053
http://www.tcia.org

US NATIONAL ARBORETUM
3501 New York Avenue, NE
Washington, DC 20002
(202) 245-2726
http://www.usna.usda.gov

Turf

THE LAWN INSTITUTE
2 East Main Street
East Dundee, IL 60118
(800) 405-8873
http://www.thelawninstitute.org

TURF PRODUCERS INTERNATIONAL
2 East Main Street
East Dundee, IL 60118
(847) 649-5555
http://www.turfgrasssod.org

TURFGRASS INFORMATION CENTER
Michigan State University Libraries
366 W. Circle Drive (10)
East Lansing, MI 48824
(517) 353-1975
http://www.lib.msu.edu

Wildflowers

LADY BIRD JOHNSON WILDFLOWER CENTER
4801 LaCrosse Avenue
Austin, TX 78739
(512) 232-0200
http://www.wildflower.org

Appendix C

Tools of the Trade

SPECIALIZED HAND TOOLS AND THEIR FUNCTIONS

TOOL AND NAME	FUNCTION	TOOL AND NAME	FUNCTION
Grass Shears 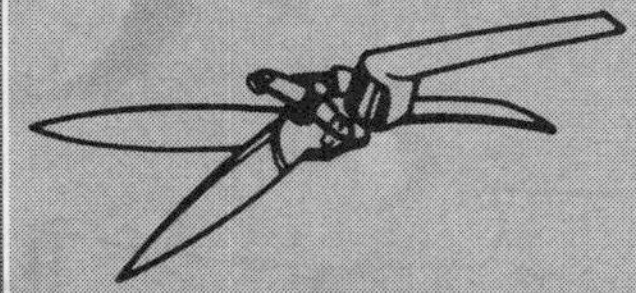	Used to trim grass along walks, roadways, the edge of planting beds, and around trees, posts, and so on.	Pruning Saw 	Removes any tree or shrub part that cannot be easily cut with the lopping shears. Usually parts are 1 inch or more in diameter.
Anvil Pruning Shears 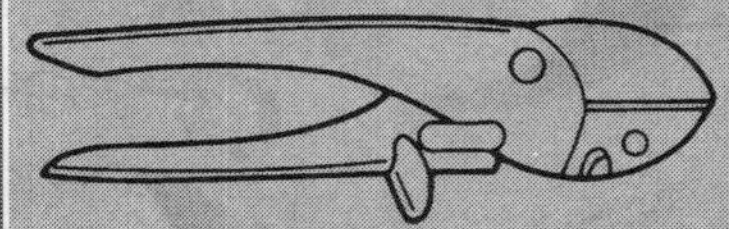	Used to trim tree and shrub twigs up to ½ inch in diameter.	Crosscut Saw 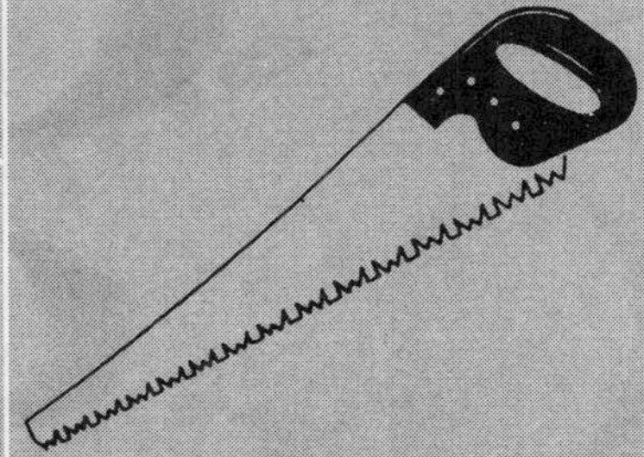	Removes large limbs and small trees. The saw has additional general uses.
Lopping Shears 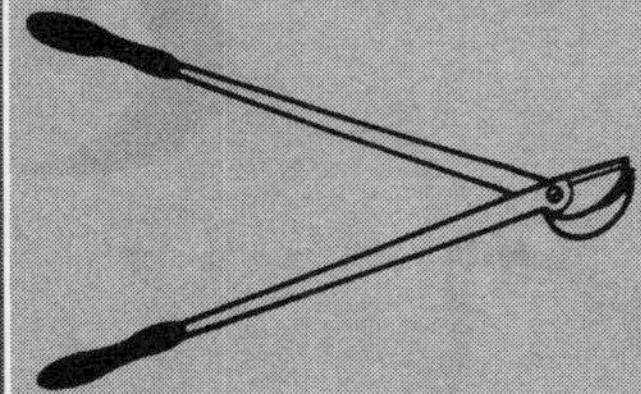	Used to trim tree and shrub twigs from 1 inch to 1½ inches in diameter.	Grass Hook 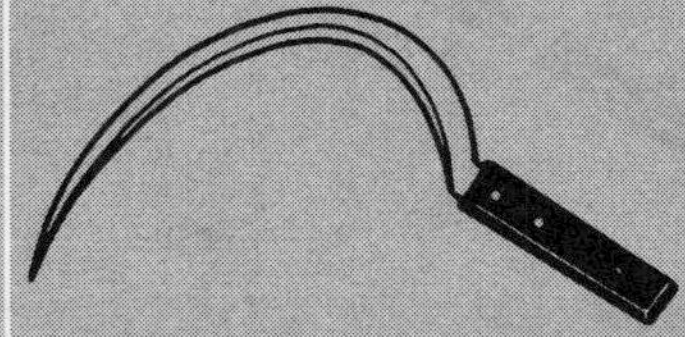	For reducing the height of overgrown grass areas. It requires the user to bend over.
Hedge Shears 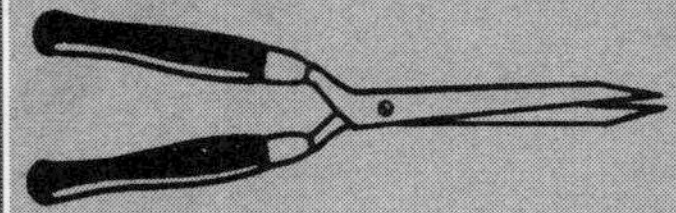	Prunes shrubs grown closely spaced as hedges. These shears are only used on young, tender new growth.		

SPECIALIZED HAND TOOLS AND THEIR FUNCTIONS (*CONTINUED*)

TOOL AND NAME	FUNCTION
Grass Whip 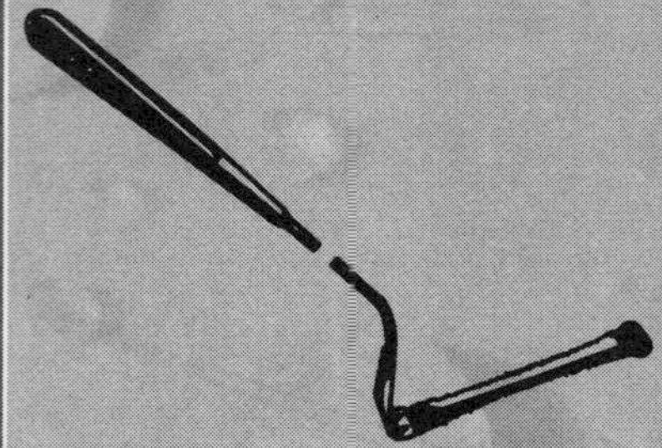	For reducing grass height without bending over. (Once reduced in height, a lawn mower can be used on the grass.)
Spading Fork 	Used for turning over the soil when it is not too hard or compacted. Also used for lifting bulbs in the fall.
Spades 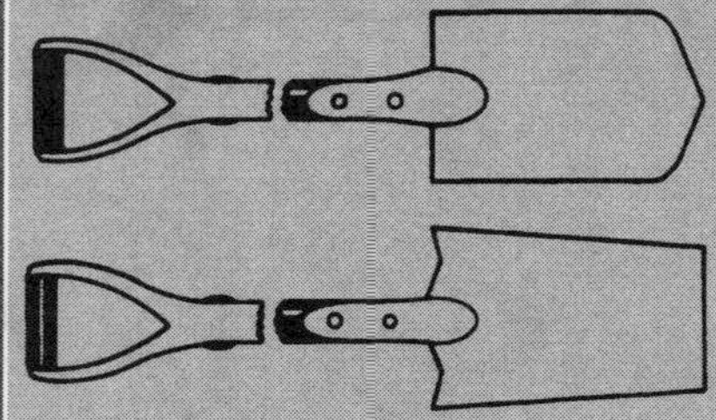	Obvious general uses in digging. Spades have flatter shapes than shovels. They penetrate the soil more easily but have less scooping capability.
Flat Shovel 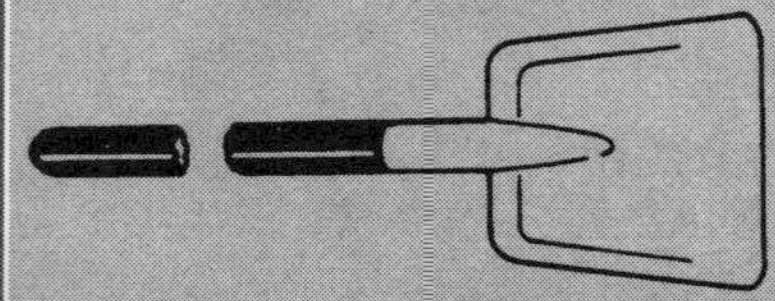	Used for cleaning loose soil from planting holes and other scooping uses. A shovel has sides that a spade does not have.
Round Shovel 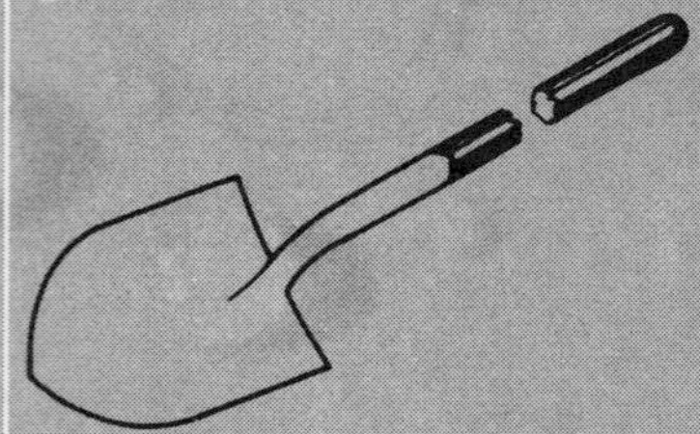	A combination tool with uses similar to both spades and shovels. It can be used for digging as well as scooping.
Scoop Shovel 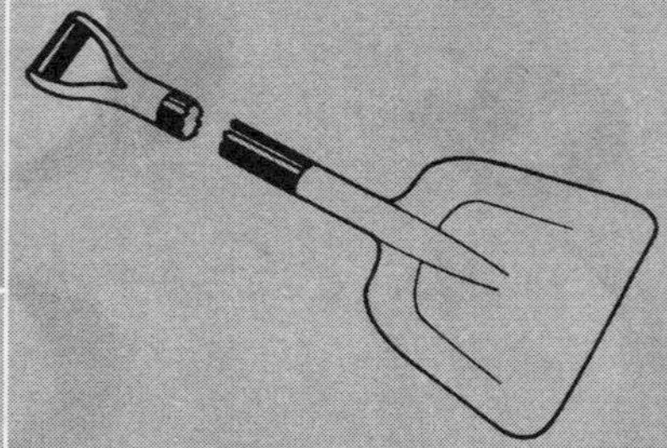	Good for moving loose materials such as crushed stone, peat moss, soil, and so on. Scoops have high sides. They are not used for digging.
Manure Fork 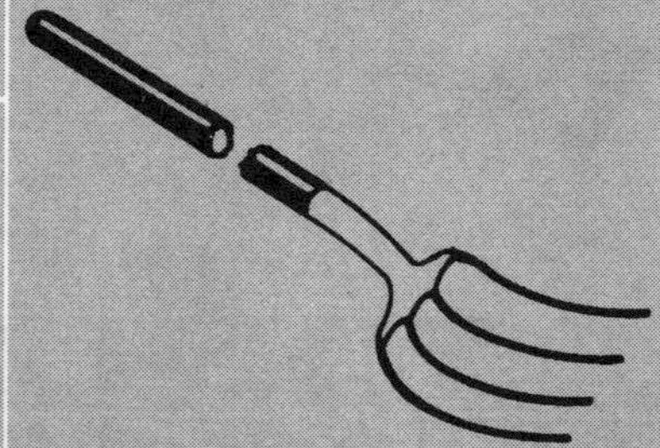	The best tool for moving coarse, lightweight materials such as straw, wood chips, and so on.

Single-Bit and Double-Bit Axes 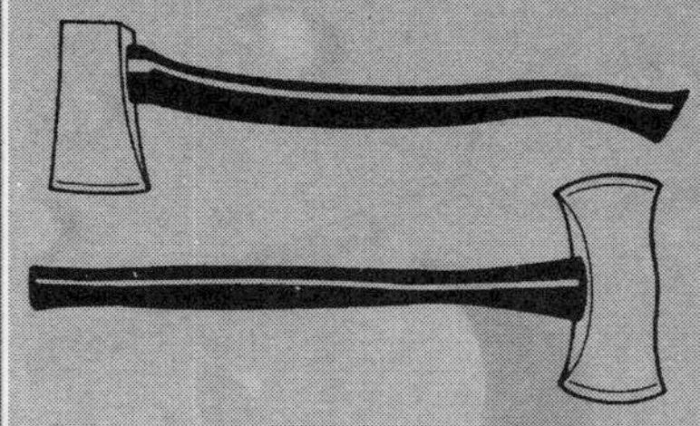	Obvious chopping uses. Especially useful in tree removal and for cutting up fallen timber.	Lawn Comb 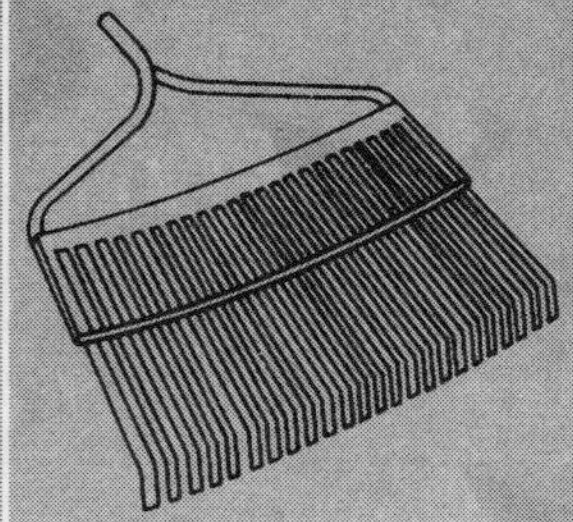	An excellent rake for collection of leaves and coarse debris from lawn surface.
Weed Cutter 	Removes annual weeds by cutting them off at ground level. Not very effective against biennial and perennial weeds.	Shrub Comb 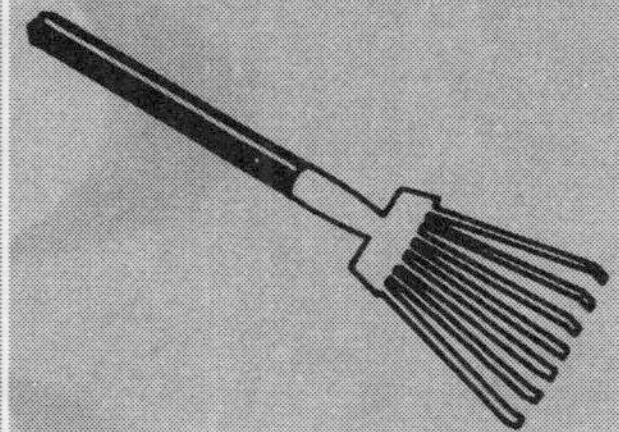	Used for raking debris from small areas between shrubs.
Toothed Rakes 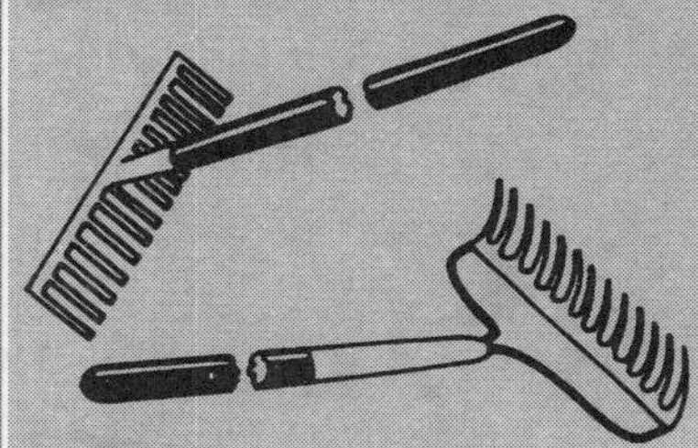	Used for heavy-duty raking that requires a strong tool. Commonly used in preparation of lawn seed-beds and cultivation of planted beds.	Bulb Planter 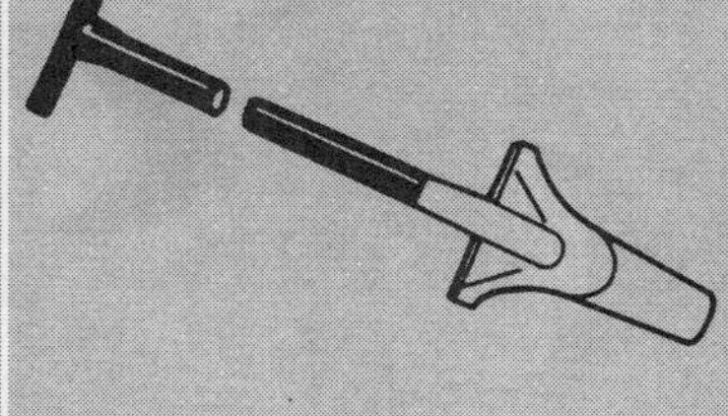	Used to install flowering bulbs.
Broom Rake 	Very useful in places where a lightweight, springy rake is needed. Very good for collecting debris and clippings from lawn surface.	Push Hoe 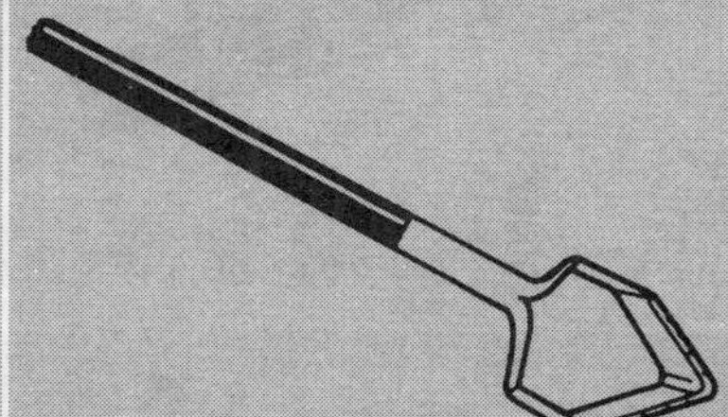	Similar to a scuffle hoe. It is good for rooting out weeds.

SPECIALIZED HAND TOOLS AND THEIR FUNCTIONS (*CONTINUED*)

TOOL AND NAME	FUNCTION	TOOL AND NAME	FUNCTION
Hand Trowel 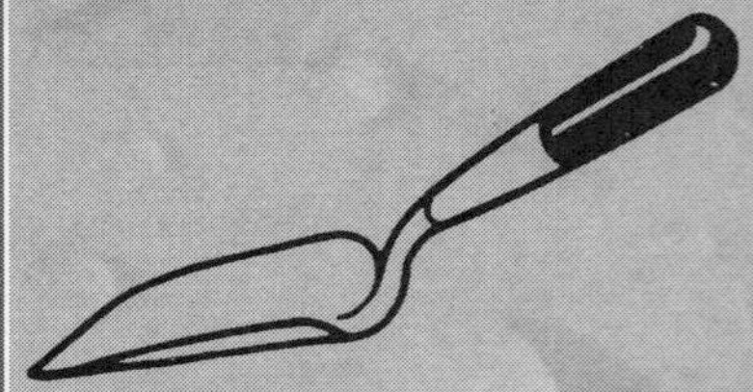	Used for transplanting bedding plants into flower beds, borders, and boxes.	Post Hole Digger 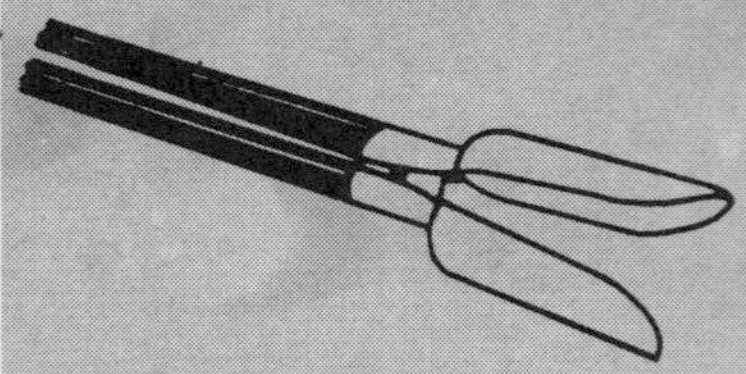	Prepares holes for the support posts of fences.
Transplanting Hoe 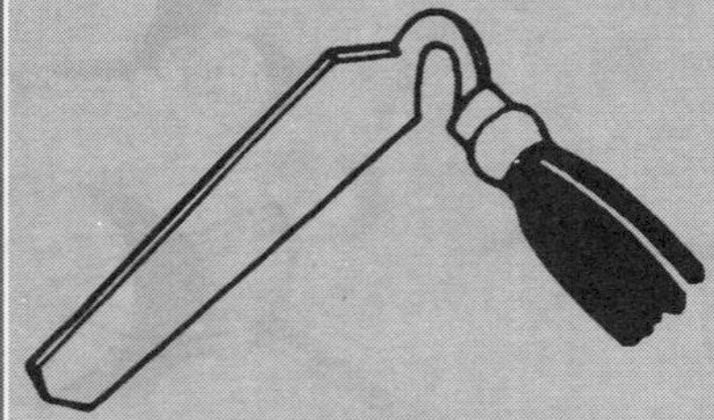	Uses are similar to those of a hand trowel. It has less adaptability for other types of digging.	Garden Hoe 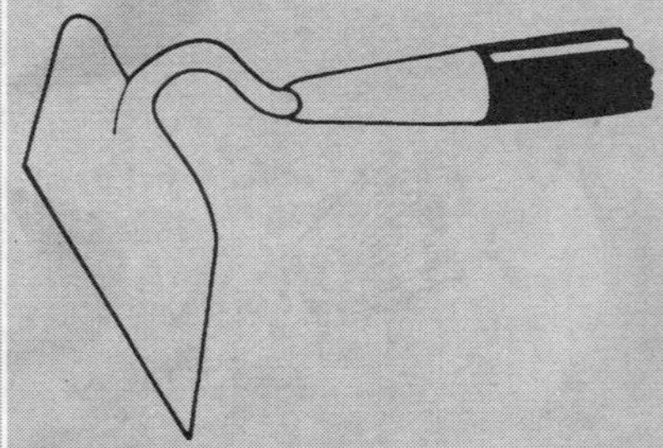	Widely used for breaking up the soil prior to planting. It is also good for cultivating planted beds and for weed removal.
Scuffle Hoe 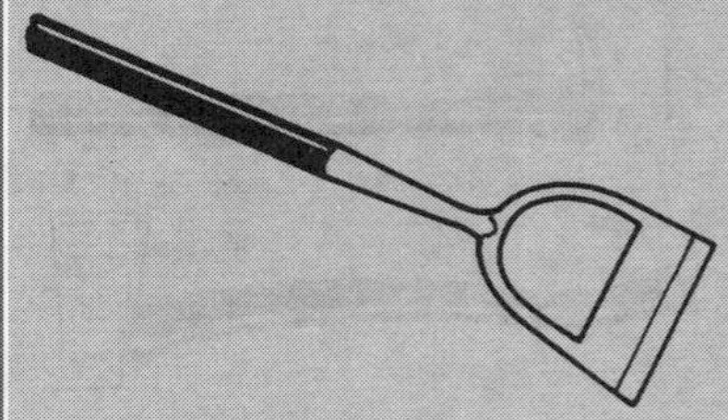	Useful in weeding and cultivating in planted beds. It cuts off weeds and loosens surface soil.	Hand Cultivator 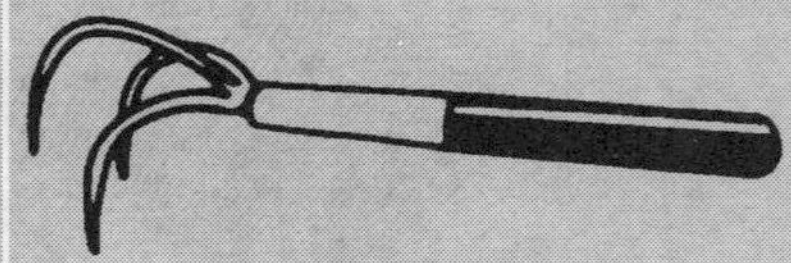	Loosens the surface soil in flower beds and around shrubs.

Pick 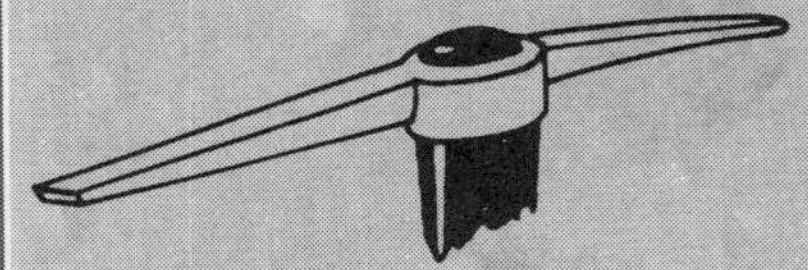	Used for breaking up hard, rocky soil. It has two pointed ends for gouging into the soil.	Cutter Mattock Hoe 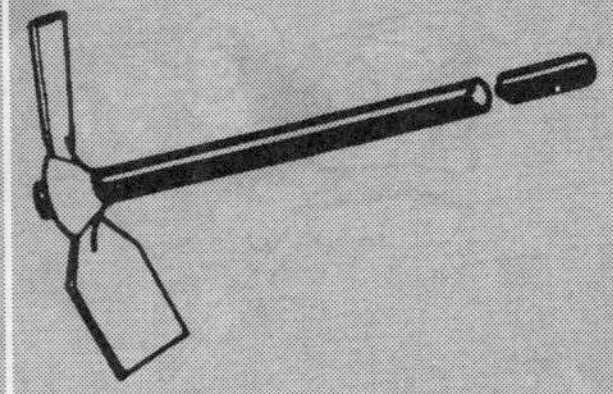	Stronger than a grading hoe but its uses are similar. It has two flat ends.
Compressed Sprayer 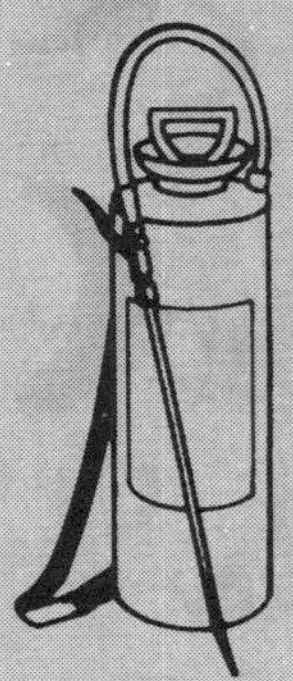	Needed to apply pesticides, antitranspirants, and other chemicals in liquid form. Sprayers are available in a wide range of sizes.	Cyclone Spreader 	Used for the application of fertilizer, seed, and other dry turf products.
Grape Hoe 	Loosens hard or compacted soil during preparation for planting. Has a sharpened flat end.	Drop Spreader 	Used for the application of fertilizer, seed, and other dry turf products.

POWER TOOLS AND THEIR FUNCTIONS

TOOL AND NAME	FUNCTION	TOOL AND NAME	FUNCTION
Walk-Behind Mower 	Professional lawn mowing equipment.	Lawn Edger 	To trim the grass along the edge of the sidewalk.
Backpack Blower 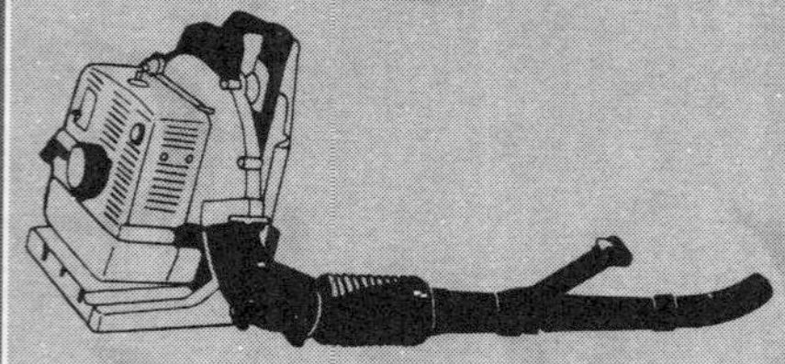	Removes leaves and trash from walks and lawn areas.	Zero-Turn Mower 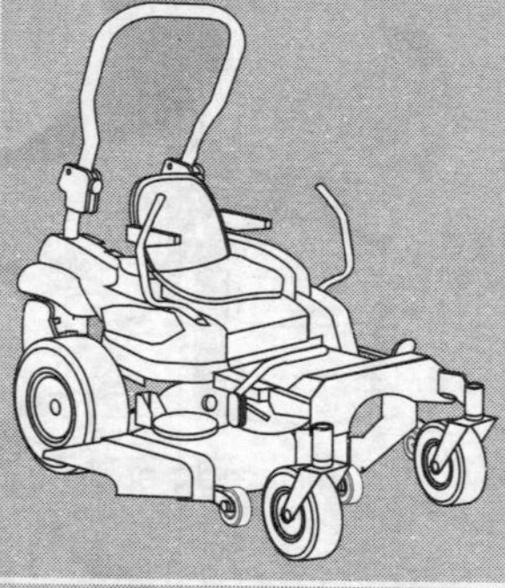	Used for professional lawn mowing.
Line and Blade Trimmer 	To trim grass and plant material from areas unable to be mown with the walk-behind mower.	Power Bed Edger 	Used for professional bed edging.

Equipment Maintenance

All equipment requires preventive maintenance to ensure that it is in proper operating condition. Preventive maintenance lengthens the useful life span of machinery and results in reduced equipment repair and replacement costs. The manager, mechanic, or equipment operator must keep records of when maintenance services are performed on a machine and how many hours the machine is in use. The equipment manual contains a preventive maintenance schedule based on the number of hours of equipment operation. The person responsible for equipment maintenance should rigidly adhere to this schedule.

The list of maintenance services necessary after 50, 100, 300, and 600 hours of equipment use is fairly typical and will help the reader understand the importance of equipment maintenance records. However, the manager should consult the manual for each piece of equipment under his or her supervision to determine exact maintenance schedule specifications. The oil level must always be checked before a machine is used.

Generalized Preventive Maintenance Schedule

AT 50 HOURS

- Check tire pressure and water level in battery.
- Clean air filter (wash or replace if necessary).
- Remove corrosion from battery terminal connections.
- Check oil level in transmission and hydraulic system.
- Grease lubrication points.

AT 100 HOURS

- Clean engine.
- Drain oil and refill engine crankcase.
- Lubricate clutch and throttle linkage.
- Tighten loose screws and nuts.

AT 300 HOURS (OR EVERY 6 MONTHS)

- Inspect and clean spark plugs.
- Check ignition point gap.
- Change hydraulic oil filters.
- Check fan and drive belts.

AT 600 HOURS (OR YEARLY)

- Change air cleaner and spark plugs.
- Touch up with paint.

Hand Tool Maintenance

Remove rust from metal tools and apply rust-inhibiting materials to metal surfaces; repair split or broken wood handles; reshape the heads of driving or driven tools; and sharpen blades or cutting tools. The sharpening frequency for pruners and shears depends on the usage and the hardness of the steel. Saw sharpening should be done by a professional service and involves the tooth sharpness and tooth set. An indication of the need to have a saw sharpened is being able to run your fingers up and down the saw teeth without getting cut. If the saw binds and becomes stuck in the groove, the teeth may need to be set.

Storing Hand Tools and Small Power Equipment

- Store equipment and fuel in dry, ventilated areas. Storage spaces within a garage area, a storage building, or a good shed are all workable options.
- Remove batteries and fully charge them before storage.
- Clean all dirt, grass, and debris from engines.
- Disconnect spark plug wires or remove spark plugs based on the manufacturer's recommendations.
- Inspect power equipment—belts, blades, and so on—and make any needed repairs.
- Drain all gasoline from tanks into a container for disposal.
- Do not blend gasoline, gasohol, or alcohol with diesel fuel, as this is a potential for an explosive hazard.
- Do not store equipment in the same location as fertilizers or swimming

pool chemicals. These chemicals are extremely corrosive to metal parts on equipment.

- Start and run engines until they quit to be sure the carburetor is dry to prevent diaphragms from sticking together.
- Before storing hand tools, be sure to clean them and apply a light coat of oil to all metal surfaces to prevent rust from forming.
- For hand tools, check to see if any repairs are needed.

Appendix

D Deer-Resistant Plants

Rarely Damaged

Aesculus parviflora—Bottlebrush Buckeye
Amelanchier—Serviceberry
Berberis—Barberry
Betula nigra—Heritage Birch
Buddleia—Butterfly Bush
Buxus—Boxwood
Calyacnathus floridus—Sweet Shrub
Caryopteris—Blue Mist Shrub
Cephalotaxus harringtonia—Japanese Plum Yew
Chamaecyparis pisifera—Gold Mop Cypress
Cryptomeria japonica—Japanese Cryptomeria
Fothergila—Fothergila
Ilex—Dragon Lady Holly
Jasminum nudiflorum—Winter Jasmine
Picea pungens glauca—Colorado Blue Spruce
Pieris japonica—Andromeda
Pseudotsuga menziesii—Douglas Fir
Saroccoca hookeriana—Dwarf Sweet Box

Seldom Damaged

Buddleia—Butterfly Bush
Callicarpa americana—American Beautyberry
Cornus florida—Flowering Dogwood
Cornus kousa—Korean Dogwood
Enkianthus campanulatus—Redvein Enkianthus
Fagus sylvatica—European Birch
Forsythia—Forsythia
Gleditsia triacanthos—Honey Locust
Ilex glabra—Inkberry
Ilex opaca—American Holly
Juniperus chinensis—Chinese Juniper
Kerria japonica—Japanese Kerria
Mahonia—Grape Holly
Myrica pennsylvanica—Bayberry
Nandina—Heavenly Bamboo
Osmanthus heterophyllus—Osmanthus
Picea abies—Norway Spruce
Pinus mugo—Mugo Pine
Prunus laurocerasus—Cherry Laurel
Prunus serrulata—Japanese Flowering Cherry
Spirea—Spirea
Syringa vulgaris—Common Lilac
Vibrunum plicatum—Doublefile Viburnum
Viburnum carlesii—Koreanspice Viburnum
Viburnum juddii—Judd Viburnum
Viburnum rhytidophyllum—Leatherleaf Viburnum
Weigela florida—Weigela

Occasionally Damaged

Acer griseum—Paperbark Maple
Acer palmatum/green-leaved only—Japanese Maple
Acer rubrum—Red Maple
Acer saccharum—Sugar Maple
Cotoneaster—Cotoneaster
Hamamelis—Witch Hazel
Hydrangea anomala petiolaris—Climbing Hydrangea
Hydrangea paniculata—Panicle Hydrangea
Ilex crenata—Japanese Holly
Liquidambar styraciflua—Sweet Gum
Pyrus calleryana—Flowering Pear
Quercus rubra—Red Oak
Salix—Willow
Spirea x bulmada—Anthony Waterer Spirea

Syringa reticulata—Japanese Lilac Tree
Tilia cordata—Littleleaf Linden
Viburnum opulus—European Cranberry Bush

Frequently Damaged

Acer palmatum/red-leaved varieties—Japanese Maple
Cedrus atlantica—Blue Atlas Cedar
Cercis canadensis—Redbud
Chionanthus virginicus—Fringe Tree
Euonymus alatus—Winged Euonymus Burning Bush
Hydrangea macrophylla—Bigleaf Hydrangea
Hydrangea quercifolia—Oakleaf Hydrangea
Ilex cornuta—Chinese Holly
Ilex verticillata—Common Winterberry
Ilex x merserveae—Merserve Holly
Ilex—Nellie Stevens
Magnolia soulangiana—Saucer Magnolia
Malus—Crabapple
Pinus strobus—White Pine
Rhododendron hybrids—Evergreen Azalea and Rhododendron
Taxus—Yew
Thuja occidentalis—Arborvitae
Thuja sp.—Arborivitae 'Green Giant'
Tsuga—Hemlock
x Cupressocyparis leylandii—Leyland Cypress

Glossary/Glosario

4-H network of youth clubs under the direction of Cooperative Extension System personnel to enhance personal development and provide skill development in many areas

4-H un colectivo de clubes juveniles bajo la dirección del Servicio Cooperativo de Extensión para fomentar el desarrollo personal y proveer adiestramiento en muchas áreas

8-week variety a variety of chrysanthemums that will flower in 8 weeks after short day length and long night lighting is started

variedad de 8 semanas una variedad de crisantemos que florecerán en 8 semanas a partir de comenzar un ciclo de días cortos y noches largas

A

accent an item used to create interest in a landscape

acento tipos de plantas que se utilizan para crear interés en un jardín

accounts receivable money owed to a business from customers

cuentas por cobrar dinero que le adeudan clientes a una empresa

acidic a condition that has a pH less than 7.0

ácido una condición que tiene un nivel de pH menor que 7.0

activated charcoal a product that absorbs harmful fumes

carbon activado un producto que absorbe los gases dañinos

actuating putting a plan into action

activar poner un plan en marcha

acute toxicity toxic or poisonous in a very short time; is used as a measure of how toxic a pesticide is after a single exposure

toxicidad aguda tóxico o venenoso en muy poco tiempo; se utiliza para medir cuán tóxico es un pesticida luego de ser aplicada una sola vez

adaptability ability to change

flexibilidad la habilidad para cambiar

adhesion a force that causes a substance to stick to things other than itself

adherencia una fuerza que hace que una sustancia se pegue a otras cosas u otras sustancias

adjourn a motion used to close a meeting

cierre de sesión una moción presentada para finalizar una reunión

aeration movement of air into and through the soil

ventilación el movimiento del aire hacia y a través de la tierra

aerosols material dispensed by air pressure

aerosoles material que se emite por spray o rociador

aesthetics the beauty and attractiveness of the landscape

estética la belleza y lo atractivo de un paisaje

aggregation the clinging together of soil particles to form larger, crumblike particles

agregación del suelo el pegarse entre sí las partículas en el suelo para forma partículas más grandes, tipo migajas

air layering a process where stems still attached to the mother plant are rooted in the air instead of being covered with soil

acodo aéreo un proceso en que los tallos, aún siendo parte de la planta madre, quedan acepado en el aire en vez de ser cubiertas por tierra

alkaline a condition that has a pH more than 7.0

alcalina una condición que tiene un nivel de pH de más de 7.0

allelopathy the toxic effect of one plant on another

alelopatía el efecto tóxico de una planta sobre otra

amend a type of motion used to add to, subtract from, or strike out words in a main motion

enmendar un tipo de moción para añadir o quitar contenido o eliminar palabras de una moción inicial o principal

annual a plant that grows, flowers, produces seed, and dies in 1 year

anual un planta que crece, florece, produce semillas y muere en un año

anvil pruner type of hand pruner with a single blade that cuts against a flat brass anvil

tijera de podar tipo yunque un tipo de podador de mano con una cuchilla que corta al dar contra un yunque llano

apical dominance a form of regulating growth in plants in which the terminal bud secretes chemicals that inhibit or prevent the growth of lateral buds on the same shoot

dominancia apical una forma de regular el crecimiento de las plantas en que la yema terminal secreta químicos que inhiben o impiden el crecimiento de yema laterales en el mismo tallo

architectural science of architecture

arquitectónico la ciencia de la arquitectura

arrangement a completed floral design

arreglo un diseño floral completado

asexually without the union of male and female sex cells

de forma asexual sin la unión de las células sexuales masculinas and femeninas

asset anything the business owns

activos cualquier propiedad mueble e inmueble de la compañía

atrium enclosed court garden area

patio interior área de jardín cerrado o cercado

attitude advertising advertising designed to build the reputation of a business

la publicidad basada en el reconocimiento la publicidad diseñada para desarrollar la reputación de la empresa

attractants chemicals or natural hormones (pheromones) that attract insects

atrayentes químicos y hormonas naturales (feromonas) que atraen a los insectos

auxins plant hormones that accelerate growth by stimulating cell enlargement

auxinas hormonas de la planta que aceleran el crecimiento al estimular la expansión y ampliación de las células

axis the point at which flower stems come together to create a floral design

axis el punto en que los tallos de las flores se unen para crear un diseño floral

B

bacteria a large group of typically single-celled microorganisms; some are beneficial in the process of fermentation and others cause diseases

bacteria un grupo grande de microorganismos que típicamente son de una sola célula; algunas son beneficiosas en el proceso de fermentación, y otras causan enfermedades

bag garden the process of growing plants in bags of plant media

jardín de bolsas de cultivo el proceso de cultivar las plantas con sus raíces en bolsas lleno de un medio para plantas

balance a basic principle of design

balance un principio básico del diseño

balance sheet a statement of the assets and liabilities of a business on a specific date

hoja de balance un informe de los activos y los pasivos o deudas de un negocio en una fecha específica

balled and burlapped (b&b) a process of forming a root ball on the plant and covering it with burlap to hold the soil around the roots

cepellón proceso en que las raíces de un árbol rodeadas de tierra forman una bola y se envuelve en yute o arpillera para mantener la tierra intacta alrededor de las raíces

balling a process of forming a root ball on the plant for transplanting

cepellón el proceso de formar una bola de las raíces rodeadas y cubiertas de tierra para facilitar el trasplante

banding fertilizer applying fertilizer in trenches on each side of a row of plants

fertilizar en franjas el colocar los fertilizantes en zanjas en cada lado de una hilera de plantas

bare root plants with no soil or soilless mix on the roots; roots are washed or shaken clean of all soil

raíz desnuda plantas cuyas raíces están sin tierra o en un medio o turba que no contiene tierra; las raíces se lavan o se sacuden para quitarles la tierra

berm a ridge of soil placed around a newly planted tree to retain water

berma un montículo de tierra colocado alrededor del árbol recién sembrado para retener el agua

biannually in plants, biannual means twice a year; for example, encore azaleas blooming in the spring and in the fall

bianual al referirse a las plantas, bianual significa dos veces al año; por ejemplo, el florecer de las azaleas en el otoño y en la primavera

binomial system a two-name system for naming plants

sistema binomial un sistema para darle dos nombres a las plantas

biochar charcoal that can increase soil fertility of acidic soils (low pH soils), increase horticultural productivity, and provide protection against some foliar and soil-borne diseases; also a rich resource of carbon

carbon biológico, biochar carbón vegetal que puede incrementar la fertilidad en los terrenos ácidos (tierra con niveles de pH bajos), aumentar la productividad hortícola, y proteger contra algunas enfermedades que provienen de las hojas o del suelo; también es una buena fuente de carbón

biochemical a chemical formed by a living organism

bioquímico un químico formado por un organismo vivo

biocontrol biological control of insects, disease, and other plant pests

control biológico control biológico de los insectos, las enfermedades y otras plagas que afectan a las plantas

biological descriptive of natural, living organisms, or from living organisms

biológico descriptivo de organismos vivos y naturales

biological agent a bacterium used to control pests

agente biológico una bacteria utilizada para controlar pestes y plagas

biological control the control of insects and diseases using natural organisms or products

control ecológico el control de los insectos y las enfermedades utilizando organismos o productos naturales

biopesticide a pesticide that came from a living organism, such as a plant, fungus, or bacteria

biopesticida un pesticida que proviene de un organismo vivo, tal como una planta, un hongo, o una bacteria

biorational pest control materials that are nontoxic with little side effects, for example, horticultural oils

pesticidas biorracionales materiales no tóxicos con pocos efectos secundarios, por ejemplo, aceites hortícolas

blossom set flower set of the plant

la floración cuando las flores de una planta brotan

bolt a full roll of ribbon

rollo un rollo entero de cinta

bolting occurs when the plant sends out a seed stalk and stops growing edible leaves

crecimiento acelerado ocurre cuando la planta saca un tallo con semillas y deja de producir hojas comestibles

bonsai the art of dwarfing plants by growing them in shallow containers

bonsái el arte de empequeñecer a las plantas al cultivarlas en envases angostos o estrechos

botanical nomenclature the international set of rules in naming plants

nomenclatura botánica las reglas internacionales para otorgarle nombre a las plantas

boutonniere a flower and/or flowers worn on a man's lapel

flor en el ojal una flor o flores que portan los hombres en la solapa

Boy Scouts youth organizations that provide opportunities for leadership development and skill development for boys

Boy Scouts organizaciones para los jóvenes que proveen oportunidades para desarrollar su liderazgo y sus destrezas

bracts specialized leaves that change to become part of the flower; see poinsettia flowers in Unit 13

brácteas hojas especializadas que cambian para convertirse en parte de la flor; véase nochebuena o flor de Pascua en la Unidad 13

brambles fruiting plants including raspberries, blackberries, boysenberries, dewberries, loganberries, and bayberries

zarzamora plantas fruteras incluyendo frambuesa, arándaro azul, bayas de Boysen,, morera roja y, claro está, la zarzamora

breaks new shoots that develop from buds in the axis of leaves or nodes

breaks (término en inglés) brotes nuevos que provienen de yemas en el axis de las hojas o en los nodos

broadleaf a large leaf of a broadleaf evergreen

latifolio una hoja grande de un árbol tropical o perenne latifoliado o de hojas anchas

broadleaf evergreen ground covers a large-leaf needlelike structure low to the ground

cubresuelo perenne latifoliado una estructura formada por plantas verdes con hojas anchas que se colocan cerca del suelo

bud shield the bud and piece of wood and bark cut to be inserted under the bark of the rootstock

injerto de escudete o de yema la yema y un pedazo de madera y de corteza se cortan y se insertan debajo de la corteza del portainjerto o planta patrón

budding a type of grafting where a single bud is used to produce the top part of the plant

injerto de yema una forma de cultivar con injerto en que se utiliza una sola yema para producir la parte superior de la planta

budwood current season growth of wood selected from plants chosen to be propagated by budding

esqueje la parte de la madera crecida en la temporada corriente que se selecciona para usar como injerto y así intentar su propagación

bulb a vegetative structure that consists of layers or fleshy scales overlapping each other, such as the onion

bulbo una estructura vegetal que consiste de capas o escamas carnosas sobreponiéndose uno al otro, como en la cebolla

bulblet immature bulb that develops at the base of the bulb

bulbillo un bulbo inmaduro que se desarrolla en la base del bulbo

business meeting a gathering of people working together to make decisions

junta una reunión de personas que trabajan juntos para tratar asuntos del negocio y tomar decisiones

buying function selection of a product or service to be marketed or sold for a profit

función de comprador de la empresa la selección de un producto o servicio a ser vendido para obtener una ganancia

C

CAD (computer-aided design) an accelerated drafting system

CAD (diseño asistido por computadora) sistema acelerado para el dibujo técnico

calcium one of the essential plant foods supplied primarily by lime, and is also used to raise soil pH

calcio uno de los alimentos esenciales para las plantas, provisto primordialmente por el cal (lime), y además se utiliza para aumentar el nivel de pH en la tierra

calibration adjusting the sprayer so that the exact amount of chemical is applied to a particular area

calibración el ajustar el rociador para que la cantidad exacta de un agente químico se aplique a un área particular

callus mass of cells that form around the wounded area of a plant to start the healing process

callo una masa de células que se forman alrededor de una herida de una planta para comenzar a sanar

calyx the floral structure at the base of flower petals to hold the flower

cáliz la estructura floral en la base de los pétalos de las flores para atarse a la flor

cambium thin, green, actively growing tissue located between the bark and wood of a plant; in grafting, the cambium of the scion must touch the cambium of the stock

cámbium tejido en crecimiento activo verde y fino que se encuentra entre la corteza y la madera de una planta; al injertar, el cámbium del vástago debe tocar el cámbium de la otra planta

cane mature shoot of a vine's previous season's growth

caña el brote maduro que surge de lo que ha crecido una enredadera en la temporada pasada

canopy a tree cover of the area

follaje una cubierta del área creada por árboles

capillary water water held against the force of gravity

agua capilar agua sostenida en contra de la fuerza de gravedad

career success creating positive habits, and setting and accomplishing goals

éxito professional el crear hábitos positivos, y establecer y cumplir metas

carrier material mixed into granular fertilizers to dilute the strength and make the material easier to spread; a 10-10-10 fertilizer is 30% fertilizer material and 70% carrier

transportador material mezclado con fertilizantes granulados para diluir su potencia y lograr que sea más fácil de esparcir; un fertilizante 10-10-10 se compone de 30 porciento fertilizante y 70 porciento material transportador

cash flow statement a financial document that describes the availability of operating funds for a business at different points in time

estado de flujo de efectivo un documento financiero que describe la disponibilidad de fondos para operar el negocio

centerpieces floral arrangements used for accents in the home

centros de mesa arreglos florales utilizados para realzar decoraciones en el hogar

centrosome a small region of cytoplasm adjacent to the nucleus that contains the centrioles and serves to organize microtubules in the mitosis stage

centrosoma una región de citoplasma adyacente al núcleo que contiene los centriolos y que sirve para organizar el huso acromático en la etapa de la mitosis

Certified Landscape Professional (CLP) program certifies individuals in the nursery and *landscape* industry who possess professional excellence in educational, scientific, and service activities in the landscape and nursery industry

Certified Landscape Professional (CLP) un programa que ofrece una certificación a los individuos en la industria de los viveros y del paisajismo por su excelencia profesional en las actividades educativas, científicas y de servicio en esa industria

Certified Professional Horticulturists (CPH) a program that certifies qualified horticultural professionals for excellence in educational, scientific, and service activities in the horticultural industry

Certified Professional Horticulturists (CPH) un programa que ofrece una certificación a los horticultores profesionales por su excelencia en las actividades educativas, científicas y de servicio en la industria hortícola

chemical retardant a chemical used to slow down, shorten, or dwarf plant growth

retardante químico una sustancia química que se utiliza para desacelerar o reducir el crecimiento de las plantas

chilling hours hours in which the temperature is below 40°F (5°C); some blueberries require a cold period to break the dormant period and resume growth

horas de frío las horas en que la temperatura es más baja que 40°F (5°C); algunos arándanos azules necesitan un tiempo en el frío para romper con ese periodo inactivo o latente y comenzar a crecer de nuevo

chloric yellowing leaves of a plant

clórico el proceso en que las hojas de las plantas se tornan amarillas

chloroplast small green particle containing chlorophyll found in leaves

cloroplasto la partícula pequeña verde que contiene clorofila y que se encuentra en las hojas

chronic toxicity a measure of how poisonous a pesticide is over a period of time
toxicidad crónica una medida de cuán venonoso es un pesticida durante un periodo de tiempo
citizenship functioning and interacting in a society in a positive way
ciudadanía el funcionar e interactuar en una sociedad de forma positiva
classic partnership when two parties share in the risk and yields of the business
sociedad clásica o típica cuando dos personas o entidades comparten los riesgos y las ganancias del negocio
clone an exact duplicate of the parent plant
clon un copia exacta de la planta madre
closed corporation does not sell its stock publicly; often owned by members of a single family
sociedad anónima limitada no vende sus acciones públicamente; frecuentemente los dueños son miembros de una sola familia
clutchwire a wiring method to support the flowers for arranging as a boutonniere, nosegay, or corsage
clutchwire (término en inglés) un método usando cables y/o alambre para proveer el soporte necesario para los arreglos florales en un ramillete para el traje, un ramillete de manos, o una flor en el ojal
cohesion a force that causes molecules of substances, such as water, to stick to each other
cohesión una fuerza que hace que las moléculas de las sustancias, como el agua, se adhieran uno al otro
cold frame an outside structure covered on top with glass or plastic used to harden off plants or protect tender plants during the winter
cajón vivero una estructura a usarse en el exterior con una tapa de cristal o plástico que se utiliza para incrementar la resistencia de las plantas o para proteger las plantas tiernas durante el invierno
cold hardiness the ability of a plant to withstand cold temperatures
resistencia al frío la habilidad de una planta de soportar las temperaturas frías
color wheel a visual representation of colors arranged according to primary, secondary, and tertiary colors
rueda de colores una representación visual de los colores, organizándolos de acuerdo a si son colores primarios, secundario o terciarios
collateral something of value that is guaranteed as security for a loan
garantía algo de valor que sirve para garantizar la seguridad de un préstamo
commercial the business world operation
comercial la manera en que opera el mundo de negocios
common name the English name of a plant that may differ in various localities
nombre común el nombre en inglés de una planta que puede ser diferente en varios lugares
compatibility ability to unite with, grow, and live together, like the scion and rootstock in the grafting and budding processes
compatibilidad la habilidad para unirse, crecer y vivir juntos, como el vástago o el esqueje y el patrón durante los procesos de crear y sembrar injertos
competitors beneficial microorganisms, especially in the soil, that colonize soil space and plant roots so that disease-producing organisms have no space to live
competidores microorganismos, especialmente en la tierra, que colonizan el espacio en el terreno y en las raíces de las plantas para que otros organismos que causan enfermedades no tengan donde vivir ni desarrollarse
complement to accent related objects and space of the design
complementar el acentuar los objetos y los espacios relacionados en el diseño
compost an organic material that is partially decomposed and used as a soil fertilizer or amendment
abono un material orgánico parcialmente descompuesto que se utiliza para fertilizar o cambiar las características del terreno
compound layering a system where a stem or vine is covered at more than one place along its length for rooting, otherwise the same as simple layering
acodo compuesto o múltiple un sistema en que a lo largo del tallo, se entierra en varios puntos para lograr enraizarse varias veces, por lo demás es igual que el acodo simple
consultation a designer that gives the client constructive advice
consulta un diseñador que le ofrece a un cliente crítica constructiva
container growing growing nursery crops to marketable size in containers. May also be referred to as container stock or containerized.
cultivos en contenedores el cultivar plantas de viveros a un tamaño vendible en contenedores
contorted plant that grows in a twisted fashion
contorsionado una planta que crece de forma torcida
Cooperative Extension System an educational agency of the US Department of Agriculture and an arm of land-grant state universities
Sistema Cooperativo de Extensión una agencia educativa del Departamento de Agricultura de los

Estados Unidos y una rama de las universidades estatales en terrenos cedidos por el gobierno, muchos de los cuales ofrecen programas de estudios agrícolas

corm swollen underground stem that grows upright; is a food storage organ and a means of reproduction

cormo un tallo hinchado bajo tierra que crece recto; es un órgano para almacenar alimento y que sirve como un medio de reproducción

cormel a new corm produced by a larger corm

cormillo un cormo nuevo producido por un cormo más grande

corporation the most public method of doing business and more governed by state and federal laws than sole proprietorships or partnerships

corporación la forma más pública de hacer negocios; más gobernada y reglamentada por leyes estatales y federales que las sociedades o negocios con un solo dueño

corsage an arrangement of flowers worn by women

ramillete un arreglo floral que lleva puesto una mujer

corsage pin one source of mechanics to hold a boutonniere or corsage in position

imperdible objeto que provee la manera de sostener en su lugar un ramillete o una flor en el ojal

cotton a ribbon fabric

algodón tela que se utiliza para las cintas

cotton wad a floral mechanism used to support trumpet florets in a boutonniere, nosegay, or corsage

bolita de algodón un mecanismo en la floristería utilizado para ofrecer soporte a los cogollos en forma de trompeta en el ramillete, el ojal de la solapa, o el ramillete de manos

courage willingness to proceed under difficult conditions

valentía voluntad para proceder a pesar de condiciones difíciles

crop rotation planting different crops each year on a 3- or 4-year cycle

rotación de los cultivos el sembrar cultivos diferentes anualmente en el mismo lugar durante un ciclo de 3 o 4 años

crown the point on the strawberry plant where the stems and roots join; very important to plant with the crown level with the soil surface

corona el punto en la planta de fresa donde la raíz se encuentra con el tallo; es muy importante sembrar con la corona al mismo nivel que la superficie del terreno

cultivars another name for a specific plant; same as variety

cultivares otro nombre para una planta específica; el mismo nombre que la variedad de plantas

cultivation working up the plant media around the plant

cultivo el trabajar con la tierra o el medio en que está sembrada la planta

customer service contributing to the ultimate satisfaction of the person who will receive the product or service being offered by a business

servicio al cliente contribuye a la satisfacción máxima de la persona que recibirá el producto o el servicio que ofrece el negocio

cytokinins plant hormones that stimulate cell division and work along with auxins (will not work without auxins present)

citoquininas las hormonas de la planta que estimulan la división celular y que trabaja conjuntamente con la auxina

cytoplasm the liquid that gives the plant cells their shape and fosters the other cell structures

citoplasma el líquido que le da su forma a las células en la planta y fomenta las otras estructuras celulares

D

damping-off a fungus disease that kills young and tender plants under wet conditions

marchitamiento fúngico enfermedades de hongos que matan las plantas jóvenes y tiernas en condiciones de mucha lluvia o mucha agua

day neutral flowering and fruiting not affected by day length

día neutral el florecimiento y la maduración de las frutas no se afectan por lo largo del día o la temporada

deciduous ground covers ground covers that lose their leaves in the winter

cubresuelos de hoja caduca plantas bajas que cubren el suelo y pierden sus hojas en el invierno

deciduous plant a plant that loses its leaves during certain seasons

planta de hoja caduca una planta que pierde sus hojas durante ciertas temporadas

decomposed to break down

podrido el descomponerse

decorator's bow an accent bow for a holiday arrangement

lazo del decorador un lazo de realce para un arreglo navideño

Della Robbia wreath a wreath made of seedpods of different plants

corona Della Robbia una corona fabricada con vainas de distintas plantas

dermal contact a compound absorbed through the skin

contacto dérmico un compuesto que se absorbe a través de la piel

desiccation loss of water from the leaf structure through the stomata

desecación la pérdida de agua de la estructura de una hoja por medio de las estomas

determinate tomatoes Tomatoes are produced on the terminal end of stems, ripen at roughly the same time, and are more compact

tomates determinados Tomates se producen en las puntas terminales de los tallos, maduran aproximadamente al mismo tiempo, y son más compactos

dexterity one's ability to move his or her fingers to manipulate floral material into a design

destreza la habilidad para mover los dedos para manipular el material floral en forma de diseño

digging a process of moving plants

excavando el proceso de mover las plantas

digitizer a device that permits direct measurement of lengths, areas, volumes, and perimeters from drawings

digitalizador un aparato que mide las longitudes, las áreas, los volúmenes y los perímetros al analizar los dibujos

direct seeding planting seeds into a planting media

siembra directa el sembrar semillas en el terreno o en un medio de cultivo

disbudding the removal of some flower buds so those remaining will produce larger flowers

desyemar el quitar algunos capullos de la planta para que los que queden produzcan flores más grandes

distribution function physical organization and delivery of product or service

función de distribución el proceso de organizar y llevar a cabo la entrega de un producto o un servicio

dividends a profit paid to corporate shareholders

dividendos una ganancia que se le paga a los accionistas de la corporación

division a method of propagation requiring the cutting and dividing of plants

división un método de propagación que requiere que se corten y se dividan las plantas

dolomitic limestone Mg_2CO limestone with a high magnesium content

caliza dolomítica Mg_2CO caliza con un contenido alto de magnesio

dormant/dormancy in a resting or non-growing state

latente en un estado de descanso o de no crecimiento

double cropping planting a second crop after an early-maturing crop has matured and been harvested

doble cultivo el sembrar en el mismo lugar un segundo cultivo luego de que otro cultivo diferente se haya madurado y se haya cosechado

double-nose bulbs that produce two flower stocks to bloom

dobles bulbos que producen dos al florecer

drench to water the soil thoroughly with a chemical, fertilizer, or just water

empapar ensopar el terreno con un químico, un fertilizante o simplemente con agua

drift blowing of spray dusts or mist, by wind, away from the target plant to other areas

deriva, desvío al viento soplar polvos o sustancias nebulizadas en spray lejos de las plantas indicadas o hacia otras áreas

drip line the imaginary circle that indicates the outer edge or farthest extension of a tree's branches

línea de goteo el círculo imaginario que indica la extensión más lejana de las ramas de un árbol

drones/Unmanned Aerial Vehicles (UAVs) used to take aerial photos for orchards and vineyards to identify problem areas throughout growing areas

drones/Vehículos Aéreos No Tripulados utilizado para tomar fotos aéreos de los huertos de árboles frutales y las viñas para identificar problemas en el cultivo

durability length of use or how long something like a container will last

durabilidad el periodo de tiempo de uso o cuánto tiempo un artículo como un contenedor durará

E

embryo live immature embryo in plant structures, seeds, bulbs, and corms

embrión el embrión vivo inmaduro en las semillas, los bulbos, y los cormos

embryonic plant the entire plant before germination; embryo

planta embriónica es la planta entera antes de su germinación; embrión

emulsifiable concentrate a concentrate that will readily mix with water

concentrado emulsionante un concentrado que fácilmente se puede mezclar con agua

endophytes small organisms, fungi or bacteria usually, that live in or on other plants and protect them from insects and disease

endofitos organismos pequeños, usualmente hongos o bacteria, que viven dentro de o sobre otras plantas y los protegen de los insectos y de las enfermedades

endosperm the food supply for the young, developing seedling that is contained in the seed

endospermo la fuente de alimentos contenido en la semilla para la planta de semillero joven y en desarrollo

enthusiasm energy to do a job and inspiration to encourage others

entusiasmo la energía para completar un trabajo y la inspiración para animar a otros

entomopathogenic (beneficial) nematodes organic control of insects, grubs, beetles, termites, and webworms
nemátodos (beneficiales) entomopatógenos el control orgánico de los insectos, escarabajos, termitas, larva, y gusanos telarañeros

Epsom salts organic salt of Magnesium sulfate (MgSO2); source Mg to correct soil pH
sal de Epsom sal orgánica de sulfato de magnesio (MgSO2); fuente de Mg para corregir el nivel de pH en la tierra

erect straight up
erecto derecho hacia arriba, vertical

erosion washing or blowing away of the soil caused by water or strong winds
erosión el desplazamiento de la tierra causado por el agua o los vientos fuertes

espalier a plant trained to grow flat against a wall or trellis
plantas en espaldera una planta "entrenada" a crecer contra una pared o un enrejado

executive meeting meeting of the officers to conduct business of the organization between regular meetings
reunión ejecutiva una reunión de los funcionarios u oficiales de más alto nivel, citada entre las reuniones pautadas regularmente, para tratar sobre los negocios y asuntos de la organización

exposure deals with light, temperature, and other environmental factors
exposición referente a la luz, la temperatura y otros factores ambientales

extemporaneous speaking a form of public speaking wherein some preparation is made, but the speech is not written out or memorized
oratoria improvisada una forma de oratoria pública donde hay cierto grado de preparación de antemano, pero no un discurso escrito ni memorizado

F

fascicles a plant structure that holds the needles together on the stem
fascículos una estructura en la planta que sujeta las agujas en el tallo

fastigiated plants that grow tall and narrow
fastigiado plantas que crecen de forma alta y estrecha

fertility an available plant food resource in the plant media
fertilidad una fuente de alimentos para las plantas que se encuentra en el medio de cultivo

FFA (the National FFA Organization) an intercurricular student group at the middle school, high school, and college level. It makes a positive difference in the lives of students by developing their potential for premier leadership, personal growth, and career success through agricultural education.
FFA (la National FFA Organization) un grupo estudiantil intercurricular en las escuelas intermedias, escuelas superiores o de bachillerato, o las universidades. Tiene una influencia positiva en las vidas de los estudiantes al desarrollar su potencial para el liderazgo de alto nivel, el desarrollo personal, y el éxito profesional a través de la educación agrícola.

field capacity water left after capillary movement stops
capacidad de retención de agua del terreno el agua remanente una vez cesa el movimiento capilar del agua

financing function obtaining capital for a business
función de financiamiento el obtener capital para un negocio

fixed assets assets owned by a company, such as real estate, furniture, or farm equipment
activos fijos los activos de una compañía, tales como los bienes raíces, los muebles, o el equipo de cultivo o de granja

floating row cover a lightweight material placed over rows of crops for protection against cold and insects; light and water can get through
cubiertas flotantes para hileras un material liviano que se coloca sobre las hileras de cultivos para protegerlos del frío y de los insectos; el agua y la luz puede penetrar la cubierta

flocked to cover the original plant with snowlike material
escarchado cubrir la planta original con material parecido a la nieve o la escarcha

floral palette array of floral blooms
paleta floral un variedad de flores de muchos colores

Floraltape tape used to fasten plant stems together
Floraltape una cinta adhesiva que se utiliza para pegar juntos los tallos de las plantas

floriculture cultivation of ornamental flowering plants
floricultura el cultivo de flores y plantas ornamentales

florist shears specialized scissor-blades that are longer than ordinary scissors
tijeras de floristas tijeras especializadas que son más largas que las tijeras comunes

florist tape a floral mechanism of a paraffin-coated paper used to cover the stem and other mechanics of the arrangement; as it is stretched it will stick and bind together to hold the arrangement
cinta de floristería un mecanismo floral de papel cubierto de parafina que se utiliza para forrar el tallo y otras piezas de la mecánica de un arreglo floral; como se estira, se pegará y se juntará para mantener el arreglo en la forma del diseño

focal point an accent point in a flower arrangement where attention is focused

foco de atención un acento en un arreglo de flores donde el que lo ve concentra o dirige su atención

foot-candles a foot-candle is the amount of light a candle would produce at a distance of 1 foot from the candle

pie-candela un pie-candela es la cantidad de luz que emite una candela a una distancia de un pie de la llama

forage radish to enhance the soil structure to improve soil aeration and moisture availability

rábano forrajero para mejorar la estructura del suelo, y así mejorar la ventilación y aumentar la humedad del terreno

forcing growing plants to flower at other than their normal season

forzar cultivar las plantas de forma que florezcan de forma acelerada o fuera de la época usual para su florecimiento o su cosecha

formaldehyde a chemical used to sterilize the soil; it is being phased out and will soon not be available

aldehído fórmico un químico que se utiliza para esterilizar la tierra; se está disminuyendo su uso y pronta no estará disponible

foundation a planting around the base of a building

foundation planting (término en inglés) la siembra de plantas dirigida a cubrir los cimientos de un edificio

fragrance smell and/or scent of the plant

fragancia olor de una planta

functional a design that is useful

funcional un diseño que es útil

fungicide any substance that destroys or prevents the growth of fungi; usually applied to plants as a spray or dust to control fungus diseases

fungicida cualquier sustancia que destruye o evita el crecimiento de los hongos; usualmente se le echa a las plantas como un aerosol o un polvo para controlar las enfermedades ocasionadas por los hongos

fungus microscopic plants that can cause plant diseases

hongos plantas microscópicas que pueden causar enfermedades en las plantas

G

gamete a mature sex cell

gameto una célula sexual reproductiva madura

garland machine a machine used to make garland

máquina para guirnaldas una máquina utilizada para fabricar guirnaldas

gavel a small wooden hammer-like tool that is used by the presiding officer to direct a meeting

martillo una herramienta tipo martillo pequeño de madera que utiliza el que dirige una reunión

genetic engineering moving or changing the genes of plants

ingeniería genética el mover o cambiar los genes de las plantas

genetically engineered plants moving or changing the genes of plants

plantas manipuladas genéticamente plantas cuyos genes han sido movidos o cambiados

genetics the study of heredity

genética el estudio de la herencia

genus a classification that includes all plants that are similar in physical structure; the genus name is the first word listed in a plant's name

género una clasificación que incluye todas las plantas que son similares en estructura física; el nombre del género es la primera palabra que aparece en el nombre de la planta

germinate to sprout or begin to grow

germinar el brotar o empezar a crecer

gibberellic acid an important growth regulator used in horticultural businesses to increase plant growth; known in the business as B9

ácido giberélico un regulador de crecimiento importante que se utiliza en los negocios de horticultura para incrementar el crecimiento de las plantas; conocido en la industria como B9

gibberellins plant hormones that stimulate growth in stem and leaf by cell elongation; also stimulate premature flowering, growth of young fruits, and breaking of dormancy

giberelinas hormonas en las plantas que estimulan el crecimiento en el tallo y en la hoja al alargar las células; además estimula el florecimiento prematuro, el crecimiento en las frutas jóvenes, y el fin de la inactividad o del estado latente

girdle to remove all the cambium and bark from a stem for a short distance to aid in rooting

girdle (término en inglés) el quitar todo el cámbium y la corteza de una parte pequeña de un tallo para facilitar el que eche raíces

Girl Scouts youth organizations that provide opportunities for leadership development and skill development for girls

Girl Scouts organizaciones de jóvenes que le proveen a las niñas oportunidades para desarrollar destrezas y liderazgo

gluten corn gluten meal is used as a pre-emergent weed killer

gluten la harina de gluten de maíz se utiliza de forma pre-emergente para matar la maleza

good salesperson a person who is positive and has an aggressive personality

buen vendedor una persona con una actitud positiva y una personalidad agresiva

grafting uniting two different plants so that they grow as one

injertar el unir dos plantas distintas de forma que crezcan como una sola

granular small particles in the pellet form
granulado partículas pequeñas en forma de bolitas o perdigones

gravitational water water that the soil is unable to hold against the force of gravity
agua gravitatorio el agua que la tierra no puede retener contra la fuerza de gravedad

green manure plants plowed or mixed into the soil to rot and add organic matter to the soil
abono verde las plantas aradas y mezcladas en el terreno para que se pudran y le añadan material orgánico a la tierra

greenery any foliage material used to create a design
vegetación verde cualquier material o follaje verde para crear un diseño

ground covers low grounding plant that covers the bare soil
cubresuelos plantas bajas que cubren el suelo pelado

growth regulator a chemical that, when applied to plants, changes growth rate, usually to shorten the plant
regulador de crecimiento un agente químico que, cuando se le aplica a las plantas, cambia el nivel de crecimiento, usualmente para que la planta sea más baja

guttation a system plants use to remove excess moisture from plant leaves
gutación un sistema de las plantas para remover la humedad y el agua excesiva de sus hojas

H

hairpin a u-shaped florist wire used to push into the base mechanics to fasten plant material into place
horquilla un alambre en forma de U que utilizan los floristas para atar el material de la planta en su lugar en los arreglos y los diseños

hairy vetch a legume that adds nitrogen to the soil and when cut off can be used as a mulch, which furnishes nutrients resulting in higher yields
veza vellosa un legumbre que le añade oxígeno a la tierra y cuando se recorta se puede usar como abono o mantillo, lo cual provee nutrientes que estimulan un rendimiento mayor

hardening off gradually subjecting plants to more difficult growing conditions by withholding water and decreasing temperature; prepares plants for transplanting
incrementar la resistencia las plantas se someten gradualmente a condiciones más difíciles para su crecimiento al privarlas de agua y bajando la temperatura; prepara las plantas para el trasplante

hardiness ability of a plant to withstand the minimum temperature of an area
resistencia la habilidad de una planta para resistir la temperatura mínima de un área

hardscape paving surface area created with interlocking pavers
pavimento hardscape el área de los superficies sólidos que se construyen con ladrillos o adoquines u otros materiales no vegetativos

hardwood cutting a cutting made from the current season's stem tissue that is mature or hard
estaquillas semileñosas una estaquilla hecha del tejido de un tallo de la temporada actual que está dura o madura

harmony creates drifts between related objects and space of the arrangement
armonía crea complementos entre objetos relacionados y los espacios en un arreglo floral o un arreglo de plantas

harvesting removal of the fruit of the plant
cosechar el remover la fruta de la planta

heading back removal of all plant growth on top of the plant
desmoche de ramas al podar, es un corte en que se quita todo el crecimiento vegetal que está sobre la planta

heat islands environmental area that retains warmth in a close area
"islas de calor" urbano puntos en el medioambiente que retienen el calor en un área cercana o concentrada

heat tolerance plants' ability to withstand extreme warm temperatures
tolerancia al calor la habilidad de las plantas de resistir temperaturas calientes extremas

heeling-in a practice of setting bareroot (br) stock and/or b&b plants into a media to keep the plant alive
heeling-in (término en inglés) la práctica de sembrar un árbol cuyas raíces no están cubiertas de tierra y/o un cepellón, un árbol cuyas raíces cubiertas de tierra forman una bola tapada con yute, en un medio temporalmente para mantenerlo vivo

heirloom tomato A variety of tomato that has been passed down from one generation to the next by harvesting the seeds from the fruit of the tomato and will produce the same tomato; original tomatoes known as open-pollinated varieties that were introduced more than 100 years ago
tomates de herencia o de reliquia Una variedad del tomate que pasa de una generación a la siguiente al cosechar las semillas de la fruta del tomate y producirá el mismo tomate; los tomates originales conocidos como las variedades de polinización abierta surgieron hace más de 100 años atrás

herbaceous refers to succulent plants, usually annuals, that are not as woody as perennials or shrubs
herbáceo se refiere a las plantas suculentas, usualmente las que florecen una vez al año, que no

tienen tanta madera como los árboles perennes y los arbustos

herbaceous perennials perennials that are less winter hardy
perennes herbáceos plantas perennes que resisten menos el invierno

herbicide a substance that destroys weeds
herbicida una sustancia que destruye la maleza

herbs plants grown primarily to be used in cooking to add flavor to foods
especies plantas que se cultivan primordialmente para usarlas al cocinar y para añadirle sabor a las comidas

hierarchy the order in which plants are classified
jerarquía el orden en que las plantas son clasificadas

high tunnels higher, plastic-covered tunnels used to extend the growing season of vegetable plants
túneles altos túneles altos y cubiertos de plástico que sirven para extender la temporada de cultivo para los vegetales

highbush the classification of a group of blueberries that are most often used in commercial production; they reach heights of 6 feet (1.8 meters)
arándaro highbush, arándaro alto la clasificación de un grupo de arándaros azules que comúnmente se producen comercialmente; llegan a medir 6 pies de alto (1.8 metros)

hill system a strawberry planting method where no runners are allowed to grow
sistema de lomita un método para cultivar fresas en que no se permite que crezcan los estolones

Hillman wreath-making machine a wreath-making machine invented by Richard Bader
Máquina para coronas Hillman una máquina para fabricar coronas, inventado por Richard Bader

hook a method used to wire mums by inserting a florist wire up through the plant stem and bringing it out into the center floret of the composite flower
gancho un método utilizado para alambrar a los crisantemos al insertar un alambre de floristería en el tallo de la planta y sacarlo en el centro de la flor

hook and blade type of pruner with a bypass scissor cut
gancho y cuchilla tipo de tijera de podar con cuchilla de paso

hormones growth-regulating substances in plants
hormonas sustancias que regulan el crecimiento en las plantas

horticulture a field that includes growing of fruits, nuts, vegetables, ornamental plants, and flowers, and the sale and processing of these items; from the Latin word meaning "garden cultivation"
horticultura un área de especialización que incluye el cultivo de frutas, nueces, vegetales, plantas ornamentales, y flores, y la venta y el procesamiento de éstos; viene de la palabra en latín que significa "cultivo de jardines"

horticulturist those individuals who cultivate gardens, orchards, nurseries, flowers, vegetables, herbs, and other ornamental plants
horticultor aquellos individuos que cultivan jardines, huertos de árboles frutales, viveros, flores, vegetales, hierbas y especies y otras plantas ornamentales

hulls covering over the seed
vaina, cáscara la cubierta sobre la semilla

humus organic layer found on the forest floor
humus capa de materia orgánica regada por los suelos de una selva

hybrid an offspring of two different varieties of one plant that possesses certain characteristics of each parent
híbrido el retoño de dos variedades distintas de una planta que posee ciertas características de cada planta ascendente

hybrid tomato A variety of tomato that results from crossing two inbred lines between two other varieties and is developed for certain qualities such as disease resistance, insect resistance, high-quality taste, and flavor; seeds from a hybrid tomato will not reproduce that same tomato
tomate híbrido una variedad del tomate que surge al cruzar dos líneas de tomates endogámicos con dos otras variedades. Se cultivan de esta forma por ciertas cualidades como la resistencia a la enfermedad y a los insectos, y un sabor de alta calidad; las semillas de un tomate híbrido no reproducirán ese mismo tomate

hybridized to crossbreed two different species or varieties
hibridado el cruzar dos especies o variedades distintas

hydroscopic water water in the soil profile that is unavailable to plant roots
agua higroscópica agua en la tierra que no puede llegar a las raíces de las plantas

I

immediate response advertising advertising designed to make customers reach for a phone or come to a store soon after reading or hearing an advertisement
publicidad para respuesta inmediata la publicidad diseñada para hacer que los clientes llamen por teléfono o vayan a la tienda pronto

indeterminate tomatoes Tomatoes are produced all along the stems and will continue to grow and produce fruit for one growing season until the first fall frost
tomates indeterminados los tomates se producen por todos los tallos y seguirán creciendo y produciendo fruta durante una temporada hasta la primera helada del otoño

inhalation the process of taking in air through the lungs; breathing in

inhalación el proceso de captar aire a través de los pulmones

inhibitors (abscisic acid), plant hormones that inhibit seed germination, stem elongation, and hasten ripening of fruit (ethylene gas)

inhibidores (ácido abscísico), hormonas en las plantas que inhiben la germinación de las semillas, el alargamiento de los tallos, y acelera la maduración de la frutas (gas etileno)

insect a six legged organism, such as the fly, ladybug, and so on

insecto un organismo con seis patas, como la mosca, la mariquita, etc.

insecticides materials used to control insects

insecticidas materiales utilizados para controlar los insectos

insert to place into an opening

insertar colocar en una apertura

insulation value this is known as the R-value; to prevent the movement or loss of heat

valor de aislamiento esto se conoce como el valor R; para prevenir el movimiento o la pérdida del calor

integrity capable of personally upholding a high moral standard

integridad capaz de mantener normas éticas de alto nivel

interiorscapes decorating with plants on the inside of a building (i.e., lobby of a hotel)

paisajes o jardinería de interiores el decorar con plantas en el interior de un edificio (por ejemplo, en el lobby o vestíbulo de un hotel)

inventory report units received and sold, unit cost, total sales and profit

informe de inventario las unidades recibidas y vendidas, el costo por unidad, las ventas totales y las ganancias

irrigation the addition of water to plants to supplement natural rainfall

irrigación el ofrecer más agua a las plantas para complementar la lluvia

K

knowledge familiarity, awareness, and understanding

conocimiento familiaridad, conciencia, entendimiento

L

lacquer a clear spray sealer to give a shiny enhancement to the arrangement

laca un sellador en aerosol sin color que le da un realce brilloso a un arreglo floral

lateral side branch

lateral rama que surge de una de las partes laterales de la planta

Latin the language often used to name plants; Latin is no longer a spoken language

latín el idioma comúnmente usado para nombrar a las plantas; el latín ya no es un idioma hablado

lay on the table a subsidiary motion that refers to temporarily setting aside a pending motion

poner sobre la mesa una moción subsidiaria que busca dejar hacia un lado temporalmente una moción pendiente

lead to show the way by going in advance or to guide the action or opinion of

dirigir el ser el primero en señalar el camino o dirigir las acciones u opinión de un grupo o una empresa

leadership the capacity or ability to lead

liderazgo la capacidad o la habilidad para dirigir

legumes a plant in the family *Fabaceae* that enriches the soil with nitrogen through the root structure of nodules that remove nitrogen from the atmosphere and fix it in the soil

legumbres una planta de la familia *Fabaceae* que enriquece el terreno con nitrógeno a través de sus raíces cuando sus nódulos toman el nitrógeno de la atmósfera y lo transporta y lo fija en la tierra

letterhead portion of a business letter that contains the name, address, and telephone number of the company

encabezamiento la parte de una carta comercial que contiene el nombre, la dirección y el número telefónico de la compañía

Lillium longiflorum the scientific name for the Easter lily

Lillium longiflorum el nombre científico para el lirio de Pascua

limestone a natural rock used to reduce soil acidity

piedra caliza una piedra natural que se utiliza para reducir la acidez en la tierra

limestone natural ground limestone (calcium carbonate—CaCo3) that raises the pH level

caliza la caliza natural de la tierra (calcio carbonato -- CaCo3) que aumenta los niveles de pH

limited partnership business where one or more partners perform some function but limit their claim on the profits or liabilities to a share agreed to in advance

sociedad limitada negocio en que uno o más de los socios llevan a cabo alguna función pero limita(n) su parte de las ganancias y su responsabilidad por los gastos y las deudas a una cantidad acordada de antemano

line flowers a spike-appearing flower, such as gladiolus, delphinium, or snapdragon

flores de línea flores tipo pico, usualmente altas, como el gladíolo, la espuela de caballero, o el dragón

lining out planting cuttings in rows in the field
lining out (término en inglés) el sembrar esquejes en hileras en el campo

loam equal mixture of sand, silt, and clay
loam mezcla con cantidades iguales de arena, cieno y barro

loam soil with approximately equal amounts of sand, silt, and clay; generally considered the best type of soil for plant growth
marga tierra con cantidades aproximadamente iguales de arena, limo y barro; generalmente se considera el mejor tipo de tierra para el cultivo de las plantas

locality surroundings nearby
localidad área alrededor

long-term plan plans accomplished over months or years
plan de largo plazo planes a cumplirse en un periodo de meses o años

lopper pruners with long handles to cut branches of 3/40 to 20 in diameter
cortaramas lopper herramienta para podar con mangos largos utilizado para cortar ramas cuyos diámetros varían desde 3/40 a 20

lotus pod fruit pod of the lotus plant
vaina de loto la vaina de la fruta de la planta loto

lowbush the most cold-hardy blueberries growing no more than 8 to 18 inches (20 cm to 45 cm) tall; they are usually found growing in the wild and are harvested commercially in some areas
arándano enano, o tipo arbusto bajo el arándano azul más resistente al frío que crece a no más de entre 8 a 18 pulgadas (20 a 45 cm) de altura; usualmente se encuentran en campos silvestres y en algunas áreas los cosechan con fines comerciales

loyalty reliable support for an individual, group, or cause
lealtad respaldo fiable para un individuo, un grupo o una causa

M

macroelements major elements of fertilizer: nitrogen, phosphorous, potassium—NPK
macroelementos los elementos primordiales en el fertilizante: nitrógeno, fósforo, potasio—NPK

macrophylla plants with big leaves
macrophylla plantas con hojas grandes

main motion a basic motion used to present a proposal for the first time
moción principal una moción básica para presentar una propuesta por primera vez

manage to use people, resources, and processes to reach a goal
administrar como gerente el utilizar personas, recursos y procesos para alcanzar una meta

mass grouping plants in a landscape setting
amasar el agrupar o juntar plantas en un jardín o un paisaje

massing flower a flower that has a rounded appearance, such as a carnation, rose, or chrysanthemum
flor de aglomeración una flor de apariencia redonda, como un clavel, una rosa o un crisantemo

matted row system a strawberry planting method where all runners are allowed to grow to form many plants in a solid row
sistema de hileras apelmazadas un método de sembrar fresas en que a todos los estolones se les permite crecer para que formen hileras sólidas con muchas plantas

medium a material that is used to start and grow seeds and plants (plural: media)
medio un material que se utiliza para hacer que las semillas y las plantas germinen y crezcan

meiosis The reproductive cell division that results in the formation of gametes
meiosis la división celular reproductiva que resulta en la formación de los gametos

microclimate the climate in a small area
microclima el clima en un área pequeño o reducido

microelements plant nutrients needed in small quantities
microelementos los nutrientes que necesitan las plantas en cantidades muy pequeñas

microorganisms microscopic (very small) organisms; ones that inhabit the soil are of most interest here
microorganismos organismos microscópicos (muy pequeños); los que se encuentran en la tierra son de mayor interés en esta publicación

microphylla plants with small leaves
microphylla plantas con hojas pequeñas

minutes the official written record of a business meeting
actas el record oficial escrito de lo acontecido en una reunión de negocios

mistletoe a parasitic plant with white berries used in holiday arrangements
muérdago una planta parásita con bayas blancas que se utiliza en los arreglos navideños

mitosis simple cell division for growth
mitosis la división celular simple necesaria para el crecimiento

molluscicide a chemical used to kill snails and slugs
molusquicida un agente químico utilizado para matar caracoles y babosas

monotony repetitiousness of plants in a design

monotonía el uso repetitivo o redundante de plantas en un diseño

motion a proposal for group action that is presented in a meeting to be acted on by an organization

moción una propuesta que se presenta en una reunión de una organización para que un grupo tome acción

mulch any material used to cover the soil for weed control and moisture retention

mantillo cualquier material utilizado para cubrir la tierra para controlar la maleza y retener la humedad

multifaceted many different aspects of the subject

polifacético muchos aspectos diferentes del tema

N

narrowleaf evergreen ground covers needlelike evergreens that are prostrate on the ground

cubresuelos o tapizantes perennes narrowleaf plantas verdes perennes que quedan prostradas sobre el suelo

natural products produced by nature

natural productos producidos por la naturaleza

nematocide a substance used to kill nematodes (small worms)

nematicida una sustancia que se utiliza para matar nemátodos (lombrices pequeños)

nematodes multicellular roundworms of the animal phylum

nemátodos lombriz intestinal multicelular del filo animal

net worth the value of an owner's assets minus liabilities (debts)

patrimonio neto el valor de los activos de un individuo menos sus deudas y obligaciones

nitrogenous a material containing nitrogen fertilizer

nitrogenado un material que contiene un fertilizante de nitrógeno

node the joint of a stem; the swollen place where leaves and buds are generally attached

nodo la coyuntura de un tallo; punto inflado donde las hojas y los capullos se conectan

nonselective herbicide an herbicide (weed killer) that will kill all plants to which it is applied

herbicida no discriminatoria un herbicida (matador de maleza) que matará todas las plantas

noxious weeds plants that are required to be controlled by state law

maleza nociva plantas que son controladas por ley

nutrient availability plant food available for plant roots

disponibilidad de nutrientes alimento disponible para las raíces de las plantas

nuts a fruit composed of a hard shell that is edible

nueces una fruta comestible compuesta de una corteza o cáscara

O

Oasis® (floral foam) spongelike material used as a base for flower arrangements

Oasis® (espuma floral) material tipo esponja utilizado como una base para arreglos florales

offset bulbs young bulbs that develop at the basal plate of the bulb splits or slabs

bulbo reproductor bulbos jóvenes que se desarrollan en el plano basal de donde se dividen los bulbos

olericulture the science of vegetable production

cultivo de hortalizas la ciencia de la producción de vegetales

oral contact a compound that enters the body through the nostrils and mouth

contacto oral un compuesto que entra al cuerpo a través de la boca y las fosas nasales

order of business the items and sequence of activities conducted at a meeting

órden del día los puntos a considerarse y la secuencia de actividades en una reunión

organic nature's natural products

orgánico productos de la naturaleza

organic fertilizers fertilizers derived from animals, plant matter, or minerals

abonos orgánicos fertilizantes derivados de animales, de plantas o de minerales

P

Parafilm a wax that is used to seal and support composite flower petals

Parafilm una cera que se utiliza para sellar y sostener los compuestos de pétalos florales

parasites organisms that search out and kill other insects or microorganisms; many small wasps are parasites

parásitos organismos que buscan y matan otros insectos o microorganismos; muchas avispas pequeñas son parásitos

parasitoids organisms that spend a portion of their life cycle on a host organism that will kill the host

parasitoides organismos que pasan una parte de su ciclo de vida en un organismo huésped y que matarán al huésped

parliamentary procedure a system of guidelines or rules for making group decisions in business meetings

procedimientos parlamentarios un sistema de normas o reglamentos para tomar decisiones de grupo en reuniones

part-time employees whose career goals usually do not involve advancement in a company

de tiempo parcial empleados cuyas metas profesionales usualmente no incluyen el ascenso en una compañía

partnership a business owned by two or more individuals who are co-owners

sociedad una empresa que cuenta con dos o más individuos que son copropietarios

pasteurize to condition soils with heat to control pests, diseases, and weed seeds

pasteurizar condicionar los terrenos utilizando el calor para controlar las plagas, los insectos, las enfermedades y las semillas de maleza

pathogen a disease organism

agente patógeno un organismo que causa enfermedad

peat moss partially decomposed vegetation that has been preserved underwater

esfagno vegetación parcialmente descompuesto que se ha preservado bajo agua

pelleted seeds small seeds coated with a clay covering to improve proper seed distribution in the media

semillas cubiertas semillas pequeñas cubiertas de barro para mejorar la distribución de ellas en el medio de cultivo

pelletized fertilizer produced as pellets to assist in easy spreading of the material

en forma de bolita o perdigón fertilizante producido en forma de bolitas o perdigones para que sea más fácil regarlas en el terreno

perennial a plant that grows year after year without replanting; plant whose roots live from year to year

perenne una planta que crece año tras año sin tener que replantarla; planta cuyas raíces se mantienen vivos año tras año

perimeter total distance around the outside of the object

perímetro la distancia alrededor del exterior de un objeto

perlite a gray-white material of volcanic origin used to improve aeration of the media

perlita un material grisáceo blancuzco de origen volcánico que se utiliza para mejorar la ventilación del medio

personal growth a lifelong goal to maximize one's individual potential

desarrollo personal una meta de toda la vida para maximizar su potencial como individuo

personal protective equipment (PPE) protective clothing worn for pesticide protection

ropa de seguridad ropa usada para protegerse de los pesticidas

personnel management the direction of workers in a manner that brings out their best efforts and attitudes on behalf of the business

administración de personal el dirigir a trabajadores de forma efectiva para estimular sus mejores esfuerzos y actitudes en pro del negocio o la compañía

pesticide an agent, chemical or living, that is known to control insects and diseases considered pests

pesticida un agente, químico o viviente, que controla insectos y enfermedades considerados como plagas

petal one type of leaf on flowers; usually considered the most striking part of the flower

pétalo un tipo de hoja en las flores; usualmente se considera la parte más llamativa de la flor

PGR Mix Master plant growth regulator app for smartphones to determine the correct concentration of the plant growth

PGR Mix Master (nombre propio en inglés) una aplicación informática para los smartphones que se utiliza para regular y determinar la concentración apropiada del crecimiento de la planta

pH level the measure of a soil's acidity or alkalinity on a scale of 1 to 14; 1 to 6 indicates an acid, 7 is neutral, and 8 to 14 indicates a base

nivel de pH la medida de la acidez o de la alcalinidad de la tierra en escala del 1 al 14; del 1 al 6 indica acidez, el 7 es neutral, y del 8 al 14 indica una base

phalaenopsis the genus of Moth Orchid; it is very popular with beginning hobbyists

phalaenopsis el género de la orquídea conocida como el Moth Orchid; es muy popular con los aficionados principiantes

pharming the process of genetically modifying plants so that they produce substances that can be used as pharmaceuticals

pharming, el fabricar fármacos el proceso en que se modifican las plantas genéticamente para que produzcan sustancias que pueden utilizarse como fármacos

phenology the relationship between climate and plant flowering; can be used as an indicator of when to plant crops or spray for insects and diseases

fenología la relación entre el clima y el florecimiento de las plantas; se puede usar para indicar cuándo sembrar los cultivos o rociar pesticidas para matar insectos y enfermedades

pheromones materials used to attract insects; often made of naturally occurring or synthetic hormones that attract adult insects, usually males

feromonas materiales utilizados para atraer a los insectos; hechos frecuentemente con hormonas naturales o sintéticas que atraen a los insectos adultos, usualmente a los insectos masculinos

phloem the tubes in plant stems that conduct food from leaves through the stem to the roots

floema los tubos en los tallos de las plantas que conducen la comida desde las hojas hacia las raíces

photoperiodism response of plants to different periods of light and darkness in terms of flowering, reproduction cycle, and dormancy

fotoperiodicidad la reacción de las plantas a la luz y a la oscuridad relacionado a su florecimiento, su ciclo reproductivo, y su estado latente o inactivo

photorespiration movement of gases and water vapor from the plant leaf during photosynthesis

respiración de la fase oscura el movimiento de gases y el vapor de agua de las hojas en las plantas durante la fotosíntesis

photosynthesis manufacture of food by green plants in which carbon dioxide and water are combined in the presence of light and chlorophyll to form sugar and oxygen

fotosíntesis la producción de alimento por parte de las plantas verdes al combinarse el dióxido de carbono y el agua en la presencia de la luz y la clorofila para formar azúcar y oxígeno

piercing a method of wiring roses through the calyx

perforación un método de atar rosas en grupo pasando alambres a través del cáliz

pinching breaking off tips from plant stems to cause side shoots to develop and form more flowers

pinzamiento quitarle las puntas a los tallos de las plantas para que los brotes laterales se desarrollen y formen más flores

pine bark organic matter of the outer bark of pine trees used as plant media and mulch

corteza de pinos material orgánico de la corteza de los árboles de pino que se utiliza como medio para las plantas o como mantillo

pinning machine a machine to affix metal to design-ing material to hold in the design

máquina de fijación una máqina para fijar metal a un material o tela para sostener el diseño

pistil the female reproductive part of a flower; contains the female sex cells in the ovary

pistil la parte reproductiva femenina de la flor; contiene las células sexuales femeninas en el ovario

plan the think-through

plan el pensar sobre y planificar lo que se va a llevar a cabo

plant growth retardants (PGRs) methods of controlling the length of the plant internode

retardantes del crecimiento de las plantas métodos para controlar la longitud del entrenudo de las plantas

plant stimulants materials (hormones or auxins) that cause plants to grow faster

estimulantes de plantas materiales (hormonas o auxinas) que hacen las plantas crecer más rápidamente

plantscaping using plants in the landscape

plantscaping (término en inglés) el usar las plantas en los jardines y los paisajes

plastic type of ribbon used to make bows or cover box wreath frames

plástico tipo de cinta que se utiliza para hacer lazos o para cubrir los marcos para las coronas

podocarpus an evergreen plant material used for accents

podocarpus un material de planta perenne que se utiliza como realce

point of order a matter is raised during consideration of a motion concerning the rules of parliamentary procedure

cuestión de órden un punto relacionado a las reglas del procedimiento parlamentario que se trae a colación durante la discusión sobre una moción

pollination the process in which the pollen cell is transferred to the female plant part and the eggs are fertilized

polinización el proceso mediante el cual la célula del polen se transfiere de la parte femenina de la planta y se fertilizan los óvulos

pollutants a substance released into the environment that has an undesirable effect

contaminantes una sustancia liberada en el medioambiente que tiene un efecto indeseable

pomology the study of fruits

pomología el estudio de las frutas

postemergence herbicides that are applied after the crop plant has come up; careful selection of herbicide used is important

posemergente herbicidas que se aplican una vez que la planta de cosecha ha brotado del terreno

pot-bound roots growing in a tight pattern around the walls of the container

pot-bound (término en inglés) las raíces que han crecido de forma muy apiñada en las paredes del envase

pot-in-pot a nursery practice of growing trees in containers by setting one container into a larger one in the ground to hold the plant and container

envase-en-envase una práctica en los viveros de cultivar árboles en envases al poner un envase dentro de otro más grande en el suelo

pottery a container made of natural materials (clay) to grow plants

alfarería un envase o una maceta hecha de materiales naturales (barro) para cultivar plantas

potting medium the porous media for growing bonsai plants

medio de macetas el medio poroso para cultivar plantas bonsai

predators living organisms that repel or prey on and search out and kill other organisms; ladybugs, green lacewings, and spiders are predators

depredadores organismos vivos que repelen o que buscan, cazan y matan otros organismos; las mariquitas o catarinas, crisopas verdes y las arañas son depredadores

preemergence herbicides applied before crop plants emerge; these chemicals prevent weed seeds from germinating and kill small, tender weeds

pre-emergente herbicidas que se usan antes de que broten las plantas; estos agentes químicos impiden que las semillas de la maleza germinen y maten las semillas y los brotes más pequeños y tiernos

premier leadership top of the line leadership
liderazgo excelente un liderazgo de lo mejor

preplanting application of an herbicide before the crop is planted

presiembra el aplicar un herbicida antes de sembrar el cultivo

presiding officer a president, vice president, or chairperson who is designated to lead a business meeting

presidente de sesión un presidente, vice presidente, o director quien está designado para dirigir una reunión

privacy being secluded from the presence of others
privacidad el estar aislado de la presencia de otros

private corporation sells its stock publicly and can be owned by many people

corporación privada vende sus acciones públicamente y muchas personas pueden ser dueños

profit and loss statement projection of costs and expenses against sales and revenue over time

estado de ganancias y pérdidas la proyección de los costos y los gastos comparado con las ventas y el ingreso durante un periodo de tiempo

promoting function plan to make potential customers aware of the product or service

plan de promoción plan para dar a conocer el producto o el servicio a posibles clientes

propagate to increase in number; to reproduce
propagar aumentar en número; reproducirse

proportion to maintain the relative size and space in the design

proporción el mantener el tamaño y el espacio relativo en un diseño

prostrate growth pattern of a plant grows close to the ground

prostrado el patrón de crecimiento de una planta que crece cerca del suelo

purity a measure of how many weed seeds or inert materials are in a grass seed mixture

pureza tomar medida de cuántas semillas de maleza o materiales inertes se encuentran en una mezcla de semillas de grama

Pythium one of the fungus diseases that causes root rot in lilies

Pythium una de las enfermedades de hongo que causa que las raíces de los lirios se pudran

Q

QR code Quick Response code
código QR código de respuesta rápida

R

rabbiteye these blueberries are native to the Southeast; bushes grow 10 or more feet (5 meters) tall

arándano rabbiteye este arándano es originario del sureste de los Estados Unidos; los arbustos crecen a 10 pies o más de altura (5 metros)

rain garden an established garden that will control surface water runoff

jardín de lluvia un jardín establecido que controlará la escorrentía

raised beds planting beds that are elevated above the original ground terrain

arriates elevados arriates para sembrar elevados sobre el terreno original

recreational an area where sports or other leisure time activities occur

recreativo un área donde se llevan a cabo los deportes u otras actividades de ocio o esparcimiento

reel mower a mower with many long, horizontal blades that cut against a bottom bar; clips grass with a clean cut

podadora de grama tipo carrete una podadora con cuchillas largas y horizontales que cortan al dar contra una barra fija al fondo; corta la grama con corte nítido y uniforme

refer a motion used to refer some other motion to a committee or person for finding more information and/or taking action on the motion on behalf of the members

referir una moción utilizada para referir alguna otra moción a un comité o a una persona para buscar más información o tomar acción a nombre de los miembros o afiliados

refrigerator equipment used for storage of products at a low temperature

refrigerador equipo para almacenar productos en temperaturas frías

relative humidity the amount of moisture in the air compared with what it can hold at that temperature; expressed as a percentage of the amount of moisture the air could hold if saturated

humedad relativa la cantidad de humedad en el aire comparado con lo que puede retener a esa temperatura; expresado como un porcentaje de la cantidad de humedad el aire pudiese aguantar al estar saturado

renewal spur a cane of a vine pruned to two buds and left to grow new canes the following year
cepa de renovación una rama de una cepa podada hasta quedar con dos capullos, y dejarla para que crezcan ramas nuevas el próximo año

repellents do not kill insects, but usually drive them away
repelentes no mata a los insectos, pero usualmente hace que se alejen

repetition to repeat the same object and space throughout the design
repetición el repetir el mismo objeto y colocarlo a través de todo el diseño

repotting removal of the bonsai plant from the present container, pruning the roots, and planting into another container
trasplantar el sacar la planta bonsái del envase actual, podar las raíces, y sembrarla en otra maceta o envase

residential the individual home in which one lives
residencial el hogar individual en que una persona vive

respiration a breathing process in which plants and animals consume oxygen and release carbon dioxide
respiración un proceso en que las plantas y los animales consumen oxígeno y emiten dióxido de carbono

response time length of time after short days (long nights) are started and plants flower
periodo de reacción periodo de tiempo luego de que comiencen los días cortos (noches largas) y que las plantas empiezan a florecer

résumé a written listing of your education and skills qualifying you for a job
curriculum vitae una presentación por escrito de su preparación académica y de sus destrezas que hace que usted cualifique para un empleo

retard to slow the growth of the plant
retrasar hacer que la planta se demore más en crecer

retardants chemicals that control the growth of the plant
retardantes agentes químicos que controlan el crecimiento de la planta

Rhizoctonia one of the fungus diseases that cause root rot in lilies
Rhizoctonia una de las enfermedades de hongos que hacen que las raíces de los lirios se pudran

rhizome underground stem that produces roots on the lower surface and extends leaves and flowering shoots above the ground
rizoma un tallo bajo tierra que produce raíces en la superficie and extiende sus hojas y sus tallos florecientes sobre tierra

rhythm movement created in the design
ritmo o armonía movimiento creado en el diseño

ribbon a nonplant accessory to create a positive color flow throughout the design
cinta un accesorio no vegetativo usado para crear un despliegue positivo de colores a través de todo el diseño

rodenticide a substance used to kill rodents
raticida un sustancia utilizada para matar ratas y roedores

root ball roots of the plant when balling and burlapping (B&B) the plant
bola de raíces las raíces de la planta cuando se apelotonan y se cubren de tierra en forma de bola. Esta, a su vez, se forra con yute, convirtiéndose así en cepellón.

root-bound when plant roots are overcrowded in the container; it usually causes the roots to grow into a circle in the container
root-bound (término en inglés) cuando las raíces de las plantas les falta espacio y están muy apiñadas en el envase; usualmente hace que las raíces crezcan en forma de círculo en el envase

rootstock the lower portion of a graft that becomes the stem and roots of the new plant
portainjerto o patrón la parte más baja de un injerto que se convierte en el tallo y las raíces de una nueva planta

rotary mower a mower that cuts with a blade that spins at a high rate of speed
cortacéspedes rotatorios un cortacéspedes que corta con un filo que gira con alta velocidad

runners long stems growing and spreading out from strawberry plants; each runner roots on the end and starts a new plant
estolones tallos que crecen y se extienden de las matas de fresas; cada estolón en su punta se adhiere al terreno como raíz y forma una planta nueva

S

safety data sheet (SDS) gives all the details about the pesticide being used
ficha de datos de seguridad ofrece todos los detalles sobre algún pesticida que se esté utilizando

salal floral greenery used as filler foliage and accent in the design
salal (nombre en inglés) vegetación floral verde que se utiliza como un follaje de relleno y para proveer realce en partes de un diseño

salutation greeting portion of a business letter
saludo el saludo al comienzo de una carta

satin type of ribbon used to make bows
satén tipo de cinta usada para hacer lazos

scale the correct size of the object and space in the design
escala el tamaño correcto de un objeto y de los espacios en un diseño

scion a short piece of shoot containing several buds that becomes the new top of a grafted plant

vástago un pedazo corto de un tallo con varios capullos que se convierte en la parte superior nueva de un injerto

seasonal workers who are hired during peak times of the year

de temporada trabajadores que se contratan para el momento cúspide de la cosecha

secretary a person elected or appointed to take notes and prepare minutes of the meeting

secretario una persona elegida o nombrada para tomar notas y preparar las actas de una reunión

seedlings young plants that have been germinated several days

plantas de semillero plantas jóvenes que germinaron hace unos días

selective herbicide a herbicide that controls certain plants

herbicida selectivo un herbicida que selecciona y controla ciertas plantas

selflessness placing the desires and welfare of others above oneself

abnegación el poner los deseos y el bienestar de otros por encima de los de uno

selling function market research, sales plans, and sales closures

función de venta estudio de mercado, planes para las ventas, finalización de las ventas

sepal green, leaflike part of the flower that covers and protects the flower bud before it opens

sépalo una parte verde de la planta parecido a una hoja que cubre y proteje el capullo antes de que se abra

separation method of propagation that occurs naturally in which reproductive organs of a plant detach from the parent plant to become new plants

separación método de propagación que ocurre de forma natural cuando órganos reproductivos de una planta se separan o se desconectan de la planta patrón para convertirse en plantas nuevas

short-term plan plan accomplished in days or weeks

plan a corto plazo un plan que se finaliza en días o semanas

side-dress to fertilize along the side of rows near plants

side-dress (término en inglés) aplicar el fertilizante en bandas a un lado del surco en que se sembrarán las plantas en hileras

simple layering a lower branch from a plant is bent to the ground where it is partially covered with soil for rooting of new shoots

acodo simple un tallo en la parte inferior de la planta se dobla hacia el suelo y se entierra parcialmente para enraizarse con tallos nuevos

site location or growing area

sitio, lugar la localización o el área para el crecimiento y el cultivo

sleeves protective sleeves to protect poinsettias in shipping

cubiertas fundas para proteger la flor de Pascua (o la flor de nochebuena) al transportarla

slow-release fertilizer fertilizer that releases plant nutrients over a long period of time

fertilizante de difusión lenta fertilizante que libera los nutrientes durante un periodo de tiempo largo

softscape paving using wood chips for walkways

pavimento de jardín utilizar virutas de madera para construir senderos de jardín

soilless mix contains combinations of perlite, sphagnum moss, peat moss, vermiculite, limestone, tree bark, and slow-release fertilizers

medio de cultivo sin tierra contiene una mezcla y combinación de perlita, musgo esfagnáceo, turba, vermiculita, caliza, corteza, y fertilizantes de difusión lenta

sole proprietorship individuals operate and fund the business, make decisions, and earn a profit; simplest form of doing business

propietario único el individuo opera y financia el negocio, toma las decisiones, y obtiene una ganancia; es la forma más sencilla de hacer negocios

soluble fertilizers fertilizers that are water soluble

fertilizantes solubles fertilizantes que se disuelven en agua

spark plug the part of an engine that delivers the electrical spark to ignite fuel in the engine cylinder

bujía la parte de un motor que provee la chispa eléctrica para encender el combustible en el cilindro del motor

species a group of plants that evolved together and have similar characteristics

especie un grupo de plantas que evolucionaron juntos y tienen características similares

sphagnum moss dehydrated remains of acid bog plants, used in shredded form for starting seeds

musgo esfagnáceo los restos deshidratados de plantas de los pantanos, usado en forma triturada para comenzar el proceso de germinación de las semillas

spongelike the texture of the floral oasis to hold the flowers

como esponja la textura de la espuma floral para sostener las flores o el arreglo floral

spreader sticker a material used to cause pesticides to evenly spread over and stick to leaves

esparcidor adherente un material que se utiliza para hacer que el pesticida se esparza de forma uniforme y se pegue a las hojas

stamen the male reproductive part of the flower containing the male sex cells or pollen

estambre la parte reproductiva masculina de la flor que contiene las células sexuales masculinas o el polen

stenting grafting a short piece of stem to another stem that roots easier to produce a larger root system and plant in a shorter time
implantar un stent injertar una pieza pequeña de un tallo a otro tallo que echa raíces más fácilmente para así producir un sistema de raíces más amplio y una planta en menos tiempo

stewardship ability to plan and manage resources
administración la habilidad para planificar y manejar los recursos

sticky cards yellow or blue cards that are used to attach insects to determine the infestation in a crop
trampas pegajosas tarjetas amarillas o azules que se utilizan para atrapar a los insectos y determinar el nivel de infestación en una cosecha

stolonized use of shredded stolons (part of a living plant) to start a grassed area instead of using seed
estolonizado el uso de estolones triturados (parte de una planta viva) para comenzar el cultivo de un área con grama en vez de utilizar semillas

stomata small pores or holes in the leaf that allow the plant to breathe and give off moisture
estoma poros u hoyos pequeños en la hoja que permiten que la planta respire y libere la humedad

stool layering the parent plant is cut back to near soil level and covered with soil; new shoots root as they emerge from the parent plant and rise above soil level
acodo en montículo proceso en que la planta matriz se recorta hasta casi el nivel de la superficie y se cubre con tierra; nuevos tallos echan raíces y brotan del terreno

stormwater the excess water that runs off the structures on a property
agua de lluvia el exceso de agua que se desplaza de las estructuras en una propiedad

subcanopy a tree cover formed under the upper tree canopy
fronda baja un follaje que forma una cubierta localizada a una altura menor que el follaje más alto en las selvas o las cosechas

subcontractor a business that carries out work for a company as a part of a larger project
subcontratista un negocio que lleva a cabo trabajos para una compañía como parte de un proyecto más grande

sulfur coated refers to a coating of sulfur on fertilizer pellets that results in slow removal of the plant nutrients to the soil over a longer period of time
cubierta de azufre se refiere a una cubierta de azufre que se le pone a las píldoras de fertilizantes para que los nutrientes en las plantas tarden más tiempo en desplazarse a la tierra

superphosphate chemical compound of monocalcium phosphate
superfosfato compuesto químico de fosfato de monocalcio

survival to remain alive
supervivencia el permanecer vivo

susceptible weak, unable to resist (disease)
susceptible débil, incapaz de resistir (la enfermedad)

swag a grouping of greenery to accent a design
swag (término en inglés) el agrupar plantas verdes para realzar un diseño

symmetrical having equal visual weight on each side of the axis
simétrico, armónico el tener la misma presencia visual y balanceada en cada lado del axis

synthetic compounds material resources from nature used for plant food
compuestos sintéticos recursos materiales provenientes de la naturaleza que se utilizan para alimento para las plantas

systemic inside and moving throughout a plant
sistémico el moverse internamente a través de la planta

T

tact skill of encouraging others in positive ways
tacto destreza efectiva al animar de manera positiva a otras personas

taxonomist a person who identifies and classifies plants, often providing the proper name for the plant
taxonomista persona que identifica y clasifica las plantas y que frecuentemente provee el nombre correcto para la planta

taxonomy the study of plant names and the identification of plants
taxonomía el estudio de los nombres y la identificación de las plantas

terrarium a covered or closed container for growing plants
terrario un envase cubierto o cerrado para el crecimiento de las plantas

textural triangle to determine the textural name of a soil
triángulo textural aparato para determinar el nombre y la clasificación textural de la tierra

texture the coarseness or fineness of plant leaves
textura la tosquedad o la finura de las hojas de las plantas

thatch dead grass clippings and roots that build up in lawns, causing growth problems

capa de esteras hierba cortada y raíces muertas que se acumulan en los céspedes, causando problemas de crecimiento para las plantas

thinning removal of lateral branches of the plant

entresacar el cortar y quitar las ramas laterales de la planta

tissue culture and micropropagation new method for rapidly reproducing plants using terminal bud tissue

cultivo de tejidos y micropropagación un método nuevo para reproducir las plantas rápidamente usando el tejido de los capullos

topiary art of creating formal design of a plant

arte topiario el arte de crear un diseño formal de una planta

toxicity the degree to which a substance can damage an organism

toxicidad el punto en que una sustancia puede causarle daño a un organismo

trailing refers to the canes of boysenberries, which are allowed to run or trail along the ground or wire and are not cut back

trailing (término en inglés) se refiere a cuando a los tallos leñosos de las bayas de Boysen se les permite rastrearse por el suelo o en los marcos de alambre y no son recortados

transpiration loss of water through the leaves or stems of plants

transpiración la pérdida de agua mediante las hojas o los tallos de las plantas

tree bark bark of pine or hardwood trees ground in small pieces

corteza corteza de pino y árboles de hoja caduca triturado en pedazos pequeños

tree wrap paper fiber to cover the tree trunk to protect it from sunscald

envoltura de árboles un especie de papel fibroso para cubrir el tronco de un árbol para protegerlo del sol y de rajaduras

trellis a supporting structure made of posts and wire that is used to tie up grapes and some of the bramble fruits

espaldar una estructura hecho de postes y de alambre para amarrar y sostener a las uvas y las zarzamoras

trench layering a plant is bent over and the branches of the entire plant are buried with soil in a trench; roots develop as new shoots form and grow up above the soil level

acodo de trinchera proceso en que se dobla una planta para que todas sus ramas queden enterradas en una zanja o trinchera; surgen raíces al formarse tallos nuevos y crecen hasta salir del terreno

tropism growth or turning movement of a plant, in response to the sun's rays, such as photosynthesis

tropismo el crecimiento o el movimiento de una planta al reaccionar a los rayos del sol, tal como la fotosíntesis

tuber fleshy root that reproduces by growing roots from an eye, or bud

tubérculo raíz carnosa que se reproduce al surgir raíces de la yema o el ojo

tuberous a fleshy enlarged portion of the root

tuberoso una parte agrandada y carnosa de la raíz

U

unglazed a container without any outer coating

no vidriado un envase sin cubierta exterior

unity pulls together the flow between scale, rhythm, repetition, harmony, and proportion in design

unidad reúne la relación entre varios elementos de un diseño como la escala, el ritmo, la repetición, la armonía y la proporción

upright canes on brambles are cut back to shorten them so they freely stand upright or are tied to wire so they remain upright

posición vertical los tallos leñosos en la zarzamora se recortan para que se puedan sostener libremente en posición vertical o se atan a las estructuras de alambre para mantenerse en posición vertical

V

variegated two different colors on the same leaf

abigarrado dos colores distintos en la misma hoja

vegetative active, living plant parts such as stems, stolons, or sprigs

vegetativo partes vivas y activas de las plantas, tales como los tallos, los estolones, o los ramitos

vegetative bud a bud that will grow into a stem or shoot, not a flower

yema vegetativa una yema o capullo que crecerá para llegar a ser un tallo o un brote, no una flor

vegetative period the time during which mums grow under long days (not more than 7 hours of continuous darkness) and grow vegetatively, but set no flower buds

periodo vegetativo el periodo de tiempo en que los crisantemos crecen durante días largos (no más de 7 horas de oscuridad contínua) y que crecen en forma vegetativa, pero no producen capullos

velvet a soft, smooth ribbon texture

terciopelo una textura suave y delicada en las cintas

vendor a supplier of materials needed in the operation of a business

vendedor un suplidor de materiales necesarios para operar un negocio

vermiculite a light mineral with a neutral pH used to increase the moisture-holding capacity of media

vermiculita un mineral liviano con un nivel del pH neutral que se utiliza para aumentar la capacidad del medio para retener la humedad

vernalized precooled bulbs that will start growing immediately after planting (some bulbs must have a period of cold temperature before growing)

vernalizado bulbos enfriados de antemano que comenzarán a crecer tan pronto se siembren (para algunos bulbos, es necesario que pasen por un periodo de tiempo en temperaturas frías antes de que comiencen a crecer)

visualize to imagine a project

visualizar imaginar un proyecto

W

website Internet site developed to provide the viewer with a positive first impression of a business, explain its products and services, and offer examples of past accomplishments

sitio web sitio en la red creado para proveerle al lector una primera impresión positiva del negocio, explicar los productos y los servicios que ofrece, y detallar ejemplos de logros en el pasado

weeping plants that have branches flowing toward the ground

weeping (término en inglés) plantas cuyas ramas, en su postura natural, "fluyen" o se dirigen hacia el suelo

wind exposure plants that are exposed to wind in the landscape

exposición al viento las plantas que son expuestas al viento en un paisaje o jardín

windbreak any natural or man-made screen to slow wind speed

rompevientos cualquier pantalla u obstáculo natural o fabricado para reducir la velocidad del viento

wiring a practice to train branches with copper wire wrapped around the branches to train the plant into a desired shape

alambrar la práctica de "entrenar" a las ramas cubriéndolas con alambre de cobre para lograr que tomen una forma específica deseada

working capital funds used by a business to pay wages, purchase supplies, and pay utility bills

capital de trabajo fondos que utiliza un negocio para pagar sueldos, comprar suministros, y pagar por servicios públicos, como la electricidad y el agua

X

xeric (waterwise) using plants that require less water

xerófilo (interesado en asuntos de preservación de agua) usar plantas que necesitan menos agua

xeriscaping a landscaping technique used to practice water conservation

xeriscape, xerojardinería una técnica de paisajismo y jardinería que se utiliza para conservar el agua

xylem conducting tissue (tubes) in plants that transports water and nutrients from the roots to the stem and leaves

xilema el tejido (tubos) en las plantas que transportan el agua y los nutrientes desde las raíces hacia el tallo y las hojas

xylem tubes tubes in a plant stem that conduct water and minerals from the roots upward

tubos xilemas tubos en el tallo de la planta que conduce el agua y los minerales desde las raíces hacia arriba

Index

D

G